Bayesian Hierarchical Models

With Applications Using R

Second Edition

Bayesian Hierarchical Models

With Applications Using R
Second Edition

By
Peter D. Congdon
University of London, England

CRC Press is an imprint of the
Taylor & Francis Group, an **informa** business
A CHAPMAN & HALL BOOK

CRC Press
Taylor & Francis Group
6000 Broken Sound Parkway NW, Suite 300
Boca Raton, FL 33487-2742

First issued in paperback 2021

ISBN-13: 978-1-4987-8575-4 (hbk)
ISBN-13: 978-1-03-217715-1 (pbk)
DOI: 10.1201/9780429113352

Publisher's Note

The publisher has gone to great lengths to ensure the quality of this reprint but points out that some imperfections in the original copies may be apparent.

Visit the Taylor & Francis Web site at
http://www.taylorandfrancis.com

and the CRC Press Web site at
http://www.crcpress.com

Contents

Preface

My gratitude is due to Taylor & Francis for proposing a revision of *Applied Bayesian Hierarchical Methods*, first published in 2010. The revision maintains the goals of presenting an overview of modelling techniques from a Bayesian perspective, with a view to practical data analysis. The new book is distinctive in its computational environment, which is entirely R focused. Worked examples are based particularly on rjags and jagsUI, R2OpenBUGS, and rstan. Many thanks are due to the following for comments on chapters or computing advice: Sid Chib, Andrew Finley, Ken Kellner, Casey Youngflesh, Kaushik Chowdhury, Mahmoud Torabi, Matt Denwood, Nikolaus Umlauf, Marco Geraci, Howard Seltman, Longhai Li, Paul Buerkner, Guanpeng Dong, Bob Carpenter, Mitzi Morris, and Benjamin Cowling. Programs for the book can be obtained from my website at https://www.qmul.ac.uk/geog/staff/congdonp.html or from https://www.crcpress.com/Bayesian-Hierarchical-Models-With-Applications-Using-R-Second-Edition/Congdon/p/book/9781498785754. Please send comments or questions to me at p.congdon@qmul.ac.uk.

QMUL, London

Preface

[illegible]

1

Bayesian Methods for Complex Data: Estimation and Inference

1.1 Introduction

The Bayesian approach to inference focuses on updating knowledge about unknown parameters θ in a statistical model on the basis of observations y, with revised knowledge expressed in the posterior density $p(\theta|y)$. The sample of observations y being analysed provides new information about the unknowns, while the prior density $p(\theta)$ represents accumulated knowledge about them before observing or analysing the data. There is considerable flexibility with which prior evidence about parameters can be incorporated into an analysis, and use of informative priors can reduce the possibility of confounding and provides a natural basis for evidence synthesis (Shoemaker et al., 1999; Dunson, 2001; Vanpaemel, 2011; Klement et al., 2018). The Bayes approach provides uncertainty intervals on parameters that are consonant with everyday interpretations (Willink and Lira, 2005; Wetzels et al., 2014; Krypotos et al., 2017), and has no problem comparing the fit of non-nested models, such as a nonlinear model and its linearised version.

Furthermore, Bayesian estimation and inference have a number of advantages in terms of its relevance to the types of data and problems tackled by modern scientific research which are a primary focus later in the book. Bayesian estimation via repeated sampling from posterior densities facilitates modelling of complex data, with random effects treated as unknowns and not integrated out as is sometimes done in frequentist approaches (Davidian and Giltinan, 2003). For example, much of the data in social and health research has a complex structure, involving hierarchical nesting of subjects (e.g. pupils within schools), crossed classifications (e.g. patients classified by clinic and by homeplace), spatially configured data, or repeated measures on subjects (MacNab et al., 2004). The Bayesian approach naturally adapts to such hierarchically or spatio-temporally correlated effects via conditionally specified hierarchical priors under a three-stage scheme (Lindley and Smith, 1972; Clark and Gelfand, 2006; Gustafson et al., 2006; Cressie et al., 2009), with the first stage specifying the likelihood of the data, given unknown random individual or cluster effects; the second stage specifying the density of the random effects; and the third stage providing priors on parameters underlying the random effects density or densities.

The increased application of Bayesian methods has owed much to the development of Markov chain Monte Carlo (MCMC) algorithms for estimation (Gelfand and Smith, 1990; Gilks et al., 1996; Neal, 2011), which draw repeated parameter samples from the posterior distributions of statistical models, including complex models (e.g. models with multiple or nested random effects). Sampling based parameter estimation via MCMC provides a full posterior density of a parameter so that any clear non-normality is apparent, and hypotheses about parameters or interval estimates can be assessed from the MCMC samples without the assumptions of asymptotic normality underlying many frequentist tests. However, MCMC methods may in practice show slow convergence, and implementation of some MCMC methods (such as Hamiltonian Monte Carlo) with advantageous estimation features, including faster convergence, has been improved through package development (rstan) in R.

As mentioned in the Preface, a substantial emphasis in the book is placed on implementation and data analysis for tutorial purposes, via illustrative data analysis and attention to statistical computing. Accordingly, worked examples in R code in the rest of the chapter illustrate MCMC sampling and Bayesian posterior inference from first principles. In subsequent chapters R based packages, such as jagsUI, rjags, R2OpenBUGS, and rstan are used for computation.

As just mentioned, Bayesian modelling of hierarchical and random effect models via MCMC techniques has extended the scope for modern data analysis. Despite this, application of Bayesian techniques also raises particular issues, although these have been alleviated by developments such as integrated nested Laplace approximation (Rue et al., 2009) and practical implementation of Hamiltonian Monte Carlo (Carpenter et al., 2017). These include:

a) Propriety and identifiability issues when diffuse priors are applied to variance or dispersion parameters for random effects (Hobert and Casella, 1996; Palmer and Pettit, 1996; Hadjicostas and Berry, 1999; Yue et al., 2012);
b) Selecting the most suitable form of prior for variance parameters (Gelman, 2006) or the most suitable prior for covariance modelling (Lewandowski et al., 2009);
c) Appropriate priors for models with random effects, to avoid potential overfitting (Simpson et al., 2017; Fuglstad et al., 2018) or oversmoothing in the presence of genuine outliers in spatial applications (Conlon and Louis, 1999);
d) The scope for specification bias in hierarchical models for complex data structures where a range of plausible model structures are possible (Chiang et al., 1999).

1.2 Posterior Inference from Bayes Formula

Statistical analysis uses probability models to summarise univariate or multivariate observations $y = (y_1,\ldots,y_n)$ by a collection of unknown parameters of dimension (say d), $\theta = (\theta_1,\ldots,\theta_d)$. Consider the joint density $p(y,\theta) = p(y\,|\,\theta)p(\theta)$, where $p(y|\theta)$ is the sampling model or likelihood, and $p(\theta)$ defines existing knowledge, or expresses assumptions regarding the unknowns that can be justified by the nature of the application (e.g. that random effects are spatially distributed in an area application). A Bayesian analysis seeks to update knowledge about the unknowns θ using the data y, and so interest focuses on the posterior density $p(\theta\,|y)$ of the unknowns. Since $p(y,\theta)$ also equals $p(y)p(\theta\,|y)$ where $p(y)$ is the unconditional density of the data (also known as the marginal likelihood), one may obtain

$$p(y,\theta) = p(y\,|\,\theta)p(\theta) = p(y)p(\theta\,|\,y). \tag{1.1}$$

This can be rearranged to provide the required posterior density as

$$p(\theta \mid y) = \frac{p(y \mid \theta)p(\theta)}{p(y)}. \tag{1.2}$$

The marginal likelihood $p(y)$ may be obtained by integrating the numerator on the right side of (1.2) over the support for θ, namely

$$p(y) = \int p(y \mid \theta)p(\theta)d\theta.$$

From (1.2), the term $p(y)$ therefore acts as a normalising constant necessary to ensure $p(\theta \mid y)$ integrates to 1, and so one may write

$$p(\theta \mid y) = kp(y \mid \theta)p(\theta), \tag{1.3}$$

where $k=1/p(y)$ is an unknown constant. Alternatively stated, the posterior density (updated evidence) is proportional to the likelihood (data evidence) times the prior (historic evidence or elicited model assumptions). Taking logs in (1.3), one has

$$\log\left[p(\theta \mid y)\right] = \log(k) + \log\left[p(y \mid \theta)\right] + \log\left[p(\theta)\right]$$

and $\log[p(y \mid \theta)] + \log[p(\theta)]$ is generally referred to as the log posterior, which some R programs (e.g. rstan) allow to be directly specified as the estimation target.

In some cases, when the prior on θ is conjugate with the posterior on θ (i.e. has the same density form), the posterior density and marginal likelihood can be obtained analytically. When θ is low-dimensional, numerical integration is an alternative, and approximations to the required integrals can be used, such as the Laplace approximation (Raftery, 1996; Chen and Wang, 2011). In more complex applications, such approximations are not feasible, and integration to obtain $p(y)$ is intractable, so that direct sampling from $p(\theta \mid y)$ is not feasible. In such situations, MCMC methods provide a way to sample from $p(\theta \mid y)$ without it having a specific analytic form. They create a Markov chain of sampled values $\theta^{(1)},\ldots,\theta^{(T)}$ with transition kernel $K(\theta_{cand} \mid \theta_{curr})$ (investigating transitions from current to candidate values for parameters) that have $p(\theta \mid y)$ as their limiting distribution. Using large samples from the posterior distribution obtained by MCMC, one can estimate posterior quantities of interest such as posterior means, medians, and highest density regions (Hyndman, 1996; Chen and Shao, 1998).

1.3 MCMC Sampling in Relation to Monte Carlo Methods; Obtaining Posterior Inferences

Markov chain Monte Carlo (MCMC) methods are iterative sampling methods that can be encompassed within the broad class of Monte Carlo methods. However, MCMC methods must be distinguished from conventional Monte Carlo methods that generate independent simulations $\{u^{(1)}, u^{(2)} \ldots, u^{(T)}\}$ from a target density $\pi(u)$. From such simulations, the expectation of a function $g(u)$ under $\pi(u)$, namely

$$E_\pi[g(u)] = \int g(u)\pi(u)du,$$

is estimated as

$$\bar{g} = \sum_{t=1}^{T} g(u^{(t)})$$

and, under independent sampling from $\pi(u)$, $\bar{g}$ tends to $E_\pi[g(u)]$ as $T \to \infty$. However, such independent sampling from the posterior density $p(\theta|y)$ is not usually feasible.

When suitably implemented, MCMC methods offer an effective alternative way to generate samples from the joint posterior distribution, $p(\theta|y)$, but differ from conventional Monte Carlo methods in that successive sampled parameters are dependent or autocorrelated. The target density for MCMC samples is therefore the posterior density $\pi(\theta) = p(\theta|y)$ and MCMC sampling is especially relevant when the posterior cannot be stated exactly in analytic form e.g. when the prior density assumed for θ is not conjugate with the likelihood $p(y|\theta)$. The fact that successive sampled values are dependent means that larger samples are needed for equivalent precision, and the effective number of samples is less than the nominal number.

For the parameter sampling case, assume a preset initial parameter value $\theta^{(0)}$. Then MCMC methods involve repeated iterations to generate a correlated sequence of sampled values $\theta^{(t)}$ $(t = 1, 2, 3, \ldots)$, where updated values $\theta^{(t)}$ are drawn from a transition distribution

$$K(\theta^{(t)} | \theta^{(0)}, \ldots, \theta^{(t-1)}) = K(\theta^{(t)} | \theta^{(t-1)})$$

that is Markovian in the sense of depending only on $\theta^{(t-1)}$. The transition distribution $K(\theta^{(t)} | \theta^{(t-1)})$ is chosen to satisfy additional conditions ensuring that the sequence has the joint posterior density $p(\theta|y)$ as its stationary distribution. These conditions typically reduce to requirements on the proposal and acceptance procedure used to generate candidate parameter samples. The proposal density and acceptance rule must be specified in a way that guarantees irreducibility and positive recurrence; see, for example, Andrieu and Moulines (2006). Under such conditions, the sampled parameters $\theta^{(t)}$ $\{t = B, B+1, \ldots, T\}$, beyond a certain burn-in or warm-up phase in the sampling (of B iterations), can be viewed as a random sample from $p(\theta|y)$ (Roberts and Rosenthal, 2004).

In practice, MCMC methods are applied separately to individual parameters or blocks of more than one parameter (Roberts and Sahu, 1997). So, assuming θ contains more than one parameter and consists of C components or blocks $\{\theta_1, \ldots, \theta_C\}$, different updating methods may be used for each component, including block updates.

There is no limit to the number of samples T of θ which may be taken from the posterior density $p(\theta|y)$. Estimates of the marginal posterior densities for each parameter can be made from the MCMC samples, including estimates of location (e.g. posterior means, modes, or medians), together with the estimated certainty or precision of these parameters in terms of posterior standard deviations, credible intervals, or highest posterior density intervals. For example, the 95% credible interval for θ_h may be estimated using the 0.025 and 0.975 quantiles of the sampled output $\{\theta_h^{(t)}, t = B+1, \ldots, T\}$. To reduce irregularities in the histogram of sampled values for a particular parameter, a smooth form of the posterior density can be approximated by applying kernel density methods to the sampled values.

Monte Carlo posterior summaries typically include estimated posterior means and variances of the parameters, obtainable as moment estimates from the MCMC output, namely

$$\hat{E}(\theta_h) = \bar{\theta}_h = \sum_{t=B+1}^{T} \theta_h^{(t)}/(T-B)$$

$$\hat{V}(\theta_h) = \sum_{t=B+1}^{T} (\theta_h^{(t)} - \bar{\theta}_h)^2/(T-B).$$

This is equivalent to estimating the integrals

$$E(\theta_h \mid y) = \int \theta_h p(\theta \mid y) d\theta,$$

$$V(\theta_h \mid y) = \int \theta_h^2 p(\theta \mid y) d\theta - [E(\theta_h \mid y)]^2$$

$$= E(\theta_h^2 \mid y) - [E(\theta_h \mid y)]^2.$$

One may also use the MCMC output to derive obtain posterior means, variances, and credible intervals for functions $\Delta = \Delta(\theta)$ of the parameters (van Dyk, 2003). These are estimates of the integrals

$$E[\Delta(\theta) \mid y] = \int \Delta(\theta) p(\theta \mid y) d\theta,$$

$$V[\Delta(\theta) \mid y] = \int \Delta^2 p(\theta \mid y) d\theta - [E(\Delta \mid y)]^2$$

$$= E(\Delta^2 \mid y) - [E(\Delta \mid y)]^2.$$

For $\Delta(\theta)$, its posterior mean is obtained by calculating $\Delta^{(t)}$ at every MCMC iteration from the sampled values $\theta^{(t)}$. The theoretical justification for such estimates is provided by the MCMC version of the law of large numbers (Tierney, 1994), namely that

$$\sum_{t=B+1}^{T} \frac{\Delta[\theta^{(t)}]}{T-B} \to E_\pi[\Delta(\theta)],$$

provided that the expectation of $\Delta(\theta)$ under $\pi(\theta) = p(\theta \mid y)$, denoted $E_\pi[\Delta(\theta)]$, exists. MCMC methods also allow inferences on parameter comparisons (e.g. ranks of parameters or contrasts between them) (Marshall and Spiegelhalter, 1998).

1.4 Hierarchical Bayes Applications

The paradigm in Section 1.2 is appropriate to many problems, where uncertainty is limited to a few fundamental parameters, the number of which is independent of the sample size n – this is the case, for example, in a normal linear regression when the independent variables are known without error and the units are not hierarchically structured. However, in more complex data sets or with more complex forms of model or response, a more general perspective than that implied by (1.1)–(1.3) is available, and also implementable, using MCMC methods.

Thus, a class of hierarchical Bayesian models are defined by latent data (Paap, 2002; Clark and Gelfand, 2006) intermediate between the observed data and the underlying parameters (hyperparameters) driving the process. A terminology useful for relating hierarchical models to substantive issues is proposed by Wikle (2003) in which y defines the data stage, latent effects b define the process stage, and ξ defines the hyperparameter stage. For example, the observations $i=1,\ldots,n$ may be arranged in clusters $j=1, \ldots, J$, so that the observations can no longer be regarded as independent. Rather, subjects from the same cluster will tend to be more alike than individuals from different clusters, reflecting latent variables that induce dependence within clusters.

Let the parameters $\theta=[\theta_L,\theta_b]$ consist of parameter subsets relevant to the likelihood and to the latent data density respectively. The data are generally taken as independent of θ_b given b, so modelling intermediate latent effects involves a three-stage hierarchical Bayes (HB) prior set-up

$$p(y,b,\theta)=p(y\,|\,b,\theta_L)p(b\,|\,\theta_b)p(\theta_L,\theta_b), \tag{1.4}$$

with a first stage likelihood $p(y\,|\,b,\theta_L)$ and a second stage density $p(b|\theta_b)$ for the latent data, with conditioning on higher stage parameters θ. The first stage density $p(y|b,\theta_L)$ in (1.4) is a conditional likelihood, conditioning on b, and sometimes called the complete data or augmented data likelihood. The application of Bayes' theorem now specifies

$$p(\theta,b\,|\,y)=\frac{p(y\,|\,b,\theta_L)p(b\,|\,\theta_b)p(\theta)}{p(y)},$$

and the marginal posterior for θ may now be represented as

$$p(\theta\,|\,y)=\frac{p(\theta)p(y\,|\,\theta)}{p(y)}=\frac{p(\theta)\int p(y\,|\,b,\theta_L)p(b\,|\,\theta_b)db}{p(y)},$$

where

$$p(y\,|\,\theta)=\int p(y,b\,|\,\theta)db=\int p(y\,|\,b,\theta_L)p(b\,|\,\theta_b)db,$$

is the observed data likelihood, namely the complete data likelihood with b integrated out, sometimes also known as the integrated likelihood.

Often the latent data exist for every observation, or they may exist for each cluster in which the observations are structured (e.g. a school specific effect b_j for multilevel data y_{ij} on pupils i nested in schools j). The latent variables b can be seen as a population of values from an underlying density (e.g. varying log odds of disease) and the θ_b are then population hyperparameters (e.g. mean and variance of the log odds) (Dunson, 2001). As examples, Paap (2002) mentions unobserved states describing the business cycle and Johannes and Polson (2006) mention unobserved volatilities in stochastic volatility models, while Albert and Chib (1993) consider the missing or latent continuous data $\{b_1, \ldots, b_n\}$ which underlie binary observations $\{y_1, \ldots, y_n\}$. The subject specific latent traits in psychometric or educational item analysis can also be considered this way (Fox, 2010), as can the variance scaling factors in the robust Student t errors version of linear regression (Geweke, 1993) or subject specific slopes in a growth curve analysis of panel data on a collection of subjects (Oravecz and Muth, 2018).

Typically, the integrated likelihood $p(y|\theta)$ cannot be stated in closed form and classical likelihood estimation relies on numerical integration or simulation (Paap, 2002, p.15). By contrast, MCMC methods can be used to generate random samples indirectly from the posterior distribution $p(\theta,b|y)$ of parameters and latent data given the observations. This requires only that the augmented data likelihood be known in closed form, without needing to obtain the integrated likelihood $p(y|\theta)$. To see why, note that the marginal posterior of the parameter set θ may alternatively be derived as

$$p(\theta \mid y) = \int p(\theta, b \mid y) db = \int p(\theta \mid y, b) p(b \mid y) db,$$

with marginal densities for component parameters θ_h of the form (Paap, 2002, p.5)

$$p(\theta_h \mid y) = \int_{\theta[h]} \int_b p(\theta, b \mid y) db d\theta_{[h]},$$

$$\propto \int_{\theta[h]} p(\theta \mid y) p(\theta) d\theta_{[h]} = \int_{\theta[h]} \int_b p(\theta) p(y \mid b, \theta) p(b \mid \theta) db d\theta_{[h]},$$

where $\theta_{[h]}$ consists of all parameters in θ with the exception of θ_h. The derivation of suitable MCMC algorithms to sample from $p(\theta,b|y)$ is based on Clifford–Hammersley theorem, namely that any joint distribution can be fully characterised by its complete conditional distributions. In the hierarchical Bayes context, this implies that the conditionals $p(b|\theta,y)$ and $p(\theta|b,y)$ characterise the joint distribution $p(\theta,b|y)$ from which samples are sought, and so MCMC sampling can alternate between updates $p(b^{(t)} \mid \theta^{(t-1)}, y)$ and $p(\theta^{(t)} \mid b^{(t)}, y)$ on conditional densities, which are usually of simpler form than $p(\theta,b|y)$. The imputation of latent data in this way is sometimes known as data augmentation (van Dyk, 2003).

To illustrate the application of MCMC methods to parameter comparisons and hypothesis tests in an HB setting, Shen and Louis (1998) consider hierarchical models with unit or cluster specific parameters b_j, and show that if such parameters are the focus of interest, their posterior means are the optimal estimates. Suppose instead that the ranks of the unit or cluster parameters, namely

$$R_j = \text{rank}(b_j) = \sum_{k \neq i}^{n} I(b_j \geq b_k),$$

(where $I(A)$ is an indicator function which equals 1 when A is true, 0 otherwise) are required for deriving "league tables". Then the conditional expected ranks are optimal, and obtained by ranking the b_j at each MCMC iteration, and taking the means of these ranks over all samples. By contrast, ranking posterior means of the b_j themselves can perform poorly (Laird and Louis, 1989; Goldstein and Spiegelhalter, 1996). Similarly, when the empirical distribution function of the unit parameters (e.g. to be used to obtain the fraction of parameters above a threshold) is required, the conditional expected EDF is optimal.

A posterior probability estimate that a particular b_j exceeds a threshold τ, namely of the integral $Pr(b_j > \tau | y) = \int_{\tau}^{\infty} p(b_j | y) db_j$, is provided by the proportion of iterations where $b_j^{(t)}$ exceeds τ, namely

$$\widehat{Pr}(b_j > \tau | y) = \sum_{t=B+1}^{T} I(b_j^{(t)} > \tau)/(T-B).$$

Thus, one might, in an epidemiological application, wish to obtain the posterior probability that an area's smoothed relative mortality risk b_j exceeds unity, and so count iterations where this condition holds. If this probability exceeds a threshold such as 0.9, then a significant excess risk is indicated, whereas a low exceedance probability (the sampled relative risk rarely exceeded 1) would indicate a significantly low mortality level in the area.

In fact, the significance of individual random effects is one aspect of assessing the gain of a random effects model over a model involving only fixed effects, or of assessing whether a more complex random effects model offers a benefit over a simpler one (Knorr-Held and Rainer, 2001, p.116). Since the variance can be defined in terms of differences between elements of the vector $(b_1,...,b_J)$, as opposed to deviations from a central value, one may also consider which contrasts between pairs of b values are significant. Thus, Deely and Smith (1998) suggest evaluating probabilities $Pr(b_j \leq \tau b_k | k \neq j, y)$ where $0 < \tau \leq 1$, namely, the posterior probability that any one hierarchical effect is smaller by a factor τ than all the others.

1.5 Metropolis Sampling

A range of MCMC techniques is available. The Metropolis sampling algorithm is still a widely applied MCMC algorithm and is a special case of Metropolis–Hastings considered in Section 1.8. Let $p(y|\theta)$ denote a likelihood, and $p(\theta)$ denote the prior density for θ, or more specifically the prior densities $p(\theta_1),...p(\theta_C)$ of the components of θ. Then the Metropolis algorithm involves a symmetric proposal density (e.g. a Normal, Student t, or uniform density) $q(\theta_{\text{cand}} | \theta^{(t)})$ for generating candidate parameter values θ_{cand}, with acceptance probability for potential candidate values obtained as

$$\alpha^{(t)} = \min\left(1, \frac{\pi(\theta_{\text{cand}})}{\pi(\theta^{(t)})}\right) = \min\left(1, \frac{p(\theta_{\text{cand}} | y)}{p(\theta^{(t)} | y)}\right) = \min\left(1, \frac{p(y | \theta_{\text{cand}})p(\theta_{\text{cand}})}{p(y | \theta^{(t)})p(\theta^{(t)})}\right). \quad (1.5)$$

So one compares the (likelihood * prior), namely, $p(y | \theta)p(\theta)$, for the candidate and existing parameter values. If the (likelihood * prior) is higher for the candidate value, it is automatically accepted, and $\theta^{(t+1)} = \theta_{\text{cand}}$. However, even if the (likelihood * prior) is lower for the candidate value, such that $\alpha^{(t)}$ is less than 1, the candidate value may still be accepted. This is decided by random sampling from a uniform density, $U^{(t)}$ and the candidate value is accepted if $\alpha^{(t)} \geq U^{(t)}$. In practice, comparisons involve the log posteriors for existing and candidate parameter values.

The third equality in (1.5) follows because the marginal likelihood $p(y) = 1/k$ in the Bayesian formula

$$p(\theta | y) = p(y | \theta)p(\theta) / p(y) = kp(y | \theta)p(\theta),$$

cancels out, as it is a constant. Stated more completely, to sample parameters under the Metropolis algorithm, it is not necessary to know the normalised target distribution, namely, the posterior density, $\pi(\theta|y)$; it is enough to know it up to a constant factor.

So, for updating parameter subsets, the Metropolis algorithm can be implemented by using the full posterior distribution

$$\pi(\theta) = p(\theta | y) = kp(y | \theta)p(\theta),$$

as the target distribution – which in practice involves comparisons of the unnormalised posterior $p(y|\theta)p(\theta)$. However, for updating values on a particular parameter θ_h, it is not just $p(y)$ that cancels out in the ratio

$$\pi(\theta_{\text{cand}})/\pi(\theta^{(t)}) = \frac{p(y | \theta_{\text{cand}})p(\theta_{\text{cand}})}{p(y | \theta^{(t)})p(\theta^{(t)})},$$

but any parts of the likelihood or prior not involving θ_h (these parts are constants when θ_h is being updated).

When those parts of the likelihood or prior not relevant to θ_h are abstracted out, the remaining part of $p(\theta | y) = kp(y | \theta)p(\theta)$, the part relevant to updating θ_h, is known as the full conditional density for θ_h (Gilks, 1996). One may denote the full conditional density for θ_h as

$$\pi_h(\theta_h | \theta_{[h]}) \propto p(y | \theta_h)p(\theta_h),$$

where $\theta_{[h]}$ denotes the parameter set excluding θ_h. So, the probability for updating θ_h can be obtained *either* by comparing the full posterior (known up to a constant k), namely

$$\alpha = \min\left(1, \frac{\pi(\theta_{h,\text{cand}}, \theta_{[h]}^{(t)})}{\pi(\theta^{(t)})}\right) = \min\left(1, \frac{p(y | \theta_{h,\text{cand}}, \theta_{[h]}^{(t)})p(\theta_{h,\text{cand}}, \theta_{[h]}^{(t)})}{p(y | \theta^{(t)})p(\theta^{(t)})}\right),$$

or by using the full conditional for the hth parameter, namely

$$\alpha = \min\left(1, \frac{\pi_h(\theta_{h,\text{cand}} | \theta_{[h]}^{(t)})}{\pi_h(\theta_h^{(t)} | \theta_{[h]}^{(t)})}\right).$$

Then one sets $\theta_h^{(t+1)} = \theta_{h,\text{cand}}$ with probability α, and $\theta_h^{(t+1)} = \theta_h^{(t)}$ otherwise.

1.6 Choice of Proposal Density

There is some flexibility in the choice of proposal density q for generating candidate values in the Metropolis and other MCMC algorithms, but the chosen density and the parameters incorporated in it are relevant to successful MCMC updating and convergence (Altaleb and Chauveau, 2002; Robert, 2015). A standard recommendation is that the proposal density for a particular parameter θ_h should approximate the posterior density $p(\theta_h|y)$ of that parameter. In some cases, one may have an idea (e.g. from a classical analysis) of what the posterior density is, or what its main defining parameters are. A normal proposal is often justified, as many posterior densities do approximate normality. For example, Albert (2007) applies a Laplace approximation technique to estimate the posterior mode, and uses the mean and variance parameters to define the proposal densities used in a subsequent stage of Metropolis–Hastings sampling.

The rate at which a proposal generated by q is accepted (the acceptance rate) depends on how close θ_{cand} is to $\theta^{(t)}$, and this in turn depends on the variance σ_q^2 of the proposal density. A higher acceptance rate would typically follow from reducing σ_q^2, but with the risk that the posterior density will take longer to explore. If the acceptance rate is too high, then autocorrelation in sampled values will be excessive (since the chain tends to move in a restricted space), while a too low acceptance rate leads to the same problem, since the chain then gets locked at particular values.

One possibility is to use a variance or dispersion estimate, σ_m^2 or Σ_m, from a maximum likelihood or other mode-finding analysis (which approximates the posterior variance) and then scale this by a constant $c > 1$, so that the proposal density variance is $\sigma_q^2 = c\sigma_m^2$. Values of c in the range 2–10 are typical. For θ_h of dimension d_h with covariance Σ_m, a proposal density dispersion $2.38^2\Sigma_m/d_h$ is shown as optimal in random walk schemes (Roberts et al., 1997). Working rules are for an acceptance rate of 0.4 when a parameter is updated singly (e.g. by separate univariate normal proposals), and 0.2 when a group of parameters are updated simultaneously as a block (e.g. by a multivariate normal proposal). Geyer and Thompson (1995) suggest acceptance rates should be between 0.2 and 0.4, and optimal acceptance rates have been proposed (Roberts et al., 1997; Bedard, 2008).

Typical Metropolis updating schemes use variables W_t with known scale, for example, uniform, standard Normal, or standard Student t. A Normal proposal density $q(\theta_{\text{cand}} \mid \theta^{(t)})$ then involves samples $W_t \sim N(0,1)$, with candidate values

$$\theta_{\text{cand}} = \theta^{(t)} + \sigma_q W_t,$$

where σ_q determines the size of the jump from the current value (and the acceptance rate). A uniform random walk samples $W_t \sim \text{Unif}(-1,1)$ and scales this to form a proposal $\theta_{\text{cand}} = \theta^{(t)} + \kappa W_t$, with the value of κ determining the acceptance rate. As noted above, it is desirable that the proposal density approximately matches the shape of the target density $p(\theta \mid y)$. The Langevin random walk scheme is an example of a scheme including information about the shape of $p(\theta \mid y)$ in the proposal, namely $\theta_{\text{cand}} = \theta^{(t)} + \sigma_q[W_t + 0.5\nabla \log(p(\theta^{(t)} \mid y)]$ where ∇ denotes the gradient function (Roberts and Tweedie, 1996).

Sometimes candidate parameter values are sampled using a transformed version of a parameter, for example, normal sampling of a log variance rather than sampling of a variance (which has to be restricted to positive values). In this case, an appropriate Jacobean adjustment must be included in the likelihood. Example 1.2 below illustrates this.

1.7 Obtaining Full Conditional Densities

As noted above, Metropolis sampling may be based on the full conditional density when a particular parameter θ_h is being updated. These full conditionals are particularly central in Gibbs sampling (see below). The full conditional densities may be obtained from the joint density $p(\theta, y) = p(y \mid \theta)p(\theta)$ and in many cases reduce to standard densities (Normal, exponential, gamma, etc.) from which direct sampling is straightforward. Full conditional densities are derived by abstracting out from the joint model density $p(y|\theta)p(\theta)$ (likelihood times prior) only those elements including θ_h and treating other components as constants (George et al., 1993; Gilks, 1996).

Consider a conjugate model for Poisson count data y_i with means μ_i that are themselves gamma-distributed; this is a model appropriate for overdispersed count data with actual variability var(y) exceeding that under the Poisson model (Molenberghs et al., 2007). Suppose the second stage prior is $\mu_i \sim Ga(\alpha,\beta)$, namely,

$$p(\mu_i|\alpha,\beta) = \mu_i^{\alpha-1}e^{-\beta\mu_i}\beta^{\alpha}/\Gamma(\alpha),$$

and further that $\alpha \sim E(A)$ (namely, α is exponential with parameter A), and $\beta \sim$ Ga(B,C) where A, B, and C are preset constants. So the posterior density $p(\theta|y)$ of $\theta = (\mu_1,..\mu_n,\alpha,\beta)$, given y, is proportional to

$$e^{-A\alpha}\beta^{B-1}e^{-C\beta}\left\{\prod_i e^{-\mu_i}\mu_i^{y_i}\right\}\left[\beta^{\alpha}/\Gamma(\alpha)\right]^n\left\{\prod_i \mu_i^{\alpha-1}e^{-\beta\mu_i}\right\} \tag{1.6}$$

where all constants (such as the denominator $y_i!$ in the Poisson likelihood, as well as the inverse marginal likelihood k) are combined in a proportionality constant.

It is apparent from inspecting (1.6) that the full conditional densities of μ_i and β are also gamma, namely,

$$\mu_i \sim Ga(y_i+\alpha, \beta+1),$$

and

$$\beta \sim Ga\left(B+n\alpha, C+\sum_i \mu_i\right),$$

respectively. The full conditional density of α, also obtained from inspecting (1.6), is

$$p(\alpha|y,\beta,\mu) \propto e^{-A\alpha}\left[\beta^{\alpha}/\Gamma(\alpha)\right]^n\left\{\prod_i \mu_i^{\alpha-1}\right\}.$$

This density is non-standard and cannot be sampled directly (as can the gamma densities for μ_i and β). Hence, a Metropolis or Metropolis–Hastings step can be used for updating it.

Example 1.1 Estimating Normal Density Parameters via Metropolis

To illustrate Metropolis sampling in practice using symmetric proposal densities, consider $n=1000$ values y_i generated randomly from a N(3,25) distribution, namely a Normal with mean $\mu=3$ and variance $\sigma^2=25$. Note that, for the particular set.seed used, the average sampled y_i is 2.87 with variance 24.87. Using the generated y, we seek to estimate the mean and variance, now treating them as unknowns. Setting $\theta=(\mu,\sigma^2)$, the likelihood is

$$p(y|\theta) = \prod_{i=1}^{n}\frac{1}{\sigma\sqrt{2\pi}}\exp\left(-\frac{(y_i-\mu)^2}{2\sigma^2}\right).$$

Assume a flat prior for μ, and a prior $p(\sigma) \propto 1/\sigma$ on σ; this is a form of noninformative prior (see Albert, 2007, p.109). Then one has posterior density

$$p(\theta | y) \propto \frac{1}{\sigma^{n+1}} \prod_{i=1}^{n} \exp\left(-\frac{(y_i - \mu)^2}{2\sigma^2}\right).$$

with the marginal likelihood and other constants incorporated in the proportionality sign.

Parameter sampling via the Metropolis algorithm involves σ rather than σ^2, and uniform proposals. Thus, assume uniform $U(-\kappa,\kappa)$ proposal densities around the current parameter values $\mu^{(t)}$ and $\sigma^{(t)}$, with $\kappa=0.5$ for both parameters. The absolute value of $\sigma^{(t)} + U(-\kappa,\kappa)$ is used to generate σ_{cand}. Note that varying the lower and upper limit of the uniform sampling (e.g. taking $\kappa=1$ or $\kappa=0.25$) may considerably affect the acceptance rates.

An R code for $\kappa=0.5$ is in the Computational Notes [1] in Section 1.14, and uses the full posterior density (rather than the full conditional for each parameter) as the target density for assessing candidate values. In the acceptance step, the log of the ratio $\frac{p(y | \theta_{\text{cand}})p(\theta_{\text{cand}})}{p(y | \theta^{(t)})p(\theta^{(t)})}$ is compared to the log of a random uniform value to avoid computer over/underflow. With $T=10000$ and $B = 1000$ warmup iterations, acceptance rates for the proposals of μ and σ are 48% and 35% respectively, with posterior means 2.87 and 4.99. Other posterior summary tools (e.g. univariate and bivariate kernel density plots, effective sample sizes) are included in the R code (see Figure 1.1 for a plot of the posterior bivariate density). Also included is a posterior probability calculation to assess $Pr(\mu<3|y)$, with result 0.80, and a command for a plot of the changing posterior expectation for μ over the iterations. The code uses the full normal likelihood, via the dnorm function in R.

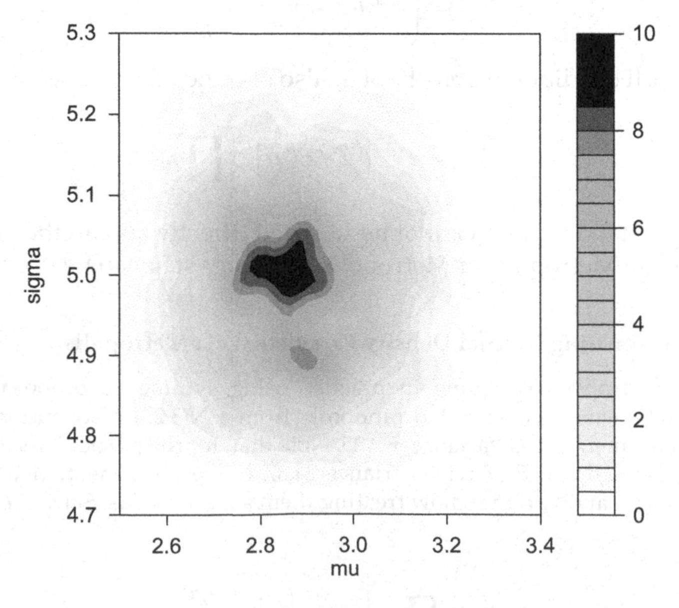

FIGURE 1.1
Bivariate density plot, normal density parameters.

Example 1.2 Extended Logistic with Metropolis Sampling

Following Carlin and Gelfand (1991), consider an extended logistic model for beetle mortality data, involving death rates π_i at exposure dose w_i. Thus, for deaths y_i at six dose points, one has

$$y_i \sim \text{Bin}(n_i, \pi(w_i)),$$

$$\pi(w_i) = [\exp(z_i)/(1+\exp(z_i)]^{m_1},$$

$$z_i = (w_i - \mu)/\sigma,$$

where m_1 and σ are both positive. To simplify notation, one may write $V = \sigma^2$.

Consider Metropolis sampling involving log transforms of m_1 and V, and separate univariate normal proposals in a Metropolis scheme. Jacobian adjustments are needed in the posterior density to account for the two transformed parameters. The full posterior $p(\mu, m_1, V \mid y)$ is proportional to

$$p(m_1)p(\mu)p(V)\prod_i [\pi(w_i)]^{y_i}(1-\pi(w_i)]^{n_i - y_i}$$

where $p(\mu)$, $p(m_1)$ and $p(V)$ are priors for μ, m_1 and V. Suppose the priors $p(m_1)$ and $p(\mu)$ are as follows:

$$m_1 \sim Ga(a_0, b_0),$$

$$\mu \sim N(c_0, d_0^2),$$

where the gamma has the form

$$Ga(x \mid \alpha, \beta) = \frac{\beta^\alpha}{\Gamma(\alpha)} x^{\alpha-1} e^{-\beta x}.$$

Also, for $p(V)$ assume

$$V \sim IG(e_0, f_0),$$

where the inverse gamma has the form

$$IG(x \mid \alpha, \beta) = \frac{\beta^\alpha}{\Gamma(\alpha)} x^{-(\alpha+1)} e^{-\beta/x}.$$

The parameters $(a_0, b_0, c_0, d_0, e_0, f_0)$ are preset. The posterior is then proportional to

$$(m_1^{a_0-1}e^{-b_0 m_1})\exp\left(-0.5\left[\frac{\mu - c_0}{d_0}\right]^2\right)V^{-(e_0+1)}e^{-f_0/V}\prod_i [\pi(w_i)]^{y_i}(1-\pi(w_i)]^{n_i-y_i}.$$

Suppose the likelihood is re-specified in terms of parameters $\theta_1 = \mu, \theta_2 = \log(m_1)$ and $\theta_3 = \log(V)$. Then the full posterior in terms of the transformed parameters is proportional to

$$\left(\frac{\partial m_1}{\partial \theta_2}\right)\left(\frac{\partial V}{\partial \theta_3}\right)p(\mu)p(m_1)p(V)\prod_i [\pi(w_i)]^{y_i}(1-\pi(w_i)]^{n_i-y_i}.$$

One has $(\partial m_1/\partial\theta_2)=e^{\theta_2}=m_1$ and $(\partial V/\partial\theta_3)=e^{\theta_3}=V$. So, taking account of the parameterisation $(\theta_1,\theta_2,\theta_3)$, the posterior density is proportional to

$$(m_1^{a_0}e^{-b_0m_1})\exp\left(-0.5\left[\frac{\mu-c_0}{d_0}\right]^2\right)V^{-e_0}e^{-f_0/V}\prod_i[\pi(w_i)]^{y_i}(1-\pi(w_i))^{n_i-y_i}.$$

The R code (see Section 1.14 Computational Notes [2]) assumes initial values for $\mu=\theta_1$ of 1.8, for $\theta_2=\log(m_1)$ of 0, and for $\theta_3=\log(V)$ of 0. Preset parameters in the prior densities are ($a_0=0.25$, $b_0=0.25$, $c_0=2$, $d_0=10$, $e_0=2.000004$, $f_0=0.001$). Two chains are run with $T=100000$, with inferences based on the last 50,000 iterations. Standard deviations in the respective normal proposal densities are set at 0.01, 0.2, and 0.4. Metropolis updates involve comparisons of the log posterior and logs of uniform random variables $\{U_h^{(t)},h=1,\ldots,3\}$.

Posterior medians (and 95% intervals) for $\{\mu,m_1,V\}$ are obtained as 1.81 (1.78, 1.83), 0.36 (0.20,0.75), 0.00035 (0.00017, 0.00074) with acceptance rates of 0.41, 0.43, and 0.43. The posterior estimates are similar to those of Carlin and Gelfand (1991). Despite satisfactory convergence according to Gelman–Rubin scale reduction factors, estimation is beset by high posterior correlations between parameters and low effective sample sizes. The cross-correlations between the three hyperparameters exceed 0.75 in absolute terms, effective sample sizes are under 1000, and first lag sampling autocorrelations all exceed 0.90.

It is of interest to apply rstan (and hence HMC) to this dataset (Section 1.10) (see Section 1.14 Computational Notes [3]). Inferences from rstan differ from those from Metropolis sampling estimation, though are sensitive to priors adopted. In a particular rstan estimation, normal priors are set on the hyperparameters as follows:

$$\mu\sim N(2,10),$$

$$\log(m_1)\sim N(0,1),$$

$$\log(\sigma)\sim N(0,5).$$

Two chains are applied with 2500 iterations and 250 warm-up. While estimates for μ are similar to the preceding analysis, the posterior median (95% intervals) for m_1 is now 1.21 (0.21, 6.58), with the 95% interval straddling the default unity value. The estimate for the variance V is lower. As to MCMC diagnostics, effective sample sizes for μ and m_1 are larger than from the Metropolis analysis, absolute cross-correlations between the three hyperparameters in the MCMC sampling are all under 0.40 (see Figure 1.2), and first lag sampling autocorrelations are all under 0.60.

1.8 Metropolis–Hastings Sampling

The Metropolis–Hastings (M–H) algorithm is the overarching algorithm for MCMC schemes that simulate a Markov chain $\theta^{(t)}$ with $p(\theta|y)$ as its stationary distribution. Following Hastings (1970), the chain is updated from $\theta^{(t)}$ to θ_{cand} with probability

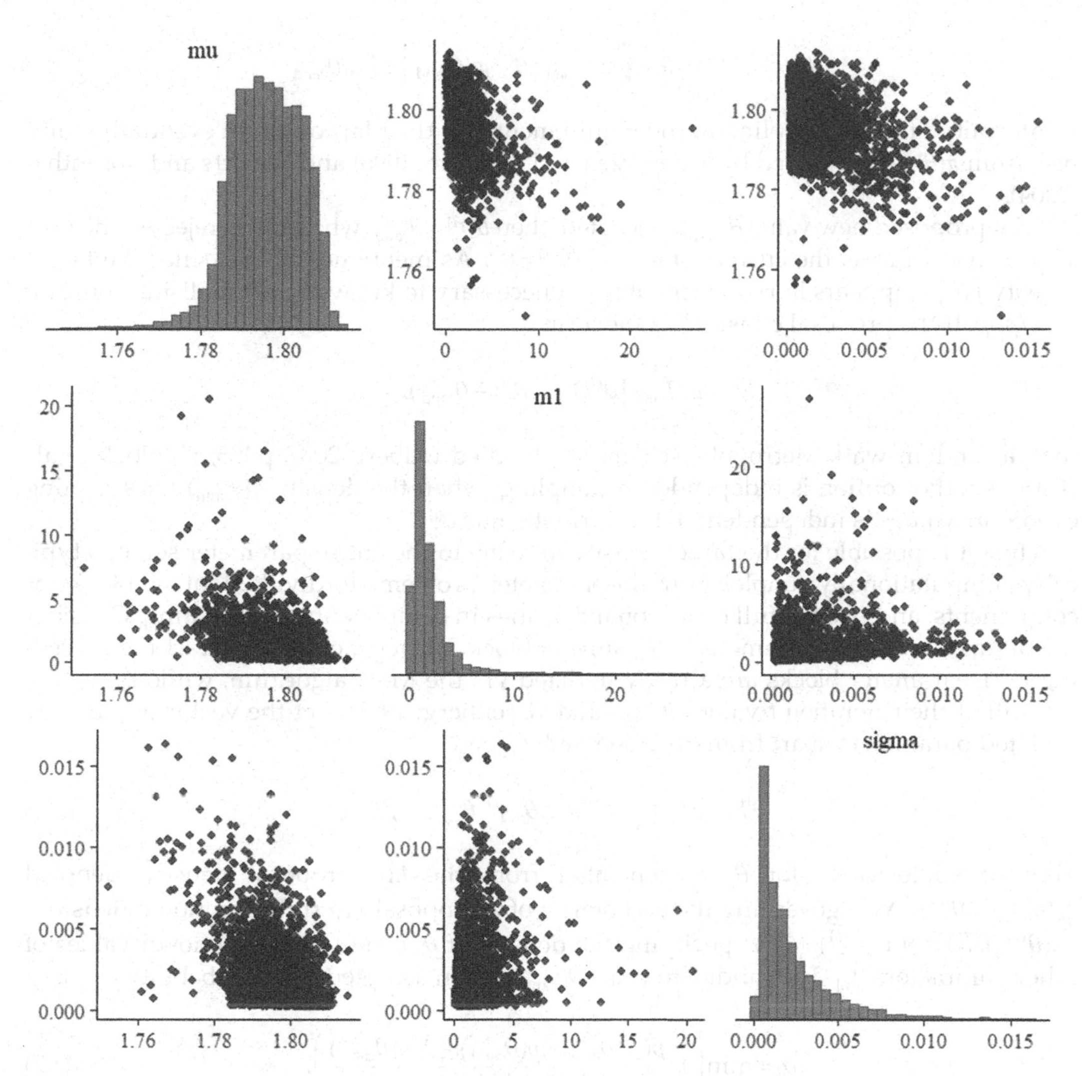

FIGURE 1.2
Posterior densities and MCMC cross-correlations, rstan estimation of beetle mortality data.

$$\alpha(\theta_{\text{cand}} \mid \theta^{(t)}) = \min\left(1, \frac{p(\theta_{\text{cand}} \mid y)q(\theta^{(t)} \mid \theta_{\text{cand}})}{p(\theta^{(t)} \mid y)q(\theta_{\text{cand}} \mid \theta^{(t)})}\right),$$

where the proposal density q (Chib and Greenberg, 1995) may be non-symmetric, so that $q(\theta_{\text{cand}} \mid \theta^{(t)})$ does not necessarily equal $q(\theta^{(t)} \mid \theta_{\text{cand}})$. $q(\theta_{\text{cand}} \mid \theta^{(t)})$ is the probability (or density ordinate) of θ_{cand} for a density centred at $\theta^{(t)}$, while $q(\theta^{(t)} \mid \theta_{\text{cand}})$ is the probability of moving back from θ_{cand} to the current value. If the proposal density is symmetric, with $q(\theta_{\text{cand}} \mid \theta^{(t)}) = q(\theta^{(t)} \mid \theta_{\text{cand}})$, then the Metropolis–Hastings algorithm reduces to the Metropolis algorithm discussed above. The M–H transition kernel is

$$K(\theta_{\text{cand}} \mid \theta^{(t)}) = \alpha(\theta_{\text{cand}} \mid \theta^{(t)})q(\theta_{\text{cand}} \mid \theta^{(t)}),$$

for $\theta_{\text{cand}} \neq \theta^{(t)}$, with a nonzero probability of staying in the current state, namely

$$K(\theta^{(t)} \mid \theta^{(t)}) = 1 - \int \alpha(\theta_{\text{cand}} \mid \theta^{(t)}) q(\theta_{\text{cand}} \mid \theta^{(t)}) d\theta_{\text{cand}}.$$

Conformity of M–H sampling to the requirement that the Markov chain eventually samples from $\pi(\theta)$ is considered by Mengersen and Tweedie (1996) and Roberts and Rosenthal (2004).

If the proposed new value θ_{cand} is accepted, then $\theta^{(t+1)} = \theta_{\text{cand}}$, while if it is rejected the next state is the same as the current state, i.e. $\theta^{(t+1)} = \theta^{(t)}$. As mentioned above, since the target density $p(\theta \mid y)$ appears in ratio form, it is not necessary to know the normalising constant $k = 1/p(y)$. If the proposal density has the form

$$q(\theta_{\text{cand}} \mid \theta^{(t)}) = q(\theta^{(t)} - \theta_{\text{cand}}),$$

then a random walk Metropolis scheme is obtained (Albert, 2007, p.105; Sherlock et al., 2010). Another option is independence sampling, when the density $q(\theta_{\text{cand}})$ for sampling candidate values is independent of the current value $\theta^{(t)}$.

While it is possible for the target density to relate to the entire parameter set, it is typically computationally simpler in multi-parameter problems to divide θ into C blocks or components, and use the full conditional densities in componentwise updating. Consider the update for the hth parameter or parameter block. At step h of iteration $t+1$ the preceding $h-1$ parameter blocks are already updated via the M–H algorithm, while $\theta_{h+1}, \ldots, \theta_C$ are still at their iteration t values (Chib and Greenberg, 1995). Let the vector of partially updated parameters apart from θ_h be denoted

$$\theta_{[h]}^{(t)} = (\theta_1^{(t+1)}, \theta_2^{(t+1)}, \ldots, \theta_{h-1}^{(t+1)}, \theta_{h+1}^{(t)}, \ldots, \theta_C^{(t)}),$$

The candidate value for θ_h is generated from the hth proposal density, denoted $q_h(\theta_{h,\text{cand}} \mid \theta_h^{(t)})$. Also governing the acceptance of a proposal are full conditional densities $\pi_h(\theta_h^{(t)} \mid \theta_{[h]}^{(t)}) \propto p(y \mid \theta_h^{(t)}) p(\theta_h^{(t)})$ specifying the density of θ_h conditional on known values of other parameters $\theta_{[h]}$. The candidate value $\theta_{h,\text{cand}}$ is then accepted with probability

$$\alpha = \min\left(1, \frac{p(y \mid \theta_{h,\text{cand}}) p(\theta_{\text{cand}}) q(\theta_h^{(t)} \mid \theta_{\text{cand}})}{p(y \mid \theta_h^{(t)}) p(\theta_h^{(t)}) q(\theta_{\text{cand}} \mid \theta_h^{(t)})}\right). \tag{1.7}$$

Example 1.3 Normal Random Effects in a Hierarchical Binary Regression

To exemplify a hierarchical Bayes model involving a three-stage prior, consider binary data $y_i \sim \text{Bern}(p_i)$ from Sinharay and Stern (2005) on survival or otherwise of n = 244 newborn turtles arranged in $J = 31$ clutches, numbered in increasing order of the average birthweight of the turtles. A known predictor is turtle birthweight x_i. Let C_i denote the clutch that turtle i belongs to. Then to allow for varying clutch effects, one may specify, for cluster $j = C_i$, a probit regression with

$$p_i \mid b_j = \Phi(\beta_1 + \beta_2 x_i + b_j),$$

where $\{b_j \sim N(0, 1/\tau_b), j = 1, \ldots, J\}$. It is assumed that $\beta_k \sim N(0,10)$ and $\tau_b \sim Ga(1, 0.001)$.

A Metropolis–Hastings step involving a gamma proposal is used for the random effects precision τ_b, and Metropolis updates for other parameters; see Section 1.14 Computational Notes [3]. Trial runs suggest τ_b is approximately between 5 and 10, and a gamma proposal $Ga(\kappa, \kappa/\tau_{b,\text{curr}})$ with $\kappa=100$ is adopted (reducing κ will reduce the M–H acceptance rate for τ_b).

A run of $T=5000$ iterations with warm-up $B=500$ provides posterior medians (95% intervals) for $\{\beta_1, \beta_2, \sigma_b = 1/\sqrt{\tau_b}\}$ of −2.91 (−3.79, −2.11), 0.40 (0.28, 0.54), and 0.27 (0.20, 0.43), and acceptance rates for $\{\beta_1, \beta_2, \tau_b\}$ of 0.30, 0.21, and 0.24. Acceptance rates for the clutch random effects (using normal proposals with standard deviation 1) are between 0.25 and 0.33. However, none of the clutch effects appears to be strongly significant, in the sense of entirely positive or negative 95% credible intervals. The effect b_9 (for the clutch with lowest average birthweight) has posterior median and 95% interval, 0.36 (−0.07, 0.87), and is the closest to being significant, while for b_{15} the median (95%CRI) is −0.30 (−0.77,0.10).

1.9 Gibbs Sampling

The Gibbs sampler (Gelfand and Smith, 1990; Gilks et al., 1993; Chib, 2001) is a special componentwise M–H algorithm whereby the proposal density q for updating θ_h equals the full conditional $\pi_h(\theta_h \mid \theta_{[h]}) \propto p(y \mid \theta_h)p(\theta_h)$. It follows from (1.7) that proposals are accepted with probability 1. If it is possible to update all blocks this way, then the Gibbs sampler involves parameter block by parameter block updating which, when completed, forms the transition from $\theta^{(t)} = (\theta_1^{(t)}, \ldots, \theta_C^{(t)})$ to $\theta^{(t+1)} = (\theta_1^{(t+1)}, \ldots, \theta_C^{(t+1)})$. The most common sequence used is

1. $\theta_1^{(t+1)} \sim f_1(\theta_1 \mid \theta_2^{(t)}, \theta_3^{(t)}, \ldots, \theta_C^{(t)})$;
2. $\theta_2^{(t+1)} \sim f_2(\theta_2 \mid \theta_1^{(t+1)}, \theta_3^{(t)}, \ldots, \theta_C^{(t)})$;

   ...

3. $\theta_C^{(t+1)} \sim f_C(\theta_C \mid \theta_1^{(t+1)}, \theta_2^{(t+1)}, \ldots, \theta_{C-1}^{(t+1)})$.

While this scanning scheme is the usual one for Gibbs sampling, there are other options, such as the random permutation scan (Roberts and Sahu, 1997) and the reversible Gibbs sampler which updates blocks 1 to C, and then updates in reverse order.

Example 1.4 Gibbs Sampling Example Schools Data Meta Analysis

Consider the schools data from Gelman et al. (2014), consisting of point estimates y_j ($j=1, \ldots, J$) of unknown effects θ_j, where each y_j has a known design variance σ_j^2 (though the listed data provides σ_j, not σ_j^2). The first stage of a hierarchical normal model assumes

$$y_j \sim N(\theta_j, \sigma_j^2),$$

and the second stage specifies a normal model for the latent θ_j,

$$\theta_j \sim N(\mu, \tau^2).$$

The full conditionals for the latent effects θ_j, namely $p(\theta_j \mid y, \mu, \tau^2)$ are as specified by Gelman et al. (2014, p.116). Assuming a flat prior on μ, and that the precision $1/\tau^2$ has a $Ga(a,b)$ gamma prior, then the full conditional for μ is $N(\bar{\theta}, \tau^2/J)$, and that for $1/\tau^2$ is gamma with parameters $(J/2 + a, 0.5\sum_j (\theta_j - \mu)^2 + b)$.

TABLE 1.1

Schools Normal Meta-Analysis Posterior Summary

	μ	τ	ϑ_1	ϑ_2	ϑ_3	ϑ_4	ϑ_5	ϑ_6	ϑ_7	ϑ_8
Mean	8.0	2.5	9.0	8.0	7.6	8.0	7.1	7.5	8.8	8.1
St devn	4.4	2.8	5.6	4.9	5.4	5.1	5.0	5.2	5.2	5.4

For the R application, the setting $a=b=0.1$ is used in the prior for $1/\tau^2$. Starting values for μ and τ^2 in the MCMC analysis are provided by the mean of the y_j and the median of the σ_j^2. A single run of $T=20000$ samples (see Section 1.13 Computational Notes [4]) provides the posterior means and standard deviations shown in Table 1.1.

1.10 Hamiltonian Monte Carlo

The Hamiltonian Monte Carlo (HMC) algorithm is implemented in the rstan library in R (see Chapter 2), and has been demonstrated to improve effective search of the posterior parameter space. Inefficient random walk behaviour and delayed convergence that may characterise other MCMC algorithms is avoided by a greater flexibility in proposing new parameter values; see Neal (2011, section 5.3.3.3), Gelman et al. (2014), Monnahan et al. (2017), and Robert et al. (2018). In HMC, an auxiliary momentum vector ϕ is introduced with the same dimension $D=\dim(\theta)$ as the parameter vector θ. HMC then involves an alternation between two forms of updating. One updates the momentum vector leaving θ unchanged. The other updates both θ and ϕ using Hamiltonian dynamics as determined by the Hamiltonian

$$H(\theta,\phi)=U(\theta)+K(\phi),$$

where $U(\theta)=-\log[p(y\,|\,\theta)p(\theta)]$ (the negative log posterior) defines potential energy, and $K(\phi)=\sum_{d=1}^{D}\theta_d^2/m_d$ defines kinetic energy (Neal, 2011, section 5.2). Updates of the momentum variable include updates based on the gradients of $U(\theta)$,

$$g_d(\theta)=\frac{dU(\theta)}{d\theta_d},$$

with $g(\theta)$ denoting the vector of gradients.

For iterations $t=1,\ldots,T$, the updating sequence is as follows:

1. sample $\phi^{(t)}$ from $N(0,I)$, where I is diagonal with dimension D;
2. relabel $\phi^{(t)}$ as ϕ_0, and $\theta^{(t)}$ as θ_0 and with stepsize ε, carry out L "leapfrog" steps, starting from $i=0$
   a) $\phi_{i+0.5}=\phi_i-0.5\varepsilon g(\theta_i)$
   b) $\theta_{d,i+1}=\theta_i+\varepsilon\phi_{i+0.5}/m_d$
   c) $\phi_{i+1}=\phi_{i+0.5}-0.5\varepsilon g(\theta_i)$;
3. set candidate parameter and momentum variables as $\theta^*=\theta_L$ and $\theta^*=\theta_L$;

4. obtain the potential and kinetic energies $U(\theta^*)$ and $K(\phi^*)$;
5. accept the candidate values with probability min(1,r) where

$$\log(r) = U(\theta^{(t)}) + K(\phi^{(t)}) - U(\theta^*) - K(\phi^*).$$

Practical application of HMC is facilitated by the No U-Turn Sampler (NUTS) (Hoffman and Gelman, 2014) which provides an adaptive way to adjust the stepsize ε, and the number of leapfrog steps L. The No U-Turn Sampler seeks to avoid HMC making backwards sampling trajectories that get closer to (and hence more correlated) with the last sample position. Calculation of the gradient of the log posterior is part of the NUTS implementation, and is facilitated by reverse-mode algorithmic differentiation (Carpenter et al., 2017).

1.11 Latent Gaussian Models

Latent Gaussian models are a particular variant of the models considered in Section 1.4, and can be represented as a hierarchical structure containing three stages. At the first stage is a conditionally independent likelihood function

$$p(y \mid x, \phi),$$

with a response y (of length n) conditional on a latent field x (usually also of length n), depending on hyperparameters θ, with sparse precision matrix Q_θ, and with ϕ denoting other parameters relevant to the observation model. The hierarchical model is then

$$y_i \mid x_i \sim p(y_i \mid x_i, \phi),$$

$$x_i \mid \theta \sim \pi(x \mid \theta) = N(., Q_\theta^{-1}),$$

$$\theta, \phi \sim \pi(\theta)\pi(\phi),$$

with posterior density

$$\pi(x, \theta, \phi \mid y) \propto \pi(\theta)\pi(\phi)\pi(x \mid \theta)\prod_i p(y_i \mid x_i, \phi).$$

For example, consider area disease counts, $y_i \sim \text{Poisson}(E_i \eta_i)$, with

$$\log(\eta_i) = \mu + u_i + s_i,$$

where $u_i \sim N(0, \sigma_u^2)$, the s_i follow an intrinsic autoregressive prior (expressing spatial dependence) with variance σ_s^2, and $s \sim \text{ICAR}(\sigma_s^2)$ and u_i are iid (independent and identically distributed) random errors. Then $x = (\eta, u, s)$ is jointly Gaussian with hyperparameters $(\mu, \sigma_s^2, \sigma_u^2)$.

Integrated nested Laplace approximation (or INLA) is a deterministic algorithm, unlike stochastic algorithms such as MCMC, designed for estimating latent Gaussian models. The algorithm is implemented in the R-INLA package, which uses R syntax throughout. For large samples (over 5,000, say), it provides an effective alternative to MCMC estimation, but with similar posterior outputs available.

The INLA algorithm focuses on the posterior density of the hyperparameters, $\pi(\theta|y)$, and on the conditional posterior of the latent field $\pi(x_i|\theta,y)$. A Laplace approximation for the posterior density of the hyperparameters, denoted $\tilde{\pi}(\theta|y)$, and a Taylor approximation for the conditional posterior of the latent field, denoted $\tilde{\pi}(x_i|\theta,y)$, are used. From these approximations, marginal posteriors are obtained as

$$\tilde{\pi}(x_i|y) = \int \tilde{\pi}(\theta|y)\tilde{\pi}(x_i|\theta,y)d\theta,$$

$$\tilde{\pi}(\theta_j|y) = \int \tilde{\pi}(\theta|y)d\theta_{[j]},$$

where $\theta_{[j]}$ denotes θ excluding θ_j, and integrations are carried out numerically.

1.12 Assessing Efficiency and Convergence; Ways of Improving Convergence

It is necessary in applying MCMC sampling to decide how many iterations to use to accurately represent the posterior density, and also necessary to ensure that the sampling process has converged. Nonvanishing autocorrelations at high lags mean that less information about the posterior distribution is provided by each iterate, and a higher sample size is necessary to cover the parameter space. Autocorrelation will be reduced by "thinning", namely, retaining only samples that are $S>1$ steps apart $\{\theta_h^{(t)},\theta_h^{(t+S)},\theta_h^{(t+2S)},\ldots\}$ that more closely approximate independent samples; however, this results in a loss of precision. The autocorrelation present in MCMC samples may depend on the form of parameterisation, the complexity of the model, and the form of sampling (e.g. block or univariate sampling for collections of random effects). Autocorrelation will reduce the effective sample size $T_{\text{eff},h}$ for parameter samples $\{\theta_h^{(t)}, t=B+1,\ldots,B+T\}$ below T. The effective number of samples (Kass et al., 1998) may be estimated as

$$T_{\text{eff},h} = T \bigg/ \left[1+2\sum_{k=0}^{\infty}\rho_{hk}\right],$$

where

$$\rho_{hk} = \gamma_{hk}/\gamma_{h0},$$

is the kth lag autocorrelation, γ_{h0} is the posterior variance $V(\theta_h|y)$, and γ_{hk} is the kth lag autocovariance $\text{cov}[\theta_h^{(t)},\theta_h^{(t+k)}|y]$. In practice, one may estimate $T_{\text{eff},h}$ by dividing T by $1+2\sum_{k=0}^{K^*}\rho_{hk}$, where K^* is the first lag value for which $\rho_{hk}<0.1$ or $\rho_{hk}<0.05$ (Browne et al., 2009).

Also useful for assessing efficiency is the Monte Carlo standard error, which is an estimate of the standard deviation of the difference between the true posterior mean $E(\theta_h \mid y) = \int \theta_h p(\theta \mid y) d\theta$, and the simulation-based estimate

$$\bar{\theta}_h = \frac{1}{T} \sum_{t=B+1}^{T+B} \theta_h^{(t)}.$$

A simple estimator of the Monte Carlo variance is

$$\frac{1}{T}\left[\frac{1}{T-1}\sum_{t=1}^{T}(\theta_h^{(t)} - \bar{\theta}_h)^2\right]$$

though this may be distorted by extreme sampled values; an alternative batch means method is described by Roberts (1996). The ratio of the posterior variance in a parameter to its Monte Carlo variance is a measure of the efficiency of the Markov chain sampling (Roberts, 1996), and it is sometimes suggested that the MC standard error should be less than 5% of the posterior standard deviation of a parameter (Toft et al., 2007).

The effective sample size is mentioned above, while Raftery and Lewis (1992, 1996) estimate the iterations required to estimate posterior summary statistics to a given accuracy. Suppose the following posterior probability

$$Pr[\Delta(\theta \mid y) < b] = p_\Delta,$$

is required. Raftery and Lewis seek estimates of the burn-in iterations B to be discarded, and the required further iterations T to estimate p_Δ to within r with probability s; typical quantities might be $p_\Delta = 0.025$, $r = 0.005$, and $s = 0.95$. The selected values of $\{p_\Delta, r, s\}$ can also be used to derive an estimate of the required minimum iterations $T_{\min}$ if autocorrelation were absent, with the ratio

$$I = T / T_{\min},$$

providing a measure of additional sampling required due to autocorrelation.

As to the second issue mentioned above, there is no guarantee that sampling from an MCMC algorithm will converge to the posterior distribution, despite obtaining a high number of iterations. Convergence can be informally assessed by examining the time series or trace plots of parameters. Ideally, the MCMC sampling is exploring the posterior distribution quickly enough to produce good estimates (this property is often called "good mixing"). Some techniques for assessing convergence (as against estimates of required sample sizes) consider samples $\theta^{(t)}$ from only a single long chain, possibly after excluding an initial $t = 1, \ldots, B$ burn-in iterations. These include the spectral density diagnostic of Geweke (1992), the CUSUM method of Yu and Mykland (1998), and a quantitative measure of the "hairiness" of the CUSUM plot (Brooks and Roberts, 1998).

Slow convergence (usually combined with poor mixing and high autocorrelation in sampled values) will show in trace plots that wander, and that exhibit short-term trends, rather than fluctuating rapidly around a stable mean. Failure to converge is typically a feature of only some model parameters; for example, fixed regression effects in a general linear mixed model may show convergence, but not the parameters relating to the random components. Often measures of overall fit (e.g. model deviance) converge, while component parameters do not.

Problems of convergence in MCMC sampling may reflect problems in model identifiability, either formal nonidentification as in multiple random effects models, or poor empirical identifiability when an overly complex model is applied to a small sample ("over-fitting"). Choice of diffuse priors tends to increase the chance that models are poorly identified, especially in complex hierarchical models for small data samples (Gelfand and Sahu, 1999). Elicitation of more informative priors and/or application of parameter constraints may assist identification and convergence.

Alternatively, a parameter expansion strategy may also improve MCMC performance (Gelman et al., 2008; Ghosh, 2008; Browne et al., 2009). For example, in a normal-normal meta-analysis model (Chapter 4) with

$$y_j \sim N(\mu + \theta_j, \sigma_y^2); \theta_j \sim N(0, \sigma_\theta^2), \ j = 1, \ldots, J$$

conventional sampling approaches may become trapped near $\sigma_\theta = 0$, whereas improved convergence and effective sample sizes are achieved by introducing a redundant scale parameter $\lambda \sim N(0, V_\lambda)$

$$y_j \sim N(\mu + \lambda\xi_j, \sigma_y^2),$$

$$\xi_j \sim N(0, \sigma_\xi^2).$$

The expanded model priors induce priors on the original model parameters, namely

$$\theta_j = \lambda\xi_j,$$

$$\sigma_\theta = |\lambda| \sigma_\xi.$$

The setting for V_λ is important; too much diffuseness may lead to effective impropriety.

Another source of poor convergence is suboptimal parameterisation or data form. For example, convergence is improved by centring independent variables in regression applications (Roberts and Sahu, 2001; Zuur et al., 2002). Similarly, delayed convergence in random effects models may be lessened by sum to zero or corner constraints (Clayton, 1996; Vines et al., 1996), or by a centred hierarchical prior (Gelfand et al., 1995; Gelfand et al., 1996), in which the prior on each stochastic variable is a higher level stochastic mean – see the next section. However, the most effective parameterisation may also depend on the balance in the data between different sources of variation. In fact, non-centred parameterisations, with latent data independent from hyperparameters, may be preferable in terms of MCMC convergence in some settings (Papaspiliopoulos et al., 2003).

1.12.1 Hierarchical Model Parameterisation to Improve Convergence

While priors for unstructured random effects may include a nominal mean of zero, in practice, a posterior mean of zero for such a set of effects may not be achieved during MCMC sampling. For example, the mean of the random effects can be confounded with the intercept, especially when the prior for the random effects does not specify the level (global mean) of the effects. One may apply a corner constraint by setting a particular random effect (say, the first) to a known value, usually zero (Scollnik, 2002). Alternatively, an empirical sum to zero constraint may be achieved by centring the sampled random effects at each iteration (sometimes known as "centring on the fly"), so that

$$u_i^* = u_i - \bar{u}$$

and inserting u_i^* rather than u_i in the model defining the likelihood. Another option (Vines et al., 1996; Scollink, 2002) is to define an auxiliary effect $u_i^a \sim N(0, \sigma_u^2)$ and obtain the u_i, following the same prior $N(0, \sigma_u^2)$, but now with a guaranteed mean of zero, by the transformation

$$u_i = \sqrt{\frac{n}{n-1}}(u_i^a - \bar{u}^a).$$

To illustrate a centred hierarchical prior (Gelfand et al., 1995; Browne et al., 2009), consider two way nested data, with $j = 1, \ldots, J$ repetitions over subjects $i = 1, \ldots, n$

$$y_{ij} = \mu + a_i + u_{ij},$$

with $a_i \sim N(0, \sigma_a^2)$ and $u_{ij} \sim N(0, \sigma_u^2)$. The centred version defines

$$\kappa_i = \mu + a_i$$

$$y_{ij} = \kappa_i + u_{ij},$$

so that

$$y_{ij} \sim N(\kappa_i, \sigma_u^2),$$

$$\kappa_i \sim N(\mu, \sigma_a^2).$$

For three-way nested data, the standard model form is

$$y_{ijk} = \mu + a_i + \beta_{ij} + u_{ijk},$$

with $a_i \sim N(0, \sigma_a^2)$, and $\beta_{ij} \sim N(0, \sigma_\beta^2)$. The hierarchically centred version defines

$$\zeta_{ij} = \mu + a_i + \beta_{ij},$$

$$\kappa_i = \mu + a_i,$$

so that

$$y_{ijk} \sim N(\zeta_{ij}, \sigma_u^2),$$

$$\zeta_{ij} \sim N(\kappa_i, \sigma_\beta^2),$$

and

$$\kappa_i \sim N(\mu, \sigma_a^2).$$

Roberts and Sahu (1997) set out the contrasting sets of full conditional densities under the standard and centred representations and compare Gibbs sampling scanning schemes.

Papaspiliopoulos et al. (2003) compare MCMC convergence for centred, noncentred, and partially non-centred hierarchical model parameterisations according to the amount of information the data contain about the latent effects $\kappa_i = \mu + a_i$. Thus for two-way nested data the (fully) non-centred parameterisation, or NCP for short, involves new random effects $\tilde{\kappa}_i$ with

$$y_{ij} = \tilde{\kappa}_i + \mu + \sigma_u e_{ij},$$

$$\tilde{\kappa}_i = \sigma_a z_i,$$

where e_{ij} and z_i are standard normal variables. In this form, the latent data $\tilde{\kappa}_i$ and hyperparameter μ are independent a priori, and so the NCP may give better convergence when the latent effects κ_i are not well identified by the observed data y. A partially non-centred form is obtained using a number $w \in [0,1]$, and

$$y_{ij} = \tilde{\kappa}_i^w + w\mu + u_{ij},$$

$$\tilde{\kappa}_i^w = (1-w)\mu + \sigma_a z_i,$$

or equivalently,

$$\tilde{\kappa}_i^w = (1-w)\kappa_i + w\tilde{\kappa}_i.$$

Thus $w=0$ gives the centred representation, and $w=1$ gives the non-centred parameterisation. The optimal w for convergence depends on the ratio σ_u/σ_α. The centred representation performs best when σ_u/σ_α tends to zero, while the non-centred representation is optimal when σ_u/σ_α is large.

1.12.2 Multiple Chain Methods

Many practitioners prefer to use two or more parallel chains with diverse starting values to ensure full coverage of the sample space of the parameters (Gelman and Rubin, 1996; Toft et al., 2007). Diverse starting values may be based on default values for parameters (e.g. precisions set at different default values such as 1, 5, 10 and regression coefficients set at zero) or on the extreme quantiles of posterior densities from exploratory model runs. Online monitoring of sampled parameter values $\{\theta_k^{(t)}, t=1,\ldots,T\}$ from multiple chains $k=1, \ldots, K$ assists in diagnosing lack of model identifiability. Examples might be models with multiple random effects, or when the mean of the random effects is not specified within the prior, as under difference priors over time or space that are considered in Chapters 5 and 6 (Besag et al., 1995). Another example is factor and structural equation models where the loadings are not specified, so as to anchor the factor scores in a consistent direction, since otherwise the "name" of the common factor may switch during MCMC updating (Congdon, 2003, Chapter 8). Single runs may still be adequate for straightforward problems, and single chain convergence diagnostics (Geweke, 1992) may be applied in this case. Single runs are often useful for exploring the posterior density, and as a preliminary to obtain inputs to multiple chains.

Convergence for multiple chains may be assessed using Gelman–Rubin scale reduction factors that measure the convergence of the between chain variance in $\theta_k^{(t)} = (\theta_{1k}^{(t)}, \ldots, \theta_{dk}^{(t)})$

to the variance over all chains $k=1, \ldots, K$. These factors converge to 1 if all chains are sampling identical distributions, whereas for poorly identified models, variability of sampled parameter values between chains will considerably exceed the variability within any one chain. To apply these criteria, one typically allows a burn-in of B samples while the sampling moves away from the initial values to the region of the posterior. For iterations $t=B+1,\ldots,T+B$, a pooled estimate of the posterior variance $\sigma^2_{\theta_h|y}$ of θ_h is

$$\sigma_{\theta_h|y} = V_h/T + TW_h/(T-1),$$

where variability within chains W_h is defined as

$$W_h = \frac{1}{(T-1)K}\sum_{k=1}^{K}\sum_{t=B+1}^{B+T}(\theta_{hk}^{(t)} - \bar{\theta}_{hk})^2,$$

with $\bar{\theta}_{hk}$ being the posterior mean of θ_h in samples from the kth chain, and where

$$V_h = \frac{T}{K-1}\sum_{k=1}^{K}(\bar{\theta}_{hk} - \bar{\theta}_{h.})^2,$$

denotes between chain variability in θ_h, with $\bar{\theta}_{h.}$ denoting the pooled average of the $\bar{\theta}_{hk}$. The potential scale reduction factor compares $\sigma^2_{\theta_h|y}$ with the within sample estimate W_h. Specifically, the scale factor is $\hat{R}_h = (\sigma^2_{\theta_h|y}/W_h)^{0.5}$ with values under 1.2 indicating convergence. A multivariate version of the PSRF for vector θ is mentioned by Brooks and Gelman (1998) and Brooks and Roberts (1998) and involves between and within chain covariances V_θ and W_θ, and pooled posterior covariance $\Sigma_{\theta|y}$. The scale factor is defined by

$$R_\theta = \max_b \frac{b'\Sigma_{\theta|y}b}{b'W_\theta b} = \frac{T-1}{T} + \left(1+\frac{1}{K}\right)\lambda_1$$

where λ_1 is the maximum eigenvalue of $W_\theta^{-1}V_\theta/T$.

An alternative multiple chain convergence criterion also proposed by Brooks and Gelman (1998), which avoids reliance on the implicit normality assumptions in the Gelman–Rubin scale reduction factors based on analysis of variance over chains. Normality approximation may be improved by parameter transformation (e.g. log or logit), but problems may still be encountered when posterior densities are skewed or possibly multimodal (Toft et al., 2007).

The alternative criterion uses a ratio of parameter interval lengths: for each chain, the length of the $100(1-\alpha)\%$ interval for a parameter is obtained, namely the gap between 0.5α and $(1-0.5\alpha)$ points from T simulated values. This provides K within-chain interval lengths, with mean L_U. From the pooled output of TK samples, an analogous interval L_P is also obtained. The ratio L_P/L_U should converge to 1 if there is convergent mixing over the K chains.

1.13 Choice of Prior Density

Choice of an appropriate prior density, and preferably a sensitivity analysis over alternative priors, is fundamental in the Bayesian approach; for example, see Gelman (2006), Daniels (1999) and Gustafson et al. (2006) on priors for random effect variances. Before the advent of MCMC methods, conjugate priors were often used in order to reduce the burden of numeric integration. Now non-conjugate priors (e.g. finite range uniform priors on standard deviation parameters) are widely used. There may be questions of sensitivity of posterior inference to the choice of prior, especially for smaller datasets, or for certain forms of model; examples are the priors used for variance components in random effects models, the priors used for collections of correlated effects, for example, in hierarchical spatial models (Bernardinelli et al., 1995), priors in nonlinear models (Millar, 2004), and priors in discrete mixture models (Green and Richardson, 1997).

In many situations, existing knowledge may be difficult to summarise or elicit in the form of an "informative prior". It may be possible to develop suitable priors by simulation (e.g. Chib and Ergashev, 2009), but it may be convenient to express prior ignorance using "default" or "non-informative" priors. This is typically less problematic – in terms of posterior sensitivity – for fixed effects, such as regression coefficients (when taken to be homogenous over cases) than for variance parameters. Since the classical maximum likelihood estimate is obtained without considering priors on the parameters, a possible heuristic is that a non-informative prior leads to a Bayesian posterior estimate close to the maximum likelihood estimate. It might appear that a maximum likelihood analysis would therefore necessarily be approximated by flat or improper priors, but such priors may actually be unexpectedly informative about different parameter values (Zhu and Lu, 2004).

A flat or uniform prior distribution on θ, expressible as $p(\theta)=1$ is often adopted on fixed regression effects, but is not invariant under reparameterisation. For example, it is not true for $\phi=1/\theta$ that $p(\phi)=1$ as the prior for a function $\phi=g(\theta)$, namely

$$p(\phi)=\left|\frac{d}{d\phi}g^{-1}(\phi)\right|,$$

demonstrates. By contrast, on invariance grounds, Jeffreys (1961) recommended the prior $p(\sigma)=1/\sigma$ for a standard deviation, as for $\phi=g(\sigma)=\sigma^2$ one obtains $p(\phi)=1/\phi$. More general analytic rules for deriving noninformative priors include reference prior schemes (Berger and Bernardo, 1992), and Jeffreys prior

$$p(\theta)\propto|I(\theta)|^{0.5},$$

where $I(\theta)$ is the information matrix, namely

$$I(\theta)=-E\left(\frac{\partial^2 l(\theta)}{\delta l(\theta_g)\delta l(\theta_h)}\right),$$

and $l(\theta)=\log(L(\theta\,|\,y))$ is the log-likelihood. Unlike uniform priors, a Jeffreys prior is invariant under transformation of scale since $I(\theta)=I(g(\theta))(g'(\theta))^2$ and $p(\theta)\propto I(g(\theta))^{0.5}g'(\theta)=p(g(\theta))g'(\theta)$ (Kass and Wasserman, 1996, p.1345).

1.13.1 Including Evidence

Especially for establishing the intercept (e.g. the average level of a disease), or regression effects (e.g. the impact of risk factors on disease) or variability in such impacts, it may be possible to base the prior density on cumulative evidence via meta-analysis of existing studies, or via elicitation techniques aimed at developing informative priors. This is well established in engineering risk and reliability assessment, where systematic elicitation approaches such as maximum-entropy priors are used (Siu and Kelly, 1998; Hodge et al., 2001). Thus, known constraints for a variable identify a class of possible distributions, and the distribution with the greatest Shannon–Weaver entropy is selected as the prior. Examples are $\theta \sim N(m,V)$, if estimates m and V of the mean and variance are available, or an exponential with parameter $-q/\log(1-p)$ if a positive variable has an estimated pth quantile of q.

Simple approximate elicitation methods include the histogram technique, which divides the domain of an unknown θ into a set of bins, and elicits prior probabilities that θ is located in each bin. Then $p(\theta)$ may be represented as a discrete prior or converted to a smooth density. Prior elicitation may be aided if a prior is reparameterised in the form of a mean and prior sample size. For example, beta priors $Be(a,b)$ for probabilities can be expressed as $Be(m\tau,(1-m)\tau)$, where $m=a/(a+b)$ and $\tau=a+b$ are elicited estimates of the mean probability and prior sample size. This principle is extended in data augmentation priors (Greenland and Christensen, 2001), while Greenland (2007) uses the device of a prior data stratum (equivalent to data augmentation) to represent the effect of binary risk factors in logistic regressions in epidemiology.

If a set of existing studies is available providing evidence on the likely density of a parameter, these may be used in a form of preliminary meta-analysis to set up an informative prior for the current study. However, there may be limits to the applicability of existing studies to the current data, and so pooled information from previous studies may be downweighted. For example, the precision of the pooled estimate from previous studies may be scaled downwards, with the scaling factor possibly an extra unknown. When a maximum likelihood (ML) analysis is simple to apply, one option is to adopt the ML mean as a prior mean, but with the ML precision matrix downweighted (Birkes and Dodge, 1993).

More comprehensive ways of downweighting historical/prior evidence have been proposed, such as power prior models (Chen et al., 2000; Ibrahim and Chen, 2000). Let $0 \leq \delta \leq 1$ be a scale parameter with beta prior that weights the likelihood of historical data y_h relative to the likelihood of the current study data y. Following Chen et al. (2000, p.124), a power prior has the form

$$p(\theta,\delta \mid y_h) \propto p(y_h \mid \theta)]^{\delta}[\delta^{a_\delta - 1}(1-\delta)^{b_\delta - 1}]p(\theta),$$

where $p(y_h|\theta)$ is the likelihood for the historical data, and (a_δ,b_δ) are pre-specified beta density hyperparameters. The joint posterior density for (θ,δ) is then

$$p(\theta,\delta \mid y,y_h) \propto p(y \mid \theta)[p(y_h \mid \theta)]^{\delta}[\delta^{a_\delta - 1}(1-\delta)^{b_\delta - 1}]p(\theta).$$

Chen and Ibrahim (2006) demonstrate connections between the power prior and conventional priors for hierarchical models.

1.13.2 Assessing Posterior Sensitivity; Robust Priors

To assess sensitivity to prior assumptions, the analysis may be repeated over a limited range of alternative priors. Thus Sargent (1998) and Fahrmeir and Knorr-Held (1997, section 3.2) suggest a gamma prior on inverse precisions $1/\tau^2$ governing random walk effects (e.g. baseline hazard rates in survival analysis), namely $1/\tau^2 \sim \text{Ga}(a,b)$, where a is set at 1, but b is varied over choices such as 0.05 or 0.0005. One possible strategy involves a consideration of both optimistic and conservative priors, with regard, say, to a treatment effect, or the presence of significant random effect variation (Spiegelhalter, 2004; Gustafson et al., 2006).

Another relevant principle in multiple effect models is that of uniform shrinkage governing the proportion of total random variation to be assigned to each source of variation (Daniels, 1999; Natarajan and Kass, 2000). So, for a two-level normal linear model with

$$y_{ij} = x_{ij}\beta + \eta_j + e_{ij},$$

with $e_{ij} \sim N(0,\sigma^2)$ and $\eta_j \sim N(0,\tau^2)$, one prior (e.g. inverse gamma) might relate to the residual variance σ^2, and a second conditional $U(0,1)$ prior relates to the ratio $\tau^2/(\tau^2+\sigma^2)$ of cluster to total variance. A similar effect is achieved in structural time series models (Harvey, 1989) by considering different forms of signal to noise ratios in state space models including several forms of random effect (e.g. changing levels and slopes, as well as season effects). Gustafson et al. (2006) propose a conservative prior for the one-level linear mixed model

$$y_i \sim N(\eta_i, \sigma^2),$$

$$\eta_i \sim N(\mu, \tau^2),$$

namely a conditional prior $p(\tau^2 \mid \sigma^2)$ aiming to prevent over-estimation of τ^2. Thus, in full,

$$p(\sigma^2, \tau^2) = p(\sigma^2)p(\tau^2 \mid \sigma^2)$$

where $\sigma^2 \sim IG(e,e)$ for some small $e > 0$, and

$$p(\tau^2 \mid \sigma^2) = \frac{a}{\sigma^2}\left[1 + \tau^2/\sigma^2\right]^{-(a+1)}.$$

The case $a = 1$ corresponds to the uniform shrinkage prior of Daniels (1999), where

$$p(\tau^2 \mid \sigma^2) = \frac{\sigma^2}{[\sigma^2 + \tau^2]^2},$$

while larger values of a (e.g. $a = 5$) are found to be relatively conservative.

For covariance matrices Σ between random effects of dimension k, the emphasis in recent research has been on more flexible priors than afforded by the inverse Wishart (or Wishart priors for precision matrices). Barnard et al. (2000) and Liechty et al. (2004) consider a separation strategy whereby

$$\Sigma = \text{diag}(S).R.\text{diag}(S),$$

where S is a $k \times 1$ vector of standard deviations, and R is a $k \times k$ correlation matrix. With the prior sequence, $p(R,S) = p(R|S)p(S)$, Barnard et al. suggest $\log(S) \sim N_k(\xi, \Lambda)$, where Λ is usually diagonal. For the elements r_{ij} of R, constrained beta sampling on $[-1,1]$ can be used subject to positive definitiveness constraints on Σ. Daniels and Kass (1999) consider the transformation $\eta_{ij} = 0.5\log[(1-r_{ij})/(1+r_{ij})]$ and suggest an exchangeable hierarchical shrinkage prior, $\eta_{ij} \sim N(0,\tau^2)$, where

$$p(\tau^2) \propto (c + \tau^2)^{-2};$$

$$c = 1/(k-3).$$

A separation strategy is also facilitated by the LKJ prior of Lewandowski et al. (2009) and included in the rstan package (McElreath, 2016). While a full covariance prior (e.g. assuming random slopes on all k predictors in a multilevel model) can be applied from the outset, MacNab et al. (2004) propose an incremental model strategy, starting with random intercepts and slopes but without covariation between them, in order to assess for which predictors there is significant slope variation. The next step applies a full covariance model only for the predictors showing significant slope variation.

Formal approaches to prior robustness may be based on "contamination" priors. For instance, one might assume a two group mixture with larger probability $1-r$ on the "main" prior $p_1(\theta)$, and a smaller probability such as $r=0.1$ on a contaminating density $p_2(\theta)$, which may be any density (Gustafson, 1996). More generally, a sensitivity analysis may involve some form of mixture of priors, for example, a discrete mixture over a few alternatives, a fully non-parametric approach (see Chapter 4), or a Dirichlet weight mixture over a small range of alternatives (e.g. Jullion and Lambert, 2007). A mixture prior can include the option that the parameter is not present (e.g. that a variance or regression effect is zero). A mixture prior methodology of this kind for regression effects is presented by George and McCulloch (1993). Increasingly also, random effects models are selective, including a default allowing for random effects to be unnecessary (Albert and Chib, 1997; Cai and Dunson, 2006; Fruhwirth-Schnatter and Tuchler, 2008).

In hierarchical models, the prior specifies both the form of the random effects (fully exchangeable over units or spatially/temporally structured), the density of the random effects (normal, mixture of normals, etc.), and the third stage hyperparameters. The form of the second stage prior $p(b|\theta_b)$ amounts to a hypothesis about the nature and form of the random effects. Thus, a hierarchical model for small area mortality may include spatially structured random effects, exchangeable random effects with no spatial pattern, or both, as under the convolution prior of Besag et al. (1991). It also may assume normality in the different random effects, as against heavier tailed alternatives. A prior specifying the errors as spatially correlated and normal is likely to be a working model assumption, rather than a true cumulation of knowledge, and one may have several models for $p(b|\theta_b)$ being compared (Disease Mapping Collaborative Group, 2000), with sensitivity not just being assessed on the hyperparameters.

Random effect models often start with a normal hyperdensity, and so posterior inferences may be sensitive to outliers or multiple modes, as well as to the prior used on the hyperparameters. Indications of lack of fit (e.g. low conditional predictive ordinates for particular cases) may suggest robustification of the random effects prior. Robust hierarchical models are adapted to pooling inferences and/or smoothing in data, subject to outliers or other irregularities; for example, Jonsen et al. (2006) consider robust space-time state-space models with Student t rather than normal errors in an analysis of travel rates of migrating leatherback turtles. Other forms of robust analysis involve discrete mixtures of random effects (e.g. Lenk and Desarbo, 2000), possibly under Dirichlet or Polya process models (e.g. Kleinman and Ibrahim, 1998). Robustification of hierarchical models reduces the chance of incorrect inferences on individual effects, important when random effects approaches are used to identify excess risk or poor outcomes (Conlon and Louis, 1999; Marshall et al., 2004).

1.13.3 Problems in Prior Selection in Hierarchical Bayes Models

For the third stage parameters (the hyperparameters) in hierarchical models, choice of a diffuse noninformative prior may be problematic, as improper priors may induce improper posteriors that prevent MCMC convergence, since conditions necessary for convergence (e.g. positive recurrence) may be violated (Berger et al., 2005). This may apply even if conditional densities are proper, and Gibbs or other MCMC sampling proceeds apparently straightforwardly. A simple example is provided by the normal two-level model with subjects $i = 1, \ldots, n$ nested in clusters $j = 1, \ldots, J$,

$$y_{ij} = \mu + \theta_j + u_{ij},$$

where $\theta_j \sim N(0, \tau^2)$ and $u_{ij} \sim N(0, \sigma^2)$. Hobert and Casella (1996) show that the posterior distribution is improper under the prior $p(\mu, \tau, \sigma) = 1/(\sigma^2\tau^2)$, even though the full conditionals have standard forms, namely

$$p(\theta_j \mid y, \mu, \sigma^2, \tau^2) = N\left(\frac{n(\bar{y}_j - \mu)}{n + \frac{\sigma^2}{\tau^2}}, \frac{1}{\frac{n}{\sigma^2} + \frac{1}{\tau^2}}\right),$$

$$p(\mu \mid y, \sigma^2, \tau^2, \theta) = N\left(\bar{y} - \bar{\theta}, \frac{\sigma^2}{nJ}\right),$$

$$p(1/\tau^2 \mid y, \mu, \sigma^2, \theta) = Ga\left(\frac{J}{2}, 0.5\sum_j \theta_j^2\right),$$

$$p(1/\sigma^2 \mid y, \mu, \tau^2, \theta) = Ga\left(\frac{nJ}{2}, 0.5\sum_{ij}(y_{ij} - \mu - \theta_j)^2\right),$$

so that Gibbs sampling could in principle proceed.

Whether posterior propriety holds depends on the level of information in the data, whether additional constraints are applied to parameters in MCMC updating, and the nature of the improper prior used. For example, Rodrigues and Assuncao (2008) demonstrate propriety in the posterior of spatially varying regression parameter models under a class of improper priors. More generally, Markov random field (MRF) priors such as random walks in time, or spatial conditional autoregressive priors (Chapters 5 and 6), may have joint forms that are improper, with a singular covariance matrix – see, for example, the discussion by Sun et al. (2000, pp.28–30). The joint prior only identifies differences between pairs of effects, and unless additional constraints are applied to the random effects, this may cause issues with posterior propriety.

It is possible to define proper priors in these cases by introducing autoregression parameters (Sun et al., 1999), but Besag et al. (1995, p.11) mention that "the sole impropriety in such [MRF] priors is that of an arbitrary level and is removed from the corresponding posterior distribution by the presence of any informative data". The indeterminacy in the level is usually resolved by applying "centring on the fly" (at each MCMC iteration) within each set of random effects, and under such a linear constraint, MRF priors become proper (Rodrigues and Assunção, 2008, p.2409). Alternatively, "corner" constraints on particular effects, namely, setting them to fixed values (usually zero), may be applied (Clayton, 1996; Koop, 2003, p.248), while Chib and Jeliazkov (2006) suggest an approach to obtaining propriety in random walk priors.

Priors that are just proper mathematically (e.g. gamma priors on $1/\tau^2$ with small scale and shape parameters) are often used on the grounds of expediency, and justified as letting the data speak for themselves. However, such priors may cause identifiability problems as the posteriors are close to being empirically improper. This impedes MCMC convergence (Kass and Wasserman, 1996; Gelfand and Sahu, 1999). Furthermore, using just proper priors on variance parameters may in fact favour particular values, despite being supposedly only weakly informative. Gelman (2006) suggests possible (less problematic) options including a finite range uniform prior on the standard deviation (rather than variance), and a positive truncated t density.

1.14 Computational Notes

[1] In Example 1.1, the data are generated ($n = 1000$ values) and underlying parameters are estimated as follows:

```
library(mcmcse)
library(MASS)
library(R2WinBUGS)
# generate data
set.seed(1234)
y = rnorm(1000,3,5)
# initial vector setting and parameter values
T = 10000; B = T/10; B1=B+1
mu = sig = numeric(T)
# initial parameter values
mu[1] = 0
sig[1] = 1
u.mu = u.sig = runif(T)
# rejection counter
REJmu = 0; REJsig = 0
# log posterior density (up to a constant)
logpost = function(mu,sig){
loglike = sum(dnorm(y,mu,sig,log=TRUE))
return(loglike - log(sig))}
# sampling loop
for (t in 2:T) {print(t)
mut = mu[t-1]; sigt = sig[t-1]
# uniform proposals with kappa = 0.5
mucand = mut + runif(1,-0.5,0.5)
sigcand = abs(sigt + runif(1,-0.5,0.5))
alph.mu = logpost(mucand,sigt)-logpost(mut,sigt)
if (log(u.mu[t]) <= alph.mu) mu[t] = mucand
else {mu[t] = mut; REJmu = REJmu+1}
alph.sig = logpost(mu[t],sigcand)-logpost(mu[t],sigt)
if (log(u.sig[t]) <= alph.sig) sig[t] = sigcand
else {sig[t] <- sigt; REJsig <- REJsig+1}}
# sequence of sampled values and ACF plots
plot(mu)
```

```
plot(sig)
acf(mu,main="acf plot, mu")
acf(sig,main="acf plot, sig")
# posterior summaries
summary(mu[B1:T])
summary(sig[B1:T])
# Monte Carlo standard errors
D=data.frame(mu[B1:T],sig[B1:T])
mcse.mat(D)
# acceptance rates
ACCmu=1-REJmu/T
ACCsig=1-REJsig/T
cat("Acceptance Rate mu =",ACCmu,"n ")
cat("Acceptance Rate sigma = ",ACCsig, "n ")
# kernel density plots
plot(density(mu[B1:T]),main= "Density plot for mu posterior")
plot(density(sig[B1:T]),main= "Density plot for sigma posterior ")
f1=kde2d(mu[B1:T], sig[B1:T], n=50, lims=c(2.5,3.4,4.7,5.3))
filled.contour(f1,main="Figure 1.1 Bivariate Density", xlab="mu",
ylab="sigma",
color.palette=colorRampPalette(c('white','blue','yellow','red','dark
red')))
filled.contour(f1,main="Figure 1.1 Bivariate Density",xlab="mu",
ylab="sigma",
color.palette=colorRampPalette(c('white','lightgray','gray','darkgra
y','black')))
# estimates of effective sample sizes
effectiveSize(mu[B1:T])
effectiveSize(sig[B1:T])
ess(D)
multiESS(D)
# posterior probability on hypothesis μ < 3
sum(mu[B1:T] < 3)/(T-B)
```

[2] The R code for Metropolis sampling of the extended logistic model is library(coda)

```
# data
w = c(1.6907, 1.7242, 1.7552, 1.7842, 1.8113, 1.8369, 1.8610, 1.8839)
n = c(59, 60, 62, 56, 63, 59, 62, 60)
y = c(6, 13, 18, 28, 52, 53, 61, 60)
# posterior density
f = function(mu,th2,th3) {
# settings for priors
a0=0.25; b0=0.25; c0=2; d0=10; e0=2.004; f0=0.001
V = exp(th3)
m1 = exp(th2)
sig = sqrt(V)
x = (w-mu)/sig
xt = exp(x)/(1+exp(x))
h = xt94m1;
loglike = y*log(h)+(n-y)*log(1-h)
# prior ordinates
logpriorm1 = a0*th2-m1*b0
logpriorV = -e0*th3-f0/V
```

```
logpriormu = -0.5*((mu-c0)/d0)942-log(d0)
logprior = logpriormu+logpriorV+logpriorm1
# log posterior
f = sum(loglike)+logprior}
# main MCMC loop
runMCMC = function(samp,mu,th2,th3,T,sd) {
for (i in 2:T+1) {
# candidates for mu
mucand = mu[i-1]+sd[1]*rnorm(1,0,1)
f.cand = f(mucand,th2[i-1],th3[i-1])
f.curr = f(mu[i-1], th2[i-1],th3[i-1])
if (log(runif(1)) <= f.cand-f.curr) mu[i] = mucand else
{mu[i] = mu[i-1]}
# candidates for log(m1)
th2cand = th2[i-1]+sd[2]*rnorm(1,0,1)
f.cand = f(mu[i],th2cand,th3[i-1])
f.curr = f(mu[i],th2[i-1], th3[i-1])
if (log(runif(1)) <= f.cand-f.curr) th2[i] = th2cand else
{th2[i] = th2[i-1]}
# candidates for log(V)
th3cand = th3[i-1]+sd[3]*rnorm(1,0,1)
f.cand = f(mu[i],th2[i],th3cand)
f.curr = f(mu[i],th2[i],th3[i-1])
if (log(runif(1)) <= f.cand-f.curr) th3[i] = th3cand else
{th3[i] = th3[i-1]}
samp[i-1.1] = mu[i]; samp[i-1.2] = exp(th2[i]); samp[i-1.3] =
exp(th3[i])}
return(samp)}
# number of iterations
T=100000
# warm-up samples
B=50000
B1=B+1
R=T-B
mu=th3=th2=numeric(T)
sd=acc=numeric(3)
# metropolis proposal standard devns
sd[1] = 0.01; sd[2] = 0.2; sd[3] = 0.4
# accumulate samples
samp = matrix(,T,3)
# initial parameter values
mu[1] = 0; th2[1]= 0; th3[1] =0
samp[1,1] = mu[1]; samp[1,2] = exp(th2[1]); samp[1,3] = exp(th3[1])
# first chain
chain1=runMCMC(samp,mu,th2,th3,T,sd)
chain1=chain1[B1:T,]
# posterior summary
quantile(chain1[1:R,1], probs=c(.025,0.5,0.975))
quantile(chain1[1:R,2], probs=c(.025,0.5,0.975))
quantile(chain1[1:R,3], probs=c(.025,0.5,0.975))
# second chain
chain2=runMCMC(samp,mu,th2,th3,T,sd)
chain2=chain2[B1:T,]
# posterior summary
```

```
quantile(chain2[1:R,1], probs=c(.025,0.5,0.975))
quantile(chain2[1:R,2], probs=c(.025,0.5,0.975))
quantile(chain2[1:R,3], probs=c(.025,0.5,0.975))
# combine chains
chain1=as.mcmc(chain1)
chain2=as.mcmc(chain2)
combchains = mcmc.list(chain1, chain2)
gelman.diag(combchains)
crosscorr(combchains)
accsum = "Acceptance rates: mu, m1, and sigma942"
print(accsum)
1 - rejectionRate(combchains)
effectiveSize(combchains)
autocorr.diag(combchains)
```

[3] The rstan code for the beetle mortality example is

```
library(rstan)
library(bayesplot)
library(coda)
# data
w = c(1.6907, 1.7242, 1.7552, 1.7842, 1.8113, 1.8369, 1.8610, 1.8839)
n = c(59, 60, 62, 56, 63, 59, 62, 60)
y = c(6, 13, 18, 28, 52, 53, 61, 60)
D=list(y=y,n=n,w=w,N=8)
# rstan code
model ="
data {
int<lower=0> N;
int n[N];
int y[N];
real w[N];
}
parameters {
real <lower=0> mu;
real log_sigma;
real log_m1;
}
transformed parameters {
real<lower=0> sigma;
real<lower=0> sigma2;
real<lower=0> m1;
real x[N];
real pi[N];
sigma=exp(log_sigma);
sigma2=sigma942;
m1=exp(log_m1);
for (i in 1:N) {x[i]=(w[i]-mu)/sigma;}
for (i in 1:N) {pi[i]=pow(exp(x[i])/(1+exp(x[i])),m1);}
}
model {
log_sigma ~normal(0,5);
mu ~normal(2,3.16);
log_m1 ~normal(0,1);
```

```
for (i in 1:N) {y[i] ~binomial_logit(n[i], pi[i]);}
}
"
fit=stan(model_code = model,data=D, iter = 2500,warmup =
250,chains=2,seed=10)
# posterior summary
print(fit,digits=6)
# bivariate density plots
color_scheme_set("gray")
afit= as.array(fit)
mcmc_pairs(afit, pars = c("mu", "m1", "sigma"), off_diag_args =
list(size = 1.5))
# MCMC diagnostics
samps <- as.matrix(fit,pars= c("mu", "m1", "sigma"))
samps <- mcmc.list(lapply(1:ncol(samps), function(x) mcmc(as.
array(samps)[,x,])))
crosscorr(samps)
effectiveSize(samps)
autocorr.diag(as.mcmc(samps))
```

[4] The R code for analysis of the turtle survival data is

```
library(bridgesampling)
options(scipen=999)
data("turtles")
y=turtles$y
x=turtles$x
C=turtles$clutch
N = length(y)
J = length(unique(C))
# posterior density function
f = function(beta,alpha,tau,e) {sig = 1/sqrt(tau)
# survival model
for (i in 1:N){p[i] = pnorm(alpha+beta*x[i]+e[C[i]])
LL[i] = y[i]*log(p[i])+(1-y[i])*log(1-p[i])}
# prior ordinates
logpr[1] = -0.5*alpha942/10
logpr[2] = -0.5*beta942/10
logpr[3] = -0.001*tau
for (j in 1:J){LLr[j] = -0.5*e[j]942/sig942-log(sig)}
# log-posterior
f = sum(LL[1:N])+sum(LLr[1:J])+sum(logpr[1:3])}
# MCMC settings
T = 5000
# warm up
B=T/10
# accumulate M-H rejections for hyperparameters
k1 = 0; k2 = 0; k3=0
# gamma parameter for precision updates
kappa=100
# uniform samples for use in hyperparameter updates
U1 = U2 = U3 = log(runif(T))
# define arrays
alpha = numeric(T); beta = numeric(T); tau = numeric(T); logpr =
numeric(3)
```

```
s = numeric(T); p = numeric(N); e = numeric(J); LL = numeric(N);
LLr = numeric(J); ec = matrix(0,T,J); en = matrix(0,T,J);
kran = numeric(J)
# initial parameter values
beta[1]= 0.35; alpha[1]= -2.6; tau[1]= 5; for (j in 1:J) {ec[1,j]= 0;
kran[j]= 0}
# main loop
# update beta
for (t in 2:T) {bstar = beta[t-1]+0.05*rnorm(1,0,1)
tn = f(bstar,alpha[t-1],tau[t-1],ec[t-1,]); tf =
f(beta[t-1],alpha[t-1],tau[t-1],ec[t-1,])
if (U1[t] <= tn-tf) beta[t] = bstar
else {beta[t] = beta[t-1]; k1 = k1+1}
# update intercept
astar = alpha[t-1]+0.5*rnorm(1,0,1)
tn = f(beta[t],astar,tau[t-1],ec[t-1,]); tf =
f(beta[t],alpha[t-1],tau[t-1],ec[t-1,])
if (U2[t] <= tn-tf) alpha[t] = astar
else {alpha[t] = alpha[t-1]; k2 = k2+1}
# update precision
taustar = rgamma(1,kappa,kappa/tau[t-1])
s[t-1] = 1/sqrt(tau[t-1])
tn = f(beta[t],alpha[t],taustar,ec[t-1,])+log(dgamma(tau[t-1],
kappa,kappa/taustar))
tc = f(beta[t],alpha[t],tau[t-1],ec[t-1,])+log(dgamma(taustar,kappa,
kappa/tau[t-1]))
if (U3[t] <= tn-tf) tau[t] = taustar
else {tau[t] = tau[t-1]; k3 = k3+1}
# update cluster effects
for (j in 1:J) {en[j] = ec[t-1,j]
ec[t,j] = ec[t-1,j]}
for (j in 1:J) {en[j] = ec[t-1,j]+rnorm(1,0,1)
tn = f(beta[t],alpha[t],tau[t],en[]); tf = f(beta[t],alpha[t],tau[t]
,ec[t,])
if (log(runif(1)) <= tn-tf) ec[t,j] = en[j]
else {en[j] = ec[t-1,j]
kran[j] = kran[j]+1}}}
# hyperparameter summaries
quantile(alpha[B:T], probs=c(.025,0.5,0.975))
quantile(beta[B:T], probs=c(.025,0.5,0.975))
quantile(tau[B:T], probs=c(.025,0.5,0.975))
quantile(s[B:T], probs=c(.025,0.5,0.975))
# random effects posterior medians and quantiles
eff.mdn = apply(ec[B:T,], 2, quantile, probs = c(0.50))
eff.q975=apply(ec[B:T,], 2, quantile, probs = c(0.975))
eff.q025=apply(ec[B:T,], 2, quantile, probs = c(0.025))
eff.q90=apply(ec[B:T,], 2, quantile, probs = c(0.90))
eff.q10=apply(ec[B:T,], 2, quantile, probs = c(0.10))
# number of significant 80% credible intervals for random effects
sum(eff.q90>0 & eff.q10 >0)+ sum(eff.q90<0 & eff.q10 <0)
# acceptance rates for hyperparameters (beta, alpha, tau.b)
1-k1/T; 1-k2/T; 1-k3/T
# acceptance rates for cluster effects
1-kran/T
```

[5] There are J+2 unknowns in the R code (N.B. the σ_j^2 are not unknowns) for implementing these Gibbs updates. There are T=20000 MCMC samples to be accumulated in the matrix samples. With $a=b=0.1$ in the prior for $1/\tau^2$, and calling on coda routines for posterior summaries, one has

```
library(coda)
# data
y=c(28,8,-3,7,-1,1,18,12)
sigma=c(15,10,16,11,9,11,10,18)
sigma2 = sigma942
J = 8
# total MCMC iterations
T = 20000
# ten unknowns (eight effects, plus their mean and variance)
samps = matrix(, T, 10)
colnames(samps) <- c("mu","tau","Sch1","Sch2","Sch3","Sch4","Sch5","
Sch6","Sch7","Sch8")
# starting values
mu=mean(y)
tau2=median(sigma2)
# sampling loop
for (t in 1:T) {th.mean=(y/sigma2+mu/tau2)/(1/sigma2+1/tau2)
th.sd=sqrt(1/(1/sigma2+1/tau2))
theta=rnorm(J,th.mean,th.sd)
mu=rnorm(1,mean(theta),sqrt(tau2/J))
# prior on random effects precision
invtau2=rgamma(1,J/2+0.1,sum((theta-mu)942)/2+0.1)
tau2 = 1/invtau2
tau = sqrt(tau2)
# accumulate samples
samps[t,3:10] = theta
samps[t,1] =mu
samps[t,2] =tau}
# posterior summary
summary(as.mcmc(samps))
post.mn = apply(samps,2,mean)
post.sd = apply(samps,2,sd)
post.median = apply(samps,2,median)
post.95=apply(samps, 2, quantile, probs = c(0.95))
post.05=apply(samps, 2, quantile, probs = c(0.05))
# trace and density plots
plot(as.mcmc(samps))
```

References

Albert J (2007) *Bayesian Computation with R*. Springer.

Albert J, Chib S (1993) Bayesian analysis of binary and polychotomous response data. *Journal of the American Statistical Association*, 88, 669–679.

Albert J, Chib S (1997) Bayesian tests and model diagnostics in conditionally independent hierarchical models. *Journal of the American Statistical Association*, 92, 916–925.

Altaleb A, Chauveau D (2002) Bayesian analysis of the logit model and comparison of two Metropolis–Hastings strategies. *Computational Statistics & Data Analysis*, 39, 137–152.
Andrieu C, Moulines E (2006) On the ergodicity properties of some adaptive MCMC algorithms. *Annals of Applied Probability*, 16(3), 1462–1505.
Barnard J, McCulloch R, Meng X (2000) Modeling covariance matrices in terms of standard deviations and correlations, with applications to shrinkage. *Statistica Sinica*, 10, 1281–1311.
Bedard M (2008) Optimal acceptance rates for Metropolis algorithms: Moving beyond 0.234. *Stochastic Processes and their Applications*, 118(12), 2198–2222.
Berger J, Bernardo J (1992) On the development of reference priors, in *Bayesian Statistics 4*, pp 35–60, eds J Bernardo, J Berger, A Dawid, A Smith. Clarendon Press, Oxford.
Berger J, Strawderman W, Tang D (2005) Posterior propriety and admissibility of hyperpriors in normal hierarchical models. *Annals of Statistics* 33, 606–646.
Bernardinelli L, Clayton D, Montomoli C (1995) Bayesian estimates of disease maps: How important are priors? *Statistics in Medicine* 14, 2411–2431.
Besag J, Green P, Higdon D, Mengerson K (1995) Bayesian computation and stochastic systems. *Statistical Science*, 10(1),103–166.
Besag J, York J, Mollie A (1991) Bayesian image restoration, with two applications in spatial statistics. *Annals of the Institute of Statistical Mathematics*, 43, 1–21.
Birkes D, Dodge Y (1993) *Alternative Methods of Regression*. John Wiley.
Brooks S, Gelman A (1998) Alternative methods for monitoring convergence of iterative simulations. *Journal of Computational and Graphical Statistics*, 7, 434–456.
Brooks S, Roberts G (1998) Convergence assessment techniques for Markov chain Monte Carlo. *Statistics and Computing*, 8, 319–335.
Browne W, Steele F, Golalizadeh M (2009) The use of simple reparameterizations to improve the efficiency of Markov chain Monte Carlo estimation for multilevel models with applications to discrete time survival models. *Journal of the Royal Statistical Society: Series A*, 172, 579–598.
Cai B, Dunson D (2006) Bayesian covariance selection in generalized linear mixed models. *Biometrics*, 62, 446–457.
Carlin B, Gelfand A (1991) An iterative Monte Carlo method for nonconjugate Bayesian analysis. *Statistics and Computing*, 1(2), 119–128.
Carpenter B, Gelman A, Hoffman M, Lee D, Goodrich B, Betancourt M, Brubaker M, Guo J, Li P, Riddell A (2017) Stan: A probabilistic programming language. *Journal of Statistical Software*, 76(1), 1–32
Chen M, Wang X (2011) Approximate predictive densities and their applications in generalized linear models. *Computational Statistics & Data Analysis*, 55(4), 1570–1580.
Chen M-H, Ibrahim J (2006) The relationship between the power prior and hierarchical models. *Bayesian Analysis*, 1, 551–574.
Chen M-H, Ibrahim J, Shao Q-M (2000) Power prior distributions for generalized linear models. *Journal of Statistical Planning and Inference*, 84, 121–137.
Chen M-H, Shao Q-M (1998) Monte Carlo estimation of Bayesian credible and HPD intervals. *Journal of Computational & Graphical Statistics*, 8(1), 69–92.
Chiang J, Chib S, Narasimhan C (1999) Markov chain Monte Carlo and models of consideration set and parameter heterogeneity. *Journal of Econometrics*, 89, 223–248.
Chib S (2001) Monte Carlo methods and Bayesian computation: Overview, in *International Encyclopedia of the Social & Behavioral Sciences*. https://doi.org/10.1016/B0-08-043076-7/00467-8
Chib S, Ergashev B (2009) Analysis of multifactor affine yield curve models. *Journal of the American Statistical Association*, 104(488), 1324–1337.
Chib S, Greenberg E (1995) Understanding the Metropolis-Hastings algorithm. *The American Statistician*, 49, 327–335.
Chib S, Jeliazkov I (2006) Inference in semiparametric dynamic models for binary longitudinal data. *Journal of the American Statistical Association*, 101, 685–700.
Clark J, Gelfand A (eds) (2006) *Hierarchical Modelling for the Environmental Sciences: Statistical Methods and Applications*. Oxford University Press.

Clayton D (1996) Generalized linear mixed models, in: *Markov Chain Monte Carlo in Practice*, eds W Gilks, S Richardson, D Spiegelhalter. Chapman & Hall, London, UK.
Congdon P (2003) *Applied Bayesian Modelling*. Wiley, Chichester, UK.
Conlon E, Louis T (1999) Addressing multiple goals in evaluating region-specific risk using Bayesian methods, pp 31–47, in *Disease Mapping and Risk Assessment for Public Health*, eds A Lawson, A Biggeri, D Bohning, E Lesaffre, J Viel, R Bertollini. Wiley.
Cressie N, Calder C A, Clark J S, Hoef J M V, Wikle C K (2009) Accounting for uncertainty in ecological analysis: The strengths and limitations of hierarchical statistical modeling. *Ecological Applications*, 19(3), 553–570.
Daniels M (1999) A prior for the variance in hierarchical models. *Canadian Journal of Statistics*, 27, 569–580.
Daniels M, Kass R (1999) Nonconjugate Bayesian Estimation of Covariance matrices and its use in hierarchical models. *Journal of the American Statistical Association*, 94, 1254–1263.
Davidian M, Giltinan D M (2003) Nonlinear models for repeated measures data: An overview and update. *Journal of Agricultural, Biological, and Environmental Statistics*, 8, 387–419.
Deely J, Smith A (1998) Quantitative refinements for comparisons of institutional performance. *Journal of the Royal Statistical Society, Series A*, 161, 5–12.
Disease Mapping Collaborative Group (2000) Disease mapping models: An empirical evaluation. *Statistic in Medicine*, 19, 2217–2241.
Dunson D (2001) Commentary: Practical advantages of Bayesian analysis of epidemiologic data. *American Journal of Epidemiology*, 153, 1222–1226.
Fahrmeir L, Knorr-Held L (1997) Dynamic discrete-time duration models. *Sociological Methodology*, 27, 417–452.
Fox J-P (2010) *Bayesian Item Response Modeling: Theory and Applications*. Springer.
Fruhwirth-Schnatter S, Tuchler R (2008) Bayesian parsimonious covariance estimation for hierarchical linear mixed models. *Statistics & Computing*, 18, 1–13.
Fuglstad G, Simpson D, Lindgren F, Rue H (2018) Constructing priors that penalize the complexity of Gaussian random fields. *Journal of the American Statistical Association*, 114(525), 445–452.
Gelfand A, Sahu S (1999) Identifiability, improper priors, and Gibbs sampling for generalized linear models. *Journal of the American Statistical Association*, 94, 247–253.
Gelfand A, Sahu S, Carlin B (1995) Efficient parameterization for normal linear mixed models. *Biometrika*, 82, 479–488.
Gelfand A, Sahu S, Carlin B (1996) Efficient parameterizations for generalised linear models, in *Bayesian Statistics 5*, pp 165–180, eds J Bernardo, J Berger, A Dawid, A Smith. Clarendon Press, Oxford, UK.
Gelfand A, Smith A (1990) Sampling-based approaches to calculating marginal densities. *Journal of the American Statistical Association*, 85, 398–409.
Gelman A (2006) Prior distributions for variance parameters in hierarchical models. *Bayesian Analysis*, 1, 515–533.
Gelman A, Rubin D (1996) Markov chain Monte Carlo methods in biostatistics. *Statistical Methods in Medical Research*, 5, 339–355.
Gelman A, Stern H, Carlin J, Dunson D, Vehtari A, Rubin D (2014) *Bayesian Data Analysis*, 3rd Edition. Chapman and Hall/CRC.
Gelman A, van Dyk D, Huang Z, Boscardin J (2008) Using redundant parameterizations to fit hierarchical models. *Journal of Computational and Graphical Statistics*, 17, 95–122.
George E, Makov U, Smith A (1993) Conjugate likelihood distributions. *Scandinavian Journal of Statistics*, 20, 147–156.
George E, McCulloch R (1993) Variable selection via Gibbs sampling. *Journal of the American Statistical Association*, 88(423), 881–889.
Geweke J (1992) Evaluating the accuracy of sampling-based approaches to calculating posterior moments, in *Bayesian Statistics*, Volume 4. eds J Bernardo, J Berger, A Dawid, A Smith. Oxford University Press, New York.
Geweke J (1993) Bayesian treatment of the Student's-t linear model. *Journal of Applied Economics*, 8, S19–S40.

Geyer C, Thompson E (1995) Annealing Markov chain Monte Carlo with applications to ancestral inference. *Journal of the American Statistical Association*, 90, 909–920.

Ghosh J (2008) Efficient Bayesian Computation and Model Search in Linear Hierarchical Models. PhD Thesis ISDS, Duke University.

Gilks W (1996) Full conditional distributions, in *Markov Chain Monte Carlo in Practice*, pp 75–88, eds W Gilks, S Richardson, D Spiegelhalter. Chapman and Hall, London, UK.

Gilks W, Richardson S, Spielgelhalter D (1996) Introducing Markov chain Monte Carlo, in *Markov Chain Monte Carlo in Practice*, pp 1–19, eds W Gilks, S Richardson, D Spiegelhalter. Chapman and Hall, London, UK.

Gilks W, Wang C, Yvonnet B, Coursaget P (1993) Random-effects models for longitudinal data using Gibbs sampling. *Biometrics*, 38, 963–974.

Goldstein H, Spiegelhalter D (1996) League tables and their limitations: Statistical issues in comparisons of institutional performance. *Journal of the Royal Statistical Society: Series A (Statistics in Society)*, 159(3), 385–409.

Green P, Richardson S (1997) On Bayesian analysis of mixtures with an unknown number of components. *Journal of the Royal Statistical Society: Series B*, 59, 731–792.

Greenland S (2007) Bayesian perspectives for epidemiological research. II. Regression analysis. *International Journal of Epidemiology*, 36, 195–202.

Greenland S, Christensen R (2001) Data augmentation priors for Bayesian and semi-Bayes analyses of conditional-logistic and proportional-hazards regression. *Statistics in Medicine*, 20, 2421–2428.

Gustafson P. (1996) Local sensitivity of inferences to prior marginals. *Journal of the American Statistical Association*, 91, 774–781.

Gustafson P, Hossain S, MacNab Y (2006) Conservative priors for hierarchical models. *Canadian Journal of Statistics*, 34, 377–390.

Hadjicostas P, Berry S (1999) Improper and proper posteriors with improper priors in a Poisson-gamma hierarchical model. *Test*, 8, 147–166.

Harvey A (1989) *Structural Time Series Models and the Kalman Filter*. Cambridge University Press.

Hastings, W (1970) Monte-Carlo sampling methods using Markov Chains and their applications. *Biometrika*, 57, 97–109.

Hobert J, Casella G (1996) The effect of improper priors on Gibbs sampling in hierarchical linear mixed models. *Journal of the American Statistical Association*, 91, 1461–1473.

Hodge R, Evans M, Marshall J, Quigley J, Walls L (2001) Eliciting engineering knowledge about reliability during design-lessons learnt from implementation. *Quality and Reliability Engineering International*, 17, 169–179.

Hoffman M, Gelman A (2014) The No-U-turn sampler: Adaptively setting path lengths in Hamiltonian Monte Carlo. *Journal of Machine Learning Research*, 15(1), 1593–1623.

Hyndman R (1996) Computing and graphing highest density regions. *American Statistician*, 50, 361–365.

Ibrahim J, Chen M-H (2000) Power prior distributions for regression models. *Statistical Science*, 15, 46–60.

Jeffreys H (1961) *Theory of Probability*, 3rd Edition. Oxford University Press, Clarendon Press, Oxford, UK.

Johannes M, Polson N (2006) MCMC methods for continuous-time financial econometrics, in *Handbook of Financial Econometrics*, eds Y Ait-Sahalia, L Hansen. North Holland, Amsterdam.

Jonsen I, Myers R, James M (2006) Robust hierarchical state–space models reveal diel variation in travel rates of migrating leatherback turtles. *Journal of Animal Ecology*, 75, 1046–1057.

Jullion A, Lambert P (2007) Robust specification of the roughness penalty prior distribution in spatially adaptive bayesian p-splines models. *Computational Statistics and Data Analysis*, 51, 2542–2558.

Kass R, Carlin B, Gelman A, Neal R (1998) Markov chain Monte Carlo in practice: A round table discussion. *The American Statistician*, 52, 93–100.

Kass R, Wasserman L (1996) The selection of prior distributions by formal rules. *Journal of the American Statistical Association*, 91, 1343–1370.

Kleinman K, Ibrahim J (1998) A semiparametric Bayesian approach to the random effects model. *Biometrics*, 54, 921–938.
Klement, R, Bandyopadhyay, P, Champ, C, Walach, H (2018) Application of Bayesian evidence synthesis to modelling the effect of ketogenic therapy on survival of high grade glioma patients. *Theoretical Biology and Medical Modelling*, 15(1), 12.
Knorr-Held L, Rainer E (2001) Projections of lung cancer mortality in West Germany: A case study in Bayesian prediction. *Biostatistics*, 2, 109–129.
Koop G (2003) *Bayesian Econometrics*. John Wiley.
Krypotos A, Blanken T, Arnaudova I, Matzke D, Beckers T (2017) A primer on Bayesian analysis for experimental psychopathologists. *Journal of Experimental Psychopathology*, 8(2), jep-057316.
Laird N, Louis T (1989) Empirical Bayes confidence intervals for a series of related experiments. *Biometrics*, 45(2), 481–495.
Lenk P, DeSarbo W (2000) Bayesian inference for finite mixture models of generalized linear models with random effects. *Psychometrika*, 65, 475–496.
Lewandowski D, Kurowicka D, Joe H (2009) Generating random correlation matrices based on vines and extended onion method. *Journal of Multivariate Analysis*, 100(9), 1989–2001.
Liechty J, Liechty M, Muller P (2004) Bayesian correlation estimation. *Biometrika*, 91, 1–14.
Lindley D, Smith A (1972) Bayes estimates for the linear model. Journal of the Royal Statistical Society, B34, 1–41.
MacNab Y, Qiu Z, Gustafson P, Dean C, Ohlsson A, Lee S (2004) Hierarchical Bayes analysis of multilevel health services data: A Canadian neonatal mortality study. *Health Services and Outcomes Research Methodology*, 5, 5–26.
Marshall C, Best N, Bottle A, Aylin P (2004) Statistical issues in the prospective monitoring of health outcomes across multiple units. *Journal of the Royal Statistical Society: Series A*, 167, 541–559.
Marshall E, Spiegelhalter D (1998) Comparing institutional performance using Markov chain Monte Carlo methods, pp 229–249, in *Statistical Analysis of Medical Data: New Developments*, eds B Everitt, G Dunn. Arnold, London, UK.
McElreath R (2016) *Statistical Rethinking: A Bayesian Course with Examples in R and Stan*. CRC Press.
Mengersen K, Tweedie R (1996) Rates of convergence of the Hastings and Metropolis algorithms. *The Annals of Statistics*, 24, 101–121.
Millar R (2004) Sensitivity of Bayes estimators to hyper-parameters, with an application to maximum yield from fisheries. *Biometrics*, 60, 536–542.
Molenberghs G, Verbeke G, Demetrio, C (2007) An extended random-effects approach to modelling repeated, overdispersed count data. *Lifetime Data Analysis*, 13, 513–531.
Monnahan C C, Thorson J T, Branch T A (2017) Faster estimation of Bayesian models in ecology using Hamiltonian Monte Carlo. *Methods in Ecology and Evolution*, 8(3), 339–348.
Natarajan R, Kass R (2000) Reference Bayesian methods for generalized linear mixed models. *Journal of the American Statistical Association*, 95, 227–237.
Neal R (2011) MCMC using Hamiltonian dynamics, Chapter 5, in *Handbook of Markov Chain Monte Carlo*, eds S Brooks, A Gelman, G Jones, X-L Meng. CRC Press.
Oravecz Z, Muth C (2018) Fitting growth curve models in the Bayesian framework. *Psychonomic Bulletin and Review*, 25(1), 235–255.
Paap R (2002) What are the advantages of MCMC based inference in latent variable models? *Statistica Neerlandica*, 56, 2–22.
Palmer J, Pettit L (1996) Risks of using improper priors with Gibbs sampling and autocorrelated errors. *Journal of Computational and Graphical Statistics*, 5, 245–249.
Papaspiliopoulos O, Roberts G, Skold M (2003) Non-centered parameterisations for hierarchical models and data augmentation, pp 307–326, in *Bayesian Statistics 7*, eds J Bernardo, S Bayarri, J Berger, A Dawid, D Heckerman, A Smith, M West. Oxford University Press.
Raftery A (1996) Approximate Bayes factors and accounting for model uncertainty in generalized linear models. *Biometrika*, 83, 251–266.
Raftery A, Lewis S (1992) One long run with diagnostics: Implementation strategies for Markov chain Monte Carlo. *Statistical Science*, 7, 493–497.

Raftery A, Lewis S (1996) The number of iterations, convergence diagnostics and generic Metropolis algorithms, in *Practical Markov Chain Monte Carlo*, eds W Gilks, D Spiegelhalter, S Richardson. Chapman & Hall, London, UK.

Robert C (2015) *The Metropolis–Hastings Algorithm*. Wiley StatsRef: Statistics Reference Online, pp 1–15. https://onlinelibrary.wiley.com/doi/full/10.1002/9781118445112.stat07834

Robert C, Elvira V, Tawn N, Wu C (2018) Accelerating MCMC algorithms. *WIRES Computational Statistics*, 10, e1435.

Roberts G, Gelman A, Gilks W (1997) Weak convergence and optimal scaling of random walk metropolis algorithms. *The Annals of Applied Probability*, 7, 110–120.

Roberts G, Rosenthal J (2004) General state space Markov chains and MCMC algorithms. *Probability Surveys*, 1, 20–71.

Roberts G, Sahu S (1997) Updating schemes, correlation structures, blocking and parameterization of the Gibbs sampler. *Journal of the Royal Statistical Society B*, 59, 291–317.

Roberts G, Sahu S (2001) Approximate predetermined convergence properties of the Gibbs sampler. *Journal of Computational and Graphical Statistics*, 10, 216–229.

Roberts G, Tweedie R (1996) Geometric convergence and central limit theorems for multidimensional Hastings and Metropolis algorithms. *Biometrika*, 83, 95–110.

Rodrigues A, Assuncao R (2008) Propriety of posterior in Bayesian space varying parameter models with normal data. *Statistics & Probability Letters*, 78, 2408–2411.

Rue H, Martino S, Chopin N (2009) Approximate Bayesian inference for latent Gaussian models using integrated nested Laplace approximations. *Journal of the Royal Statistical Society, Series B*, 71(2), 319–392.

Sargent D (1998) A general framework for random effects survival analysis in the Cox proportional hazards setting. *Biometrics*, 54(4), 1486–1497.

Scollnik D (2002) Implementation of four models for outstanding liabilities in WinBUGS: A discussion of a paper by Ntzoufras and Dellaportas (2002). *North American Actuarial Journal*, 6, 128–136.

Shen W, Louis T (1998) Triple-goal estimates in two-stage hierarchical models. *Journal of the Royal Statistical Society: Series B*, 60, 455–471.

Sherlock C, Fearnhead P, Roberts G (2010) The random walk Metropolis: Linking theory and practice through a case study. *Statistical Science*, 25(2), 172–190.

Shoemaker J, Painter I, We B (1999) Bayesian statistics in genetics: A guide for the uninitiated. *Trends in Genetics*, 15, 354–358.

Simpson D, Rue H, Riebler A, Martins T, Sørbye S (2017) Penalising model component complexity: A principled, practical approach to constructing priors. *Statistical Science*, 32(1), 1–28.

Sinharay S, Stern H (2005) An empirical comparison of methods for computing Bayes factors in generalized linear mixed models. *Journal of Computational and Graphical Statistics*, 14, 415–435.

Siu N, Kelly D (1998) Bayesian parameter estimation in probabilistic risk assessment. *Reliability Engineering and System Safety*, 62, 89–116.

Spiegelhalter D (2004) Incorporating Bayesian Ideas into Health-Care evaluation. *Statistical Science*, 19, 156–174.

Sun D, Speckman P, Tsutakawa R (2000) Random effects in generalized linear mixed models (GLMMs), pp 23–39, in *Generalized Linear Models: A Bayesian Perspective*, eds D Dey, S Ghosh, B Mallick. Dekker, New York.

Sun D, Tsutakawa R, Speckman P (1999) Posterior distribution of hierarchical models using CAR(1) distributions. *Biometrika*, 86, 341–350.

Tierney L (1994) Markov Chains for exploring posterior distributions. *Annals of Statistics*, 21, 1701–1762.

Toft N, Innocent G, Gettinby G, Reid S (2007) Assessing the convergence of Markov Chain Monte Carlo methods: An example from evaluation of diagnostic tests in absence of a gold standard. *Preventive Veterinary Medicine*, 79, 244–256.

van Dyk D (2003) Hierarchical models, data augmentation, and Markov chain Monte Carlo, pp 41–56, in *Statistical Challenges in Modern Astronomy III*, eds G Babu, E Feigelson. Springer, New York.

Vanpaemel W (2011) Constructing informative model priors using hierarchical methods. *Journal of Mathematical Psychology*, 55(1), 106–117.

Vines S, Gilks W, Wild P (1996) Fitting bayesian multiple random effects models. *Statistics and Computing*, 6, 337–346.

Wetzels R, van Ravenzwaaij D,Wagenmakers E (2014) Bayesian analysis, in *The Encyclopedia of Clinical Psychology*, eds R Cautin, S Lilienfeld. Wiley-Blackwell, Hoboken, NJ.

Wikle C (2003) Hierarchical models in environmental science. *International Statistical Review*, 71, 181–199.

Willink R, Lira I (2005) A united interpretation of different uncertainty intervals. *Measurement*, 38, 61–66.

Yu B, Mykland P (1998) Looking at Markov samplers through cusum path plots: A simple diagnostic idea. *Statistics and Computing*, 8(3), 275–286.

Yue Y, Speckman P, Sun D (2012) Priors for Bayesian adaptive spline smoothing. *Annals of the Institute of Statistical Mathematics*, 64(3), 577–613.

Zhu M, Lu A (2004) The counter-intuitive non-informative prior for the Bernoulli family. *Journal of Statistics Education [Online]*, 12(2).

Zuur G, Garthwaite P, Fryer R (2002) Practical use of MCMC methods: Lessons from a case study. *Biometrical Journal*, 44, 433–455.

2

Bayesian Analysis Options in R, and Coding for BUGS, JAGS, and Stan

2.1 Introduction

R, available at https://cran.r-project.org/, is an integrated suite of software facilities for data manipulation, statistical analysis, and graphical display (R Core Team, 2016). The advantages of the R environment for Bayesian analysis are considerable, including access to extensive graphical capabilities (e.g. ggplot) and data manipulation facilities; a range of posterior diagnostic and summarisation tools; and the ability to obtain classical estimates in tandem with a full Bayesian analysis. A full list of packages in R is available at https://cran.r-project.org/web/packages/available_packages_by_name.html and www.onlinetoolz.com/tools/r-packages.php, while Bayesian analysis packages are listed at https://cran.r-project.org/web/views/Bayesian.html.

Worked examples in subsequent chapters focus primarily on three options for generic Bayesian analysis in R, based on user-defined program code. Implementation in R uses interfaces for BUGS such as R2OpenBugs and R2MultiBUGS, for JAGS (e.g. rjags, runjags, jagsUI), and for Stan (rstan). The LaplacesDemon package (CRAN, 2018) also offers Bayesian estimation options, with entirely R based user code. A number of packages use one or more of BUGS, JAGS, or Stan as a basis for coding and computation, but provide extra compilation checks, posterior summarisation, or data analysis options. Thus, the rube package (Seltman, 2016) interfaces with BUGS and JAGS to provide additional compilation details to assist with code debugging, while MCMCvis (Youngflesh, 2017) provides tools for posterior summarisation and visualisation which can be applied across all three generic options. The Nimble package aims to update BUGS and retain its functionality in the R environment (de Valpine et al., 2017), while R2MultiBUGS is a recently developed alternative to R2OpenBUGS and links to MultiBUGS (Goudie et al., 2019). Comparative analyses of some of these packages include Li et al. (2018) and Monnahan et al. (2017).

A range of application packages not requiring user-defined code adapted to the application is available. These have a different design philosophy to the generic coding options, using MCMC algorithms that are model-specific and hence likely to be more efficient (Martin and Quinn, 2006). As one example, bamlss (Bayesian Additive Models for Location, Scale, and Shape) enables Bayesian estimation of generalised linear models, additive regression, and spatial models (Umlauf et al., 2018). MCMCpack (Martin et al., 2011) allows estimation of generalised linear models, change-point models, quantile linear regression, and certain latent variable models. The rstanarm package (Gabry and Goodrich, 2018) uses Stan as a basis for estimation, but using simplified functions: for example, the stan_glm function to represent generalised linear models. The R-INLA package uses the Integrated

Nested Laplace Approximation as a computationally effective alternative to MCMC estimation and is illustrated in later examples. It is applicable to a range of generalised mixed models which can be represented in hierarchical form with latent Gaussian effects.

2.2 Coding in BUGS and for R Libraries Calling on BUGS

R interfaces to BUGS include R2OpenBUGS, rube, R2WinBUGS, and R2MultiBUGS. These use unmodified BUGS coding principles, as used in the standalone WINBUGS and OPENBUGS packages, and in the more recent MultiBUGS. For a review of BUGS, see Lykou and Ntzoufras (2011).

As also applies to JAGS, a BUGS program is declarative, namely, a description of the model and of the parameters or other stochastic nodes that may be monitored at each MCMC step. Like JAGS (but unlike Stan), there is no prescribed order for particular code elements. Thus, the code specifying prior densities may precede or follow the code specifying the model and likelihood. A wide variety of worked examples, including program codes, suggested initial value settings (for unknown parameters) are available at www.openbugs.net/Examples/Volumei.html and www.openbugs.net/Examples/Volumeii.html.

Many coding elements in BUGS are R-like, such as for-loops. However, unlike R, the specification of the univariate normal density (in both BUGS and JAGS) is in terms of mean and inverse variance (precision), with the multivariate normal being parameterised in terms of the precision matrix.

As an example, consider a BUGS code for a normal linear regression model with responses $y = (y_1,\ldots,y_n)$, and a single predictor $x = (x_1,\ldots,x_n)$. We consider in particular the ereturns data from the R package heavy (for heavy tailed regression, and related techniques), for which a normal linear regression may provide poorly fitted cases. The response (m.marietta in the dataset ereturns) is for excess returns from the Martin Marietta company, and the predictor (CRSP, an index for the excess rate returns for the New York stock exchange, in ereturns) is an index for the excess rate returns for the New York Stock Exchange.

The BUGS code includes predictions (replicates) at observed predictor values, and comparisons between these and the actual response data. It also includes a calculation of the residual precision from the standard deviation, which is assigned a uniform prior. Log-likelihood calculations are included to enable a subsequent call to the loo library (Vehtari et al., 2017). The code is included within a call to R2OpenBUGS as follows:

```
options(scipen=999)
library(R2OpenBUGS)
library(heavy)
library(loo)
data(ereturns)
x=as.vector(ereturns[[4]])
y=as.vector(ereturns[[3]])
# Data
D=list(y=y,x=x,n=60,x.new=0.13)
# Model Code
model <- function() { for (i in 1:n) {y[i] ~dnorm(mu[i],tau)
mu[i] <- beta[1] + beta[2]*(x[i]-mean(x[]))
```

```
# log-likelihood
LL[i] <- -0.92+0.5*log(tau)-0.5*tau*pow(y[i]-mu[i],2)
# replicates (predictions) at observed x[i]
yrep[i] ~dnorm(mu[i],tau)
# check replicate against actual observation
check[i] <- step(yrep[i]-y[i])}
# priors
for (j in 1:2) {beta[j] ~dnorm(0,0.001)}
# calculate precision
tau <- 1/(sigma*sigma)
sigma ~dunif(0,100)
# prediction at new x value
mu.new <- beta[1]+beta[2]*(x.new-mean(x[]))
y.new ~dnorm(mu.new,tau)}
inits1 = list(beta=rep(0,2), sigma=1)
inits2 = list(beta=rep(0,2), sigma=2)
inits = list(inits1,inits2)
pars = c("beta","sigma","check","y.new","LL")
n.iters=10000; n.burnin=500; n.chains=2
R=bugs(D,inits,pars,n.iters,model,n.chains,n.burnin,debug=T,
codaPkg = F,bugs.seed=10)
R$summary
LOO=loo(R$sims.list$LL)
LOO.PW=LOO$pointwise[,3]
```

As expected, a number of cases, particularly 8, 15, 34, and 58 have extreme posterior predictive checks, and these cases also have the most extreme pointwise LOO-IC values.

This example could also be run using R2MultiBUGS, with the second line now library(R2MultiBUGS), and the bugs command being:

```
R=bugs(D, inits, pars, model,n.chains=2, n.workers = 2, n.iter=10000,
MultiBUGS.pgm = "C:/Program Files/MultiBUGS/MultiBUGS.exe").
```

MultiBUGS (www.multibugs.org/) parallelises the MCMC algorithm with resulting shorter computing times.

2.3 Coding in JAGS and for R Libraries Calling on JAGS

While it is an adaptation of BUGS, JAGS has the advantages of more parsimonious coding. For example, if the prior on a linear regression residual variance is specified as a uniform on the standard deviation named as sigma (as in the example above), then one can directly specify:

```
y[i] ~ dnorm(mu[i],1/(sigma^2)).
```

Drawbacks of JAGS code relative to BUGS are that loop limits cannot involve any calculation, and the inability to take sub-samples at each MCMC iteration (see Example 3.5).

The JAGS code for the above regression example emphasises its essential similarity with the BUGS code, but also coding flexibility, in that equality rather than assignment signs are allowed, and extra facilities such as the logdensity.norm function to obtain log-likelihoods. The JAGS code also includes a function to generate suitable initial parameter values and calls on the jagsUI package. The jagsUI package has the benefit of repeatedly checking convergence and thus avoiding unnecessary computing. The calling sequence is as follows:

```
library(jagsUI)
library(heavy)
library(loo)
data(ereturns)
x=as.vector(ereturns[[4]])
y=as.vector(ereturns[[3]])
# Data
D=list(y=y,x=x,n=60,x.new=0.13)
cat("
model {for (i in 1:n) {y[i] ~dnorm(mu[i], 1/sigma^2)
mu[i] = beta[1] + beta[2]*(x[i]-mean(x[]))
# log-likelihood
LL[i] = logdensity.norm(y[i],mu[i],1/sigma^2)
# replicates at observed x[i]
yrep[i] ~dnorm(mu[i],1/sigma^2)
# check replicate against actual observation
check[i] = step(yrep[i]-y[i])}
# priors
for (j in 1:2) {beta[j] ~dnorm(0,0.001)}
sigma ~dunif(0,100)
# prediction at new x value
mu.new = beta[1]+beta[2]*(x.new-mean(x[]))
y.new ~dnorm(mu.new, 1/sigma^2)}
", file="model.jag")
# Estimation
inits <- function(){list(sigma=runif(0,5), beta=rnorm(2,0,0.1))}
pars = c("beta","sigma","check","y.new","LL")
R=autojags(D,inits,pars,model.file="model.jag",2,iter.increment=1000,
n.burnin=100,Rhat.limit=1.025, max.iter=5000, seed=1234, codaOnly=
c('LL'))
# Posterior Summary
R$summary
# Fit
LOO=loo(as.matrix(R$sims.list$LL))
LOO.PW.JAGS=LOO$pointwise[,3]
order(LOO.PW.JAGS)
```

Automatic convergence checking is also included in the package runjags (Denwood, 2016) and a calling sequence for this is as follows:

```
model ="model {for (i in 1:n) {y[i] ~dnorm(mu[i], 1/sigma^2)
mu[i] = beta[1] + beta[2]*(x[i]-mean(x[]))
# log-likelihood
LL[i] = logdensity.norm(y[i],mu[i],1/sigma^2)
# replicates at observed x[i]
yrep[i] ~dnorm(mu[i],1/sigma^2)
# check replicate against actual observation
check[i] = step(yrep[i]-y[i])}
```

```
# priors
for (j in 1:2) {beta[j] ~dnorm(0,0.001)}
sigma ~dunif(0,100)
# prediction at new x value
mu.new = beta[1]+beta[2]*(x.new-mean(x[]))
y.new ~dnorm(mu.new, 1/sigma^2 )} "
inits <- function(){list(sigma=runif(0,5), beta=rnorm(2,0,0.1))}
pars = c("beta","sigma","check","y.new","LL")
R = autorun.jags(model,data=D,startburnin=500,startsample=4000,
inits=inits,
monitor=pars ,n.chains=2)
add.summary(R)
# MCMC output for log-likelihoods
LLsamps=as.matrix(as.mcmc.list(R, vars = "LL"))
LOO=loo(LLsamps)
LOO.PW.JAGS=LOO$pointwise[,3]
order(LOO.PW.JAGS)
```

2.4 Coding for rstan

2.4.1 Hamiltonian Monte Carlo

Stan differs from BUGS and JAGS in using the no-U-turn sampler (Hoffman and Gelman, 2014) based on the Hamiltonian Monte Carlo (HMC) scheme (Neal, 2011; Duane et al., 1987; Betancourt and Girolami, 2015). HMC typically explores the posterior parameter space faster and more efficiently than Metropolis–Hastings or related algorithms, and so reaches convergence earlier, especially for high-dimensional models (Hoffman and Gelman, 2014), avoiding delays associated with random walk samplers. As mentioned in Stan Development Team (2017, p.593), "Stan might work fine with 1000 iterations with an example where BUGS would require 100,000 for good mixing."

2.4.2 Stan Program Syntax

Stan codes break the model specification into blocks. First is the data block, specifying all the data that is supplied in the input dataset, and referred to in later blocks in the code. This includes the number of observations. Integer and real data items are distinguished. The parameters block names the parameters, on which priors are usually explicit, and whose estimation is sought. Priors on these parameters are usually specified in the subsequent model block, which also specifies the data likelihood. Note that priors may be omitted, so that one has a 'flat' prior. In that case, parameter bounds (lower, upper, or both) should be stipulated, and this is good practice anyway. A transformed parameters block is for obtaining functions of parameters, for example, if the parameter block names the residual standard error σ, specified in the parameters block as real

```
<lower=0> sigma,
```

then one may be interested in estimating the precision as $\tau=1/\sigma^2$. The transformed parameters block may also specify limits, and this may facilitate particular types of model.

For example, one may specify a log-link in binomial regression by stipulating that probabilities π_i are between 0 and 1, with a log-link obtained as

```
real <lower=0,upper=1> pi[n];
for (i in 1:n) {pi[i]=exp(x[i]*beta);}
```

The generated quantities block specifies names and derivation of any quantities, such as log-likelihoods, resulting from the calculations during estimation. All distinct statements in all blocks must be terminated by a semicolon, which in for-loops precedes the closing } of the loop.

Flexibility in rstan coding is provided by opportunities to vectorise prior and likelihood statements in the model block (and hence avoid for-loops); see Section 9 on Regression Models in the Example Models section of the Stan User Guide (Stan Development Team, 2017).

We continue the regression example above, now specifying the predictors (intercept and CRSP) as a regression matrix. The generated quantities block includes log-likelihoods, generation of replicate data, and posterior predictive checks comparing replicate and actual data. Vectorisation is illustrated in the code by the statement

```
y ~normal(eta,sigma);
```

in the model block.
The calling sequence is:

```
options(scipen=999)
library(loo)
library(rstan)
# Regression Data
library(heavy)
data(ereturns)
x=as.vector(ereturns[[4]])
y=as.vector(ereturns[[3]])
x_new=0.13
K=2
X=matrix(,60,K)
X[,1]=1
X[,2]=x-mean(x)
x_new=x.new-mean(x)
D=list(y=y,X=X,n=60,K=2,x_new=x_new)
model="
data {
int n;    // number of observations
real y[n]; // response
real x_new; // new predictor value
int K; // number of predictors
matrix[n,K] X; // predictor matrix
}
parameters {
vector[K] beta; // regression coefficients
real <lower=0> sigma; // residual standard deviation
}
transformed parameters {
vector[n] eta; // linear regression term
eta = X*beta;
}
```

```
model {
sigma ~uniform(0,100);
beta ~normal(0,31.6);
y ~normal(eta,sigma);
}
generated quantities { real LL[n];
real y_rep[n];
real y_new;
real check[n];
for (i in 1:n) {LL[i]= normal_lpdf(y[i] eta[i],sigma); }
for (i in 1:n) {y_rep[i] =normal_rng(eta[i],sigma);}
for (i in 1:n) {check[i] =step(y_rep[i]-y[i]);}
y_new = normal_rng(beta[1]+beta[2]*x_new,sigma); // prediction at new x
value
}
"
# Estimation
fit=stan(model_code = model,data=D, iter = 1500,warmup = 250,chains=2)
# Posterior Summary
print(fit,digits=3)
# plot of posterior densities
# stan_dens(fit)
# Fit
LLsamps <- as.matrix(fit,pars="LL")
LOO=loo(LLsamps)
LOO.PW.STAN= LOO$pointwise[,3]
order(LOO.PW.STAN)
```

2.4.3 The Target + Representation

An advantage of rstan is that the log posterior can be explicitly specified using the target + representation. Thus, in the regression example, the prior samples sigma ~uniform(0,100) and beta ~normal(0,31.6) can be expressed as target +=uniform_lpdf(sigma | 0,100) and target += normal_lpdf(beta | 0, 31.6) respectively, and the likelihood y ~normal(eta,sigma) can be expressed as target += normal_lpdf(y | eta, sigma). So the model block becomes:

```
model {
target += uniform_lpdf(sigma | 0,100);
target += normal_lpdf(beta | 0, 31.6);
target += normal_lpdf(y | eta, sigma);
}
```

This is relevant in, say, marginal likelihood estimation, if one seeks to scale the contribution of the log-likelihood to the log-posterior (see Example 3.1); in regression using weighted log-likelihoods or regression using frequency tabulations; or in fitting distributions with custom likelihoods (not available among the standard densities included in rstan).

Using the target + format, rstan can accommodate improper priors as long as the posteriors are proper. Whereas BUGS and JAGS code specify a formal graphical model, for rstan, the code simply specifies a joint density function needed for HMC. Thus, Jeffreys prior on a variance σ^2, namely

$$p(\sigma) = 1/\sigma,$$

can be coded

```
target += -log(sigma);
```

As an illustration of a weighted log-likelihood estimated using the target + option, consider the api dataset from the R package survey (Carnes, 2017). This dataset has 200 observations with sample weights w_i. Unweighted logistic regression of the binary response yr.rnd.numeric on predictors' meals and mobility gives respective β coefficients (and s.e.) as 0.041 (0.010) and 0.041 (0.015). By contrast, weighted logistic regression using the Zelig package gives respective β coefficients (and s.e.) as 0.034 (0.012) and 0.086 (0.027), with a much-amplified effect of the mobility predictor and a diminished effect of the meals predictor.

For rstan estimation, the target + option includes survey weights w[i] to scale the log-likelihood contributions, as shown by the code in the following calling sequence:

```
data(api, package = "survey")
library(Zelig)
library(rstan)
apistrat$yr.rnd.numeric <- as.numeric(apistrat$yr.rnd == "Yes")
w.logit = zelig(yr.rnd.numeric ~meals + mobility, model = "logit.survey",
weights = apistrat$pw, data = apistrat)
unw.logit = glm(yr.rnd.numeric ~meals + mobility, data = apistrat, family
= "binomial")
attach(apistrat)
# Data for STAN, weights w scaled to average 1.
D=list(N=200,y=yr.rnd.numeric,x1=meals,x2=mobility,w=pw/mean(pw))
model ="
data { int N;
int<lower=0, upper=1> y[N]; // outcomes
real x1[N];
real x2[N];
real w[N]; // weights
}
parameters { real beta[3];
}
model { beta ~normal(0,5);
for (i in 1:N) {target += w[i]*bernoulli_lpmf(y[i] 1/
(1+exp(-beta[1]-beta[2]*x1[i]-beta[3]*x2[i])));}
}
"
fit=stan(model_code = model,data=D, iter = 1500,warmup =
250,chains=2,seed=10)
# Posterior Summary
print(fit,digits=3)
```

The rstan estimation gives respective posterior means (sd) for the coefficients on meals and mobility of 0.037 (0.09) and 0.064 (0.020). Again, the effect of mobility is amplified as compared to unweighted logistic regression (though less so than under the Zelig approach), while the effect of meals is attenuated.

The target + option can also be used with frequency data. Suppose housing tenants are grouped into 72 groups (with frequencies FREQ) according to an ordinal satisfaction response (three categories) and three categorical predictors, one with four categories, one with three, and one binary. Then an ordinal logistic regression can be applied via the

```
model {for (i in 1:72) { target += FREQ[i]*ordered_logistic_lpmf(y[i] |
x[i] * beta, tau);}}
```

This scenario is in fact applicable to the housing dataset in the R MASS library.

To demonstrate the target + option applied to a non-standard density, consider the Kumaraswamy distribution, obtained by sampling $y \sim \text{Beta}(1,b)$ and then $x = y^{1/a}$. The density is $p(x|a,b) = abx^{a-1}(1-x^a)^{b-1}$. We generate 1000 observations with $a=3$ and $b=2$.

The code sequence below provides posterior means (sd) for a and b of 3.00 (0.11) and 1.92 (0.09).

```
N =1000; a = 3; b = 2
# Kumaraswamy density
x = rbeta(N, 1, b)^(1/a)
library(rstan)
model ="
data {
int<lower=1> N;
real<lower=0,upper=1> x[N];
}
transformed data {
real sum_log_x;
sum_log_x = 0.0;
for (i in 1:N) {sum_log_x = sum_log_x + log(x[i]);}
}
parameters {
real<lower=0> a;
real<lower=0> b;
}
model {
target += N * (log(a) + log(b)) + (a - 1) * sum_log_x;
for (i in 1:N) { target += (b - 1) * log1m(pow(x[i],a)); }
}
"
D = list(N = N, x = x)
fit=stan(model_code = model,data=D, iter = 2500,warmup =
250,chains=2,seed=10)
# Posterior Summary
print(fit,digits=3)
```

2.4.4 Custom Distributions through a Functions Block

Although the target + specification may be used for non-standard densities, these can also be implemented using a functions block; see Chapter 24 in Stan Development Team (2017). The functions block (if used) should precede other blocks in an rstan code. The functions command will provide a log-likelihood term and the function name will include a _log suffix. However, the function call in the subsequent likelihood block will omit the _log suffix (Annis et al., 2017, p.872). Function names should not duplicate existing functions in rstan.

We consider the generalised Poisson (Consul, 1989):

$$p(x \mid \omega, \mu) = (1-\omega)\mu \frac{[(1-\omega)\mu + \omega x]^{x-1}}{x!} \exp\left(-[(1-\omega)\mu + \omega x]\right)$$

and the application by Joe and Zhu (2005). Joe and Zhu (2005, Table 3) consider data for n = 158 tumour count observations and provide estimates (mean, se) for ω and $\vartheta = \mu(1-\omega)$, namely 0.79 (0.04) and 0.91 (0.10).

An rstan implementation involves the sequence:

```
library(rstan)
# Tumour count data from Joe and Zhu (2005)
x=c(0,0,0,0,0,0,0,0,0,0,0,0,0,0,0,0,0,0,0,0,0,0,0,0,0,0,0,0,0,0,0,0,0
,0,0,0,0,0,0,0,0,0,0,0,0,0,0,0,0,0,
0,0,0,0,0,0,0,0,0,0,0,0,0,0,0,0,0,0,0,1,1,1,1,1,1,1,1,1,1,1,1,1,2,2,2,2,2
,2,2,2,2,2,2,2,2,2,2,3,3,3,3,3,3,
4,4,4,4,4,4,4,5,5,5,5,5,6,6,6,6,6,6,6,6,6,7,7,7,7,7,7,7,7,7,8,9,9,10,10,1
0,10,10,11,13,14,15,16,20,20,20,
21,24,24,24,26,30,50,50)
D=list(x=x,N=158)
model ="
functions {
real generalized_poisson_log(int x, real theta, real omega) {
return log(theta) + (x - 1) * log(theta + x*omega) - lgamma(x + 1)
- x * omega - theta ; }
}
data {
int<lower=0> N;
int x[N];
}
parameters {
real<lower=0> mu;
real<lower=-1, upper=1> omega;
}
transformed parameters {
real<lower=0> theta;
theta=mu*(1-omega);
}
model {
for (i in 1:N) {x[i] ~generalized_poisson(theta, omega);}
}
"
fit=stan(model_code = model,data=D, iter = 2500,warmup =
250,chains=2,seed=10)
# Posterior Summary
print(fit,digits=3)
```

We obtain estimates (posterior mean (sd)) for ω and ϑ of 0.797 (0.037) and 0.919 (0.095). Note that this model can be extended to better account for the zero inflation present in the data.

2.5 Miscellaneous Differences between Generic Packages (BUGS, JAGS, and Stan)

We consider the detailed availability of particular analysis options in the R-based implementations of the generic coding packages, namely BUGS, JAGS, and Stan; see also Appendix B in the Stan User's Guide and Reference Manual (Stan Development Team, 2017).

Missing Data: An advantage of BUGS and JAGS (and their R versions) is simple handling of missing data, either as missing at random, or informatively missing, if a selection mechanism is modelled as well. If response values for certain cases are specified as NA, and no specification regarding the origin of missingness is included, then values are imputed on a missing at random assumption. If predictor values are missing (again coded as NA), a specific generating density for that predictor needs to be included in the code. Note that Stan may provide estimates of the missing data values in the generated quantities block, possibly after data rearrangement (see Example 6.5). Missing data may also be included in the model likelihood, after appropriate subdividing of the response data: see Section 11.1 in Stan Development Team (2014).

Predictor Retention and Discrete Mixture Models: BUGS and JAGS allow predictor selection in regression using binary retention or exclusion indicators, whereas this is not possible, at least directly, in Stan, because HMC requires continuous parameter spaces. This is an illustration of a broader issue of "discrete assignment indicators," which also arises in discrete mixture analysis. Stan handles such analysis by focusing on the marginal likelihood, for example, using the likelihood assignment target += log_mix; see Chapters 13 and 15 in the Stan manual (Stan Development Team, 2017) and McElreath (2018).

Spatial Modelling: Compared to the other main languages, BUGS has the advantage of including priors to represent conditional spatial dependence for both univariate and multivariate outcomes, via the carnormal and mvcar priors. These priors can also be used in time series modelling. JAGS does not have these options. Stan can estimate spatially correlated effects (for SAR as well as CAR models) using the full multivariate priors (e.g. normal, Student t) of dimension N, where N is the number of units or points (Joseph, 2016; Morris, 2018), but this may become computationally intensive for large N. Multivariate normal prior options for CAR and SAR models are included in the brms package (Bürkner, 2017).

Augmented Data Representations: As compared to JAGS and Stan, BUGS can directly estimate augmented data versions of binary and multinomial regression, with latent utility interpretations (e.g. Albert and Chib, 1993). Stan can, however, estimate latent utilities in the generated quantities block.

Priors on Precision and Covariance Matrices: A limitation common to BUGS and JAGS (and their R implementations) is the limited choice in specifying the prior on the precision matrix (inverse covariance matrix) for the multivariate normal or multivariate t densities. For a univariate normal, there is more flexibility in these languages regarding the prior on the precision (e.g. gamma, lognormal). One may also obtain a univariate precision parameter from the corresponding standard deviation or variance, with a prior (e.g. uniform, lognormal) on the latter. For the multivariate normal, rstan (and associated R libraries such as rethinking and arms) offers extra flexibility in allowing (a) LKJ prior distributions for a correlation matrix, combined with priors on the standard deviations; (b) Cholesky decomposition expressions of the covariance matrix; see Sections 61 and 63 of Stan Development Team (2017); and (c) offering multivariate normal sampling using either the covariance or precision matrix (e.g. y ~multi_normal_prec).

References

Albert J, Chib S (1993) Bayesian analysis of binary and polychotomous response data. *Journal of the American Statistical Association*, 88, 669–679.

Annis J, Miller B, Palmeri T (2017) Bayesian inference with Stan: A tutorial on adding custom distributions. *Behavior Research Methods*, 49(3), 863–886.

Betancourt M, Girolami M. (2015) Hamiltonian Monte Carlo for hierarchical models. Chapter 4, pp 79–102, in U. Singh, S. Upadhyay, D. Dey (eds) *Current Trends in Bayesian Methodology with Applications*. CRC, Boca Raton, FL.

Bürkner P (2017) brms: An R package for Bayesian multilevel models using Stan. *Journal of Statistical Software*, 80(1), 1–28.

Carnes N (2017) Logistic Regression for Survey Weighted Data. http://docs.zeligproject.org/articles/zelig_logitsurvey.html

Consul P (1989) *Generalized Poisson Distribution: Properties and Applications*. Marcel Decker, New York.

CRAN (2018) Laplaces Demon: Complete Environment for Bayesian Inference. https://cran.r-project.org/web/packages/LaplacesDemon/LaplacesDemon.pdf

Denwood M (2016) runjags: An R package providing interface utilities, model templates, parallel computing methods and additional distributions for MCMC models in JAGS. *Journal of Statistical Software*, 71(9), 1–25.

de Valpine P, Turek D, Paciorek C, Anderson-Bergman C, Lang D, Bodik R (2017) Programming with models: Writing statistical algorithms for general model structures with NIMBLE. *Journal of Computational and Graphical Statistics*, 26(2), 403–413.

Duane S, Kennedy AD, Pendleton BJ, Roweth D (1987) Hybrid Monte Carlo. *Physics Letters B*, 195, 216–222.

Gabry J, Goodrich B (2018) How to Use the rstanarm Package. https://cran.r-project.org/web/packages/rstanarm/vignettes/rstanarm.html

Goudie R, Turner R, De Angelis D, Thomas A (2019) MultiBUGS: A parallel implementation of the BUGS modelling framework for faster Bayesian inference. *Journal of Statistical Software*. arXiv:1704.03216

Hoffman M, Gelman A (2014) The No-U-turn sampler: adaptively setting path lengths in Hamiltonian Monte Carlo. *Journal of Machine Learning Research*, 15(1), 1593–1623.

Joe H, Zhu R (2005) Generalized Poisson distribution: The property of mixture of Poisson and comparison with negative binomial distribution. *Biometrical Journal*, 47(2), 219–229.

Joseph M (2016) Exact sparse CAR models in Stan. http://mc-stan.org/documentation/case-studies/mbjoseph-CARStan.html

Li M, Dushoff J, Bolker B (2018) Fitting mechanistic epidemic models to data: A comparison of simple Markov chain Monte Carlo approaches. *Statistical Methods in Medical Research*, 27(7), 1956–1967.

Lykou A, Ntzoufras I (2011) WinBUGS: A tutorial. *WIRES: Wiley Interdisciplinary Reviews*, 3(5), 385–396.

Martin A, Quinn K (2006) Applied Bayesian inference in R using MCMCpack. *R News*, 6(1), 2–7.

Martin A, Quinn K., Park J (2011) MCMCpack: Markov chain Monte Carlo in R. *Journal of Statistical Software*, 42(9), 1–21. www.jstatsoft.org/v42/i09/

McElreath R (2018) Algebra and the Missing Oxen. http://elevanth.org/blog/2018/01/29/algebra-and-missingness/

Monnahan C, Thorson J, Branch T (2017) Faster estimation of Bayesian models in ecology using Hamiltonian Monte Carlo. *Methods in Ecology and Evolution*, 8(3), 339–348.

Morris M (2018) Spatial Models in Stan: Intrinsic Auto-Regressive Models for Areal Data. http://mc-stan.org/users/documentation/case-studies/icar_stan.html

Neal R (2011) MCMC Using Hamiltonian Dynamics, Chapter 5, in S Brooks, A Gelman, G Jones, X-L Meng (eds) *Handbook of Markov Chain Monte Carlo*. CRC Press, Boca Raton, FL, pp 113–162.

R Core Team (2016) R: A language and environment for statistical computing. R Foundation for Statistical Computing, Vienna. www.r-project.org/

Seltman H (2016) R Package rube (Really Useful WinBUGS (or JAGS) Enhancer). Version 0.3-8. http://www.stat.cmu.edu/~hseltman/rube/
Stan Development Team (2014) Stan Modeling Language: User's Guide and Reference Manual. https://github.com/stan-dev/stan/releases/download/v2.4.0/stan-reference-2.4.0.pdf
Stan Development Team (2017) Modeling Language User's Guide and Reference Manual, Version 2.17.0. https://mc-stan.org/users/documentation/
Umlauf N, Klein N, Zeileis A (2018) BAMLSS: Bayesian additive models for location, scale, and shape (and beyond). *Journal of Computational and Graphical Statistics*, 27(3), 612–627.
Vehtari A, Gelman A, Gabry J (2017) Practical Bayesian model evaluation using leave-one-out cross-validation and WAIC. *Statistics and Computing*, 27(5), 1413–1432.
Youngflesh C (2017) MCMCvis: Tools to Visualize, Manipulate, and Summarize MCMC Output. https://mran.microsoft.com/snapshot/2017-04-22/web/packages/MCMCvis/index.html

3

Model Fit, Comparison, and Checking

3.1 Introduction

Model assessment involves choices between competing models in terms of best fit, and checks to ensure model adequacy. For example, even if one model has a superior fit, it still needs to be established whether predictions from that model check with, namely, reproduce satisfactorily, the observed data. Checking may also seek to establish whether model assumptions (e.g. normality of random effects) are justified, whether the model reproduces particular aspects of the data, and whether particular observations are poorly fit (Sinharay and Stern, 2003; Berkhof et al., 2000; Kelly and Smith, 2011; Lucy, 2018; Conn et al., 2018; Park et al., 2015).

Once adequacy is established for a set of candidate models, one may seek to choose a particular best fitting model to base inferences on, or average over two or more adequate models with closely competing fit. This chapter focuses on three main strategies to assess model fit and carry out model checks: the formal approach; approaches based on posterior analysis of the likelihood; and predictive methods based on samples of replicate data. Particular emphasis is placed on their application in hierarchical models. Hierarchical indicator priors for selecting predictors are considered here (Section 3.4), and more extensively in Chapter 7.

R packages focusing particularly on Bayesian model selection or other aspects of model comparison include loo (Vehtari et al., 2017); mombf for regression and mixture analyses (Rossell, 2018; Johnson and Rossell, 2012); AICcmodavg (for deviance information criterion (DIC) calculation) (https://rdrr.io/cran/AICcmodavg/); BayesFactor (https://rdrr.io/cran/BayesFactor/), and the bridgesampling package (Gronau et al., 2017a). Packages focusing on predictor selection include BayesVarSel (https://rdrr.io/cran/BayesVarSel/), and BMA (https://rdrr.io/cran/BMA/).

3.2 Formal Model Selection

What is termed the formal approach to Bayes model selection is based on integration over the model parameter space to estimate marginal likelihoods and posterior model probabilities, leading on to possible model averaging. The canonical situation is provided by a "model-closed" or M-closed scenario (Key et al., 1999; Bernardo and Smith, 1994) where the set of models under consideration are judged to include the correct model, then formal model choice strategies are directed towards finding which model is most likely given the data.

Let prior model probabilities be denoted $p(m=k)$, where $m \in (1,\ldots,K)$ is a model indicator. Then posterior model probabilities are obtained as

$$p(m=k \mid y) = \frac{p(y \mid m=k)p(k)}{p(y)}$$

where

$$p(y \mid m=k) = \int p(y \mid \theta_k)p(\theta_k)d\theta_k$$

is the marginal likelihood for model k, with parameter θ_k of dimension d_k. This section considers approximations to marginal likelihoods and to Bayes factors

$$BF_{jk} = Pr(y \mid m=j)/Pr(m=k \mid y)$$

that compare such likelihoods. In simple models, such as normal linear regressions with regression coefficients and residual variance as the only unknowns, the formal approach is relatively simple to implement, and marginal likelihoods are available analytically under certain priors (Bos, 2002).

Approximate methods (Tierney and Kadane, 1986) for obtaining summary fit measures (e.g. marginal likelihoods) or posterior densities of parameters are also reliable in simple models. A large sample approximation for the log marginal likelihood is provided by the Bayesian Information Criterion (BIC) (Schwarz, 1978; Myung and Pitt, 2004) defined as

$$\text{BIC} = \log[p(y \mid \hat{\theta}_k)] - 0.5\, d_k \log(n)$$

where $\hat{\theta}_k$ is the maximum likelihood estimator, d_k is a known model dimension, and n is the sample size. The BIC is consistent for a wide set of problems, meaning that the probability of selecting the most parsimonious true model tends to 1 as the sample size tends to infinity. However, for singular model selection problems (discrete mixtures, factor models where the true number of factors is unknown, etc.), the asymptotic justification for the BIC no longer applies: considering the case of discrete parametric mixtures (Chapter 4), the Fisher information matrix with K components is singular at a distribution based on K-1 components. An alternative for such problems, the singular BIC, or sBIC, has been proposed (Drton and Plummer, 2017) and implemented in the R package sBIC (Weihs and Plummer, 2016). The widely applicable Bayesian information criterion (WBIC) can also be applied for nonsingular models (Watanabe, 2013; Friel et al., 2017).

Posterior model probabilities on nested models may also be obtainable by adding model selection indicators, as illustrated by Bayesian variable selection algorithms (Mitchell and Beauchamp, 1988; Fernandez et al., 2001) for choosing predictors in regression. Such selection has been extended to variance hyperparameters in hierarchical models (e.g. Cai and Dunson, 2006; Chen and Dunson, 2003; Fruhwirth-Schnatter and Tuchler, 2008; Kinney and Dunson, 2008), enabling selection which avoids the complex issues involved in marginal likelihood estimation for random effects models. Section 3.4 considers variance selection in hierarchical models.

However, in more complex random effect applications with discrete responses or hierarchically structured data, there remain issues which impede the straightforward application of the formal approach (Han and Carlin, 2001). For example, in approximating marginal likelihoods, there is a choice whether or not to integrate over random effects (Sinharay and Stern, 2005). The more commonly advocated approach of integrating out random effects becomes impractical when there are multiple possibly correlated random effects. The formal approach is also sensitive to priors adopted on parameters, which in the case of random effect models include the form of prior on variance components (e.g. inverse gamma or uniform), as well as the degree of prior informativeness. As priors become more diffuse, the formal approach tends to select the simplest least parameterised models, in line with the so-called Lindley or Bartlett paradox (Bartlett, 1957). Finally, the formal approach to model averaging requires both posterior densities $p(\theta_k \mid y, m = k)$, and posterior model probabilities $p(m = k \mid y)$. Estimates of posterior densities $p(\theta_k \mid y, m = k)$ may be difficult to obtain in complex random effects models with large numbers of parameters.

Straightforward and pragmatic approaches to model comparison, which are also applicable to complex hierarchical models, are available as alternatives to formal methods. The two main approaches are based on posterior densities of fit measures (log-likelihood, deviance) and on predictive assessment using samples of replicate data. Section 3.3 considers the posterior deviance as a fit measure and the related measure of model complexity (effective dimension) that are of utility in comparing hierarchical models. Bayesian fit measures such as the DIC or LOO-IC (Vehtari et al., 2017) are analogous to information theoretic approaches in frequentist statistics (Burnham and Anderson, 2002), but more widely applicable (e.g. to non-nested models). The components of the overall fit deriving from each observation (e.g. the deviance contributions from particular observations) may be used in model checking (Plummer, 2008).

The predictive approach to model choice and diagnosis (Section 3.5) has also been simplified by MCMC (Gelfand, 1996). Predictive methods shift the focus onto observables away from parameters (Geisser and Eddy, 1979) and seek to alleviate the impact on model comparison of factors such as specification of priors. The predictive approach is particularly advantageous in model checking, namely ensuring that a model actually reproduces the data satisfactorily (e.g. Kacker et al., 2008), but is also applied to model choice, for example, under posterior predictive loss criteria (Gelfand and Ghosh, 1998).

Predictive model checking typically involves repeated sampling of replicate data y_{new} from a model's parameters at each MCMC iteration (Gelfand et al., 1992). For a satisfactory model this process generates data like the observed data such that (y, y_{new}) are exchangeable draws from the joint density (Stern and Sinharay, 2005, pp.176–177).

$$p(y_{\text{new}}, y, \theta) = p(y_{\text{new}} \mid \theta, y) p(y \mid \theta) p(\theta) = p(y_{\text{new}} \mid \theta) p(y \mid \theta) p(\theta).$$

When all the data is used in model estimation, such sampling provides estimates of the posterior predictive density of model k, $p(y_{\text{new}} \mid y, m = k)$. However, predictive comparisons based on models using all the data in estimation may be overly favourable to the model being fitted (i.e. be conservative in terms of detecting model discrepancies) (Bayarri and Berger, 1999). An alternative involves cross-validation (Alqallaf and Gustafson, 2001) where the model predicts values for certain observations (the test sample) on the basis of a model estimated using the remaining observations (the learning sample). Key et al. (1999) argue that cross-validation is approximately optimal in an M-open scenario, where none of the models being considered is believed to be the true model.

3.2.1 Formal Methods: Approximating Marginal Likelihoods

As mentioned above, the global fit of a model with parameter vector θ under the formal Bayes paradigm is provided by the marginal likelihood $p(y|m=k)$, obtained by integrating the likelihood

$$p(y) = \int p(y \mid \theta)p(\theta)d\theta.$$

The marginal likelihood is also a component in Bayes formula, such that at any parameter value θ

$$p(\theta \mid y) = \frac{p(y \mid \theta)p(\theta)}{p(y)}.$$

Consider models 1 and 2 with equal prior model probabilities $p(m=1)=p(m=2)=0.5$. Then the ratio of posterior model probabilities is obtained as

$$\frac{p(m=2 \mid y)}{p(m=1 \mid y)} = \frac{p(y \mid m=2)}{p(y \mid m=1)} = B_{21}$$

where B_{21} is the Bayes factor. Kass and Raftery (1995) provide guidelines for interpreting B_{21}. If $2\log_e B_{21}$ is larger than 10, the evidence for model 2 is very strong, while values of $2\log_e B_{21} < 2$ are inconclusive as evidence in favour of one model or another. Note that such criteria are influenced by the prior adopted. In general, diffuse priors (whether on fixed effect parameters or variances) are to be avoided, as they tend to favour the selection of the simpler model.

Estimating the marginal likelihood by direct integration is generally infeasible in multi-parameter applications. Hence, a range of approximations have been proposed for estimating marginal likelihoods or associated model choice criteria, such as the Bayes factor. For example, on suitable rearrangement (Chib, 1995), the Bayes formula implies that the marginal likelihood may be approximated by estimating the posterior ordinate $p(\theta|y)$ in the relation

$$\log[p(y)] = \log[p(y \mid \theta_h)] + \log[p(\theta_h)] - \log[p(\theta_h \mid y)]$$

where θ_h is a point with high posterior density (e.g. posterior mean or median). One may estimate $p(\theta|y)$ by kernel density methods or by moment approximations based on MCMC output – see Lenk and DeSarbo (2000) for a discussion of such estimates. Let $g(\theta)$ denote an estimated density that approximates $p(\theta|y)$. One may then evaluate $g(\theta)$ at θ_h (Sinharay and Stern, 2005; Bos, 2002), so providing an estimate of the log marginal likelihood as

$$\log[p(y \mid \theta_h)] + \log[p(\theta_h)] - \log[g(\theta_h)]$$

The relation $\log[p(y)] = \log[p(y \mid \theta_h)] + \log[p(\theta_h)] - \log[p(\theta_h \mid y)]$ also implies a sampling-based estimator of the log marginal likelihood. Since this relation applies for all samples $\theta^{(r)}$, one may average over values

$$H^{(r)} = \log[p(y \mid \theta^{(r)})] + \log[p(\theta^{(r)})] - \log[g(\theta^{(r)})]$$

to estimate the log of the marginal likelihood, $\log[p(y)]$. Using log transforms is likely to be the most suitable approach for larger samples, to avoid numeric overflow. For small samples, one may set $L^{(r)} = p(y\,|\,\theta^{(r)})$, $\pi^{(r)} = p(\theta^{(r)})$, and $g^{(r)} = g(\theta^{(r)})$. Then an estimator of the marginal likelihood is provided by the simple average of the ratios $L^{(r)}\pi^{(r)}/g^{(r)}$.

Alternatively, suppose θ contains B parameter sub-blocks. When the full conditionals of each sub-block are available in closed form, Chib (1995) considers a marginal/conditional decomposition of $p(\theta_h|y)$ as follows

$$p(\theta_h\,|\,y) = p(\theta_{1h}\,|\,y)p(\theta_{2h}\,|\,\theta_{1h},y)p(\theta_{3h}\,|\,\theta_{1h},\theta_{2h},y)\ldots p(\theta_{Bh}\,|\,\theta_{1h},..\theta_{B-1,h},y)$$

with $p(\theta_h|y)$, and thus $p(y)$, estimated by using $B-1$ sampling sequences subsidiary to the main scheme. If $B=2$, namely $\theta_h = (\theta_{1h},\theta_{2h})$, the posterior ordinate $p(\theta_h|y)$ is then $p(\theta_{1h}\,|\,y)p(\theta_{2h}\,|\,y,\theta_{1h})$, where $p(\theta_{1h}\,|\,y)$ is estimated from the output of the main sample e.g. as

$$p(\theta_{1h}\,|\,y) = \sum_{r=1}^{R} p(\theta_{1h}\,|\,y,\theta_2^{(r)})$$

or by an approximation technique (e.g. assuming univariate/multivariate posterior normality of θ_1, or a kernel method). The second ordinate is available by inserting θ_{1h} and θ_{2h} in the relevant full conditional density. Chib and Jeliazkov (2001) extend this method to cases where full conditionals do not have a known normalising constant and have to be updated by Metropolis–Hastings steps.

3.2.2 Importance and Bridge Sampling Estimates

Let θ_k be the parameter vector for model k, denote the marginal likelihoods $p(y|m=k)$ as M_k, and let

$$p^*(\theta_k\,|\,y,m=k) = p(y\,|\,\theta_k,m=k)p(\theta_k\,|\,m=k)$$

denote the un-normalised posterior density of θ_k with

$$p^*(\theta_k\,|\,y,m=k)\,/\,c_k = p(\theta_k\,|\,y).$$

Then by definition

$$p(y\,|\,m=k) = \int p^*(\theta_k\,|\,y,m=k)d\theta.$$

Consider a function $g(\theta)$ with known normalising constants, often termed an importance function, and one that should ideally approximate the posterior $p(\theta|y)$. Then one has

$$p(y\,|\,m=k) = \int p^*(\theta_k\,|\,y,m=k)d\theta = \int \frac{p^*(\theta_k\,|\,y,m=k)}{g(\theta_k)}g(\theta_k)d\theta_k.$$

This suggests that an estimator for the marginal likelihood may be obtained using samples $\tilde{\theta}_k^{(r)}$ $(r=1,\ldots R)$ from $g(\theta_k)$, namely

$$M_k = \sum_r \frac{p^*(\tilde{\theta}_k^{(r)} \mid y, m=k)}{g(\tilde{\theta}_k^{(r)})}.$$

Let $\tilde{L}_k^{(r)} = p(y \mid \tilde{\theta}_k^{(r)})$, $\tilde{\pi}_k^{(r)} = p(\tilde{\theta}_k^{(r)})$ and $\tilde{g}_k^{(r)} = g(\tilde{\theta}_k^{(r)})$. Then, the importance sample estimator may be written in terms of weights $w_k^{(r)} = \tilde{\pi}_k^{(r)} / \tilde{g}_k^{(r)}$ comparing the prior and importance function, namely

$$M_k = \sum_r \tilde{L}_k^{(r)} w_k^{(r)}.$$

Bridge sampling estimators of marginal likelihoods use the fact that the marginal likelihood of model k is the normalising constant $c_k = p(y \mid m=k)$ in the relation

$$p(\theta_k \mid y, m=k) = \frac{p(y \mid \theta_k, m=k) p(\theta_k \mid m=k)}{p(y \mid m=k)} = \frac{p^*(\theta_k \mid y, m=k)}{c_k}.$$

The Bayes factor $B_{jk} = p(y \mid m=j) / p(y \mid m=k)$ is then a ratio c_j/c_k of normalising constants. Let $g(\theta)$ be an approximation to $p(\theta|y)$ with known normalising constant (e.g. suppose g consists of a multivariate normal density and a gamma density). Then one has

$$1 = \frac{\int \alpha(\theta_k) p(\theta_k \mid y) g(\theta_k) d\theta_k}{\int \alpha(\theta_k) g(\theta_k) p(\theta_k \mid y) d\theta_k} = \frac{E_g[\alpha(\theta_k) p(\theta_k \mid y)]}{E_p[\alpha(\theta_k) g(\theta_k)]}$$

where $\alpha(\theta)$ is a bridge function linking the densities $g(\theta)$ and $p(\theta|y)$ (Meng and Wong, 1996; Gronau et al., 2017b), $E_g[]$ denotes expectation with regard to the density $g(\theta)$, and $E_p[]$ denotes expectation with regard to the density $p(\theta|y)$. Substituting $p^*(\theta_k \mid y, m=k)/c_k$ for $p(\theta|y)$ in $1 = E_g[\alpha(\theta_k) p(\theta_k \mid y)] / E_p[\alpha(\theta_k) g(\theta_k)]$ gives the result

$$c_k = \frac{E_g[\alpha(\theta_k) p^*(\theta_k \mid y, m=k)]}{E_p[\alpha(\theta_k) g(\theta_k)]}.$$

For simplicity, omit conditioning on model k. Then with samples $\theta^{(r)}$ $(r=1,\ldots,S)$ and $\tilde{\theta}^{(r)}$ $(r=1,\ldots,R)$ from $p(\theta|y)$ and $g(\theta)$ respectively, one may estimate the marginal likelihood $p(y)$ of a particular model as

$$\left\{ \frac{1}{R} \sum_{r=1}^{R} \left[\alpha(\tilde{\theta}^{(r)}) p^*(\tilde{\theta}^{(r)} \mid y) \right] \right\} \Big/ \left\{ \frac{1}{S} \sum_{r=1}^{S} \left[\alpha(\theta^{(r)}) g(\theta^{(r)}) \right] \right\}$$

Setting $\alpha(\theta) = 1/g(\theta)$ then gives a marginal likelihood estimator

$$M = \frac{1}{R} \sum_{r=1}^{R} \frac{p^*(\tilde{\theta}^{(r)} \mid y)}{g(\tilde{\theta}^{(r)})}$$

that uses only samples from the approximate posterior (or importance) density $g(\theta)$.

Setting $\alpha(\theta) = 1/p^*(\theta \mid y)$ gives an estimator based on the harmonic mean of the ratios $p^*(\theta^{(r)} \mid y)/g(\theta^{(r)})$, and using parameters sampled from $p(\theta|y)$ rather than $g(\theta)$ (Gelfand and Dey, 1994). So

$$\frac{1}{M} = \frac{1}{S}\sum_{r=1}^{S}\frac{g(\theta^{(r)})}{p^*(\theta^{(r)} \mid y)}.$$

The choice $\alpha(\theta) = 1/[g(\theta)p^*(\theta \mid y)]$ leads to the geometric estimator of Lopes and West (2004), namely

$$M = \frac{\frac{1}{R}\sum_{r=1}^{R}\left[p^*(\tilde{\theta}^{(r)} \mid y)/g(\tilde{\theta}^{(r)} \mid y)\right]^{0.5}}{\frac{1}{S}\sum_{r=1}^{S}\left[g(\theta^{(r)} \mid y)/p^*(\theta^{(r)} \mid y)\right]^{0.5}}.$$

A recursive scheme for obtaining an optimal estimate of $\alpha(\theta)$ is also available, and mentioned by Lopes and West (2004, p.54) and Frühwirth-Schnatter (2004, equation 8). This simplifies if $R = S$, as in the first illustrative worked application below. With $R \neq S$, $s_1 = S/(S+R)$ and $s_2 = 1 - s_1$, one has an updated estimate for M at recursion j

$$M_j = A(M_{j-1})/B(M_{j-1})$$

where

$$A(u) = \sum_r W_{2r}/(s_1 W_{2r} + s_2 u), \quad B(u) = \sum_s 1/(s_1 W_{1s} + s_2 u),$$

$$W_{2r} = p(y \mid \tilde{\theta}^{(r)})p(\tilde{\theta}^{(r)})/g(\tilde{\theta}^{(r)}),$$

and

$$W_{1s} = p(y \mid \theta^{(s)})p(\theta^{(s)})/g(\theta^{(s)}).$$

3.2.3 Path Sampling

Another approximation may be obtained by a technique known as path sampling (Gelman and Meng, 1998; Xie et al., 2011; Friel and Pettitt, 2008). Consider a path variable t ranging from 0 to 1, and define the power posterior based on various levels of weighted likelihood, namely

$$p_t(\theta \mid y) \propto p(y \mid \theta)]^t p(\theta).$$

Define the posterior expectation

$$z(y \mid t) = \int \left[p(y \mid \theta)\right]^t p(\theta)d\theta$$

so that $z(y|t=0)$ is the integral of the prior, namely 1 for proper priors, while $z(y|t=1)$ is the marginal likelihood, $p(y) = \int p(y \mid \theta)p(\theta)d\theta$.

To derive an estimate of $z(y|t=1)$, one may use the identity

$$\log(p(y)) = \log\left(\frac{z(y|t=1)}{z(y|t=0)}\right) = \int_0^1 E_{\theta|y,t} \log[p(y|\theta)]dt$$

which states that the log marginal likelihood is the expected log-likelihood with respect to the power posterior at temperature t, with t ranging from 0 to 1. This follows (Friel and Pettitt, 2008) because

$$\frac{d}{dt}\log\left[z(y|t)\right] = \frac{1}{z(y|t)}\frac{d}{dt}z(y|t) = \frac{1}{z(y|t)}\frac{d}{dt}\left[\int \{p(y|\theta)\}^t p(\theta)d\theta\right]$$

$$= \frac{1}{z(y|t)}\int \{p(y|\theta)\}^t \log[p(y|\theta)]p(\theta)d\theta$$

$$= \int \frac{\{p(y|\theta)\}^t p(\theta)}{z(y|t)} \log[p(y|\theta)]d\theta$$

$$= E_{\theta|y,t} \log[p(y|\theta)].$$

One may numerically evaluate the integral over t using the trapezoid rule over T intervals defined using T+1 temperature functions

$$q_s = a_s^c \tag{3.1}$$

defined at cutpoints $\{a_0, \ldots a_L\}$ in [0,1], where c is a specified positive power. So, the estimate $\log(M_c)$ of the log marginal likelihood at that power is obtained by summing over T grid points that combine information from successive expected log likelihoods,

$$\log(M_c) = \sum_{s=0}^{T-1}(q_{s+1} - q_s)\frac{1}{2}\left[E_{\theta|y,q_{s+1}} \log[p(y|\theta)] + E_{\theta|y,q_s} \log[p(y|\theta)]\right].$$

Friel and Pettitt (2008) take $c=4$ in (3.1), while Xie et al. (2011) recommend values of c between 2.5 and 5. So with $T=40$ intervals, equally spaced cutpoints $\{a_0 = 0, a_1 = 0.025, a_2 = 0.05, \ldots a_{40} = 1\}$, and setting $c=4$, one has $q_0=0$, $q_1 = (0.025)^4$, …, $q_{39} = 0.975^4$, $q_{40}=1$. The Monte Carlo standard error of $\log(M_c)$ is obtained as the square root of the summed variances of the contributions to $\log(M_c)$ at each of T grid points. Thus let $\delta_s = (1/2)(q_{s+1} - q_s)$ and let ν_s be the Monte Carlo variance of $E_{\theta|y,q_{s+1}} \log[p(y|\theta)]$. Then the variance at each grid point is $\delta_s^2 \nu_s$ and the Monte Carlo variance of $\log(M_c)$ is $\sum_{s=0}^{T-1} \delta_s^2 \nu_s$.

To illustrate estimation in path sampling consider the online vignette demonstrating the use of the R package bridgesampling (Gronau et al., 2017a; https://cran.r-project.org/web/packages/bridgesampling/vignettes/bridgesampling_example_jags.html). This example assumes a normal-normal two stage hierarchy (see Chapter 4), as often used in meta-analysis, with known first level variance σ^2

$$y_i \sim N(\theta_i, \sigma^2),$$

$$\theta_i \sim N(\mu, \tau^2).$$

The comparison is between a model with $\mu=0$ and a model with μ unknown. The data ($n=20$ cases) are generated under the first option, namely with $\mu=0$, and also with $\sigma^2=1$ and $\tau^2=0.5$.

Path sampling as in Friel and Pettitt (2008) is applied, with $q_s = a_s^4$ where $\{a_s\} = \{a_0, 1/T, 2/T, \ldots, T-1/T, T\}$ and $T=30$. For numeric stability, a_0 is taken as 0.00001 rather than 0, so that $q_0 = 1E-20$. Estimates are made using jagsUI. The parameters and likelihoods at each of the T+1 points are estimated using the device from Barry (2006). The likelihood is specified as in (4.4.3), namely:

$$y_i \sim N(\theta_i, \sigma^2 + \tau^2),$$

with $\sigma^2 = 1$ known. Using the code listed in the Computational Notes [1] in Section 3.7, the estimated marginal likelihoods are closely similar to those reported in the bridgesampling vignette, namely −37.53 for the zero-mean model, and −37.81 for the model with μ unknown.

3.2.4 Marginal Likelihood for Hierarchical Models

For conjugate hierarchical models (e.g. Poisson-gamma mixtures) the marginal likelihood can be obtained analytically (Albert, 1999). However, general linear mixed models (Clayton, 1996) are widely used for handling multiple random effects, with regression terms

$$\eta_i = X_i\beta + W_i b_i,$$

where X_i and W_i are predictors, and b_i are latent data. For such non-conjugate schemes, the marginal likelihood is not obtainable analytically, and one possible approach to evaluating marginal likelihoods is to work with the integrated likelihood

$$p(y|\theta) = \int p(y,b|\theta)db = \int p(y|b,\theta)p(b|\theta)db,$$

where the random effects or latent data have been integrated out, and where θ includes hyperparameters ψ (e.g. covariances) governing the b, as well as parameters φ (e.g. fixed regression effects) not relevant to the random effect hyperdensity (Sinharay and Stern, 2005; Fruhwirth-Schnatter, 1999). This can be done in practice in MCMC sampling by applying importance sampling, the Laplace approximation, or numeric integration methods to the complete data likelihood $p(y,b|\theta)$.

However, it may be argued that under a Bayesian approach, the distinction between fixed and random regression coefficients is less relevant, and so use of the integrated likelihood approach and implied numerical complexity may be avoided. For example, one may (e.g. Clayton, 1996) adopt a unified perspective on the parameters in the joint precision matrix for the fixed effects (and other parameters not in the hyperdensity) φ, and the random effects hyperparameters ψ. Chib (2008) proposes marginal likelihood estimation for different classes of panel model by marginalisation over the random effects. Sinharay and Stern (2005) also mention obtaining the marginal likelihood by considering the expanded parameter set $\omega = (b,\varphi,\psi)$, so that

$$p(y) = \iint p(y|\varphi,\psi)p(\varphi,\psi)d\varphi d\psi,$$

$$= \iiint p(y|\varphi,b)p(b|\psi)p(\psi,\varphi)dbd\psi d\varphi.$$

The advantage of working with the expanded likelihood $p(y|b,\varphi)$ is the avoidance of repeated integration, but this comes at the expense of an often considerably increased dimension of the parameter space (namely by the number of components in b). Marginal likelihood approximation retaining the expanded likelihood is considered in real examples by Nandram and Kim (2002) and Gelfand and Vlachos (2003).

Let $g(\theta|b)$ be a density subject to $\int g(\theta|b)d\theta=1$, where $\theta=(\psi,\varphi)$, and let θ^* be an appropriate fixed point (e.g. a posterior mean). Chen (2005) mentions an estimator for the log marginal likelihood $M=p(y)$ in a hierarchical modelling situation based on the identity

$$p(y|\theta^*)=\int p(y|\theta^*,b)p(b|\theta^*)g(\theta|b)d\theta db,$$

$$=\int \frac{g(\theta|b)}{p(\theta)}\frac{p(y|\theta^*,b)p(b|\theta^*)}{p(y|\theta,b)p(b|\theta)}p(y|\theta,b)p(b|\theta)p(\theta)d\theta db,$$

$$=p(y)E\left[\frac{g(\theta|b)}{p(\theta)}\frac{p(y|\theta^*,b)p(b|\theta^*)}{p(y|\theta,b)p(b|\theta)}|y\right],$$

where the expectation is over samples from $p(\theta|y)$. Taking logarithms provides

$$\log[M]=\log[p(y|\theta^*)]-\log[E\left[\frac{g(\theta|b)}{p(\theta)}\frac{p(y|\theta^*,b)p(b|\theta^*)}{p(y|\theta,b)p(b|\theta)}|y\right].$$

So, with samples $\{\theta^{(r)},b^{(r)}\}$ from $p(\theta,b|y)$, an estimator for log(M) is

$$\log[M]=\log[p(y|\theta^*)]-\log\left[\frac{1}{R}\sum_{r=1}^{R}\frac{g(\theta^{(r)}|b^{(r)})}{p(\theta^{(r)})}\frac{p(y|\theta^*,b^{(r)})p(b^{(r)}|\theta^*)}{p(y|\theta^{(r)},b^{(r)})p(b^{(r)}|\theta^{(r)})}\right].$$

One option is then to set $g(\theta|b)=p(\theta)$, leading to

$$\log[M]=\log\left[p(y|\theta^*)\right]-\log\left[\frac{1}{R}\sum_{r=1}^{R}\frac{p(y|\theta^*,b^{(r)})p(b^{(r)}|\theta^*)}{p(y|\theta^{(r)},b^{(r)})p(b^{(r)}|\theta^{(r)})}\right],$$

The component $\log[p(y|\theta^*)]=\log(L^*)$ may be estimated from the Monte Carlo average

$$L^*=\frac{1}{R}\sum_{r=1}^{R}p(y|\theta^*,b^{(r)}).$$

Chen (2005) shows that a variance minimising estimator is, however, obtained by setting $g(\theta|b)=p(\theta|b,y)$, namely the conditional posterior density of θ given b.

Example 3.1 Marginal Likelihood and Bayes Factors, Turtle Mortality Data

This example applies approximations to the marginal likelihood to data from Sinharay and Stern (2005). These are nested binary data y_{ij} on $n=244$ newborn turtles $i=1,\ldots,m_j$ clustered into clutches $j=1,\ldots,J$, with responses $y_{ij}=1$ or 0 according to survival or death. The known predictor is turtle birthweight x_{ij} so there are $p=2$ regression parameters, including an intercept. Graphical analysis suggests that heavier turtles have better survival chances, but also suggests extraneous variability in survival rates across clutches.

Sinharay and Stern (2005) compare several methods of deriving formal model fit measures, namely marginal likelihoods or Bayes factors. Here two alternative models are evaluated for the probability $\pi_{ij} = Pr(y_{ij} = 1)$ using the temperature path approach of Friel and Pettitt (2008) and jagsUI. One involves a fixed effects only regression on birthweight with a probit link. The other assumes additional random effects based on clutch membership. So model 1 specifies $\pi_{ij1} = \Phi(\beta_1 + \beta_2 x_{ij})$, while model 2 has

$$\pi_{ij2} = \Phi(\beta_1 + \beta_2 x_{ij} + b_j),$$

where $b_j \sim N(0, \sigma_b^2)$. The predictor x_{ij} is standardised, and unlike Sinharay and Stern (2005), N(0,1) priors are assumed on the fixed effects $\{\beta_1, \beta_2\}$. A gamma prior on τ_b, namely $\tau_b \sim Ga(0.1, 0.1)$, is assumed, as the shrinkage prior

$$p(\sigma_b^2) \propto \frac{1}{(1+\sigma_b^2)^2}, \tag{3.2}$$

used by Sinharay and Stern cannot be implemented in jagsUI.

There are some possible sources of sensitivity: formal model measures may depend on informativeness in the priors and on the form of prior, for example, the prior density adopted on the random effects variance σ_b^2 or the precision $\tau_b = 1/\sigma_b^2$. For example, the value $\sigma_b^2 = 1$ has a quarter of the prior weight for $\sigma_b^2 = 0$ under the prior used by Sinharay and Stern (2005). With this prior, they obtain an inconclusive Bayes factor of 1.27 in favour of the simpler fixed effects only model. Under particular methods, additional sensitivity issues occur. Using the temperature path approach, estimates of the marginal likelihood may be affected by the number of sequence points T (Drummond and Bouckaert, 2015) and by the location of the points, especially points near zero.

The temperature path a_s has T = 10, and $q_s = a_s^4$ where a = (0.00025,0.0005,0.001,0.005,0.05,0.075,0.1,0.25,0.5,0.75,0.99). The parameters and likelihoods at each of the T + 1 points are estimated using the device from Barry (2006). For numeric stability, the initial point in the path is taken as 0.00025 rather than 0. Although formally, the estimate of $\log[p(y)]$ is obtained by piecing together the separate posterior estimates $E_{\theta_k|y,t_s} p(y \mid \theta_k)$, an essentially identical estimate is obtained by applying the trapezoid rule at each iteration and monitoring the composite log marginal likelihood node.

The marginal likelihood estimate for model 1 is thus obtained as −150.4 as compared to −152.56 for the random effects alternative, model 2, giving B_{12} = 8.7. σ_b^2 is estimated at 0.152. Relatively large clutch effects (mean, posterior sd) are obtained under model 2 for clutches 9 and 15, namely 0.46 (0.27) and −0.37 (0.27).

As an alternative option for the random effects approach, defining model 3, a shrinkage prior is implemented by taking a uniform prior on $U = 1/(1+\sigma_b^2)^2$, namely

$$U \sim \text{Unif}(0,1),$$

$$\sigma_b^2 = (1 - U^{0.5}) / U^{0.5}.$$

This option produces a clutch variance estimate $\sigma_b^2 = 0.149$. The marginal likelihood is −150.4, so that BF_{13} is 1. Thus varying Bayes factors illustrate the impacts of different priors on the variance or precision. This approach can be extended to allow uncertainty in the shrinkage prior power, allowing potentially more pronounced shrinkage (Gustafson et al., 2006). Thus, one takes a uniform prior on

$$U = 1/(1+\sigma_b^2)^P,$$

where P is unknown with a minimum of 1. So with $P = 1 + P_1$, where $P_1 \sim Ga(0.1, 0.1)$, one has

$$\sigma_b^2 = (1 - U^{1/P}) / U^{1/P}.$$

This leads to an estimate for P of the default value 1, but with an estimated marginal likelihood of −151.4 (and so BF_{13}=2.7), while the posterior mean for σ_b^2 is now 0.155.

An advantage with rstan is that the prior (3.2) can be represented using the expression (within the model segment)

```
target += -2 * log(1 + sigma2);
```

where sigma2 is the unknown variance. Using rstan in combination with the bridgesampling package provides respective marginal likelihoods for models 1 and 2 of −156.48 and −156.71, and a Bayes factor BF_{12}=1.26 (Gronau et al., 2017a). This is close to that reported by Sinharay and Stern (2005). The clutch variance for model 2 (with prior as in equation 3.2) is $\sigma_b^2 = 0.153$.

This option is also used to compare the fixed effects regression model with a variable slopes model (model 4), namely

$$\pi_{ij4} = \Phi(\beta_1 + (\beta_2 + b_j)x_{ij}),$$

where $b_j \sim N(0, \sigma_b^2)$. A gamma prior on $1/\sigma_b^2$ is taken. The resulting marginal likelihood is −160.13, and so a more decisive advantage for the simpler model, with BF_{14}=38.66. This counts as strong evidence in favour of the simpler model according to the schedules of Jeffreys (1961), and of Kass and Raftery (1995).

One may also apply rstan to direct path sampling, namely to estimating a sequence of models with varying temperatures t,

$$p_t(\theta | y) \propto [p(y | \theta)]^t p(\theta)$$

with t ranging from 0 to 1. If $U[t_i]$ denotes the actual log likelihood at an ascending temperature sequence $t_i \in [0,1]$, for $i = 1, \ldots, T$, then the marginal likelihood estimate is $\sum_{i=1}^{T} U(t_i)/T$. Alternatively, one may generate T temperatures randomly from the uniform $U(0,1)$. For a selected temperature t, the code for the fixed effects model is

```
model="data {
int<lower = 1> N;
int<lower = 0, upper = 1> y[N];
real<lower = 0> x[N];
real<lower=0, upper=1> t;//parameter for path sampling
}
parameters {
real alpha;
real beta;
}
transformed parameters {
real U_case[N];
for (i in 1:N) {U_case[i]= bernoulli_lpmf(y[i] |
Phi(alpha+beta*x[i]));}
}
```

```
model {
target += normal_lpdf(alpha 0, 3.16);
target += normal_lpdf(beta 0, 3.16);
for (i in 1:N) {target += t*bernoulli_lpmf(y[i] |
Phi(alpha+beta*x[i]));}
}
generated quantities {
real U;
U = sum(U_case);
}
"
For example, a calling sequence with T=1000 randomly generated
temperatures is
T=1000
temps=runif(T,0,1)
U=c()
sink("sink.txt")
for (i in 1:length(temps)) {D=list(y = turtles$y,x = turtles$x,N
=244,t=temps[i])
fit=stan(model_code=model,data=D,iter=1250,warmup=250,chains=1,refre
sh=-1,seed=100)
U[i]= summary(fit, pars = c("U"))$summary[1]}
sink()
# marginal likelihood estimate
mean(U)
```

where the likelihood at temperature t_i is U[i]. In practice, this approach is computationally intensive, requiring high T, and producing different results each time. The above call led to an estimated marginal likelihood of –157.15.

3.3 Effective Model Dimension and Penalised Fit Measures

Classical model choice is frequently based on penalised likelihood criteria, such as the Akaike Information Criterion or AIC (Akaike, 1973), and the Bayesian Information Criterion or BIC (Schwarz, 1978). Such criteria are applicable in comparing fixed effects models with known dimension d, and with models assumed nested within one another. With L denoting a log likelihood, and $D=-2L$ denoting the deviance, log likelihood ratio tests comparing maximised log likelihoods of models 1 and 2 are obtained with

$$C = -2(\log L_1 - \log L_2) = D_1 - D_2$$

where C is approximately chi-square, with degrees of freedom $d_2 - d_1$ equal to the number of additional parameters in the more complex model 2. The AIC is defined as $2d - 2L = D + 2d$ and the difference in AICs between models 1 and 2 is $\Delta AIC = C + 2(d_1 - d_2)$. However, classical likelihood ratio testing is not possible in random effects models or models with parameter constraints (e.g. order or size constraints) that make the effective number of estimated parameters itself a random variable so that the asymptotic distribution of the log likelihood ratio is unknown.

3.3.1 Deviance Information Criterion (DIC)

Spiegelhalter et al. (2002) provide a penalised fit criterion analogous to the AIC and BIC, called the deviance information criterion or DIC. This is applicable to comparing non-nested models and also to models including random effects where the true model dimension is another unknown. The DIC is based on the posterior distribution of the deviance statistic

$$D(\theta \mid y) = -2\log[p(y \mid \theta)] + 2\log[h(y)]$$

where $p(y|\theta)$ is the likelihood of the data y given parameters θ, and $h(y)$ is a standardising function of the data only (and so does not affect model choice).

Suppose the deviance is monitored during an MCMC run, providing samples $\{D^{(1)},\ldots,D^{(R)}\}$. The overall fit of a model is measured by the posterior expected deviance obtained by averaging over the posterior density of the parameters,

$$\bar{D} = E_{\theta|y}[D],$$

while the effective model dimension, d_e, is estimated as

$$d_e = E_{\theta|y}[D] - D(E_{\theta|y}[\theta]) = \bar{D} - D(\bar{\theta} \mid y), \tag{3.3}$$

namely the expected deviance minus the deviance at the posterior means of the parameters; the latter is also known as the plug-in deviance (Plummer, 2008). In hierarchical random effects models, the effective number of parameters in total is typically lower than the nominal number of parameters, due to borrowing of strength under the hyperdensity (e.g. Zhu et al., 2006; Buenconsejo et al., 2008).

The DIC is then obtainable as the expected deviance plus the effective model dimension,

$$\text{DIC} = \bar{D} + d_e = D(\bar{\theta}) + 2d_e. \tag{3.4}$$

So the DIC will prefer models with lower values of $\bar{D}$, combined with smaller values of d_e (which indicate a relatively parsimonious model). A possible disadvantage with the DIC is that it can be affected by reparameterisation of θ or by the form of link in general linear models, with this applying in particular to the "plug-in" deviance $D(\bar{\theta} \mid y)$; hence the value of d_e may be sensitive to parameterisation.

The deviance $D(\bar{\theta} \mid y)$ at the posterior mean $\bar{\theta}$ of the parameters may also be estimated by using posterior means of quantities involved in defining the deviance, such as case means (Poisson likelihood), means and overdispersion parameter (negative binomial likelihood), means and variance (normal likelihood), and so on (Spiegelhalter et al., 2002, p.596). Thus, let μ_i denote case specific means and ξ denote any other parameters needed to derive the deviance. Then an estimate $D(\bar{\mu},\bar{\xi} \mid y)$ may be more easily obtainable than $D(\bar{\theta} \mid y)$ in complex (e.g. discrete mixture) models, or in models with many random effects, where the number of nominal parameters may considerably exceed the number of cases. This type of procedure is also mentioned by Spiegelhalter (2006) in terms of monitoring the "direct parameters" that appear in the distributional syntax and plugging these into the deviance; it was adopted in the paper by Ohlssen et al. (2006, section 2).

The DIC and d_e can be disaggregated to individual observations, and provide a measure of local complexity, namely of observations that are more problematic under the model relative to others. Spiegelhalter et al. (2002, p.602) mention that the local complexity measures

$$d_{ei} = \bar{D}_i - D_i(\bar{\theta})$$

measure the leverage of observation i, defined as the relative influence that each observation has on its own fitted value. Unusually large observation specific DIC measures, namely

$$\text{DIC}_i = \bar{D}_i + d_{ei}$$

are used by Spiegelhalter et al. (2002) as indicators of outlier status – observations inconsistent with the model. The DIC can be seen as a Bayesian version of AIC and may under-penalise model complexity, as pointed out by discussants to Spiegelhalter et al. (2002). By contrast, it is well established (Burnham and Anderson, 2002) that the BIC tends to select overly parsimonious models. A fit criterion analogous to the BIC may be defined as

$$\text{DIC}^* = D(\bar{\theta}) + d_e \log(n)$$

and was used by Pourahmadi and Daniels (2002, p.228) for panel data with repeated observations over n subjects.

Note that the model with the lowest DIC or DIC* will not necessarily be a suitable model if it does not reproduce the data adequately. Hence, model checks are required to assess consistency of predictions from the model with the actual observations.

Just as there are alternative approaches to marginal likelihood derivation in hierarchical models, Spiegelhalter et al. (2002) point that for such models, one cannot uniquely define the likelihood or model complexity without specifying the level of the hierarchy that is the model focus. Thus one might analyse count data using a complete data likelihood (with unknown latent data b as well as hyperparameters θ) using a Poisson-gamma or Poisson-lognormal model, or alternatively apply a negative binomial likelihood with the random effects integrated out (Fahrmeir and Osuna, 2003), and the complexity measures will obviously differ.

Model choice may be affected by the focus, as shown by Plummer (2008, p.530) in an analysis of a discrete mixture model, with one approach considering a complete data likelihood $p_C(y \mid b, \theta)$ (with the parameters including missing component indicators), and the other considering the integrated likelihood $p_I(y \mid \theta)$. Ando (2007) considered DICs based on both conditional and integrated likelihoods, namely DIC^C and DIC^I, and showed that both tend to select overfitted (i.e. non-parsimonious) models.

3.3.2 Alternative Complexity Measures

Plummer (2008) confirms that the DIC tends to under-penalise complex models, particularly when the ratio of the sample size to the effective number of parameters is relatively low. Plummer (2008) proposes an alternative effective dimension penalty based on cross-validation considerations. Thus, suppose Z constitutes training data and Y are test data, and consider a loss function $L(Y,Z)$. Assume that Y and Z are conditionally independent given θ. Then $p(Y \mid \theta, Z) = p(Y \mid \theta)$, and the log-scoring rule (Gneiting and Raftery, 2007) for

Y is then the log-likelihood $\log\{P(Y\,|\,\theta)\}$ of θ. The corresponding loss function is the deviance $D(\theta) = -2\log\{p(Y\,|\,\theta)\}$.

As estimates of the loss function, one may consider either the plug-in deviance

$$L^p(Y\,|\,Z) = -2\log[\{p(Y\,|\,\bar{\theta}(Z)\}]$$

where $\theta(Z) = E(\theta\,|\,Z)$, or the expected deviance

$$L^e(Y\,|\,Z) = -2\int \log\{P(Y\,|\,\theta)P(\theta\,|\,Z)d\theta,$$

with the test data considered fixed. Whereas the plug-in deviance is sensitive to reparameterisation and does not take account of the precision of $\theta(Z)$, the expected deviance is coordinate-free and takes account of precision.

When there are no training data, Y must be used to estimate θ and assess model fit. However, $L(Y,Y)$ is optimistic as a measure of model adequacy, as it uses the data twice. Consider the corresponding function for observation i, namely $L(Y_i,Y)$. This can be compared with the cross-validation loss $L(Y_i,Y_{[i]})$, where $Y_{[i]}$ is Y with observation i excluded. The excess of $L(Y_i,Y)$ over $L(Y_i,Y_{[i]})$ provides a measure of optimism from using the data twice. The expected decrease in loss due to using $L(Y_i,Y)$ instead of $L(Y_i,Y_{[i]})$ is obtained as

$$d_{\text{opt},i} = E\{L(Y_i, Y_{[i]}) - L(Y_i, Y)\,|\,Y_{[i]}\}. \tag{3.5}$$

Summing over observations provides a complexity measure d_{opt}, with a corresponding penalised fit measure

$$\text{DIC}_{\text{opt}} = D + d_{\text{opt}}. \tag{3.6}$$

Issues of focus, as well as the derivation of the complexity measure d_e, are also considered by Celeux et al. (2006). In general terms, a complexity measure or effective parameter count is obtained by comparing the mean deviance with the deviance at the pseudo-true parameter values θ^t (Spiegelhalter et al., 2002, section 3.2). There are various estimators $\tilde{\theta}$ of the pseudo-true parameter values θ^t, apart from the element wise posterior means. Another possibility is to consider the posterior mode posterior value, $\hat{\theta}$, that generates the maximum posterior density $p(\theta\,|\,y) \propto p(y\,|\,\theta)p(\theta)$ (Celeux et al., 2006, p.654), namely $\hat{\theta} = \underset{\theta}{\operatorname{argmax}}\; p(\theta\,|\,y)$. In applications (e.g. discrete mixture models and random effect models) with missing data b, this extends to considering the pair $(\hat{\theta},\hat{b})$ that generates the maximum posterior density (Celeux et al., 2006, p.656). Celeux et al. (2006) mention other possibilities for $\tilde{\theta}$, such as the EM maximum likelihood estimate.

They state different DIC definitions under three alternative foci (observed data likelihood, complete data likelihood, and conditional likelihood) and under different options for $\tilde{\theta}$. For the observed data focus with likelihood $p(y\,|\,\theta)$, obtained possibly after integrating out random effects, one has

$$\text{DIC} = \bar{D} + d_e = D(\tilde{\theta}) + 2d_e = 2\bar{D} - D(\tilde{\theta}) = -4E_\theta[\log\{p(y\,|\,\theta)\}\,|\,y] + 2\log[p(y\,|\,\tilde{\theta})].$$

It can be seen that taking $\tilde{\theta}$ as the posterior mean amounts to assuming

$$\text{DIC} = -4E_\theta\left[\log\{p(y\,|\,\theta)\}\,|\,y\right] + 2\log\left[p\{y\,|\,E_\theta(\theta\,|\,y)\}\right],$$

whereas taking $\tilde{\theta}$ as the posterior mode $\hat{\theta}$ amounts to an alternative DIC definition, denoted DIC_2 by Celeux et al., namely

$$\text{DIC} = -4E_\theta\left[\log\{p(y|\theta)\}|y\right] + 2\log\left[p(y|\hat{\theta})\right].$$

For a complete data focus, with likelihood $p(y,b\,|\,\theta) = p(y\,|\,b,\theta)p(b\,|\,\theta)$ including the second stage likelihood model $p(b|\theta)$ for the missing data (e.g. Kuhn and Lavielle, 2005), one obtains

$$\bar{D} = -2E_{\theta,b}\left[\log\{p(y,b\,|\,\theta)\}\,|\,y\right].$$

Taking b as additional parameters, one may define $\tilde{\theta}$ on the basis of joint modal or maximum a posteriori parameters, $(\hat{\theta},\hat{b})$, with d_e obtained by comparing the average deviance $\bar{D}$ with

$$D(\tilde{\theta}) = -2\log\left[p(y,\hat{b}|\hat{\theta})\right].$$

The joint mode $(\hat{\theta},\hat{b})$ may be estimated by monitoring the posterior density over an MCMC sequence, and finding that set of values $\{\theta^{(r)},b^{(r)}\}$ associated with the maximum value, $p_{\max}(\theta|\,y)$, of the posterior density. The DIC may then be defined as

$$\text{DIC} = -4E_{\theta,b}\left[\log\{p(y,b|\theta)\}|y\right] + 2\log\left[p(y,\hat{b}|\hat{\theta})\right],$$

with complexity estimated as

$$d_e = -2E_{\theta,b}\left[\log\{p(y,b|\theta)\}|y\right] + 2\log\left[p(y,\hat{b}|\hat{\theta})\right].$$

3.3.3 WAIC and LOO-IC

The WAIC (widely applicable information criterion) and the LOO-IC (leave one out information criterion) are more recently developed measures of complexity penalised fit, and are based on averaging over the posterior distribution, rather than using posterior means $\bar{\theta}$ or other point estimates of θ (Watanabe, 2010; Vehtari et al., 2017). The WAIC is obtained as

$$\text{WAIC} = -2(\text{LPPD}(y\,|\,\theta) - d_e)$$

where

$$\text{LPPD}(y\,|\,\theta) = \sum_{i=1}^{n} \log\int p(y\,|\,\theta)p(\theta\,|\,y)d\theta$$

is the log posterior predictive density (LPPD) for y (Gelman et al., 2014), and d_e is the estimated effective model dimension (complexity). The LLPD is an estimate, albeit a biased overestimate, of the expected log posterior predictive density (ELPD) for (unobserved) new data $\tilde{y}$ generated from the same density as the observed data y, and the complexity measure is a measure of the bias.

To estimate the LPPD for a particular observation, one obtains the likelihood for that observation at each MCMC iteration (i.e. conditioning on $\theta^{(r)}$ at iteration r). The resulting vector of likelihoods, for observation i and samples $r=1,\ldots,R$, can be denoted $L_i=(L_{i1},\ldots,L_{iR})$. The log of the mean of L_{ir} over iterations r provides the LPPD for observation i, namely $\text{LPPD}(y_i \mid \theta)=\log(\bar{L}_i)$. The total of these over observations is the estimate of the LPPD.

The estimated complexity for the WAIC is obtained by monitoring log-likelihoods during MCMC sampling, namely $LL_{ir}=\log(L_{ir})$. Then the variance of $LL_i=(LL_{i1},\ldots,LL_{iR})$ provides an estimate of complexity d_{ei} for that observation, $d_{ei}=\text{var}(LL_i)$. The total of the d_{ei} is the total complexity d_e. The estimated piecewise WAIC can be obtained as $-2(\log(\bar{L}_i)-d_{ei})$, and the total WAIC as the sum of the piecewise WAIC.

If the R package loo is used to obtain the LOO-IC, then it is more convenient to monitor log-likelihoods (which are the input to loo), and then obtain sampled likelihoods by exponentiation. For example, using rjags or jagsUI (for example), and with R an object containing model results (including sampled log-likelihoods, LL), WAIC calculations are as follows:

```
LL = as.matrix(R$sims.list$LL)
L=exp(LL)
waic1=log(apply(L,2,mean))
waic2=apply(LL,2,sd)
# casewise waic
waic.pw=-2*(waic1-waic2)
elpd_waic=sum(waic1)-sum(waic2)
# total waic
waic=-2*elpd_waic.
```

The LOO-IC uses an estimate of the leave-one-out predictive fit (or ELPD)

$$\sum_{i=1}^{n}\log[p(y_i \mid y_{[i]})],$$

where $y_{[i]}$ is the set of observations omitting y_i, and

$$p(y_i \mid y_{[i]})=\int p(y_i \mid \theta)p(\theta \mid y_{[i]})d\theta.$$

The latter may be estimated using samples θ^r from the full data posterior $p(\theta|y)$ using importance ratios

$$IR_i^r=\frac{1}{p(y_i \mid \theta^r)}\propto\frac{p(\theta \mid y_{[i]})}{p(\theta \mid y)},$$

with the estimator for $p(y_i \mid y_{[i]})$ then being

$$\sum_{r=1}^{R}IR_i^r p(\tilde{y}_i \mid \theta^r)/\sum_{r=1}^{R}IR_i^r \approx 1/[\frac{1}{R}\sum_{r=1}^{R}\frac{1}{p(y_i \mid \theta^r)}.$$

This estimator may be unstable due to high variances of the importance ratios for certain observations.

Vehtari et al. (2017) use a smoothed version of the importance ratios based on fitting a generalised Pareto density to the upper tail of the importance ratios, leading to Pareto smoothed importance sampling (PSIS) estimates of the LOO-IC. Let w_i^r denote the smoothed importance weights. Then the estimate of the ELPD is

$$\text{ELPD}_{\text{PSIS-LOO}} = \sum_{i=1}^{n} \log\left(\frac{\sum_{r=1}^{R} w_i^r p(y_i \mid \theta^r)}{\sum_{r=1}^{R} w_i^r}\right),$$

with the LOO-IC estimated as $-2 \times \text{ELPD}_{\text{PSIS-LOO}}$. The estimate of the effective parameter total is then

$$d_{\text{PSIS-LOO}} = \text{LPPD} - \text{ELPD}_{\text{PSIS-LOO}}.$$

The LOO-IC may be obtained directly as follows:

```
LL = as.matrix(R$sims.list$LL)
L=exp(LL)
library(resample)
S = nrow(LL)
n = ncol(LL)
lpd_pw = log(colMeans(L))
w = 1/exp(LL-max(LL))
w_n = w/matrix(colMeans(w),S,n,byrow=TRUE)
w_r = pmin (w_n, sqrt(S))
elpd_loo_pw = log(colMeans(L*w_r)/colMeans(w_r))
p_loo_pw = lpd_pw - elpd_loo_pw
# Complexity
sum(p_loo_pw)
# LOO-IC
-2*sum(elpd_loo_pw)
```

Though the WAIC and LOO-IC provide an estimate of predictive ability, both are subject to stochastic variability which can be considerable for smaller datasets (Piironen and Vehtari, 2017). There may also be cautions regarding the estimates of WAIC and LOO-IC, provided in the loo package and discussed by Vehtari et al. (2017, p.1416). For the LOO-IC, these are based on the estimated shape parameter of the generalized Pareto, values of which indicate whether the variance of the importance ratios is effectively infinite.

3.3.4 The WBIC

The BIC is a penalised fit measure, and the widely applicable Bayesian information criterion or WBIC (Watanabe, 2013) is therefore included here, though it is essentially based on an estimator of the marginal likelihood. Thus following Friel et al. (2017), and referring to path sampling ideas, there exists a unique temperature t^* such that

$$p(y) = E_{\theta|y,t^*} p(y \mid \theta).$$

Watanabe (2013) shows that asymptotically, as the sample size n tends to ∞, $t^* \approx 1/\log(n)$. Friel et al. (2017) show for a number of worked examples that the optimal t^* is smaller than

$1/\log(n)$, but that the latter approximation may be a useful practical option, except when weakly informative priors are used.

Example 3.2 Turtle Survival and Penalised Fit Measures

This example compares penalised fit measures for the data of Example 3.1. As before, a fixed effects model is compared with random clutch effects alternatives. Models are compared first using rjags and jagsUI. Normal N(0,10) priors are adopted on the fixed effects, and in the random effects models, a shrinkage prior is implemented by taking a uniform prior on $U = 1/(1+\sigma_b^2)^2$, so that $\sigma_b^2 = (1-U^{0.5})/U^{0.5}$. The WBIC is for these models (Watanabe, 2013) is also estimated using rstan, and using the prior on σ_b^2 as in Sinharay and Stern (2005).

The DICs of models 1 and 2 (fixed effects and random clutch intercepts) are estimated with JAGS using different penalties on the mean deviance, as proposed by Plummer (2008). The usual DIC estimates, following Spiegelhalter et al. (2002), as in (3.4), can then be compared with optimism adjusted DIC estimates, as in (3.6). For model 1, these are denoted DIC1.pD and DIC1.popt in the code.

On this basis, the effective parameter total d_{e2} for model 2 is 12.3 compared to 2.2 under model 1, a difference of approximately 10, whereas there are 32 extra nominal parameters (the random effects variance and the 31 cluster effects). The DIC is then 3.4 lower under model 2, as the mean deviance is lessened from 299.8 to 286.3.

Comparing the DICs, as defined by (3.4), suggests a small advantage for the random effects model, though the small DIC difference is unlikely to be significant according to the rule of thumb in Spiegelhalter et al. (2002, section 9.2.4). The WAIC and LOO-IC also both show a slight advantage for model 2 as against model 1, with the respective WAIC being 299.4 and 301.7, and the respective LOO-IC being 299.6 and 301.7.

The optimism adjusted DICs (DIC.popt), by contrast, show a considerably better fit for model 1. There are now more effective parameters under model 2, namely $d_{opt,2} = 25.4$, as compared to $d_{opt,1} = 4.0$, and this offsets the reduction in mean deviance. The WBIC also prefers model 1, with respective values for models 1 and 2 of 308.4 and 312.9.

Other criteria may be considered. For example, the 5% worst fitting cases under model 1 (without clutch effects) account for 14% of the WAIC, suggesting some issues in fit for individual cases, especially in clutches 9 and 10. Additionally, estimates from model 2 shows the density of σ_b to be relatively symmetric and to have its mass away from zero (Figure 3.1), supporting the presence of at least some random variability. In this connection, MacNab et al. (2004) illustrate how – when a form of random variation is not supported by the data – the density of the random effects standard deviation can be heavily skewed to the left or "spiked," with the posterior mass piled up against zero. In this regard, it is relevant to consider the significance of individual clutch effects under model 2. In fact, there are three clutch effect (clutches 9, 10, and 26) with a 0.85 chance of exceeding zero, while the clutch 15 effect has posterior probability of 0.07 of exceeding 0.

It is of interest also to compare the fixed effects model to a random slopes model (now denoted model 3), as the Bayes factor showed strong positive support for the simpler model. In this regard, the optimism adjusted DIC has the same preference for the simpler model, with $DIC_{opt,3} = 314.3$ compared to $DIC_{opt,1} = 303.3$. Similarly, the WBIC for model 3 is estimated as 314.7, higher than 308.4 for the simplest model 1. By contrast, the DIC of (3.4), the WAIC and the LOO-IC all show a slight preference for the random slopes model over model 1. Such findings suggest that in certain applications, penalised fit measures and formal model comparisons may prefer different models. Figure 3.2 plots out the varying slopes under model 3.

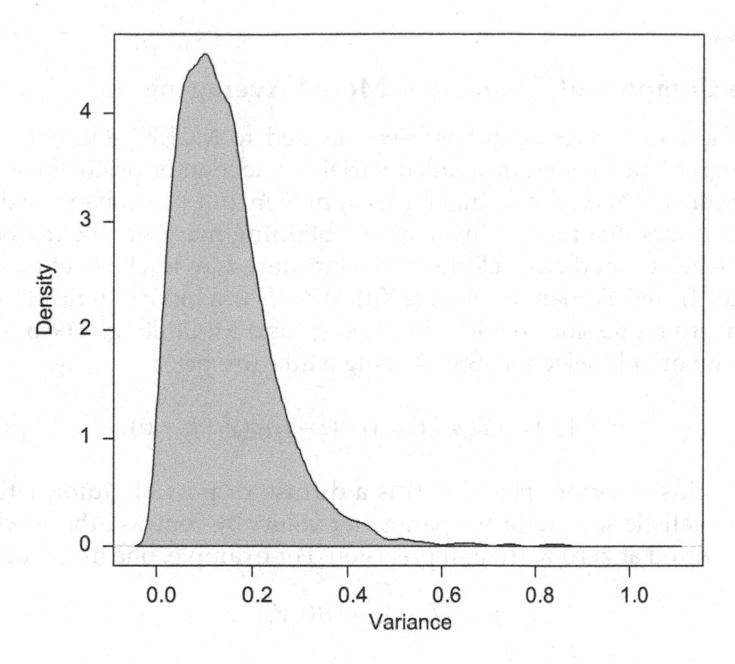

FIGURE 3.1
Density of clutch variance.

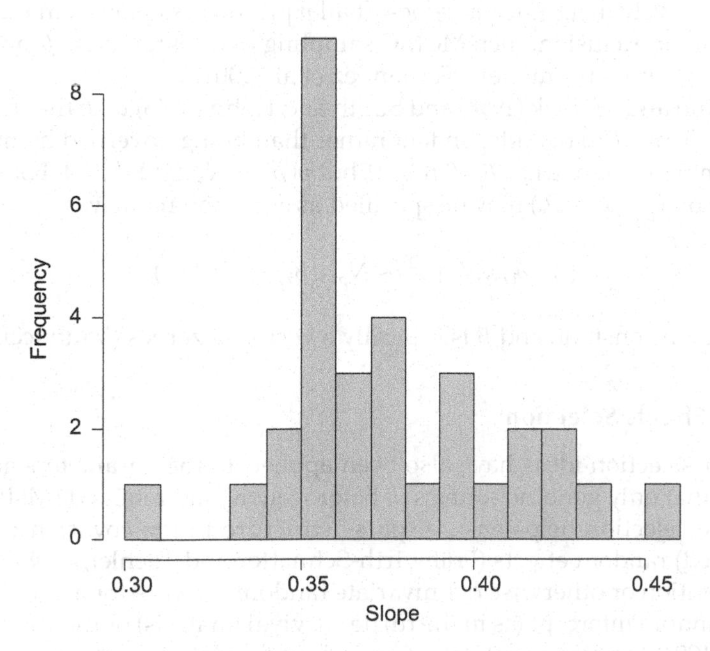

FIGURE 3.2
Random slopes on birth weight.

3.4 Variance Component Choice and Model Averaging

A considerable amount of research has been devoted to MCMC selection of significant predictors in regression – sometimes called variable selection or predictor selection. Such techniques are consistent with a formal Bayes approach, but less constrained by the complex integration issues that may be involved in obtaining marginal likelihoods. Different possible approaches to predictor selection are considered by Rockova et al. (2012), Sala-i-Martin et al. (2004), and Fernandez et al. (2001). With J_j as a binary indicator for retaining or excluding the jth regression coefficient β_j, George and McCullough (1993, 1997) develop stochastic search variable selection (SSVS) using a mixture prior

$$p(\beta_j \mid J_j) = J_j p(\beta_j \mid J_j = 1) + (1 - J_j) p(\beta_j \mid J_j = 0)$$

in which the "inclusion prior" $p(\beta_j \mid J_j = 1)$ is a diffuse or possibly informative prior, but one that allows realistic search for the parameter value. By contrast, the "exclusion prior" $p(\beta_j \mid J_j = 0)$ is centred at zero with high precision. For example, one might have

$$p(\beta_j \mid J_j = 1) \sim N(0, V_j)$$

with V_j large, but

$$p(\beta_j \mid J_j = 0) \sim N(0, V_j / K_j) \quad K_j >> 1$$

with K_j chosen so that the sampling from the prior is constrained to values around zero, that is, to substantively insignificant values. If all p predictors apart from the intercept are open to inclusion or exclusion, then MCMC sampling over parameters β_j and indicators J_j is averaging over 2^p possible models (Fernandez et al., 2001).

By contrast, Kuo and Mallick (1998) and Smith and Kohn (1996) take the selection indicators J_j and coefficients β_j to be independent rather than being governed by mixture priors. Assuming normal priors, one has $\beta_j = 0$ if $J_j = 0$, but $p(\beta_j) \sim N(0, V_j)$ if $J_j = 1$. Following Zellner (1986), the prior on $(\beta_0, \beta_1, \ldots \beta_p)$ may be specified as a g-prior, namely

$$(\beta_0, \beta_1, \ldots \beta_p \mid \sigma^2) \sim N_{p+1}(B, g\sigma^2 (X'X)^{-1})$$

where g is a known constant, and B is typically a vector of zeroes (Vannucci, 2000).

3.4.1 Random Effects Selection

Model indicator selection ideas have also been applied to the parameters governing random effects, so that only genuine sources of heterogeneity are retained (Müller et al., 2013). Such covariance selection helps ensure sparse structure in the covariance matrix of the selected (retained) random effects (Frühwirth-Schnatter and Tüchler, 2008). Selection may relate to the retention or otherwise of univariate random effects – for example, a multilevel model with a random intercept (as in the turtle survival analysis) or the convolution model of Besag et al. (1991) for area count data. For multivariate random effects, such as random cluster intercepts b_{0j} and slopes $\{b_{1j}, \ldots, b_{pj}\}$ in a multilevel analysis

$$y_{ij} = b_{0j} + b_{1j} x_{1ij} + \ldots + b_{pj} x_{pij} + u_{ij},$$

one can consider retaining covariances Σ_{bgh} subject to variances in both effects b_{gj} and b_{hj} being retained. Thus, Smith and Kohn (2002) identify zero off-diagonal elements in the inverse $\Pi_b = \Sigma_b^{-1}$ of the variance-covariance matrix. Alternatively, one may also allow the exclusion of variance components (diagonal terms in Σ_b), which necessarily leads to exclusion of associated covariances.

Selection schemes applicable to both diagonal and off-diagonal elements in covariance matrices for random effects have been developed by Fruhwirth-Schnatter and Tuchler (2008), Chen and Dunson (2003), Kinney and Dunson (2008), and Cai and Dunson (2006); for applications, see Yang (2012), Saville et al. (2011), and Harun and Cai (2014). Note that these methods may be relatively difficult to implement, with Saville and Herring (2009) finding "these methods are generally time consuming to implement, require special software, and rely on subjective choice of hyperparameters."

Consider a general linear mixed model for nested responses y_{ij} (as in longitudinal data with repetitions i over subjects j) with means μ_{ij}. These means are linked to a $P \times 1$ vector of regressors X_{ij} and $Q \times 1$ vector of regressors Z_{ij} via the model

$$g(\mu_{ij}) = X_{ij}'\beta + Z_{ij}'b_j,$$

where g is an appropriate link, $\beta = (\beta_1,\ldots\beta_Q)'$ denotes the central fixed effects, and $b_j = (b_{j1},\ldots,b_{jQ})'$ are zero mean random effects with covariance $\Sigma_b = \{\sigma_{bkl}\}$. For continuous data – and discrete outcomes subject to overdispersion – an observation level residual is also present, so that

$$g(\mu_{ij}) = X_{ij}'\beta + Z_{ij}'b_j + u_{ij}$$

with u_{ij} usually taken as iid.

Following Cai and Dunson (2006), one possible Cholesky decomposition of the covariance matrix for $b_j = (b_{j1},\ldots,b_{jQ})'$ has the form

$$\Sigma_b = \Lambda\Gamma\Gamma'\Lambda,$$

where $\Lambda = \text{diag}(\lambda_1,\ldots\lambda_Q)$ and Γ is a lower triangular matrix

$$\Gamma = \begin{pmatrix} 1 & 0 & \ldots & 0 \\ \gamma_{21} & 1 & \ldots & 0 \\ \ldots & \ldots & \ddots & 0 \\ \gamma_{Q1} & \gamma_{Q2} & \ldots & 1 \end{pmatrix},$$

implying

$$\sigma_{bkl} = \lambda_k\lambda_l\left(\gamma_{r_2 r_1} + \sum_{s=1}^{r_1-1}\gamma_{ks}\gamma_{ls}\right),$$

where $r_2 = \max(k,l)$, $r_1 = \min(k,l)$. Then one has

$$g(\mu_{ij}) = X_{ij}'\beta + Z_{ij}'\Lambda\Gamma c_j + u_{ij},$$

where $\{c_{jq} \sim N(0,1), q = 1,\ldots,Q\}$ are uncorrelated standard normal variables.

The selection indicators for retaining variances and covariances are $J_q \sim \text{Bern}(\pi_\Lambda)$, governing the diagonal terms in Λ, and $H_{kl} \sim \text{Bern}(\pi_\Gamma)$ governing the terms in Γ. Note that retaining γ_{kl} requires not only $H_{kl}=1$, but $J_k = J_l = 1$. If either J_k or J_l is zero, then γ_{kl} is necessarily excluded. Cai and Dunson (2006) suggest positive truncated normal priors with variance 10 for the diagonal terms λ_q, namely

$$\lambda_q \sim N(0,10)\ \ I(0,)\quad \text{if } J_q = 1$$

$$\lambda_q = 0 \quad \text{if } J_q = 0$$

Diffuse priors are not recommended (Cai and Dunson, 2008, p.72), as they may favour the null model. There may also be a case for interlinked priors for λ_q and the variances of the u_{ij} effects (if present).

Fruhwirth-Schnatter and Tuchler (2008) consider the covariance matrix decomposition

$$\Sigma_b = CC',$$

with C a lower triangular matrix of dimension Q including unknown diagonal terms C_{qq}. To illustrate the covariance selection procedure, a hierarchical linear normal model with varying cluster regression effects $\beta_j = \beta + b_j$ of dimension Q would be reframed as

$$y_{ij} = X_{ij}(\beta + b_j) + u_{ij}$$
$$= X_{ij}\beta + X_{ij}Cz_j + u_{ij},$$

where $u_{ij} \sim N(0,1/\tau_u)$, and $z_j = (z_{j1}, z_{j2} \ldots, z_{jQ})'$ is a $Q \times 1$ vector distributed as $N_Q(0,I)$. Consider binary indicators J_{kl} for retention or otherwise of each of the $Q(Q+1)/2$ elements of C. Then

$$C_{kl} \neq 0 \text{ if } J_{kl} = 1 \text{ (for } k \geq l),$$

$$C_{kl} = 0 \text{ if } J_{kl} = 0,$$

and b_{jk} is 0 at a particular iteration if all C_{kl} in the kth row of C are zero. A possible prior for the J_{kl} indicators is Bernoulli with probability π_J, where π_J follows a beta density,

$$\pi_J \sim Be(T_J + 1, Q(Q+1)/2 - T_J + 1),$$

based on the total free covariance parameters, and the number T_J of J_{kl} taking the value 1 (i.e. the number of non-zero elements in C). For $Q=1$ in a model where a cluster level random intercept is to be tested for inclusion, one would have

$$\mu_{ij} = \beta_0 + X_{ij}\beta + b_j + u_{ij}$$
$$= \beta_0 + X_{ij}\beta + cz_j + u_{ij},$$

where $z_j \sim N(0,1)$, and $c \neq 0$ if $J=1$ and $c=0$ if $J=0$. The (model averaged) estimate of the covariance matrix Σ_b of the b_j over $r = 1,\ldots,R$ iterations of a chain is obtained as

$$\hat{\Sigma}_b = \frac{1}{R}\sum_{r=1}^{R} C^{(r)}(C')^{(r)}.$$

Methods for selecting the entire random effect term extend to selection of individual random effects. For selecting the entire term, consider a spike and slab prior with the spike component having considerably lower variance:

$$b_i \sim (1-\delta)N(0, r\sigma_b^2) + \delta N(0, \sigma_b^2),$$

where $r << 1$. This extends to selection of individual random effects, for example using Lasso random effect models (Fruhwirth-Schnatter and Wagner, 2010) involving component-specific indicators δ_i and a hierarchical prior on the variances. For example, a mixture of Laplace densities is obtained under

$$b_i \sim (1-\delta_i)N(0, \zeta_{1i}) + \delta_i N(0, \zeta_{2i}),$$

$$\zeta_{1i} \sim E(1/(2rQ)),$$

$$\zeta_{2i} \sim E(1/(2Q)),$$

with r set small, so that $\zeta_{1i} \approx 0$. The δ_i are binary indicators with unknown probability π_δ, the prior proportion of subjects with non-zero random effects. If Q is also unknown, there may be identification issues under independent priors, as different combinations of π_δ and Q can give similar b_i.

Example 3.3 Seeds Data

The widely analysed seeds data from Crowder (1978) may be considered an example where not all random effects may be necessary. The binomial data $\{y_i, n_i\}$ over N=21 plates refer to germinations y_i among n_i seeds, and are subject to a binomial logit analysis with random normal plate effects to account for binomial overdispersion. Predictors are x_{1i} and x_{2i} (respectively seed type and root extract), and their interaction $x_{1i}x_{2i}$. Thus

$$y_i \sim N(n_i, \pi_i),$$

$$\text{logit}(\pi_i) = \beta_1 + b_i + \beta_2 x_{1i} + \beta_3 x_{2i} + \beta_4 x_{1i} x_{2i},$$

$$b_i \sim N(0, \sigma_b^2).$$

Fitting this baseline model, without any random effects selection, suggests not all the plate effects are needed. Posterior mean probabilities for $Pr(b_i > 0 \mid y)$ are inconclusive, ranging from 0.34 to 0.63.

As one approach to selection, the method of Fruhwirth-Schnatter and Wagner (2010) seeks to classify units as either close to average (with $\delta_i \approx 0$, with b_i close to zero, and effectively unnecessary), above average with $\delta_i \approx 1$, and high $Pr(b_i > 0 \mid y)$, or below average, also with $\delta_i \approx 1$ but high $Pr(b_i < 0 \mid y) = 1 - Pr(b_i > 0, y)$. A Laplace mixture density for the plate effects is used, namely

$$b_i \sim (1-\delta_i)N(0, \zeta_{1i}) + \delta_i N(0, \zeta_{2i}),$$

$$\zeta_{1i} \sim E(1/(2rQ)),$$

$$\zeta_{2i} \sim E(1/(2Q)),$$

$$\delta_i \sim \text{Bern}(\omega),$$

$$\omega \sim \text{Beta}(1,1),$$

with $r = 0.00001$ and $1/Q \sim Ga(0.5, 0.2275)$, the latter as suggested by Fruhwirth-Schnatter and Wagner (2010).

Estimated retention probabilities $Pr(\delta_i = 1 \mid y)$ range from 0.48 to 0.70, while the probabilities of high effects $Pr(b_i > 0 \mid y)$ range from 0.18 to 0.82. The most distinctive $Pr(b_i > 0 \mid y)$ are for plates 10 and 17, with probabilities $Pr(b_i < 0 \mid y)$ around 0.80, and plates 4 and 15 with probabilities $Pr(b_i > 0 \mid y)$ exceeding 0.80 (cf. Fruhwirth-Schnatter and Wagner, 2010, Table 7). Figure 3.3 plots out the probabilities $Pr(b_i > 0 \mid y)$. The probabilities of high effects $Pr(b_i > 0 \mid y)$ are relatively stable as less informative $Ga(1, 0.05)$ and $Ga(1, 0.01)$ priors are assumed for $1/Q$.

We also consider a horseshoe prior for the plate effects, namely

$$b_i \sim N(0, \lambda_i^2 \sigma_b^2),$$

with half Cauchy $C(0,1)^+$ priors on both the λ_i and σ_b^2. As mentioned by Carvalho et al. (2009), $\varphi_i = 1/(1 + \lambda_i^2)$ is interpretable as the amount of weight that the posterior mean for b_i places on zero. We consider instead $\kappa_i = \lambda_i^2/(1 + \lambda_i^2)$ as an indicator for non-zero posterior mean b_i, analogous to a probability that $b_i \neq 0$. The estimated κ_i range from 0.35 to 0.61, with κ_i greater than 0.5 for plates 4,10, 15, 16 and 17 (see Figure 3.4). Despite the extra parameters in this extended model as compared to the baseline model, a formal comparison shows similar marginal likelihoods for the extended and baseline models.

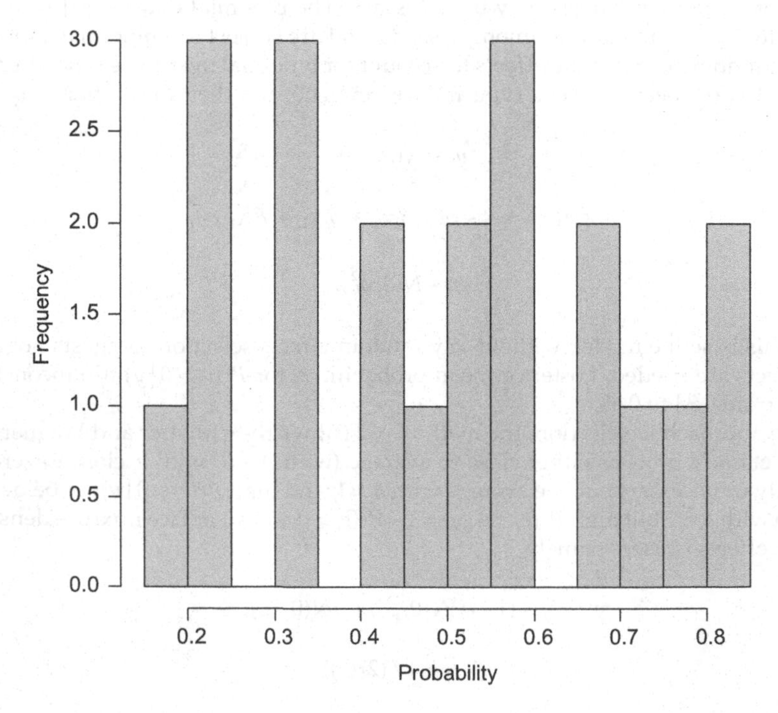

FIGURE 3.3
Probabilities of high random effects.

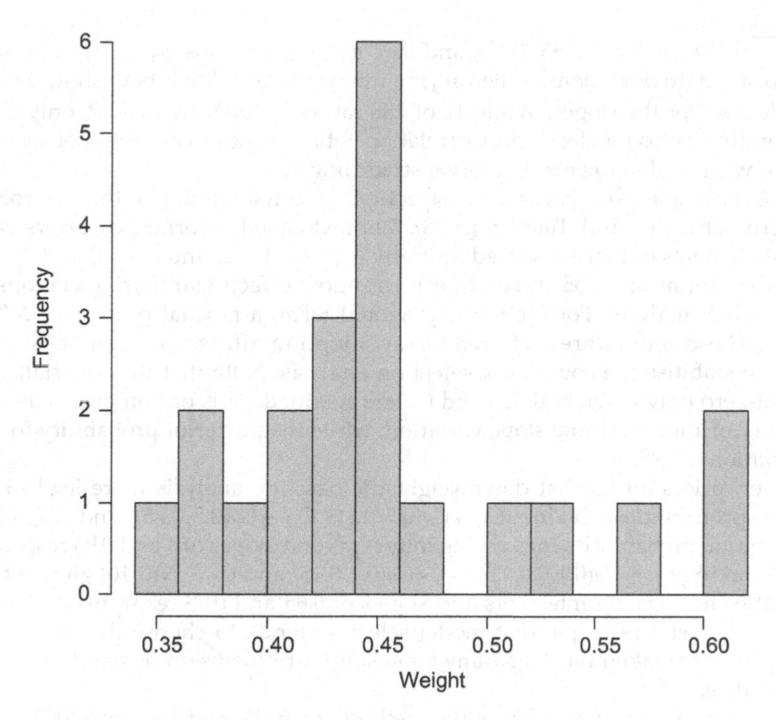

FIGURE 3.4
Histogram of weights for non-zero effects, horseshoe prior.

Example 3.4 Hypertension Trial

To illustrate covariance selection for potentially correlated multiple random effects, this example considers clinical trial data from Brown and Prescott (1999). In this trial, 288 patients are randomly assigned to one of three drug treatments for hypertension, 1 = Carvedilol, 2 = Nifedipine, and 3 = Atenolol. The data consist of a baseline reading BP_i of diastolic blood pressure (these are centred), and four post-treatment blood pressure readings y_{it} at two weekly intervals (weeks 3,5,7 and 9 after treatment). Some patients are lost to follow up (there are 1092 observations rather than $4\times288=1152$), but for simplicity, their means are modelled for all $T=4$ periods.

A baseline analysis includes random patient intercepts, and random slopes on the blood pressure readings. Additionally, the new treatment Carvedilol is the reference in the fixed effects comparison vector $\eta=(\eta_1,\eta_2,\eta_3)$, leading to the corner constraint $\eta_1=0$. Then for patients $i=1,\ldots,288$, with treatments Tr_i and waves $t=1,\ldots 4$,

$$y_{it}=\beta_1+b_{1i}+(\beta_2+b_{2i})BP_i+\eta_{Tr_i}+u_{it},$$

$$u_{it}\sim N(0,1/\tau_u),$$

with errors taken to be uncorrelated through time. In line with a commonly adopted methodology, the b_{qi} are taken to be bivariate normal with mean zero and covariance Σ_b. The precision matrix Σ_b^{-1} is assumed to be Wishart with 2 degrees of freedom and identity scale matrix, S. The observation level precision is taken to have a gamma prior, $\tau_u\sim Ga(1,0.001)$.

A two-chain run of 5000 iterations in jagsUI give posterior means (sd) for $\beta=(\beta_1,\beta_2)$ of 92.6 (0.7) and 0.41 (0.10). Posterior means (sd) for the random effect standard deviations

$\sigma_{b_j}=\sqrt{\Sigma_{b_{jj}}}$ of $\{b_{1i},b_{2i}\}$ are 5.55 (0.42), and 0.64 (0.13). The ratios $|b_{ji}/sd(b_{ji})|$ of posterior means to standard deviations of the varying intercepts and slopes both show variation, though less so for the slopes. While 42 of 288 ratios $|b_{1i}/sd(b_{1i})|$ exceed 2, only 2 of the corresponding ratios for slopes do. Correlation between the effects does not seem to be apparent, with $\sigma_{b_{12}}$ having a 95% interval straddling zero.

In a second analysis, covariance selection is considered via the approach of Fruhwirth-Schnatter and Tuchler (2008). Context-based informative priors for the diagonal elements of C are assumed. Initially $C_{11}\sim Ga(1,0.2)$ and $C_{22}\sim Ga(1,1.5)$, based on the posterior means 5.55 and 0.64 for the random effects standard deviations from the preceding analysis. For the lower diagonal term, a normal prior $C_{21}\sim N(0,1)$ is assumed. These options are preferred to, say, adopting diffuse priors on the C_{jk} terms, in order to stabilise the covariance selection analysis. Note that the covariance term Σ_{21} is non-zero only when both C_{11} and C_{21} are retained. This option gives a posterior probability of 1 for retaining slope variation, while the posterior probability for intercept variation is 0.98.

However, priors on C_{jk} that downweight the baseline analysis more lead to lower retention probabilities. Taking $C_{11}\sim Ga(0.5,0.1)$, $C_{22}\sim Ga(0.5,0.75)$ and $C_{21}\sim N(0,2)$ gives retention probabilities for varying intercepts and slopes of 1 and 0.93 respectively. Similarly, taking $C_{11}\sim Ga(0.1,0.0.02)$, $C_{22}\sim Ga(0.1,0.15)$ and $C_{21}\sim N(0,10)$ gives retention probabilities for varying intercepts and slopes of 0.65 and 0.98 respectively. This is in line with a general principle that model selection tends to choose the null model if diffuse priors are taken on the parameter(s) subject to inclusion or rejection (Cai and Dunson, 2008).

We also consider an adaptation of the method of Saville and Herring (2009) for continuous nested outcomes, which involves scaling factors $\exp(\phi_j)$ premultiplying random effects (e.g. cluster intercepts and slopes) taken to have the same variance as the main residual term. This allows Bayes factor calculation using Laplace methods. In the current application, and allowing for correlated slopes and intercepts, one has

$$y_{it}=\beta_1+\beta_2 BP_i+\exp(\phi_1)b_{1i}+\exp(\phi_2)(\gamma_{12}b_{1i}+b_{2i})BS_i+\eta_{Tr_i}+u_{it},$$

$$u_{it}\sim N(0,1/\tau_u),$$

$$b_{1i}\sim N(0,1/\tau_u),$$

$$b_{2i}\sim N(0,1/\tau_u),$$

$$\gamma_{12}\sim N(0,1).$$

For the ϕ_j discrete mixture, priors are adopted, with one option corresponding to $\lambda_j=\exp(\phi_j)$ being close to zero, while in the other, the prior on ϕ_j allows unrestricted sampling. Here

$$\phi_j\sim(1-J_j)N(-5,0.1)+J_jN(0,10),$$

$$J_j\sim \text{Bern}(0.5).$$

This provides posterior probabilities $Pr(J_j=1)$ of 1 and 0.41 respectively for random intercepts and slopes. Posterior means (sd) for the random effect standard deviations $\sigma_{bj}=\sqrt{\Sigma_{bjj}}$ of $\{b_{1i},b_{2i}\}$ are 6.18 (0.39), and 0.16 (0.17).

3.5 Predictive Methods for Model Choice and Checking

A number of studies have pointed to drawbacks in focusing solely on the marginal likelihood or Bayes factor, as a single global assessment measure of the performance of complex models, and point out computational and inferential difficulties with the Bayes factor when priors are diffuse, as well as the need to examine fit for individual observations to make sense of global criteria (e.g. Gelfand, 1996; Johnson, 2004). While formal Bayes methods can be extended to assessing the fit of single observations (Pettit and Young, 1990), it may be argued that predictive likelihood methods offer a more flexible approach to assessing the role of individual observations. In fact, predictive methods have a role both in model choice and model checking.

3.5.1 Predictive Model Checking and Choice

The formal predictive likelihood approach assumes only part of the observations are used in estimating a model. On this basis, one may obtain cross-validation predictive densities (Vehtari and Lampinen, 2002), $p(y_s \mid y_{[s]})$, where y_s denotes a subset of y (the "validation data"), and $y_{[s]}$ is the complementary "test data" formed by excluding y_s from y. If $[i]$ is defined to contain all the data $\{y_1,\ldots y_{i-1}, y_{i+1},\ldots,y_n\}$ except for a single observation i, then the densities

$$p(y_i \mid y_{[i]}) = \int p(y_i \mid \theta, y_{[i]}) p(\theta \mid y_{[i]}) d\theta,$$

are called conditional predictive ordinates or CPOs (e.g. Chaloner and Brant, 1988; Geisser and Eddy, 1979), and sampling from them shows what values of y_i are likely when a model is applied to all the data points except the ith, namely to the data $y_{[i]}$. The predictive distribution $p(y_i \mid y_{[i]})$ can be compared to the actual observation in various ways (Gelfand et al., 1992).

For example, to assess whether the observation is extreme (not well fitted) in terms of the model being applied, replicate data $y_{i,\text{rep}}$ may be sampled from $p(y_i \mid y_{[i]})$ and their concordance with the data may be represented by probabilities (Marshall and Spiegelhalter, 2003),

$$Pr(y_{i,rep} \leq y_i \mid y_{[i]}).$$

These are estimated in practice by counting iterations r where the constraint $y_{i,\text{rep}}^{(r)} \leq y_i$ holds. For discrete data, this assessment is based on the probability

$$Pr(y_{i,\text{rep}} < y_i \mid y_{[i]}) + 0.5Pr(y_{i,\text{rep}} = y_i \mid y_{[i]}).$$

Gelfand (1996) recommends assessing concordance between predictions and actual data by a tally of how many actual observations y_i are located within the 95% interval of the corresponding model prediction $y_{i,\text{rep}}$. For example, if 95% or more of all the observations are within 95% posterior intervals of the predictions $y_{i,\text{rep}}$, then the model is judged to be reproducing the observations satisfactorily.

The collection of predictive ordinates $\{p(y_i \mid y_{[i]}), i = 1, n\}$ is equivalent to the marginal likelihood $p(y)$ when $p(y)$ is proper, in that each uniquely determines the other. A pseudo

Bayes factor is obtained as a ratio of products of leave one out cross-validation predictive densities (Vehtari and Lampinen, 2002) under models M_1 and M_2, namely

$$PsBF(M_1,M_2) = \prod_{i=1}^{n} \{p(y_i \mid y_{[i]}, M_1) / p(y_i \mid y_{[i]}, M_2)\}$$

In practical data analysis, one typically uses logs of CPO estimates, and totals the log(CPO) to derive log pseudo marginal likelihoods and log pseudo Bayes factors (Sinha et al., 1999, p.588).

Monte Carlo estimates of conditional predictive ordinates $p(y_i \mid y_{[i]})$ may be obtained without actually omitting cases, so formal cross-validation based on n separate estimations (the 1st omitting case 1, the 2nd omitting case 2, etc) may be approximated by using a single estimation run. For parameter samples $\{\theta^{(1)},\ldots,\theta^{(R)}\}$ from an MCMC chain, an estimator for the CPO, $p(y_i \mid y_{[i]})$, is

$$\frac{1}{p(y_i \mid y_{[i]})} = \frac{1}{R}\sum_{r=1}^{R} \frac{1}{p(y_i \mid \theta^{(r)})},$$

namely the harmonic mean of the likelihoods for each observation (Aslanidou et al., 1998; Silva et al., 2006; Sinha, 1993) In computing terms, an inverse likelihood needs to be calculated for each case at each iteration, the posterior means of these inverse likelihoods obtained, and the CPOs are the inverse of those posterior mean inverse likelihoods. Denoting the inverse likelihoods as $H_i^{(r)} = 1/p(y_i \mid \theta^{(r)})$, one would in practice take minus the logarithms of the posterior means of H_i as an estimate of $\log(\text{CPO})_i$. The sum over all cases of these estimates provides a simple estimate of the log pseudo marginal likelihood. In the turtle data example (Example 3.1), the fixed effects only model 1 has a *PsBF* of –151.8, while the random intercepts model 2 has a PsBF of –149.6 under a *Ga*(0.1,0.1) prior for τ_b. So the pseudo Bayes factors tends to weakly support the random effects option.

Model fit (and hence choice) may also be assessed by comparing samples y_{rep} from the posterior predictive density based on all observations, though such procedures may be conservative since the presence of y_i influences the sampled $y_{i,\text{rep}}$ (Marshall and Spiegelhalter, 2003). Laud and Ibrahim (1995) and Meyer and Laud (2002) propose model choice based on minimisation of the criterion

$$C = E\left[c(y_{\text{rep}}, y) \mid y\right] = \sum_{i=1}^{n} \left\{\text{var}(y_{i,\text{rep}}) + [y_i - E(y_{i,\text{rep}})]^2\right\},$$

where for y continuous, $c(y_{\text{rep}}, y)$ is the predictive error sum of squares

$$c(y_{\text{rep}}, y) = (y_{\text{rep}} - y)'(y_{\text{rep}} - y).$$

The C measure can be obtained from the posterior means and variances of sampled $y_{i,rep}^{(r)}$ or from the posterior average of $\sum_{i=1}^{n} (y_{i,\text{rep}}^{(r)} - y_i)^2$. Carlin and Louis (2000) and Buck and Sahu (2000) propose related model fit criteria appropriate to both metric and discrete outcomes.

Posterior predictive loss (PPL) model choice criteria allow varying trade-offs in the balance between bias in predictions and their precision (Gelfand and Ghosh, 1998; Ibrahim et al., 2001). Thus for k positive and y continuous, one possible criterion has the form

$$\text{PPL}(k) = \sum_{i=1}^{n} \left\{ \text{var}(y_{i,\text{rep}}) + \left(\frac{k}{k+1}\right)\left[y_i - E(y_{i,\text{rep}})\right]^2 \right\}.$$

This criterion would be compared between models at selected values of k, typical values being $k=0$, $k=1$, and $k=10{,}000$, where higher k values put greater stress on accuracy in predictions, and less on precision. One may consider calibration of such measures, namely expressing the uncertainty of C or PPL in a variance measure (Laud and Ibrahim, 1995; Ibrahim et al., 2001). De la Horra and Rodríguez-Bernal (2005) suggest predictive model choice based on measures of distance between the two densities that can potentially be used for predicting future observations, namely sampling densities and posterior predictive densities.

To assess poorly fitted cases, the CPO values may be scaled (dividing by their maximum) and low values for particular observations (e.g. under 0.001) will then show observations which the model does not reproduce effectively (Weiss, 1994). If there are no very small scaled CPOs, then a relatively good fit of the model to all data points is suggested, and is likely to be confirmed by other forms of predictive check. The ratio of extreme percentiles of the CPOs is useful as an indicator of a good fitting model e.g. the ratio of the 99th to the 1st percentile.

An improved estimate of the CPO may be obtained by weighted resampling from $p(\theta|y)$ (Smith and Gelfand, 1992; Marshall and Spiegelhalter, 2003). Samples $\theta^{(r)}$ from $p(\theta|y)$ can be converted (approximately) to samples from $p(\theta|y_{[i]})$ by resampling the $\theta^{(r)}$ with weights

$$w_i^{(r)} = G(y_i|\theta^{(r)}) / \sum_{r=1}^{R} G(y_i|\theta^{(r)}),$$

where

$$G(y_i|\theta^{(r)}) = 1/p(y_i|\theta^{(r)}),$$

is the inverse likelihood of case i at iteration r. Using the resulting re-sampled values $\tilde{\theta}^{(r)}$, corresponding predictions $\tilde{y}_{\text{rep}}$ can be obtained which are a sample from $p(y_i | y_{[i]})$.

3.5.2 Posterior Predictive Model Checks

A range of model checks can also be applied using samples from the posterior predictive density without actual case omission. To assess predictive performance, samples of replicate data y_{rep} from

$$p(y_{\text{rep}}|y) = \int p(y_{\text{rep}}|\theta)p(\theta|y)d\theta,$$

may be taken, and checks made against the data, for example, whether the actual observations y are within 95% credible intervals of y_{rep}. Formally, such samples are obtained by the method of composition (Chib, 2008), whereby if $\theta^{(r)}$ is a draw from $p(\theta|y)$, then $y_{\text{rep}}^{(r)}$ drawn from $p(y_{\text{rep}}|\theta^{(r)})$ is a draw from $p(y_{\text{rep}}|y)$. In a satisfactory model, namely one that adequately reproduces the data being modelled, predictive concordance (accurate reproduction of the actual data by replicate data) is at least 95% (Gelfand, 1996, p.158).

Other comparisons of actual and predicted data can be made, for example by a chi-square comparison (Gosoniu et al., 2006). Johnson (2004) proposes a Bayesian chi-square approach based on partitioning the cumulative distribution into K bins, usually of equal probability. Thus, one chooses quantiles

$$0 \equiv a_0 < a_1 < \ldots < a_{K-1} < a_K \equiv 1,$$

with corresponding bin probabilities

$$p_k = a_k - a_{k-1}, k = 1, \ldots, K.$$

Then using model means μ_i for subject $i \in (1,\ldots,n)$ one obtains the implied cumulative density q_i, say $a_{k^*-1} < q_i < a_{k^*}$, and allocates the fitted point to a bin randomly chosen from bins $1,\ldots,k^*$.

For example, suppose there are $K=5$ equally probable intervals, with $p_k = 0.2$. If $\mu_i = 1.4$, the probability assigned to an observation $y_i=1$ by the cumulative density function falls in the interval (0.247,0.592), which straddles bins 2 and 3. To allocate a bin, a U(0.247,0.592) variable is sampled, and the predicted bin is 2 or 3, according to whether the sampled uniform variable falls within (0.247,0.4), or (0.4, 0.592). The totals so obtained accumulating over all subjects define predicted counts $m_k(\tilde{\theta})$ which are compared (at each MCMC iteration) to actual counts np_k, as in formula (3) in Johnson (2004). This provides the Bayesian chi-square criterion

$$R^B(\tilde{\theta}) = \sum_{k=1}^{K} \frac{[m_k(\tilde{\theta}) - np_k]^2}{np_k},$$

where $R^B(\tilde{\theta})$ is asymptotically χ^2_{K-1}, regardless of the parameter dimension of the model being fitted [2]. One can assess the posterior probability that $R^B(\tilde{\theta})$ exceeds the 95th percentile of the χ^2_{K-1} density. Poor fit will show in probabilities considerably exceeding 0.05.

Analogues of classical significance tests are obtained using the posterior predictive p-value (Kato and Hoijtink, 2004). This was originally defined (Rubin, 1984; Meng, 1994) as the probability that a test statistic $T(y_{\text{rep}})$ of future observations y_{rep} is larger than or equal to the observed value of $T(y)$, given the adopted model M, the response data y, and any ancillary data x,

$$p_{\text{post}} = Pr[T(y_{\text{rep}}, x) \geq T(y, x) \mid y, x, M], \tag{3.7}$$

where x would typically be predictors measured without error. The probability is calculated over the posterior predictive distribution of y_{rep} conditional on M and x. By contrast, the classical p-test integrates over y, as in

$$p_c = Pr[T(y_{\text{rep}}, x) \geq T(y, x) \mid x, M].$$

The formulation of Meng (1994) is extended by Gelman et al. (1996) to apply to discrepancy criteria $D(y,\theta)$ based on data and parameters, as well as to observation-based functions $T(y)$. So the posterior predictive check is

$$p_{\text{post}} = Pr[D(y_{\text{rep}}, x, \theta) \geq D(y, x, \theta) \mid y, x, M],$$

where the probability is taken over the joint posterior distribution of y_{rep} and θ given M and x. In estimating the corresponding p_{post}, the discrepancy is calculated at each MCMC iteration. This is done both for the observations, giving a value $D(y,x,\theta^{(r)})$, and for the replicate data $y_{\text{rep}}^{(r)}$, sampled from $p(y_{\text{rep}}^{(r)} \mid \theta^{(r)},x)$ resulting in a value $D(y_{\text{rep}}^{(r)},x,\theta^{(r)})$ for each sampled parameter $\theta^{(r)}$. The proportion of samples where $D(y_{\text{rep}}^{(r)},x,\theta^{(r)})$ exceeds $D(y,x,\theta^{(r)})$ is then the Monte Carlo estimate of p_{post} For example, Kato and Hoijtink (2004) show good performance of p_{post} using both statistics T and discrepancies D in a normal multilevel model context with subjects $i = 1,\ldots m_j$ in clusters $j = 1,\ldots,J$

$$y_{ij} = b_{1j} + b_{2j}x_{ij} + u_{ij},$$

where $u_{ij} \sim N(0,\sigma_j^2)$. The hypotheses considered (i.e. in the form of reduced models) are $b_{kj} = \beta_k$ and $\sigma_j^2 = \sigma^2$.

Posterior predictive checks may be used to assess model assumptions. For instance, in multilevel and general linear mixed models, assumptions of normality regarding random effects are often made by default, and a posterior check against such assumptions is sensible. A number of classical tests have been proposed such as the Shapiro–Wilk W statistic (Royston, 1993) and the Jarque–Bera test (Bera and Jarque, 1980). These statistics can be derived at each iteration for actual and replicate data, and the comparison $D(y_{\text{rep}},x,\theta) \geq D(y,x,\theta)$ applied over MCMC iterations to provide a posterior predictive p-value.

3.5.3 Mixed Predictive Checks

The posterior predictive check makes double use of the data and so may be conservative as a test (Bayarri and Berger, 1999; Bayarri and Berger, 2000), since the observation y_i has a strong influence on the replicate $y_{i,\text{rep}}$. For example, Sinharay and Stern (2003) show posterior predictive checks may fail to detect departures from normality in random effects models. However, also in the context of random effect and hierarchical models, Marshall and Spiegelhalter (2003) and Marshall and Spiegelhalter (2007) mention a mixed predictive scheme, which uses a predictive prior distribution $p(y_{i,\text{rep}} \mid b_{i,\text{rep}},\theta)$ for a new set of random effects. The associated model check is called a mixed predictive p-test, whereas the (conservative) option of sampling from $p(y_{i,\text{rep}} \mid b_i,\theta)$ results in what Marshall and Spiegelhalter (2007, p.424) term full-data posterior predictive p-values. Mixed predictive replicates for each case seek to reduce dependence on the observation for that case, as the replicate data is sampled conditional only on global hyperparameters. Therefore, mixed predictive p-values are expected to be less conservative than posterior predictive p-values.

Let b denote random effects for cases $i = 1,\ldots,n$, or for clusters j in which individual cases are nested. To generate a replicate $y_{i,\text{rep}}$ for the ith case under the mixed scheme involves sampling (θ,b) from the usual posterior $p(\theta,b \mid y)$ conditional on all observations, but the sampled b are ignored and instead replicate b_{rep} values taken. A fully cross-validatory method would require that $b_{i,\text{rep}}$ be obtained by sampling from $p(b_{i,\text{rep}} \mid y_{[i]})$, or $b_{j,\text{rep}}$ sampled from $p(b_{j,\text{rep}} \mid y_{[j]})$. In fact, Green et al. (2009) compare mixed predictive assessment schemes with full cross-validation based on omitting single observations.

As full cross-validation is computationally demanding when using MCMC methods, approximate cross-validatory procedures are proposed by Marshall and Spiegelhalter (2007), in which the replicate random effect is sampled from $p(b_{i,\text{rep}} \mid y)$, followed by sampling $y_{i,\text{rep}}$ from $p(y_{i,\text{rep}} \mid b_{i,\text{rep}},\theta)$. This is called "ghosting" by Li et al. (2017). A discrepancy measure T^{obs} based on the observed data is then compared to its reference distribution

$$p(T \mid y) = \int p(T \mid b) p^M(b \mid y) db,$$

where $p^M(b \mid y) = \int p(b \mid \theta) p(\theta \mid y) d\theta$ may be termed the 'predictive prior' for b (Marshall and Spiegelhalter, 2007, p.413). This contrasts with more conservative posterior predictive checks based on replicate sampling from $p(y_{i,\text{rep}} \mid b_i, \theta)$, under which T_{obs} is compared to the reference distribution

$$p(T \mid y) = \int p(T \mid b) p(b \mid y) d\theta.$$

Marshall and Spiegelhalter (2003) confirm that a mixed predictive procedure reduces the conservatism of posterior predictive checks in relatively simple random effects models, and is more effective in reproducing $p(y_i \mid y_{[i]})$ than weighted importance sampling. However, this procedure may be influenced by the informativeness of the priors on the hyperparameters θ, and also by the presence of multiple random effects.

Marshall and Spiegelhalter (2007) also consider full cross-validatory mixed predictive checks to assess conflict in evidence regarding random effects b between the likelihood and the second stage prior; see also Bayarri and Castellanos (2007). Consider nested data $\{y_{ij}, i = 1, \ldots, n_j; j = 1, \ldots, J\}$ with likelihood

$$y_{ij} \sim N(b_j, \sigma^2),$$

and second stage prior on random cluster effects

$$b_j \sim N(\mu, \tau^2).$$

Under a cross-validatory approach, the discrepancy measure T_j^{obs} for cluster j would be based on the remaining data $y_{[j]}$ with cluster j excluded, and its reference distribution is then

$$p(T_j^{\text{rep}} \mid y_{[j]}) = \int p(T_j^{\text{rep}} \mid b_j, \sigma^2) p(b_j \mid \mu, \tau^2) p(\sigma^2, \tau^2, \mu \mid y_{[j]}) db_j d\sigma^2 d\tau^2 d\mu.$$

Marshall and Spiegelhalter (2007) also propose a conflict p-test based on comparing a predictive prior replicate $b_{j,\text{rep}} \mid y_{[j]}$ with a fixed effect estimate or "likelihood replicate" $b_{j,\text{fix}}$ for b_j based only on the data. The latter is obtained using a highly diffuse fixed effects prior on the b_j, rather than a borrowing strength hierarchical prior, for example, $b_j \sim Be(1,1)$ or $b_j \sim Be(0.5,0.5)$. Defining

$$b_{j,\text{diff}} = b_{j,\text{rep}} - b_{j,\text{fix}},$$

the conflict p-value for cluster j is obtained as

$$p_{j,\text{conf}} = Pr(b_{j,\text{diff}} \leq 0 \mid y).$$

This can be compared to a mixed predictive p-value, based on sampling $y_{j,\text{rep}}$ from a cross-validatory model using only the remaining cases $y_{[j]}$ to estimate parameters, and then comparing $y_{j,\text{rep}}$, or some function $T_j^{\text{rep}} = T(y_{j,\text{rep}})$, with $y_{j,\text{obs}}$ or with $T_j^{\text{obs}} = T(y_{j,\text{obs}})$.

Thus, depending on the substantive application, one may define lower or upper tail mixed p-values

$$p_{j,\text{mix}} = Pr(T_j^{\text{rep}} \leq T_j^{\text{obs}} \mid y),$$

or

$$p_{j,\text{mix}} = Pr(T_j^{\text{rep}} \geq T_j^{\text{obs}} \mid y),$$

with the latter being relevant in (say) assessing outliers in hospital mortality comparisons. If $T(y)=y$, and y is a count, then a mid p-value is relevant instead, with the upper tail test being

$$p_{j,\text{mix}} = Pr(y_{j,\text{rep}} > y_{j,\text{obs}} \mid y_{[j]}) + 0.5Pr(y_{j,\text{rep}} = y_{j,\text{obs}} \mid y_{[j]}). \quad (3.8)$$

Li et al. (2016, 2017) combine the principle of mixed predictive tests with that of importance sampling, with the intention of further correcting for optimism present in standard posterior predictive tests. Consider a particular MCMC iteration t. Sub-samples of random effects are obtained conditional on hyperparameters $\theta^{(t)}$ and the random effects $b^{(t)}$. One set of sub-samples $b_{j,s,\text{rep}}^{A}$ (for observations j and sub-samples $s=1,\ldots,S$) are obtained, along with the corresponding $y_{j,s,\text{rep}}$ conditional on $b_{j,s,\text{rep}}$. One then obtains the corresponding $p_{j,s,\text{mix}}$ as per (3.8), assuming the data are binomial or Poisson. Integrated importance weights for the $p_{j,\text{mix}}$ are based on an independent set of replicate random effects, say $b_{j,s,\text{rep}}^{B}$. Equations (38) to (40) in Li et al. (2017) set out the procedure more completely. Integrated importance weights are obtained as

$$W_i^{(t)} = 1 \Big/ \left(\sum_s p(y_i \mid \theta^{(t)}, b_{j,s,\text{rep}}^{B}) / S \right),$$

and can be used to provide WAIC estimates (denoted iWAIC); see equations (26)–(27) in Li et al. (2016).

Example 3.5 Seeds Data, Predictive Assessment of Logit-Binomial Model

This example considers case specific predictive assessment of the standard binomial logit model for the seeds data (the baseline model in Example 3.3). A first analysis obtains estimated log(CPO) values for each plate, and mixed predictive checks, as in Marshall and Spiegelhalter (2007). Mixed predictions are assessed by sampling replicate random effects $\{b_{i,\text{rep}}\}$, the corresponding $\{y_{i,\text{rep}} \mid b_{i,\text{rep}}\}$, and then deriving p-values

$$p_{i,\text{mix}} = Pr[y_{i,\text{rep}} > y_i \mid y) + 0.5Pr[y_{i,\text{rep}} = y_i \mid y).$$

Note that replicating this calculation in JagsUI or R2OpenBUGS needs to account for the fact that the step function is a greater than or equals calculation.

We also include posterior predictive checks (Section 3.5.2) based on comparing deviances for actual data and replicate data. Replicate data can be drawn from the model in an unmodified form (which may provide conservative posterior checks), or with replicates obtained using the mixed sampling approach.

With estimation using jagsUI, the mixed p-tests $p_{i,\text{mix}}$ and log(CPO) statistics are found to imply similar inferences regarding less well-fitted cases. The lowest $p_{i,\text{mix}}$ is for slide 4, which also has the second lowest log(CPO), while the second highest $p_{i,\text{mix}}$ is for slide 10, which has the lowest log(CPO). Regarding the posterior predictive checks, as in (3.7), taking replicates from the original model leads to a relatively low probability of 0.10, while using mixed replicates provides a probability of 0.08. Both these indicate possible model failure.

For this small sample, it is relatively straightforward to carry out a full (leave one out) cross-validation based on omitting each observation in turn. This shows slides 4, 15, and 20 as underpredicted (low probabilities that replicates exceed actual), and slides 10 and 17 as overpredicted.

Integrated importance cross-validation probabilities based on $S=10$ subsamples are also obtained; see the code used in [3]. These are very similar to the full cross-validation probabilities (see Table 3.1, which highlights slides with full cross-validation probabilities over 0.95 or under 0.05). We also use subsampling to obtain estimated iWAIC, following the notation of Li et al. (2016). Thus, the total iWAIC is 121.9, as compared to a LOO-IC of 121.6 and a WAIC of 119.8. Casewise iWAIC confirm the poor fit to slides 4 and 10. For this example, log(CPO) and casewise iWAIC statistics correlate closely, namely 0.9976.

In an attempt to improve fit, we replace the single intercept by a three-group discrete mixture intercept. Thus

$$y_i \sim N(n_i, \pi_i),$$

$$\text{logit}(\pi_i) = \beta_{1,G_i} + b_i + \beta_2 x_{1i} + \beta_3 x_{2i} + \beta_4 x_{1i} x_{2i},$$

TABLE 3.1

Seeds Data. Comparing Cross-Validation Probabilities, log(CPO), and Casewise iWAIC

Plate	Mixed Cross-Validation	IIS Cross-Validation	Full Cross-Validation	log(CPO)	Casewise iWAIC
1	0.883	0.925	0.922	−3.15	−3.19
2	0.473	0.460	0.454	−2.58	−2.51
3	0.894	0.949	0.946	−3.97	−3.98
4	*0.050*	**0.015**	*0.013*	*−4.86*	**−4.88**
5	0.226	0.178	0.176	−2.66	−2.65
6	0.250	0.240	0.242	−1.28	−1.29
7	0.312	0.266	0.258	−2.70	−2.76
8	0.117	0.058	0.053	−3.62	−3.75
9	0.735	0.808	0.811	−2.72	−2.75
10	*0.925*	**0.980**	*0.978*	*−5.03*	**−4.79**
11	0.271	0.267	0.267	−1.64	−1.66
12	0.263	0.188	0.194	−2.11	−2.15
13	0.726	0.769	0.772	−2.32	−2.37
14	0.856	0.899	0.902	−2.87	−2.84
15	*0.128*	*0.028*	*0.034*	*−4.08*	*−4.12*
16	0.934	0.936	0.937	−2.08	−2.06
17	*0.959*	**0.973**	*0.974*	*−3.52*	**−3.48**
18	0.442	0.451	0.456	−2.28	−2.35
19	0.606	0.625	0.630	−2.13	−2.18
20	*0.130*	*0.047*	*0.049*	*−3.70*	*−3.79*
21	0.681	0.692	0.692	−1.42	−1.41

$$G_i \sim \text{Categorical}(\phi[1:3]),$$

$$\phi \sim \text{Dirichlet}(5,5,5),$$

$$b_i \sim N(0,\sigma_b^2).$$

The posterior predictive checks, whether or not based on mixed replicates, are now satisfactory, both around 0.48. There are now no casewise predictive exceedance probabilities exceeding 0.95 or under 0.05. The LOO-IC and WAIC now stand at 116.6 and 110.4 respectively.

3.6 Computational Notes

[1] The code for the bridgesampling vignette is as follows:

```
require(jagsUI)
# generate data
set.seed(12345)
mu = 0
tau2 = 0.5
sigma2 = 1
# number of observations
n = 20
theta = rnorm(n, mu, sqrt(tau2))
y = rnorm(n, theta, sqrt(sigma2))
# define w according to length, T=30, of bridge-sampling schedule
T=30
T1=T+1
D= list(T=T,T1=T1, w=matrix(1,n,T1),n=n,y=y, path.pow=4)
# Model 1, mu=0
cat("model {for (h in 1:n) {for (s in 1:T1) {
L.tem[h,s] <- pow(L[h,s],q[s])
w[h,s] ~dunif(a1[h,s],b1[h,s])
a1[h,s] <- -1/L.tem[h,s]
b1[h,s] <- 1/L.tem[h,s]
LL[h,s] <- log(L[h,s])
# log-likelihood
log(L[h,s]) <- 0.5*log(phi[s]/(1+phi[s]))-0.919-0.5*phi[s]/
(1+phi[s])*y[h]*y[h]}}
# precision parameters
for (s in 1:T1) {phi[s] ~dgamma(1,1)}
phi.est <- phi[T1]
# path sampling calculations
for (s in 1:T1) {q[s] <- pow(a[s],path.pow)
expLL[s] <- sum(LL[1:n,s])}
a[1] <- 0.00001
for (s in 1:T) {a[s+1] <- s/T
mc[s] <- (q[s+1]-q[s])*(expLL[s+1]+expLL[s])*0.5}
logML <- sum(mc[])}
```

```
", file="model1.jag")
inits1 = list(phi=rep(1,T1))
inits2 = list(phi=rep(2,T1))
inits=list(inits1,inits2)
pars = c("logML","phi.est")
R1 = autojags(D, inits, pars,model.file="model1.jag",2,iter.
increment=1000,
n.burnin=100,Rhat.limit=1.025, max.iter=5000, seed=1234)
R1$summary
# Model 2, mu unknown
cat("model {for (h in 1:n) {for (s in 1:T1) {
L.tem[h,s] <- pow(L[h,s],q[s])
w[h,s] ~dunif(a1[h,s],b1[h,s])
a1[h,s] <- -1/L.tem[h,s]
b1[h,s] <- 1/L.tem[h,s]
LL[h,s] <- log(L[h,s])
# log-likelihood
log(L[h,s])<-0.5*log(phi[s]/(1+phi[s]))-0.919-0.5*phi[s]/
(1+phi[s])*(y[h]-mu[s])*(y[h]-mu[s])}}
# mean and precision parameters
for (s in 1:T1) {phi[s] ~dgamma(1,1)
mu[s] ~dnorm(0,1)}
phi.est <- phi[T1]
mu.est <- mu[T1]
# path sampling calculations
for (s in 1:T1) {q[s] <- pow(a[s],path.pow)
expLL[s] <- sum(LL[1:n,s])}
a[1] <- 0.00001
for (s in 1:T) {a[s+1] <- s/T
mc[s] <- (q[s+1]-q[s])*(expLL[s+1]+expLL[s])*0.5}
logML <- sum(mc[])}
", file="model2.jag")
inits1 = list(phi=rep(1,T1),mu=rep(0,T1))
inits2 = list(phi=rep(2,T1),mu=rep(0,T1))
inits=list(inits1,inits2)
pars = c("logML","phi.est","mu.est")
R2= autojags(D, inits, pars,model.file="model2.jag",2,iter.
increment=1000,
n.burnin=100,Rhat.limit=1.025, max.iter=5000, seed=1234)
R2$summary
# Marginal Likelihoods and Bayes Factor
ML=c()
ML[1]=R1$summary[1]
ML[2]=R2$summary[1]
BF12=exp(ML[1]-ML[2])
```

[2] The Bayesian chi-square method is illustrated using model 5 for the Scottish lip cancer incidence, as considered in Johnson (2004, pp.2374–2376). Thus, with E_i denoting expected incidence counts,

$$y_i \sim Po\left(E_i \exp\left(\rho_i\right)\right),$$

where the ρ_i are modelled as diffuse fixed effects. The BUGS code is as follows:

```
model {for (i in 1:n) {y[i] ~dpois(mu[i])
mu[i] <- E[i]*exp(rho[i])
rho[i] ~dnorm(0,0.001)
# Poisson probs (up to maximum count 50), ym[i]=y[i]-1 unless y=0
for (j in 1:51) {cdf[i,j] <- exp(-mu[i])*pow(mu[i],j-1)/
exp(logfact(j-1))*step(y[i]-j+1)}
for (j in 1:51) {cdfm[i,j] <- exp(-mu[i])*pow(mu[i],j-1)/
exp(logfact(j-1))*step(ym[i]-j+1)}
# cdf probs for y[i] and (y[i]-1)
t[i] <- sum(cdf[i,1:51])
tm[i] <- sum(cdfm[i,1:51])
# lower limit of interval from which bin randomly chosen
s[i] <- (1-equals(y[i],0))*tm[i]
u[i] ~dunif(0,1)
a[i] <- s[i]+u[i]*(t[i]-s[i])
ybin[i,1] <- step(0.2-a[i])
ybin[i,5] <- step(a[i]-0.8)
ybin[i,2] <- step(a[i]-0.2)*step(0.4-a[i])
ybin[i,3] <- step(a[i]-0.4)*step(0.6-a[i])
ybin[i,4] <- step(a[i]-0.6)*step(0.8-a[i])}
for (k in 1:K) {mhat[k] <- sum(ybin[,k])
m[k] <- n*p[k]
r.B[k] <- pow(mhat[k]-m[k],2)/m[k]}
# compare R.B with 95th quantile of the chi² distribution for K-1 df
R.B <- sum(r.B[])
P <- step(R.B-9.49)}
```

From iterations 5–100 thousand of a single chain run, the probability that R^B exceeds the 95% point of a χ_4^2 density is 0.157, and the posterior means of the number (mhat[] in the code) of the $n=56$ counts assigned to the five bins are (8.6,9.9,10.9,12.1,14.5).

[3] The code used for the IIS cross-validation probability estimates (seeds data) is

```
model3 <- function() {for(i in 1: N) {y[i] ~dbin(p[i],n[i])
  b[i] ~dnorm(0,tau)
  logit(p[i]) <- beta[1]+beta[2]*x1[i]+beta[3]*x2[i]+beta[4]*x1[i]*x2[i]
  +b[i]
  # IPD and log(IPD) based on averages over sub-samples
  IPD[i] <- mean(L.new[i,])
  # IIS-CV predictive estimates are posterior mean of A div'd by
  posterior mean of Awt
  A[i] <- mean(a[i,])*Awt[i]
  Awt[i] <- 1/IPD[i]
  # subsamples
  for (s in 1:S) {b.new.1[i,s]~dnorm(0,tau)
  logit(p.new.1[i,s])<- beta[1]+beta[2]*x1[i]+beta[3]*x2[i]+beta[4]*x1[i
  ]*x2[i]+b.new.1[i,s]
  y.new[i,s]~dbin(p.new.1[i,s],n[i])
  a[i,s]<- step(y.new[i,s]-y[i])-0.5*equals(y.new[i,s],y[i])
  b.new.2[i,s]~dnorm(0,tau)
  logit(p.new.2[i,s])<- beta[1]+beta[2]*x1[i]+beta[3]*x2[i]+beta[4]*x1[i
  ]*x2[i]+b.new.2[i,s]
```

```
log(L.new[i,s])<- logfact(n[i])-logfact(y[i])-logfact(n[i]-y[i])
+y[i]*log(p.new.2[i,s])+(n[i]-y[i])*log(1-p.new.2[i,s])}}
# priors
for (j in 1:P) {beta[j] ~dnorm(0.0,1.0E-6)}
tau ~dgamma(1,0.001)}
```

References

Albert J (1999) Criticism of a hierarchical model using Bayes factors. *Statistics in Medicine*, 18, 287–305.

Alqallaf F, Gustafson P (2001) On cross-validation of Bayesian models. *Canadian Journal of Statistics*, 29, 333–340.

Akaike H (1973) Information theory and an extension of the maximum likelihood principle, in *The Second International Symposium on Information Theory*, eds B Petrov, F Csaki. Akademiai Kiado, Budapest.

Ando T (2007) Bayesian predictive information criterion for the evaluation of hierarchical Bayesian and empirical Bayes models. *Biometrika*, 94, 443–458.

Aslanidou H, Dey D, Sinha D (1998) Bayesian analysis of multivariate survival data using Monte Carlo methods. *Canadian Journal of Statistics*, 26, 33–48.

Barry R (2006) An alternative to the 'ones' trick? BUGS Archive, 09/11/2006. https://www.jiscmail.ac.uk/cgi-bin/webadmin?A1=ind06&L=BUGS#13

Bartlett M (1957) A comment on D.V. Lindley's statistical paradox. *Biometrika*, 44, 533–534.

Bayarri M, Berger J (1999) Quantifying surprise in the data and model verification, pp 53–82, in *Bayesian Statistics 6*, eds J Bernardo, J Berger, A Dawid, A Smith. Oxford University Press, London, UK.

Bayarri M, Berger J (2000) P-values for composite null models. *Journal of the American Statistical Association*, 95, 1127–1142.

Bayarri M, Castellanos M (2007) Bayesian checking of the second levels of hierarchical models. *Statistical Science*, 22, 363–367.

Bera A, Jarque C (1980) Efficient tests for normality, homoscedasticity and serial independence of regression residuals. *Economics Letters*, 6, 255–259.

Berkhof J, van Mechelen I, Hoijtink H (2000) Posterior predictive checks: Principles and discussion. *Computational Statistics*, 3, 337–354.

Bernardo J, Smith A (1994) *Bayesian Theory*. Wiley.

Besag J, York J, Mollié A (1991) Bayesian image restoration, with two applications in spatial statistics. *Annals of the Institute of Statistical Mathematics*, 43(1), 1–20.

Bos C (2002) A comparison of marginal likelihood computation methods, pp 111–117, in *COMPSTAT 2002: Proceedings in Computational Statistics*, eds W Härdle, B Ronz. Springer, Berlin.

Brown H, Prescott R (1999) *Applied Mixed Models in Medicine*. John Wiley & Sons.

Buck C, Sahu S (2000) Bayesian models for relative archaeological chronology building. *Applied Statistics*, 49, 423–444.

Buenconsejo J, Fish D, Childs J, Holford T (2008) A Bayesian hierarchical model for the estimation of two incomplete surveillance data sets. *Statistics in Medicine*, 27, 3269–3285.

Burnham K, Anderson D (2002) *Model Selection and Multimodel Inference: A Practical Information-Theoretic Approach*, 2nd Edition. Springer-Verlag, New York.

Cai B, Dunson D (2006) Bayesian covariance selection in generalized linear mixed models. *Biometrics*, 62, 446–457.

Cai B, Dunson D (2008) Bayesian variable selection in generalized linear mixed models, in *Random Effect and Latent Variable Model Selection*, ed D Dunson. Springer.

Carlin B, Louis T (2000) *Bayes and Empirical Bayes Methods for Data Analysis*, 2nd Edition. Chapman and Hall, London, UK.

Carvalho C, Polson N, Scott J (2009) Handling sparsity via the horseshoe. *Proceedings of Machine Learning Research*, 5, 73–80.
Celeux G, Forbes F, Robert C, Titterington M (2006) Deviance information criteria for missing data models. *Bayesian Analysis*, 1, 651–674.
Chaloner K, Brant R (1988) A Bayesian approach to outlier detection and residual analysis. *Biometrika*,75, 651–660.
Chen M-H (2005) Computing marginal likelihoods from a single MCMC output. *Statistica Neerlandica*, 59, 16–29.
Chen Z, Dunson D (2003) Random effects selection in linear mixed models. *Biometrics*, 59, 762–769.
Chib S (1995). Marginal likelihood from the Gibbs output. *Journal of the American Statistical Association*, 90(432), 1313–1321.
Chib S (2008) Panel data modeling and inference: A Bayesian primer, pp 479–515, in *The Econometrics of Panel Data*, 3rd Edition, eds L Matyas, P Sevestre. Springer-Verlag, Berlin, Germany.
Chib S, Jeliazkov I (2001) Marginal likelihood from the Metropolis–Hastings output. *Journal of the American Statistical Association*, 96(453), 270–281.
Clayton D (1996) Generalized linear mixed models, in *Markov Chain Monte Carlo in Practice*, eds W Gilks, S Richardson, D Spiegelhalter. Chapman & Hall, London, UK.
Conn, P, Johnson D, Williams P, Melin S, Hooten M (2018) A guide to Bayesian model checking for ecologists. *Ecological Monographs*, 88(4), 526–542.
Crowder MJ (1978) Beta-binomial ANOVA for proportions. *Applied Statistics*, 27, 34–37.
de la Horra, J, Rodrguez-Bernal M (2005) Bayesian model selection: A predictive approach with losses based on distances. *Statistics & Probability Letters*, 71, 257–265.
Drton M, Plummer M (2017) A Bayesian information criterion for singular models. *Journal of the Royal Statistical Society: Series B (Statistical Methodology)*, 79(2), 323–380.
Drummond A, Bouckaert R (2015) *Bayesian Evolutionary Analysis with BEAST*. Cambridge University Press.
Fahrmeir L, Osuna L (2003) Structured count data regression. *Sonderforschungsbereich*, 386, Discussion Paper 334, University of Munich.
Fernandez C, Ley E, Steel M (2001) Benchmark priors for Bayesian model averaging. *Journal of Econometrics*, 100, 381–427.
Friel N, McKeone J, Oates C, Pettitt A (2017) Investigation of the widely applicable Bayesian information criterion. *Statistics and Computing*, 27(3), 833–844.
Friel N, Pettitt A (2008) Marginal likelihood estimation via power posteriors. *Journal of the Royal Statistical Society: Series B*, 70, 589–607.
Fruhwirth-Schnatter S (1999) Bayes Factors and Model Selection for Random Effect Models. Working Paper, Department of Statistics, University of Business Administration and Economics, Vienna.
Fruhwirth-Schnatter S (2004) Estimating marginal likelihoods for mixture and Markov switching models using bridge-sampling techniques. *The Econometrics Journal*, 7, 143–167.
Fruhwirth-Schnatter S, Tuchler R (2008) Bayesian parsimonious covariance estimation for hierarchical linear mixed models. *Statistics & Computing*, 18, 1–13.
Frühwirth-Schnatter S, Wagner H (2010) Stochastic model specification search for Gaussian and partial non-Gaussian state space models. *Journal of Econometrics*, 154(1), 85–100.
Geisser S, Eddy W (1979) A predictive approach to model selection. *Journal of the American Statistical Association*, 74, 153–160.
Gelfand A (1996) Model determination using sampling based methods, Chapter 9, in *Markov Chain Monte Carlo in Practice*, eds W Gilks, S Richardson, D Spiegelhalter. Chapman & Hall/CRC, Boca Raton.
Gelfand A, Dey D (1994) Bayesian model choice: Asymptotics and exact calculations. *Journal of the Royal Statistical Society, Series B*, 56, 501–514.
Gelfand A, Dey D, Chang H (1992) Model determination using predictive distributions with implementations via sampling-based methods, pp 147–168, in *Bayesian Statistics 4*, eds J Bernardo et al. Oxford University Press.
Gelfand A, Ghosh S (1998) Model choice: A minimum posterior predictive loss approach. *Biometrika*, 85, 1–11.

Gelfand A, Vlachos P (2003) On the calibration of Bayesian model choice criteria. *Journal of Statistical Planning and Inference*, 111, 223–234.

Gelman A, Carlin J, Stern H, Dunson D, Vehtari A, Rubin D (2014) *Bayesian Data Analysis*. CRC, Boca Raton, FL.

Gelman A, Meng XL (1998) Simulating normalizing constants: From importance sampling to bridge sampling to path sampling. *Statistical Science*, 13(2), 163–185.

Gelman A, Meng XL, Stern H (1996) Posterior predictive assessment of model fitness via realized discrepancies. *Statistica Sinica*, 6, 733–807.

George E, McCulloch R (1993) Variable selection via Gibbs sampling. *Journal of the American Statistical Association*, 88(423), 881–889.

George E, McCulloch R (1997) Approaches for Bayesian variable selection. *Statistica Sinica*, 7, 339–373.

Gneiting T, Raftery AE (2007) Strictly proper scoring rules, prediction, and estimation. *Journal of the American Statistical Association*, 102(477), 359–378.

Gosoniu L, Vounatsou P, Sogoba N, Smith T (2006) Bayesian modelling of geostatistical malaria risk data. *Geospatial Health*, 1, 127–139.

Green M, Medley G, Browne W (2009) A comparison of methods of posterior predictive assessment in multilevel logistic regression using an example from veterinary medicine. *Veterinary Research*. 40(4), 1–10.

Gronau Q, Sarafoglou A, Matzke D, Ly A, Boehm U, Marsman M (2017b) A tutorial on bridge sampling. *Journal of Mathematical Psychology*, 81, 80–89.

Gronau Q, Singmann H, Wagenmakers E (2017a). Bridgesampling: An R package for estimating normalizing constants. arXiv preprint arXiv:1710.08162

Gustafson P, Hossain S, Macnab Y (2006) Conservative prior distributions for variance parameters in hierarchical models. *Canadian Journal of Statistics*, 34(3), 377–390.

Han C, Carlin B (2001) Markov chain Monte Carlo methods for computing Bayes factors: A comparative review. *Journal of the American Statistical Association*, 96, 1122–1132.

Harun N, Cai B (2014) Bayesian random effects selection in mixed accelerated failure time model for interval-censored data. *Statistics in Medicine*, 33(6), 971–984.

Ibrahim J, Chen M, Sinha D (2001) Criterion-based methods for Bayesian model assessment. *Statistica Sinica*, 11, 419–443.

Jeffreys H. (1961) *The Theory of Probability*, 3rd edn. Oxford, UK, Clarendon Press.

Johnson V (2004) A Bayesian χ^2 test for goodness-of-fit. *Annals of Statistics*, 32, 2361–2384.

Johnson V, Rossell D (2012) Bayesian model selection in high-dimensional settings. *Journal of the American Statistical Association*, 107(498), 649–660.

Kacker R, Forbes A, Kessel R, Sommer K-D (2008) Bayesian posterior predictive p-value of statistical consistency in interlaboratory evaluations. *Metrologia*, 45, 512–523.

Kass R, Raftery A (1995) Bayes factors. *Journal of the American Statistical Association*, 90, 773–795.

Kato B, Hoijtink H (2004) Testing homogeneity in a random intercept model using asymptotic, posterior predictive and plug-in p-values. *Statistica Neerlandica*, 58, 179–196.

Kelly D, Smith C (2011) Bayesian model checking, pp 39–50, in *Bayesian Inference for Probabilistic Risk Assessment*. eds D Kelly, C Smith. Springer, London, UK.

Key J, Pericchi L, Smith A (1999) Bayesian model choice: What and why?, pp 343–370, in *Bayesian Statistics 6*, eds J Bernardo, J Berger, A Dawid, A Smith. Oxford Science Publications, Oxford, UK.

Kinney, S, Dunson D (2008) Bayesian model uncertainty in mixed effects models, in *Random Effect and Latent Variable Model Selection*, ed D Dunson. Springer.

Kuhn E, Lavielle M (2005) Maximum likelihood estimation in nonlinear mixed effects models. *Computational Statistics & Data Analysis*, 49, 1020–1038.

Kuo L, Mallick B (1998) Variable selection for regression models. *Sankhyā: The Indian Journal of Statistics, Series B*, 60(1), 65–81.

Laud P, Ibrahim J (1995) Predictive model selection. *Journal of The Royal Statistical Society: Series B*, 57, 247–262.

Lenk P, DeSarbo W (2000) Bayesian inference for finite mixture models of generalized linear models with random effects. *Psychometrika*, 65, 475–496.
Li L, Qiu S, Zhang B, Feng C (2016) Approximating cross-validatory predictive evaluation in Bayesian latent variable models with integrated IS and WAIC. *Statistics and Computing*, 26(4), 881–897.
Li L, Feng C, Qiu S (2017) Estimating cross-validatory predictive p-values with integrated importance sampling for disease mapping models. *Statistics in Medicine*, 36(14), 2220–2236.
Lopes HF, West M (2004) Bayesian model assessment in factor analysis. *Statistica Sinica*, 14(1), 41–68.
Lucy L (2018) Bayesian model checking: A comparison of tests. *Astronomy & Astrophysics*, 614, A25.
MacNab Y, Qiu Z, Gustafson P, Dean C, Ohlsson A, Lee S (2004) Hierarchical Bayes analysis of multilevel health services data: A Canadian neonatal mortality study. *Health Services and Outcomes Research Methodology*, 5, 5–26.
Marshall C, Spiegelhalter D (2003) Approximate cross-validatory predictive checks in disease mapping models. *Statistics in Medicine*, 22, 1649–1660.
Marshall C, Spiegelhalter D (2007) Identifying outliers in Bayesian hierarchical models: A simulation-based approach. *Bayesian Analysis*, 2, 1–33.
Meng X (1994) Posterior predictive p-values. *The Annals of Statistics*, 22, 1142–1160.
Meng XL, Wong WH (1996) Simulating ratios of normalizing constants via a simple identity: A theoretical exploration. *Statistica Sinica*, 6(4), 831–860.
Meyer M, Laud P (2002) Predictive variable selection in generalized linear models. *Journal of the American Statistical Association*, 97, 859–871.
Mitchell TJ, Beauchamp JJ (1988) Bayesian variable selection in linear regression. *Journal of the American Statistical Association*, 83(404), 1023–1032.
Müller S, Scealy J, Welsh A (2013) Model selection in linear mixed models. *Statistical Science*, 28(2), 135–167.
Myung J, Pitt M (2004) Model comparison methods, pp 351–366, in *Methods in Enzymology*, Vol. 383, eds L Brand, M Johnson. Elsevier, Amsterdam.
Nandram B, Kim H (2002) Marginal likelihoods for a class of Bayesian generalized linear models. *Journal of Statistical Computation and Simulation*, 73, 319–340.
Ohlssen D, Sharples L, Spiegelhalter D (2006) Flexible random-effects models using Bayesian semi-parametric models: Applications to institutional comparisons. *Statistics in Medicine*, 26, 2088–2112.
Park J Y, Johnson M, Lee Y-S (2015) Posterior predictive model checks for cognitive diagnostic models. *International Journal of Quantitative Research in Education*, 2(3–4), 244–264.
Pettit L,Young K (1990) Measuring the effect of observations on Bayes factors. *Biometrika*, 77, 455–466.
Piironen J, Vehtari A (2017) Comparison of Bayesian predictive methods for model selection. *Statistics and Computing*, 27(3), 711–735.
Plummer M (2008) Penalized loss functions for Bayesian model comparison. *Biostatistics*, 9, 523–539.
Pourahmadi M, Daniels M (2002) Dynamic conditionally linear mixed models for longitudinal data. *Biometrics*, 58, 225–231.
Rockova V, Lesaffre E, Luime J, Löwenberg B (2012). Hierarchical Bayesian formulations for selecting variables in regression models. *Statistics in Medicine*, 31(11–12), 1221–1237.
Rossell P (2018) Bayesian Model Selection and Averaging with mombf. https://cran.r-project.org/web/packages/mombf/vignettes/mombf.pdf
Royston P (1993) A toolkit for testing for non-normality in complete and censored samples. *The Statistician*, 42, 37–43.
Rubin DB (1984) Bayesianly justifiable and relevant frequency calculations for the applied statistician. *The Annals of Statistics*, 12(4), 1151–1172.
Sala-i-Martin X, Doppelhofer G, Miller RI (2004) Determinants of long-term growth: A Bayesian averaging of classical estimates (BACE) approach. *American Economic Review*, 94(4), 813–835.
Saville B, Herring A (2009) Testing random effects in the linear mixed model using approximate Bayes factors. *Biometrics*, 65, 369–376.

Saville B, Herring A, Kaufman J (2011) Assessing variance components in multilevel linear models using approximate Bayes factors: A case-study of ethnic disparities in birth weight. *Journal of the Royal Statistical Society: Series A*, 174(3), 785–804.

Schwarz G (1978) Estimating the dimension of a model. *Annals of Statistics*, 6, 461–464.

Silva R, Lopes H, Migon H (2006) The extended generalized inverse Gaussian distribution for log-linear and stochastic volatility models. *Brazilian Journal of Probability and Statistics*, 20, 67–91.

Sinha D (1993) Semiparametric Bayesian analysis of multiple event time data. *Journal of the American Statistical Association*, 88(423), 979–983.

Sinha D, Chen M-H, Ghosh S (1999) Bayesian analysis and model selection for interval-censored survival data. *Biometrics*, 55, 585–590.

Sinharay S, Stern H (2003) Posterior predictive model checking in hierarchical models. *Journal of Statistical Planning and Inference*, 111, 209–221.

Sinharay S, Stern H (2005) An empirical comparison of methods for computing bayes factors in generalized linear mixed models. *Journal of Computational and Graphical Statistics*, 14, 415–435.

Smith AF, Gelfand AE (1992) Bayesian statistics without tears: A sampling–resampling perspective. *The American Statistician*, 46(2), 84–88.

Smith M, Kohn R (1996) Nonparametric regression using Bayesian variable selection. *Journal of Econometrics*, 75(2), 317–343.

Smith M, Kohn R (2002) Parsimonious covariance matrix estimation for longitudinal data. *Journal of the American Statistical Association*, 97(460), 1141–1153.

Spiegelhalter D (2006) Two brief topics on modelling With WinBUGS. Presented at ICEBUGS Conference, Helsinki 2006.

Spiegelhalter D, Best N, Carlin B, van der Linde A (2002) Bayesian measures of model complexity and fit. *Journal of the Royal Statistical Society, Series B*, 64, 583–639.

Stern H, Sinharay S (2005) Bayesian model checking and model diagnostics, pp 171–192, in *Bayesian Thinking: Modeling and Computation, Handbook of Statistics*, Vol. 25, eds D Dey, C Rao. Elsevier, Amsterdam, Netherlands.

Tierney L, Kadane J (1986) Accurate approximations for posterior moments and marginal densities. *Journal of the American Statistical Association*, 81, 82–86.

Vannucci M (2000) Matlab code for Bayesian variable selection. *ISBA Bulletin*, 7(3), 1–3.

Vehtari A, Gelman A, Gabry J (2017) Practical Bayesian model evaluation using leave-one-out cross-validation and WAIC. *Statistics and Computing*, 27(5), 1413–1432.

Vehtari A, Lampinen J (2002) Expected utility estimation via cross-validation, in *Bayesian Statistics 7*, eds J Bernardo, M Bayarri, J Berger, A Dawid, D Heckerman, A Smith, M West. Clarendon Press.

Watanabe S (2010) Asymptotic equivalence of Bayes cross validation and widely applicable information criterion in singular learning theory. *Journal of Machine Learning Research* 11, 3571–3594.

Watanabe S (2013) A widely applicable Bayesian information criterion. *Journal of Machine Learning Research*, 14, 867–897.

Weihs C, Plummer M (2016) Package sBIC. Computing the singular BIC for multiple models. https://cran.r-project.org/web/packages/sBIC/sBIC.pdf

Weiss R (1994) Pediatric pain, predictive inference and sensitivity analysis. *Evaluation Review*, 18, 651–678.

Xie W, Lewis P, Fan Y, Kuo L, Chen M-H (2011) Improving marginal likelihood estimation for Bayesian phylogenetic model selection. *Systematic Biology*, 60(2), 150–160.

Yang M (2012) Bayesian variable selection for logistic mixed model with nonparametric random effects. *Computational Statistics & Data Analysis*, 56(9), 2663–2674.

Zellner A (1986) On assessing prior distributions and Bayesian regression analysis with g-prior distributions, pp 233–243, in *Bayesian Inference and Decision Techniques: Essays in Honor of Bruno de Finetti*. North-Holland/Elsevier.

Zhu L, Gorman D, Horel S (2006) Hierarchical Bayesian spatial models for alcohol availability, drug "hot spots" and violent crime. *International Journal of Health Geographics*, 5, 54.

4

Borrowing Strength via Hierarchical Estimation

4.1 Introduction

What is sometimes termed ensemble estimation, or borrowing strength, refers to inferences for collections of similar (exchangeable) units $i = 1,\ldots,n$ (schools, health agencies, etc.) using Bayesian hierarchical methods (Burr and Doss, 2005; Clark and Gelfand, 2006; Rounder et al., 2013; Rhodes et al., 2016). Among possible examples are surgical outcome rates (Kuhan et al., 2002; Bayman et al., 2013), drug development (Gupta, 2012), baseball batting averages (Kruschke and Vanpaemel, 2015), health quality measures (Staggs and Gajewski, 2017), or oviposition preference data (Fordyce et al., 2011). Fixed effects models for such collections are problematic (Marshall and Spiegelhalter, 1998), whereas hierarchical random effects approaches pool information across units to obtain more reliable estimates for each unit, identify units with unusually high or low values, and enable comparisons between units. Borrowing strength may need to be modified to account for, or accommodate, unusual observations (Baker and Jackson, 2016; Farrell et al., 2010). Rankings of the units may often be required, or probabilities of significant difference between units or against a threshold (Deely and Smith, 1998; Staggs and Gajewski, 2017).

Implementations for hierarchical methods in R include Bayesian applications, as in bayesPref (Gompert and Fordyce, 2015), LearnBayes (Albert, 2015), bmeta (Ding and Baio, 2016), bamdit (Verde, 2018), meta4diag (Guo and Riebler, 2016), and frequentist applications, such as metaplus (Beath, 2016) and metafor (Viechtbauer, 2010; Viechtbauer, 2017); see also https://cran.r-project.org/web/views/MetaAnalysis.html. For semiparametric and discrete mixture models, packages include DPpackage (Jara et al., 2011), bspmma (Burr, 2012), bayesmix (Gruen and Plummer, 2015), and label.switching (Papastamoulis, 2016).

A prototypical Bayesian hierarchical model for interrelated units specifies an outcome model (first stage likelihood) $p(y_i \mid b_i, \Phi)$, and a process model involving unobserved effects b_i, with density $p(b_i \mid \Psi)$, conditional on hyperparameters Ψ. In a longitudinal linear regression, the Φ might be regression coefficients and the residual regression variance, while Ψ could include the variance of unit random intercepts b_i. Similarly, in a Poisson-gamma mixture, the likelihood $p(y_i \mid b_i)$ conditions on latent gamma effects b_i. At the second stage, the gamma density $p(b_i \mid \Psi)$ for the b_i conditions on gamma shape and scale parameters Ψ, while prior densities for the gamma parameters form the third stage.

The procedures considered in this chapter are typically based on an exchangeability principle: that units are similar enough to justify being modelled by a common density and that the units are not configured in ways (e.g. over time or space) that implies higher correlations between some units than others (Spiegelhalter et al., 2004; Lindley and Smith, 1972, p.4). Structuring of units in space, time, or other forms of non-exchangeability does not preclude borrowing strength, but a prior reflecting that structuring is required (see Chapters 5, 6). Exchangeability means that there is no prior basis for supposing some units have higher true effects than others, or that certain subgroups of units are more similar between themselves than other subgroups (e.g. that mortality in hospitals i and j is more similar than between hospitals i and k). For units of the same type and observations generated under similar conditions, exchangeability means all possible permutations of the sequence of units have the same probability: random variables $\{y_1,\ldots,y_n\}$ are exchangeable if their joint distribution $P(y_1,\ldots,y_n)$ is invariant under permutation of its arguments, so that

$$P(y_1^*,\ldots,y_n^*) = P(y_1,\ldots,y_n)$$

where $\{y_1^*,\ldots,y_n^*\}$ is any permutation of $\{y_1,\ldots,y_n\}$ (Greenland and Draper, 1998). Sometimes units are better considered exchangeable within subgroups of the data; a UK example relates to mortality in cardiac surgery units, with exchangeability within "closed" procedures involving no use of heart bypass during anaesthesia, and "open" procedures where the heart is stopped and heart bypass needed (Spiegelhalter, 1999). Sometimes exchangeability can only be supported for residual effects, b_i, obtained after controlling for known differences between studies, for example, as represented by covariates in meta-regression.

Hierarchical smoothing methods result in shrinkage of estimates for each unit towards the average outcome rate in the population within which exchangeability is assumed; shrinkage will be greater for units with observations based on small samples (Staggs and Gajewski, 2017). When the single population hierarchical model is appropriate, pooling of strength results in more precise estimates, and may provide better out of sample predictive performance – see Deely and Smith (1998) for an application of such predictions to performance indicators. However, borrowing strength may increase the risk of bias, as compared to unadjusted fixed effect estimates. The increase in precision but possible bias inherent in hierarchical estimation provides a dilemma known as the "bias-variance trade-off." In some applications, inferences are over more than one variable as well as over a collection of similar units (Everson and Morris, 2000; van Houwelingen et al., 2002). Inferences will typically be improved for related outcomes over similar units (e.g. surgical and non-surgical mortality in different hospitals).

While smoothing is the leading motivation for hierarchical models, a related theme is to achieve smoothed estimates that allow appropriately for heterogeneity between sample units – that is, they do not oversmooth, and show some robustness or flexibility to individual units, or to clusters of units, that are somewhat discrepant or outlying from the rest of the population (Baker and Jackson, 2008, 2016; Zhang et al., 2015; Beath, 2014). Such heterogeneity will typically be associated with overdispersion in Poisson or binomial data, or with heavy tails or skew in the case of departures from normality in continuous outcomes. One way to modify the standard densities to take account of heterogeneity greater than postulated under that density is to allow adaptive continuous mixing at unit level. Examples of such mixing are the scale mixture approach to the t-density discussed in Section 4.3. Another option is discrete mixing (see Sections 4.8 et seq.), in which a single population assumption is replaced by an assumption of two or more subpopulations. Shrinkage will then be towards the subgroup characteristics that each unit has the highest posterior probability of belonging to.

Undershrinkage (undersmoothing) also raises issues: this will lead to over-estimation of random effect variability and is to be avoided when a type I error has worse consequences than a type II error (Gustafson et al., 2006). Similarly, Spiegelhalter (2005) points out that there is a danger in performance indicator analysis that the units (e.g. institutions) that one is trying to detect could be accommodated by a random effects approach, and it is therefore important that robust methods are used to estimate the standard deviation of the random effects distribution.

4.2 Hierarchical Priors for Borrowing Strength Using Continuous Mixtures

Observations for related units are typically considered in aggregate form, such as means y_i for a metric variable, or numbers of successes for a binomial variable, even though originally collected in disaggregated form for repetitions j within each unit of observation i. For example, consider a normal first stage model $y_i \sim N(b_i, s_i^2)$, with known variances s_i^2. Latent means b_i vary according to a stage 2 density $p(b_i \mid \Psi)$, such as $b_i \sim N(\mu, \tau^2)$. Stage 3 specifies priors on the population parameters $\Psi = (\mu, \tau^2)$.

Inferences of interest include the posterior densities, such as $p(b_i \mid y)$ and $p(\mu \mid y)$, and posterior probabilities that b_i and μ are in specified intervals, such as the probability of a positive effect $Pr(\mu > 0 \mid y)$, when the y_i measure clinical treatment benefit. Interest may also be in predictions for hypothetical future units (e.g. for a new clinical trial, or for the next year in a performance ranking application), $p(y_{\text{new}} \mid y)$ (Friede et al., 2017). If $p(\Psi \mid y)$ can be obtained analytically, or samples $\Psi^{(1)}, \Psi^{(2)}, \ldots \Psi^{(t)}$ obtained directly, then $p(b_i \mid y)$, $p(\mu|y)$ and $p(y_{\text{new}} \mid y)$ can be obtained by Monte Carlo simulation, as in

$$p(b_i \mid y) = \int p(b_i \mid y, \Psi) p(\Psi \mid y) d\Psi,$$

leading to the estimate $\hat{p}(b_i \mid y) = \sum_{t=1}^{T} p(b_i \mid y, \Psi^{(t)})$.

An alternative to direct simulation is to simulate the full posterior $p(\Psi, b \mid y)$ using MCMC methods, by obtaining samples $\{b_i^{(t)}, \Psi^{(t)}\}$ from the full conditional posteriors $p(b_i \mid b_{[i]}, \Psi, y)$ and $p(\psi_q \mid \Psi_{[q]}, b, y)$. For example, often the first stage density $p(y|b)$ is in the full exponential family, so that

$$p(y_i \mid b_i) = \exp\left(\frac{y_i b_i - B(b_i)}{A(\phi_i)} + C(y_i, \phi_i) \right) \tag{4.1}$$

where ϕ_i is a scale parameter. Assuming a conjugate second stage prior, the conditional posterior of each b_i follows the same density. For example, assume (Frees, 2004; Das and Dey, 2006; Das and Dey, 2007; Ferreira and Gamerman, 2000) that

$$p(b_i \mid \Psi) = k_1 \exp(b_i g_1(\Psi) - B(b_i) g_2(\Psi)) \tag{4.2}$$

where k_1 is a normalising constant. Then the posterior density of b_i and Ψ given y is of exponential form

$$p(b_i, \Psi \mid y) = k_2 \exp\left(\left[g_1(\Psi) + \frac{y_i}{A(\phi_i)} \right] b_i - B(b_i) \left[g_2(\Psi) + \frac{1}{A(\phi_i)} \right] \right). \tag{4.3}$$

With proper log-concave priors $p(\Psi)$, the full conditionals $p(\psi_q \mid \Psi_{[q]}, b, y)$ are logconcave, and can be sampled using methods such as those of Gilks and Wild (1992). By contrast, if improper priors are assumed on hyperparameters $\{\psi_1,...,\psi_Q\}$, then the full posterior $p(b,\Psi \mid y)$ is not necessarily proper (George et al., 1993; George and Zhang, 2001; Browne and Draper, 2006), and empirical convergence of the MCMC sequence $\{b^{(t)},\Psi^{(t)}\}$ may be problematic even if the posterior is proper analytically. George and Zhang (2001) consider posterior propriety results for the Poisson-gamma, the binomial-beta, and multinomial-Dirichlet models in terms of conditions on the hyperparameter prior tail behaviour. For the latter two hierarchical schemes, no improper prior can guarantee a proper posterior. Similar convergence and identification issues apply to the general linear mixed model formulation.

4.3 The Normal-Normal Hierarchical Model and Its Applications

A widely applied conjugate hierarchical scheme assumes normal sampling of observations and normally distributed latent effects. A typical borrowing strength or meta-analysis template is for continuous observations for unit level effects y_i, and intra-unit variation $a(\phi_i) = s_i^2$, even though the underlying data might have involved two-way nesting with $j = 1,\dots,J_i$ replications for units $i = 1,\dots,n$. This is often the case in clinical meta-analysis where patient level results are summarised as treatment or risk factor effect measures (e.g. change in a clinical measure between treatment and control groups, or the slope of a dose-response curve) along with moment estimates of sampling variances. Assuming the observed summary measures are exchangeable, obtained from similar study designs and relating to similar types of unit (Spiegelhalter et al., 2004, p.92), they may be regarded as draws from an underlying common density for the unknown true means b_i.

Often the normal-normal model is applied to originally discrete data using normal approximations for the effect measures (e.g. Bakbergenuly and Kulinskaya, 2017; Friede et al., 2017). Suppose r_{iT} of N_{iT} treated subjects in study i exhibit a particular response (e.g. disease or death), as compared to r_{iC} of N_{iC} control subjects. Define log odds

$$\omega_{iT} = \log(r_{iT}/(N_{iT} - r_{iT}),$$

and

$$\omega_{iC} = \log(r_{iC}/(N_{iC} - r_{iC}),$$

in the treated and control arms. Then the log odds ratios form the unit level response,

$$y_i = \omega_{iT} - \omega_{iC},$$

which may be assumed approximately normal with variance

$$s_i^2 = \frac{1}{r_{iT}} + \frac{1}{N_{iT} - r_{iT}} + \frac{1}{r_{iC}} + \frac{1}{N_{iC} - r_{iC}},$$

(see Example 4.1). It is also possible to take y_i as a log relative risk between treatment and control groups, namely

$$y_i = \log\left(\frac{r_{iT}}{N_{iT}}\right) - \log\left(\frac{r_{iC}}{N_{iC}}\right),$$

with variance

$$\frac{1}{r_{iT}} + \frac{1}{r_{iT}} - \frac{1}{N_{iT}} - \frac{1}{N_{iC}}.$$

Another option is to take the risk difference

$$y_i = \frac{r_{iT}}{N_{iT}} - \frac{r_{iC}}{N_{iC}},$$

as approximately normal with variance

$$\frac{r_{iT}(N_{iT} - r_{iT})}{N_{iT}^3} + \frac{r_{iC}(N_{iC} - r_{iC})}{N_{iC}^3}.$$

Unless heavy tails, skewness, or multiple modes are suspected, an appropriate hierarchical model then has a normal first stage, and a second stage normal density for the b_i with variance constant over units. So

$$y_i \sim N(b_i, s_i^2), \tag{4.4.1}$$

and

$$b_i \sim N(\mu, \tau^2). \tag{4.4.2}$$

Integrating out the b_i, the marginal likelihood for y_i (Guolo and Varin, 2017) is then

$$y_i \mid \mu, \tau^2 \sim N(\mu, s_i^2 + \tau^2). \tag{4.4.3}$$

Often the summary measures are unit or trial means and different observational variances are associated with differing sample sizes N_i, so that $s_i^2 = \sigma^2/N_i$, where σ^2 is an additional unknown. While clinical meta-analysis applications are common, a similar scenario occurs in small area estimation from multiple surveys where s_i^2 are sampling variances obtained according to the survey design.

More complex situations can be fitted into this framework. For example, Abrams et al. (2000) consider the effect of testing positive or negative in a screening test on subsequent levels of anxiety; see also Abrams et al. (2005). Let x_{ik} be baseline anxiety in study i, with $k=1$ (tested positive) and $k=2$ (tested negative), and with N_{i1} and N_{i2} subjects in different arms. Let z_{ik} be follow-up anxiety according to screening result, and let $d_{ik} = z_{ik} - x_{ik}$ denote change in anxiety. Then the measure of interest is the contrast between anxiety growth according to screening result, namely $y_i = d_{i1} - d_{i2}$, with variance

$$s_i^2 = \frac{(N_{i1} - 1)V(d_{i1}) + (N_{i2} - 1)V(d_{i2})}{(N_{i1} + N_{i2} - 2)},$$

where

$$V(d_{ik}) = V(x_{ik}) + V(z_{ik}) - 2\rho\sqrt{V(x_{ik})V(z_{ik})},$$

and ρ is a within-subject correlation taken constant across studies and arms. Studies may not report all the relevant statistics: they may report the d_{ik} and their variances, or the separate baseline and follow-up measures in each arm $\{x_{ik}, z_{ik}\}$ and their variances. In either case, meta-analysis requires a prior on ρ.

In (4.4), assume independent priors on the hyperparameters

$$p(\tau^2, \mu) = p(\tau^2)p(\mu),$$

with a commonly adopted option being

$$\tau^2 \sim IG\left(\frac{\nu}{2}, \frac{\nu\lambda}{2}\right),$$

$$\mu \sim N(m_\mu, V_\mu),$$

where ν, λ, m_μ, V_μ are assumed known. The full posterior conditional for b_i is then (Browne and Draper, 2006; George et al., 1993; Silliman, 1997, p.927)

$$p(b_i \mid b_{[i]}, \mu, \tau, y) = p(b_i \mid \mu, \tau, y) = N([1 - w_i]y_i + w_i\mu, D_i),$$

where

$$D_i = \left(\frac{1}{s_i^2} + \frac{1}{\tau^2}\right)^{-1} = \frac{\tau^2 s_i^2}{\tau^2 + s_i^2},$$

$$w_i = \frac{s_i^2}{s_i^2 + \tau^2},$$

and the first equality is by virtue of conditional independence of the b_i. The full conditional for τ^2 is

$$\tau^2 \sim IG\left(0.5[n + \nu], 0.5\left[\nu\lambda + \sum_{i=1}^{n}(y_i - \mu)^2\right]\right),$$

while that for μ involves a precision weighted average of m_μ, and the average of the b_i, namely

$$\mu \sim N\left(\bar{b}\left(\frac{nV_\mu}{nV_\mu + \tau^2}\right) + m_\mu\left(\frac{\tau^2}{nV_\mu + \tau^2}\right), \frac{\tau^2 V_\mu}{nV_\mu + \tau^2}\right).$$

Allowing interrelatedness between units leads to inferences about underlying unit means that are different from those obtained under alternative scenarios sometimes used, namely, (a) the "independent units" case, with b_i taken as unknown and mutually unrelated fixed effects, with $\tau^2 \to \infty$, and (b) the complete pooling model of classical meta-analysis where the studies are regarded as effectively interchangeable and $\tau^2 = 0$.

By contrast, the intermediate "exchangeable units" Bayes model leads to a posterior mean for b_i,

$$E[b_i \mid y] = w_i \mu + [1 - w_i] y_i,$$

that averages over the prior mean μ and the data mean y_i with weights $w_i = s_i^2/(s_i^2 + \tau^2)$ and $1 - w_i = \tau^2/(s_i^2 + \tau^2)$ respectively, as is apparent from the Gibbs sampling full conditionals. The b_i under an exchangeability scenario have narrower posterior intervals than under an independent units assumption, with precision related to the confidence about the prior mean and the prior assumed for τ^2 (see also Section 4.4). Assume the intra-study variances can be expressed as $s_i^2 = \sigma^2/N_i$ and then set $\tau^2 = \sigma^2/N_\mu$, where N_μ is the sample size assigned to the prior mean. Then the weights w_i become $N_\mu/(N_i + N_\mu)$ demonstrating that shrinkage to the prior mean increases as the confidence about the prior mean increases.

The normal-normal model may be robustified against skewness, heavy tails, and outlier studies in either the sampling density or the latent effects density. If non-normality is suspected at the second stage, a heavy-tailed prior can be used to accommodate possibly outlying studies. A normal-t approach involves study-specific scale adjustments at the second stage (West, 1984), downplaying the influence of atypical studies on posterior estimates of the overall effect μ, and avoiding over-shrinkage of individual study effects b_i. The scaling factors are gamma with shape and rate $\nu/2$:

$$b_i \sim N(\mu, \tau^2/\lambda_i)$$

$$\lambda_i \sim Ga\left(\frac{\nu}{2}, \frac{\nu}{2}\right).$$

Skewness in the observed data can often be reduced or eliminated by transformation. However, continuous data (e.g. cost data or data resulting from psychometric tests) will sometimes have more unusual departures from normality that render transformation inapplicable, such as clumping of zero values as well as positive skewness in positive responses (Delucchi and Bostrom, 2004). Skewness in the latent effects b_i may be handled by more specific parametric adaptations (e.g. Fernandez and Steele, 1998; Sahu et al., 2003; Lee and Thompson, 2008), multivariate versions of which are considered in Section 4.5.

Robustness may also be achieved by shrinkage priors (see Chapter 7) such as the horseshoe, Lasso, or scaled beta2 priors (Zollinger et al., 2015; Pérez et al., 2017), by median regression at the second stage, or a discrete mixture over two or more normal densities (Marshall and Spiegelhalter, 1998). For example, Beath (2014) develops a two-group mixture for the second stage variance. Baker and Jackson (2016) point out possible identifiability issues with extended models (e.g. discrete mixture models), and investigate instead estimation using the marginal likelihood formulations. Moreno et al. (2018) consider problems in estimating the second stage mean μ when there is clustering in the latent means, with $b_i = b_j$ (for $i \neq j$), so that the distribution of the treatment effect is not fully heterogenous across units. This could potentially be approached using a discrete mixture over cluster-specific second stage means, with the population-wide mean obtained by averaging over cluster means.

Example 4.1 Local Anaesthesia

Guolo and Varin (2017) report inferential difficulties when there are a small number of studies, comparing significance tests on μ for different classical procedures. In particular, they compare five studies of the effectiveness of local anaesthesia in controlling pain during intrauterine pathological examinations. The procedure of DerSimonian and Laird (1986) obtains a 95% confidence interval for μ of (−2.22, −0.35) with an associated p-value of 0.007. By contrast, the application of six alternative procedures provides p-values between 0.056 and 0.17.

Here we compare the normal-normal (with Bayesian estimation) with three other approaches: a normal-t hierarchy, a 2nd stage shrinkage prior, and a hierarchical approach using (intercept only) median regression at the second stage. The latter option is a particular case of quantile regression (see Chapter 12). A normal-t is applied with preset degrees of freedom ($df=5$), as the data would provide little information to identify an unknown parameter. As a posterior predictive check (PPC), the Q statistic for detecting study heterogeneity is compared between the original and replicated data. The mixed exceedance procedure of Marshall and Spiegelhalter (2007) is also used to detect poorly fitted cases.

A normal-normal model with a Ga(1,0.001) prior on $1/\tau^2$ provides a 95% CRI for μ of (−2.76, 0.15), with an estimated posterior probability $Pr(\mu<0\,|\,y)=0.033$ that the treatment reduces pain. The PPC check is unsatisfactory, and there is a 95% exceedance probability $Pr(y_{\text{new},i}>y_i\,|\,y)$ for study 5, which has a much stronger observed treatment effect than the other studies. A normal-t model provides a more precise estimate of μ, with 95% CRI (−2.3, 0.2), and with the estimated scale adjustment λ_5 having mean 0.43, so downweighting the impact of the fifth study on inferences. However, the estimate $Pr(\mu<0\,|\,y)=0.049$ remains under 0.05.

A horseshoe prior on the second stage effects is more conservative, leading to an estimate $Pr(\mu<0\,|\,y)=0.080$, with the estimated κ_i values (see Equation 7.2) seemingly downweighting study 2 as well as study 5. The overall treatment effect has a much narrower 95% CRI, namely (−0.86, −0.04). The scaled beta2 prior with settings p=q=1, as in Perez et al. (2017) provides a 95% CRI for μ of (−2.38, 0.41), with a conservative estimate also for $Pr(\mu<0\,|\,y)$ of 0.11. The exceedance probability for study 5 is raised to over 0.99.

Median regression using the Asymmetric Laplace distribution as in Yu and Moyeed (2001) leads to an estimate $Pr(\mu<0\,|\,y)=0.068$. In view of the sensitivity evident in this application, it may also be relevant to note sensitivity to the prior on $1/\tau^2$ or τ^2, even if the same stage 2 model is used (see Section 4.4). Lambert et al. (2005) provide a sensitivity analysis of inferences under simulations with a small number of studies.

4.3.1 Meta-Regression

Sometimes it is necessary to control explicitly for trial design, study location, and other design features in order to justify an exchangeability assumption (Marshall and Spiegelhalter, 1998; Pauler and Wakefield, 2000; Prevost et al., 2000). Similarly, in survey-based small area estimation, the estimate of b_i may incorporate information from administrative area data X_i (Rao, 2003; Jiang and Lahiri, 2006). So, with centred predictors X_i of dimension p (excluding a constant term), the normal-normal model becomes

$$y_i \sim N(b_i, s_i^2),$$

$$b_i \sim N(\mu + X_i\beta, \tau^2),$$

with marginal likelihood then (DuMouchel, 1996)

$$y_i \,|\, \beta, \tau^2 \sim N(\mu + X_i\beta, s_i^2 + \tau^2).$$

Recent Bayesian applications of meta-regression including Markham et al. (2017) and Druyts et al. (2017). Writing the model as

$$y_i = \mu + X_i\beta + \delta_i + \varepsilon_i,$$

$$\delta_i \sim N(0, \tau^2),$$

$$\varepsilon_i \sim N(0, s_i^2),$$

the true effect for unit i is then $\mu + X_i\beta + \delta_i$.

4.4 Prior for Second Stage Variance

The prior assumed for the second stage variance τ^2, or precision $1/\tau^2$, plays an important role in governing the degree of shrinkage or pooling strength (Lambert et al., 2005), with diffuse priors leading to lesser shrinkage (Conlon et al., 2007), and convenient choices prone to overfitting. As discussed in Chapter 1, improper or highly diffuse priors may also lead to identification or propriety problems. For example, the prior

$$p(\tau^2) \propto \frac{1}{\tau^2},$$

equivalent to taking $\tau^2 \sim IG(0,0)$ and to a flat prior on $\log(\tau)$ over $(0,\infty)$, can lead to improper posteriors in random-effects models (DuMouchel and Waternaux, 1992). A just proper risk-averse alternative, such as $\tau^2 \sim IG(c,c)$ with c small is often used (Simpson et al., 2016). However, this prior has a spike near zero (Browne and Draper, 2006), and different values of c can influence posterior influences despite the supposedly diffuse nature of the prior (Gelman, 2006). The prior $1/\tau^2 \sim Ga(1,c)$ similarly may lead to overfitting (Simpson et al., 2016).

One might carry out a sensitivity analysis over a range of proper but diffuse $Ga(c,d)$ priors for the precision $1/\tau^2$, such as $\{c = 0.1, d = 0.001\}$ or $c = d = 0.0001$ (Fahrmeir and Lang, 2001; van Dongen, 2006). An alternative scheme is to compare alternative values of c in $Ga(1,c)$ priors for $1/\tau^2$ (Besag et al., 1995), possibly using a mixture prior over M possible values for $c = (c_1,\ldots,c_M)$ in the prior $1/\tau^2 \sim Ga(1,c)$, such as $c_m = 1, 0.1, 0.01$ and 0.001 (Jullion and Lambert, 2007). Then

$$c \mid p \sim \sum_{m=1}^{M} p_m Ga(1, c_m)$$

$$p \sim \text{Dirichlet}(\omega)$$

where $(\omega_1,\ldots,\omega_M)$ are prior weights.

Introducing some degree of prior information may be relevant, and is natural under the inverse chi-squared density (sometimes called the scaled inverse chi-squared) with parameters $\{\nu,\lambda\}$. For τ^2 a variance, taking

$$\tau^2 \sim \chi^{-2}(\nu, \lambda),$$

is equivalent to assuming $\tau^2 \sim IG(\nu/2, \nu\lambda/2)$, where λ is a prior guess at the mean variance, and ν is a prior sample size (or level of confidence) parameter. Conlon et al. (2007) consider informative inverse gamma priors on τ^2 for inter-study variability in logexpression ratios in a microarray data application; for example, they use relatively large prior sample sizes ν.

Smith et al. (1995) discuss elicitation of informative inverse gamma priors for τ^2 based on anticipated variation in the underlying rates b_i, and the fact that assuming normality, 95% of the b_i will lie between $\mu - 1.96\tau$ and $\mu + 1.96\tau$. Assume the b_i are measured on a log scale (e.g. log relative risks or log odds ratios), and suppose the expected ratio of the 97.5th and 2.5th percentiles of risks (or odds) between centres or studies is 5, then the gap between the 97.5th and 2.5th percentiles for b_i is log(5) = 1.61. For normal b_i, the prior mean for τ^2 is then $(0.5 \times 1.61/1.96)^2 = 0.17$, and the prior mean for $1/\tau^2$ is 5.93. If the upper limit for the ratio of the 97.5th and 2.5th percentile of rates or odds is set at 10, this defines the 97.5th percentile of τ^2 namely $(0.5 \times 2.3/1.96)^2 = 0.34$. The expectation and variability are then used to define an inverse gamma prior on τ^2 or a gamma prior on $1/\tau^2$. Another procedure based on expected contrasts in relative risk (RR) or relative odds (RO) is mentioned by Marshall and Spiegelhalter (2007, p.422): 95% of units will have RRs or ROs in the range $\exp(\pm 1.96\tau)$, and an expectation of reasonable homogeneity might correspond to values of τ less than $\tau_h = 0.2$. Setting $\psi = 0.5\tau_h = 0.1$, these expectations are expressed via a half normal prior on τ, with $\tau = |T|$ where

$$T \sim N(0, \psi^2),$$

with prior 95% point at $1.96 \times \psi = 0.2$.

As another way to use prior evidence on variability, Marshall and Spiegelhalter (1998) mention a hyperprior for the scale parameter ϕ in a gamma prior for $1/\tau^2$, namely

$$1/\tau^2 \sim Ga(\gamma, \phi),$$

$$\phi \sim Ga(c, d),$$

where d is a small multiple of $1/R^2$ and R is the range of the observed centre effects, and with γ and c constrained according to $\gamma > 1 > c$. When the first stage sampling density involves an unknown variance, Gustafson et al. (2006) suggest a conditional prior sequence adapted to avoiding undersmoothing, namely

$$p(\sigma^2, \tau^2) = p(\sigma^2)p(\tau^2 \mid \sigma^2),$$

where $\sigma^2 \sim IG(e, e)$ for some small $e > 0$, and

$$p(\tau^2 \mid \sigma^2) \propto \left[\frac{1}{\tau^2 + \sigma^2}\right]^{a+1} \exp\left[-\frac{b}{\tau^2 + \sigma^2}\right].$$

This corresponds to a truncated inverse Gaussian prior on τ^2, with $Z \sim IG(a, b)$ or $(1/Z) \sim Ga(a, b)$ where $Z = \tau^2 + \sigma^2$. The case $\{a = 1, b = 0\}$ corresponds to the uniform shrinkage prior while larger values of a (e.g. $a = 5$) are "conservative" in the sense of guarding against over-estimation of τ^2.

4.4.1 Non-Conjugate Priors

Among non-conjugate strategies (for normal-normal meta-analysis) an effective choice in terms of being genuinely non-informative (Gelman, 2006) is a bounded uniform prior on the random effects standard deviation $\tau \sim U(0, H)$ with H large. However, this prior may be biased towards relatively large variances when the number of units (trials, studies, etc.) is small (van Dongen, 2006, p.92).

A prior selection strategy based on the principles of penalising complexity, and of preferring simpler models when more complex models are not strongly supported (Occam's razor), may be adopted. Thus Simpson et al. (2016) propose that the prior $\pi(\xi)$ on a flexibility parameter (hyperparameter), such as the level 2 standard deviation in a normal-normal model, be set so as to prefer the simpler base model in which $\xi = 0$. A penalising complexity (PC) prior has density decreasing at high values and maximum at $\xi = 0$ in order to prevent overfitting; that is, the mode of the PC prior is always at the base model. A suitable value for the prior on ξ can be obtained via a user-defined condition $Pr(Q(\xi) > U) = a$. This specifies an upper value U for a function $Q(\xi)$ of ξ, and the associated probability α. For the τ standard deviation parameter in a normal-normal hierarchy, the PC prior is an exponential with rate λ, $\tau \sim \text{Exp}(\lambda)$, and if one specifies $Pr(\tau > \tau_U) = a$, the resulting exponential rate is $\lambda = -\ln(a)/\tau_U$. The PC prior for the precision $1/\tau^2$ is a Gumbel type 2 density.

Variations on the uniform shrinkage prior, suggested by Christiansen and Morris (1997) and Daniels (1999), may also be used. One is a uniform prior on the shrinkage weights $w_i = s_i^2/(s_i^2 + \tau^2)$, or on the shrinkage weight $w = \sigma^2/(\sigma^2 + \tau^2)$, when σ^2 is unknown. Alternatively, one might represent different shades of opinion (sceptical, neutral, enthusiastic with regard to meta-analytic shrinkage) via the shrinkage weight. One might set a prior probability of 1/3 on the value $w = 0.9$, or on values $w > 0.9$, corresponding to nearly complete shrinkage to μ as under classical meta-analysis. A prior probability of 1/3 would also be set on $w = 0.1$, or values $w < 0.1$, corresponding to a sceptical view on exchangeability. Finally, a prior probability of 1/3 could be set on neutral values $w \sim U(0.1, 0.9)$.

Another possibility when the s_i^2 are provided as part of study summaries, is a uniform prior on the average shrinkage (Spiegelhalter et al., 2004, Chapter 5), namely

$$w = \frac{s_0^2}{s_0^2 + \tau^2},$$

where

$$\frac{1}{s_0^2} = \frac{1}{n} \sum_{i=1}^{n} \frac{1}{s_i^2},$$

is the harmonic mean of the study sampling variances. DuMouchel (1996) proposes a uniform prior on $s_0/(s_0 + \tau)$ which is equivalent to a Pareto prior, namely

$$p(\tau) = \frac{s_0}{(s_0 + \tau)^2}.$$

This prior is proper but with $E(\tau) = \infty$, and with (0.01,0.25,0.5,0.75,0.99) percentile points at $(s_0/99, s_0/3, s_0, 3s_0, 99s_0)$. Note that the Pareto can also be parameterised as

$$p(u) = b s_0^b u^{-b-1},$$

with $\tau = u - s_0$ when $b=1$.

Other robust options are half normal, half Student t or half Cauchy priors on the second-stage standard deviation τ (Lambert et al., 2005; Burke et al., 2016; Williams et al., 2018; Spittal et al., 2015). If $T \sim N(0,V)$ and $\tau = |T|$, then τ is half-normal with variance V (Spiegelhalter et al., 2004). One then has $E(\tau|V) = \sqrt{2V/\pi}$ and $\text{var}(\tau|V) = V(1-2/\pi)$. If τ_U represents a likely upper value for τ, then one may take $V = (\tau_U/1.96)^2$ as in Pauler and Wakefield (2000). A value such as $\tau_U=1$ is often suitable (Spiegelhalter et al., 2004).

Note that if $T \sim N(m,V)$ (i.e. the normal has an unknown mean), then $\tau = |T|$ is folded-normal with

$$E(\tau|V) = \sqrt{2V/\pi}\exp(-m^2/2V) - m[1-2\Phi(m/V^{0.5})],$$

and variance $m^2 + V$. Gelman (2006) and Zhao et al. (2006) adopt folded non-central t-densities for τ, obtained by dividing the absolute value of a normal variable by the square root of a gamma variable.

If the normal variable has mean zero, then the folded non-central t becomes a half t variable. With degrees of freedom in the t density set to 1, this leads to a half-Cauchy for τ, exemplified by

$$\Delta \sim N(0, \sigma_\Delta^2),$$

$$\sigma_\Delta \sim U(0,K),$$

$$\lambda \sim Ga(0.5, 0.5),$$

$$\tau = |\Delta| / \lambda^{0.5}.$$

Setting $\sigma_\Delta^2 = 1$ leads to a $C^+(0,1)$ prior, as in the horseshoe prior (Chapter 7). The half Cauchy prior on τ is included in rstan and runjags libraries in R.

Half t and half Cauchy priors for the second stage parameter τ may also be achieved by a reparameterisation of the second-stage prior on the latent trial means which strictly involves parameter redundancy. Such over-parameterisation may improve MCMC convergence (Gelman, 2006, section 3.2). With preset parameters ν and A (degrees of freedom and prior scale respectively) one has, for $y_i \sim N(b_i, s_i^2)$,

$$b_i = \mu + \xi\eta_i,$$

$$\xi \sim N(0, A)$$

$$\eta_i \sim N(0, \sigma_\eta^2)$$

$$1/\sigma_\eta^2 \sim \chi_\nu^2$$

with the standard deviation of the b_i then obtained as

$$\tau = |\xi|\sigma_\eta.$$

Applications are provided by van Dongen (2006) and Chelgren et al. (2011). Setting $\nu = 1$ leads to a half Cauchy prior

$$p(\sigma_b) \propto (\tau^2 + A)^{-1},$$

where Gelman (2006, p.524) uses a value $A = 25$ in a meta-analysis with small n, based on a prior belief that τ is well below 100.

Example 4.2 Nicotine Replacement Therapies

To illustrate approximately normal responses based on discrete (binomial) data, this example considers $n = 90$ studies of the benefits of nicotine replacement therapy (NRT) (Cepeda-Benito et al., 2004). These data also raise issues of potential outliers.

The data are supplied as the numbers r_{iT} quitting smoking among those under therapy N_{iT}, and numbers of quitters r_{iC} in control or placebo groups of size N_{iC}. Then the empirical log odds ratios measuring treatment effects, namely

$$y_i = \log\left(\frac{r_{iT}}{N_{iT} - r_{iT}}\right) - \log\left(\frac{r_{iC}}{N_{iC} - r_{iC}}\right),$$

are taken as approximately normal with known variances

$$s_i^2 = \frac{1}{r_{iT}} + \frac{1}{N_{iT} - r_{iT}} + \frac{1}{r_{iC}} + \frac{1}{N_{iC} - r_{iC}}.$$

A normal higher stage is assumed with $y_i \sim N(b_i, s_i^2)$ and $b_i \sim N(\mu, \tau^2)$. A uniform shrinkage prior on

$$w = \frac{s_0^2}{s_0^2 + \tau^2},$$

as considered above, is assumed for the second-stage variance, with the half-Cauchy also considered. Additionally, a $N(0, 100)$ prior on μ is adopted. Various kinds of prediction may be considered. Here, the predicted treatment effect in a new trial is sampled according to

$$b_{\text{new}} \sim N(\mu, \tau^2)$$

$$y_{\text{new}} \sim N(b_{\text{new}}, s_0^2).$$

Early convergence in a two-chain run of 5000 iterations with jagsUI is obtained. τ^2 is estimated as 0.085 (mean) and 0.080 (median). A clear benefit of NRT is seemingly apparent, with the odds ratio $\exp(\mu)$ having a posterior mean (and 95% CrI) of 1.93 (1.73,2.17). On the other hand, the predicted odds ratio for a new trial (OR.new in the rjags code) includes null values for the benefit from NRT, having mean (95% CrI) of 2.2 (0.7,5.1). Some deficiencies against model assumptions are evident: although the Shapiro–Wilk normality test of the posterior mean b_j is inconclusive (a p-value of 0.07), the Jarque–Bera test (Jarque and Bera,1980) shows a significant departure from normality.

Similarly, evaluating individual components of the total WAIC (widely applicable information criterion) shows studies 4, 36, and 59 as having distinctively high values. Trial 4 has an exceptionally high empirical log odds ratio in support of NRT, while trial 36 shows unusually low NRT benefit. Mixed predictive exceedance checks (Marshall and Spiegelhalter, 2007) show aberrant values (0.001 and 0.992) for these two trials, with study 59 also having an extreme value. A reanalysis using the normal-normal scheme uses a half-Cauchy prior, with the setting on the Cauchy scale parameter as in Gelman et al. (2008). τ^2 is now estimated as 0.081 (mean) and 0.078 (median). The LOO-IC (leave-one-out information criterion) is reduced slightly, but model checks show similar features to the analysis using the uniform shrinkage prior.

To allow for potential outlier trials and downweight their effect, an alternative analysis adopts a second-stage Student density with

$$b_i \sim N(\mu, \tau^2 / \lambda_i)$$

$$\lambda_i \sim G\left(\frac{\nu}{2}, \frac{\nu}{2}\right).$$

Less typical trial results will have values of λ_i considerably under 1 and a test for the posterior probability that λ_i is less than 1 can be included. The prior on ν is specified in two steps as $\nu \sim E(\kappa)$ and $\kappa \sim U(0.01, 0.5)$. Evidence in support of a heavy tailed second stage is equivocal. From a two-chain run of 10000 iterations with jagsUI, ν has a posterior mean of 9.3, suggesting departure from normality. The posterior mean and median for τ^2 are reduced to 0.051 and 0.045 respectively. Two trials (4 and 36) have posterior probability that $\lambda_i < 1$ in excess of 0.8, namely, trials 4 and 36. These trials also have extreme mixed predictive exceedance p-values. On the other hand, gain in goodness of fit is not obtained: the marginal density likelihood, uncorrected for complexity, is unchanged, and the WAIC increases.

The skew t model (Lee and Thompson, 2008; Fernandez and Steel, 1998) is also estimated using rube. This involves asymmetric scaling of the second-stage variance according to whether the residual $e_j = y_j - b_j$ is negative or positive. For positive residuals, τ^2 is scaled by a factor $\gamma^2 > 0$, while for negative residual terms, the scaling is by $1/\gamma^2$. The value $\gamma = 1$ corresponds to a symmetric t density, while $\gamma > 1$ ($\gamma < 1$) corresponds to positive (negative) skew. Applied to the NRT data, a two-chain run of 5,000 iterations shows no gain in fit, or any evidence that the 95% CRI for γ excludes 1. Mixed predictive exceedance checks for studies 4, 36, and 59 are still extreme, with values 0.016, 0.97 and 0.014.

Finally, a two-category discrete mixture is assumed on the second-stage variance (Beath, 2014), with an outlier group ($G_j = 2$) posited to have higher variance. To improve identifiability, prior probabilities for the outlier and main groups, $Pr(G_j = 2)$ and $Pr(G_j = 1)$ are set at 0.05 and 0.95 respectively. The default (main group) variance is assigned a uniform shrinkage prior as above, while the increment in the outlier group variance is assigned an informative $E(10)$ prior. The posterior probability that $Pr(G_j = 2|y)$ is then 0.25 for trial 4 (a marginal Bayes factor of 6.3), while the corresponding marginal Bayes factor for trial 36 is 3.6. Again, fit is not improved against the standard normal-normal model, and mixed predictive exceedance checks for studies 4, 36, and 59 remain extreme.

4.5 Multivariate Meta-Analysis

Multivariate meta-analysis may adopt a normal-normal strategy, albeit often with originally binary, count, or time to event data (Mavridis and Salanti, 2013). A multivariate analysis for metric outcomes may arise in different ways. These include clinical applications involving treatment and control arms; studies where multiple outcomes are reported; in meta-analysis of diagnostic test studies, where sensitivity and specificity are reported (Guo and Riebler, 2016; Guo et al., 2017); in multiple treatments meta-analysis; and in network meta-analysis (Greco et al., 2016).

In the first scenario, the event rate in the control arm may be taken as indicating baseline risk, and there is interest in whether the treatment effect is related in any way to baseline risk (Arends, 2006). Suppose r_{iT} of N_{iT} treated subjects in trial i exhibit a particular response (e.g. disease or death), as compared to r_{iC} of N_{iC} control subjects, and define log odds $y_{iT} = \log(r_{iT}/(N_{iT} - r_{iT}))$ and $y_{iC} = \log(r_{iC}/(N_{iC} - r_{iC}))$. Often the outcome may be taken as the log of the odds ratio, $y_{iT} - y_{iC}$, assumed normal (see Example 4.2). However, to separate out baseline risk, one may model $\{y_{iT}, y_{iC}\}$ as (approximately) bivariate normal. If the trial is randomised, it is legitimate to assume that $\{y_{iT}, y_{iC}\}$ are independent at the first stage (van Houwelingen et al., 2002). So

$$\begin{pmatrix} y_{iT} \\ y_{iC} \end{pmatrix} \sim N\left(\begin{pmatrix} b_{iT} \\ b_{iC} \end{pmatrix}, \begin{pmatrix} s_{iT}^2 & 0 \\ 0 & s_{iC}^2 \end{pmatrix}\right),$$

$$\begin{pmatrix} b_{iT} \\ b_{iC} \end{pmatrix} \sim N\left(\begin{pmatrix} \mu_T \\ \mu_C \end{pmatrix}, \Sigma_b\right),$$

where $\Sigma_b = \begin{pmatrix} \tau_T^2 & \tau_{TC} \\ \tau_{TC} & \tau_C^2 \end{pmatrix}$, with $\tau_{TC} = \rho \tau_T \tau_C$, and diagonal terms τ_T^2 and τ_C^2 represent variability in the true treatment and control event rates. Then $\gamma = \mu_T - \mu_C$ defines the underlying treatment effect with variance $\tau_T^2 + \tau_C^2 - 2\tau_{TC}$. The conditional variance of the treatment effect, given the true control group rate, is $\tau_T^2 - (\tau_{TC}^2/\tau_C^2)$. So, baseline risk explains a portion

$$\frac{\tau_C^2 - 2\tau_{TC} + \dfrac{\tau_{TC}^2}{\tau_C^2}}{\tau_T^2 + \tau_C^2 - 2\tau_{TC}}$$

of the treatment effect variance.

Most commonly, a multivariate analysis is generated when more than one outcome is associated with a specific unit (Everson and Morris, 2000; Wei and Higgins, 2013). In this case, suppose there are K outcomes such that

$$\begin{pmatrix} y_{i1} \\ y_{i2} \\ \cdot \\ y_{iK} \end{pmatrix} \sim N\left(\begin{pmatrix} b_{i1} \\ b_{i2} \\ \cdot \\ b_{iK} \end{pmatrix}, S_i\right),$$

where

$$S_i = \begin{pmatrix} s_{i1}^2 & s_{i12} & \cdot & s_{i1K} \\ s_{i21} & s_{i2}^2 & \cdot & s_{i2K} \\ \cdot & \cdot & \cdot & \cdot \\ s_{iK1} & s_{iK2} & \cdot & s_{iK}^2 \end{pmatrix},$$

is the known covariance matrix between outcomes for trial i, with $s_{ijk} = r_{ijk}(s_{ij}^2 s_{ik}^2)^{0.5}$. A multivariate normal second-level prior for $(b_{i1}, \ldots b_{iK})$ involves means $\{\mu_1, \ldots \mu_K\}$, and $K \times K$ covariance matrix

$$\Sigma_b = \begin{pmatrix} \tau_1^2 & \tau_1\tau_2\rho_{12} & . & \tau_1\tau_K\rho_{1K} \\ \tau_1\tau_2\rho_{12} & \tau_2^2 & . & \tau_2\tau_K\rho_{2K} \\ . & . & . & . \\ \tau_1\tau_K\rho_{1K} & \tau_2\tau_K\rho_{2K} & . & \tau_K^2 \end{pmatrix}.$$

There may well be sensitivity to the priors adopted for Σ_b, especially when there are a small number of trials, or some missingness in outcomes, with results from the inverse Wishart potentially sensitive to the prior scale matrix (Wei and Higgins, 2013). Alternatives involve decomposition approaches to the covariance matrix, so that separate priors are specified on variances and correlations (Barnard et al., 2000; Lu and Ades, 2009; Burke et al., 2016; Guo et al., 2017; Hurtado Rua et al., 2015). Incorporating evidence into multivariate priors on variances and correlations leads to stabilised inferences (Burke et al., 2016; Guo et al., 2017). Alternatives to a $U(-1, 1)$ prior on correlations include a normal prior on the Fisher z-transformed correlation $\text{logit}((\rho+1)/2)$, a uniform prior U(0,1), constrained to positive correlations (Burke et al., 2016), or penalised complexity priors, as included in the R program meta4diag (Guo et al., 2017). Alternative to gamma priors on precisions $1/\tau_k^2$, which may lead to relatively high estimated τ_k, are half-normal priors (Burke et al., 2016; Lambert et al., 2005). Alternative methods are available if some, or all, within study correlations are not observed (i.e. only standard errors of treatment effects are available), for example, specifying an informative prior (Mavridis and Salanti, 2013). For the bivariate case, an alternative model may be specified (Riley et al., 2008), entirely avoiding the need for observed intra-study correlations.

Multivariate normality is often a simplification, and one may wish to allow both for heavier tails, skewness, or multi-modality; see Genton (2004) and Lee and Thompson (2008) regarding use of skew-elliptical densities as a route to greater robustness. These models build on the principle (Azzalini, 1985) that if f and g are symmetric densities with parameters μ and σ, with G the cumulative density corresponding to g, then the new density defined by

$$h(x \mid \mu,\sigma,\delta) = \frac{2}{\sigma} f\left(\frac{x-\mu}{\sigma}\right) G\left(\delta\frac{x-\mu}{\sigma}\right)$$

is skew for non-zero δ.

Following Sahu et al. (2003), a multivariate skew-normal model is a particular type of skew-elliptical model (of dimension K) obtained by considering errors $\varepsilon_{K\times 1} \sim N_K(0,\Sigma)$, positive variables $Z_{K\times 1} \sim N_K(0,I)$ and taking $y = DZ + \varepsilon$ where D is a diagonal matrix, $\text{diag}(\delta_1,\ldots\delta_K)$. In a regression setting with a K dimensional mean μ, one has

$$y \mid Z = z \sim N_K(\mu + Dz, \Sigma).$$

Values $\delta_k > 0$ correspond to positive skew in the kth outcome while a negative δ_k arises from negative skew. A multivariate skew-t model (allowing for both heavier tails than the normal, and also for skewness) is obtained by sampling $Z_{K\times 1} \sim t_{K,\nu}(0,I)$, where ν is a degrees of freedom parameter, and then

$$y \mid Z = z \sim t_{K,\nu+K}\left(\mu + Dz, \frac{\nu + z^T z}{\nu + K}\Sigma\right).$$

Example 4.3 BCG Vaccine Trials; Bivariate Normal Model

Following van Houwelingen et al. (2002), an example of a bivariate meta-analysis involves data from 13 trials regarding the effectiveness of the BCG vaccine against tuberculosis. We consider, in particular, sensitivity on priors for the second stage covariance and relevance for assessing potential outliers.

Each trial compares vaccinated and non-vaccinated groups of size $\{N_T, N_C\}$, with the outcome being counts of tuberculosis $\{r_T, r_C\}$, and with the infection rate in the control arm taken as indicating the baseline risk. The response variables are the log odds in each trial arm, $y_{iT} = \log(r_{iT}/(N_{iT} - r_{iT}))$ and $y_{iC} = \log(r_{iC}/(N_{iC} - r_{iC}))$. Predictive fit is assessed by comparing replicate data(obtained using mixed predictions) (Marshall and Spiegelhalter, 2007) with actual observations in a sum of squares criterion.

Here an initial analysis assumes normality at both levels, with

$$\begin{pmatrix} y_{iT} \\ y_{iC} \end{pmatrix} \sim N\left(\begin{pmatrix} b_{iT} \\ b_{iC} \end{pmatrix}, \begin{pmatrix} s_{iT}^2 & 0 \\ 0 & s_{iC}^2 \end{pmatrix}\right),$$

$$\begin{pmatrix} b_{iT} \\ b_{iC} \end{pmatrix} \sim N\left(\begin{pmatrix} \mu_T \\ \mu_C \end{pmatrix}, \Sigma_b\right),$$

where

$$s_{iT}^2 = \frac{1}{r_{iT}} + \frac{1}{N_{iT} - r_{iT}}, \; s_{iC}^2 = \frac{1}{r_{iC}} + \frac{1}{N_{iC} - r_{iC}}.$$

It is also assumed that the precision matrix Σ_b^{-1} of the latent effects is Wishart with identity scale matrix and 2 degrees of freedom, while the $\{\mu_l, \mu_C\}$ parameters have N(0, 1000) priors.

Using jagsUI, posterior means for (μ_T, μ_C) are estimated as (−4.87, −4.07), with mean vaccination effect $\gamma = \mu_T - \mu_C$ of −0.79 (−1.27, −0.32), slightly more negative than the estimate of −0.74 found by van Houwelingen et al. (2002) using classical methods (in the SAS package). The posterior mean for the second-stage covariance matrix is

$$\Sigma_b = \begin{pmatrix} 1.83 & 2.21 \\ 2.21 & 3.29 \end{pmatrix},$$

with correlation between treatment and control effects (where effects are log-odds), obtained from monitoring the components of Σ_b, as 0.90.

Similarly, the slope of the regression to predict the vaccination group log-odds from the control group log-odds, obtained by averaging $\Sigma_{b12}^{(t)}/\Sigma_{b22}^{(t)}$ over iterations t, is 0.67 (slope.TC in the code). The variance of the true treatment effects $b_{iT} - b_{iC}$ is obtained by monitoring $V_t = \Sigma_{b,11} + \Sigma_{b,22} - 2\Sigma_{b,12}$, while the conditional variance of the vaccination log-odds effects b_{iT} given b_{iC} (and hence the variance of $b_{iT} - b_{iC}$ given b_{iC}) is obtained by monitoring $V_c = \Sigma_{b,11} - \Sigma_{b,12}^2/\Sigma_{b,22}$. Finally, the proportion of treatment effect variation explained by baseline risk (i.e. the true log-odds in the control group), obtained by monitoring $1 - V_c/V_t$, has a posterior mean of 0.51 (r2.base in the code).

To assess possible outliers, a mixed predictive exceedance check (Marshall and Spiegelhalter, 2007) is carried out by sampling replicate random effects $(b_{iT,\text{new}}, b_{iC,\text{new}})$ and then sampling replicate data $y_{ij,\text{new}}$. There are four observations out of the 26 (13 pairs) with predictive exceedance checks

$$Pr(y_{ij,\text{new}} > y_{ij} \mid y) \quad (j = T \text{ or } C)$$

under 0.10 or over 0.90, with the most extreme being an exceedance probability of 0.024 in the vaccination arm of trial 6 (pred.exc[1,6] in the code). As a summary fit measure, a predictive criterion (Laud and Ibrahim, 1995) based on comparing y_{new} and y is derived.

Accordingly, a second analysis (using WINBUGS via rube) adopts a bivariate Student t model at stage 2. The degrees of freedom parameter is set at 4, providing a robust analysis (Gelman et al., 2014, section 17.2). This leads to only two observations, (6,T) and (6,C), having predictive exceedance checks under 0.10. Extension of the model to a skew bivariate t (Sahu et al., 2003), namely model 3 in the code, provides no further gain in fit.

Model fit using the predictive criterion (PFC.mix in the code), in fact, is worse for these two extensions, illustrating that improved model fit does not always follow measures to counteract adverse model checks. The mean vaccination effect $\gamma = \mu_T - \mu_C$ is less precise under these models, namely a mean (95% CRI) of −0.78 (−1.34, −0.23) under the second model, and −0.80 (−1.54, −0.04) under the third.

A final analysis uses a Cholesky decomposition (Wei and Higgins, 2013) for the second-stage covariance matrix in an MVN-MVN analysis,

$$\Sigma_b = V_b R_b V_b$$

$$R_b = L'L$$

where V_b is a diagonal matrix of standard deviations, and R_b is a correlation matrix. For a bivariate analysis, $L_{11} = L_{22} = 1$, while a $U(-1, 1)$ prior is assumed for $L_{12} = r_{b12}$, and the standard deviations τ_T and τ_C have $U(0,5)$ priors. This analysis produces slightly larger variances in Σ_b and a slightly larger correlation, 0.9, between treatment and control effects.

A more pronounced effect on estimates is obtained if the Wishart scale matrix in the initial MVN-MVN analysis is a diagonal with elements 0.1. This option increases the correlation between treatment and control effects to 0.95.

Example 4.4 Hypertension Treatments

The data for this example are from ten studies into the effectiveness of hypertension treatment (Jackson et al., 2013), included in the R library mvmeta. Data from each study consists of two treatment effects: differences in systolic blood pressure (SBP) and diastolic blood pressure (DBP) between treatment and control groups (adjusted for baseline blood pressure). A larger reduction in blood pressure indicates greater effectiveness. Within-study correlations are known for all studies. In the absence of covariates, one would represent the two-stage model as

$$\begin{pmatrix} y_{1i} \\ y_{2i} \end{pmatrix} \sim N\left(\begin{pmatrix} b_{1i} \\ b_{2i} \end{pmatrix}, \begin{pmatrix} s_{1i}^2 & \rho s_{1i} s_{2i} \\ \rho s_{1i} s_{2i} & s_{2i}^2 \end{pmatrix}\right),$$

$$\begin{pmatrix} b_{1i} \\ b_{2i} \end{pmatrix} \sim N\left(\begin{pmatrix} \mu_1 \\ \mu_2 \end{pmatrix}, \Sigma_b\right),$$

where $\Sigma_b = \begin{pmatrix} \tau_1^2 & \tau_1\tau_2\rho_b \\ \tau_1\tau_2\rho_b & \tau_2^2 \end{pmatrix}$. In fact, three studies contain subjects with isolated systolic hypertension (namely high SBP, but normal DBP). So the treatment effect may be smaller in these trials. To represent this effect, we introduce a second-stage regression (i.e. multivariate meta-regression):

$$\begin{pmatrix} b_{1i} \\ b_{2i} \end{pmatrix} \sim N\left(\begin{pmatrix} \nu_{1i} \\ \nu_{2i} \end{pmatrix}, \Sigma_b\right),$$

$$\nu_{1i} = \mu_1 + \beta_1 \text{ISH}_i$$

$$\nu_{2i} = \mu_2 + \beta_2 \text{ISH}_i.$$

Following Mavridis and Salanti (2013), a Cholesky decomposition with spherical parameterisation on the correlation is used as a prior for the second-stage covariance, and Ga(1,0.001) priors on the between study precisions. The jags code uses conditional expectations and variances for y_{2i} and b_{2i}. A two-chain run using jagsUI provides posterior mean (sd) for (β_1,β_2) of 0.47 (1.61) and 1.43 (0.78), similar to Jackson et al. (2013, Table 4). The second stage correlation ρ is estimated as 0.46 (0.26) (rho.tau in the code). As to between study standard deviations, posterior mean (sd) for (τ_1,τ_2) are 1.85 (0.66) and 0.98 (0.32). Mixed exceedance checks throw some doubt on study 1, which has relatively small treatment effects, especially for DBP.

A U(0,1) prior on the between study correlation may be sensible given the correlated blood pressure outcomes (Burke et al., 2016, p.23) and this is combined with half normal priors HN(0,2) on τ_1 and τ_2. This provides posterior mean (sd) for (β_1,β_2) of 0.63 (1.26) and 1.46 (0.76), with the estimated correlation similar to the first analysis, namely 0.45 (0.21). Estimated study standard deviations are affected more, with posterior mean (sd) for (τ_1,τ_2) of 1.51 (0.29) and 1.00 (0.25).

4.6 Heterogeneity in Count Data: Hierarchical Poisson Models

The adoption of higher stage densities for count data is often linked to apparent departures from the Poisson mean-variance assumption. The most common departure is that count data show more variability than expected under the Poisson, so that the coefficient of variation $V(y)/\bar{y}$ exceeds 1. Overdispersion may reflect unobserved subject frailties, multiple modes, non-random sampling (Efron, 1986), or widely different exposures o_i (e.g. when a count outcome y_i is surgical deaths for hospitals, with means $\mu_i o_i$ where o_i are patient totals). The conjugate continuous mixture models in the presence of excess heterogeneity is the Poisson-gamma, though greater flexibility in more complex models (e.g. multilevel or multivariate) is generally obtained by mixing with non-conjugate links.

The Poisson-gamma model allows for unit mean rates μ_i to vary according to a gamma density $\mu_i \sim Ga(a,\beta)$, which is unimodal, but flexibly shaped. Thus for count data y_i assumed Poisson with means μ_i, set

$$B(b_i) = e^{b_i}$$

in (4.1), where $\mu_i = e^{b_i}$, $a(\phi_i) = 1$ and $c(y_i,\phi_i) = \log y_i!$. Then equation (4.2) has the form

$$\begin{aligned} p(b_i \mid \psi) &= k_1 \exp(b_i g_1(\psi) - e^{b_i} g_2(\psi)) \\ &= k_1 (e^{b_i})^{g_1(\psi)} \exp(-e^{b_i} g_2(\psi)) \end{aligned}$$

Namely, a gamma density for $\mu_i = e^{b_i}$ with parameters $\alpha = g_1(\psi)$ and $\beta = g_2(\psi)$. The conditional posterior is

$$p(b_i \mid y, \alpha, \beta) = k_2(e^{b_i})^{\alpha+y_i} \exp(-e^{b_i}[\beta+1])$$

namely a gamma for μ_i with parameters $\alpha+y_i$ and $\beta+1$. Denoting the mean of the μ_i as $\xi=\alpha/\beta$, one obtains $V(\mu_i) = \alpha/\beta^2 = \xi^2/\alpha$. Then

$$V(y_i) = E[V(y_i \mid \mu_i)] + V[E(y_i \mid \mu_i)] = \xi + \xi^2/\alpha$$

so that overdispersion is present when $\phi > 0$, where $\phi = 1/\alpha$.

Different parameterisations of the Poisson-gamma mixture can be used. For example, one may set $\mu_i = \xi\omega_i$, with overall mean parameter ξ, and multiplicative random effects ω_i having mean 1 for identifiability, namely $\omega_i \sim Ga(\alpha, \alpha)$ with $V(\omega_i) = 1/\alpha$. Integrating the ω_i out, as in

$$p(y_i \mid \xi, \alpha) = \int p(y_i \mid \omega_i, \xi) p(\omega_i \mid \alpha) d\omega_i,$$

leads to a marginal negative binomial density for the y_i, namely

$$p(y_i \mid \xi, \alpha) = \frac{\Gamma(\alpha+y_i)}{\Gamma(\alpha)\Gamma(y_i+1)} \left(\frac{\alpha}{\alpha+\xi}\right)^{\alpha} \left(\frac{\xi}{\alpha+\xi}\right)^{y_i}.$$

If predictors X_i are present, negative binomial regression is obtained with $\xi_i = \exp(X_i\beta)$. Note that in many applications, there may be forms of truncation, as when zero counts do not enter the analysis (Larson and Soule, 2009). So a zero truncated negative binomial has

$$p(y_i \mid \xi, \alpha, y_i > 0) = \frac{\frac{\Gamma(\alpha+y_i)}{\Gamma(\alpha)\Gamma(y_i+1)} \left(\frac{\alpha}{\alpha+\xi}\right)^{\alpha} \left(\frac{\xi}{\alpha+\xi}\right)^{y_i}.}{1-\left(\frac{\alpha}{\alpha+\xi}\right)^{\alpha}}.$$

Alternatively, one may assume $y_i \sim Po(\mu_i), \mu_i \sim Ga(\alpha, \beta)$, with

$$E(\mu_i) = m_\mu = \alpha/\beta$$

$$\text{var}(\mu_i) = V_\mu = \alpha/\beta^2$$

(e.g. Clayton and Kaldor, 1987). When this parameterisation includes offsets o_i, the posterior $p(\mu_i, \alpha, \beta \mid y)$ has the form

$$L(\alpha, \beta, \mu \mid y) p(\alpha, \beta) = \left[\prod_{i=1}^{n} \frac{\exp(-\mu_i o_i)(\mu_i o_i)^{y_i}}{y_i!} \left(\frac{\beta^\alpha}{\Gamma(\alpha)}\right)^n \left(\prod_{i=1}^{n} \mu_i\right)^{\alpha-1} \exp\left(-\beta \sum_{i=1}^{n} \mu_i\right)\right] p(\alpha, \beta),$$

with conditional posterior for μ_i now $Ga(\alpha + y_i, \beta + o_i)$. Hence the posterior mean is

$$E(\mu_i \mid y_i, a, \beta) = \frac{y_i + a}{o_i + \beta}.$$

One may define reliabilities (Staggs and Gajewski, 2017) using V_μ and the conditional variances $\text{var}((y_i/o_i) \mid \mu_i, o_i) = \mu_i/o_i$. Reliabilities in unit rates are estimated as:

$$\frac{V_\mu}{V_\mu + \dfrac{\mu_i}{o_i}}.$$

So higher reliabilities attach to units with larger offsets.

The conditional likelihoods (George et al., 1993, p.191) for α and β under this structure are obtained from $L(a, \beta, \mu \mid y)$, namely

$$L(\alpha \mid \beta, \mu) = k_\alpha \left(\frac{\beta^\alpha}{\Gamma(\alpha)} \right)^n \left(\prod_{i=1}^{n} \mu_i \right)^{\alpha-1},$$

and

$$L(\beta \mid \alpha, \mu) = k_\beta \beta^{n\alpha} \exp\left(-\beta \sum_{i=1}^{n} \mu_i \right),$$

where k_α and k_β are normalising constants. Hence $L(\beta|\alpha,\mu)$ is gamma with parameters $n\alpha+1$ and $\sum_{i=1}^{n} \mu_i$. The conditional posteriors $p(a \mid \beta, \mu) = L(a \mid \beta, \mu)p(a)$ and $p(\beta \mid a, \mu) = L(\beta \mid a, \mu)p(\beta)$ are log-concave when the priors $p(\alpha)$ and $p(\beta)$ are log-concave. Assuming a gamma prior $p(\beta) = Ga(c, d)$, the full conditional for β is $Ga(n\alpha + 1 + c, \sum_{i=1}^{n} \mu_i + d)$. However, the full conditional for α is non-standard, whatever form for $p(\alpha)$ is adopted.

Another Poisson-gamma mixture formulation (e.g. Albert, 1999; Christiansen and Morris, 1996) assumes

$$y_i \mid \lambda_i \sim Po(o_i \lambda_i),$$

$$\lambda_i \sim Ga\left(\zeta, \frac{\zeta}{\mu_i} \right),$$

where $V(\lambda_i) = \mu_i^2/\zeta$ and the Poisson corresponds to $\zeta \to \infty$. If $\mu_i = \mu$ and a gamma prior is assumed for μ, then the posterior mean for λ_i conditional on μ and ζ is

$$E(\lambda_i \mid y, \zeta, \mu) = \frac{y_i + \zeta}{o_i + \zeta/\mu} = B_i \mu + (1 - B_i) \frac{y_i}{o_i},$$

where

$$B_i = \frac{\zeta}{\zeta + o_i \mu},$$

measures the level of shrinkage towards the overall mean μ. Thus, shrinkage will be greater when o_i (e.g. the population at risk in a mortality application) is small, or when ζ is large. As for the second-stage variance in the normal-normal model, the prior on ζ

influences the degree of shrinkage that is obtained. Let $r_i = y_i/o_i$. Then Christiansen and Morris (1996) suggest a uniform prior based on the average shrinkage factor

$$B_0 = \frac{\zeta}{\zeta + \min(o_i)\bar{r}} \sim U(0,1),$$

with the prior value of ζ then obtained as $B_0 \min(o_i)\bar{r}/(1-B_0)$.

Extended parameterisations of the negative binomial have been suggested (Liu and Dey, 2007). Winkelmann and Zimmermann (1991) suggest a variance function

$$V(y_i) = E[V(y_i \mid \mu_i)] + V[E(y_i \mid \mu_i)] = \xi + \phi\xi^{k+1}$$

with $k \geq -1$, and obtained by taking $\mu_i \sim Ga(\xi^{1-k}/\phi, \xi^{-k}/\phi)$. Setting $k=0$ and $k=1$ leads to what are called NB1 and NB2 forms of the negative binomial, under which the variances are linear and quadratic in ξ, namely $V(y_i) = \xi + \phi\xi$ and $V(y_i) = \xi + \phi\xi^2$ respectively.

4.6.1 Non-Conjugate Poisson Mixing

Alternatives to the conjugate model are the Poisson lognormal model, and models such as the generalised Poisson density, zero-inflated Poisson, and the hurdle model adapted to different types of departure from the typical Poisson frequency pattern. The Poisson lognormal (PLN) model has been suggested as more appropriate than the conjugate mixture in certain applications, such as species abundance – see Bulmer (1974) and Diserud and Engen (2000). The Poisson lognormal representation may be beneficial in terms of robustness to contamination or outliers, as the tails of the lognormal are heavier than for the gamma distribution (Connolly et al., 2009; Wang and Blei, 2017).

The PLN model is obtained for $y_i \sim Po(\mu_i)$ when μ_i are lognormally distributed, or equivalently when the logarithms $w_i = \log(\mu_i)$ of the Poisson means are assumed normal with mean M and variance V (Aitchison and Ho, 1989). The marginal density under lognormal mixing is obtained by integrating the sampling density over the domain of the log mean, namely

$$p(y_i \mid M, V) = \frac{(2\pi V)^{-0.5}}{y_i!} \int_0^\infty \mu_i^{y_i - 1} e^{-\mu_i} \exp\left[\frac{-(\log \mu_i - M)^2}{2V}\right] d\mu_i$$

with marginal mean and variance respectively $e^{M+V/2}$ and $e^{2M+V}[e^V - 1]$. As $V \to 0$, this reduces to a Poisson density. An alternative parameterisation (Weems and Smith, 2004) has $y_i \sim Po(\mu_i U_i)$ with $\log(\mu_i) = \beta_0 + \beta_1 x_{1i} + \ldots \beta_p x_{pi}$, and $\log(U_i) \sim N(1, V)$.

The Poisson lognormal generalises readily to multivariate count data (Chib and Winkelmann, 2001) or to mixing with heavier tails than available under the lognormal; for example, the log Student t with a low degrees of freedom parameter for a heavy tailed, albeit symmetric, mixing density. Skew normal and skew Student t mixing can also be used, since in some applications, extremes of frailty tend to be above rather than below the centre of the density (Sahu et al., 2003).

The exchangeable Poisson lognormal model is quite widely applied to pooling inferences over sets of units (e.g. hospitals) when health event totals y_i such as surgical deaths are obtained and there are o_i expected events; the Poisson lognormal is also widely applied in modelling for spatially structured disease count data (Chapter 6). The o_i might be based on multiplying the patient total for hospital i by an average event rate and are usually assumed known (i.e. not to be subject to measurement error). If the average rate is based on the total set of n hospitals then one has $\sum y_i = \sum o_i$, and with $\mu_i = o_i \rho_i$ one has

$$y_i \sim Po(o_i \rho_i),$$

with the ρ_i interpretable as relative risks averaging 1 over all units. However, this feature is not always present, and allowing for mean risk other than 1 (e.g. if a national surgical mortality rate is applied to a particular set of hospitals), the Poisson lognormal then assumes

$$\log(\rho_i) = \beta_0 + w_i$$

where the $w_i \sim N(0, V_w)$ are exchangeable normal random effects, with relative risks ρ_i pooled towards a global average rate $\exp(\beta_0)$ according to the size of V_w. Equivalently $v_i = \exp(w_i)$ are lognormal with mean $\mu = \exp(0.5V_w)$ and variance $\mu^2(\exp^{V_w} - 1)$.

Generalised Poisson and Poisson process models are also often useful in particular settings, including underdispersion (Consul, 1989; Scollnik, 1995; Podlich et al., 2004). The generalised Poisson density (Consul, 1989) specifies

$$p(y \mid \lambda, \rho) = \frac{\lambda(\lambda + y\rho)^{y-1}}{y!} e^{-\lambda - \rho y}$$

with mean $\lambda/(1-\rho)$, variance $\lambda/(1-\rho)^3$ and hence coefficient of variation $1/(1-\rho)^2 \geq 1$. This reduces to a Poisson density as $\rho \to 0$.

Example 4.5 Hospital Mortality

To exemplify the Poisson-gamma methodology, consider counts of patient deaths following heart transplant surgery in 131 hospitals in the US between October 1987 and December 1989. These were analysed by Christiansen and Morris (1996, 1997). Let o_i be expected deaths (calculated by a logit regression on patient characteristics).

The first model considered is the usual Poisson-gamma mixture,

$$y_i \sim Po(o_i \mu_i),$$

$$\mu_i \sim Ga(\alpha, \beta),$$

where the μ_i are relative risks, since actual and expected deaths are equal. The hyperparameters α and β are assigned diffuse gamma priors. The DIC for this model is 475. To assess variations in the extent of shrinkage, one may plot the lengths of 90% credible intervals for percentile ranks against posterior mean reliabilities $V_\mu/(V_\mu + (\mu_i/o_i))$ (Staggs and Gajewski, 2017). As expected, more precise estimates of percentile ranks are associated with higher reliability. A mixed predictive exceedance check (Marshall and Spiegelhalter, 2007) shows 13 observations with exceedance probabilities under 0.05 or over 0.95.

A second model adopts the scheme of Christiansen and Morris (1996), which includes data-based priors. Thus

$$y_i \mid \mu_i \sim Po(o_i \mu_i),$$

$$\mu_i \sim Ga\left(\zeta, \frac{\zeta}{M_\mu}\right),$$

with shrinkage factors

$$B_i = \frac{\zeta}{\zeta + o_i M_\mu}.$$

The prior on ζ is indirect, via a uniform prior on $B_0 = \zeta/(\zeta + \min(o_i)\bar{r})$. A two-chain run of 5,000 iterations provides a DIC of 530, and high values for both B_0 and ζ, namely 0.986 and 7.81. The mixed predictive exceedance check now shows seven observations with exceedance probabilities under 0.05 or over 0.95.

Christiansen and Morris (1996) argue that exchangeability between all 131 units might not be applicable, since hospitals with larger patient totals have lower crude death rates. As one remedy for such a pattern, one might take

$$y_i \sim Po(\nu_i),$$

$$\nu_i \sim Ga\left(\zeta, \frac{\zeta}{\rho_i}\right),$$

where

$$\log(\rho_i) = \beta_1 + \beta_2 \log(o_i),$$

now includes a regression on $\log(o_i)$. So expected deaths is no longer an offset with implicit coefficient $\beta_2 = 1$. Here, we instead split the hospitals into two groups with indicator G_i, one group (with $G_i = 1$) containing 37 hospitals with under 10 patients, the other (with $G_i = 2$) containing the remaining 94 hospitals. Different means and variance parameters are assumed in the two groups. So

$$y_i \mid \mu_i \sim Po(o_i \mu_i),$$

$$\mu_i \sim Ga\left(\zeta_{G_i}, \frac{\zeta_{G_i}}{\mu_{G_i}}\right)$$

with uniform priors on group-specific average shrinkage factors ($k = 1, 2$)

$$B_{0k} = \frac{\zeta_k}{\zeta_k + \min(o_i; G_i = k)\bar{r}_k}.$$

This extension to partial exchangeability produces a deviance reduction to 506. The mean mortality relative risk (with 95% interval) is found to be 2.05 (1.33, 2.95) in the low workload hospitals, but lower, namely 0.96 (0.83,1.10), in the higher workload hospitals (m.mu[] in the code). The variance factor ζ_k is higher in the low workload hospitals, but the average shrinkages B_{0k} are similar, at 0.95 and 0.93 respectively. Six observations now have exceedance probabilities under 0.05 or over 0.95.

4.7 Binomial and Multinomial Heterogeneity

Heterogeneity in binary and categoric outcomes is commonly found in consumer and demographic data. Among possible approaches are the beta-binomial, the logistic-normal and generalisations of the binomial (e.g. Alanko and Duffy, 1996). Analogous methods apply for categoric data ($M > 2$ categories) with the conjugate model being the multinomial-Dirichlet. Although the Poisson-gamma mixture is widely applied to health and disease events, the beta-binomial may also be used if populations are relatively small, and has different implications for shrinkage: shrinkage is greater under the Poisson-gamma (Howley and Gibberd, 2003). Binomial and multinomial mixture methods have recently become popular in the analysis of ecologic problems where marginals of a contingency table are available, often from different sources such as census and voting data, but the internal cells are unobserved (King, 1997; King et al., 2004). They may also be applied in meta-analysis, avoiding normal approximations (Bakbergenuly and Kulinskaya, 2017; Kulinskaya and Olkin, 2014).

For binomial data $y_i \sim \text{Bin}(N_i, \pi_i), i = 1, \ldots, n$, the exponential family parameterisation sets

$$B(b_i) = N_i \log(1 + e^{b_i}),$$

in (4.1), where $\pi_i = e^{b_i}/(1+e^{b_i})$, $a(\phi_i) = 1$, and $c(y_i, \phi_i) = \log\binom{N_i}{y_i}$. Then equation (4.2) has the form

$$\begin{aligned} p(b_i \mid \psi) &= k_1 \exp(b_i g_1(\psi) - N_i \log(1+e^{b_i}) g_2(\psi)) \\ &= k_1 \left(\frac{e^{b_i}}{1+e^{b_i}}\right)^{g_1(\psi)} \left(1+e^{b_i}\right)^{-N_i g_2(\psi) + g_1(\psi)} \end{aligned}$$

namely a beta density for π_i with parameters $g_1(\psi)$ and $N_i g_2(\psi) - g_1(\psi)$. The conditional posterior of π_i is then also a beta with parameters $g_1(\psi) + y_i$ and $N_i[g_2(\psi)+1] - g_1(\psi) - y_i$. The marginal density is the beta-binomial with

$$p(y_i \mid g_1, g_2) = \binom{N_i}{y_i} \frac{Be(g_1 + y_i, N_i(g_2+1) - (g_1 + y_i))}{Be(g_1, N_i g_2 - g_1)}.$$

Shrinkage effects are apparent under the beta mixing parameterisation

$$\pi_i \sim Be(\gamma\rho, \gamma(1-\rho)),$$

with mean $\rho \in (0,1)$, and where $\gamma > 0$, termed the spread parameter by Howley and Gibberd (2003), is inversely related to the prior variance of the proportions $\rho(1-\rho)/(1+\gamma)$. The conditional posterior for π_i is

$$\pi_i \sim Be(\gamma\rho + y_i, \gamma(1-\rho) + N_i - y_i),$$

and the posterior mean is

$$E(\pi_i \mid y, \gamma, \rho) = \frac{\gamma}{\gamma + N_i}\rho + \frac{N_i}{\gamma + N_i}\left(\frac{y_i}{N_i}\right),$$

namely a weighted average of the observed rate and the prior mean rate. Shrinkage to the prior mean is greater when γ is large and for small populations N_i. The marginal density is

$$p(y_i \mid \gamma, \rho) = \binom{N_i}{y_i} \frac{Be(\gamma\rho + y_i, \gamma(1-\rho) + N_i - y_i)}{Be(\gamma\rho, \gamma(1-\rho))}$$

$$= \binom{N_i}{y_i} \frac{\Gamma(\gamma\rho + y_i)\Gamma(\gamma(1-\rho) + N_i - y_i)\Gamma(\gamma)}{\Gamma(\gamma\rho)\Gamma(\gamma(1-\rho))\Gamma(\gamma + N_i)},$$

with expectation $E(y_i) = E[E(y_i \mid \pi_i)] = E(N_i\pi_i) = N_i\rho$, and variance

$$V(y_i) = V[E(y_i \mid \pi_i)] + E[V(y_i \mid \pi_i)] = \rho(1-\rho)\left(\frac{\gamma + N_i}{\gamma + 1}\right).$$

so that $\gamma \to \infty$ corresponds to the binomial density.

Quintana and Tam (1996) consider both marginal and conditional likelihood MCMC estimation approaches to the beta-binomial. With beta mixing according to $\pi_i \sim Be(a,b)$, and prior $p(a,b)$, they apply Hastings sampling to the joint marginal likelihood (with π_i integrated out)

$$L(a,b,y) \propto \left[\frac{\Gamma(a+b)}{\Gamma(a)\Gamma(b)}\right]^n \left[\prod_i \frac{\Gamma(a+y_i)\Gamma(b+N_i-y_i)}{\Gamma(a+b+N_i)}\right] p(a,b),$$

and mixed Gibbs–Hastings sampling to the joint conditional likelihood

$$L(a,b,\pi,y) \propto \left[\frac{\Gamma(a+b)}{\Gamma(a)\Gamma(b)}\right]^n \left[\prod_i \pi_i^{a+y_i-1}(1-\pi_i)^{b+N_i-y_i+1}\right] p(a,b).$$

They also consider implications for posterior parameter correlation of the reparameterisation (Lee and Sabavala, 1987)

$$\pi_i \sim Be(\mu, \eta),$$

$$\mu = a/(a+b),$$

$$\eta = 1/(1+a+b),$$

where η is a measure of heterogeneity.

4.7.1 Non-Conjugate Priors for Binomial Mixing

Alternatives to conjugate beta mixing are the binomial with normal errors in the link, generalised binomial models (Makuch et al., 1989), generalised beta-binomial models (Rodriguez-Avi et al., 2007), and models adapted to departures from the typical binomial frequency pattern, such as zero-inflated binomial models. The logistic-normal model with normal random effects in the logit link specifies

$$y_i \mid \pi_i \sim \text{Bin}(N_i, \pi_i),$$

$$\text{logit}(\pi_i) = b_i,$$

$$b_i \mid \mu, \tau \sim N(\mu, \tau^2).$$

Here π_i then follows a logistic-normal density,

$$p(\pi_i \mid \mu, \tau^2) = \frac{1}{\tau\sqrt{2\pi}} \exp\left(-\frac{1}{2\tau^2}\left[\log\frac{\pi_i}{1-\pi_i} - \mu\right]^2\right)\frac{1}{\pi_i(1-\pi_i)}.$$

The logistic-normal prior with $\tau=2.67$ and $\mu=0$ matches a Jeffreys prior on π_i in the first two moments, and setting $\tau=1.69$ matches the uniform prior in the first two moments (Agresti and Hitchcock, 2005). As for the Poisson lognormal, one may generalise to heavier tailed or skewed mixing densities. Teather (1984) proposes a family of symmetric prior densities for logit(π_i) that includes the normal and double exponential as special cases. Alternative links (e.g. probit) or mixing over links are possible.

In many applications (e.g. studies with patients allocated to multiple treatment), the random effect variation is representing differential frailty in the patient population of the study, so that for studies $i = 1,\ldots,n$ with $k = 1,\ldots,K$ treatment categories

$$y_{ik} \sim Bin(N_{ik}, \pi_{ik}),$$

$$\text{logit}(\pi_{ik}) = b_i + \beta_k,$$

$$b_i \sim N(0, \tau^2),$$

where the β_k are fixed treatment effects, while the b_i can be interpreted as between study variation in treatment effects. For example, Gao (2004) considers this structure for data from Winship (1978) on a meta-analysis of eight randomised clinical trials comparing healing rates in duodenal ulcer patients. For trials with treatment and control arms only, with patient totals $\{N_{iT}, N_{iC}\}$, the logistic-normal model is often applied in meta-analysis when trial totals are small, rather than adopting a normal approximation (Warn et al., 2002; Parmigiani, 2002). In fact, other links (combined with binomial sampling) may be more useful in clinical interpretability.

The prior structure often focuses on the control arm probabilities π_{iC}, and on differences between trial and control group probabilities. Thus assume

$$y_{iT} \sim \text{Bin}(N_{iT}, \pi_{iT}),$$

$$y_{iC} \sim \text{Bin}(N_{iC}, \pi_{iC}).$$

Then analysis of treatment-control differences δ_i on the log odds ratio scale would involve transforms $\omega_{iT} = \text{logit}(\pi_{iT})$, and $\omega_{iC} = \text{logit}(\pi_{iC})$, and taking

$$\delta_i = \omega_{iT} - \omega_{iC},$$

one might assume

$$\delta_i \sim N(\Delta, \sigma_\delta^2).$$

For the π_{iC}, random effect options might be to take $\omega_{iC} \sim N(\mu_C, \tau_C^2)$, with $\{\mu_C, \tau_C^2\}$ as additional unknowns, or $\pi_{iC} \sim Be(a_C, b_C)$ with $\{a_C, b_C\}$ additional unknowns.

Consider instead a log link, so that $\omega_{iT} = \log(\pi_{iT})$, and $\omega_{iC} = \log(\pi_{iC})$, again with $\delta_i \sim N(\Delta, \sigma_\delta^2)$. The δ_i now measure log relative risks, which are often more clinically useful than log odds ratios, and exp(Δ) will measure the relative risk of (say) recurrence or mortality under the treatment. In practice, sampling has to be constrained to ensure δ_i is less than $-\log(\pi_{iC})$, so that

$$\omega_{iT} = \omega_{iC} + \min(\delta_i, -\log(\pi_{iC})),$$

$$\delta_i \sim N(\Delta, \sigma_\delta^2).$$

Similarly, for a risk difference analysis

$$\omega_{iC} = \pi_{iC},$$

$$\pi_{iT} = \omega_{iT} = \delta_i + \omega_{iC},$$

$$\delta_i \sim N(\Delta, \sigma_\delta^2),$$

sampling has to be constrained to ensure that $\pi_{iT} \in [0,1]$. This involves confining δ_i to the interval $[-\pi_{iC}, 1-\pi_{iC}]$ with the actually sampled model specifying

$$\omega_{iT} = \omega_{iC} + \min(\max(\delta_i, -\pi_{iC}), 1-\pi_{iC}).$$

If the control group probabilities are regarded as proxies for the underlying risk of subjects in a study, then the model involves a regression on centred control group effects, namely

$$\omega_{iT} = \omega_{iC} + \delta_i + \beta(\omega_{iC} - \bar{\omega}_C),$$

$$\delta_i \sim N(\Delta, \sigma_\delta^2),$$

where $\bar{\omega}_C$ is the average of the control arm effects (calculated at each iteration), and β is an extra unknown.

4.7.2 Multinomial Mixtures

For representing overdispersion in multinomial data with M categories

$$(y_{i1}, \ldots y_{iM}) \sim \text{Mult}(N_i, [\pi_{i1}, \ldots \pi_{iM}]),$$

$$N_i = \sum_m y_{im},$$

the beta prior generalises to a Dirichlet prior with parameters $(a_1, \ldots, a_M)$. With $\pi_i = [\pi_{i1}, \ldots \pi_{iM}]$ and $A = \sum_m a_m$, one has

$$p(\pi_i | a) = \frac{\Gamma(A)}{\prod_{m=1}^{M} \Gamma(a_m)} \prod_{m=1}^{M} \pi_{im}^{a_m - 1},$$

so that prior means for π_{im} are α_m/A, with variances $a_m(K-a_m)/A^2(A+1)$. The posterior density for $[\pi_{i1},\ldots,\pi_{iM}]$ is Dirichlet with parameters $(y_{i1}+a_1,\ldots,y_{iM}+a_M)$. Assuming equal prior mass is assigned to all categories, namely $a_1=a_2=\ldots=a_M$, there is greater shrinkage or flattening towards an equal prior cell probability across the M categories as A increases.

Greater flexibility may be provided by a multivariate generalisation of the logistic-normal prior (Aitchison and Shen, 1980; Hoff, 2003). Thus with $(y_{i1},\ldots,y_{iM})\sim \text{Mult}(N_i,[\pi_{i1},\ldots,\pi_{iM}])$,

$$\pi_{ij}=\frac{\exp(b_{ij})}{\sum_{m=1}^{M}\exp(b_{im})},$$

where the vector $(b_{i1},\ldots b_{i,M-1})$ of the first M−1 effects is multivariate normal with mean $\mu_i=(\mu_{i1},\ldots,\mu_{i,M-1})$ and covariance matrix Σ of dimension M−1. For the reference category, one sets $b_{iM}=0$. If the categories are ordered and similarity of probabilities in adjacent categories is expected on substantive grounds, the covariance matrix or its inverse may be stipulated in line with a low order autoregressive form; this is known as "histogram smoothing" (Leonard, 1973).

Another generalisation is to add a higher stage prior on the Dirichlet parameters, for example, on the total mass A. Thus, Albert and Gupta (1982) consider a two-stage prior in multinomial-Dirichlet analysis of contingency tables. With the reparameterisation $a_i=A\rho_i$ where $\sum_m \rho_m=1$, one possible hierarchical prior generalises the binomial-beta with

$$\pi_i=[\pi_{i1},\ldots\pi_{iM}]\sim \text{Dir}(A\rho_1,\ldots,A\rho_M),$$

$$A\sim Gu(u_A,b_A),$$

$$(\rho_1,\ldots,\rho_M)\sim \text{Dir}(w_1,\ldots,w_M),$$

where the w_m and $\{a_A,b_A\}$ are known.

4.7.3 Ecological Inference Using Mixture Models

Binomial-beta and multinomial-Dirichlet models (or non-conjugate alternatives) have recently found wide application in ecological inference. Much of the impetus for this research has come from political science, and may involve counts of a behaviour or event for unit i (e.g. constituency) with $m=1,\ldots,M$ outcomes (e.g. party voting affiliation) by demographic attribute with $c=1,\ldots,C$ levels (e.g. social class, ethnic group). The underlying data are the totals N_{imc}. What is observed in practice are the marginals N_{im+} (e.g. constituency voting data by party voted for), and information from another source (e.g. from the census) on the relative distribution of the voting age population across levels of the demographic attribute. This is proxy information regarding the ratios $x_{ic}=N_{i+c}/N_{i++}$ (which might be census-based percentages of the voting population in different ethnic groups).

Consider the simplest case, ecological inference in 2×2 tables. Suppose the observations are the total electorate N_i, and the number who turn out V_i. So $M=2$, for voting and not voting. Also available from census data is the proportion x_i of the voting age population who are black. Given this information, the goal of ecological inference is to estimate parameters governing the internal table cells, namely the proportions r_{i1} and r_{i2} of black and white voters who turned out. Since $M=2$, the data are binomial, and the overall turnout rate in area i is modelled as $V_i \sim \text{Bin}(N_i, p_i)$. Modelling the turnout rates in terms of ethnic-specific voting rates proceeds using the probabilistic statement

$$Pr(\text{Turnout}) = Pr(\text{Turnout} \mid \text{Black})Pr(\text{Black}) + Pr(\text{Turnout} \mid \text{White})Pr(\text{White}),$$

with the corresponding relation in area i being

$$p_i = r_{i1}x_i + r_{i2}(1 - x_i).$$

Among possible priors for the unknown r_{i1} and r_{i2} in a 2×2 ecological problem are:

a) Independent beta densities $r_{i1} \sim Be(a_1, b_1)$, $r_{i2} \sim Be(a_2, b_2)$;
b) A bivariate normal for $w_{ij} = \text{logit}(r_{ij})$, with mean $\mu = (\mu_1, \mu_2)$ and covariance Σ, allowing $\{r_{i1}, r_{i2}\}$ to be correlated; and
c) A trivariate normal for $w_{i1} = \text{logit}(r_{i1})$, $w_{i2} = \text{logit}(r_{i2})$, and $w_{i3} = \text{logit}(x_i)$.

Imai et al. (2008) typify ecological missing data as data "coarsening," and the first two priors above are consistent with coarsening at random. By contrast, the final option amounts modelling the joint density $p(x,r)$ of racial composition x and turnout behaviour $r = (r_1, r_2)$ via the sequence $p(x \mid r)p(r)$. This is similar to joint modelling of missingness and observed data in non-random models for missing data (Pastor, 2003) and hence may be termed "coarsening not at random." If predictors of turnout rates are available, then the means μ_{i1} and μ_{i2} include regression terms.

Example 4.6 Breast Cancer Recurrence; Binomial Meta-Analysis

Parmigiani (2002, p.127) considers 14 trials concerning the impact of tamoxifen on breast cancer recurrence rates. The trials are mostly large and a normal approximation might well be applied, though one trial involved only 20 patients. A binomial analysis is adopted with

$$y_{iT} \sim \text{Bin}(N_{iT}, \pi_{iT}),$$

$$y_{iC} \sim \text{Bin}(N_{iC}, \pi_{iC}).$$

A beta density is assumed for the control group rates, namely

$$\pi_{iC} \sim Be(a_C, \beta_C),$$

with uniform priors on the unknowns, $a_C \sim U(1,100)$ and $\beta_C \sim U(1,100)$. Different comparison scales can be defined. For example, on the log odds ratio scale

$$\omega_{iT} = \text{logit}(\pi_{iT}),$$

$$\omega_{iC} = \text{logit}(\pi_{iC}),$$

$$\delta_i = \omega_{iT} - \omega_{iC},$$

with treatment effects assumed normal

$$\delta_i \sim N(\Delta, \sigma_\delta^2),$$

and diffuse normal and inverse gamma priors on Δ and σ_δ^2 respectively. Under an absolute risk difference scale, one has instead

$$\omega_{iT} = \pi_{iT},$$

$$\omega_{iC} = \pi_{iC},$$

$$\delta_i = \omega_{iT} - \omega_{iC},$$

with appropriate constraints (Warn et al., 2002) to ensure $\pi_{iT} \in [0,1]$.

To provide a summary index of treatment benefit, the treatment gain δ_{new} for a hypothetical new trial is sampled and added to a predicted baseline recurrence rate $\pi_{\text{new},C}$ (transformed on the appropriate scale) to give a predicted new trial treatment rate, $\pi_{T,\text{new}}$. Then the probability that the predictive relative risk $RR_{new} = \pi_{\text{new},T} / \pi_{\text{new},C}$ exceeds 1 is obtained.

Treatment and placebo groups are compared on three different effect scales, namely the log odds ratio (LOR), the log relative risk (LRR), and the absolute risk difference (ARD). On the LRR scale, the predictive density for RR_{new} has 95% interval (0.79, 1.01) with a 3.5% chance that RR_{new} exceeds 1. The ARD scale admits a larger element of doubt, with a 95% interval (0.65,1.06) and $Pr(RR_{\text{new}} > 1 \mid y) = 0.092$. By contrast, under a LOR scale, RR_{new} has 95% interval (0.77,0.97), with only a 1.1% chance of exceeding 1. The lowest DIC is for the LOR scale, namely 218. Mixed predictive checks show all observations with exceedance probabilities $Pr(y_{iT,\text{new}} > y_{iT} \mid y) + 0.5Pr(y_{iT,\text{new}} = y_{iT} \mid y)$ between 0.1 and 0.9 (exc.mx in the code).

A considerably different result (providing a DIC of 222) is obtained using beta-binomial sampling for both treatment and control arms. Consider the reparameterised beta density $Be(\theta S,(1-\theta)S)$ where S is the prior sample size, and θ is the prior probability. Then $1/(S+1)$ is an estimator of the beta-binomial intra-class correlation. Assuming a common prior sample size (Bakbergenuly and Kulinskaya, 2017), one has

$$y_{iT} \sim \text{Bin}(N_{iT}, \pi_{iT}),$$

$$y_{iC} \sim \text{Bin}(N_{iC}, \pi_{iC}),$$

$$\pi_{iT} \sim Be(\theta_T S,(1-\theta_T)S),$$

$$\pi_{iC} \sim Be(\theta_C S,(1-\theta_C)S),$$

with S assigned a gamma prior, and the θ parameters themselves assumed beta distributed. Under these assumptions, RR_{new} has a 95% interval (0.3,2.2), with a 38% chance of exceeding 1.

Example 4.7 Adverse Effects from Terbinafine

We consider overdispersed binomial data from Young-Xu and Chan (2008) on patients with adverse effects in treatment with an oral anti-fungal agent (terbinafine) for onychomycosis and dermatophytosis. These data raise issues regarding the accommodation (i.e. satisfactory representation within the model) of extreme values, as against identifying poorly fitted cases (Marshall and Spiegelhalter, 2007). Another distinction can be made, namely, between extreme values and outliers, in areas such as pharmaceutical testing. Thus Walfish (2006) suggests that

> an outlier is defined as an observation that appears to be inconsistent with other observations in the data set. An outlier has a low probability that it originates from the same statistical distribution as the other observations in the data set. On the other hand, an extreme value is an observation that might have a low probability of

occurrence but cannot be statistically shown to originate from a different distribution than the rest of the data.

There are 41 studies in the terbinafine analysis, and each study has N_i patients and y_i patients with adverse reactions. The binomial logit-normal (BLN) representation

$$y_i \mid \pi_i \sim \text{Bin}(N_i, \pi_i),$$

$$\text{logit}(\pi_i) = \mu + b_i,$$

$$b_i \mid \tau \sim N(0, \tau^2),$$

is compared with the beta-binomial,

$$y_i \mid \pi_i \sim \text{Bin}(N_i, \pi_i)$$

$$\pi_i \sim Be(\alpha, \beta).$$

The latter can be also be represented directly in rstan using the beta_binomial density. Pooling across all studies, 111 of 3002 patients have adverse effects (around 3.7%).

The mixed replicate checking scheme is used to identify poorly fitted cases. In the BLN representation, this is implemented by sampling replicate normal random effects $b_{\text{rep},i}$, and then the corresponding predicted totals of adverse reactions. Poorly fitted cases are identified by extreme exceedance probabilities, namely $p.exc_i = Pr(y_{\text{rep},i} > y_i \mid Y) + 0.5Pr(y_{\text{rep},i} = y_i \mid Y)$ under 0.05 or over 0.95. Two studies (19, 38) are identified as poorly fitted (a potential outlier), with study 19, containing 186 patients but 0 adverse effects, having $p.exc_i = 0.96$.

In the hierarchical beta-binomial representation, we sample replicate $\pi_{\text{rep},i}$ and then the corresponding predicted adverse reactions. Now three studies are identified as problematic: study 19 with $p.exc_i = 0.96$, and two studies (33 and 38) with relatively high adverse reaction totals. Inferences regarding the mean adverse rate are similar between the two approaches: the beta-binomial adverse mean rate is 3.44%, compared to 3.45% under the binomial logitnormal (based on averaging over all samples of all π_i). However, sampled π_i under the binomial logit-normal show greater positive skew than the beta-binomial (1.61 vs 1.22), reflecting accommodation for the higher rates for some studies. There is a similar contrast in outlier accommodation between the conjugate Poisson-gamma mixture and the Poisson lognormal, as the tails of the lognormal are heavier than for the gamma distribution (Connolly et al., 2009; Wang and Blei, 2017).

4.8 Discrete Mixtures and Semiparametric Smoothing Methods

Hierarchical models for pooled inferences or density estimation based on a single underlying population with a specific parametric form are often a simplification. Pooling strength applications such as meta-analysis and density estimation are often seeking to identify the main features of the data, or to predict further observations y_{new} via the predictive distribution $p(y_{\text{new}} \mid y)$, and a single population model may not be appropriate for data exhibiting asymmetry, multiple modes, isolated outliers, or outlier clusters (Mohr, 2006). While the standard densities can be extended (e.g. to reflect asymmetry), mixtures of standard densities (normal and t densities) can be used to represent a wide variety of density shapes (Everitt and Hand, 1981). Use of a single population density model in such circumstances will provide improper pooling and poor predictions for a new unit (Hoff, 2003). For example, a normal random-effects analysis of hospital mortality rates may shrink extreme rates considerably, and this might mask potentially unusual results for units with smaller totals of patients at risk (Ohlssen et al., 2007).

Among the principles that govern robust smoothing and regression methods for non-standard densities are discrete mixing of densities over $K>1$ subpopulations (Bohning, 1999) and various types of local regression based on kernel or smoothness priors (Muller et al., 1996). In this chapter, the focus is on discrete mixture modelling, where the Bayesian approach has been coupled with many recent advances. These include the Bayesian analogue to non-parametric maximum likelihood estimation, with MCMC implementation as set out by Diebolt and Robert (1994), and Richardson and Green (1997), and numerous developments of the Dirichlet process methodology, as reviewed by Hanson et al. (2005). The Bayesian approach is flexible in terms of prior structures that can be imposed in estimation, either grounded in substantive theory, or to improve definition of the subgroups (e.g. Robert and Mengersen, 1999). On the other hand, repeated sampling without appropriate parameter constraints is subject to "label switching," since labelling of the subgroups is arbitrary (Fruhwirth-Schattner, 2001; Chung et al., 2004).

4.8.1 Finite Mixtures of Parametric Densities

In a discrete parametric mixture model, a single parametric density is typically assumed in each subpopulation $k \in (1,\ldots,K)$, but a different hyperparameter ψ_k, so that within the kth subpopulation $y \sim p(y \mid \psi_k)$. Unobserved subgroup or allocation indicators $S_i \in (1,\ldots K)$ describe how the units are distributed over subpopulations. These are also known as configuration indicators (Gopalan and Berry, 1998). The joint or complete data density p(y,S) can be written

$$p(y_i,S_i) = p(y_i \mid S_i)p(S_i) = p(y_i \mid \psi_{S_i})\pi_{S_i}$$

where $p(y \mid S) = p(y \mid \psi_S)$ is the density for y_i conditional on S_i, and $\{\pi_1,\ldots,\pi_K\}$ are the prior subgroup probabilities, with $\sum_{k=1}^{K} \pi_k = 1$. The unconditional or marginal density for a single y_i is

$$p(y_i \mid \pi_1,\ldots \pi_K,\psi_1,..\psi_K) = \sum_{k=1}^{K} \pi_k p(y_i \mid \psi_k),$$

with the total marginal likelihood being

$$p(y \mid \pi,\psi) = \prod_{i=1}^{n}\sum_{k=1}^{K} \pi_k p(y_i \mid \psi_k).$$

Classical analysis via non-parametric maximum likelihood estimation involves maximisation of the log of this marginal density – for example, see Rattanasiri et al. (2004) for a disease-mapping application where y_i are malaria counts and $p(y|\psi)$ is a Poisson density.

In MCMC applications discrete mixture models can be represented hierarchically using the latent subpopulation indicators (Marin et al., 2005, p.462). Thus, at the highest stage or level are the parameters $\varphi = (\pi_1,\ldots,\pi_K,\psi_1,\ldots,\psi_K)$, then the missing configuration data the distribution of which depends on φ,

$$S_i \sim P(S_i \mid \varphi)$$

and at the lowest (first) stage the distribution of the observations $p(y \mid \varphi, S)$ depends on both φ and $S=(S_1,\ldots,S_n)$. The joint distribution is therefore

$$p(y,S,\varphi)=p(y \mid S,\varphi)p(S \mid \varphi)p(\varphi).$$

4.8.2 Finite Mixtures of Standard Densities

There is a considerable literature on univariate and multivariate normal mixtures for continuous data, and on Poisson and binomial mixtures for discrete data, with Bayesian references including Richardson and Green (1997), Roberts et al. (1998), Militino et al. (2001), and Hurn et al. (2003). Overdispersed or skew alternatives to the major densities can be used in discrete mixtures instead: for continuous data, the Student t distribution involves an additional tuning parameter useful for outlier accommodation, and greater robustness to such points may be obtained by discrete mixtures over univariate and multivariate Student t densities with varying degrees of freedom (Lin et al., 2004). Discrete mixtures of skew normal and skew Student t densities are considered by Lin et al. (2007a,b). Lin et al. (2007a) argue that a simple normal discrete mixture model tends to overfit when additional components are added to capture skewness in continuous data.

Parameter sampling via MCMC is facilitated by conjugate prior choices for the mixing density. For example, consider a univariate normal mixture with $\psi_k=(\mu_k,\sigma_k^2)$, and

$$p(y \mid \pi,\mu,\sigma)=\sum_{k=1}^{K}\pi_k\phi(y \mid \mu_k,\sigma_k),$$

where $\phi(y \mid \mu,\sigma)$ is the normal density, $N(\mu,\sigma^2)$. The conjugate prior for $\psi_k=(\mu_k,\sigma_k)$ takes $\sigma_k^2 \sim IG(V_k/2,V_k/2)$, namely

$$p(\sigma_k^2)\propto\sigma_k^{-v_k-1}\exp(-V_k/2\sigma_k^2),$$

and

$$p(\mu_k \mid \sigma_k^2)=N\left(\xi_k,\frac{\sigma_k^2}{\kappa_k}\right).$$

Also assume a Dirichlet prior for the unknown mixture probabilities

$$(\pi_1,\ldots,\pi_K)\sim Dir(a,\ldots,a),$$

with a preset or possibly an extra unknown. Gibbs sampling then samples the missing data (the allocation indicators) according to a multinomial density with probabilities at iteration t,

$$p(S_i^{(t)}=k \mid \pi^{(t)},\mu^{(t)},\sigma^{(t)})=\rho_{ik}^{(t)}=\frac{\pi_k^{(t)}\phi(y_i \mid \mu_k^{(t)},\sigma_k^{(t)})}{\sum_{k=1}^{K}\pi_k^{(t)}\phi(y_i \mid \mu_k^{(t)},\sigma_k^{(t)})}.$$

Let $d_{ik}^{(t)}=1$ if $S_i^{(t)}=k$ and $d_{ik}^{(t)}=0$ otherwise. Suppose $N_k^{(t)}=\#\{S_i^{(t)}=k\}$ is the total number of cases with $S_i^{(t)}=k$, that $m_k^{(t)}=\sum d_{ik}^{(t)}y_i/N_k^{(t)}$ is the average response for these cases, and that

$E_k^{(t)} = \sum d_{ik}^{(t)}(y_i - m_k^{(t)})^2$ is the sum of squared errors for this subgroup. Then, with conditioning on remaining parameters understood, the π_k are updated according to a Dirichlet with

$$(\pi_1^{(t)}, \pi_2^{(t)}, \ldots, \pi_K^{(t)}) \sim D(a + N_1^{(t)}, a + N_2^{(t)}, \ldots, a + N_K^{(t)}),$$

the subgroup variances are sampled from an updated inverse gamma,

$$\sigma_k^{2(t)} \sim IG\left(0.5[\nu_k + N_k^{(t)}], 0.5\left[V_k + E_k^{(t)} + \frac{N_k^{(t)}\kappa_k}{\kappa_k + N_k^{(t)}}(\xi_k - m_k^{(t)})\right]\right)$$

and the subgroup means are updated according to

$$\mu_k^{(t)} \sim N\left(\frac{\kappa_k \xi_k + N_k^{(t)} m_k^{(t)}}{\kappa_k + N_k^{(t)}}, \frac{\sigma_k^{2(t)}}{\kappa_k + N_k^{(t)}}\right).$$

Diebolt and Robert (1994) suggest stabilising adjustments to these updates to improve convergence. A refinement is to take the mixture proportions as subject specific as in $(\pi_{i1}, \ldots, \pi_{iK}) \sim \text{Dir}(a, \ldots, a)$, and in the updates for $\pi_{ik}^{(t)}$, the $N_k^{(t)}$ are replaced by binary indicators according to which class subject i is allocated to at a particular iteration.

4.8.3 Inference in Mixture Models

Parametric mixture models, such as the univariate normal just considered, are subject to identification issues due to the arbitrariness of the sub-population labels. Other forms of identifiability relate to potential overfitting (e.g. K taken too large), so that some groups are overlapping and difficult to distinguish (Betancourt, 2017; Frühwirth-Schnatter, 2006, p.107). An additional issue is whether all parameters in the separate densities need to be taken to vary between groups. For example, the mclust package (Scrucca et al., 2016) compares models with different K, and with group-specific variances as against common variances, $\sigma_k^2 = \sigma^2$. Identification and relative fit may also be affected by the setting for the prior Dirichlet weights (as shown in Example 4.8, using the galaxy data).

Assessing the fit of discrete mixture models also raises distinct problems. Compared with random effects models, there is the benefit that the number of parameters is known, so estimating information criteria such as the BIC or the AIC (Akaike information criterion) is straightforward (McLachlan and Rathnayake, 2014). On the other hand, the asymptotic justification for such criteria is affected by singularities at the boundaries of the parameter space in moving from a $K-1$ group solution to a K group solution (Biernacki et al., 2000). Recently developed information criteria, such as the sBIC may be considered instead (Drton and Plummer, 2017).

To illustrate label identifiability, in the absence of parameter constraints or other prior information to distinguish the components, the likelihood is invariant under permutation of the components, and there are $K!$ possible labelling schemes. It is essential to produce MCMC draws with a unique labelling, if interest lies in the estimation of group-specific parameters or classification probabilities π_k (Frühwirth-Schnatter et al., 2004). Note though that inferences on some aspects of the model are unaffected by group labelling – for example, the unit means that pool over the subpopulation category means. Cluster labelling issues are also not generally considered in the Dirichlet process approach (Section 4.9), where the emphasis is on the smoothed unit means.

Identifying (usually ordering) constraints may be imposed on parameters to avoid label-switching (Roeder and Wasserman, 1997, Richardson and Green 1997), providing what may be termed "non-exchangeable priors" (Betancourt, 2017). Label switching or labelling degeneracy refers to permuting the mixture component subscripts without altering the likelihood (Redner and Walker, 1984). However, Celeux et al. (2000), Marin et al. (2005), and Geweke (2007) consider drawbacks to such identifiability constraints (e.g. distortions of the posterior distribution of the parameters). For example, in a normal mixture, constraints may be imposed on prior masses π_k (e.g. $\pi_1 > \pi_2 > \ldots > \pi_K$), or on the subpopulation parameters, μ_k or on the scale parameters σ_k. A preliminary MCMC sampling analysis without parameter constraints may be used to assess the most suitable form of constraint (Fruhwirth-Schattner, 2001). Another possibility is to use maximum likelihood solutions (e.g. using the R package flexmix) to set constraints and/or relatively informative priors that are sensible for the dataset. Re-analysis of the posterior output to impose a consistent labelling is another possibility (Frühwirth-Schnatter, 2001), as are data-based priors, albeit not fully Bayesian (Wasserman, 2000). For example, in a two-group model without regression on predictors, the unit with the maximum y value could be pre-labelled as belonging to one or other subpopulation.

Particular types of parameterisation may be used to improve identification, such as introducing dependence between the parameters ψ_k in different components so that they are perturbations of one another (Robert and Mengersen, 1999). For example, a normal mixture model with $\psi_k = (\mu_k, \sigma_k^2)$ would be based on taking $\{\theta_1, \sigma_1^2\}$ as reference parameters and adopting the parameterisation

$$\sigma_2 = \sigma_1\omega_1,$$

$$\sigma_3 = \sigma_2\omega_2,$$

$$\sigma_4 = \sigma_3\omega_3,$$

$$\ldots$$

$$\sigma_K = \sigma_{K-1}\omega_{K-1} = \sigma_1\omega_1\omega_2\ldots\omega_{K-1},$$

where $\omega_k \sim U(0,1)$. With $\theta_1 = \mu_1$, the prior on the series of normal means takes a perturbation form

$$\mu_2 = \theta_1 + \sigma_1\theta_2,$$

$$\mu_3 = \theta_1 + \sigma_1\theta_2 + \sigma_1\sigma_2\theta_3,$$

$$\ldots$$

$$\mu_K = \theta_1 + \sigma_1\theta_2 + \sigma_1\sigma_2\theta_3 + \ldots + (\sigma_1\sigma_2\ldots\sigma_{K-1})\theta_K.$$

The mixture weights have the form

$$\pi_1 = p_1,$$

$$\pi_2 = (1 - p_1)p_2,$$

$$\pi_3 = (1-p_1)(1-p_2)p_3,$$

$$\ldots$$

$$\pi_{K-1} = (1-p_1)(1-p_2)\ldots(1-p_{K-2})p_{K-1},$$

$$\pi_K = (1-p_1)(1-p_2)\ldots(1-p_{K-1})$$

with $p_k \sim U(0,1)$. This prior is still invariant under permutation of the cluster indices and an identifying constraint is placed on the variances by taking $1 \geq \omega_1 \geq \ldots \geq \omega_{K-1}$. An advantage of this representation is that an improper prior on $\{\mu_1, \sigma_1^2\}$ can be used (Robert and Titterington, 1998). For the two group case, Basu (1996) presents the parameterisation $\nu = \sigma_1^2 / \sigma_2^2$ and $\Delta = (\mu_2 - \mu_1)/\sigma_1$ to test for normal or Student t unimodality as against bimodality; posterior probabilities of unimodality are obtained using the results of Robertson and Fryer (1968).

Celeux et al. (2000) and others apply post-processing to the MCMC output resulting from a discrete mixture analysis without parameter constraints; the goal is to reconfigure the output with a consistent labelling. Suppose there are p parameters in any subpopulation. If MCMC convergence is assumed, one may select a short run of iterations (say $S = 100$ iterations) where there is no label switching to provide a reference labelling. The initial run of parameter samples provides a base reference label sequence $1, 2, \ldots, K$ (one among the $K!$ possible), and K means of dimension p, $\bar{\theta}_k = \{\bar{\theta}_{1k}, \bar{\theta}_{2k}, \ldots, \bar{\theta}_{pK}\}$, that can be permuted to include all other remaining $K! - 1$ possible labelling schemes. In a subsequent run of R iterations where label switching might occur, iteration r is assigned to that scheme (among the $K!$) closest to it in distance terms and a relabelling applied if there has been a switch away from the base reference label. Additionally, the means under the schemes are recalculated at each iteration $S + r$ (Celeux et al., 2000, p.965).

Schemes for gaining identifiability can be applied within the MCMC sampling, as illustrated in the rjags online code for the BUGS example concerning peak sensitivity wavelengths (the Eyes example) (https://sourceforge.net/p/mcmc-jags/examples/ci/3765ddfd606e96c5de12818b50ef1b807f77af53/tree/classic-bugs/vol2/eyes/eyes.bug). Assume an unconstrained analysis, with no constraints on the mixture parameters. Then, assuming relabelling based on sampled means, processing resorts these sampled means, named say m0[1:K] in the code, with identifiable means mu[K], mu[K–1],...,mu[1] defined according to which of the m0[1:K] has the maximum value, the second highest, etc. Other mixture parameters (weights and variances for each group in a normal mixture) are reassigned using the same relabelling rule. This procedure corresponds to adopting a standard set of labels or standard ordering to obtain an identified solution (Betancourt, 2017).

The rjags online code is for the case K = 2. For K = 3, assume a normal univariate mixture, with reassignment based on the means, but applied also to resorting weights, from the sampled P0[1:3] to the identified P[1:3]. Then one possible rjags code fragment is

```
rank <- rank(m0)
for (j in 1:K) {
J1[j] <- equals(rank[j],1)
J2[j] <- equals(rank[j],2)
J3[j] <- equals(rank[j],3)}
P[1] <- P0[1] * J3[1] + P0[2] *J3[2] + P0[3] *J3[3]
P[2] <- P0[1] * J2[1] + P0[2] *J2[2] + P0[3] *J2[3]
```

```
P[3] <- P0[1] * J1[1] + P0[2] *J1[2] + P0[3] *J1[3]
mu[1] <- m0[1] * J3[1] + m0[2] *J3[2] + m0[3] *J3[3]
mu[2] <- m0[1] * J2[1] + m0[2] *J2[2] + m0[3] *J2[3]
mu[3] <- m0[1] * J1[1] + m0[2] *J1[2] + m0[3] *J1[3]
```

Assume precisions tau0[1:K] are to be reassigned as well. A general code for larger K can be written more compactly as follows:

```
rank <- rank(m0)
for (j in 1:K) {P[j] <- sum(P0prod[j,])
mu[j] <- sum(m0prod[j,])
tau[j] <- sum(tau0prod[j,])
for (k in 1:K) {P0prod[j,k] <- P0[k]*equals(rank[k],j)
m0prod[j,k] <- m0[k]* equals(rank[k],j)
tau0prod[j,k] <- tau0[k]* equals(rank[k],j)}}}
```

This procedure is illustrated in Example 4.8. Which parameter is selected as the basis for resorting (e.g. means or weights) may partly be decided using measures of fit.

We illustrate this procedure with jagsUI applied to the randomly generated dataset used in Betancourt (2017), consisting of a two-group Gaussian mixture with means (–2.75, 2.75), prior weights P = (0.6,0.4), and variances 1 in both groups. Prior Dirichlet sample sizes of 2 are assumed. The code assumes a conditional likelihood (conditional on allocation indicators) and is:

```
mu <- c(-2.75, 2.75)
sigma <- c(1, 1)
lambda <- 0.4
set.seed(689934)
N <- 1000
z <- rbinom(N, 1, lambda) + 1
y <- rnorm(N, mu[z], sigma[z])
D <- list(N= N, y = y, K = 2)
require(jagsUI)
K=2
cat("model { for (i in 1:N){ # conditional likelihood
y[i] ~dnorm(m0[S[i]], tau0[S[i]])
# latent allocation indicators (conditional likelihood)
S[i] ~dcat(P0[1:K])
ynew[i] ~dnorm(mu[S[i]],tau0[S[i]])
exc[i] <- step(ynew[i]-y[i])
LL[i] <- log(sum(L[i,]))}
P0 ~ddirch(alpha[]); # prior for weights
for (j in 1:K) { m0[j] ~dnorm(0, 0.01)
alpha[j] <- 2        # prior Dirichlet sample sizes
tau0[j] ~dgamma(1,0.001)
for (i in 1:N) { L[i,j] <- exp(log(P[j])+0.5*log(tau[j])-0.919-0.5*tau[j]
*pow(y[i]-mu[j],2)) }}
tLL <- sum(LL[])
# Processing to obtain identifiable groups
rank <- rank(m0)
for (j in 1:K) {P[j] <- sum(P0prod[j,])
mu[j] <- sum(m0prod[j,])
tau[j] <- sum(tau0prod[j,])
```

```
s2[j] <- 1/tau[j]
for (k in 1:K) {P0prod[j,k] <- P0[k]*equals(rank[k],j)
m0prod[j,k] <- m0[k]* equals(rank[k],j)
tau0prod[j,k] <- tau0[k]* equals(rank[k],j)}}}
", file="discmix.jag")
# initial values and estimation
inits <- function(){list(m0=rnorm(K,0,0.01),tau0=rexp(K,1))}
pars <- c("P","mu","tau","s2","tLL")
summary(autojags(D, inits, pars, model.file="discmix.jag",2, n.adapt=100,
iter.increment=1000, n.burnin=500,Rhat.limit=1.1, max.
iter=50000,seed=1234))
```

We obtain a solution with μ_2 as the larger mean, with mean (sd) of 2.87 (0.05), and with corresponding estimated weight $p_2 = 0.38$. In this solution μ_1 is the smaller mean, with posterior mean (sd) of −2.73 (0.04), and with corresponding estimated weight $p_1 = 0.62$. The estimated weights reflect the actually sampled assignment indicator totals at line 7 of the code, respectively sum(z==1) = 622 and sum(z==2) = 378. Convergence was attained at under 2000 iterations.

A less satisfactory result is obtained under the alternative scenario investigated by Betancourt (2017) where the means are (−0.75,0.75), separated by less than a standard deviation. As before prior weights are P = (0.6,0.4) and variances are 1 in both groups. This time prior Dirichlet sample sizes of 5 are assumed. Convergence is obtained by under 5000 iterations with this more informative prior, but the estimated means are not fully reproducing the simulation, namely −0.50 (0.21) and 0.44 (0.38) with estimated weights of p = (0.57,0.43). This demonstrates the identifiability issues present when components are not widely separated.

4.8.4 Particular Types of Discrete Mixture Model

Heterogeneity within classes can be accommodated using discrete mixtures for unit level conjugate or non-conjugate random effects (Lenk and DeSarbo, 2000; Fruhwirth-Schnatter et al., 2004). For example, the standard discrete mixture to account for heterogeneity in count data involves $K < n$ homogenous subpopulations with means $\mu_1, \ldots, \mu_K$

$$y_i \sim \sum_{k=1}^{K} \pi_k Po(\mu_k),$$

where π_k is the prior probability that a unit belongs to sub-population k, with $\sum_{k=1}^{K} \pi_k = 1$. Alternatively accounting for heterogeneity within subpopulations would involve K Poisson-gamma subgroups

$$y_i \sim Po(\mu_i),$$

$$\mu_i \sim \sum_{k=1}^{K} \pi_k Ga(a_k, b_k),$$

or K Poisson lognormal subgroups

$$y_i \sim Po(\mu_i),$$

$$\mu_i \sim \sum_{k=1}^{K} \pi_k LN(\mu_k, \sigma_k^2),$$

where $LN(m,V)$ denotes a lognormal density with mean m and variance V.

Discrete mixtures can also be used to modify the shape of standard densities such as the Poisson or binomial. For example, a manufacturing process may move between different regimes, one where faults are essentially unknown and another where they occur according to a Poisson process. This will generate excess zeroes as compared to the standard Poisson, leading to a zero-inflated Poisson (ZIP). One may introduce a binary regime indicator S_i with marginal probability $\pi = Pr(S_i = 1)$ that the fault-free regime applies, and $(1-\pi)$ that sampling is from a Poisson density with mean μ. In more generality, with $p(y|\psi)$ as a density for count data (e.g. Poisson, negative binomial, binomial), the corresponding zero-inflated density is

$$p(y=0 \mid \pi, \psi) = \pi + (1-\pi)p(y=0 \mid \psi) \quad y=0$$

$$p(y \mid \pi, \psi) = (1-\pi)p(y \mid \psi) \quad y > 0.$$

Conditionally $Pr(S_i = 1 \mid y > 0) = 0$, while

$$Pr(S_i = 1 \mid y = 0) = \frac{\pi}{\pi + (1-\pi)p(y=0 \mid \psi)}.$$

The process generating the S_i needs only to be considered for zero observations $y_i=0$, and the complete data likelihood (assuming S_i to be given) is

$$L(\pi, \psi \mid y, S) = \prod_{y_i>0}(1-\pi_i)p(y_i \mid \psi)\prod_{y_i=0}\pi_i^{S_i}[(1-\pi_i)p(0 \mid \psi_i)]^{1-S_i}.$$

For example, if $p(y|\psi)$ is taken to be Poisson with mean $\psi=\mu$ then $E(y \mid \pi, \mu) = (1-\pi)\mu$ and

$$V(y \mid \pi, \mu) = (1-\pi)\mu(1+\pi\mu) > E(y \mid \pi, \mu)$$

so that the ZIP model is necessarily overdispersed.

4.8.5 The Logistic-Normal Alternative to the Dirichlet Prior

A generalisation of the logistic-normal to multivariate contexts has been applied to non-parametric analysis and by authors such as Aitchison and Shen (1980), Lenk (1988), and Hoff (2003). The goal is to replace the restrictive Dirichlet prior for the unknown mixture probabilities π_k with a multinomial logistic framework. Consider the case where units are exchangeable, and there are no covariates relevant to allocation between subpopulations. Then for subjects or units $i = 1,\ldots,n$, and assuming

$$y_i \sim \sum_{k=1}^{K} \pi_{ik} p_k(y_i \mid \psi_k),$$

the mixing probabilities are obtained as

$$\pi_{ik} = \frac{e^{z_{ik}}}{1+\sum_{k=1}^{K} e^{z_{ik}}}, \quad k=1,\ldots,K-1$$

$$\pi_{iK} = \frac{1}{1+\sum_{k=1}^{K} e^{z_{ik}}},$$

where the $\{z_{ik}, k=1,\ldots,K-1\}$ are multivariate normal with mean ν and variance Σ_z. For example, Hoff (2003) argues for the use of normal mixtures in density smoothing and, in this case, the $p_k(y \mid \psi_k)$ would be univariate or multivariate normal themselves. This approach generalises to multivariate skewnormal or multivariate Student t densities, and can be adapted to allow non-exchangeable mixture priors, as in histogram smoothing (Leonard, 1973).

Instead of subject-specific z_{ik}, one may also assume a single vector $\{z_1,\ldots,z_{K-1}\}$ to be multivariate normal. For unique identification of the subgroups one may impose order constraints on the parameters in ψ_k or on those underlying $\{z_1,\ldots,z_{K-1}\}$. In the univariate normal case with $\psi_k = \{\mu_k, \sigma_k^2\}$, one might assume an ordering either on the means μ_k, or on the means ν_k of the z_{ik}.

Example 4.8 Galaxy Data

The number of clusters detected in the much-analysed galaxy data has varied over different studies, under the model

$$y_i \sim \sum_{k=1}^{K} p_k N(\mu_k, \sigma_k^2),$$

with y being measured in thousands of kilometres per second. Classical analysis using the flexmix package (Leisch, 2004) in R shows a better AIC and BIC for 5 clusters. The mclust program selects K=4 as optimal, and for the K=6 solution selects an equal variance solution. The K=4 and K=5 solutions have the drawback of a large variance in the group with the largest mean.

Bayesian studies such as Ishwaran and James (2002) find at least 5–6 clusters with a Dirichlet process approach, and under an inverse gamma prior for the σ_k^2. They do, however, find only four clusters when a uniform prior U(0,20.83) is used for σ_k^2, with 20.83 being the observed variance, $V(y)$. Ando (2007) reports six clusters (assuming a monotonic constraint on the μ_k) via several model fit criteria, and $K=6$ is also the best fitting using the sBIC criterion of Drton and Plummer (2017, p.350).

Here we compare solutions with $K=4$, $K=5$ and $K=6$. First of all, the rstan ordered vector parameterisation will be used, following Betancourt (2017) and Savage (2016). A half-Cauchy(0,2) is assumed for the group standard deviations. Prior Dirichlet sample sizes α of 2 and 4 are also compared. For $K=5$ and $K=6$, estimation is with 2 chains and 10,000 iterations. For $K=4$, a higher number of iterations (50,000) is needed for convergence.

With $\alpha=2$, respective posterior mean total log-likelihoods, namely $\log[\sum_{k=1}^{K} p_k \phi(y \mid \mu_k, \sigma_k)]$, are –206.0, –205.8 and –206.5, with respective LOO-IC 421.7, 422.4 and 424.2. So there is little to separate these solutions in terms of fit. With $\alpha=4$, the posterior mean log-likelihoods are –206.4, –205.7 and –206.2, with respective LOO-IC being 421.2, 421.4 and 422.8. The rstan solutions generally show the lowest mean group with a mean lower than the minimum, namely 9.17, of the observed data points. This can be taken as generalising beyond the observed data.

We also implement jagsUI with the latent means constrained to lie between the minimum and maximum of the observations. MCMC convergence is focused on relabelled parameters, using the standard labelling approach set out above [1]. Convergence is problematic with independent priors on the precisions $\psi_k = 1/\sigma_k^2$ when $K>4$. Improved convergence is obtained if a hierarchical prior is adopted instead, namely $\psi_k \sim Ga(a_\psi, b_\psi)$, where a_ψ and b_ψ are assigned $E(1)$ priors. This is an intermediate option between independent priors and assuming the same variance across all groups. As noted by Baudry et al. (2010), the most appropriate number of mixture components may not guarantee well-separated groups. To assess cluster overlap, we use the entropy measure $-2\sum_i \sum_k d_{ik} \log(\rho_{ik})$ (Scrucca et al., 2016, p.297); another form, with effective numerical equivalence, is $-2\sum_i \sum_k \rho_{ik} \log(\rho_{ik})$.

For $K=4$, 5 and 6 respective posterior mean log-likelihoods are –205.6, –204.4, and –204.4, so BIC-type penalised fit measures (with respective penalties 48.5, 61.7, and 74.9) would favour K=4. Respective posterior mean entropies are 59, 87, and 109, so penalisation by entropy (Biernacki et al., 2000) would also decisively favour K=4. The LOO-IC measures also favour K=4, with the values for K=4, K=5, and K=6 being respectively 357, 370, and 376. A solution with K=3 was also run, which gave a mean log-likelihood of –216.4 and an entropy of 68.3. The estimated group means under K=4 are 9.7, 19.9, 22.4, and 28.0, with group probabilities 0.10, 0.33, 0.42, and 0.15 via jagsUI.

By comparison, mclust provides estimated means of 9.7, 19.8, 22.9, and 24.5 with respective probabilities 0.08, 0.39, 0.37, and 0.16, and bayesmix (Gruen and Plummer, 2015) provides estimated means of 10.3, 20.4, 22.5, and 30.5 with respective probabilities 0.09, 0.45, 0.39, and 0.06. The bayesmix run used the code:

```
variables <- c("mu","tau","eta")
M4<-BMMmodel(k=4,priors=list(kind="independence",parameter="priorsFi
sh",hierarchical="tau"))
C <- JAGScontrol(variables = c(variables, "S"), n.iter = 5000,burn.
in = 500)
R4 <- JAGSrun(y, model = M4, initialValues = list(S0 = 2),control =
C, cleanup = T, tmp = F)
Sort4 <- Sort(R4, "mu")
Sort4
```

Predictive checks for K=4 under a hierarchical prior for ψ_k show only one exceedance probabilities under 0.1 or over 0.9.

4.9 Semiparametric Modelling via Dirichlet Process and Polya Tree Priors

In applications of hierarchical models, inferences may depend on the assumed forms (e.g. normal, gamma) for higher stage priors, and will be distorted if there are unrecognised features such as multiple modes in the underlying second stage effects. Instead of assuming a known prior distribution G for second stage latent effects, such as b_i in the normal-normal model of Section 4.3, the Dirichlet process (DP) prior involves a distribution on G itself, so acknowledging uncertainty about its form (Carvalho and Branscum, 2017; Gill and Casella, 2009). The DP prior involves a baseline or base prior G_0, the expectation of G, and a precision or mass parameter α governing the concentration of the prior for G about its mean G_0. For any partition $A_1,\ldots,A_M$ on the support of G_0, the vector $\{G(A_1),\ldots,G(A_M)\}$ of probabilities $G(A_m)$ contained in the set $\{A_m, m=1,\ldots,M\}$ follows a Dirichlet distribution $D(\alpha G_0(A_1),\ldots,\alpha G_M(A_M))$. Such an approach may be termed semiparametric as it involves a parametric model at the first stage for the observations, but a non-parametric model at the second stage (Basu and Chib, 2003).

Original forms of the DP prior assumed G_0 to be known (fixed). One problem with a Dirichlet process when G_0 is known is that it assigns a probability of 1 to the space of discrete probability measures (Hanson et al., 2005, p.249). An alternative is to take the parameters in G_0 to be unknown, and to follow a set of parametric distributions, with possibly unknown hyperparameters, resulting in a mixture of Dirichlet process or MDP model (Walker et al., 1999, p.489). Computational procedures for such models are discussed by Jara (2007), Ohlssen et al. (2007), Jara et al. (2011), Burr (2012), Karabatsos (2016), Karabatsos (2017), with associated R packages including DPpackage (Jara et al., 2011), and bspmma (Burr, 2012).

Following West et al. (1994), assume conventional first-stage sampling densities $y_i \sim p(y_i \mid b_i, \psi)$, with distributions $P(y_i \mid b_i, \psi)$. The uncertainty about the appropriate form of prior arises about the distribution G for the latent effects b_i. Under a DP prior, any set of unitspecific parameters $\{b_1,\ldots,b_n\}$ generated from G lies in a set of $K \leq n$ distinct values $\{\zeta_1,\ldots,\zeta_K\}$ which are sampled from G_0. The concentration parameter α governing the closeness of G to G_0 can be taken as an unknown, or assigned a preset value (e.g. $\alpha=1$) (Da Silva, 2009). The number of distinct values or clusters K is stochastic, with an implicit prior determined by α, with limiting mean $\alpha\log(1+n/\alpha)$. Note that the posterior mean of K is not necessarily a reliable guide to the number of components in the data or effects (e.g. components with substantive meaning), though it can be interpreted as an upper bound on the number of components (Ishwaran and Zarepour, 2000, pp.381–382).

Given the realised number of clusters K (at any particular MCMC iteration), the b_i are sampled from the set $\{\zeta_1,\ldots,\zeta_K\}$ according to a multinomial distribution. Define cluster indicators $S=\{S_1,\ldots S_n\}$, where $S_i=k$ if $b_i=\zeta_k$, and denote $N_k = \#\{S_i = k\}$ as the total number of units with $S_i=k$ (i.e. units in the same cluster with a common value ζ_k for the second stage latent effect). If α is taken as unknown, its prior is important in determining the number of clusters. Taking $\alpha \sim Ga(\eta_1,\eta_2)$ where η_1 and η_2 are relatively large will tend to discourage unduly small or large values for α. Typical values are $\eta_1=\eta_2=1$ or $\eta_1=\eta_2=2$, though taking $\eta_2>\eta_1$ as in $\{\eta_1=2,\eta_2=4\}$ tends to encourage repetitions in the ζ_k, and can be used to assess the number of components present in the data (Ishwaran and Zarepour, 2000, p.377). It is clear that the parameters used in the prior for α may affect the number of components, but typically there is less concern with this aspect in non-parametric mixture modelling (Leslie et al., 2007).

Consider the assignment of a latent effect b_i to a particular unit, given that the remaining $n-1$ latent effects $b_{[i]}=\{b_1,\ldots,b_{i-1},b_{i+1},\ldots,b_n\}$ are already assigned. Also let $S_{[i]}$ be a particular configuration of the remaining $n-1$ effects $b_{[i]}$ into $K_{[i]}$ distinct values, with $N_{[i]k} = \#\{S_j = k, j \neq i\}$ denoting the total of those $n-1$ units having a common value $\zeta_{[i]k}$. Then the conditional prior for b_i follows a Polya urn scheme (West et al., 1994; Hanson et al., 2005, p.252; Dunson et al., 2007, p.165)

$$(b_i \mid b_{[i]}, S_{[i]}, K_{[i]}, \alpha) \sim \frac{\alpha}{\alpha+n-1} G_0 + \frac{1}{\alpha+n-1}\sum_{k \neq i} \delta(b_k),$$

$$\sim \frac{a}{a+n-1} G_0 + \frac{1}{a+n-1} \sum_{k=1}^{K_{[i]}} N_{[i]k} \delta(\zeta_{[i]k}), \quad (4.5)$$

where $\delta(u)$ denotes a degenerate distribution having a single value at u. So b_i is distinct from the remaining latent values with probability $a/(a+n-1)$, in which case it is drawn from the base prior G_0. Alternatively, it is selected from the existing distinct effects $\zeta_{[i]k}$ according to a multinomial with probabilities proportional to $N_{[i]k}/(a+n-1)$. This selection scheme extends to the predictive scenario i.e. to the latent effect b_{n+1} for a hypothetical new unit $n+1$, with

$$(b_{n+1} \mid b, S, K, a) \sim \frac{a}{a+n} G_0 + \frac{1}{a+n} \sum_{k=1}^{K} N_k \delta(\zeta_k).$$

Predictions of the first stage response for unit $n+1$ are obtained as

$$(y_{n+1} \mid b, S, K, a) \sim \frac{a}{a+n} P_{n+1}(\mid \zeta_{n+1}) + \frac{1}{a+n} \sum_{k=1}^{K} N_k P_{n+1}(\mid \zeta_k),$$

where ζ_{n+1} is an extra draw from G_0. Predictions beyond $n+1$ may be relevant in panel or time series applications (Hirano, 1998).

In terms of Gibbs sampling, (4.5) implies conditional posteriors (West et al., 1994, p.367; Ishwaran and James, 2001, p.166)

$$(b_i \mid y, b_{[i]}, S_{[i]}, K_{[i]}, a) \sim a q_{i0} g_0(b_i \mid y) p(y_i \mid b_i) + \sum_{k=1}^{K_{[i]}} q_{ik} \delta(\zeta_{[i]k}),$$

where $g_0(b_i \mid y)$ is the density corresponding to G_0 evaluated at b_i, and where

$$q_{i0} = \int p(y_i \mid b_i) g_0(b_i) db_i \quad (4.6.1)$$

$$q_{ik} = N_{[i]k} p(y_i \mid \zeta_{[i]k}) \quad k > 0 \quad (4.6.2)$$

Normalising the values aq_{i0} and q_{ik} to probabilities $\{r_{i0}, r_{i1}, \ldots r_{iK_{[i]}}\}$ summing to 1, the conditional posteriors for the subgroup indicators are then

$$Pr(S_i = k \mid y, b_{[i]}, S_{[i]}, K_{[i]}) = r_{ik}$$

where $S_i = 0$ corresponds to drawing a new sample from G_0 under the Polya urn scheme.

4.9.1 Specifying the Baseline Density

An important aspect of the MDP framework is the specification of G_0. Assume there are p parameters $(\psi_1, \ldots, \psi_p)$ in G_0, then one has

$$y_i \mid b_i \sim p(y_i \mid b_i),$$

$$b_1, \ldots b_n \mid G,$$

$$G \mid a, G_0 \sim DP(\alpha G_0),$$

$$G_0 = \{p_{01}(\psi_1 \mid \xi_1), \ldots p_{0p}(\psi_p \mid \xi_p)\},$$

where $\{\psi_1, \ldots, \psi_p\}$, are unknown, and also possibly some of the defining ξ parameters. Consider a normal mixture with both means and variances possibly differing for each unit (Cao and West, 1996; Hirano, 2002), namely

$$y_i \sim N(\mu_i, \sigma_i^2).$$

The appropriate prior G for $b_i = (\mu_i, \sigma_i^2)$ is not certain, and so

$$(\mu_i, \sigma_i^2) \sim G,$$

$$G \sim DP(\alpha G_0),$$

where G_0 involves the priors

$$\mu_i \sim p_{01}(\mu_i \mid \xi_1),$$

$$\sigma_i^2 \sim p_{02}(\sigma_i^2 \mid \xi_2),$$

with ξ_1 and ξ_2 possibly including further unknowns. For example, Hirano (2002) takes

$$1/\sigma_i^2 \sim \chi^2(s)/(sQ),$$

and

$$\mu_i \sim N(m, c\sigma_i^2),$$

where s, Q, m and c are specified, but may be varied in a sensitivity analysis.

The marginal distribution of the y_i (averaged over all possible G) in this case is a mixture of normal distributions, with the number of subgroups K randomly varying between 1 and n. The n unit specific parameter pairs $b_i = (\mu_i, \sigma_i^2)$ are selected under G from the set of $K_{[i]}$ possible values $\zeta_k = (\mu_k, \sigma_k^2)$ already drawn from G_0, or by fresh sampling from G_0. The q_{ih} in (4.6) are then obtained as

$$q_{i0} = \int \frac{1}{\sigma_i \sqrt{2\pi}} e^{-(y_i - \mu_i)^2 / 2\sigma_i^2} g_0(\mu_i, \sigma_i^2) d\mu_i d\sigma_i^2,$$

$$q_{ik} = N_{[i]k} \frac{1}{\sigma_k \sqrt{2\pi}} e^{-(y_i - \mu_k)^2 / 2\sigma_k^2} \quad k > 0.$$

As other examples, Chib and Hamilton (2002) consider a potential outcomes model for panel data with DP errors, while Kleinman and Ibrahim (1998) consider Gibbs updates in an MDP framework for parameters in general linear mixed models for nested data. For example, let X_i and Z_i be predictors of dimension q and r (possibly overlapping), and consider repeated data y_{it} over subjects i, with observation vectors $y_i = (y_{i1}, \ldots y_{iT})$, and first stage model

$$y_i \sim N(X_i\beta + Z_i b_i, \sigma^2),$$

where one may assume conventional normal and inverse gamma priors for β and σ^2. However, for $b_i = (b_{i1}, \ldots b_{ir})$, greater flexibility is obtained by taking

$$b_i \sim G,$$

$$G \sim DP(a, G_0),$$

where G_0 is multivariate normal of dimension r, with mean 0, but unknown covariance D. The Wishart distribution in the Gibbs update for D^{-1} is modified for clustering of values among the sampled b_i (Kleinman and Ibrahim, 1998, p.94).

4.9.2 Truncated Dirichlet Processes and Stick-Breaking Priors

Implementation may be simplified if an alternative way to generate the DP prior is adopted. The basis of this alternative scheme is to regard the density of the unit level effects b_i as an infinite mixture of point masses or continuous densities (Ohlssen et al., 2007; Hirano, 1998), with

$$b_i \sim \sum_{k=1}^{\infty} \pi_k h(b_i \mid \psi_k).$$

This approach is called a Dirichlet process mixture by Hanson et al. (2005, p.250), and a dependent Dirichlet process by Dunson et al. (2007, p.164). For practical application, Ishwaran and Zarepour (2000) and Ishwaran and James (2002) suggest the infinite representation be approximated by one truncated at $M \leq n$ components with

$$g(b) = \sum_{m=1}^{M} \pi_m h(b \mid \psi_m),$$

where the π_m are sampled by introducing M − 1 beta distributed random variables,

$$V_m \sim Be(c_m, d_m),$$

with $V_M = 1$ to ensure the random weights π_m sum to 1 (Ishwaran and James, 2001; Sethuraman, 1994). Then

$$\pi_1 = V_1,$$

$$\pi_m = (1 - V_1)(1 - V_2)\ldots(1 - V_{m-1})V_m \quad m > 1.$$

This method of generation is known as stick-breaking, since at each stage, the procedure randomly breaks what is left of a stick of unit length and assigns the length of the break to the current π_m. Griffin (2016) proposes an adaptive technique for selecting the truncation point in truncated DP priors. Recent applications include Prabhakaran et al. (2016) and Hu et al. (2018). It may be noted that rstan can use the TDP principle to estimate mixtures, but taking M as a known rather than maximum number of components [2].

Following Pitman and Yor (1997), the beta parameters $\{c_m, d_m\}$ in the prior for V_m can be written $c_m = 1 - C$, $d_m = D + mC$, where $C \in [0,1)$ and $D > -C$. For an infinite dimensional mixture, the Dirichlet process is obtained by taking $C = 0$ and $D = \alpha$, so that $V_m \sim Be(1, \alpha)$. When a finite (truncated) mixture is used, setting

$$c_m = 1 + \frac{a}{M},$$

$$d_m = a - \frac{ma}{M} = a\left(1 - \frac{m}{M}\right)$$

is asymptotically equivalent to the DP process (Ishwaran and Zarepour, 2002; Taylor-Rodriguez et al., 2017).

However, using an approximate DP scheme with

$$V_m \sim Be(1, a)$$

and M large is equivalent to the infinite DP process for practical purposes (Ishwaran and James, 2002; Ishwaran and Zarepour, 2000, p.383). If a $Ga(\eta_1, \eta_2)$ prior is used for α, its full conditional is $a \sim Ga(M + \eta_1 - 1, \eta_2 - \log(\pi_M))$ (Ishwaran and Zarepour, 2000, p.387). The realised number of clusters is $K \leq M$ as above, and (Ishwaran and James, 2002) suggest AIC and BIC penalties based on K that can be used for model selection.

Taking $V_m \sim Be(a, 1)$ rather than $V_m \sim Be(1, a)$ in the truncated stick-breaking scheme means that larger values of α now imply greater clustering into a few sub-populations. This is an example of the beta process priors considered by Ishwaran and Zarepour (2000). Other truncated mixture sampling schemes that start with a prior on α to give an implicit prior on a stochastic K are available. For example, Ishwaran and Zarepour (2000, p.376) consider taking α as an unknown in

$$(\pi_1, \ldots \pi_M) \sim D\left(\frac{a}{M}, \frac{a}{M}, \ldots, \frac{a}{M}\right).$$

Alternatively, Green and Richardson (2001, p.357) start off with a prior on K and then select the cluster indicators from a multinomial vector with probabilities $p(S_i = k) = \pi_i$, where $(\pi_1, \ldots, \pi_K)$ follow a Dirichlet density $D(\delta, \ldots, \delta)$. They refer to this as an explicit allocation prior and show how the DP prior is obtained as $K \to \infty$ and $\delta \to 0$ in such a way that $K\delta \to a > 0$.

4.9.3 Polya Tree Priors

The Polya tree is a more general class than the Dirichlet process, and has the benefit that it can place probability 1 on the space of continuous densities (Hanson et al., 2005; Walker et al., 1999). In essence, if the support of a parameter ω is denoted Γ, then the Polya Tree (PT) prior chooses the most appropriate value for ω by successive binary partitioning of Γ. The first partition splits Γ into 2 disjoint sets $\{B_0, B_1\}$; the probabilities of moving into B_0 and B_1 are C_{00} and $C_{01} = 1 - C_{00}$, with C_{00} set to 0.5. At the second partition B_0 is split into $\{B_{00}, B_{01}\}$ and B_1 is split into $\{B_{10}, B_{11}\}$ so there are 2^2 sets. At the third partition, B_{00} is split into $\{B_{000}, B_{001}\}$, B_{01} into $\{B_{010}, B_{011}\}$, B_{10} into $\{B_{100}, B_{101}\}$, and B_{11} into $\{B_{110}, B_{111}\}$, so there are 2^3 sets. The number of sets at the mth partition is generally 2^m.

The partition probabilities at second and subsequent stages are unknown. Let ε denote a sequence of 0s and 1s. For example, suppose B_1 is selected at step 1, and B_{11} is selected at step 2, then $\varepsilon=[1,1]$. The choice at the next stage between sets $B_{\varepsilon 0}$ and $B_{\varepsilon 1}$ (i.e. between B_{110} and B_{111}) is governed by probabilities $(C_{\varepsilon 0}, C_{\varepsilon 1})$, with a beta prior for $C_{\varepsilon 0}$, and $C_{\varepsilon 1} = 1 - C_{\varepsilon 0}$. The canonical form for the prior on the partition probabilities at partition m is

$$C_{\varepsilon 0} \sim Be(c_m, c_m)$$

$$c_m = dm^2$$

where d may be taken as an extra unknown. The Dirichlet process occurs when $c_m = d/2^m$, so that $c_m \to 0$ as $m \to \infty$, whereas $c_m \to \infty$ as $m \to \infty$ is appropriate if the underlying distribution G is expected to be continuous.

While theoretically the completely continuous case corresponds to $m \to \infty$, in practice the partitioning is truncated at a finite value M. Hanson and Johnson (2002) recommend $M = \log_2(n)$ where n is the sample size. The partitions can be taken to coincide with percentiles of G_0, so for example

$$B_0 = (-\infty, G_0^{-1}(0.5)], \quad B_1 = [G_0^{-1}(0.5), \infty);$$

$$B_{00} = (-\infty, G_0^{-1}(0.25)], \quad B_{01} = [G_0^{-1}(0.25), G_0^{-1}(0.5)];$$

$$B_{10} = [G_0^{-1}(0.5), G_0^{-1}(0.75)], \quad B_{11} = [G_0^{-1}(0.75), \infty);$$

and so on.

Let d_{ki} at partition k, and option i, be a re-expression of the B_ε (e.g. for $k=3$, $d_{31}=B_{000}$, $d_{32}=B_{001}$, $d_{33}=B_{010}$, $d_{34}=B_{011}$, $d_{35}=B_{100}$, $d_{36}=B_{101}$, $d_{37}=B_{110}$, $d_{38}=B_{111}$). Then at partition k, for $i = 1, \ldots 2^k$, the interval boundaries are

$$d_{ki} = \left[G_0^{-1}\left(\frac{i-1}{2^k}\right), G_0^{-1}\left(\frac{i}{2^k}\right) \right],$$

with appropriate modifications for the extreme tails.

For example, consider a PT prior on unstructured errors in a Poisson lognormal mixture, with

$$y_i \sim Po(\mu_i),$$

$$\log(\mu_i) = \beta + \sigma b_i.$$

Then G_0 for $v_i = \sigma b_i$ is a $N(0, \sigma^2)$ density, with G_0 for b_i being a $N(0,1)$ density. So with $M=3$ levels, the relevant ordinates from G_0 for defining the 8 intervals are (–1.15,–0.67,–0.32,0,0.32,0.67,1.15).

Example 4.9 Nicotine Replacement Therapy

We re-analyse the NRT trials data using truncated DP priors (Section 4.9.2). Thus, second stage random trial effects are obtained as $b_i = \zeta_k$ conditional on latent group indicator $S_i=k$, and with $\{\zeta_1,\ldots,\zeta_M\}$ sampled from G_0. The realised number of clusters is $K \le M$, where a maximum of $M=50$ possible normal clusters $N(\mu_m,\tau_m^2)$ are assumed as potential second stage priors. The M potential parameter pairs $\{\mu_m,\tau_m^2\}$ defining G_0 are respectively sampled from normal densities with means $\mu_m \sim N(m_\mu,1)$, where m_μ is itself unknown, and from exponential densities, with $1/\tau_m^2 \sim E(1)$.

Then $V_m \sim Be(1,\alpha)$, with an exponential $E(1)$ prior assumed on the concentration parameter α, and with a lower sampling limit of 0.25 for numeric stability. A mixed predictive check is based on sampling replicate $\{\zeta_{\text{rep},1},\ldots,\zeta_{\text{rep},M}\}$ from G_0, and taking $b_{\text{rep},i}=\zeta_{\text{rep},k}$.

A two-chain run of 5,000 iterations using rube shows convergence in α, K, and the realised latent effects b. The posterior mean and median of K are respectively 3.9 and 4, supporting a relatively small number of components in the second-stage prior of NRT effects; α has a posterior mean of 0.85. Mixed predictive checks are satisfactory, with none exceeding 0.9 or being under 0.1.

A plot of the posterior means of the b_i does not show sharply distinct subgroups (Figure 4.1), though outlier random effects can be seen, such as trials 4, 36, and 59. However, the effects show more peakedness than under a normal density (superimposed plot).

The analysis is also run using a Pitman–Yor prior, with $V_m \sim Be(1-C,D+mC)$, where $C\in[0,1)$ and $D>-C$, and with a maximum of $M=20$ clusters. This is implemented using R2OpenBUGS with a two-chain run of 20,000 iterations. A uniform $U(0,1)$ prior is adopted on C, with D obtained as $D=D_1-C$, where $D_1 \sim Ga(1,0.01)$ is assigned a gamma prior. This analysis provides posterior means (sd) for C and D of 0.55 (0.25) and 132 (101), with the mean number of clusters being 4.7. Posterior means for b_i are similar to those of the first analysis, the correlation between them exceeding 0.95, while exceedance probabilities again show no model failure.

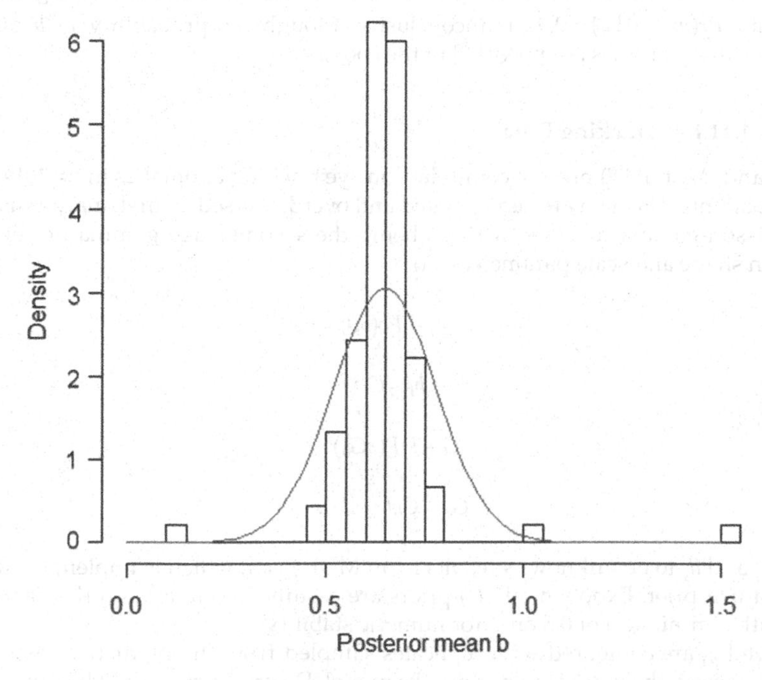

FIGURE 4.1
Nicotine replacement. Estimated random effects.

Example 4.10 Digestive Tract Decontamination

Data on log odds ratios y_i and their variances s_i^2 in $n=14$ trials are considered by Burr and Doss (2005), and relate to mortality after treatment vs control comparison for decontamination of the digestive tract. Assumptions under the normal-normal model (equations 4.4.1 to 4.4.3) are cast into doubt by quantile plots of the y_i. We consider a truncated DP prior (with $M=n=14$), with second-stage effects $b_i=\zeta_k$ when allocation indicators $S_i=k$, and

$$\zeta_k \sim N(\mu_k, \tau^2),$$

with $\{\mu_1,\ldots,\mu_M\}$, themselves sampled from a normal density

$$\mu_m \sim N(m_\mu, \tau_\mu^2).$$

Both m_μ and τ_μ^2 are unknowns, assigned N(0,100) and Ga(1,0.01) priors. The second-stage variance parameter τ^2 is also assigned a $Ga(1,0.01)$ prior.

Analysis compares the DPMmeta option in the R library DPpackage, and a BUGS code estimated using rube in R, with $V_m \sim Be\left(1+\frac{\alpha}{M}, \alpha\left[1-\frac{m}{M}\right]\right)$ in the stick-breaking prior. Either computing option suggests α is not strongly identified by the data: alternative settings for a_0 and b_0 in $\alpha \sim Ga(a_0,b_0)$ tend to carry over to the estimated α. So alternative preset values such as $\alpha=1$, $\alpha=10$, etc. may be adopted instead (Burr, 2012).

With the setting $\alpha=10$, DPMmeta shows a mean of around 9 realised clusters, as against 6.8 under the TDP prior. Treatment benefit can be measured by the probability that m_μ is negative, or the probability that the mean of the realised b_i is negative. The probability $Pr(m_\mu < 0 \mid y) = 0.92$ is inconclusive, though the probability $Pr(\bar{b} < 0 \mid y)$ is 0.97 (their two quantities are pben[1:2] in the code).

Example 4.11 Eye Tracking Data

Escobar and West (1998) present count data on eye tracking anomalies in $n=101$ schizophrenic patients. The data are zero-inflated and overdispersed. A first analysis assumes a DP Poisson-gamma mixture, with G_0 being the second-stage gamma density with unknown shape and scale parameters. So

$$y_i \sim Po(b_i),$$

$$b_i \sim G,$$

$$G \sim DP(\alpha G_0),$$

$$G_0 = Ga(c_g, d_g).$$

Taking c_g and d_g to be unknowns results in an MDP prior, which is implemented using the Polya urn prior. Exponential $E(1)$ priors are assumed on α, and on the parameters (c_g,d_g), with a minimum of 0.5 on c_g for numeric stability.

Observed y_i are compared with replicates sampled from the predictive distribution $p(y_{\text{rep}} \mid y)$ to see if the y_i are at odds with the model. Discrepancies could be due to genuine outlier status, or to model failures. For discrete data, the relevant p-value is

$$Pr(y_{\text{rep},i} < y_i) + 0.5Pr(y_{\text{rep},i} = y_i).$$

A related check is whether the 95% intervals for $y_{\text{rep},i}$ include y_i (Gelfand, 1996).

A two-chain run of 20,000 iterations in R2OPENBUGS provides an estimated mean of $K=12$ clusters, with posterior means (95% CRI) for α, c_g, and d_g of 2.83 (0.74,6.45), 0.82 (0.51,1.61), and 0.13 (0.04,0.29). Figure 4.2 shows the prediction y_{new} for a new case, and demonstrates that the main source of overdispersion is skewness in the latent frailties b_i rather than multiple modes. The predictive checks based on replicate samples are satisfactory. Note that the same does not apply if the gamma mixing density parameters are set, e.g. $c_g = d_g = 1$. In this case, bimodal posteriors are obtained on some b_i (e.g. b_{92}), and predictive checks for $y_{101} = 34$ suggest it to be an extreme observation.

A second analysis involves a Polya tree prior, and a Poisson-lognormal model, namely

$$y_i \sim Po(\mu_i),$$

$$\log(\mu_i) = \beta + \sigma b_i$$

where G_0 for $v_i = \sigma b_i$ is a $N(0,\sigma^2)$ density. The number of stages is set at $M=4$, and an $E(1)$ prior is assumed on $1/\sigma^2$. Once an interval B_{em} is selected, uniform sampling to generate b_i takes place within the interval defined by G_0, except in the tails where the sampling is from a $N(0,1)$.

As for the Polya urn model, both types of predictive check indicate no major discrepancies. σ has posterior mean (and 95% interval) 2.06 (1.65, 2.51). If σ is taken to equal 1 so that G_0 is assumed known, then predictive discrepancies do occur. Taking $\sigma=1$ also leads to bimodal posteriors for individual b_i indicating a clash between prior and data, such that the prior cannot accommodate certain values. A plot of the estimated b_i shows the distinct zero inflation combined with positive skew (Figure 4.3).

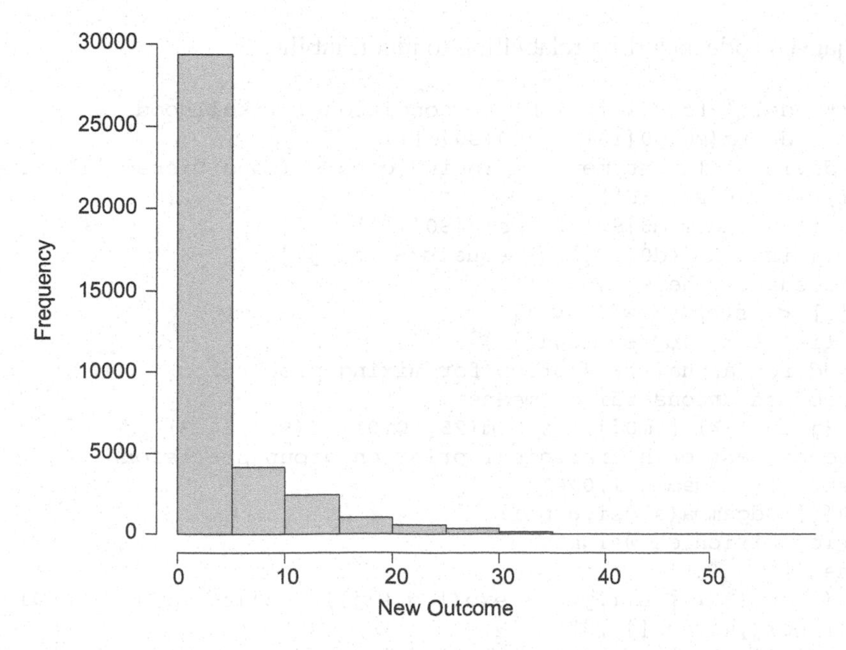

FIGURE 4.2
Predictive samples, new outcome, eye tracking data.

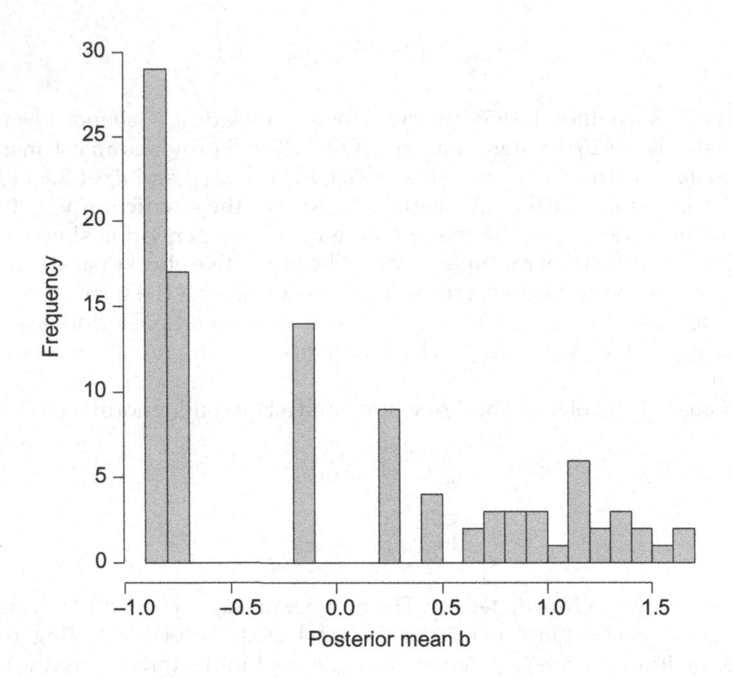

FIGURE 4.3
Estimated random effects, eye tracking data.

4.10 Computational Notes

1. The jagsUI code including relabelling to identifiability is

```
cat("model { for (i in 1:N){ # conditional likelihood
y[i] ~ dnorm(m0[S0[i]], psi0[S0[i]])
# individual latent membership indicators (conditional likelihood)
S0[i] ~ dcat(P0[1:K])
ynew[i] ~ dnorm(m0[S0[i]],psi0[S0[i]])
for (j in 1:K) {d0[i,j] <- equals(S0[i],j)}
# exceedance check
exc[i] <- step(ynew[i]-y[i])
log_lik[i] <- log(sum(L[i,]))}
P0 ~ddirch(alpha[]); # prior for mixing proportion
# prior on unconstrained means
for (j in 1:K) { m0[j] ~dnorm(25, 0.01) T(9.2,34.3)
# independent or hierarchical prior on group precisions
# psi0[j] ~ dgamma(1,0.01)
psi0[j] ~dgamma(a.psi,b.psi)
# prior Dirichlet weights
alpha[j] <- 2.5
for (i in 1:N) { L[i,j] <- exp(log(P[j])+0.5*log(psi[j])-0.919-0.5*p
si[j]*pow(y[i]-mu[j],2))
# conditional allocation probabilities
rho[i,j] <- L[i,j]/sum(L[i,])
```

```
# entropy
ent1[i,j] <- equals(S[i],j)*log(rho[i,j])
ent2[i,j] <- rho[i,j]*log(rho[i,j])}}
Ent[1] <- -2*sum(ent1[1:N,1:K])
Ent[2] <- -2*sum(ent2[1:N,1:K])
tLL <- sum(log_lik[])
# hyperparameters, hierarchical prior on precisions
a.psi ~dexp(1)
b.psi ~dexp(1)
# Processing to obtain identifiable groups, using ranks of
unconstrained means
rank <- rank(m0)
# relabelled weights, means, precisions, variances
for (j in 1:K) {P[j] <- sum(P0prod[j,])
mu[j] <- sum(m0prod[j,])
psi[j] <- sum(psi0prod[j,])
s2[j] <- 1/psi[j]
for (k in 1:K) {P0prod[j,k] <- P0[k]*equals(rank[k],j)
m0prod[j,k] <- m0[k]* equals(rank[k],j)
psi0prod[j,k] <- psi0[k]* equals(rank[k],j)}}
# relabelled allocation indicators
for (i in 1:N) { S[i] <- sum(dcat[i,])
for (j in 1:K) {d[i,j] <- sum(d0prod[i,j,])
dcat[i,j] <- j*d[i,j]
for (k in 1:K) {d0prod[i,j,k] <- d0[i,k]*equals(rank[k],j)}}}}
", file="mixnorm.jag")
```

2. Consider the galaxy data and suppose M=4 is the number of mixture components. Unknown means are centred at the observed mean of the data. Then a truncated DP prior can be implemented as

```
stan_model <- "
data{
int<lower=0> M;// number of clusters
int<lower=0> N;// number of observations
real y[N];
}
parameters {
positive_ordered[M]mu; //cluster means
real <lower=0,upper=1> v[M];
real<lower=0> sigma[M]; // cluster scales
real<lower=0> alpha; // concentration parameter
}
transformed parameters{
simplex [M] pi;
pi[1] = v[1];
// stick-break process
for(j in 2:(M-1)){ pi[j]= v[j]*(1-v[j-1])*pi[j-1]/v[j-1]; }
pi[M]=1-sum(pi[1:(M-1)]);
}
model { real comp[M];
sigma ~exponential(1);
alpha ~ gamma(2,4);
mu ~ normal(21,5);
```

```
v ~ beta(1,alpha);
for(i in 1:N){ for(c in 1:M){
comp[c]=log(pi[c])+normal_lpdf(y[i]mu[c],sigma[c]); }
target += log_sum_exp(comp); }}
"
D=list(y=y,N=82,M=4)
fit = stan(model_code = stan_model, data =D, iter = 2000, chains = 2)
summary(fit,pars = c("mu","pi","alpha"),probs=
c(0.025,0.975))$summary
```

The estimated parameters are in Table 4.1 and are similar to those estimated in Example 4.8.

TABLE 4.1

Galaxy Data Discrete Mixture, Galaxy Data, TDP Prior

	Mean	St devn	2.5%	97.5%
μ_1	9.76	0.24	9.33	10.33
μ_2	20.02	0.54	19.42	21.48
μ_3	22.45	1.08	21.20	26.01
μ_4	28.65	4.18	22.59	33.96
π_1	0.09	0.03	0.04	0.16
π_2	0.36	0.20	0.08	0.86
π_3	0.44	0.23	0.02	0.79
π_4	0.12	0.11	0.01	0.39
α	0.78	0.36	0.23	1.61

References

Abrams K, Gillies C, Lambert P (2005) Meta-analysis of heterogeneously reported trials assessing change from baseline. *Statistics in Medicine*, 24, 3823–3844.

Abrams K, Lambert P, Sanso B, Shaw S, Marteau T (2000) Meta-analysis of heterogeneously reported study results: A Bayesian approach, pp 29–64, in *Meta-Analysis in Medicine and Health Policy*, eds D Berry, D Stangl. Marcel Dekker.

Agresti A, Hitchcock D (2005) Bayesian inference for categorical data analysis. *Statistical Methods and Applications*, 14, 297–330.

Aitchison J, Ho C (1989) The multivariate Poisson-log normal distribution. *Biometrika*, 76, 643–653.

Aitchison J, Shen S (1980) Logistic-normal distributions: Some properties and uses. *Biometrika*, 67, 261–272.

Alanko T, Duffy J (1996) Compound Binomial distributions for modeling consumption data. *The Statistician*, 45, 269–286.

Albert J (1999) Criticism of a hierarchical model using Bayes factors. *Statistics in Medicine*, 18, 287–305.

Albert J (2015) Package 'LearnBayes': Functions for Learning Bayesian Inference. https://cran.r-project.org/web/packages/LearnBayes/LearnBayes.pdf

Albert JH, Gupta AK (1982) Mixtures of Dirichlet distributions and estimation in contingency tables. *The Annals of Statistics*, 10(4), 1261–1268.

Ando T (2007) Bayesian predictive information criterion for the evaluation of hierarchical Bayesian and empirical Bayes models. *Biometrika*, 94, 443–458.

Arends L (2006) Multivariate meta-analysis: Modelling the heterogeneity. Repub/EUR Repository. http://repub.eur.nl/publications/med_hea

Azzalini A (1985) A class of distributions which includes the normal ones. *Scandinavian Journal of Statistics*, 12, 171–178.
Bakbergenuly I, Kulinskaya E (2017) Beta-binomial model for meta-analysis of odds ratios. *Statistics in Medicine*, 36, 1715–1734.
Baker R, Jackson D (2008) A new approach to outliers in meta-analysis. *Health Care Management Science*, 11(2), 121–131.
Baker R, Jackson D (2016) New models for describing outliers in meta-analysis. *Research Synthesis Methods*, 7, 314–328.
Barnard J, McCulloch R, Meng XL (2000) Modeling covariance matrices in terms of standarddeviations and correlations, with applications to shrinkage. *Statistica Sinica*, 10, 1281–311.
Basu S (1996) Bayesian tests for unimodality, pp 77–82, in *Proceedings of the Section on Bayesian Statistical Science*. American Statistical Association.
Basu S, Chib S (2003) Marginal likelihood and Bayes factors for Dirichlet process mixture models. *Journal of the American Statistical Association*, 98(461), 224–235.
Baudry J, Raftery A, Celeux G, Lo K, Gottardo R (2010) Combining mixture components for clustering. *Journal of Computational and Graphical Statistics*, 19(2), 332–353.
Bayman E, Chaloner K, Hindman B, Todd M (2013) Bayesian methods to determine performance differences and to quantify variability among centers in multi-center trials: The IHAST trial. *BMC Medical Research Methodology*, 13, 5.
Beath K (2014) A finite mixture method for outlier detection and robustness in meta-analysis. *Research Synthesis Methods*, 5(4), 285–293.
Beath K (2016) metaplus: An R package for the analysis of robust meta-analysis and meta-regression. *The R Journal*, 8(1), 5–16.
Besag J, Green P, Higdon D, Mengerson K (1995) Bayesian computation and stochastic systems. *Statistical Science*, 10(1), 103–166.
Betancourt M (2017) Identifying Bayesian Mixture Models. http://mc-stan.org/users/documentation/case-studies/identifying_mixture_models.html
Biernacki C, Celeux G, Govaert G (2000) Assessing a mixture model for clustering with the integrated completed likelihood. *IEEE Transactions on Pattern Analysis and Machine Intelligence*, 22(7), 719–725.
Bohning D (1999) *Computer-Assisted Analysis of Mixtures and Applications: Meta-Analysis, Disease Mapping and Others*. Chapman & Hall, New York.
Browne W, Draper D (2006) A comparison of Bayesian and likelihood-based methods for fitting multilevel models. *Bayesian Analysis*, 1, 473–550.
Bulmer M (1974) On fitting the Poisson log-normal distribution to species abundance data. *Biometrics*, 30, 101–110.
Burke D, Bujkiewicz S, Riley R (2016) Bayesian bivariate meta-analysis of correlated effects: Impact of the prior distributions on the between-study correlation, borrowing of strength, and joint inferences. *Statistical Methods in Medical Research*, 27(2), 428–450.
Burr D (2012) bspmma: An R package for Bayesian semiparametric models for meta analysis. *Journal of Statistical Software*, 50, 1–23.
Burr D, Doss H (2005) A Bayesian semi-parametric model for random effects meta analysis. *Journal of the American Statistical Association*, 100, 242–251.
Cao G, West M (1996) Practical Bayesian inference using mixtures of mixtures. *Biometrics*, 52, 1334–1341.
Carvalho, V, Branscum, A (2017) Bayesian nonparametric inference for the three-class Youden index and its associated optimal cutoff points. *Statistical Methods in Medical Research*, 27, 689–700.
Celeux G, Hurn M, Robert C (2000) Computational and inferential difficulties with mixture posterior distributions. *Journal of the American Statistical Association*, 95, 957–970.
Cepeda-Benito A, Reynoso N, Erath S (2004) Meta-analysis of the efficacy of nicotine replacement therapy for smoking cessation: Differences between men and women. *Journal of Consulting and Clinical Psychology*, 72, 712–722.
Chelgren N, Adams M, Bailey L, Bury, B (2011) Using multilevel spatial models to understand salamander site occupancy patterns after wildfire. *Ecology*, 92, 408–421.

Chib S, Hamilton B (2002) Semiparametric Bayes analysis of longitudinal data treatment models. *Journal of Econometrics*, 110(1), 67–89.

Chib S, Winkelmann R (2001) Markov chain Monte Carlo analysis of correlated count data. *Journal of Business & Economic Statistics*, 19(4), 428–435.

Christiansen C, Morris C (1996) Fitting and checking a two-level Poisson model: modeling patient mortality rates in heart transplant patients, pp 467–501, in *Bayesian Biostatistics*, eds D Berry, D Stangl. Marcel Dekker, New York.

Christiansen C, Morris C (1997) Hierarchical Poisson regression modeling. *Journal of the American Statistical Association*, 92, 618–632.

Chung H, Loken E, Schafer J (2004) Difficulties in drawing inferences with finite-mixture models: A simple example. *The American Statistician*, 58, 152–158.

Clark J, Gelfand A (2006) *Hierarchical Modelling for the Environmental Sciences: Statistical Methods and Applications*. Oxford University Press.

Clayton D, Kaldor J (1987) Empirical Bayes estimates of age-standardized relative risks for use in disease mapping. *Biometrics*, 43(3), 671–681.

Conlon E, Song J, Liu A (2007) Bayesian meta-analysis models for microarray data: A comparative study. *BMC Bioinformatics*, 8, 80.

Connolly S, Dornelas M, Bellwood D, Hughes T (2009) Testing species abundance models: A new bootstrap approach applied to Indo-Pacific coral reefs. *Ecology*, 90(11), 3138–3149.

Consul P (1989) *Generalized Poisson Distributions*. Marcel Dekker, New York.

Daniels M (1999) A prior for the variance in hierarchical models. *Canadian Journal of Statistics*, 27, 569–580.

Das S, Dey D (2006) On Bayesian analysis of generalized linear models using the Jacobian technique. *The American Statistician*, 60, 264–268.

Das S, Dey D (2007) On Bayesian analysis of generalized linear models: A new perspective. Technical Report 2007-8, Statistical and Applied Mathematical Sciences Institute, UNC. www.samsi.info

Da Silva, A (2009) Bayesian mixture models of variable dimension for image segmentation. *Computer Methods and Programs in Biomedicine*, 94(1), 1–14.

Deely N, Smith A (1998) Quantitative refinements for comparisons of institutional performance. *Journal of the Royal Statistical Society: Series A*, 161, 5–12.

Delucchi K, Bostrom A (2004) Methods for analysis of skewed data distributions in psychiatric clinical studies: Working with many zero values. *The American Journal of Psychiatry*, 161, 1159–1168.

DerSimonian R, Laird N (1986) Meta-analysis in clinical trials. *Controlled Clinical Trials*, 7, 177–188.

Diebolt N, Robert C (1994) Estimation of finite mixture distributions through Bayesian sampling. *Journal of the Royal Statistical Society: Series B*, 56, 363–375.

Ding T, Baio G (2016) bmeta: Bayesian Meta-analysis and Metaregression. http://www.statistica.it/gianluca/software/bmeta/

Diserud O, Engen S (2000) A general and dynamic species abundance model, embracing the lognormal and the gamma models. *The American Naturalist*, 155, 497–511.

Drton M, Plummer M (2017) A Bayesian information criterion for singular models. *Journal of the Royal Statistical Society: Series B*, 79(2), 323–380.

Druyts E, Palmer J, Balijepalli C, Chan K, Fazeli M, Herrera V (2017) Treatment modifying factors of biologics for psoriatic arthritis: A systematic review and Bayesian meta-regression. *Clinical and Experimental Rheumatology*, 35(4), 681–688.

DuMouchel W (1996) Predictive cross-validation of Bayesian meta-analyses, pp 107–127, in eds J Bernardo, J Berger, A Dawid, A Smith, *Bayesian Statistics 5*. Oxford University Press.

DuMouchel W, Waternaux C (1992) Discussion of "Hierarchical models for combining information and for meta-analysis," by C Morris and S Normand, pp 338–341, in *Bayesian Statistics*, Vol. 4, eds J Bernardo, J Berger, A Dawid, A Smith. Clarendon Press, Oxford, UK.

Dunson D, Pillai N, Park J (2007) Bayesian density regression. *Journal of the Royal Statistical Society: Series B*, 69, 163–183.

Efron B (1986) Double exponential families and their use in generalized linear regression. *Journal of the American Statistical Association*, 81(395), 709–721.

Escobar M, West, M (1998) Computing nonparametric hierarchical models, in *Practical Nonparametric and Semiparametric Bayesian Statistics*, eds D Dey, P Muller, D Sinha. Springer-Verlag.
Everitt B, Hand D (1981) *Finite Mixture Distributions*. Chapman & Hall, London, UK.
Everson P, Morris C (2000) Inference for multivariate normal hierarchical models. *Journal of the Royal Statistical Society: Series B*, 62, 399–412.
Fahrmeir L, Lang S (2001) Bayesian inference for generalized additive mixed models based on Markov random field priors. *Journal of the Royal Statistical Society: Series C (Applied Statistics)*, 50(2), 201–220.
Farrell PJ, Groshen S, MacGibbon B, Tomberlin T (2010) Outlier detection for a hierarchical Bayes model in a study of hospital variation in surgical procedures. *Statistical Methods in Medical Research*, 19(6), 601–619.
Fernandez C, Steele M (1998) On Bayesian modeling of fat tails and skewness. *Journal of the American Statistical Association*, 93, 359–371.
Ferreira M, Gamerman D (2000) Dynamic generalized linear models, pp 57–72, in *Generalized Linear Models: A Bayesian Perspective*, eds D Dey, S Ghosh, B Mallick. Marcel Dekker, New York.
Fordyce J A, Gompert Z, Forister M L, Nice C C (2011) A hierarchical Bayesian approach to ecological count data: A flexible tool for ecologists. *PLOS ONE*, 6(11), e26785.
Frees E (2004) *Longitudinal and Panel Data*. Cambridge University Press.
Friede T, Röver C, Wandel S, Neuenschwander B (2017) Meta-analysis of few small studies in orphan diseases. *Research Synthesis Methods*, 8(1), 79–91.
Fruhwirth-Schattner S (2001) Markov chain Monte Carlo estimation of classical and dynamic switching and mixture models. *Journal of the American Statistical Association*, 96, 194–209.
Fruhwirth-Schnatter S (2006) *Finite Mixture and Markov Switching Models*. Springer.
Fruhwirth-Schnatter S, Otter T, Tuchler R (2004) Bayesian analysis of the heterogeneity model. *Journal of Business & Economic Statistics*, 22, 2–15.
Gao S (2004) Combining binomial data using the logistic normal. *Journal of Statistical Computation and Simulation*, 74, 293–306.
Gelfand A (1996) Model determination using sampling-based methods, pp 145–161, in *Markov Chain Monte Carlo in Practice*, eds W Gilks, S Richardson, D Spiegelhalter. Chapman & Hall/CRC.
Gelman A (2006) Prior distributions for variance parameters in hierarchical model. *Bayesian Analysis*, 1, 515–534.
Gelman A, Carlin J, Stern H, Dunson D, Vehtari A, Rubin D (2014) *Bayesian Data Analysis*. CRC, Boca Raton, FL.
Gelman A, Jakulin A, Pittau M, Su Y (2008) A weakly informative default prior distribution for logistic and other regression models. *The Annals of Applied Statistics*, 2(4), 1360–1383.
Genton M (2004) *Skew-Elliptical Distributions and Their Applications: A Journey Beyond Normality*, Edited Volume. Chapman & Hall/CRC, Boca Raton, FL.
George E, Makov U, Smith A (1993) Conjugate likelihood distributions. *Scandinavian Journal of Statistics*, 20, 147–156.
George E, Zhang Z (2001) Posterior propriety in some hierarchical exponential family models, in *Data Analysis from Statistical Foundations: Festschrift in Honor of Donald A.S. Fraser*, ed A Saleh. Nova Science Publishers, New York.
Geweke J (2007) Interpretation and inference in mixture models: Simple MCMC works. *Computational Statistics & Data Analysis*, 51, 3529–3550.
Gilks WR, Wild P (1992) Adaptive rejection sampling for Gibbs sampling. *Journal of the Royal Statistical Society: Series C (Applied Statistics)*, 41(2), 337–348.
Gill J, Casella G (2009) Nonparametric priors for ordinal Bayesian social science models: Specification and estimation. *Journal of the American Statistical Association*, 104, 453–464.
Gompert Z, Fordyce J (2015) Package 'bayespref': Hierarchical Bayesian Analysis of Ecological Count Data. https://cran.r-project.org/web/packages/bayespref/bayespref.pdf
Gopalan R, Berry D (1998) Bayesian multiple comparisons using Dirichlet process priors. *Journal of the American Statistical Association*, 93, 1130–1139.
Green P, Richardson S (2001) Modelling heterogeneity with and without the Dirichlet process. *Scandinavian Journal of Statistics*, 28, 355–375.

Greenland S, Draper D (1998) Exchangeability, in *Entry in Encyclopedia of Biostatistics*, eds P Armitage, T Colton. Wiley, London, UK.

Greco T, Landoni G, Biondi-Zoccai G, D'Ascenzo F, Zangrillo A (2016) A Bayesian network meta-analysis for binary outcome: How to do it. *Statistical Methods in Medical Research*, 25(5), 1757–1773.

Griffin, J (2016) An adaptive truncation method for inference in Bayesian nonparametric models. *Statistics and Computing*, 26(1–2), 423–s441.

Gruen B, Plummer M (2015) BayesMix: An R Package for Bayesian Mixture Modeling. http://ifas.jku.at/gruen/BayesMix/

Guo J, Riebler A (2016) meta4diag: Bayesian Bivariate Meta-analysis of Diagnostic Test Studies for Routine Practice. https://arxiv.org/pdf/1512.06220.pdf

Guo J, Riebler A, Rue H (2017) Bayesian bivariate meta-analysis of diagnostic test studies with interpretable priors. *Statistics in Medicine*, 36(19), 3039–3058.

Guolo A, Varin C (2017) Random-effects meta-analysis: The number of studies matters. *Statistical Methods in Medical Research*, 26(3), 1500–1518.

Gupta S K (2012) Use of Bayesian statistics in drug development: Advantages and challenges. *International Journal of Applied and Basic Medical Research*, 2(1), 3–6.

Gustafson P, Hossain S, MacNab Y (2006) Conservative priors for hierarchical models. *Canadian Journal of Statistics*, 34, 377–390.

Hanson T, Branscum A, Johnson W (2005) Nonparametric Bayesian data analysis: An introduction, in *Handbook of Statistics*, Vol. 25, eds C Rao, D Dey. Elsevier.

Hanson T, Johnson W (2002) Modeling regression error with a mixture of polya trees. *Journal of the American Statistical Association*, 97(460), 1020–1033.

Hirano K (1998) Nonparametric Bayes models for longitudinal earnings data, in *Practical Nonparametric and Semiparametric Bayesian Statistics*, eds D Dey, P Muller, D Sinha. Springer-Verlag.

Hirano K (2002) Semiparametric Bayesian inference in autoregressive panel data models. *Econometrica*, 70, 781–799.

Hoff P (2003) Nonparametric modelling of hierarchically exchangeable data. Technical Report 421, Department of Statistics, University of Washington.

Howley P, Gibberd R (2003) Using hierarchical models to analyse clinical indicators: A comparison of the gamma-Poisson and beta-binomial models. *International Journal for Quality in Health Care*, 15, 319–329.

Hu J, Reiter J, Wang Q (2018) Dirichlet process mixture models for modeling and generating synthetic versions of nested categorical data. *Bayesian Analysis*, 13(1), 183–200.

Hurn M, Justel A, Robert C (2003) Estimating mixtures of regressions. *Journal of Computational and Graphical Statistics*, 12, 55–79.

Hurtado Rua S, Mazumdar M, Strawderman R (2015). The choice of prior distribution for a covariance matrix in multivariate meta-analysis: A simulation study. *Statistics in Medicine*, 34(30), 4083–4104.

Imai K, Ying L, Strauss A (2008) Bayesian and likelihood inference for 2×2 ecological tables: An incomplete data approach. *Political Analysis*, 16, 41–69.

Ishwaran H, James L (2001) Gibbs sampling methods for stick-breaking priors. *Journal of the American Statistical Association*, 96, 161–173.

Ishwaran H, James L (2002) Approximate Dirichlet process computing in finite normal mixtures: Smoothing and prior information. *Journal of Computational and Graphical Statistics*, 11, 508–532.

Ishwaran H, Zarepour M (2000) Markov chain Monte Carlo in approximate Dirichlet and beta two-parameter process hierarchical models. *Biometrika*, 87, 371–390.

Ishwaran H, Zarepour M (2002) Exact and approximate sum-representations for the Dirichlet process. *Canadian Journal of Statistics*, 30, 269–283.

Jackson D, White I, Riley R (2013) A matrix based method of moments for fitting the multivariate random effects model for meta-analysis and meta-regression. *Biometrical Journal*, 55(2), 231–245.

Jara A (2007) Applied Bayesian non- and semi-parametric Inference using Dppackage. *R News*, 7/3, 17–26.

Jara A, Hanson T, Quintana F, Müller P, Rosner G (2011) DPpackage: Bayesian semi-and nonparametric modeling in R. *Journal of Statistical Software*, 40(5), 1–30.

Jarque C, Bera A (1980) Efficient tests for normality, homoscedasticity and serial independence of regression residuals. *Econometric Letters*, 6, 255–259.

Jiang J, Lahiri P (2006) Mixed model prediction and small area estimation. *Test*, 15(1), 1.

Jullion A, Lambert P (2007) Robust specification of the roughness penalty prior distribution in spatially adaptive Bayesian P-splines models. *Computational Statistics & Data Analysis*, 51(5), 2542–2558.

Karabatsos G (2016) A menu-driven software package for Bayesian regression analysis. The ISBA Bulletin, 22(4), 13–16.

Karabatsos G (2017) A menu-driven software package of Bayesian nonparametric (and parametric) mixed models for regression analysis and density estimation. *Behavior Research Methods*, 49(1), 335–362.

King G (1997) *A Solution to the Ecological Inference Problem: Reconstructing Individual Behavior from Aggregate Data*. Princeton University Press, Princeton, NJ.

King G, Rosen O, Tanner M (eds) (2004) *Ecological Inference: New Methodological Strategies*. Cambridge University Press, New York.

Kleinman KP, Ibrahim JG (1998) A semiparametric Bayesian approach to the random effects model. *Biometrics*, 54(3), 921–938.

Kruschke J, Vanpaemel W (2015) Bayesian estimation in hierarchical models, pp 279–299, in *The Oxford Handbook of Computational and Mathematical Psychology*, eds J R Busemeyer, Z Wang, J T Townsend, A Eidels. Oxford University Press, Oxford, UK.

Kuhan G, Marshall E C, Abidia A F, Chetter I C, McCollum P (2002) A Bayesian hierarchical approach to comparative audit for carotid surgery. *European Journal of Vascular and Endovascular Surgery*, 24(6), 505–510.

Kulinskaya E, Olkin I (2014) An overdispersion model in meta-analysis. *Statistical Modelling*, 14(1), 49–76.

Lambert P, Sutton A, Burton P, Abrams K, Jones D (2005) How vague is vague? A simulation study of the impact of the use of vague prior distributions in MCMC using WinBUGS. *Statistics in Medicine*, 24(15), 2401–2428.

Larson J, Soule S (2009) Sector-level dynamics and collective action in the United States, 1965–1975. *Mobilization: An International Quarterly*, 14(3), 293–314.

Laud PW, Ibrahim JG (1995) Predictive model selection. *Journal of the Royal Statistical Society: Series B (Methodological)*, 57(1), 247–262.

Lee J, Sabavala D (1987) Bayesian estimation and prediction for the beta binomial model. *Journal of Business and Economic Statistics*, 5, 357–367.

Lee K, Thompson S (2008) Flexible parametric models for random-effects distributions *Statistics in Medicine*, 27, 418–434.

Leisch F (2004) FlexMix: A general framework for finite mixture models and latent class regression in R. *Journal of Statistical Software*, 11(8), 1–18.

Lenk P (1988) The logistic normal distribution for Bayesian nonparametric predictive densities. *Journal of the American Statistical Association*, 83, 509–516.

Lenk P, Desarbo W (2000) Bayesian inference for finite mixtures of generalized linear models with random effects. *Psychometrika*, 65, 93–119.

Leonard T (1973) A Bayesian method for histograms. *Biometrika*, 60, 297–308.

Leslie D, Kohn R, Nott D (2007) A general approach to heteroscedastic linear regression. *Statistics and Computing*, 17, 131–146.

Lin T, Lee J, Hsieh W (2007b) Robust mixture modeling using the skew t distribution. *Statistics and Computing*, 17, 81–92.

Lin T, Lee J, Ni H (2004) Bayesian analysis of mixture modelling using the multivariate t distribution. *Statistics and Computing*, 14, 119–130.

Lin T, Lee J, Yen S (2007a) Finite mixture modelling using the skew normal distribution. *Statistica Sinica*, 17, 909–927.

Lindley D, Smith A (1972) Bayes estimates for the linear model. *Journal of the Royal Statistical Society: Series B*, 34, 1–41.

Liu J, Dey D (2007) Hierarchical overdispersed Poisson model with macrolevel autocorrelation. *Statistical Methodology*, 4(3), 354–370.

Lu G, Ades AE (2009) Modeling between-trial variance structure in mixed treatment comparisons. *Biostatistics*, 10(4), 792–805.

Makuch R, Stephens M, Escobar M (1989) Generalized binomial models to examine the historical control assumption in active control equivalence studies. *The Statistician*, 38, 61–70.

Marin J, Mengersen K, Robert C (2005) Bayesian modelling and inference on mixtures of distributions, in *Handbook of Statistics*, Vol. 25, eds D Dey, C Rao. Elsevier.

Markham F, Young M, Doran B, Sugden M (2017) A meta-regression analysis of 41 Australian problem gambling prevalence estimates and their relationship to total spending on electronic gaming machines. *BMC Public Health*, 17(1), 495.

Marshall E, Spiegelhalter D (1998) Comparing institutional performance using Markov chain Monte Carlo methods, in *Statistical Analysis of Medical Data: New Developments*, eds B Everitt, G Dunn. Arnold.

Marshall E, Spiegelhalter D (2007) Simulation-based tests for divergent behaviour in hierarchical models. *Bayesian Analysis*, 2, 409–444.

Mavridis D, Salanti G (2013) A practical introduction to multivariate meta-analysis. *Statistical Methods in Medical Research*, 22(2), 133–158.

McLachlan G, Rathnayake S (2014) On the number of components in a Gaussian mixture model. *Wiley Interdisciplinary Reviews: Data Mining and Knowledge Discovery*, 4(5), 341–355.

Militino A, Ugarte M, Dean C (2001) The use of mixture models for identifying high risks in disease mapping. *Statistics in Medicine*, 20, 2035–2049.

Mohr D (2006) Bayesian identification of clustered outliers in multiple regression. *Computational Statistics & Data Analysis*, 51, 3955–3967.

Moreno E, Vázquez-Polo F, Negrn M (2018) Bayesian meta-analysis: The role of the between-sample heterogeneity. *Statistical Methods in Medical Research*, 27(12), 3643–3657.

Muller P, Erkanli A, West M (1996) Bayesian curve fitting using multivariate normal mixtures. *Biometrika*, 83, 67–79.

Ohlssen D, Sharples L, Spiegelhalter D (2007) Flexible random-effects models using Bayesian semi-parametric models: Applications to institutional comparisons. *Statistics in Medicine*, 26, 2088–2112.

Papastamoulis P (2016) label.switching: An R package for dealing with the label switching problem in MCMC outputs. *Journal of Statistical Software*, 69. https://www.jstatsoft.org/article/view/v069c01.

Parmigiani G (2002) *Modeling in Medical Decision Making: A Bayesian Approach*. Wiley, New York.

Pastor N (2003) Methods for the analysis of explanatory linear regression models with missing data not at random. *Quality and Quantity*, 37, 363–376.

Pauler D, Wakefield J (2000) Modeling and implementation issues in Bayesian meta-analysis, pp 205–230, in *Bayesian Meta-Analysis*, eds Stangl D, Berry D. Marcel Dekker.

Pérez M, Pericchi L, Ramrez I (2017) The Scaled Beta2 distribution as a robust prior for scales. *Bayesian Analysis*, 12(3), 615–637.

Pitman J, Yor M (1997) The two-parameter Poisson-Dirichlet distribution derived from a stable subordinator. *Annals of Probability*, 25, 855–900.

Podlich H, Faddy M, Smyth G (2004) Semi-parametric extended Poisson process models for count data. *Statistics and Computing*, 14, 311–321.

Prabhakaran S, Azizi E, Carr A, Pe'er D (2016) Dirichlet process mixture model for correcting technical variation in single-cell gene expression data, pp 1070–1079, in *Proceedings of the International Conference on Machine Learning*, New York.

Prevost T, Abrams K, Jones D (2000) Hierarchical models in generalized synthesis of evidence: An example based on studies of breast cancer screening. *Statistics in Medicine*, 19, 3359–3376.

Quintana F, Tam W (1996) Bayesian estimation of beta-binomial models by simulating posterior densities. *Revista de la Sociedad Chilena de Estadstica*, 13, 43–56.

Rao J (2003) *Small Area Estimation*. Wiley, New York.

Rattanasiri S, Bohning D, Roianavipart P, Athipanyakom S (2004) A mixture model application in disease mapping of malaria. *The Southeast Asian Journal of Tropical Medicine and Public Health*, 35, 38–47.

Redner R, Walker H (1984) Mixture densities, maximum likelihood, and the EM algorithm. *S1Ah-I Review*, 26, 195–239.

Rhodes K, Turner R, White I, Jackson D, Spiegelhalter D, Higgins J (2016) Implementing informative priors for heterogeneity in meta-analysis using meta-regression and pseudo data. *Statistics in Medicine*, 35(29), 5495–5511.

Richardson S, Green P (1997) On Bayesian analysis of mixtures with an unknown number of components. *Journal of the Royal Statistical Society: Series B*, 59, 731–758.

Riley RD, Dodd SR, Craig JV, Thompson JR, Williamson PR (2008) Meta-analysis of diagnostic test studies using individual patient data and aggregate data. *Statistics in Medicine*, 27(29), 6111–6136.

Robert C, Mengersen K (1999) Reparameterisation issues in mixture modelling and their bearing on the Gibbs sampler. *Computational Statistics and Data Analysis*, 29, 325–343.

Robert C, Titterington D (1998) On perfect simulation for some mixtures of distributions. *Statistics and Computing*, 8, 145–158.

Roberts S, Husmeier D, Rezek I, Penny W (1998) Bayesian approaches to Gaussian mixture modeling. *IEEE Transactions on Pattern Analysis and Machine Intelligence*, 20, 1133–1142.

Robertson C, Fryer J (1968) Some descriptive properties of normal mixtures. *Scandinavian Actuarial Journal*, 52, 137–146.

Rodrguez-Avi J, Conde-Sânchez A, Sáez-Castillo A, Olmo-Jiménez M (2007) A generalization of the beta–binomial distribution. *Journal of Applied Statistics*, 56, 51–61.

Roeder K, Wasserman L (1997) Practical Bayesian density estimation using mixtures of normals. *Journal of the American Statistical Association*, 92, 894–902.

Rouder J N, Morey R, Pratte M (2013) Hierarchical Bayesian models, in *The New Handbook of Mathematical Psychology, Volume 1: Measurement and Methodology*, eds W H Batchelder, H Colonius, E Dzhafarov, J I Myung. Cambridge University Press, London, UK.

Sahu S, Dey D, Branco M (2003) A new class of multivariate skew distributions with applications to Bayesian regression models. *The Canadian Journal of Statistics*, 31: 129–150.

Savage J (2016) Finite Mixture Models in Stan. http://modernstatisticalworkflow.blogspot.co.uk/2016/10/finite-mixture-models-in-stan.html

Scollnik DP (1995) Bayesian analysis of two overdispersed Poisson regression models. *Communications in Statistics-Theory and Methods*, 24(11), 2901–2918.

Scrucca L, Fop M, Murphy T, Raftery A (2016) mclust 5: Clustering, classification and density estimation using gaussian finite mixture models. *The R Journal*, 8(1), 289.

Sethuraman J (1994) A constructive definition of Dirichlet priors. *Statistica Sinica*, 4, 639–650.

Silliman N (1997) Hierarchical selection models with applications in meta-analysis. *Journal of the American Statistical Association*, 92, 926–936.

Simpson D P, Rue H, Martins T G, Riebler A, Sørbye S H (2016) Penalising model component complexity: A principled, practical approach to constructing priors. *Statistical Science* (Forthcoming). arXiv preprint arXiv:1403.4630.

Smith TC, Spiegelhalter DJ, Thomas A (1995) Bayesian approaches to random-effects meta-analysis: A comparative study. *Statistics in Medicine*, 14(24), 2685–2699.

Spiegelhalter D (1999) Surgical audit: Statistical lessons from Nightingale and Codman. *Journal of the Royal Statistical Society: Series A*, 162, 45–58.

Spiegelhalter D (2005) Handling over-dispersion of performance indicators. *Quality and Safety in Health Care*, 14, 347–351.

Spiegelhalter D, Abrams K, Myles J (2004) *Bayesian Approaches to Clinical Trials and Health-Care Evaluation*. Wiley, New York.

Spittal M J, Pirkis J, Gurrin L (2015) Meta-analysis of incidence rate data in the presence of zero events. *BMC Medical Research Methodology*, 15(1), 42.

Staggs V, Gajewski B (2017) Bayesian and frequentist approaches to assessing reliability and precision of health-care provider quality measures. *Statistical Methods in Medical Research*, 26(3), 1341–1349.

Taylor-Rodrguez D, Kaufeld K, Schliep E, Clark J, Gelfand A (2017) Joint species distribution modeling: Dimension reduction using Dirichlet processes. *Bayesian Analysis*, 12(4), 939–967.
Teather D (1984) The estimation of exchangeable binomial parameters. *Communications in Statistics, Part A*, 13, 671–680.
van Dongen S (2006) Prior specification in Bayesian statistics: Three cautionary tales. *Journal of Theoretical Biology*, 242: 90–100.
van Houwelingen H, Arends L, Stiinen T (2002) Advanced methods in meta-analysis: Multivariate approach and meta-regression. *Statistics in Medicine*, 21, 589–624.
Verde PE (2018) bamdit: An R package for Bayesian meta-analysis of diagnostic test data. *Journal of Statistical Software*, Articles, 86, 1–32.
Viechtbauer W (2010) Conducting meta-analyses in R with the metafor package. *Journal of Statistical Software*, 36(3), 1–48.
Viechtbauer W (2017) Package 'metafor'. https://cran.r-project.org/web/packages/metafor/metafor.pdf
Walfish S (2006) A review of statistical outlier methods. *Pharmaceutical Technology*, 30(11), 82–86.
Walker S, Damien P, Laud P, Smith A (1999) Bayesian nonparametric inference for random distributions and related functions. *Journal of the Royal Statistical Society: Series B*, 61, 485–527.
Wang C, Blei D (2017) A general method for robust Bayesian modeling. *Bayesian Analysis*, 13(4), 1163–1191.
Warn D, Thompson S, Spiegelhalter D (2002) Bayesian random effects meta-analysis of trials with binary outcomes: Methods for the absolute risk difference and relative risk scales. *Statistics in Medicine*, 21, 1601–1623.
Wasserman L (2000) Asymptotic inference for mixture models using data-dependent priors. *Journal of the Royal Statistical Society: Series B*, 62, 159–180.
Weems K, Smith P (2004) On robustness of maximum likelihood estimates for Poisson-lognormal models. *Statistics & Probability Letters*, 66, 189–196.
Wei Y, Higgins JP (2013) Bayesian multivariate meta-analysis with multiple outcomes. *Statistics in Medicine*, 32(17), 2911–2934.
West M (1984) Outlier models and prior distributions in Bayesian linear regression. *Journal of the Royal Statistical Society: Series B*, 46, 431–439.
West M, Muller P, Escobar M (1994) Hierarchical priors and mixture models, with application in regression and density estimation, pp 363–386, in *Aspects of Uncertainty: A Tribute to D. V. Lindley*, eds P Freeman, A Smith. Wiley, New York.
Williams D, Rast P, Bürkner P (2018) Bayesian Meta-Analysis with Weakly Informative Prior Distributions. PsyArXiv. https://andrewgelman.com/wp-content/uploads/2018/01/bayes_donny.pdf
Winkelmann R, Zimmermann KF (1991) A new approach for modeling economic count data. *Economics Letters*, 37(2), 139–143.
Winship DA (1978) Cimetidine in the treatment of duodenal ulcer. *Gastroenterology*, 74, 402–406.
Young-Xu Y, Chan K (2008) Pooling overdispersed binomial data to estimate event rate. *BMC Medical Research Methodology*, 8, 58.
Yu K, Moyeed R (2001) Bayesian quantile regression. *Statistics and Probability Letters*, 54(4), 437–447.
Zhang J, Fu H, Carlin B (2015) Detecting outlying trials in network meta-analysis. *Statistics in Medicine*, 34(19), 2695–2707.
Zhao Y, Staudenmayer J, Coull B, Wand M (2006) General design Bayesian generalized linear mixed models. *Statistical Science*, 21, 35–51.
Zollinger A, Davison A, Goldstein D (2015) Meta-analysis of incomplete microarray studies. *Biostatistics*, 16(4), 686–700.

5

Time Structured Priors

5.1 Introduction

A time series is a sequence of stochastic observations which are ordered in time, most often at equally spaced discrete times $t = 1,\ldots,T$, though extensions to unequally spaced intervals are relatively straightforward (Lee and Nelder, 2001). Major goals of time series analysis include modelling the interrelationship of variables evolving jointly through time, as in econometric growth models (Paap and van Dijk, 2003), forecasting future values of time series variables (Beck, 2004), and identifying the structural components of a sequence of observations (Huerta and West, 1999). In the analysis of temporal data, one generally expects positive covariation between observations that are close to each other in time, so that exchangeable priors are not appropriate. While time series are sometimes analysed exchangeably, at least within subgroups of the data, as in change-point models (Mira and Petrone, 1996), in most applications, there is a gain from modelling temporal covariation. Hence, hierarchical priors for time series modelling are typically structured in the sense of explicitly recognising adjacency in time as the basis for smoothing or prediction. Hierarchical methods also assist in identifying underlying relatively smooth or recurring features of the data, for example, underlying trends or seasonal effects.

Bayesian methods are widely applied to autoregressive moving average models, without necessarily imposing the stationarity restrictions and preliminary detrending that feature in classical estimation. However, a general scheme for specifying priors for modelling time series data is provided by the state-space approach, considered in Sections 5.3 and 5.4 (Harvey et al., 2006; West, 2013; Giordani et al., 2011; Petris et al., 2009), which includes ARMA (autoregressive moving average) models as special cases. State-space models recognise multiple underlying components in time series, with the priors governing the evolution of the components under an expectation of smoothness. The linear state-space (or dynamic linear model) specification for the changing level of a univariate continuous response y_t has the form

$$y_t = \beta_t X_t + u_t,$$

$$\beta_t = \beta_{t-1} G_t + w_t,$$

where the errors

$$u_t \sim N(0, V_t),$$

$$w_t \sim N(0, W_t),$$

are unstructured white noise, X_t is a predictor or design matrix, and G_t is a known matrix governing the evolution of the state vector β_t (Durbin, 2000; West and Harrison, 1997). The time structured latent effects β_t may include level, trend, seasonal, or cyclical effects. Taking u_t and w_t to be normal leads to the normal dynamic linear model (West, 1998), with extension to generalised linear model forms for discrete data leading to dynamic generalised linear models (West et al., 1985). State-space principles can also be applied to model stochastically evolving variances, as in stochastic volatility models (Kim et al., 1998; Jacquier et al., 2002); see Section 5.5.

While there may be benefits from borrowing strength methods that take account of correlations between units, the use of multiple random effects to represent unobserved components in time raises potential identification issues (Auger-Méthé et al., 2016; Knape, 2008). For example, priors for correlated effects in time may specify differences in effects between adjacent units without specifying the mean level of the effects. MCMC methods may then require centring of the effects during sampling to ensure identification of other parameters. Methods for smoothing or interpolation in time may also need to retain robustness to take account of regime shifts, or to accommodate temporal outliers. Structured priors assume relatively smooth variation over adjacent units, and their parameters may be distorted if mechanisms are not incorporated for accommodating extreme points.

There is a wide range of time series analysis options in R using frequentist estimation packages (https://cran.r-project.org/web/views/TimeSeries.html) which may be useful for comparative purposes. Bayesian computing options in R for time series include bsts, particularly for state-space modelling (Scott, 2017); BMR, Bayesian Macroeconometrics in R (https://github.com/kthohr/BMR/tree/master/man); stochvol for stochastic volatility analysis (Kastner and Hosszejni, 2016); and tsPI (Helske, 2017). Generic packages such as rstan and R-INLA may facilitate estimation and identification in complex random effects time series models (Monnahan et al., 2017; Betancourt and Girolami, 2015).

The chapter below considers schemes for modelling correlated observations and latent effects in time series. Sections 5.2 and 5.3 consider autoregressive and state-space priors for time series analysis, while Section 5.4 considers state-space methods for discrete time series. Section 5.5 considers Bayesian approaches to stochastic volatility and Section 5.6 considering models adaptive to temporal discontinuities.

5.2 Modelling Temporal Structure: Autoregressive Models

Many time series show evidence of serial dependence in the observations or error terms, leading to what are sometimes denoted as observation- and parameter-driven models, respectively (Oh and Lim, 2001). A widely used model for expressing such serial dependence is the lag p autoregressive or $AR(p)$ model. An $AR(p)$ scheme for dependent outcomes y_t in a normal linear framework is represented by

$$y_t = \phi_0 + \phi_1 y_{t-1} + \phi_2 y_{t-2} + \ldots + \phi_p y_{t-p} + u_t, \quad t = 1, \ldots, T$$

where the innovation errors $u_t \sim N(0, \sigma^2)$ are homoscedastic white noise, independent of each other and lagged y values $\{y_{t-1}, \ldots, y_{t-p}\}$. So $E(u_t u_{t-s}) = E(u_{t-j} u_{t-j-s}) = 0$ for all s and j. Note that a full likelihood analysis will refer to p latent preseries values (Marriott et al.,

1996), with Naylor and Marriott (1996) suggesting preseries values follow a heavy tailed version of the density assumed for the observed series, for instance $(y_0, y_{-1}, \ldots y_{1-p})$ as Student t with variance σ^2 and low degrees of freedom v. Autoregressive dependence may also be present in error terms, such that

$$y_t = \phi_0 + \phi_1 y_{t-1} + \phi_2 y_{t-2} + \cdots + \phi_{p1} y_{t-p1} + \varepsilon_t$$

$$\varepsilon_t = \rho_1 \varepsilon_{t-1} + \rho_2 \varepsilon_{t-2} + \cdots + \rho_{p2} \varepsilon_{t-p2} + u_t,$$

where the u_t are IID.

Furthermore, moving average effects may occur in the white noise errors u_t, and so have an impact on y_t of lagged disturbances u_t. A lag q moving average effect, combined with a lag p effect in the y_t series, provides the ARMA(p, q) model

$$y_t = \phi_0 + \phi_1 y_{t-1} + \cdots + \phi_p y_{t-p} + u_t + \gamma_1 u_{t-1} + \gamma_2 u_{t-2} + \cdots \gamma_q u_{t-q},$$

with assumptions as in Chib and Greenberg (1994). Assuming the y-series is centred around its mean, and defining $By_t = y_t - y_{t-1}$, one has $y_t - \phi_1 y_{t-1} - \cdots \phi_p y_{t-p} = y_t(1 - \phi_1 B - \cdots \phi_p B^p) = \Phi(B) y_t$, and the ARMA($p$, q) model can be written

$$\Phi(B) y_t = \Gamma(B) u_t.$$

As for other regressions, collinearity may occur, and parameter selection for the ARMA(p,q) may include shrinkage priors (Schmidt and Makalic, 2013) and RJMCMC (Ehlers and Brooks, 2004).

Classical estimation methods typically require stationarity and constant variances in estimating such models. Stationarity is equivalent to the roots of $\Phi(B) = 1 - B - B^2 \cdots - B^p$ being outside the unit circle, and invertibility refers to the same condition on the roots of $\Gamma(B)$. This typically involves preliminary data differencing or transformation to gain stationarity, or regression to remove trend (e.g. Abraham and Ledolter, 1983, p.225), with the actual model then applied to differenced data or to regression residuals. To assess whether stationarity has been achieved, one can consider the autocorrelation sequence of model residuals: a stationary process should show a sequence fading to zero at high lags, whereas significant values at high lags indicate nonstationarity. In Bayesian analyses, it is common to estimate parameters without presuming stationarity (or invertibility), but instead obtain the posterior probabilities of stationarity via monitoring the sampled parameters (McCulloch and Tsay, 1994; Marriott et al., 1996).

Example 5.1 Southern Oscillation Index

This example illustrates the estimation of ARMA models via the rstan package. The data is NINO3.4 index (as in the R package tseries with T=598), this being one of several El Niño/ Southern Oscillation (ENSO) indicators based on sea surface temperatures. The Nino 3.4 Region is bounded by 120W–170W and 5S–5N. Use of options aic.wge from the R package tswge suggests the best fit (under classical estimation) to be for an ARMA(4,0,1) model.

Estimates from ARMA models may be affected by the specification of the initial conditions, and we consider first the estimation of the ARMA(1,1) model. Thus

$$y_t = \phi_0 + \phi_1 y_{t-1} + u_t + \gamma_1 u_{t-1},$$

where no stationarity constraints are imposed on ϕ_1 or γ_1. The first observation y_1 is included in the estimation, and a composite fixed effect parameter is assumed for $\phi_1 y_0+\gamma_1 u_0$, referring to latent preseries data. Classical estimation via the tseries ARMA option provides estimates (mean, s.e.) of 3.67 (0.58), 0.86 (0.02), and 0.26 (0.03) for ϕ_0, ϕ_1, and γ_1 respectively. Estimation using rstan including the first observation provides corresponding posterior mean (sd) estimates of 3.76 (0.59), 0.86 (0.02), and 0.27 (0.03). The LOO-IC (leave-one-out information criterion) is 615.

Estimates may also be obtained by conditioning on the first observation (i.e. not including that point in the likelihood). The rstan estimates of ϕ_0, ϕ_1, and γ_1 on this basis are 3.67 (0.58), 0.86 (0.02) and 0.26 (0.03).

Classical estimates of the ARMA(4,0,1) model vary slightly according to the package. Note also that parameterisations of the intercept, and the way MA terms are signed, differs between R packages. The ARMA option in tseries is unable to fit this model, while the aic.wge option provides estimated AR lag parameters (0.21,0.99, −0.36, −0.14) and MA parameter 0.98. The FitARMA package provides estimated AR parameters (0.31, 0.89, −0.38, −0.10), and MA parameter 0.89.

The rstan estimation, with code as in [1], of the ARMA(4,0,1) model gives estimates for $(\phi_1, \phi_2, \phi_3, \phi_4)$ of 0.29 (sd = 0.08), 0.91 (0.08), −0.37 (0.04), and −0.10 (0.04) respectively, and 0.92 (0.07) for γ_1. The LOO-IC is reduced to 547.

5.2.1 Random Coefficient Autoregressive Models

A hierarchical generalisation of the AR(p) prior allows the lag coefficient to vary over time, as in random coefficient AR or RCAR models – see, for example, Lee (1998), Berkes et al. (2009), Wang and Ghosh (2002), and Araveeporn (2017). These are also called time-varying autoregressive or TVAR models. Thus, for a centred and univariate y, an RCAR(p) model in the observations specifies

$$y_t = \sum_{t=1}^{p} \phi_{tj} y_{t-j} + u_t,$$

$$\phi_t = \mu_\phi + \Sigma_\phi^{0.5} e_t,$$

where $u_t \sim N(0,\sigma^2)$, $e_t \sim N_p(0,I)$, $\phi_t = (\phi_{t1},\ldots,\phi_{tp})$, $\Sigma_\phi = \text{diag}(\sigma^2_{\phi 1},\ldots,\sigma^2_{\phi p})$, and $\mu_\phi = (\phi_1,\ldots,\phi_p)$. Instead of a multivariate normal prior for the ϕ_t, sequential updating of the ϕ_t may be applied, for example, via a multivariate random walk (Section 5.3),

$$\phi_t = \phi_{t-1} + w_t, \quad w_t \sim N_p(0,W_t).$$

Another possibility (Godsill et al., 2004) is to take both the AR coefficient vector and the innovation variance σ^2 to be time-varying, for example, by setting a random walk prior on $h_t = \log(\sigma_t)$, or by a second-stage autoregression, such as

$$h_t \sim N\left(\rho_h h_{t-1}, \sigma_h^2\right).$$

As in many Bayesian applications, stationarity constraints are not necessarily placed on the ϕ_{tj} at each t (Prado et al., 2000). However, if the AR parameters lie in the stationary region, then the series can be considered locally stationary. For example, for an RCAR(1) model including a latent preseries value y_0, a hierarchical scheme such as

$$y_t \sim N\left(\phi_{t1} y_{t-1}, \sigma^2\right), \quad t > 1$$
$$y_0 \sim t_2\left(m_0, \sigma^2\right),$$
$$\phi_{t1} \sim N\left(\phi_1, \sigma_\phi^2\right), \quad t > 1$$

may be applied. For this model, stationarity holds if $\phi_1^2 + \sigma_\phi^2 < 1$.

5.2.2 Low Order Autoregressive Models

Simple dependence models for observations, errors, or latent effects are obtained via first- or second-order autoregression. In the AR(1) observation model, one has

$$y_t - \mu = \phi\left(y_{t-1} - \mu\right) + u_t$$

or

$$y_t = \phi y_{t-1} + u_t$$

for centred data, where under stationarity $-1 < \phi < 1$, and y_r and y_s for $1 \le r \le s \le T$ are conditionally independent, given $\{y_{r+1}, \dots, y_{s-1}\}$ if $r - s > 1$ (Rue and Held, 2005). The AR(2) model has

$$y_t = \phi_1 y_{t-1} + \phi_2 y_{t-2} + u_t$$

where stationarity requires $\phi_1 + \phi_2 < 1, \phi_2 - \phi_1 < 1$, and $|\phi_2| < 1$.

An AR(1) error sequence $\varepsilon_t = \rho\varepsilon_{t-1} + u_t$ with $u_t \sim N(0, \sigma^2)$ similarly requires $-1 < \rho < 1$ for stationarity. The covariance for such a sequence has the form $\text{Cov}(\varepsilon) = \sigma^2 C$ with (s, t)th element in the correlation matrix $\text{corr}(\varepsilon_s, \varepsilon_t) = \rho^{|s-t|}/(1-\rho^2)$, so correlations decline as the gap between observations increases.

For the stationary AR(1) observation model $y_t = \phi y_{t-1} + u_t$, the marginal density of the first observation is $y_1 \sim N(0, \sigma^2/(1-\phi^2))$, and the joint density can also be obtained by density decomposition as

$$p\left(y_1, \dots, y_T\right) = p\left(y_1\right) p\left(y_2 \mid y_1\right) p\left(y_3 \mid y_2\right) \dots p\left(y_T \mid y_{T-1}\right)$$
$$\propto \left(1 - \phi^2\right)^{0.5} \sigma^{-n} \exp\left[-0.5H / \sigma^2\right],$$

where $H = (1-\phi^2) y_1^2 + \sum_{t=2}^{T} (y_t - \phi y_{t-1})^2$. The same sequence of marginal and conditional densities applies for AR(1) autoregressive errors.

The precision (inverse covariance) matrix of autoregressive models has interesting theoretical properties demonstrating how conditional independence structures determine the precision matrix and vice versa (Speed and Kiiveri, 1986; Rue and Held, 2005). Specifically, zeros in the precision matrix define, and are defined by, conditional independencies in the joint density. Thus, for an AR(1) prior on errors ε with lag coefficient ρ, the precision matrix Π is tridiagonal with (r, s)th cell equalling zero only if the complete conditional distribution of ε_r does not depend on ε_s, namely

$$\Pi = \sigma^{-2} C^{-1} = \sigma^{-2} \begin{bmatrix} 1 & -\rho & 0 & & & & \\ -\rho & 1+\rho^2 & -\rho & & & & \\ 0 & -\rho & 1+\rho^2 & \dots & & & \\ & & \dots & \dots & & & \\ & & & & -\rho & 1+\rho^2 & -\rho \\ & & & & 0 & -\rho & 1 \end{bmatrix}.$$

For an AR(2) error sequence with lag parameters $\{\rho_1, \rho_2\}$, the precision matrix is

$$\Pi = \sigma^{-2} \begin{bmatrix} 1 & -\rho_1 & -\rho_2 & 0 & & & & \\ -\rho_1 & 1+\rho_1^2 & -\rho_1(1-\rho_2) & -\rho_2 & & & & \\ -\rho_2 & -\rho_1(1-\rho_2) & 1+\rho_1^2+\rho_2^2 & -\rho_1(1-\rho_2) & & & & \\ 0 & -\rho_2 & -\rho_1(1-\rho_2) & 1+\rho_1^2+\rho_2^2 & \dots & & & \\ & & & \dots & \dots & & & \\ & & & & & -\rho_1(1-\rho_2) & 1+\rho_1^2 & -\rho_1 \\ & & & & & -\rho_2 & -\rho_1 & 1 \end{bmatrix}.$$

Such simplifications in structure are useful in multidimensional applications involving spatio-temporal or multiple-time scale errors. For example, if the covariance matrix of a spatio-temporal error ε_{st} is represented as a Kronecker product $\Sigma_t \otimes \Sigma_s$ of a temporal covariance Σ_t and spatial covariance Σ_s, then the corresponding precision matrix is $\Pi_t \otimes \Pi_s$ (Bijma et al., 2005).

There is considerable literature around the unit root and explosive root solutions of the AR(1) observation model $y_t = \phi y_{t-1} + u_t$. One may apply an autoregressive prior not constrained to stationarity, and a substantial posterior probability of nonstationarity would support using random walk priors, as a parsimonious autoregressive prior that allows for potential nonstationarity. For example, Lubrano (1995) considers the alternative composite hypotheses $H_0: \phi < 1$ and $H_1: \phi \geq 1$. Schotman and van Dijk (1991) consider the autoregression plus trend observation model $y_t = \phi_0 + \phi_1 y_{t-1} + \delta t + u_t$ and reframe it in equivalent AR(1) error form as

$$y_t = \delta_0 + \delta_1 t + \varepsilon_t,$$

$$\varepsilon_t = \phi \varepsilon_{t-1} + u_t,$$

while Chatuverdi and Kumar (2005) consider the unit root hypothesis under a more general polynomial trend $y_t = \delta_0 + \Sigma_j \delta_j t^j + \varepsilon_t$.

5.2.3 Antedependence Models

Structured antedependence models may offer flexibility in time series specification; they resemble autoregressions in entailing a regression over preceding observations or latent effects, but are specified in a way that avoids stationarity constraints (Nunez-Anton and Zimmerman, 2000; Pourahmadi, 2002). Observations $\{y_1, \ldots, y_T\}$ are antedependent of order s if y_t depends only on $\{y_{t-1}, \ldots, y_{t-s}\}$ for all $t \geq s$ (Gabriel, 1962). For example, Jaffrezic et al. (2003) consider a second-order antedependence model for normal longitudinal data of the form $y_{it} = \eta_{it} + g_{it} + u_{it}$, where η_{it} models fixed effects, e.g. $\eta_{it} = x_{it}\beta, u_{it}$ are unstructured white noise errors with fixed variance, and the genetic component g_{it} follows a second-order structured antedependence or AD(2) scheme. This scheme specifies

$$g_1 = \varepsilon_1$$

$$g_2 = \phi_{12} g_1 + \varepsilon_2$$

$$g_t = \phi_{1t} g_{t-1} + \phi_{2t} g_{t-2} + \varepsilon_t, \quad t > 2$$

with $\varepsilon_t \sim N(0, \omega_t)$. Because of the initial condition $g_1 = \varepsilon_1$, the antedependence parameters, such as $\{\phi_{1t}, \phi_{2t}\}$ in an AD(2) model, are unconstrained, in contrast to the stationarity constraints needed for autoregressive models.

To reduce the number of parameters being estimated, changing variances ω_t may be modelled via a parametric function of time, for example

$$\log(\omega_t) = a_1 + a_2 t + a_3 t^2,$$

while the antedependence parameters can also be modelled using time functions. For example, a Box–Cox power law can be used to parameterise time-varying AD coefficients ϕ_{kt}, namely

$$\phi_{kt} = \phi_k^{r_t - r_{t-k}}$$

where $\{r_t = t^{\lambda_k - 1} / \lambda_k, r_{t-k} = (t-k)^{\lambda_k - 1} / \lambda_k\}$ if $\lambda_k \neq 0$, and $\{r_t = \log(t), r_{t-k} = \log(t-k)\}$ if $\lambda_k = 0$ (Nunez-Anton and Zimmerman, 2000). The ϕ and ω parameters may be adjusted to account for unevenly spaced times located at points $\{a_1, \dots a_T\}$ (Jaffrezic et al., 2004).

Example 5.2 NASDAQ Daily Volume, 2017

This example considers the log transforms of the NASDAQ daily trading volume statistics during 2017 (from https://finance.yahoo.com/). There are 252 uncentred observations. The analysis compares random coefficient AR models in terms of fit, and robustness to outliers, against fixed coefficient models and stochastic volatility GARCH models (cf. Wang and Ghosh, 2002).

Thus, the first two analyses compare fixed-coefficient and random-coefficient AR1 models. For the fixed-coefficient AR1 model, no prior assumption of stationarity is made, whereby

$$y_t = \phi_0 + \phi_1 y_{t-1} + u_t,$$

$$u_t \sim N(0, \sigma^2),$$

with priors

$$\phi_1 \sim N(0,1),$$

$$\phi_0 \sim N(0,20),$$

$$\sigma \sim N^+(0,1).$$

The estimates (posterior mean and st devn) for ϕ_0, ϕ_1, and σ are respectively 12.6 (1.25), 0.41 (0.06), and 0.143 (0.006). The LOO-IC is –255, with Figure 5.1A showing the extreme pointwise LOO-IC associated with certain observations.

The random-coefficient AR1 specifies

$$y_t = \phi_0 + \phi_{1t} y_{t-1} + u_t,$$

$$u_t \sim N(0, \sigma^2),$$

$$\phi_{1t} = \mu_{\phi 1} + \sigma_{\phi 1} \varepsilon_t,$$

$$\varepsilon_t \sim N(0,1),$$

where the parameterisation of ϕ_t follows Gelman et al. (2014), and provides improved MCMC sampling via rstan [2]. For extreme outliers, such as at t=120 and t=227, the mean likelihoods are higher under this model. However, the overall LOO-IC rises to –250, with the improved fit per se (lower ELPD-LOO) offset by a higher complexity measure (113 vs 9). The WAIC (widely applicable information criterion) also favours the simpler model (–256 as against –245). The parameter $\sigma_{\phi 1}$ has a mean of 0.0034, with posterior mean ϕ_{1t} varying from 0.401 to 0.423.

A GARCH(1,1) model (section 5.5) specifies

$$y_t = \phi_0 + u_t,$$

$$u_t \sim N\left(0, \sigma_t^2\right),$$

with variance model

$$\sigma_t^2 = \alpha_0 + \alpha_1 u_{t-1}^2 + \beta_1 \sigma_{t-1}^2.$$

This provides evidence of volatility, as in Figure 5.1B, but the LOO-IC deteriorates to –205. The α_1 and β_1 coefficients have skew posterior densities with respective means (medians) of 0.17 (0.07) and 0.67 (0.87).

Finally, an AR1 lag in y is added to the GARCH(1,1), namely

$$y_t = \phi_0 + \phi_1 y_{t-1} + u_t,$$

$$u_t \sim N(0, \sigma_t^2),$$

$$\sigma_t^2 = \alpha_0 + \alpha_1 u_{t-1}^2 + \beta_1 \sigma_{t-1}^2.$$

This provides a LOO-IC of –260, a slight improvement on the basic AR(1) model. The lagged effect of u_{t-1}^2 is now virtually eliminated, with posterior means (medians) for β_1 and ϕ of 0.58 (0.65) and 0.38 (0.39).

5.3 State-Space Priors for Metric Data

Nonstationary models based on state-space priors are widely used in applications where time series parameters are evolving through time, especially in analysing separate unobserved components representing trend, cyclical, or seasonal effects (West, 2013). The

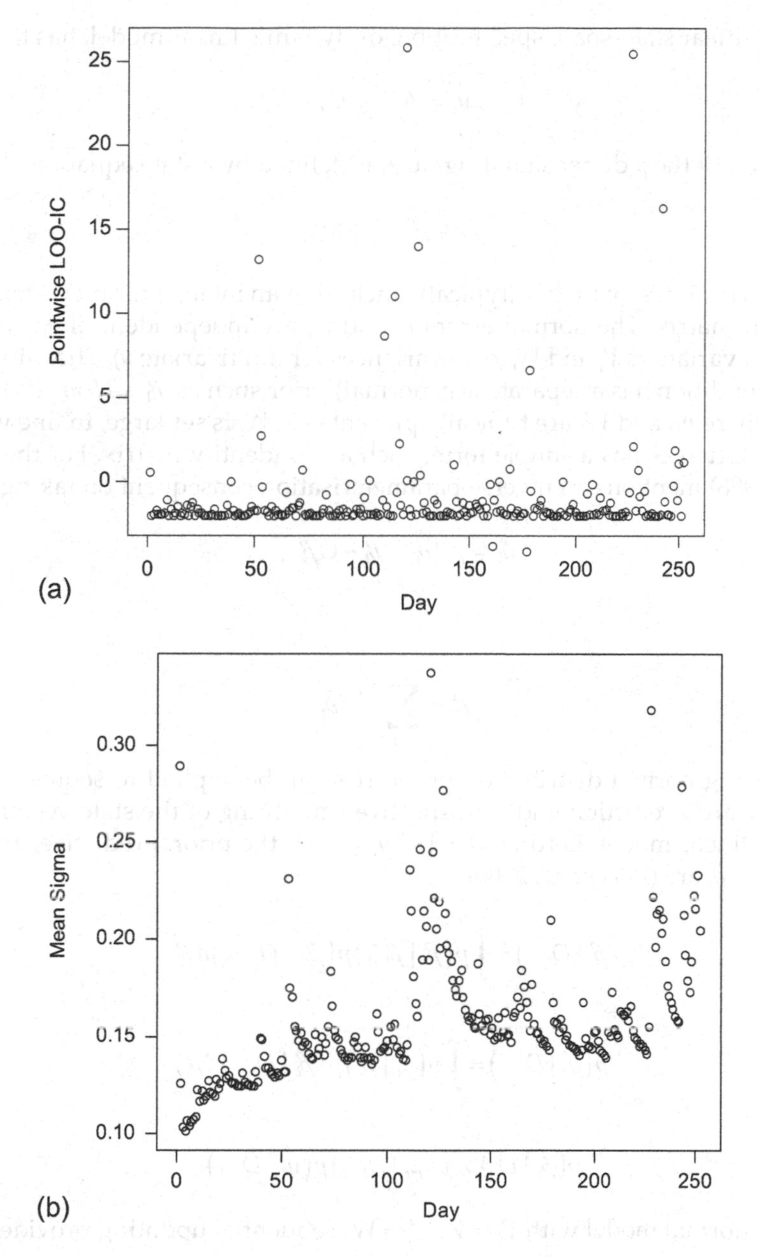

FIGURE 5.1
(A) Pointwise LOO-IC. Fixed Coefficient AR1 Model. (B) Posterior Mean σ. GARCH(1,1) Model.

idea that a time series is composed of several unobserved components contrasts with Box–Jenkins or ARMA methods that require differencing to eliminate trend or periodic effects and achieve stationary means and variances (Durbin, 2000, p.2). ARMA models are selected using autocorrelation and partial and autocorrelation functions that are subject to sampling variability, and quite different models can provide similar fits for the same series. In fact, ARMA sequences can be represented as particular instances of state-space models with implicit components. Among informative discussions on state-space vs Box–Jenkins methods, see Durbin and Koopman (2001, p.51) and Harvey and Todd (1983).

The normal linear state-space specification, or dynamic linear model, has the form

$$y_t = \beta_t X_t + u_t,$$

where evolution of the p dimensional signal β_t is defined by a state equation

$$\beta_t = \beta_{t-1} G_t + w_t,$$

with X_t being a $p \times 1$ design matrix (typically including an intercept), and G_t defining a $p \times p$ state evolution matrix. The normal errors u_t and w_t are independent of each other, with mean zero and variances V_t and W_t (or covariances for multivariate y). The initial state vector or initial condition has a separate (e.g. normal) prior such as $\beta_1 \sim N(m_1, W_1)$ (Strickland et al., 2008), where m_1 and W_1 are typically present (e.g. W_1 is set large, in line with diffuse expectations). Often G_t has a simple form, such as an identity matrix. For the case $G_t = G$, Gamerman (1998) mentions an inverse parameterisation consequent on taking

$$\delta_1 = \beta_1, \delta_t = \beta_t - G\beta_{t-1},$$

so that

$$\beta_t = \sum_{j=1}^{t} G^{t-j} \delta_j.$$

Algorithms using normal distribution properties can be applied to sequential updating (filtering), forward prediction and retrospective smoothing of the state vector in the normal dynamic linear model. Letting $D_t = (y_t, y_{t-1}, ..y_1)$, the prior, predictive, and posterior distributions of β_t are (Reis et al., 2006)

$$p(\beta_t | D_{t-1}) = \int p(\beta_t | \beta_{t-1}) p(\beta_{t-1} | D_{t-1}) d\beta_{t-1},$$

$$p(y_t | D_{t-1}) = \int p(y_t | \beta_t) p(\beta_t | D_{t-1}) d\beta_t,$$

$$p(\beta_t | D_t) \propto p(\beta_t | D_{t-1}) p(y_t | D_{t-1}).$$

For the linear normal model with $V_t = V, W_t = W$, sequential updating provides posteriors

$$\beta_t | D_t \sim N(m_t, C_t),$$

where

$$a_t = G_t m_{t-1},$$

$$e_t = y_t - X_t' a_t,$$

$$m_t = a_t + A_t e_t,$$

$$C_t = R_t - A_t A_t' q_t,$$

$$R_t = G_t C_{t-1} G_t' + W,$$

$$q_t = X_t' R_t X_t + V,$$

The one step ahead state and observation predictive densities are normal densities, namely,

$$(\beta_t \mid D_{t-1}) \sim N(a_t, R_t)$$

$$(y_t \mid D_{t-1}) \sim N(X_t' a_t, q_t).$$

5.3.1 Simple Signal Models

As an illustration of a normal state-space or dynamic linear model, assume that observations y_t are obtained with measurement error and in fact generated by a relatively smooth underlying signal β_t. This is a hierarchical model – analogous to the normal-normal model of Chapter 4 – with the first level being the observation equation, the second level being the state equation, and the priors on the variances and initial conditions defining hyperparameters at the third stage (Berliner, 1996). Assuming iid measurement errors u_t, one has an observation or measurement equation

$$y_t = \beta_t + u_t, \tag{5.1.1}$$

for $t = 1,\ldots,T$ and a state equation defining the evolution of the signal

$$\beta_t = \beta_{t-1} + w_t, \tag{5.1.2}$$

for $t = 2,\ldots,T$. This is also known as a local level model (Durbin and Koopman, 2001), or random walk plus noise model (Durbin, 2000), and the second stage is a nonstationary first order random walk or RW(1) prior, corresponding to the unit root case of an AR(1) prior.

As for the AR(1) prior, future values of the signal depend on (β_t, β_{t-1},..., β_1) only through the current value β_t. Denoting $\beta_{[t]} = (\beta_1,\ldots,\beta_{t-1})$, the conditional form of the RW(1) prior is

$$p(\beta_t \mid \beta_{[t]}, y) \propto \left\{ \begin{array}{ll} p(\beta_2 \mid \beta_1) p(\beta_1) p(y_1 \mid \beta_1) & t = 1 \\ p(\beta_{t+1} \mid \beta_t) p(\beta_t \mid \beta_{t-1}) p(y_t \mid \beta_t) & t = 2,\ldots T-1 \\ p(\beta_T \mid \beta_{T-1}) p(y_T \mid \beta_T) & t = T \end{array} \right\}$$

so that for times $t = 2,\ldots,T-1$ there is averaging over preceding and following states. The first period signal (initial condition) β_1 is typically taken as an unknown fixed effect with large variance, while the observation error u_t, and state error w_t are taken as respectively $N(0, V)$ and $N(0, W)$, and assumed uncorrelated in time, independent of one other, and also independent of the signal β_t.

Assume $\beta_1 \sim N(b_1, S_1)$, $1/V \sim \text{Ga}(a_u, b_u)$, $1/W \sim \text{Ga}(a_w, b_w)$ then the full conditionals are

$$\beta_1 \sim N\left(\left[\frac{\beta_2}{W} + \frac{b_1}{S_1} + \frac{y_1}{V} \right] \left[\frac{1}{W} + \frac{1}{S_1} + \frac{1}{V} \right]^{-1}, \left[\frac{1}{W} + \frac{1}{S_1} + \frac{1}{V} \right]^{-1} \right)$$

$$\beta_t \sim N\left(\left[\frac{(\beta_{t+1}+\beta_{t-1}}{W}+\frac{y_t}{V}\right]\left[\frac{2}{W}+\frac{1}{V}\right]^{-1},\left[\frac{2}{W}+\frac{1}{V}\right]^{-1}\right) \quad t=2,\ldots,T-1$$

$$\beta_T \sim N\left(\left[\frac{\beta_{T-1}}{W}+\frac{y_T}{V}\right]\left[\frac{1}{W}+\frac{1}{V}\right]^{-1},\left[\frac{1}{W}+\frac{1}{V}\right]^{-1}\right)$$

$$1/V \sim Ga\left(a_u+\frac{T}{2},b_u+0.5\sum_{t=1}^{T}(y_t-\beta_t)^2\right)$$

$$1/W \sim Ga\left(a_w+\frac{(T-1)}{2},b_w+0.5\sum_{t=2}^{T}(\beta_t-\beta_{t-1})^2\right).$$

Higher order random walks in the signal are another possibility, with a kth order random walk having prior

$$\Delta^k\beta_t \sim N(0,W)$$

(Berliner, 1996; Kitagawa and Gersch, 1996; Fahrmeir and Lang, 2001). For example, a second difference random walk or RW(2) prior specifies $y_t=\beta_t+u_t$ and state equation $\Delta^2\beta_t=w_t$. Hence

$$\Delta(\Delta\beta_t)=\Delta(\beta_t-\beta_{t-1})=\Delta\beta_t-\Delta\beta_{t-1}=(\beta_t-\beta_{t-1})-(\beta_{t-1}-\beta_{t-2})=w_t$$

and the RW(2) prior can be stated as

$$\beta_t \sim N(2\beta_{t-1}-\beta_{t-2},W).$$

Whereas first order random walks penalise abrupt jumps between successive values, the RW(2) prior penalises deviations from a linear trend. The RW(2) and higher order RW priors therefore lead to a smoother evolution of β_t through time. This is relevant not just to time series, but to processes operating on other time scales (e.g. age, cohort), for example, in survival analysis or in graduating (smoothing) demographic schedules (Carlin and Klugman, 1993).

5.3.2 Sampling Schemes

Different MCMC sampling schemes have been proposed for state-space models according to the form of outcome (e.g. metric or discrete) and the form of the observation-state equations (e.g. linear or nonlinear). Multi-state or joint sampling of the state vectors β_t is generally more efficient than single-state sampling that updates one state parameter vector at a time (Knorr-Held, 1999). Joint sampling for β when y is metric is discussed by Carter and Kohn (1994) and Fruhwirth-Schnatter (1994), while de Jong and Shephard (1995) focus on sampling the u_t and w_t error series, as opposed to the state effects β_t; recent overviews are provided by Reis et al. (2006) and Simpson et al. (2017). Gamerman (1998) proposes updating via the δ_t rather than the usually highly correlated β_t using the re-parameterisation mentioned above.

Knorr-Held (1999) uses properties of the penalty (inverse covariance) matrix of the joint density for the state vectors as a basis for sampling sub-blocks of the elements $(\beta_1, \ldots, \beta_T)$. Thus Gaussian state-space priors can be written in joint form as

$$p(\beta_1, \ldots, \beta_T \mid W) \propto \exp\left(-\frac{\beta' K \beta}{2W}\right),$$

where the penalty matrix K is determined by the form of autoregressive prior. For a first order random walk with $\beta_t \sim N(\beta_{t-1}, W)$, the penalty matrix is

$$K = \begin{pmatrix} 1 & -1 & & & & & \\ -1 & 2 & -1 & & & & \\ & -1 & 2 & -1 & & & \\ & \ldots & \ldots & \ldots & & & \\ & & & -1 & 2 & -1 & \\ & & & & -1 & 2 & -1 \\ & & & & & -1 & 1 \end{pmatrix},$$

while for a second order random walk with $\beta_t \sim N(2\beta_{t-1} - \beta_{t-2}, W)$, one has

$$K = \begin{pmatrix} 1 & -2 & 1 & & & & & & \\ -2 & 5 & -4 & & & & & & \\ 1 & -4 & 6 & -4 & 1 & & & & \\ & 1 & -4 & 6 & -4 & 1 & & & \\ & \ldots & \ldots & \ldots & \ldots & & & & \\ & & & 1 & -4 & 6 & -4 & 1 & \\ & & & & 1 & -4 & 6 & -4 & 1 \\ & & & & & 1 & -4 & 5 & -2 \\ & & & & & & 1 & -2 & 1 \end{pmatrix}.$$

For an RW(p) prior at equally spaced time points, the elements of the matrix K (apart from edge effects) are expressible as

$$k_{ij} = (-1)^{|i-j|} \binom{2p}{p - |i-j|} \quad \text{if } |i-j| \leq p,$$

and $k_{ij} = 0$ otherwise.

Let β_{ab} denote the subvector $(\beta_a, \beta_{a+1}, \ldots, \beta_b)$ of state effects, and K_{ab} denote the corresponding submatrix of K. Let $K_{1,a-1}$ and $K_{b+1,T}$ denote the submatrices to the left and right of K_{ab}, namely

$$K = \begin{pmatrix} & K'_{1,a-1} & \\ K_{1,a-1} & K_{ab} & K_{b+1,T} \\ & K'_{b+1,T} & \end{pmatrix}.$$

Then the conditional density for β_{ab} given $\beta_{1,a-1}$, $\beta_{b+1,T}$ and W, is normal $\beta_{ab} \sim N(\nu_{ab}, WK_{ab}^{-1})$, where

$$\nu_{ab} = \begin{array}{ll} -K_{ab}^{-1}K_{b+1,T}\beta_{b+1,T} & a=1 \\ -K_{ab}^{-1}\left[K_{1,a-1}\beta_{1,a-1} + K_{b+1,T}\beta_{b+1,T}\right] & a>1, b<T. \\ -K_{ab}^{-1}K_{1,a-1}\beta_{1,a-1} & b=T \end{array}$$

Using this density, a Metropolis-Hastings block sample may be used to update the full conditional

$$p(\beta_{ab} \mid) \propto \prod_{t=a}^{b} p(y_t \mid \beta_t) p(\beta_{ab} \mid \beta_{b+1,T}, \beta_{1,a-1}, W).$$

This involves drawing a proposal β_{ab} from $N(\nu_{ab}, WK_{ab}^{-1})$ with $\{\nu_{ab}, K_{ab}\}$ evaluated at the current sampled values β and W in a chain, with the proposal accepted or rejected according to a probability

$$\min\left(1, \prod_{t=a}^{b} p(y_t \mid \beta_t^*) \Big/ \prod_{t=a}^{b} p(y_t \mid \beta_t)\right),$$

that may be calculated by comparing likelihoods only (Knorr-Held, 1999, p.134).

5.3.3 Basic Structural Model

To allow for a trend in the mean level or signal, one may extend the state equation in (5.1) to include a stochastic increment, so that

$$y_t = \beta_t + u_t$$

$$\beta_t = \beta_{t-1} + \Delta_t + w_{1t}$$

$$\Delta_t = \Delta_{t-1} + w_{2t}$$

where Δ_t represent the changing slope of the trend. This provides the local linear trend model or dynamic trend model (Fruhwirth-Schnatter, 1994).

A constant parameter Δ provides a linear trend, as in the Carter–Lee mortality forecasting model considered by Pedroza (2006); this is sometimes known as a random walk with drift. Other variations on the local linear model in (5.1) include autoregressive rather than random walk state equations, such as

$$\beta_t = \phi\beta_{t-1} + w_t$$

as in Carlin et al. (1992, p.496). An autoregression or random walk in y itself might be added, as in Ghosh and Tiwari (2007), who assume a local linear model for common cancer deaths of the form $y_{t+1} \sim N(y_t + \beta_t, V)$.

The basic structural model (BSM) or unobserved components model (Koopman, 1993; Koopman et al., 1999) adds seasonal effects s_t to the above local linear trend model, so that with μ_t representing the level of the series, one has

$$y_t = \mu_t + s_t + u_t,$$

$$\mu_t = \mu_{t-1} + \Delta_t + w_{1t},$$

$$\Delta_t = \Delta_{t-1} + w_{2t},$$

$$s_t + s_{t-1} + \cdots s_{t-S+1} = w_{3t},$$

where S is the number of seasons, and $w_{jt} \sim N(0, W_j)$. Relevant R packages for estimating the BSM include stsm (via maximum likelihood), and bsts and dlm (via Bayesian estimation).

Fruhwirth-Schnatter (1994) sets out the full conditionals for this model under gamma priors for the precisions $1/W_j$. The last equation provides the time domain prior for seasonal effects, whereas a frequency domain prior specifies

$$s_t = \sum_{j=1}^{[S/2]} s_{jt},$$

$$s_{jt} = s_{j,t-1}\cos(\lambda_j) + v_{j,t-1}\sin(\lambda_j) + w_{3t},$$

$$v_{jt} = -s_{j,t-1}\sin(\lambda_j) + v_{j,t-1}\cos(\lambda_j) + w_{4t},$$

where $\lambda_j = 2\pi j / S$ and $[S/2]$ denotes the integer part of $S/2$.

Certain series (e.g. natural phenomena) may show unknown periodicities, so cyclical components are added as well as, or instead of, seasonal components. For example, Piegorsch and Bailer (2005, p.229) consider unknown frequencies in carbon dioxide concentrations from Mauna Loa volcano in Hawaii. So for a local linear trend model with a single unknown cycle

$$y_t = \mu_t + c_t + u_t,$$

$$\mu_t = \mu_{t-1} + \Delta_t + w_{1t},$$

$$\Delta_t = \Delta_{t-1} + w_{2t},$$

$$c_{t+1} = c_t\cos(\lambda) + d_t\sin(\lambda) + w_{3t},$$

$$d_{t+1} = -c_t\sin(\lambda) + d_t\cos(\lambda) + w_{4t},$$

where λ is an unknown frequency.

5.3.4 Identification Questions

Identification issues in state-space random effect models occur for two main reasons. One is that the mean or level of the state effects is not specified (rather the mean of pairwise or higher order differences is specified). The other is the presence of multiple confounded sources of random variation, as in the basic structural model with level and seasonal effects, whereas the data can only identify the sum of the random effects $u_t + s_t$. These questions raise issues in MCMC sampling, for example, whether effects need to be centred at each iteration, because an intercept (if included) will otherwise be confounded with the means of the random effects.

To exemplify issues occurring due to the mean of the latent series consider the measurement error with RW(1) signal model in 5.3.1. The state equation can be stated as

$$\Delta\beta_t = \beta_t - \beta_{t-1} \sim N(0,W),$$

so the prior only defines a level for differences in β_t, but the level of the (undifferenced) β_t is not defined by the prior. If the model for y_t does not have a separate intercept parameter, the level of the β_t will be identified by the level of the y_t. Suppose though that the observation equation includes a separate constant γ_0 with

$$y_t = \gamma_0 + \beta_t + u_t.$$

Then γ_0 and the mean of the β_t are confounded and for identification one may apply a centring or corner constraint to the β_t. An identifying corner constraint involves setting a single β_t to a known value; taking the initial condition β_1 to have a known value, e.g. $\beta_1=0$, is one option (Clayton, 1996). By contrast, if the initial conditions (β_1 in an RW(1) prior, β_1 and β_2 in an RW(2) prior, etc) are taken as unknowns, then a centring constraint may be applied at each MCMC iteration, so that the centred β_t satisfy $\sum_{t=1}^{T} \beta_t = 0$.

As in other models with multiple sources of random variation, priors on the variance components in state-space models may affect inferences. This is not simply a matter of selecting prior densities for scale parameters, but of also a question of how such priors influence the partitioning of total random variation. One may recognise the interdependence between variance components using devices such as uniform priors on shrinkage ratios $B=V/V+W$ combined with a prior on V or $V+W$ (Daniels, 1999). Alternatively (V, W) may be reparameterised as (V, qV), where q is a signal to noise ratio. So the prior on q might be centred on 1 in line with a prior belief that signal and observation variances are equal.

These approaches extend to models with competing sources of variation in the state equation. Consider the three errors w_{jt} (for levels, slopes, and seasonals) in the basic structural model. Denoting $W_j = \text{Var}(w_{jt})$ and $V = \text{Var}(u_t)$, one may set $W_j = q_j V$ where q_j are signal to noise ratios (Koopman, 1993; Harvey, 1989, p.33). One may then set priors on the q_j separately (e.g. separate gammas), or jointly; for example, via a multivariate normal on the $\log(q_j)$. Another option is a prior on V and uniform priors on the ratios $V/(V+W_j)$. Such devices amount to assuming prior correlation between the respective variances.

An alternative approach to ensure stable identification is to set informative priors on the variance of each random walk, possibly based on expected stochastic variation around a deterministic trend. For example, following Berzuini and Clayton (1994), for counts $y_t \sim Po(\lambda_t)$, consider a second order random walk for $\beta_t = \log(\lambda_t)$

$$\beta_t = 2\beta_{t-1} - \beta_{t-2} + w_t$$

then the value $W=0$ for $\text{Var}(w_t)$ corresponds to a log-linear deterministic relationship between the λ_t and time. To allow for stochastic variation, one may assume $\nu W^*/W \sim \chi^2_\nu$, or equivalently

$$1/W \sim Ga\left(\frac{\nu}{2}, \frac{W^*\nu}{2}\right),$$

where W^* is a prior setting for W, and higher values of ν represent stronger degrees of belief in that setting. For example, taking $W^*=0.01$ corresponds to assuming a 95% probability that λ_t will be within −18 and +22% of a log-linear extrapolation from β_{t-1}.

The single source of error approach (Ord et al., 2005) may also assist in achieving parsimony, and in resolving the partitioning of variance between multiple sources of variation in unobserved component models. Thus, the local linear trend model in multiple source of error (MSOE) form is

$$y_t = \mu_t + u_t,$$

$$\mu_t = \mu_{t-1} + \Delta_t + w_{1t},$$

$$\Delta_t = \Delta_{t-1} + w_{2t},$$

but in single source of error (SSOE) form is

$$y_t = \mu_t + u_t,$$

$$\mu_t = \mu_{t-1} + \Delta_t + \lambda_1 u_t,$$

$$\Delta_t = \Delta_{t-1} + \lambda_2 u_t,$$

where λ_1 and λ_2 are loadings. By contrast to the MSOE scheme, the state and observation errors are now correlated.

Example 5.3 Air Passenger Data

As an application of the basic structural model, consider monthly air passenger totals using London airports (Heathrow, Gatwick, etc.) from January 1999 through to March 2014, so $T = 183$). Totals are in millions. A monthly seasonal effect is assumed so there are S-1 = 11 initial conditions for the s_t sequence. Then

$$y_t = \mu_t + s_t + u_t,$$

$$\mu_t = \mu_{t-1} + \Delta_t + w_{1t},$$

$$\Delta_t = \Delta_{t-1} + w_{2t},$$

$$s_t + s_{t-1} + \ldots s_{t-S+1} = w_{3t},$$

with $w_{jt} \sim N(0,\sigma_{j+1}^2)$, and normal observation errors $u_t \sim N\left(0,\sigma_1^2\right)$. Half t(0,1) priors with 4 degrees of freedom are assumed on the σ_j. For $t=1$, the μ_t and Δ_t series refer to pre-series values which are assigned $N(0,10)$ priors.

Convergence is obtained readily in a two-chain run of 5000 iterations using rstan, with a LOO-IC of 44.1. The posterior means (medians) of the σ_j are 0.202 (0.202), 0.174 (0.173), 0.0036 (0.0025) and 0.0136 (0.0118).

Figure 5.2A–C show respectively the clear seasonal variations, the generally upward trend in the slope parameters Δ_t (though most evident in the early part of the period), and the combined level and trend. These series all include forecasts for nine extra months through to the end of 2014. A similar slope trajectory is estimated

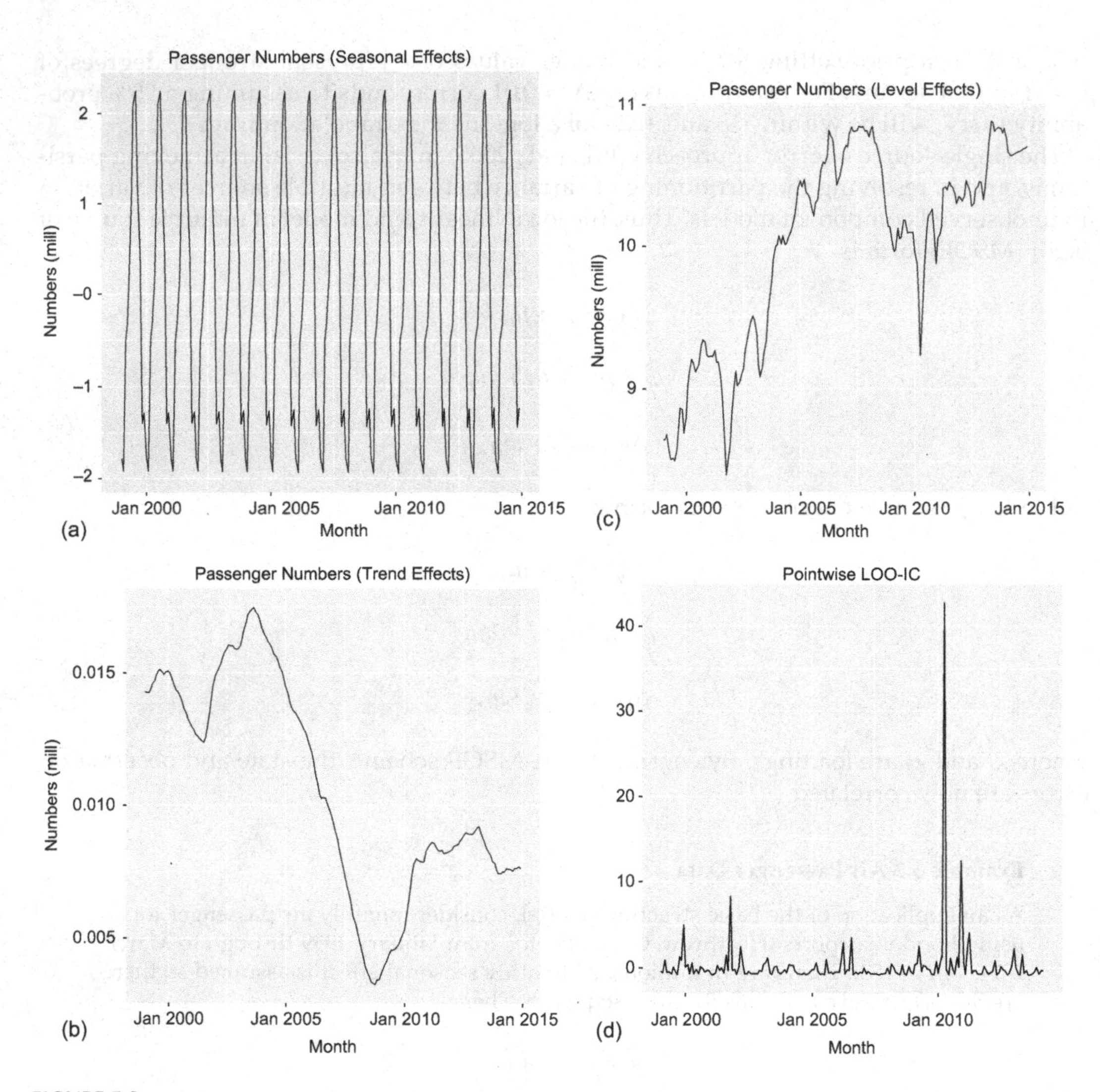

FIGURE 5.2
(A) Passenger numbers, seasonal effects. (B) Passenger numbers, trend effects. (C) Passenger numbers, level effects. (D) Passenger numbers model, pointwise LOO-IC.

using the R program rucm. Reversals to the broad upward trend in modelled passenger numbers, as in Figure 5.2C, reflect especially the recession of 2008–09, as well as more distinct outliers for individual months. An examination of the pointwise LOO-IC, as in Figure 5.2D, shows the most discrepant month ($t=136$) to be April 2010, reflecting the impact on flights of the Eyjafjallajökull volcanic eruption in Iceland.

To alleviate the impact of outlier values, a student t observation model is also estimated, namely $u_t \sim t(0, \nu, \sigma_1^2)$. The unknown degrees of freedom ν is assigned an $E(0.1)$ prior. This provides an improved LOO-IC of 16 with a posterior mean (median) of ν of 2.48 (2.33), with the posterior mean (median) for σ_1 reduced to 0.093 (0.092).

Estimation of the basic structural model is also straightforward with R-INLA, with the simplest code involving a random effect that combines level and trend. The pointwise WAIC from a normal errors-based model reproduces the extreme outlier at $t=136$.

Example 5.4 Global Sea Level Change

This example compares in-sample predictions and out of sample forecasts from a local linear model with linear trend, and a simple hierarchical model involving unit level linear trends. Identifiability issues and their resolution are discussed.

A number of studies have analysed local relative sea-level records from tide gauge observations, and considered broader inferences regarding global mean sea level change. Local confounding factors may hinder quantifying a "global" signal from such data. However, Patwardhan and Small (1992) consider a set of stations from around the world with relatively long continuous records that were representative of other stations in the same region, and seemed relatively free of local confounding factors.

The analysis here with jagsUI follows them in using records for 1900–1980 from five stations, namely San Francisco, Tonoura (Japan), Sydney, Bombay, and Cascais (Portugal), with out-of-sample forecasts to 2000. Therefore, the data file has T = 101 points, with the last 20 being recorded as NA. The data are in mm from the Permanent Service for Mean Sea Level website (www.psmsl.org/).

For station j at time t, the model used by Patwardhan and Small (1992) involves a 1st order random walk in the mean global sea level M_t plus a homogenous linear trend (common coefficient b across sites j). So model 1 has

$$y_{jt} = M_t + bt + u_{jt}$$

$$u_{jt} \sim N(0, \sigma_u^2),$$

$$M_t \sim N(M_{t-1}, \sigma_M^2) \quad (t > 1),$$

with the initial condition M_1 assigned a diffuse N(6900,10000) prior. A gamma prior is assumed for $\xi = \tau_u + \tau_M = 1/\sigma_u^2 + 1/\sigma_M^2$, so with $\kappa = \tau_u/\xi \sim U(0,1)$, one obtains $\tau_u = \kappa\xi$ and $\tau_M = (1-\kappa)\xi$. Patwardhan and Small mention that compilations of trends in relative sea level data suggest an upward trend of 0.5–3.0 mm/year, so a $N(0,1)$ prior on b seems reasonable.

For improved identification and convergence, the M_t series are differenced with respect to M_1, namely $\Delta_t = M_t - M_1$, and a level parameter β_0 is introduced. So the M_t are effectively represented as $\Delta_t + \beta_0$, and the observation model is $y_{jt} = \beta_0 + \Delta_t + bt + u_{jt}$. Convergence is much delayed without using this re-expression. An alternative device is centring, whereby $\Delta_t = M_t - \bar{M}$.

An alternative model (model 2) allowing site-specific linear trends is considered, namely

$$y_{jt} = M_t + b_j t + u_{jt},$$

$$M_t \sim N(M_{t-1}, \sigma_M^2),$$

$$u_{jt} \sim N(0, \sigma_u^2),$$

$$b_j \sim N(\mu_b, \sigma_b^2),$$

$$\mu_b \sim N(0,1).$$

A relatively informative exponential $E(1)$ prior for $1/\sigma_b^2$ is adopted, as diffuse options lead to delayed convergence. The same identification strategy as under model 1 is adopted for the M_t series.

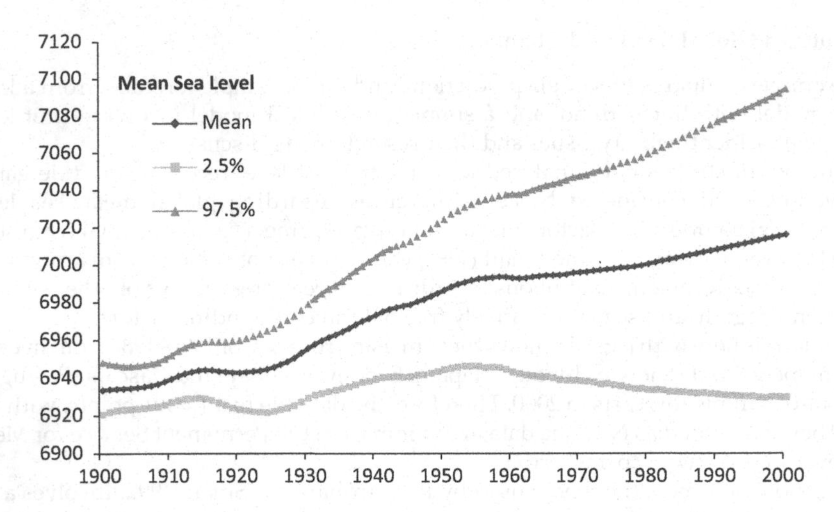

FIGURE 5.3
Modelled global sea level.

For model 1, a two-chain run using jagsUI converges after 20,000 iterations. There is a mean (95% CRI) linear growth rate b of 0.85 (−0.08, 1.54). Variation around the random level component M_t is comparatively small, with the posterior median of σ_M^2 standing at 7.6, compared to a median 2005 for σ_u^2. A posterior predictive p-test based on squared deviations is satisfactory. As to fit, let $y_{\text{rep},jt}$ be replicate data from the model. Then a posterior predictive loss criterion is calculated (within the observed data period to 1980, and with $k=1000$) as

$$\text{PPL} = \frac{1}{R}\frac{k}{k+1}\sum_{t=1}^{80}\sum_{j=1}^{5}\sum_{r=1}^{R}(y_{\text{rep},jt}^{(r)} - y_{jt})^2 + \sum_{t=1}^{80}\sum_{j=1}^{5} V(y_{\text{rep},jt}).$$

The respective components are obtained as 1629 and 1054 (in units of 1000), with the second a measure of complexity.

For model 2, a two-chain analysis with jagsUI converges after 10,000 iterations. This analysis gives a mean (95% interval) for μ_b of 0.78 (−0.15,1.68), while site level mean growth rates range from 0.30 to 1.89. Variation in the random components (M_t and b_j) is such as to reduce the median σ^2 to 1378. The respective PPL components are also much reduced, namely to 1116 and 731. Figure 5.3 shows the evolution of modelled global sea level to 1980 and forecasts thereafter.

5.3.5 Nonlinear State-Space Models for Continuous Data

Assumptions such as linearity of state transitions and additive normal errors are often not realistic, for example, in population dynamics (Maunder et al., 2015). In fisheries population dynamics, a state-space approach recognises both biological or process variation (randomness in population dynamics), and measurement error in the observations (Auger-Méthé et al., 2016). The process model may focus on the total biomass of a particular fish species (in a particular fishing region) above the minimum legal catch size, or the biomass of fish for particular age or length groups. In the former aggregate case, the state equation typically specifies the current biomass as a function of the biomass in the previous period, additions due to production (natural increase or other forms of recruitment), and removals due to fishing or natural mortality. The observations for this case are C_t, observed catch totals, and an index of abundance, A_t, such as the catch per unit effort, though such indices are often imperfect measures (Maunder et al., 2006).

A widely applied population dynamics model is the logistic function of Schaefer (1954) whereby biomass at period $t+1$, B_{t+1}, is represented as

$$B_{t+1} = [B_t + g(B_t) - C_t]\varepsilon_t^p,$$

where ε_t^p are multiplicative errors, and $g(B_t)$ represents "surplus production" as a function of biomass. Thus, one of the observation series, namely C_t, appears in the process model. The Schaefer model involves three parameters, r, K, and q, which can be interpreted respectively as the maximum intrinsic growth rate, the arithmetic mean biomass at unexploited equilibrium (or carrying capacity), and a catchability parameter (or proportionality constant). These parameters define the surplus production function, namely

$$g(B_t) = rB_t(1 - B_t/K),$$

and the abundance observation model

$$A_t = qB_t\varepsilon_t^o.$$

Additional parameters are the variances $\sigma_{2o} = \text{var}(\varepsilon_t^o)$ and $\sigma_{2p} = \text{var}(\varepsilon_t^p)$ of the observation and process models. Identification may be improved by using several abundance indices, so that

$$A_{jt} = q_j B_t \varepsilon_{jt}^o.$$

Typically, lognormal likelihoods are adopted for both the process and observation models. Derived parameters of interest include the maximum sustainable yield, $\text{MSY} = rK/4$.

Estimation of this model may well require informative priors for at least some of the parameters. The data may contain relatively little information on the parameters, so the prior may considerably influence the posterior. The literature has discussion about appropriate forms of prior, such as a uniform or lognormal prior for K, or uniform on $\log(K)$ (Punt and Hilborn, 1997; McAllister, 2014). For q, McAllister (2014) suggests a uniform density for $\log(q)$ over (−20,2), namely a diffuse prior concentrated on values under one, but including values above one. By contrast, Parent and Rivot (2013) suggest a $U(-20,20)$ prior on $\log(q)$, while Rankin and Lemos (2015) assume $\log(q) \sim U(-20,-3)$. For the rate of natural increase, r, there may be more substantive prior evidence, though Parent and Rivot (2013) adopt a $U(0.01,3)$ prior. Regarding the process and observation series variances, Parent and Rivot (2013) propose a parameter λ governing the ratio σ_p/σ_o, but in an application, assume $\lambda = 1$ and adopt a diffuse $U(-20,20)$ prior on the log of the common variance σ^2. McAllister (2014) discusses the basis for more informative priors on σ_p^2; for example, a value $\sigma_p = 0.05$ results in an interannual change in total recruited stock biomass of about 5%. Rankin and Lemos (2015) follow Parent and Rivot (2013) in adopting relatively diffuse priors except on K.

Example 5.5 North Pacific Blue Shark

This example illustrates fish population dynamics, focusing on the Pacific blue shark population in the North Pacific, as considered by the ISC Shark Working Group (2013). The observed data on catch (in units of 1000 metric tons), and an availability index, run from 1976 to 2010. Relatively diffuse priors are adopted, following recommendations in the literature (Parent and Rivot, 2013; Rankin and Lemos, 2015), except for K, where a uniform prior, $\log(K) \sim U(3.92, 7.6)$ follows the ISC Shark Working Group (2013). Thus for r, q, and σ^2, the priors are $r \sim U(0.01, 3)$, $\log(q) \sim U(-20, 2)$, and $\log(\sigma^2) \sim U(-10, 10)$.

The biomass series is expressed as

$$P_t = B_t / K$$

as in Meyer and Millar (1999). The lognormal prior on the initial condition P_1 is as in ISSWG (2013).

In the first model, the process and observation priors are related according to $\sigma_p = \lambda\sigma_o$, with the prior for λ exponential, $\lambda \sim E(1)$, centred at 1. Convergence using rstan is rapid, with the LOO-IC pooling over process and observation likelihoods, as the process likelihood involves observing the catch data. The LOO-IC is −99, with posterior mean (median) estimates of the maximum sustainable yield (MSY) of 77.6 (68.1), higher than the estimate of 52 in ISCSWG (2013). Estimates may be affected by the use in ISCSWG (2013) of the extended Fletcher–Schaefer model, and by the inclusion in ISCSWG (2013) of five earlier years when the abundance index was missing. The carrying capacity K has posterior mean (median) of 1130 (1105). The mean for the A_t series can be modified numerically by changing K or q, and samples of these parameters are negatively correlated, a feature that might be used in setting a prior.

In a second model, it is assumed that $\sigma_p = \sigma_o$, and this model has a higher LOO-IC, namely −89. Under this model, the posterior mean (median) estimates of the MSY are 73.5 (66.7). Both analyses show biomass at low levels in the late 1980s, as in Figure 8 of ISCSWG (2013). Under the first model, the posterior median biomass for 1989 is 588, compared to 1213 in 1976, and 994 in 2010.

5.4 Time Series for Discrete Responses; State-Space Priors and Alternatives

Dynamic generalised linear models extend the Gaussian state-space representation to outcomes with density $p(y_t|\zeta_t)$ belonging to the exponential family of distributions, where ζ_t is the natural parameter (Helske, 2017; Soyer et al., 2015; Davis et al., 2016). One may also condition on the history of previous observations plus previous and current predictors, $D_{t-1} = (y_{t-1}, y_{t-2}, \ldots, y_1, X_t, \ldots, X_1)$ to allow for observation driven components in the model (Fahrmeir and Tutz, 2001, p.242). So

$$p(y_t \mid \zeta_t, D_{t-1}) = \exp\left[\phi_t \left\{y_t\zeta_t - b(\zeta_t)\right\}\right] c(y_t, \phi_t)$$

with $\mu_t = E(y_t \mid \zeta_t) = b'(\zeta_t)$ and μ_t linked to a linear predictor η_t via a link function g, $g(\mu_t) = \eta_t$. Also a known scale parameter ϕ_t defines the conditional variance $\text{Var}(y_t \mid \zeta_t) = b''(\zeta_t)/\phi_t$.

Then an observation equation for design matrix X_t of dimension p would typically be of the form

$$g(\mu_t) = \eta_t = \beta_t X_t + u_t,$$

with state or system equation

$$\beta_t = \beta_{t-1}G_t + w_t,$$

where $w_t \sim N_p(0,W)$. The error $u_t \sim N(0,V)$ is not necessarily included for discrete responses, but may be necessary to represent unstructured extra-variation.

An alternative state-space approach, sometimes termed a linear Bayes approach, involves conjugate priors for the natural parameters and a guide relationship

$$h(\zeta_t) = \beta_t X_t,$$

linking the natural parameters to the state vector (West et al., 1985, p.74; Ferreira and Gamerman, 2000, p.60). So with time specific parameters (g_t,h_t), the prior for the natural parameter at time t is

$$p(\zeta_t \mid D_{t-1}, g_t, h_t) = k(g_t, h_t)\exp[g_t\zeta_t - h_t b(\zeta_t)],$$

while the updated natural parameters have density

$$p(\zeta_t \mid D_t, g_t, h_t) = k(g_t, h_t)\exp\left[(g_t + \phi_t y_t)\zeta_t - (h_t + \phi_t)b(\zeta_t)\right].$$

As for normal linear state-space models, the state vector may include level, trend and seasonal effects. For an underlying signal model (with X_t containing only an intercept), the regression and state equations become (Kitagawa and Gersch, 1996, Ch 13),

$$g(\mu_t) = \beta_t + u_t,$$

$$\Delta^k \beta_t = w_t,$$

with $u_t \sim N(0,V)$, $w_t \sim N(0,W)$. Thus Kashiwagi and Yanagimoto (1992) consider Poisson data on disease counts $y_t \sim Po(\mu_t)$, and take $k=1$ in the signal equation.

For binary data with $\pi_t = Pr(y_t = 1)$, a signal may be combined with randomly time-varying dependence on lagged responses (Cox, 1970), providing a parameter driven representation, whereas an observation-driven model would only involve fixed effect coefficients on lagged observed y_t values (Wu and Cui, 2014). For example, a time-varying level and lag 1 effect could specify

$$g(\pi_t) = \eta_t = \beta_{1t} + \beta_{2t}y_{t-1},$$

$$(\beta_{1t}, \beta_{2t}) \sim N_2([\beta_{1,t-1}, \beta_{2,t-1}], W).$$

Time series of categorical data vectors, namely $y_t = (y_{t1}, y_{t2}, \ldots y_{tJ})$ with only a single $y_{tj}=1$ if (say) diagnosis j applies, or mutually exclusive choices j made at time t, are multinomial according to

$$y_t = (y_{t1}, y_{t2}, \ldots y_{tJ}) \sim \text{Mult}(1, [p_{t1}, p_{t2}, \ldots p_{tJ}]).$$

Typically, a multiple logit link is assumed for the unknown probabilities p_{tj} (Fahrmeir and Tutz, 2001; Cargnoni et al., 1996). A signal model would then involve a $(J-1)$ dimensional state vector, though by analogy to binary Markov dependence, the regression term η_{tj} for the jth choice may also involve lags on both the same response $y_{t-k,j}$, and lagged cross-responses $y_{t-k,m}$ $(m \neq j)$. For a general predictor, possibly varying by category, X_{tj}, one has

$$p_{tj} = \exp(\beta_{tj} X_{tj}) \Big/ \sum_{j=1}^{J} \exp(\beta_{tj} X_{tj}),$$

where $\beta_{tJ} = 0$ for identifiability. Cross-series borrowing of strength via random walk priors may be applied for the $J-1$ category specific state vectors β_{tj}. Thus, for the coefficient on predictor k, X_{tjk}, one might have

$$\beta_{tk} \sim N_{J-1}(\beta_{t-1,k}, \Sigma_k),$$

where $\beta_{tk} = (\beta_{t1k}, \ldots, \beta_{t,J-1,k})$, and Σ_k is of dimension $J-1$.

An alternative for binary and multinomial responses is to introduce the augmented metric data y_t^* that underlie the observed discrete responses. Thus, for binary data, consider the scheme

$$y_t^* = \beta_t X_t + u_t,$$

where y_t^* is positive or negative according as $y_t = 1$ or $y_t = 0$, and the variance of u_t is assumed known for identifiability, usually with var($u_t = 1$). A simple signal model with $X_t = 1$ may then be expressed as

$$y_t^* \mid W, y_t, \beta_t \propto N(\beta_t, 1) \;\; I(0, \infty) \text{ if } y_t = 1$$

$$y_t^* \mid W, y_t, \beta_t \propto N(\beta_t, 1) \;\; I(-\infty, 0) \text{ if } y_t = 0$$

$$\beta_t \sim N(\beta_{t-1}, W).$$

5.4.1 Other Approaches

Other general schemes for modelling time series of exponential family data include the generalised autoregressive moving average (GARMA) representation (Benjamin et al., 2003; Li, 1994; Silveira de Andrade et al., 2015). The GARMA representation involves conditional means μ_t, link function $g(\mu_t) = \eta_t$, and regression term in the form

$$\eta_t = \gamma_t X_t + \sum_{j=1}^{p} \phi_j \left[g(y_{t-j}) - \gamma_{t-j} X_{t-j} \right] + \sum_{k=1}^{q} \gamma_k \left[g(y_{t-k}) - \eta_{t-k} \right].$$

For example, for Poisson data $y_t \sim Po(\mu_t)$ and $y_t^* = \max(y_t, m)$ for a small positive constant m, one has

$$\log(\mu_t) = \gamma_t X_t + \sum_{j=1}^{p} \phi_j \left[\log(y_{t-j}^*) - \gamma_{t-j} X_{t-j} \right] + \sum_{k=1}^{q} \gamma_k \left[\log(y_{t-k}^* / \mu_{t-k}) \right].$$

More general autoregression in the state vector (not limited to random walks) may be adopted. Thus Oh and Lim (2001) and Chan and Ledolter (1995) adopt an autocorrelated error θ_t for count data with

$$y_t \sim Po(e^{\eta_t}),$$

$$\eta_t = \theta_t + X_t\beta,$$

$$\theta_t = \rho\theta_{t-1} + w_t,$$

where ρ is constrained to stationarity, and the w_t are normally distributed. Utazi (2017) considers a variant of this model allowing a changepoint in the autoregressive parameter ρ.

Dependence on lagged counts can also be achieved by binomial thinning (Silva et al., 2009), whereby

$$\eta_t = \beta_t X_t + \rho \circ y_{t-1},$$

is equivalent to $\eta_t = \beta_t X_t + h_t$, $h_t \sim \text{Bin}(y_{t-1}, \rho)$, and by thinning schemes applicable to both count and categorical data (Angers et al., 2017; Khoo and Ong, 2014). Conjugate mixture schemes for time series counts are exemplified by Jowaheer and Sutradhar (2002), and Bockenholt (1999), with, for instance,

$$y_t \sim Po(e^{\eta_t}\kappa_t),$$

$$\kappa_t \sim Ga\left(\frac{1}{c}, \frac{1}{c}\right),$$

where marginally $\text{var}(y_t) = \exp(\eta_t) + c\exp(2\eta_t)$.

Autoregression on both past observations and means for count data is included in autoregressive conditional Poisson (ACP) models (Heinen, 2003; Fokianos et al., 2009). Classical estimation is implemented in the R package acp (https://cran.r-project.org/web/packages/acp/acp.pdf). The ACP model is a particular case of the ARMA-GLM model for counts set out by Liboschik et al. (2017). Poisson means in the ACP(p,q) model are specified as

$$\mu_t = \omega + \sum_{j=1}^{p} \phi_j y_{t-j} + \sum_{k=1}^{q} \gamma_j \mu_{t-k}$$

with all parameters positive. Under an ACP(1,1) model, one therefore has

$$\mu_t = \omega + \phi y_{t-1} + \gamma\mu_{t-1}.$$

In this model, stationarity is obtained subject to a constraint $\phi + \gamma < 1$. Defining $\Delta = 1 - (\phi + \gamma)^2$, the unconditional variance is given (Heinen, 2003, p.5) by

$$\text{Var}(y_t) = \mu_t(\Delta + \phi^2)/\Delta \geqq \mu_t,$$

so that unconditionally the ACP is overdispersed, even though the conditional distribution is equidispersed. Covariates may be introduced by defining a conditional mean $\mu_t^* = \exp(X_t\beta)\mu_t$ (Jung et al., 2006), and further overdispersion can be achieved by setting $\mu_t^* = \exp(X_t\beta)\mu_t\lambda_t$, where λ_t is lognormal.

Example 5.6 Ontario Car Fatalities

To illustrate different models for count time series and intervention analysis (Santos et al., 2010) we consider data on road fatalities in Ontario between 1931 and 2001, and the impact of a seat belt law introduced on January 1, 1976. Expected accidents, E_t, obtained as average accident rate times the number of registered drivers in a year, are used as an offset in the regression term.

The first analysis uses the ACP(1,1) model with a multiplicative lognormal error

$$\mu_t^* = E_t \exp(\beta S_t)\mu_t\lambda_t,$$

$$\mu_t = \omega + \phi y_{t-1} + \gamma\mu_{t-1}.$$

The seatbelt intervention (S_t) is represented by a binary variable with values 1 from 1976 onwards, 0 before. The estimates from a two-chain run of 25,000 iterations using jagsUI show most of the lag effect on μ_t operating through the conditional means, with γ having posterior mean (sd) of 0.98 (0.002). Predictive checks (comparing replicates from the posterior predictive distribution with actual observations), are satisfactory. The seatbelt effect β is estimated as −0.62 (0.08), but the pointwise LOO values still show 1976 (and surrounding years) as poorly fit. In that year, the accident rate per million fell to 350 compared to 433 in the previous year. It may be noted that estimates of the LOO-IC and WAIC (respectively 797 and 744) are unstable.

A second analysis adopts an antedependence approach, whereby $y_t \sim Po(\mu_t)$, $\log(\mu_t) = \log(E_t) + \beta S_t + g_t$, with g_t following a first-order antedependence scheme, whereby

$$g_1 = \varepsilon_1$$

$$g_t = \phi_t g_{t-1} + \varepsilon_t \quad t > 2$$

with $\varepsilon_t \sim N(0, \omega_t)$. The variances ω_t are modelled as a quadratic function of time, $\log(\omega_t) = a_1 + a_2 t + a_3 t^2$. This model provides rapid convergence using jagsUI. Estimates of the LOO-IC and WAIC are lowered (to 781 and 737 respectively). The seatbelt parameter has a mean (95% CRI) of −0.25 (−0.48, −0.08), while the estimated variances ω_t decline over time.

The pointwise LOO-IC values are no longer clustered in the 1970s under the antedependence model, but in an attempt to better represent the discontinuity at 1976, a decay effect in the impact of the seat belt law is implemented. The effect is constrained as negative, most pronounced in 1976, with subsequently monotonically fewer negative values, and the effect set to zero, unless the effect is negative. On this basis, we find that the year 1985 is the last year with a probability exceeding 0.5 that the seatbelt effect is negative. The LOO-IC and WAIC are now respectively 778 and 736, with the most extreme pointwise LOO-IC being for 1931 and 1982, followed by 1976. A plot of the relative risks (after controlling for the intervention) shows a continuing downward trend (Figure 5.4).

We also consider the CLAR(1) model (Grunwald et al., 2000), with form

$$\mu_t = \rho_1 y_{t-1} + \exp(\log(E_t) + \beta S_t + \eta_t),$$

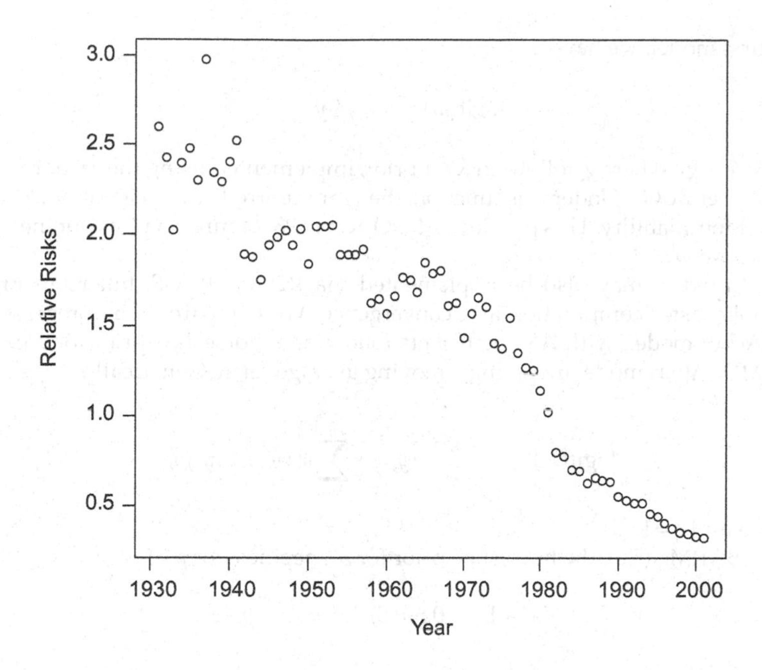

FIGURE 5.4
Annual relative risks, Ontario accidents, posterior means.

$$\eta_t \sim N(\rho_2 \eta_{t-1}, \sigma_\eta^2),$$

with $\rho_1 \sim U(0,1)$, and $\rho_2 \sim U(-1,1)$. This model has satisfactory predictive checks, and there is no significant correlation between successive errors $(y_t - \mu_t)/\mu_t^{0.5}$. However, the LOO-IC and WAIC, at 811 and 754 are higher than for other models, with performance being vitiated by discontinuities in the series, such as in 1937. The β coefficient has a mean (95%CRI) of −0.33 (−1.23, −0.01).

Finally, we use R-INLA to estimate a model with random walk level, with $y_t \sim Po(\mu_t)$,

$$\log(\mu_t) = \log(E_t) + \beta.S_t + w_t,$$

$$w_t \sim N(w_{t-1}, \sigma_w^2).$$

This model provides a posterior mean and CRI for β of −0.22 (−0.41, −0.04), and a WAIC of 751. A R-INLA model code including trend as well as level can be achieved using an augmented data representation (Ruiz-Cárdenas et al., 2012).

Example 5.7 Old Faithful Data

This example uses binary data generated from the Old Faithful geyser data in R, setting y=1 if the eruption time exceeds 3 minutes, and y=0 otherwise. There are T=272 data points. These data are characterised by several long runs of ones, but at most two zeroes in a run. We compare a parameter driven model, including random AR1 dependence, with two observation driven binary autoregressive moving average (BARMA) models (Startz, 2008).

In the first model, we have

$$\text{logit}(\pi_t) = \beta_1 + \beta_{2t} y_{t-1},$$

with $\beta_{2t} = \beta_2 + g_t$, where g_t follow an RW1 prior, implemented using the carnormal function in R2OpenBUGS. Under this function, the g_t are centred at each iteration, leading to improved identifiability. This provides a LOO-IC of 293. Figure 5.5 plots out the varying AR coefficients β_{2t}.

BARMA models may also be implemented via R2OpenBUGS, but rstan provides considerably faster computation and convergence. We compare an autoregressive lag 5 BARMA(5,0) model, with AR coefficients following a horseshoe prior for parsimony, with a BARMA(5,1) model including a moving average term. Generically

$$\text{logit}(\pi_t) = \beta_1 + \sum_{j=1}^{p} \rho_j y_{t-j} + \sum_{k=1}^{q} \theta_k (y_{t-k} - \pi_{t-k}),$$

where $p=5$ and $q=1$.

Under a BARMA(5,0), the horseshoe prior for ρ_j specifies (for $p=5$),

$$\kappa_j \sim \text{Beta}(0.5, 0.5), \quad j = 1, \ldots, p$$

$$\tau_\lambda^2 \sim IG(1, 0.001)$$

$$\lambda_j = (1/\kappa_j - 1)$$

$$\rho_j \sim N(0, \lambda_j \tau_\lambda),$$

where the posterior mean $\varphi_j = 1 - \kappa_j$ are effectively analogous to posterior selection rates. The model includes initial condition parameters so that all data points can be

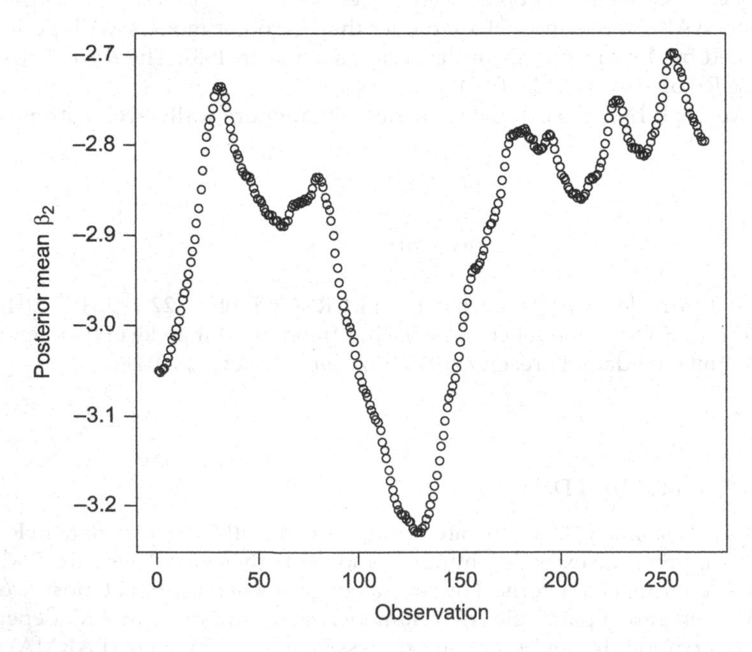

FIGURE 5.5
Old Faithful data. Varying AR1 regression coefficient

included in the likelihood. For example, the model at $t=1$ refers to unobserved preseries data points represented in the parameter $\varepsilon_1 = \sum_{j=1}^{p} \rho_j y_{1-j}$. The alternative is to condition on the first five observations. In fact, only the AR1 coefficient plays a significant role, with ρ_1 having posterior mean (sd) of −2.56 (0.44), and with $\varphi_1 = 0.98$. The LOO-IC for this model deteriorates to 332.

A BARMA(5,1) model finds ρ_2 and θ_1 (beta[3] and beta[7] in the code) to be significant with respective posterior means (sd) 1.39 (0.37) and −2.3 (0.67). The LOO-IC for this model is 331.

5.5 Stochastic Variances

Many state-space applications assume constant variances in the observation and state equations, but there is often nonstationarity in such variances (Omori and Watanabe, 2015; Broto and Ruiz, 2004). Certain types of data such as exchange rate and share price series r_t are particularly likely to demonstrate volatility clustering (Granger and Machina, 2006), with fluctuating variances $V_t = \text{var}(r_t)$. Typically, there are periods where volatility is relatively high and periods where volatility is relatively low, often with relatively smooth transition between high and low volatility regimes. In many applications, the series is transformed to have an effectively zero mean (Meyer and Yu, 2000, p.200). For example, the ratio of successive exchange rates r_t/r_{t-1} has approximate average 1, so that a response obtained as $y_t = \log(r_t/r_{t-1})$ can be taken to average zero. Hence, one may write a model without intercept (or predictor effects) as

$$y_t = V_t^{0.5} u_t,$$

where $u_t \sim N(0,1)$, but the variances V_t are unknowns.

Stochastic volatility models apply state-space techniques to model changing variances. A widely used scheme involves a state-space or autoregressive model in log scale parameters (Meyer and Yu, 2000; Jacquier et al., 2004; Kim et al., 1998; Harvey et al., 1994). With $h_t = \log(V_t)$, and stationary AR1 model for h_t, one has

$$y_t = \sqrt{V_t} u_t = \exp\left(\frac{h_t}{2}\right) u_t, \tag{5.2}$$

$$h_t = \mu + \phi(h_{t-1} - \mu) + \sigma_w w_t, \quad t > 1$$

$$h_1 \sim N\left(\mu, \frac{\sigma_w^2}{1-\phi^2}\right),$$

$$\begin{pmatrix} u_t \\ w_t \end{pmatrix} \sim N\left(\begin{pmatrix} 0 \\ 0 \end{pmatrix}, \begin{pmatrix} 1 & 0 \\ 0 & 1 \end{pmatrix}\right)$$

where $|\phi| < 1$ measures persistence in the volatility, but the u_t and w_t series are uncorrelated. This scheme can be generalised to multivariate responses subject to volatility, such as a set of exchange rates – see Chapter 7, and Yu and Meyer (2006).

As a heavy tailed alternative, one may consider a Student t likelihood for the log scale series, implemented as a scale mixture of normals (Jacquier et al., 2004). With ν degrees of freedom, one has

$$y_t = \sqrt{\lambda_t}\sqrt{V_t}u_t,$$

$$= \sqrt{\lambda_t}\exp\left(\frac{h_t}{2}\right)u_t,$$

$$1/\lambda_t \sim Ga\left(\frac{\nu}{2},\frac{\nu}{2}\right),$$

and other aspects as above. A diffuse prior on ν is not suitable, and one option is an exponential prior with prior mean 10 or 20 (Fernández and Steel, 1998). For a recent alternative prior (applicable to other types of Student t regression), see Fonseca et al. (2008). This model deals with isolated y-outliers by introducing a large λ_t, and it requires a sequence of large $|y_t|$ before V_t is increased (Jacquier et al., 2004, p.190).

By contrast, generalised autoregressive conditional heteroscedastic (GARCH) models involve autoregression in y_t^2 and/or V_t. A GARCH(p,q) model specifies

$$V_t = \gamma + \sum_{j=1}^{p} a_j y_{t-j}^2 + \sum_{j=1}^{q} \beta_j V_{t-j}$$

where coefficients $\{\gamma, a_j, \beta_j\}$ are constrained to be positive, and setting $q=0$ leads to the ARCH(p) model (Engle, 1982). Stationarity requires

$$\sum_{j=1}^{p} a_j + \sum_{j=1}^{q} \beta_j < 1$$

though is not necessarily imposed a priori. Whichever approach is used, departures from normality are frequently relevant, such that $y_t/\sqrt{V_t}$ is non-Gaussian. Among heavy tailed alternatives, one may consider a Student t, either $u_t \sim t(0,1,\nu)$, or a scale mixture of normals (Bauwens and Lubrano, 1998; Chib et al., 2002).

In case y has a non-zero mean, or there are predictors, one may widen the model for y. For example, a model with a zero mean y and lag 1 effect in y would be

$$y_t = \rho y_{t-1} + \sqrt{V_t}u_t.$$

One variant is the doubly autoregressive model (Ling, 2004)

$$y_t = \rho y_{t-1} + u_t\sqrt{\gamma + a y_{t-1}^2}.$$

For u_t normal, this can be shown equivalent to the random coefficient AR model

$$y_t = (\rho + a_t)y_{t-1} + c_t,$$

where (a_t, c_t) are bivariate normal with mean 0 and covariance matrix Diag(α,γ).

A generalisation of the state-space approach is to introduce correlation between the u_t and w_t terms, and so reflect leverage effects. Positive and negative shocks then have different impacts on future volatility (Wang et al., 2011; Asai et al., 2006; Jacquier et al., 2004; Meyer and Yu, 2000; Chen and So, 2006). So one possible scheme has

$$y_t = \sqrt{V_t} u_t = \exp\left(\frac{h_t}{2}\right) u_t,$$

$$h_t = \mu + \phi(h_{t-1} - \mu) + \sigma_w w_t,$$

$$\begin{pmatrix} u_t \\ w_t \end{pmatrix} \sim N\left(\begin{pmatrix} 0 \\ 0 \end{pmatrix}, \begin{pmatrix} 1 & \varphi \\ \varphi & 1 \end{pmatrix}\right),$$

where φ is a correlation. A heavy tailed version of the leverage model (Jacquier et al., 2004; Omori et al., 2007) may be obtained with

$$y_t = \sqrt{\lambda_t} \exp\left(\frac{h_t}{2}\right) u_t,$$

$$1/\lambda_t \sim Ga\left(\frac{\nu}{2}, \frac{\nu}{2}\right).$$

A GARCH model including leverage is obtained by setting $z_t = \sqrt{V_t} u_t$ in $y_t = \mu_y + \rho(y_{t-1} - \mu_y) + \sqrt{V_t} u_t$. Leverage is then obtained under the following asymmetric model (Glosten et al., 1994; Fonseca et al., 2016)

$$V_t = \gamma + \alpha_1 z_{t-1}^2 + \alpha_2 z_{t-1}^2 I(z_{t-1} > 0) + \beta V_{t-1}.$$

Under the model (5.2), assume priors $\mu \sim N(0, \sigma_\mu^2)$, $(\phi + 1)/2 \sim Be(r_\phi, s_\phi)$, and $\sigma_w^2 \sim IG(\kappa_w, \lambda_w)$, where $\{\sigma_\mu^2, r_\phi, s_\phi, \kappa_w, \lambda_w\}$ are known. Then with $\psi = (\mu, \phi, \sigma_w^2)$, the posterior is

$$p(\psi \mid y) \propto \left[\prod_{t=1}^{T} \exp\{-h_t/2\} \exp\left\{\frac{-y_t^2}{2e^{h_t}}\right\}\right]\left[\left(\frac{1-\phi^2}{\sigma_w^2}\right)^{0.5} \exp\left\{\frac{1-\phi^2}{2\sigma_w^2}(h_1 - \mu)^2\right\}\right]$$

$$\left[\prod_{t=2}^{T} \left(\frac{1}{\sigma_w^2}\right)^{0.5} \exp\left\{-\frac{1}{2\sigma_w^2}(h_t - \mu - \phi(h_{t-1} - \mu))^2\right\}\right] p(\mu) p(\phi) p(\sigma_w^2)$$

and Gibbs sampling from full conditionals is obtained (Kim et al., 1998). The Griddy–Gibbs technique may also be used to enable Gibbs sampling of all parameters in a GARCH(1,1) model, with normal or Student distributed u_t (Bauwens and Lubrano, 1998). Chib et al. (2002) consider more general Metropolis–Hastings techniques including particle filtering, to sample from models with discontinuities in the observations.

Example 5.8 Bitcoin Price

Consider 365 observations r_t of the Bitcoin price (in $000s) during 2017. The data are obtained as daily returns, namely $y_t = (r_t - r_{t-1})/r_t$ – see Figure 5.6 for a plot of y_t which shows several spells of high volatility. As one of several ways to represent the data, the double autoregressive model (Ling, 2004), namely

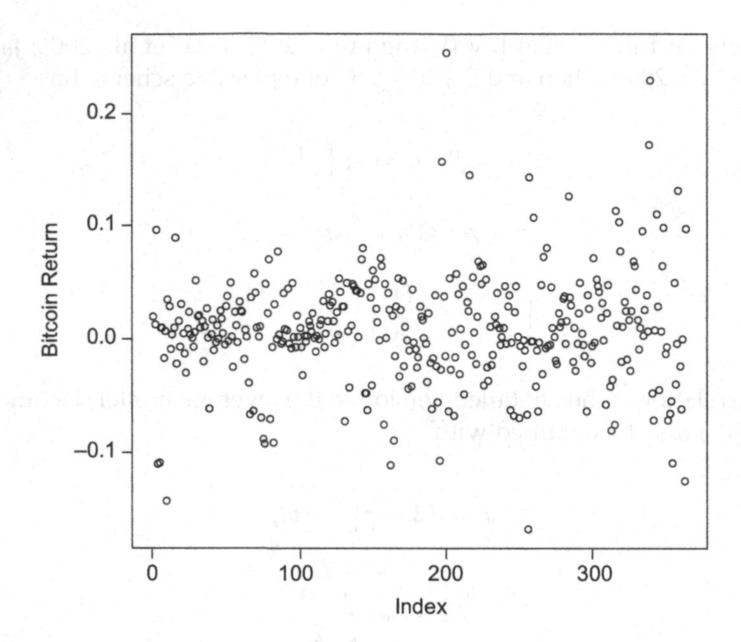

FIGURE 5.6
Fluctuations in returns $y_t = (r_t - r_{t-1})/r_t$ (r_t is Bitcoin price).

$$y_t = \rho y_{t-1} + u_t\sqrt{\gamma + a y_{t-1}^2},$$

is applied, with the constraint $\rho^2 + a < 1$ sufficient to ensure $E(y_t^2 < \infty)$. The analysis conditions on the first observation. A two-chain run of 5,000 iterations using jagsUI gives posterior means (sd) for ρ and a of 0.14 (0.05) and 0.16 (0.06), with γ estimated as 0.0021(0.0002). The LOO-IC is obtained as −1155.

A second approach is based on a stationary autoregressive stochastic volatility model (in stan), as in (5.2), with

$$y_t = \sqrt{V_t}u_t = \exp\left(\frac{h_t}{2}\right)u_t,$$

$$u_t \sim N(0,1),$$

$$h_t = \mu + \phi(h_{t-1} - \mu) + \sigma_w w_t, \quad t > 1$$

with a uniform U(−1,1) prior on ϕ, and a half Cauchy prior on σ_w. With a two-chain run of 2,000 iterations, we obtain posterior estimates (mean, sd) for μ and ϕ of −6.34 (0.35) and 0.91 (0.05), with the LOO-IC estimated as −1249. Figure 5.7 plots the evolving variance $V_t = \exp(h_t)$ under this model.

To better represent the extreme volatility in the series, a Student t (scale mixture) version of the preceding stochastic volatility model is applied. Thus

$$y_t = \sqrt{\lambda_t}\sqrt{V_t}u_t = \sqrt{\lambda_t}\exp\left(\frac{h_t}{2}\right)u_t,$$

$$1/\lambda_t \sim Ga\left(\frac{\nu}{2},\frac{\nu}{2}\right),$$

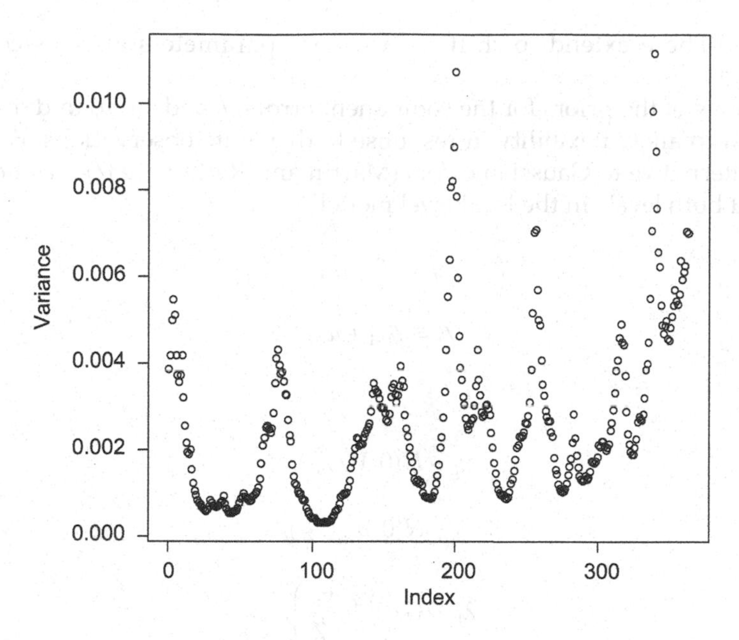

FIGURE 5.7
Stochastic volatility. Bitcoin data.

with a prior $\nu \sim E(0.1)$. This provides an improved LOO-IC of –1255, with a posterior mean (sd) for ν of 15.2 (9.6). Low values of the precision scaling parameters $\zeta_t = 1/\lambda_t$ are indicators of outlier status, and we find that cases 200, 284,10, 216, and 340 have the lowest posterior mean ζ_t. Two of these cases have return values exceeding 20%.

Outliers can also be represented by a binary shift mechanism (Wang, 2011). Thus

$$y_t = J_t N_t + \exp\left(\frac{h_t}{2}\right) u_t,$$

$$h_t = \mu + \phi(h_{t-1} - \mu) + \sigma_w w_t, \quad t > 1$$

where

$$J_t \sim \text{Bern}(\pi_J),$$

$$N_t \sim N(0, \sigma_N^2),$$

represent the shift mechanism and its potential size respectively. The probability π_J can be preset, or assigned a prior favouring a low outlier rate. Taking $\pi_J \sim \text{Beta}(2, 48)$, this model (fitted using jagsUI) provides a LOO-IC of –1256. The highest posterior probabilities $Pr(J_t = 1 \mid y)$ are for cases 284, 39, 10, 216, and 200.

5.6 Modelling Discontinuities in Time

Aberrant observations or shifts in a series can bias parameter estimates and other inferences in time series models (Chen and Liu, 1993; Tsay, 1986; Hamilton, 2007), and a variety of methods exist for modelling shifts or outliers in the observation, state, or error series. These extend to shifts in variance parameters also (as considered in Example 5.10).

Robust versions of the priors for the component errors u_t and/or w_{jt} in dynamic models may be applied to allow flexibility in response to disparate observations. For example, a heavy tailed alternative to Gaussian errors (Martin and Raftery, 1987) may be invoked by scale mixing at both levels in the local level model

$$y_t = \beta_t + u_t,$$

$$\beta_t = \beta_{t-1} + w_t,$$

with

$$u_t \sim N(0, V/\lambda_{1t}),$$

$$w_t \sim N(0, \sigma_w^2/\lambda_{2t}),$$

$$\lambda_{1t} \sim Ga\left(\frac{\nu_u}{2}, \frac{\nu_u}{2}\right),$$

$$\lambda_{2t} \sim Ga\left(\frac{\nu_w}{2}, \frac{\nu_w}{2}\right).$$

This generalisation is adapted to detecting or accommodating additive outliers (outliers in the observation errors) and innovation outliers in the state equation errors. Geweke (1993) points out problems with adopting diffuse priors for ν, and possibilities include an exponential density such as $\nu \sim E(0.1)$ (Fernandez and Steel, 1998).

Many outlier mechanisms involve discrete mixing around default normal error assumptions, as in a contaminated normal density (Verdinelli and Wasserman, 1991). Thus, let π be a given prior probability of an outlier (e.g $\pi = 0.05$). Then the observation error in a state-space model can be modified to allow innovation outliers

$$u_t \sim (1-\pi)N(0, W_1) + \pi N(0, W_2),$$

where $W_2 = KW_1$ with K large. A comprehensive generalisation of the normal errors dynamic linear model is provided by taking y_t and β_t to follow the univariate or multivariate exponential power distribution (Gomez et al., 2002).

More specialised binary switching in observation error or state error processes may be applied (Diggle and Zeger, 1989), for example, adapted to positive pulses (e.g. periods with abnormally heavy rainfall). To illustrate switching in observation errors to accommodate positive pulses, consider the $AR(1)$ observation model

$$y_t = \phi y_{t-1} + u_t,$$

such that usually $u_t = u_{1t}$, but exceptionally $u_t = u_{2t}$, where the latter error is necessarily positive, namely

$$u_{1t} \sim N(0, \sigma^2),$$
$$u_{2t} \sim Ga(g_1, g_2),$$

where $\{g_1, g_2\}$ are preset. Define latent allocation indicators $S_t \in (1,2)$, as in Chapter 3. Then $u_t = u_{2t}$ with probabilities $\pi_t = Pr(S_t = 2)$, that might be defined by a separate model, such as

$$\text{logit}(\pi_t) = \eta_0 + \eta_1 y_{t-1}.$$

One may also distinguish innovation outliers from additive outliers corresponding to isolated shifts or "gross errors" in the observation series (Tsay, 1986; Fox, 1972). This involves separate binary indicators $\{S_{At}, S_{It}\}$, or a single multinomial indicator S_t. For example, let π_A and π_I be prior probabilities of additive and innovative outliers, and consider an $AR(1)$ observation model with $AR(1)$ errors

$$y_t = \phi_0 + \phi_1 y_{t-1} + a_t S_{At} + \varepsilon_t,$$

$$\varepsilon_t = \rho \varepsilon_{t-1} + u_t,$$

where $S_{At} \sim \text{Bern}(\pi_A)$, and $a_t \sim N(0, \sigma_a^2)$ represents the sizes of the additive outliers (McCulloch and Tsay, 1994). Innovation outliers are encompassed by a variance inflation mechanism with

$$u_t \sim (1 - \pi_I) N(0, V) + \pi_I N(0, KV),$$

with $K \gg 1$, as determined by latent indicators $S_{It} \sim \text{Bern}(\pi_I)$.

The possibility of additive and innovative outliers coinciding at a single point may be discounted (Barnett et al., 1996; Gerlach et al., 1999). So with both additive and innovation outliers generated by variance inflation factors (respectively K_A and K_I), one may have a single trinomial indicator S_t governing outlier occurrence, with $S_t = 1$ if neither type of outlier is present ($K_A = 0, K_I = 1$), $S_t = 2$ if an additive outlier is present ($K_A = 10, K_I = 0$), and $S_t = 3$ if an innovation outlier is present ($K_A = 0, K_I = 10$). Then

$$S_t \sim \text{Mult}(1, [\pi_1, \pi_2, \pi_3]),$$

where π_2 and π_3 may be assigned preset values (e.g. $\pi_2 = \pi_3 = 0.025$), and

$$y_t = \phi_0 + \phi_1 y_{t-1} + a_{tS_t} + \varepsilon_t$$

$$\varepsilon_t = \rho \varepsilon_{t-1} + u_{tS_t}$$

where $a_{tS_t} \sim N(0, K_A \sigma^2)$, $u_{tS_t} \sim N(0, K_I \sigma^2)$.

Enduring, rather than temporary, shifts in the mean or variance of a series require another approach. Models with a single or small number of enduring changes in the level of the series may be handled by extending conventional discrete mixture methods (e.g. Leonte et al., 2003; Mira and Petrone, 1996; Albert and Chib, 1993; Perreault et al., 2000). To illustrate binary switching in both levels and variances in an autoregressive error model (McCulloch and Tsay, 1993), determined by binary pairs (S_{1t}, S_{2t}), consider

$$y_t = \mu_t + \varepsilon_t,$$

where a change in level, namely

$$\mu_t = \mu_{t-1} + S_{1t} \Delta_t,$$

occurs when $S_{1t}=1$, with $Pr(S_{1t}=1)=\pi_1$, and the Δ_t are random effects representing the shifts. The errors are $AR(p)$

$$\varepsilon_t = \rho_1\varepsilon_{t-1} + \rho_2\varepsilon_{t-2} + \ldots + \rho_{t-p}\,\varepsilon_{t-p} + u_t$$

where shifts in the variance of $u_t \sim N(0,V_t)$ occur when $S_{2t}=1$ with $Pr(S_{2t}=1)=\pi_2$. If there is conditioning on $(y_1,\ldots,y_p)$, then the variance sequence commences with $V_{p+1}=\sigma^2$, and subsequently,

$$V_t = V_{t-1} \text{ when } S_{2t}=0,$$

$$V_t = \kappa_t V_{t-1} \text{ when } S_{2t}=1,$$

where the κ_t are positive variables (e.g. gamma distributed) that model proportional shifts in the error variance.

Shocks in different components of the basic structural model can also be considered (De Jong and Penzer, 1998; Penzer, 2006). For example, in a three-component local linear trend model, binary shock indicators (S_{1t}, S_{2t}, S_{3t}) are invoked, such that

$$y_t = \mu_t + S_{1t}\Delta_{1t} + u_t,$$

$$\mu_t = \mu_{t-1} + S_{2t}\Delta_{2t} + \Delta_t + w_{1t},$$

$$\Delta_t = \Delta_{t-1} + S_{3t}\Delta_{3t} + w_{2t},$$

where the Δ_{1t} represent temporary additive shocks that occur when $S_{1t}=1$, the Δ_{2t} represent shifts in mean, and the Δ_{3t} represent shifts in the slope.

Regime switching models (Geweke and Terui, 1993; Lubrano, 1995) typically involve discrete switching between two or more levels, regression regimes, or variances, though smooth transition mechanisms can also be used. The choice between regimes is governed by a binary switching function S_t, or a continuous transition function ϕ_t with values between 0 and 1, such as the logit (Bauwens et al., 2000). A binary function S_t might be defined as one if time t exceeds a threshold κ and zero otherwise, as in change-point models for the mean level of a series. In self-exciting threshold autoregressive (SETAR) models, the mechanism involves a lag in y; for example, $S_t=1$ if $y_{t-1}>\kappa$. The continuous version in these two cases would be

$$\phi_t = \frac{\exp(\omega[t-\kappa])}{1+\exp(\omega[t-\kappa])},$$

$$\phi_t = \frac{\exp(\omega[y_{t-1}-\kappa])}{1+\exp(\omega[y_{t-1}-\kappa])},$$

where ω is an extra unknown. Additionally, the lag r in the comparison $y_{t-r}>\kappa$ may be unknown (Geweke and Terui, 1993).

Example 5.9 Nile Discharges

Data on Nile discharges for 1871–1970 ($T=100$) have been analysed by a variety of ARMA and other methods and illustrate possible identification issues associated with outlier and shift points. The initial analysis compares an *AR*(2) model for these data to one allowing for an intercept shift (cf Balke, 1993). Following that analysis, a Bayesian estimation of the AR(2) is applied. To facilitate prior specification for latent preseries values y_0 and y_{-1}, we centre the original data Y_t by subtracting Y_1 from all points. So $y_t = Y_t - Y_1$.

For the *AR*(2) model with no shift mechanism (and a heavy tailed Student t prior for the preseries points) is applied. Thus,

$$y_t = \phi_0 + \phi_1 y_{t-1} + \phi_2 y_{t-2} + u_t \quad t = 1,\ldots,T$$

$$u_t \sim N(0, \sigma^2),$$

$$y_0 \sim t_2(\phi_0 + \varepsilon_1, \sigma^2),$$

$$y_{-1} \sim t_2(\phi_0 + \varepsilon_2, \sigma^2),$$

where ε_j are fixed effects, and N(0,1) priors are adopted for $\{\phi_1, \phi_2\}$ so that nonstationarity is allowed. A two-chain run using jagsUI provides a LOO-IC of 1284. The posterior means (and 95% credible intervals) on the AR parameters $\{\phi_1, \phi_2\}$ are obtained as 0.45 (0.27,0.64), and 0.25 (0.06,0.44).

Suppose, however, a shift in the series level is allowed: a series plot suggests such a shift around 1895. One may also allow for coefficient selection via binary variables, namely $d_j = 1$ if ϕ_j $(j > 0)$ is to be retained, with prior probabilities $Pr(d_j = 1) = \pi_d$, with $\pi_d \sim$ Beta(1,1) So

$$y_t = \phi_{01} + \phi_{02} I(t > \kappa) + d_1 \phi_1 y_{t-1} + d_2 \phi_2 y_{t-2} + u_t$$

where κ is taken to be uniform between 3 and $T-3$. Fitting this model provides an improved LOO-IC of 1275, with posterior mean for κ of 29.8. The selection process indicates that the lag in y_{t-2} is now in doubt, with $Pr(d_2 = 1 \mid y) = 0.5$, whereas $Pr(d_1 = 1 \mid y) = 0.998$.

So an *AR*(1) model with shift mechanism is applied, namely

$$y_t = \phi_{01} + \phi_{02} I(t > \kappa) + \phi_1 y_{t-1} + u_t \quad t = 1,\ldots,T$$

The LOO-IC is reduced to 1272, with κ now having mean 28.6 (i.e. the year 1899). This is similar to the classical estimate of 28 obtained from the changepoint package (Killick and Eckley, 2014). The lag 1 coefficient estimate is now 0.43 with 95% interval (0.25,0.62).

Finally, we consider an AR(2) SETAR model (e.g. Korenok, 2009), which bases the shift threshold on the discharge value. Specifically,

$$y_t = \phi_{01} + \phi_{02} I(y_{t-1} > \kappa_y) + \phi_1 y_{t-1} + \phi_2 y_{t-2} + u_t \quad t = 2,\ldots,T$$

with κ_y assigned a uniform prior, $\kappa_y \sim U(-700, 300)$, based on actual (differenced) y values, which have minimum (maximum) of −664 and 250. This model provides a LOO-IC of 1282.6, with κ_y estimated as −336. The latter parameter is only weakly identified, as can be verified by a prior-posterior overlap plot using MCMCvis. This explains the small reduction in LOO-IC as against an AR(2) model with no shift. Figure 5.8 shows the extent of updating in κ_y.

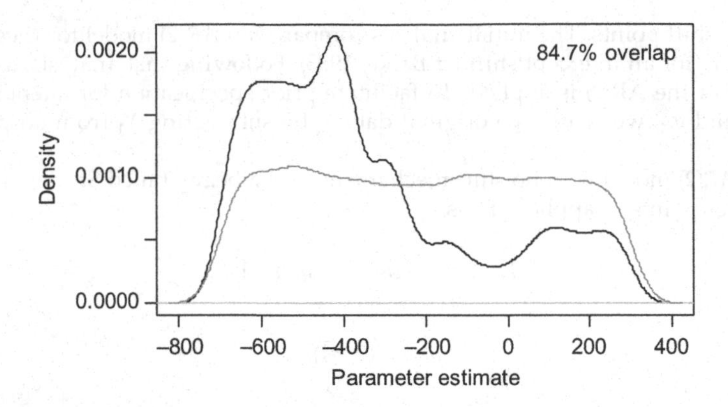

FIGURE 5.8
Density of κ_y.

Example 5.10 Box–Jenkins Series A

This example involves the Box–Jenkins series A, and demonstrates outlier modelling via variance inflation in the observation component of an autoregressive state-space model (cf. Gerlach et al., 1999). The observation model is

$$y_t \sim N(\beta_0 + \theta_t, V_{J_t}),$$

where J_t is a trinomial indicator modelling the measurement error outlier mechanism. The state equation is

$$\theta_t = \phi\theta_{t-1} + w_t,$$

where $w_t \sim N(0,W)$, and ϕ is constrained to stationarity.

As discussed above, outlier probabilities are often preset. However, if variance inflation factors are preset instead, then it is possible to take the outlier probabilities as unknowns. Thus assume $\pi_1 = Pr(J_t = 1)$ is the unknown probability of a default measurement error with variance V_1, while $\pi_2 = \pi_3$ are unknown probabilities of moderate and extreme outliers with variances $10V_1$ and $32V_1$ respectively. It is assumed that $V_1 \sim Ga(1,0.001)$, together with the parameterisation

$$\pi_1 = 1/(1+r),$$

$$\pi_2 = \pi_3 = 0.5r/(1+r),$$

where $r \sim E(9)$. Additionally, the variances of the observation and state equations are linked by taking $W = qV_1$ with an $E(1)$ prior on q.

A two-chain run using jagsUI shows early convergence with estimated probability $\pi_1 = 0.94$ (and 95% interval from 0.82 to 0.99). The observation error variance V_1 has a posterior mean of 0.031, while the state variance W has mean 0.037.

5.7 Computational Notes

[1] The code for the ARMA(4,0,1) model in Example 5.1 is

```
ARMA41.stan <- "
data {int<lower=1> T;//length of series
real y[T];
}
parameters {real phi0;//intercept
real phi[4];//autoregression coeffs
real gamma1;//moving avg coeff
real kappa[4];
real<lower=0> sigma;//residual sd
}
transformed parameters {real y_fit[T];
real error[T];
for (t in 1:T) {error[t] =y[t]-y_fit[t];}
//kappa[1] is composite parameter for effects of y[0],y[-1], etc., and
// gamma1*(y[0]-y_fit[0])
y_fit[1]=phi0+kappa[1];
y_fit[2]=phi0+phi[1]*y[1]+gamma1*(y[1]-y_fit[1])+kappa[2];

y_fit[3]=phi0+phi[1]*y[2]+phi[2]*y[1]+gamma1*(y[2]-y_
fit[2])+kappa[3];
y_fit[4]=phi0+phi[1]*y[3]+phi[2]*y[2]+phi[3]*y[1]+gamma1*(y[3
]-y_fit[3])+kappa[4];
for(tin5:T){y_fit[t]=phi0+phi[1]*y[t-1]+phi[2]*y[t-2]+phi[3]*y[t-3]+p
hi[4]*y[t-4]+
gamma1*(y[t-1]-y_fit[t-1]);}
}
model {real eps[T];
phi0 ~normal(0,10);
kappa ~normal(0,10);
phi ~normal(0,2);
gamma1 ~normal(0,2);
sigma ~cauchy(0,5);
eps[1]=y[1]-phi0-kappa[1];
eps[1]~normal(0,sigma);
eps[2]=y[2]-(phi0+phi[1]*y[1]+gamma1*(y[1]-y_fit[1])+kappa[2]);
eps[2]~normal(0,sigma);
eps[3]=y[3]-(phi0+phi[1]*y[2]+phi[2]*y[1]+gamma1*(y[2
]-y_fit[2])+kappa[3]);
eps[3]~normal(0,sigma);
eps[4]=y[4]-(phi0+phi[1]*y[3]+phi[2]*y[2]+phi[3]*y[1]+
gamma1*(y[3]-y_fit[3])+kappa[4]);
eps[4]~normal(0,sigma);
for(tin5:T){eps[t]=y[t]-(phi0+phi[1]*y[t-1]+phi[2]*y[t-2]+phi[3]*y[t-
3]+phi[4]*y[t-4]
+gamma1*eps[t-1]);
eps[t] ~normal(0,sigma);}}
generated quantities {
vector[T] log_lik;
for (t in 1:T) {log_lik[t] = normal_lpdf(y[t] y_fit[t], sigma);}}
"
# Initial Values and Estimation
sm <- stan_model(model_code=ARMA41.stan)
INI <- list(list(phi0=8,gamma1=0.8,ph
i=c(0.3,0.9,-0.4,-0.2),sigma=0.5,
```

```
kappa=c(-1,1,-1,1.5)),
list(phi0=7,gamma1=0.9,phi=c(0.4,0.8,-0.3,-0.1),sigma=0.7,kapp
a=c(-2,1.5,-1.5,2)))
fit4<-sampling(sm,data =D,pars
=c("phi0","phi","gamma1","y_fit","kappa","log_lik"),
iter = 10000,warmup=500,chains = 2,seed= 12345,init=INI)
print(fit4)
# Fit
LLsamps <- extract(fit4,"log_lik",permute=F)
LLsamps <- matrix(LLsamps, 2*9500, 598)
loo(LLsamps)
```

[2] The code for the random coefficient AR1 model in Example 5.2 is

```
RCAR.stan <- "
data {
int<lower=0> T;
vector[T] y;
}
parameters {
real mu;
real eta[T];
real y0;
real mu_phi;
real<lower=0> sigma;
real<lower=0> sigma_phi;
}
transformed parameters {
vector[T] muy;
vector[T] phi;
phi[1]=mu_phi+eta[1]*sigma_phi;
for(tin2:T){phi[t]=mu_phi+eta[t]*sigma_phi;}
muy[1]=mu+(mu_phi+eta[1]*sigma_phi)*y0;
for(tin2:T){muy[t]=mu+(mu_phi+eta[t]*sigma_phi)*y[t-1];}
}
model {
sigma ~normal(0, 1);
eta ~normal(0,1);
mu ~normal(0, 20);
mu_phi ~normal(0, 1);
y0 ~normal(0,20);
for (t in 1:T) {y[t] ~normal(muy[t], sigma);}
}
generated quantities {
vector[T] log_lik;
for (t in 1:T) {log_lik[t] = normal_lpdf(y[t] muy[t], sigma);}
}
```

[3] The code for the intervention decay effect antedependence model is as follows:

```
cat("model {for (t in 1:71) {
y[t] ~dpois(mu[t])
# Scaled deviance and likelihood terms
yts[t] <- equals(y[t],0)+(1-equals(y[t],0))*y[t]
mus[t] <- equals(y[t],0)+(1-equals(y[t],0))*mu[t]
```

```
dv[t] <- 2*(y[t]*log(yts[t]/mus[t])-(y[t]-mu[t]))
LL[t] <- -mu[t]+y[t]*log(mu[t])-logfact(y[t])
# Predictive checks
ynew[t] ~dpois(mu[t])
ch[t] <- step(ynew[t]-y[t])-0.5*equals(ynew[t],y[t])
# Regression
log(mu[t]) <- log(E[t])+beta[t]*SB[t]+g[t]
# Relative risk after control for intervention
RR[t] <- exp(g[t])}
Dv <- sum(dv[1:71])
g.m <- mean(g[])
# priors
phi ~dnorm(0,1)
g[1] ~dnorm(0,1/omega[1])
for (t in 2:71) {g[t] ~dnorm(phi*g[t-1],1/omega[t])}
# Variance model
for (t in 1:71) {log(omega[t]) <- gam[1]+gam[2]*t/100+gam[3]*
t*t/10000}
for (j in 1:3) {gam[j] ~dnorm(0,1)}
# Intervention effect
for(r in 1:26) {b[r] ~dnorm(0,tau.b)}
tau.b ~dexp(1)
# sort ascending order
bsort <- sort(b)
for (j in 1:45) {beta[j] <- 0}
# Decay in effect from year of introduction
for (j in 46:71) {betas[j] <- bsort[j-45]
# Retain negative coefficients
beta[j] <- betas[j]*step(-betas[j])
# Probability that SB effect still relevant
decay.prob[j-45] <- step(-betas[j])}}
", file="model3.jag")
# Initial values and estimation
init1= list(gam=c(-3,0,0),phi=0.8)
init2= list(gam=c(-3,0,0),phi=0.7)
inits=list(init1,init2)
pars <- c("beta","gam","LL","Dv","phi","RR","ch","decay.prob")
R <- autojags(D, inits, pars,model.file="model3.jag",2,iter.
increment=5000, n.burnin=500,Rhat.limit=1.1, max.iter=50000, seed=1234)
R$summary
samps <- as.matrix(R$samples)
# Select log-likelihood samples in samps
LL <- samps[,75:145]
LOO=loo(LL,pointwise=T)
waic(LL)
# Relative risks after controlling for intervention
RR <- samps[,148:218]
plot(apply(RR,2,mean),x=year,xlab="Year",ylab="Relative Risks")
# plots and listing, pointwise LOO
loocase <- as.vector(LOO$pointwise[,3])
plot(loocase,x=year,xlab="Year",ylab="Pointwise LOO-IC")
year=seq(1931,2001,1)
list.loocase <- data.frame(year,loocase)
list.loocase=list.loocase[order(-list.loocase$loocase),]
head(list.loocase,10)
```

References

Abraham B, Ledolter J (1983) *Statistical Methods for Forecasting*. Wiley, New York.

Albert J, Chib S (1993) Bayes inference via Gibbs sampling of autoregressive time series subject to Markov mean and variance shifts. *Journal of Business & Economic Statistics*, 11(1), 1–15.

Angers J, Biswas A, Maiti R (2017) Bayesian forecasting for time series of categorical data. *Journal of Forecasting*, 36(3), 217–229.

Araveeporn A (2017) Comparing random coefficient autoregressive model with and without autocorrelated errors by Bayesian analysis. *Statistical Journal of the IAOS*, 33(2), 537–545.

Asai M, McAleer M, Yu J (2006) Multivariate stochastic volatility: A review. *Econometric Reviews*, 25, 145–175.

Auger-Méthé M, Field C, Albertsen C M, Derocher A, Lewis M, Jonsen I, Flemming J (2016) State-space models' dirty little secrets: Even simple linear Gaussian models can have estimation problems. *Scientific Reports*, 6, 26677.

Balke N (1993) Detecting level shifts in time series. *The Journal of Business and Economic Statistics*, 11, 81–92.

Barnett G, Kohn R, Sheather S (1996) Bayesian estimation of an autoregressive model using Markov chain Monte Carlo. *Journal of Econometrics*, 74, 237–254.

Bauwens L, Lubrano M (1998) Bayesian inference on GARCH models using the Gibbs sampler. *Econometrics Journal*, 1, C23–C46.

Bauwens L, Lubrano M, Richard J (2000) *Bayesian Inference in Dynamic Econometric Models*. OUP.

Beck N (2004) Time series, in *Encyclopedia of Social Science Research Methods*, eds M Lewis-Beck, A Bryman, T Futing Liao. Sage.

Benjamin M, Rigby R, Stasinopoulos D (2003) Generalized autoregressive moving average models. *Journal of the American Statistical Association*, 98, 214–223.

Berkes I, Horvath L, Ling S (2009) Estimation in nonstationary random coefficient autoregressive models. *Journal of Time Series Analysis*, 30, 395–416.

Berliner L (1996) Hierarchical Bayesian time series models, pp 15–22, in *Maximum Entropy and Bayesian Methods*, eds K Hanson, R Silver. Kluwer Academic Publishers.

Berzuini C, Clayton D (1994) Bayesian analysis of survival on multiple time scales. *Statistics in Medicine*, 13, 823–838.

Betancourt M, Girolami M (2015) Hamiltonian Monte Carlo for hierarchical models, in *Current Trends in Bayesian Methodology with Applications*, eds S Upadhyay, U Singh, D Dey, A Loganathan. CRC.

Bijma F, De Munck J, Huizenga H, Heethaar R, Nehorai A (2005) Simultaneous estimation and testing of sources in multiple MEG data sets. *IEEE Transactions on Signal Processing*, 53, 3449–3460.

Bockenholt U (1999) An INAR(1) negative multinomial regression model for longitudinal count data. *Psychometrika*, 64, 53–68.

Broto C, Ruiz E (2004) Estimation methods for stochastic volatility models: A survey. *Journal of Economic Surveys*, 18, 613–649.

Cargnoni C, Muller P, West M (1996) Bayesian forecasting of multinomial time series through conditionally Gaussian dynamic models. *Journal of the American Statistical Association*, 92, 587–606.

Carlin BP, Klugman SA (1993) Hierarchical Bayesian Whittaker graduation. *Scandinavian Actuarial Journal*, 1993(2), 183–196.

Carlin B, Polson D, Stoffer D (1992) A Monte Carlo approach to nonnormal and nonlinear state space modelling. *Journal of the American Statistical Association*, 87, 493–500.

Carter C, Kohn R (1994) On Gibbs sampling for state space models. *Biometrika*, 81, 541–553.

Chan K, Ledolter J (1995) Monte Carlo EM estimation for time series models involving counts. *Journal of the American Statistical Association*, 90, 242–252.

Chatuverdi A, Kumar J (2005) Bayesian unit root test for model with maintained trend. *Statistics & Probability Letters*, 74, 109–115.

Chen C, Liu L (1993) Joint estimation of model parameters and outlier effects in time series. *Journal of the American Statistical Association*, 88, 284–297.

Chen C, So M (2006) On a threshold heteroscedastic model. *International Journal of Forecasting*, 22, 73–89.
Chib S, Greenberg E (1994) Bayes inference in regression models with ARMA(p, q) errors. *Journal of Econometrics*, 64(1–2), 183–206.
Chib S, Nardari F, Shephard N (2002) Markov Chain Monte Carlo methods for stochastic volatility models. *Journal of Econometrics*, 108, 281–316.
Clayton D (1996) Generalized linear mixed models, pp 275–301, in *Markov Chain Monte Carlo in Practice*, eds W Gilks, S Richardson, D Spiegelhalter. Chapman and Hall, London, UK.
Cox D (1970) *The Analysis of Binary Data*. Methuen, London, UK.
Daniels M (1999) A prior for the variance in hierarchical models. *Canadian Journal of Statistics*, 27, 569–580.
Davis R A, Holan S H, Lund R, Ravishanker N (eds) (2016) *Handbook of Discrete-Valued Time Series*. CRC Press.
De Jong P, Penzer J (1998) Diagnosing shocks in time series. *Journal of the American Statistical Association*, 93, 796–806.
De Jong P, Shephard N (1995) The simulation smoother for time series models. *Biometrika*, 82, 339–350.
Diggle P, Zeger S (1989) A non-Gaussian model for time series with pulses. *Journal of the American Statistical Association*, 84, 354–359.
Durbin J (2000) The Foreman lecture: The state space approach to time series analysis and its potential for official statistics. *Australian & New Zealand Journal of Statistics*, 42, 1–24.
Durbin J, Koopman S (2001) *Time Series Analysis by State Space Methods*. Oxford University Press, Oxford, UK.
Ehlers R, Brooks S (2004) Bayesian analysis of order uncertainty in ARIMA models. Technical Report, Federal University of Parana.
Engle RF (1982) Autoregressive conditional heteroscedasticity with estimates of the variance of United Kingdom inflation. *Econometrica: Journal of the Econometric Society*, 50(4), 987–1007.
Fahrmeir L, Lang S (2001) Bayesian inference for generalized additive mixed models based on Markov random field priors. *Applied Statistics*, 50, 201–220.
Fahrmeir L, Tutz G (2001) *Multivariate Statistical Modelling Based on Generalized Linear Models*, 2nd Edition. *Springer Series in Statistics*. Springer Verlag, New York/Berlin/Heidelberg.
Fernandez C, Steel M (1998) On Bayesian modeling of fat tails and skewness. *Journal of the American Statistical Association*, 93, 359–371.
Ferreira M, Gamerman D (2000) Dynamic generalized linear models, pp 57–72, in *Generalized Linear Models: A Bayesian Perspective*, eds D Dey, S Ghosh, B Mallick. Marcel Dekker, New York.
Fokianos K, Rahbek A, Tjøstheim D (2009) Poisson autoregression. *Journal of the American Statistical Association*, 104(488), 1430–1439.
Fonseca T, Cerqueira V, Migon H, Torres C (2016) Full Bayesian inference for asymmetric Garch models with Student-T innovations. IPEA Discussion Paper.
Fonseca T, Ferreira M, Migon H (2008) Objective Bayesian analysis for the Student-t regression model. *Biometrika*, 95, 325–333.
Fox A (1972) Outliers in time series. *Journal of the Royal Statistical Society, Series B*, 34, 350–363.
Fruhwirth-Schnatter S (1994) Data augmentation and dynamic linear models. *Journal of Time Series Analysis*, 15, 183–202.
Gabriel K (1962) Ante-dependence analysis of an ordered set of variables. *Annals of Mathematical Statistics*, 33, 201–212.
Gamerman D (1998) Markov chain Monte Carlo for dynamic generalized linear models. *Biometrika*, 85, 215–227.
Gelman A, Carlin J, Stern H, Dunson D, Vehtari A, Rubin D (2014) *Bayesian Data Analysis*. CRC, Boca Raton, FL.
Gerlach R, Carter C, Kohn R (1999) Diagnostics for time series analysis. *Journal of Time Series Analysis*, 20, 309–330.
Geweke J (1993) Bayesian treatment of the Students-t linear model. *Journal of Applied Economics*, 8, S19–S40.

Geweke J, Terui N (1993) Bayesian threshold auto-regressive models for nonlinear time series. *Journal of Time Series Analysis*, 14, 441–454.
Ghosh K, Tiwari R (2007) Prediction of U.S. cancer mortality counts using semiparametric Bayesian techniques. *Journal of the American Statistical Association*, 102, 7–15.
Giordani P, Pitt M, Kohn R (2011) Bayesian inference for time series state space models, in *The Oxford Handbook of Bayesian Econometrics*, eds J Geweke, G Koop, H Van Dijk. OUP.
Glosten L, Jagannathan R, Runkle D (1994) On the relation between the expected value and the variance of the nominal excess return on stocks. *Journal of Finance*, 48(5), 1779–1801.
Godsill S, Doucet A, West M (2004) Monte Carlo smoothing for nonlinear time series. *Journal of the American Statistical Association*, 99, 156–168.
Gómez E, Gómez-Villegas M, Marn J (2002) Continuous elliptical and exponential power linear dynamic models. *Journal of Multivariate Analysis*, 83, 22–36.
Granger C, Machina M (2006) Structural attribution of observed volatility clustering. *Journal of Econometrics*, 135, 15–29.
Grunwald G, Hyndman R, Tedesco L, Tweedie R (2000) Non-Gaussian conditional linear AR(1) models. *Australian & New Zealand Journal of Statistics*, 42, 479–495.
Grunwald S (2005) *Environmental Soil-Landscape Modeling: Geographic Information Technologies and Pedometrics*. CRC Press.
Hamilton J (2007) Regime-switching models, in *Palgrave Dictionary of Economics*, 2nd Edition, eds S Durlauf, L Blume. Palgrave MacMillan, London.
Harvey A (1989) *Structural Time Series Models and the Kalman Filter*. Cambridge University Press.
Harvey A, Ruiz E, Shepherd N (1994) Multivariate stochastic variance models. *Review of Economic Studies*, 61, 247–264.
Harvey A, Todd P (1983) Forecasting economic time series with structural and Box-Jenkins models: A case study. *Journal of Business & Economic Statistics*, 1, 299–307.
Harvey A, Trimbur T, Van Dijk H (2006) Trends and cycles in economic time series: A Bayesian approach. *Journal of Econometrics*, 140(2), 618–649.
Heinen A (2003) Modelling time series count data: An autoregressive conditional Poisson model. *SSRN Electronic Journal*. DOI:10.2139/ssrn.1117187
Helske J (2017) tsPI: Improved Prediction Intervals for ARIMA Processes and Structural Time Series. https://cran.r-project.org/web/packages/tsPI/index.html
Huerta G, West M (1999) Priors and component structurres in autoregressive time series. *Journal of the Royal Statistical Society, Series B*, 61, 881–899.
ISC Shark Working Group (2013) Stock assessment and future projections of blue shark in the North Pacific ocean. WCPFC-SC9-2013/SA-WP-11. WCPFC-SC. https://www.wcpfc.int/node/19204
Jacquier E, Polson N, Rossi P (2004) Bayesian analysis of stochastic volatility models with fat-tails and correlated errors. *Journal of Econometrics*, 122, 185–212.
Jacquier E, Polson NG, Rossi PE (2002) Bayesian analysis of stochastic volatility models. *Journal of Business & Economic Statistics*, 20(1), 69–87.
Jaffrézic F, Thompson R, Hill G (2003) Structured antedependence models for genetic analysis of repeated measures on multiple quantitative traits. *Genetics Research*, 82, 55–65.
Jaffrézic F, Venot E, Laloë D, Vinet A, Renand G (2004) Use of structured antedependence models for the genetic analysis of growth curves. *Journal of Animal Science*, 82, 3465–3473.
Jowaheer V, Sutradhar B (2002) Analysing longitudinal count data with overdispersion. *Biometrika*, 89, 389–399.
Jung R, Kukuk M, Liesenfeld R (2006) Time series of count data: modeling, estimation and diagnostics. *Computational Statistics & Data Analysis*, 51(4), 2350–2364.
Kashiwagi N, Yanagimoto T (1992) Smoothing serial count data through a state-space model. *Biometrics*, 48, 1187–1194.
Kastner G, Hosszejni D (2016) Package 'stochvol'. Efficient Bayesian Inference for Stochastic Volatility (SV) Models. https://cran.r-project.org/web/packages/stochvol/stochvol.pdf
Khoo W, Ong S (2014) A new model for time series of counts. *AIP Conference Proceedings*, 1605(1), 938–942.

Killick R, Eckley I (2014) Changepoint: An R package for changepoint analysis. *Journal of Statistical Software*, 58(3), 1–19.
Kim S, Shephard N, Chib S (1998) Stochastic volatility: Likelihood inference and comparison with ARCH models. *The Review of Economic Studies*, 65, 361–393.
Kitagawa G, Gersch W (1996) *Smoothness Priors Analysis of Time Series*. Springer, New York.
Knape J (2008) Estimability of density dependence in models of time series data. *Ecology*, 89, 2994–3000.
Knorr-Held L (1999) Conditional prior proposals in dynamic models. *Scandinavian Journal of Statistics*, 26, 129–144.
Koopman S (1993) Disturbance smoother for state space models. *Biometrika*, 80, 117–126.
Koopman S, Shephard N, Doornik J (1999) Statistical algorithms for models in state space form using SsfPack 2.2. *Econometrics Journal*, 2, 113–166.
Korenok O (2009) Bayesian methods in non-linear time series, pp 441–455, in *Encyclopedia of Complexity and Systems Science*. Springer, New York.
Lee S (1998) Coefficient constancy test in a random coefficient autoregressive model. *Journal of Statistical Planning and Inference*, 74, 93–101.
Lee Y, Nelder J (2001) Modelling and analysing correlated non-normal data. *Statistical Modelling*, 1, 3–16.
Leonte D, Nott D, Dunsmuir W (2003) Smoothing and change point detection for gamma ray count data. *Mathematical Geology*, 35, 175–194.
Li W (1994) Time series models based on generalized linear models: Some further results. *Biometrics*, 50, 506–511.
Liboschik T, Fokianos K, Fried R (2017) tscount: An R package for analysis of count time series following generalized linear models. *Journal of Statistical Software*, 82(5), 1–50.
Ling S (2004) Estimation and testing stationarity for double-autoregressive models. *Journal of the Royal Statistical Society: Series B*, 66, 63–78.
Lubrano M (1995) Testing for unit root in a Bayesian framework. *Journal of Econometrics*, 69, 81–109.
Marriott J, Ravishanker N, Gelfand A, Pai J (1996) Bayesian analysis of ARMA processes: Complete sampling based inference under full likelihoods, pp 243–256, in *Bayesian Analysis in Statistics and Econometrics*, eds D Barry, K Chaloner, J Geweke. Wiley, New York.
Martin D, Raftery A (1987) Non-Gaussian state-space modeling of nonstationary time series: Robustness, computation, and non-Euclidean models. *Journal of the American Statistical Association*, 82, 1044–1050.
Maunder M, Sibert J, Fonteneau A, Hampton J, Kleiber P, Harley S (2006) Interpreting catch per unit effort data to assess the status of individual stocks and communities. *ICES Journal of Marine Science*, 63(8), 1373–1385.
Maunder MN, Deriso RB, Hanson CH (2015) Use of state-space population dynamics models in hypothesis testing: Advantages over simple log-linear regressions for modeling survival, illustrated with application to longfin smelt (Spirinchus thaleichthys). *Fisheries Research*, 164, 102–111.
McAllister M K (2014) A generalized Bayesian surplus production stock assessment software (BSP2). *Collective Volumes of Scientific Papers ICCAT*, 70(4), 1725–1757.
McCulloch R, Tsay R (1993) Bayesian inference and prediction for mean and variance shifts in autoregressive time series. *Journal of the American Statistical Association*, 88, 968–978.
McCulloch R, Tsay R (1994) Bayesian analysis of autoregressive time series via the Gibbs sampler. *Journal of Time Series Analysis*, 15, 235–250.
Meyer R, Millar RB (1999) BUGS in Bayesian stock assessments. *Canadian Journal of Fisheries and Aquatic Sciences*, 56(6), 1078–1087.
Meyer R, Yu J (2000) BUGS for a Bayesian analysis of stochastic volatility models. *Econometrics Journal*, 3, 198–215.
Mira A, Petrone S (1996) Bayesian hierarchical nonparametric inference for change point problems, pp 693–703, in *Bayesian Statistics 5*, eds J Bernardo, J Berger, A Dawid, A Smith. OUP, Oxford.

Monnahan C, Thorson J, Branch T (2017) Faster estimation of Bayesian models in ecology using Hamiltonian Monte Carlo. *Methods in Ecology and Evolution*, 8(3), 339–348.
Naylor J, Marriott J (1996) A Bayesian analysis of non-stationary autoregressive series, pp 705–712, in *Bayesian Statistics 5*, eds J Bernardo, J Berger, A Dawid, A Smith. Clarendon Press.
Nunez-Anton V, Zimmerman D (2000) Modeling non-stationary longitudinal data. *Biometrics*, 56, 699–705.
Oh M-S, Lim Y (2001) Bayesian analysis of time series Poisson data. *Journal of Applied Statistics*, 28, 259–271.
Omori Y, Chib S, Shephard N, Nakajima J (2007) Stochastic volatility with leverage: Fast and efficient likelihood inference. *Journal of Econometrics*, 140(2), 425–449.
Omori Y, Watanabe T (2015) Stochastic volatility and realized stochastic volatility models, pp 435–456, Chapter 21, in *Current Trends in Bayesian Methodology with Applications*, eds S Upadhyay, U Singh, D Dey, A Loganathan. Chapman and Hall/CRC.
Ord J, Snyder R, Koehler A, Hyndman R, Leeds M (2005) Time series forecasting: The case for the single source of error state space approach. Working Paper 7/05, Department of Econometrics and Business Statistics, Monash University.
Paap R, van Dijk H (2003) Bayes estimation of Markov trends in possibly cointegrated series: an application to U.S. consumption and income. *Journal of Business & Economic Statistics*, 21, 547–563.
Parent E, Rivot E (2013) *Introduction to Hierarchical Bayesian Modeling for Ecological Data*. Chapman and Hall/CRC.
Patwardhan A, Small M (1992) Bayesian methods for model uncertainty analysis with application to future sea level rise. *Risk Analysis*, 12, 513–523.
Pedroza C (2006) A Bayesian forecasting model: Predicting U.S. male mortality. *Biostatistics*, 7, 530–550.
Penzer J (2006) Diagnosing seasonal shifts in time series using state space models. *Statistical Methodology*, 3, 193–210.
Perreault L, Berniera J, Bobéeb B, Parent E (2000) Bayesian change-point analysis in hydrometeorological time series: Comparison of change-point models and forecasting. *Journal of Hydrology*, 235, 242–263.
Petris G, Petrone S, Campagnoli P (2009) *Dynamic Linear Models with R*. Springer, New York.
Piegorsch W, Bailer J (2005) *Analyzing Environmental Data*. Wiley.
Pourahmadi M (2002) Graphical diagnostics for modeling unstructured covariance matrices. *International Statistical Review*, 70, 395–417.
Prado R, Huerta G, West M (2000) Bayesian time-varying autoregressions: Theory, methods and applications. *Journal of the Institute of Mathematics and Statistics of the University of Sao Paolo*, 4, 405–422.
Punt A, Hilborn R (1997) Fisheries stock assessment and decision analysis: The Bayesian approach. *Reviews in Fish Biology and Fisheries*, 7, 35–63.
Rankin P S, Lemos R T (2015) An alternative surplus production model. *Ecological Modelling*, 313, 109–126.
Reis E, Salazar E, Gamerman D (2006) Comparison of sampling schemes for dynamic linear models. *International Statistical Review*, 74, 203–214.
Rue H, Held L (2005) *Gaussian Markov Random Fields: Theory and Applications*. Chapman and Hall/CRC.
Ruiz-Cárdenas R, Krainski E T, Rue H (2012) Direct fitting of dynamic models using integrated nested laplace approximations—INLA. *Computational Statistics & Data Analysis*, 56(6), 1808–1828.
Santos T, Franco G, Gamerman D (2010) Comparison of classical and Bayesian approaches for intervention analysis. *International Statistical Review*, 78(2), 218–239.
Schaefer MB (1954) Some aspects of the dynamics of populations important to the management of the commercial marine fisheries. *Inter-American Tropical Tuna Commission Bulletin*, 1(2), 23–56.
Schmidt D, Makalic E (2013) Estimation of stationary autoregressive models with the Bayesian LASSO. *Journal of Time Series Analysis*, 34(5), 517–531.

Schotman P, Van Dijk H (1991) On Bayesian routes to unit roots. *Journal of Applied Econometrics*, 6, 387–401.
Scott S (2017) Package 'bsts'. Bayesian Structural Time Series. https://cran.r-project.org/web/packages/bsts/bsts.pdf
Silva N, Pereira I, Silva M E (2009) Forecasting in INAR (1) model. *REVSTAT*, 7(1), 119–134.
Silveira de Andrade B, Andrade M, Ehlers R (2015) Bayesian GARMA models for count data. *Communications in Statistics: Case Studies, Data Analysis and Applications*, 1(4), 192–205.
Simpson M, Niemi J, Roy V (2017) Interweaving Markov chain Monte Carlo strategies for efficient estimation of dynamic linear models. *Journal of Computational and Graphical Statistics*, 26(1), 152–159.
Soyer R, Aktekin T, Kim B (2015) Bayesian modeling of time series of counts with business applications, in *Handbook of Discrete-Valued Time Series*, eds R Davis, S Holan, R Lund, N Ravishanker. CRC.
Speed T, Kiiveri H (1986) Gaussian distributions over finite graphs. *Annals of Statistics*, 14, 138–150.
Startz R (2008) Binomial autoregressive moving average models with an application to US recessions. *Journal of Business & Economic Statistics*, 26(1), 1–8.
Strickland C, Turner I, Denham R, Mengersen K (2008) Efficient Bayesian Estimation of Multivariate State Space Models. http://eprints.qut.edu.au
Tsay R (1986) Time series model specification in the presence of outliers. *Journal of the American Statistical Association*, 81, 132–141.
Utazi C (2017) Bayesian single changepoint estimation in a parameter-driven model. *Scandinavian Journal of Statistics*, 44(3), 765–779.
Verdinelli I, Wasserman L (1991) Bayesian analysis of outlier problems using the Gibbs sampler. *Statistics and Computing*, 1, 105–117.
Wang D, Ghosh S (2002) Bayesian analysis of random coefficient autoregressive models. *Model Assisted Statistics and Applications*, 3(2), 281–295.
Wang J, Chan J, Choy S (2011) Stochastic volatility models with leverage and heavy-tailed distributions: A Bayesian approach using scale mixtures. *Computational Statistics & Data Analysis*, 55(1), 852–862.
Wang P (2011) Pricing currency options with support vector regression and stochastic volatility model with jumps. *Expert Systems with Applications*, 38(1), 1–7.
West M (1998) Bayesian forecasting, in *Encyclopedia of Statistical Sciences*, eds S Kotz, C Read, D Banks. Wiley.
West M (2013) Bayesian dynamic modelling, pp 145–166, in *Bayesian Inference and Markov Chain Monte Carlo: In Honour of Adrian FM Smith*, eds M West, P Damien, P Dellaportas, N Polson, D Stephens. Oxford University Press.
West M, Harrison P (1997) *Bayesian Forecasting and Dynamic Models*, 2nd Edition. Springer-Verlag, New York.
West M, Harrison P, Migon H (1985) Dynamic generalised linear models and Bayesian forecasting. *Journal of the American Statistical Association*, 80, 73–97.
Wu R, Cui Y (2014) A parameter-driven logit regression model for binary time series. *Journal of Time Series Analysis*, 35(5), 462–477.
Yu J, Meyer R (2006) Multivariate stochastic volatility models: Bayesian estimation and model comparison. *Econometric Reviews*, 25, 361–384.

6

Representing Spatial Dependence

6.1 Introduction

In the analysis of spatially configured data, positive covariation is typically expected between observations (areas, points) that are close to each other, so that residual spatial dependence may remain under a simple iid residual assumption (Anselin and Bera, 1998). Spatial heterogeneity in regression relationships is also common (Anselin, 2010). Spatial regression aims to represent the residual structure appropriately, or represent heterogeneity, and may also be used to obtain improved estimates, especially when applying Bayesian spatial smoothing. Consider disease counts for areas, when small event totals or small populations lead to unstable point estimates of rates or relative risks. One is then led to hierarchical regression models for borrowing strength to achieve more stable estimates (Riggan et al., 1991; Waller, 2002). If there is spatial covariation (e.g. when contiguous areas have similar disease levels), an appropriate borrowing strength mechanism would incorporate local smoothing towards the mean of adjacent areas (Clayton and Kaldor, 1987). By contrast, assuming exchangeable random effects implies global smoothing, with rates or risks smoothed towards the overall mean, and does not account for spatial dependence.

Priors for spatial covariance modelling are therefore structured in the sense of explicitly recognising the role of adjacency or proximity, and use this structure as the basis for smoothing or prediction. Often smoothing of rates is an end in itself; for example, spatial smoothing of area health data to reflect similarity of disease risks in nearby areas is a more reliable guide for health interventions (e.g. Zhu et al., 2006). However, structured priors may also be suitable when the goals of analysis include out-of-sample prediction. In geostatistical applications, a frequent goal is interpolation of a modelled surface to unsampled locations based on proximity to observed locations (Gotway and Wolfinger, 2003; Webster et al., 1994; Jiruse et al., 2004).

The R environment now offers considerable potential for analysing spatial data, as discussed, for example, in Bivand et al. (2013), Allard et al. (2017), and Brunsdon and Comber (2015). On-line R-based resources for spatial data analysis include www.rspatial.org/spatial/ and https://data.cdrc.ac.uk/tutorial/an-introduction-to-spatial-data-analysis-and-visualisation-in-r. Bayesian spatial estimation in R is facilitated by packages such as CARBayes (Lee, 2013), R-INLA (Blangiardo and Cameletti, 2015; Schrödle and Held, 2011), INLABMA (Goméz-Rubio and Bivand, 2018), geostatsp (Brown, 2015), spBayes (Finley et al., 2015), geoR (Ribeiro and Diggle, 2018), and spNNGP.

While there may be benefits from borrowing strength methods based on spatial proximity, using random effects to represent unobserved components may raise potential identification issues. For example, priors for random effects may specify differences between adjacent observations without specifying their mean, so that MCMC methods then require centring of the effects to ensure identification of other parameters. Furthermore, methods for smoothing or interpolation assuming relatively smooth variation over adjacent units may need to be adaptive to spatial discontinuities (Knorr-Held and Rasser, 2000).

As an example of spatially structured prior and its Bayesian implementation, the pairwise difference or Markov random field (MRF) prior may be specified via conditional densities, which are naturally suited for Gibbs sampling (Finley et al., 2015). For univariate effects $\theta = (\theta_1, \ldots, \theta_n)$, the conditional MRF prior takes the form (Besag et al., 1995, p.11; Rue and Tjelmeland, 2002; Furrer and Sain, 2010)

$$p(\theta_i \mid \theta_{[i]}) \propto \tau \exp\left[-\sum_{j \neq i} w_{ij} \Phi(\tau[\theta_i - \theta_j])\right],$$

where $\theta_{[i]}$ denotes values for cases other than i, w_{ij} are weights specifying spatial dependence between observations i and j, $\Phi(u)$ is an increasing function in u, subject to $\Phi(u) = \Phi(-u)$, and τ a precision parameter. Under a neighbourhood prior, where $w_{ij} = 1$ when observations (usually areas) i and j are neighbours and $w_{ij} = 0$ otherwise, an equivalent representation is

$$p(\theta_i \mid \theta_{[i]}) \propto \tau \exp\left[-\sum_{j \in \partial_i} \Phi(\tau[\theta_i - \theta_j])\right],$$

where ∂_i is the set of areas adjacent to area i. The case $w_{ij} = 1$ if $|i - j| = 1$ and $w_{ij} = 0$ otherwise leads to first order random walk priors relevant to modelling time-ordered data. The MRF prior generalises to variables θ_{ij} in two-dimensional lattices (e.g. areas i and times j), and a neighbourhood might then be defined as $\partial_{ij} = [(i+1, j), (i-1, j), (i, j+1), (i, j-1)]$ (Lavine, 1999). Taking $\Phi(u) = u^2/2$ leads to a Gaussian or L2 norm conditional prior for θ_i (Waller, 2002)

$$\theta_i \mid \theta_{[i]} \sim N\left(\sum_{j \neq i} \frac{w_{ij}\theta_j}{w_{i+}}, \frac{1}{\tau w_{i+}}\right), \tag{6.1}$$

whereas if $\phi(u) = |u|$ then

$$p(\theta_i \mid \theta_{[i]}) \propto \tau \exp\left(-\tau \sum_{j \# i} w_{ij} \left|\theta_i - \theta_j\right|\right),$$

known as the $L1$ norm prior (Richardson et al., 2004). To achieve robust smoothing, the latter form may be better suited to spatial discontinuities, since its mode is at the median rather than the mean.

6.2 Spatial Smoothing and Prediction for Area Data

Whereas exchangeable hierarchical analysis is appropriate for independently generated area or point data, such data often cannot be regarded as independent because of the presence of similarities between neighbouring areas or points (Anselin and Bera, 1998). Modelling area differences or point patterns with spatially structured effects reflects the empirical regularity that neighbouring areas or points tend to be similar, and that similarity typically diminishes as distance increases. Even if known predictors are available, it is likely that other relevant influences on the underlying process cannot be identified or measured, and this residual heterogeneity is likely (at least in part) to be spatially structured (Lawson, 2008, p.94). For example, Gelfand et al. (2005a) consider spatial modelling of residuals in the analysis of species distributions, both for areas and points as the units, where unobserved influences might include habitat and inter-species competition. Bayesian techniques have played a central role in recent developments for analysing spatial data, whether space is viewed from a discrete or continuous perspective, e.g. Banerjee et al. (2014) and Waller and Carlin (2010).

In studies with a discrete framework, the data are typically aggregated, with observations consisting of counts (e.g. of diseased subjects in spatial epidemiology) or of regional indicators (e.g. average income per head or house prices in spatial econometrics). By contrast, in geostatistical models for geochemical readings, species distribution, or disease events in relation to a pollution source, a continuous spatial framework is more relevant (Section 6.5), allowing interpolation between observed point readings.

Consider metric responses y_i for areas i, or at sites specified by grid references $g_i = (g_{1i}, g_{2i})$. To allow greater flexibility, one may assume a "convolution" prior that compromises between structured and unstructured variation; so the model includes both a spatially structured random effect s_i and a fully exchangeable effect u_i, with

$$y_i = a + u_i + s_i,$$

where $u_i \sim N(0, \sigma_u^2)$, but the s_i are spatially correlated. Alternatively, suppose y_i are counts, and that P_i are populations at risk with $y_i \sim Bin(P_i, \pi_i)$. Then one may specify

$$\text{logit}(\pi_i) = a + u_i + s_i,$$

where π_i are latent probabilities of the event. Alternatively, for rare events in relation to the risk population, a Poisson assumption is relevant with $y_i \sim Po(P_i \lambda_i)$, and

$$\log(\lambda_i) = a + u_i + s_i,$$

where λ_i are latent event rates per unit of P_i. If the offsets to the Poisson mean are expected health events E_i, such that $\Sigma_i y_i = \Sigma_i E_i$ with $y_i \sim Po(E_i \lambda_i)$, then the λ_i are interpretable as latent relative risks (Wakefield, 2007, p.160).

One way to model the correlation in the elements of the vector $s = (s_1, \ldots, s_n)$ is to directly specify a joint multivariate prior with covariance matrix that expresses spatial correlation between areas i and j or sites g_i and g_j (Richardson et al., 1992, p.541; Wakefield, 2007). Typical assumptions in such models (also considered in Section 6.5) are of stationarity and isotropy, with the latter meaning the correlation is the same in all directions. For example, a multivariate normal prior would take

$$(s_1, \ldots s_n) \sim N_n(0, \Sigma_s),$$

$$\Sigma_s = \sigma_s^2 W = \sigma_s^2 \begin{bmatrix} 1 & w_{12} & . & w_{1n} \\ w_{21} & 1 & . & w_{2n} \\ . & . & . & . \\ w_{n1} & w_{n2} & . & 1 \end{bmatrix},$$

where $w_{ij}=f(d_{ij})$ are correlation functions that decline as the spatial separation d_{ij} between areas i and j (or sites g_i and g_j) increases, and defined to ensure that W is always non-negative definite (Mardia and Watkins, 1989).

For example, one may specify exponential spatial decay,

$$w_{ij} = \exp(-\delta d_{ij}),$$

where $\delta > 0$, or for area units, allow for both inter-area distance d_{ij} and the length b_{ij} of the common border between area i and j, namely

$$w_{ij} = d_{ij}^{\gamma_1}[b_{ij} + c]^{\gamma_2},$$

where γ_1 is negative, and γ_2 is positive. Another choice is the disc model with

$$w_{ij} = \frac{2}{\pi}\left[\cos^{-1}\left(\frac{d_{ij}}{\kappa}\right) - \left\{\frac{d_{ij}}{\kappa}\left(1 - \frac{d_{ij}^2}{\kappa^2}\right)\right\}^{0.5}\right] \quad d_{ij} \leq \kappa,$$

with $w_{ij}=0$ for $d_{ij} > \kappa$, so that κ controls the decline in correlation with distance. Such choices are to some degree arbitrary, and inferences may be sensitive to the choice of spatial weights (e.g. Bhattacharjee and Jensen-Butler, 2006).

6.2.1 SAR Schemes

A widely used scheme, especially in spatial econometrics, specifies the joint density via simultaneous autoregressive or SAR effects (Richardson et al., 1992). By analogy with ARMA time series models, the autoregression may operate both for (metric) responses $y = (y_1, \ldots y_n)'$, and for the error vector $\varepsilon = (\varepsilon_1, \ldots \varepsilon_n)'$. Let $W = [w_{ij}]$ be a spatial dependence matrix as above, but with $w_{ii}=0$ rather than $w_{ii}=1$. One possible SAR scheme has the form

$$y_i = \rho_1 \sum_{h \neq i} w_{ih} y_h + X_i \beta + \varepsilon_i,$$

$$\varepsilon_i = \rho_2 \sum_{h \neq i} w_{ih} \varepsilon_h + u_i,$$

where ρ_1 and ρ_2 are measures of spatial dependence, and the $u = (u_1, \ldots, u_n)'$ are independently distributed, with diagonal covariance matrix Σ_u. The covariance matrix for $\varepsilon = (\varepsilon_1, \ldots \varepsilon_n)'$ is $(I - \rho_2 W)^{-1}\Sigma_u (I - \rho_2 W')^{-1}$. In matrix form

$$y = \rho_1 W y + X\beta + \varepsilon,$$

$$\varepsilon = \rho_2 W \varepsilon + u.$$

The ρ coefficients are constrained to lie between $1/\eta_{\min}$ and $1/\eta_{\max}$, where $\{\eta_1, \ldots, \eta_n\}$ are the eigenvalues of W, in order to ensure that $(I - \rho W)$ is invertible. If the weights matrix is standardised to have row sums of unity, so that $w_{ij}^* = w_{ij}/\Sigma_h w_{ih}$, then the maximum eigenvalue of W^* is 1 and since negative spatial correlation is unlikely, one may specify uniform or beta priors on ρ coefficients in the interval [0,1]. Wall (2004) points out that SAR priors (and also CAR priors, as considered below) may generate implausible covariance patterns when considered in terms of the joint priors.

Variants of the above scheme include the spatial errors model (SEM), with $\rho_1 = 0$ (Cressie and Wikle, 2011),

$$y = X\beta + \varepsilon, \tag{6.2}$$

$$\varepsilon = \rho W \varepsilon + u,$$

and the spatial lag model (SLM) with $\rho_2 = 0$, namely

$$y = \rho W y + X\beta + u, \tag{6.3}$$

where in both models $u \sim N(0, \sigma^2)$ are iid. The spatial errors model may be expressed as

$$y = X\beta + (I - \rho W)^{-1} u,$$

or, equivalently,

$$y \sim MVN(X\beta, \sigma^2[(I - \rho W)'(I - \rho W)]^{-1}).$$

The SEM model may also be considered as a prior for spatially correlated effects. For example, in (6.2) one may assume spatially varying β_i over units i, with

$$\beta = \beta_\mu + \varepsilon_\beta,$$

$$\varepsilon_\beta = \rho_\beta W \varepsilon_\beta + u_\beta,$$

where β_μ is the average coefficient. Another option is a spatial moving average errors representation (Hepple, 2003) with

$$y = X\beta + \varepsilon,$$

$$\varepsilon = \rho W u + u,$$

$$u \sim N(0, \sigma^2).$$

Equivalently, $y \sim MVN(X\beta, \sigma^2 (I + \rho W)'(I + \rho W))$.

Regarding the spatial errors model expressed as

$$y = X\beta + (I - \rho W)^{-1} u,$$

an alternative to assuming uncorrelated u and X, and allowing greater generality, specifies

$$u = X\gamma + v,$$

where the v are *iid*. Then

$$y = X\beta + (I - \rho W)^{-1} X\gamma + (I - \rho W)^{-1} v.$$

Expressed with *iid* errors, this leads to the spatial Durbin model or SDM (Seya et al., 2012, Lacombe and LeSage, 2015).

$$y = \rho Wy + X(\beta + \gamma) - \rho WX\beta + v,$$

which may be reparameterised as

$$y = \rho Wy + X\theta_1 + WX\theta_2 + v. \tag{6.4}$$

Example 6.1. SAR Models for Long-Term Limiting Illness Data

This example considers area data on limiting long-term illness (LLTI) for 133 electoral wards (small areas) among people aged 50–59 in East London from Congdon (2008), and also considered in Example 6.3. Here we express the illness totals T_i and the population denominators P_i as long-term illness rates per 1,000, namely $y_i = 1000 \times T_i / P_i$. Since area deprivation is typically a strong influence on area morbidity, an index of multiple deprivation (IMD) is used as a single predictor.

As discussed by Bivand et al. (2014, 2015) one may use the Integrated Nested Laplace Approximation to estimate spatial autoregressive models conditioning on particular values of the spatial autocorrelation parameter. Conditioned models can be estimated over a suitable grid, and subsequently combined using Bayesian model averaging to provide posterior marginals of parameters (Goméz-Rubio et al., 2018). In particular, the spatial Durbin representations (SDM) of the LLTI rates (and the impact on them of area deprivation) is estimated using the INLABMA package.

The spatial interaction matrix W is obtained using the spdep and maptools packages, applied to a relevant shape file (note that dbf and shx files are needed in the same directory but not explicitly referenced). Thus the R sequence:

```
library(easypackages)
libraries("INLA","spdep","INLABMA","maptools")
setwd("C:/R Files BHMRA")
# shapefile East London electoral wards
ELmap <- readShapePoly("Example_6_1")
ELnb <- poly2nb(ELmap, queen=F)
lw=nb2listw(ELnb, glist=NULL,, zero.policy=NULL)
# Sparse Adjacency matrix
W = as(as_dgRMatrix_listw(nb2listw(ELnb)), "CsparseMatrix")
```

A grid for the spatial autocorrelation parameter ρ in the SDM model (6.4) is specified with limits 0.2 and 0.9, namely grid.rho = seq(0.2, 0.9, length.out=20). This is based on an estimate of 0.48 from maximum likelihood estimation. The estimates from INLABMA are shown in Table 6.1, with mean (sd) for ρ of 0.53 (0.09). The DIC is estimated as 1242.

With rstan, we may estimate the spatial errors errors (SEM) model, using the multi_normal_prec option (Brunsdon, 2018). Thus with

$$y \sim MVN(X\beta, \sigma^2[(I - \rho W)'(I - \rho W)]^{-1}),$$

the precision is $[(I - \rho W)'(I - \rho W)]/\sigma^2$. The full code, with flat priors on hyperparameters, is

```
model="data {
int N;
vector[N] x;
vector[N] y;
matrix<lower=0>[N,N] W;
matrix<lower=0,upper=>[N,N] I;
}
parameters {
real beta;
real alpha;
real<lower = 0> sigma;
real<lower=-1,upper=1> rho;
}
model {
y ~multi_normal_prec(alpha + x * beta, crossprod(I - rho * W)/
(sigma*sigma));
}
generated quantities
{
real LL;
LL= multi_normal_prec_lpdf(y alpha + x * beta, crossprod(I - rho *
W)/(sigma*sigma));
}"
```

This leads to very similar estimates to those obtained using maximum likelihood, with posterior mean (sd) for ρ of 0.50 (0.10). The log-likelihood is estimated at −627, and the DIC (estimated as the mean deviance plus the number of parameters) is obtained as 1258.

TABLE 6.1

Spatial Autoregressive Models Compared

		Mean	St devn	2.5%	Median	97.5%
Spatial Error Model	Intercept	109.8	10.4	90.1	109.9	129.9
	IMD	6.2	0.3	5.7	6.2	6.8
	ρ	0.50	0.10	0.30	0.51	0.69
	DIC	1258.0				
Spatial Lag Model	Intercept	84.6	12.2	60.8	84.1	108.2
	IMD	5.42	0.39	4.66	5.43	6.18
	ρ	0.16	0.07	0.03	0.17	0.30
	DIC	1271.4				
Spatial Moving Average Errors Model	Intercept	109.4	8.8	92.4	109.3	126.8
	IMD	6.3	0.3	5.7	6.3	6.7
	ρ	0.47	0.12	0.23	0.46	0.71
	DIC	1262.3				
Spatial Durbin Model	Intercept	48.0	6.7	34.7	48.0	61.2
	IMD	6.2	0.4	5.4	6.2	7.0
	IMD-spatial lag	−3.2	0.5	−4.1	−3.2	−2.3
	ρ	0.53	0.09	0.36	0.53	0.71
	DIC	1241.6				

A similar approach may be applied to estimate the spatial moving average errors model, except that the likelihood is now

```
y ~multi_normal(alpha + x * beta, crossprod(I + rho * W)*sigma^2).
```

The DIC for this model is slightly higher than for the spatial autocorrelated errors model, with posterior mean (sd) for ρ of 0.47 (0.12).

The spatial lag model may be estimated using the target + representation to accommodate the likelihood. The log-likelihood is

$$\log(L) = -0.5N\log(2\pi) - 0.5N\log(\sigma^2) + \log|I - \rho W|$$
$$-(y - \rho Wy - X\beta)'(y - \rho Wy - X\beta)/(2\sigma^2),$$

where

$$|I - \rho W| = \prod_{i=1}^{n}(1 - \rho\lambda_i)$$

with $\lambda = (\lambda_1,\ldots,\lambda_n)$ being the eigenvalues of W. So the log determinant term may be written

$$\log|I - \rho W| = \sum_{i=1}^{n}\log(1 - \rho\lambda_i).$$

For simplicity, the target + calculations include the squared regression error terms and the log determinant contributions in the same summand terms, albeit with the total of these summands still being the overall log-likelihood. Discrepancies at case level might be assessed by standardised residuals. Estimates for the four hyperparameters are very similar to those from maximum likelihood, with posterior mean (sd) for ρ of 0.16 (0.07), and 5.42 (0.39) for the regression coefficient on IMD.

To illustrate the SEM as a prior for random spatial effects, we extend the above rstan code to allow the random coefficients scheme

$$\beta = \beta_\mu + \varepsilon_\beta,$$

$$\varepsilon_\beta = \rho_\beta W\varepsilon_\beta + u_\beta,$$

where β_μ is the average coefficient. This involves an extra input vector, e = rep(1,N), in the data block:

```
vector<upper=>[N] e;
```

and beta as a vector

```
vector[N] beta;
```

There are extra parameters beta_mu, sigma_b, rho_b, and a model block as follows:

```
model {
beta ~ multi_normal_prec(e * beta_mu, tcrossprod(I - rho_b * W)/
(sigma_b*sigma_b));
y ~ multi_normal_prec(alpha + x .* beta, tcrossprod(I - rho * W)/
(sigma*sigma));
}
```

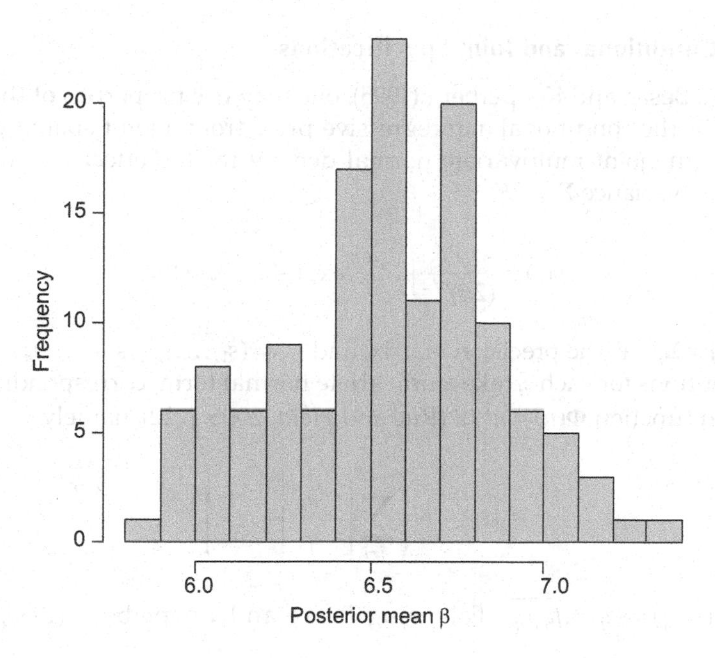

FIGURE 6.1
Histogram of spatially varying predictor effect.

This option shows an increase in the log-likelihood from −627.0 to −589.0, with Figure 6.1 showing the variation in the impacts of deprivation, and slopes varying from 5.9 to 7.3.

6.3 Conditional Autoregressive Priors

In contrast to simultaneous autoregressive spatial priors, conditional autoregressive priors for spatial errors $s = (s_1, \ldots, s_n)$ have the advantage of facilitating random effects analysis under an MCMC sampling approach, especially for large numbers of areas. Such priors are often applied to discrete outcomes, such as disease counts y_i, taken to be Poisson or binomial in relation to populations P_i or expected events E_i (e.g. Besag et al., 1991; Norton and Niu, 2009; De Oliveira, 2012). If the event is relatively infrequent (or populations at risk are small), one often seeks to estimate an underlying smooth pattern of disease risk by borrowing strength over areas taking account of spatial dependence (MacNab et al., 2006).

Disease counts typically display extra-variation which can be modelled by including random effects in the regression for disease rates. In particular, spatially correlated random effects may account for much extra-variation, serve to borrow strength in estimation, and proxy unobserved risk factors that are also spatially correlated (Richardson and Monfort, 2000). These might be shared environmental or social capital factors in neighbouring areas. While spatially correlated random effects acting alone may be assumed, this may constitute an informative prior, since there may be areas (e.g. areas of deprived social renting surrounded by affluent areas) discrepant from surrounding areas in terms of risk factors. A more general and less informative approach allows adaptive downweighting of a spatial prior.

6.3.1 Linking Conditional and Joint Specifications

As discussed by Besag and Kooperberg (1995), one may use properties of the multivariate normal to obtain the conditional autoregressive prior from a joint spatial prior and vice versa. Consider the joint multivariate normal density for the effects $s' = (s_1, \ldots, s_n)$, with mean zero and covariance Σ_s,

$$p(s) = \frac{1}{(2\pi)^{n/2}} |\Sigma_s|^{-0.5} \exp(-0.5 s' \Sigma_s^{-1} s).$$

Denote $Q = [q_{ij}] = \Sigma_s^{-1}$ as the precision matrix, and $s_{[i]} = (s_1, \ldots, s_{i-1}, s_{i+1}, \ldots, s_n)$. Then the conditional distributions for each s_i take a univariate normal form, corresponding to the pairwise interaction function $\Phi(u) = u^2/2$ (Rue and Held, 2005, p.22), namely

$$s_i \mid s_{[i]} \sim N\left(\sum_{j \neq i} \left[-\frac{q_{ij}}{q_{ii}}\right] s_j, \frac{1}{q_{ii}}\right)$$

with $\mathrm{corr}(s_i, s_j \mid s_{[i,j]}) = -q_{ij} / \sqrt{q_{ii} q_{jj}}$. Following Besag and Kooperberg (1995, p.734) define $h_{ii} = 0$, and set

$$h_{ij} = -q_{ij} / q_{ii} \quad (i \neq j).$$

Also set $q_{ii} = a_i / \delta$ with variance parameter δ, so that

$$h_{ij} = -q_{ij} \delta / a_i. \tag{6.5}$$

The above conditional density is then in the conditional autoregressive form specified by Besag (1974),

$$s_i \mid s_{[i]} \sim N\left(\sum_{j \neq i} h_{ij} s_j, \delta / a_i\right). \tag{6.6}$$

To obtain the joint density from the conditional one, symmetry of Q means $-Q_{ij} = -Q_{ji}$, so that from (6.5), the constraint

$$h_{ij} a_i = h_{ji} a_j$$

applies. Note that expressing $\delta / a_i = \tau_i^2$ or $a_i = \delta / \tau_i^2$, this constraint can also be stated (Cressie and Kapat, 2008) as

$$h_{ij} \tau_j^2 = h_{ji} \tau_i^2.$$

Letting $R = A(I - H)$, where $A = \mathrm{diag}(a_1, \ldots, a_n)$, one has that R is symmetric with diagonal elements a_i and off-diagonal elements $-a_i h_{ij}$. So the joint density (Besag and Green, 1993; Banerjee et al., 2014) implied by the conditional priors is

$$(s_1, \ldots s_n) \sim N_n(0, \delta R^{-1})$$

where $Q = \delta^{-1}R$. If R is positive definite as well as symmetric, the joint density of the spatial effects is proper. Positive definiteness of R holds under diagonal dominance (Rue and Held, 2005, p.20; Besag and Kooperberg, 1995, p.734), namely, that in at least one row (or column) of R, the diagonal element r_{ii} exceeds the absolute sum of the off-diagonal elements $|\Sigma_{j \neq i} r_{ij}|$.

6.3.2 Alternative Conditional Priors

Different schemes for defining the h_{ij} and a_i in (6.5) are possible, including options where R is not positive definite. Setting

$$h_{ij} = \rho \frac{w_{ij}}{\sum_{k \neq i} w_{ik}} \tag{6.7}$$

$$a_i = \sum_{k \neq i} w_{ik},$$

where $0 \leq \rho \leq 1$, and taking $w_{ij} = w_{ji}$, with $w_{ii} = 0$, ensures the symmetry constraint is met, with $h_{ij}a_i = \rho w_{ij} = h_{ji}a_j$. This is sometimes called the proper CAR, as the covariance matrix of the corresponding multivariate density is invertible. The most commonly applied approach is to set $w_{ij} = 1$ for adjacent areas and $w_{ij} = 0$ otherwise, and let $a_i = d_i = \Sigma_{k \neq i} w_{ik}$, where d_i is then the number of areas adjacent to area i. For example, when a region is partitioned into grid cells, then each grid cell has eight (first order) neighbours (Gelfand et al., 2005a). However, distance or common boundary length based forms for w_{ij} can be used.

In this case, $R = A(I - H)$ has diagonal elements d_i and off-diagonal elements $-\rho w_{ij}$. This provides the intrinsic conditional autoregression or ICAR(ρ) prior, with

$$s_i \mid s_{[i]} \sim N\left(\rho \bar{A}_i, \frac{\delta}{d_i}\right)$$

where $\bar{A}_i$ is the average of the s_i in locality L_i of area i, i.e.

$$\bar{A}_i = \frac{\sum_{j \in L_i} s_j}{d_i}.$$

Note that $R = A(I - H) = D - \rho W$ is positive definite, and the joint prior on $(s_1, \ldots s_n)$ is proper, only when $|\rho| < 1$. Lower values of ρ imply lesser degrees of spatial dependence between the s_i, though the limiting case when $\rho = 0$ has the disadvantage that the variance is not constant, but depends on the number of neighbours d_i.

Alternatively, in a CAR(ρ) spatial prior, as distinct from the ICAR(ρ) prior, one may set

$$h_{ij} = \rho w_{ij}, \quad a_i = 1,$$

so that

$$s_i \mid s_{[i]} \sim N\left(\rho \sum_{j \neq i} w_{ij} s_j, \delta\right),$$

with a homogenous conditional variance (Cressie and Kapat, 2008, p.729). In this case, $R=I-\rho W$ is positive definite, and so invertible (and the joint density is proper), when the correlation parameter is between $1/\eta_{\min}$ and $1/\eta_{\max}$ where $\eta_1,\ldots,\eta_n$ are the eigenvalues of W (Bell and Broemeling, 2000).

A compromise scheme for the variance deflators a_i – see MacNab et al. (2006) and Leroux et al. (1999) – sets

$$a_i = (1-\lambda)+\lambda\sum_{j\neq i} w_{ij},$$

with $0\leq\lambda\leq 1$ subject to a prior such as $\lambda\sim U(0,1)$. This representation has identifiability advantages in involving a single set of random effects (Lee, 2011) and can be estimated in R using CARBayes or the INLABMA packages, as well as R2OpenBUGS. It can also be estimated in rstan using multi_normal_prec applied to the joint distribution [1].

The symmetry condition $h_{ij}a_i = h_{ji}a_j$ is ensured by setting

$$h_{ij} = \frac{\lambda w_{ij}}{1-\lambda+\lambda\sum_{j\neq i} w_{ij}},$$

since $h_{ij}a_i = \lambda w_{ij} = \lambda w_{ji} = h_{ji}a_j$. So the joint density for $(s_1,\ldots s_n)$ has covariance δR^{-1} where

$$R = \lambda F + (1-\lambda)I,$$

$$f_{ii} = \sum_{j\neq i} w_{ij},$$

$$f_{ij} = -w_{ij} \quad i\neq j.$$

The case $\lambda=0$ corresponds to a lack of spatial interdependence, with R then reducing to an identity matrix, and borrowing strength confined to "global smoothing." By contrast, $\lambda=1$ leads to the ICAR(1) model (see 6.3.3). So

$$s_i \mid s_{[i]} \sim N\left(\frac{\lambda}{1-\lambda+\lambda\sum_{j\neq i} w_{ij}}\sum_{j\neq i} w_{ij}s_j, \frac{\delta}{1-\lambda+\lambda\sum_{j\neq i} w_{ij}}\right),$$

and when the w_{ij} are defined by contiguity, one obtains

$$s_i \mid s_{[i]} \sim N\left(\frac{\lambda}{1-\lambda+\lambda d_i}\sum_{j\in L_i} s_j, \frac{\delta}{1-\lambda+\lambda d_i}\right).$$

The scheme of Leroux et al. (1999) can be generalised to allow greater spatial adaptivity with varying λ (Congdon, 2008). The symmetry condition $h_{ij}a_i = h_{ji}a_j$ is maintained by setting $a_i = (1-\lambda_i)+\lambda_i\sum_{j\neq i} w_{ij}$, and taking

$$h_{ij} = \frac{\lambda_i \lambda_j w_{ij}}{1 - \lambda_i + \lambda_i \sum_{j \neq i} w_{ij}},$$

since this ensures the constraint

$$h_{ij} a_i = h_{ji} a_j = \lambda_i \lambda_j w_{ij}.$$

A possible borrowing strength prior for these parameters is

$$\text{logit}(\lambda_i) \sim N(\lambda_\mu, 1/\tau_\lambda),$$

where the average λ_μ and precision τ_λ are extra unknowns. Setting $\Lambda = \text{diag}(\lambda_1, \ldots, \lambda_n)$, the covariance in the joint prior is then

$$\delta[\Lambda F^* + (I - \Lambda)]^{-1},$$

where

$$f_{ii}^* = \sum_{j \neq i} w_{ij},$$

$$f_{ij}^* = -w_{ij} \lambda_j \quad i \neq j.$$

Pettitt et al. (2002) propose a scheme with

$$h_{ij} = \frac{\phi w_{ij}}{1 + |\phi| \sum_{j \neq i} w_{ij}},$$

and

$$a_i = 1 + |\phi| \sum_{j \neq i} w_{ij},$$

where ϕ measures the strength of spatial dependency, and the case $\phi = 0$ corresponds to an absence of spatial interdependence, such that $R = I$ (see also Gschlößl and Czado, 2006). Gibbs updating for ϕ can be applied. So

$$s_i \mid s_{[i]} \sim N\left(\frac{\phi}{1 + |\phi| \sum_{j \neq i} w_{ij}} \sum_{j \neq i} w_{ij} s_j, \frac{\delta}{1 + |\phi| \sum_{j \neq i} w_{ij}} \right).$$

Under both the MacNab et al. (2006) and Pettitt et al. (2002) schemes, the joint distribution of s is proper, ensuring a proper posterior when either is taken as the prior distribution. Retaining $h_{ij} = \phi w_{ij}/(1 + |\phi| \Sigma_{j \neq i} w_{ij})$, but setting $a_i = (1 + |\phi| \Sigma_{j \neq i} w_{ij})/(1 + |\phi|)$, means that $\phi \to \infty$ corresponds to the ICAR(1) prior, with the conditional variance $(1 + |\phi| \delta)/(1 + |\phi| \Sigma_{j \neq i} w_{ij})$ tending to $\delta / \Sigma_{j \neq i} w_{ij}$.

6.3.3 ICAR(1) and Convolution Priors

The ICAR(ρ) prior when $\rho=1$ is sometimes known as the ICAR(1) model, when one has

$$h_{ij} = \frac{w_{ij}}{\sum_{j \neq i} w_{ij}}, \quad a_i = \sum_{j \neq i} w_{ij},$$

and for counts $y_i \sim Po(\lambda_i P_i)$, if one assumes

$$\log(\lambda_i) = \alpha + s_i,$$

then borrowing of strength is purely spatial, with

$$s_i \mid s_{[i]} \sim N\left(\bar{A}_i, \frac{\delta}{\sum_{j \neq i} w_{ij}}\right),$$

where $\bar{A}_i = \Sigma_{j \neq i} w_{ij} s_j / \Sigma_{j \neq i} w_{ij}$. The precision matrix of the joint prior is $\delta^{-1}R$, where

$$r_{ii} = \sum_{j \neq i} w_{ij},$$

$$r_{ij} = -w_{ij} \quad i \neq j.$$

When the w_{ij} are binary indicators of adjacency ($w_{ij}=1$ for areas i and j contiguous, $w_{ij}=0$ otherwise), then $r_{ii}=d_i$ and the off-diagonal elements r_{ij} are -1 if i and j are neighbours, but zero otherwise. This case demonstrates most directly that conditional independence properties relating to spatial effects are stipulated by the matrix R and vice versa (Rue and Held, 2005, p.4). Despite the relative simplicity of this form and the wide use of the ICAR(1) conditional prior, R is not invertible under this model, and the joint prior is improper (Haran et al., 2003).

To see this in another way, for the case where the w_{ij} are binary, the joint prior can be specified in terms of pairwise comparisons between the s_i (Knorr-Held and Becker, 2000). Let $i \sim j$ denote that areas i and j are neighbours, then for a normal ICAR(1) model, the joint prior in terms of differences $s_i - s_j$ is (Hodges et al., 2003)

$$p(s_1, \ldots s_n) \propto \delta^{-0.5(n-1)} \exp\left(-\frac{1}{2\delta}\sum_{i \sim j}(s_i - s_j)^2\right).$$

Thus the prior only specifies differences between spatial effects and not their overall level. However, all linear contrasts $c's$ with $c'1=0$ have proper distributions (Besag and Kooperberg, 1995, p.740).

To tie down the effects and remove their locational invariance, one method involves centring the sampled values at every iteration to have mean zero. This is one form of linear constraint, and so the joint distribution becomes integrable and propriety is obtained (Rodrigues and Assuncao, 2008). Another possibility is a corner constraint, i.e. setting a particular effect to a known value, such as $s_1=0$ (Besag et al., 1995). Finally, one may omit the intercept so that the s_i model the level of the data. In this case, $y_i \sim Po(P_i \exp(s_i))$ with the s_i not constrained, rather than $y_i \sim Po(P_i \exp(a+s_i))$.

As mentioned above a spatial effects-only assumption is relatively informative, and the ICAR(1) spatial prior is often combined with an exchangeable prior to form a convolution prior (Richardson et al., 2004). It may be argued that an exchangeable iid effect should only be introduced in combination with an ICAR(1) spatial prior, since conditional priors including a correlation parameter, such as the ICAR(ρ) can adjust to varying mixtures of spatial and unstructured variation by varying the ρ parameter (Wakefield, 2007). Thus, for a Poisson response, $y_i \sim Po(\lambda_i P_i)$, the convolution prior of Besag et al. (1991), also called the Besag-York-Mollie (BYM) prior, specifies

$$\log(\lambda_i) = a + s_i + u_i$$

with $s_i \mid s_{[i]} \sim N(\bar{A}_i, \delta_s / d_i)$, and $u_i \sim N(0, \delta_u)$ usually homoscedastic. Note that heteroscedasticity or heavier tails than under the normal might be represented by taking $u_i \sim N(0, \psi_i)$ where

$$\psi_i = \delta_u / \kappa_i$$

where the κ_i are positive variables with mean 1 (LeSage, 1999). While only the sum $z_i = s_i + u_i$ is identifiable in this model, Norton and Niu (2009) show that the precisions δ_s and δ_u are identifiable from the distribution of z_i.

6.4 Priors on Variances in Conditional Spatial Models

As in the exchangeable hierarchical models considered in Chapter 4, the prior on the conditional spatial variance parameter δ_s, or on the pair $\{\delta_s, \delta_u\}$ in a convolution model, is important in governing the degree of smoothing towards the neighbourhood or global mean. Prior specification is important as an aspect in the general identifiability of complex random effects models for spatial variation, with potential weak identifiability of hyperparameters and sensitivity of posterior estimates to the form of prior; see Example 6.4. The same applies to the spatial smoothing parameters in the proper CAR prior and the Leroux et al. CAR (MacNab, 2014), and on hyperparameters for spatial priors based on group allocation, such as the Potts prior (Moores et al., 2015).

Regarding variance priors, some applications of conditional autoregressive priors use vague priors for δ_s, such as $p(\delta_s) \propto 1/\delta_s$ or just proper priors, with $1/\delta_s \sim Ga(\varepsilon, \varepsilon)$ with ε small. However, these may lead to effective impropriety in the posterior such that MCMC convergence is impeded (Besag and Kooperberg, 1995, p.741). The prior $1/\delta_s \sim Ga(\varepsilon, \varepsilon)$ with ε small may also put undue weight on low variances. Suppose the prior relates to a variance for unstructured random effects in a log-linear model for relative risks λ_i, with $y_i \sim Po(E_i \lambda_i)$. Wakefield (2007) mentions that a $Ga(0.001,0.001)$ prior on $1/\delta_u$ in the model $\log(\lambda_i) = a + u_i$ is equivalent to assuming relative risks $e^{\lambda_i - a}$ follow a log-t distribution with 0.002 degrees of freedom.

Prior specification is most problematic for the convolution model, since the data identify the total variation in log relative risks (under a Poisson model), but not the pair of variances $\{\delta_s, \delta_u\}$ (MacNab, 2014). Following Bernardinelli et al. (1995), the marginal standard deviation $sd(s_i)$ of the spatial effects is approximately equal to a multiple 1.43 (=1/0.7) of the conditional scale term, $(\delta_s/\bar{d})^{0.5}$, where $\bar{d}$ is the average number of neighbours. Hence a "fair" prior on $sd(u_i)=\delta_u^{0.5}$ (Banerjee et al., 2014, section 6.4.3.3) is one that ensures

$$sd(u_i) \approx sd(s_i) \approx 1.43 \times (\delta_s/\bar{d})^{0.5}.$$

Riebler et al. (2016) propose a modified BYM scheme retaining the two random effects, but with a single scale parameter δ for the composite effects

$$t_i = u_i + s_i = \sqrt{\delta}[\sqrt{1-\rho}\theta_i + \sqrt{\rho}\phi_i^*].$$

Here $\theta_i \sim N(0,1)$ are iid effects, the ϕ_i^* are scaled versions of spatial effects ϕ_i following an ICAR(1) prior, and $\rho \in [0,1]$ governs the proportion of residual variance due to spatial dependence. To ensure $\sqrt{\delta}$ is legitimate as the standard deviation of the composite effect, one requires $\text{var}(\phi_i) \approx \text{var}(\theta_i) \approx 1$. To achieve this, Riebler et al. (2016) propose a scaling whereby the geometric mean of variances of ϕ_i is 1. To obtain a scaling factor F, with $\phi_i^* = \phi_i/F$, one may apply the R-INLA function inla.scale.model to the adjacency matrix.

Example 6.2 Blood Lead in Children, Virginia Counties

The data here, considered by Schabenberger and Gotway (2004), relate to elevated blood level readings y_i among n_i children (under 72 months) tested in the $N=133$ counties of Virginia (including Independent Cities) in 2000. Numbers sampled n_i vary considerably (from 1 to 3808). Spatial proximity is binary, with $w_{ij}=1$ for intercounty distances under 50 km and $w_{ij}=0$ otherwise.

Assuming binomial sampling with $y_i \sim \text{Bin}(n_i, \pi_i)$, one option considered is the convolution, or BYM, model of Besag et al. (1991), namely

$$\text{logit}(\pi_i) = a + s_i + u_i,$$

with conditional variance δ_s for ICAR(1) spatial effect s_i, and variance δ_u for the unstructured effects. Using rstan for estimation, positive $N^+(0,25)$ priors are assumed on the standard deviations $\sigma_s = \delta_s^{0.5}$ and $\sigma_u = \delta_u^{0.5}$. In rstan, the ICAR(1) spatial prior is implemented using the pairwise difference form of the joint multivariate density (e.g. Gerber and Furrer, 2015; Morris, 2018), and in particular the target + formulation,

```
target += 0.5*(N-1)*log(tau_s) -0.5*tau_s*dot_self(s[node1]
- s[node2]);
```

where tau_s is the precision of the spatial effects.

Convergence is non-problematic, despite the default strategy for the priors on the standard deviations. Posterior means (medians) for σ_s and σ_u are obtained as 1.57 (1.62) and 0.37 (0.38) respectively. Twenty (from 133) of the s_i parameters are judged significant in terms of posterior probabilities over 0.95, or under 0.05, that $s_i>0$. By contrast, 41 composite terms $t_i = u_i + s_i$ are significant. The LOO-IC (leave-one-out information criterion) is 608, with the highest individual LOO-IC being for Winchester (county 3), which has an unusually high proportion of elevated readings, but a small population. More relevant in establishing significantly elevated readings may be the second highest LOO-IC value, namely for county 63, with a much larger population than county 3. The proportion of variance due to spatial effects can be obtained as var(s)/(var(s)+var(u)) and is estimated at 0.77.

A second analysis applies the Riebler et al. (2016), or BYM2, prior, with a single set of effects, and with the proportion of spatial variance now a parameter. This model provides an unchanged LOO-IC of 608. The proportion of spatial variance ρ is estimated at 0.82, though with a wide 95% interval from 0.32 to 1. Forty-two of the composite effects t_i are now significant.

An area spatial model may also be assessed by whether residual spatial dependence is removed, and this can be established using the moran.mc function in R. The moran.mc function uses a Monte Carlo permutation test for Moran's I statistic. Significant residual correlation shows in extreme tail p-values, either values close to zero (positive residual correlation), or p-values near 1 (negative residual correlation).

Here 100,000 permutations are taken, with the calculations using a binary adjacency spatial interaction matrix for the 133 areas, converted to listw format. We find a non-significant p-value of around 0.25 for the first model, and 0.27 for the second.

These models are also estimated by R-INLA, with the default log-gamma priors on random effect precisions. The total random effects $t_i = u_i + s_i$ under the BYM model are very similar to those from the rstan application, with a correlation of 0.99 between the two sets of posterior means. However, possibly reflecting sensitivity to priors on scale parameters, spatial effects are smaller under R-INLA, and unstructured effects larger. The BYM2 model estimated using R-INLA produces a lower DIC than the BYM model. The proportion ρ of total residual variation due to spatial effects is estimated with mean (95% CRI) of 0.69 (0.30,0.95), as against 0.82 (0.32,1.00) under rstan. The spatial effects under the two estimations are highly correlated.

6.5 Spatial Discontinuity and Robust Smoothing

Spatial pooling assuming a smoothly varying outcome over contiguous areas may not be appropriate when there are clear discontinuities in the spatial pattern of events (Adin et al., 2018). For instance, a low mortality area surrounded by high mortality areas will have a distorted smoothed rate when heterogeneity is assumed to be entirely spatially structured. More generally one may seek robustness against mis-specification of the distribution of latent event rates or risks; for example, virtually all applications of spatial conditional autoregression models assume normality by default. Finally, one may seek some degree of spatial adaptiveness. For example, under conditional autoregressive models, the conditional variance δ is constant across the region, whereas one might expect spatial correlation to be stronger in some sub-regions. In the convolution model, the variances δ_s and δ_u are global parameters, so that the relative amount of spatially structured and unstructured heterogeneity is constant across the study region (Knorr-Held and Becker, 2000; Congdon, 2007).

Robustness against spatial discrepancies or non-normality may be important when event totals are small, since then the prior structure of the latent risks has a greater effect; this is the case with the much analysed Scottish lip cancer data, where certain areas have elevated SMRs, but small counts y and expected cases E . A high relative risk apparent from a crude or moment estimate not based on a large y or E may be shrunk considerably under a spatial random effects approach, particularly if surrounded by lower morbidity areas, so that important excess risks may not be flagged up (Conlon and Louis, 1999).

One strategy is to adopt heavier tailed alternatives to the CAR normal, such as the double exponential (Laplace) or $L1$-norm version of the ICAR(1) prior, which Besag (1989, p.399) mentions as preferable when the s_i have discontinuities. For a connected graph (i.e. with no isolated areas in the region) this prior is

$$p(s_1,\ldots s_n) \propto \frac{1}{\delta^{n-1}}\exp\left(-0.5\frac{1}{\delta}\sum_{j\neq i}|s_i - s_j|\right),$$

and has its posterior mode at the median rather than mean of the neighbouring s_j. One might also apply Student t versions of the ICAR(ρ) which, if applied using scale mixtures, give a natural measure of outlier status. Thus, for a Student t with ν degrees of freedom,

$$s_i \mid s_{[i]} \sim N\left(\rho\bar{A}_i, \frac{\delta}{\gamma_i d_i}\right)$$

where $\gamma_i \sim Ga(\nu/2, \nu/2)$, and low values of γ_i correspond to spatial outliers.

Forms of discrete mixture have been proposed. Green and Richardson (2002) distinguish between clustering models and allocation models, while Knorr-Held and Rasser (2000) propose a scheme whereby at each MCMC iteration, areas are allocated to clusters of mutually contiguous areas, with identical risks within each cluster. Lawson and Clark (2002) propose a mixture of the ICAR(1) and Laplace priors for the case $y_i \sim Po(E_i\lambda_i)$, with continuous (beta) weights r_i rather than binary mixture weights, namely

$$\log(\lambda_i) = a + r_i s_{1i} + (1 - r_i)s_{2i},$$

where $r_i \sim Be(c,c)$, with c known, s_{1i} is an ICAR error, but s_{2i} follows a spatial Laplace prior.

Following Congdon (2007), analogous mixture forms can be applied to the errors in the convolution model itself, giving more emphasis to the unstructured term u_i in outlier areas:

$$\log(\lambda_i) = a + r_i s_i + (1 - r_i)u_i.$$

This type of representation may be useful for modelling edge effects, with the u effects taking a greater role on the peripheral areas where neighbours are fewer. Another possibility is a discrete mixture in a "spatial switching" model (Congdon, 2007), allowing an unstructured term only for areas where the pure spatial effects model is inappropriate. Thus, for a count response,

$$y_i \sim Po(E_i\lambda_{J_i,i})$$

$$J_i \sim \text{Categoric}(\pi_1, \pi_2)$$

$$(\pi_1, \pi_2) \sim \text{Dirichlet}(\xi_1, \xi_2)$$

$$\log(\lambda_{1i}) = a + s_i$$

$$\log(\lambda_{2i}) = a + s_i + u_i$$

where the ξ_j are extra unknowns, and the $s_i \sim$ICAR(1). The posterior estimates for the ξ_j provide overall weights of evidence in favour of a pure spatial model as compared to a convolution model, while high posterior probabilities $Pr(S_i = 2|y)$ for particular areas indicate that pure spatial smoothing is inappropriate for them.

Fernandez and Green (2002) use a discrete mixture model generated via mixing over several spatial priors. Thus, for count data, assume K possible components with area-specific probabilities π_{ik} on each component

$$y_i \sim \sum_{k=1}^{K} \pi_{ik} Po(E_i \lambda_{ik})$$

where $\log(\lambda_{ik}) = \alpha_k$ for a model without predictors. Then K sets of underlying spatial effects $\{s_{ik}\}$ are generated from separate conditional spatial priors, and used to estimate area-specific mixture weights

$$\pi_{ik} = \exp(\chi s_{ik}) \Big/ \sum_{k=1}^{K} \exp(\chi s_{ik})$$

where $\chi > 0$. As χ tends to 0, the π_{ik} tend to $1/K$ without spatial patterning, whereas large χ reduce over-shrinkage.

Another discrete mixture model for robust spatial dependence modelling uses the Potts prior (Green and Richardson, 2002). Thus let $J_i \in 1,\ldots,K$ be unknown allocation indicators with $y_i \sim Po(E_i \mu_{S_i})$ where $\{\mu_1,\ldots,\mu_K\}$ are distinct cluster means. Also let $d_{ik} = 1$ if $J_i = k$. Then the joint prior for the allocation indicators incorporates spatial dependence with

$$Pr(J_i = k) = \exp\left[\omega \sum_{j \sim i} I(d_{ik} = d_{jk})\right] \Big/ \sum_{h=1}^{K} \exp\left[\omega \sum_{j \sim i} I(d_{ih} = d_{jh})\right]$$

where $\omega > 0$ multiplies the number of same label neighbour pairs, so that lower values of ω indicating lesser spatial dependence. So pooling towards the local neighbourhood average will tend not to occur if an area's latent risk is discrepant with those of its neighbours. Richardson et al. (2004) compare this model with the convolution model under various simulated scenarios for differentiated spatial risks. Additional effects can be included by multiplying the μ_{J_i}. For example, a spatially unstructured multiplicative effect could be modelled as $\nu_i \sim Ga(b_\nu, b_\nu)$, or a log-normal prior assumed with $\nu_i = \exp(u_i)$, and $u_i \sim N(0, \delta_u)$. Then $y_i \sim Po(E_i \mu_{S_i} \nu_i)$.

Assumptions such as normality in the spatial effects can be avoided by adapting the Dirichlet process stick-breaking prior of Sethuraman (1994) to spatial settings. The stick-breaking prior specifies an unknown distribution G by a mixture

$$G = \sum_{m=1}^{M} p_m \delta(\rho_m)$$

where M may in principle be infinite, but in practical computing is taken as finite, the mixing probabilities satisfy $\Sigma_{m=1}^{M} p_m = 1$, and $\delta(\rho_m)$ has a point mass at ρ_m which may be scalar or vector values for areas (e.g. relative risks) or at grid locations. For example, the ρ_m may be drawn from a baseline borrowing-strength prior G_0 such as a stationary Gaussian process in the case of continuous point-referenced spatial data $y(g_i)$ at sites g_i. One may incorporate spatial information into either the ρ_m, as in Gelfand et al. (2005b), or into the mixture probabilities p_m, as in Griffin and Steel (2006). Such formulations are typically for point-referenced data, and allow for nonstationarity and non-Gaussian features in the response when the stationary Gaussian process is not appropriate (Duan et al., 2007).

Example 6.3 Long-Term Illness, NE London

This example compares the original Leroux et al. (1999) model with the adaptive Leroux scheme of Congdon (2008). The application is to 133 small areas in NE London, electoral wards, defined for political and administrative purposes. As well as census counts of limiting long-term illness (LLTI) among people aged 50–59, and corresponding binomial denominators, a deprivation index is used, not to model varying LLTI propensities, but to measure discrepancies between areas and their surrounding localities on this potential risk factor.

The spatially adaptive approach retains the principle of spatial borrowing of strength, but modifies it to better represent discontinuities in the outcome and/or observed risk factors. The Leroux global index of spatial dependence λ is allowed to vary between areas, with one possible prior for λ_i linking varying spatial dependence to spatial dissimilarity (or similarity) in risk factors. For example, illness is commonly linked to socio-economic deprivation, and spatial correlation in illness may be weaker when socio-economically distinct areas are adjacent, with localised dissimilarity in risk factors.

Possible priors for the λ_i include beta priors, or probit-normal or logit-normal priors, such as $\text{logit}(\lambda_i) \sim N(\mu_\lambda, 1/\tau_\lambda)$, where the average and precision $\{\mu_\lambda, \tau_\lambda\}$ are extra unknowns. However, if predictors D_i measuring dissimilarity in observed risk factors are available, and so relevant to whether there should be some attenuation of the principle of local borrowing of strength, one can use the scheme

$$\text{logit}(\lambda_i) \sim N(\gamma_1 + \gamma_2 D_i, 1/\tau_\lambda),$$

where $\gamma = (\gamma_1, \gamma_2)$ are regression parameters. One would expect lower λ_i for areas dissimilar from their neighbours on the risk factor; that is, γ_2 is anticipated to be negative. Here the discrepancy measure is based on the index z_i of socioeconomic deprivation, whereby dissimilarity may be represented as

$$D_i = \left| z_i - \bar{Z}_i \right|$$

with $\bar{Z}_i$ being the average deprivation level in the locality L_i around area i, namely $\bar{Z}_i = \Sigma_{j \in L_i} z_j / d_i$.

Estimation of the original Leroux et al. (1999) model using R2OpenBUGS provides a posterior mean for the global λ of 0.86, with a LOO-IC of 1278 and WAIC (widely applicable information criterion) of 1185. Estimation using CARBayes provides a slightly higher estimate of λ, namely 0.93, but a higher WAIC of 1190.

Improved fit is provided by the adaptive Leroux model, with the LOO-IC and WAIC respectively at 1267 and 1178. The coefficient γ_2 has mean (95% CRI) of −0.54 (−0.83,−0.30). In contrast to the estimated global λ of 0.88, there are eight local λ_i under 0.5, with the minimum being for area 133 (the City of London) with posterior mean $\lambda_{133} = 0.002$. This area has an illness rate (illness total divided by population, as percentage) of 14.5%, as compared to the rate in its locality (surrounding adjacent wards) of 38.4%. Its deprivation index is 16.4, compared to the locality average of 43.9. Figure 6.2 maps out the local λ_i.

Example 6.4 Robust Priors for London Suicides

This analysis compares the Potts prior, the convolution prior and spatial median regression for modelling the distribution of suicides in 983 middle level super output areas (MSOAs) in London over 2011–15. Expected suicides E_i are based on England wide rates, with a subsequent scaling to ensure $\Sigma_i y_i = \Sigma_i E_i$. As for Examples 6.1 and 6.3, the adjacency matrix is obtained by inputting a shapefile.

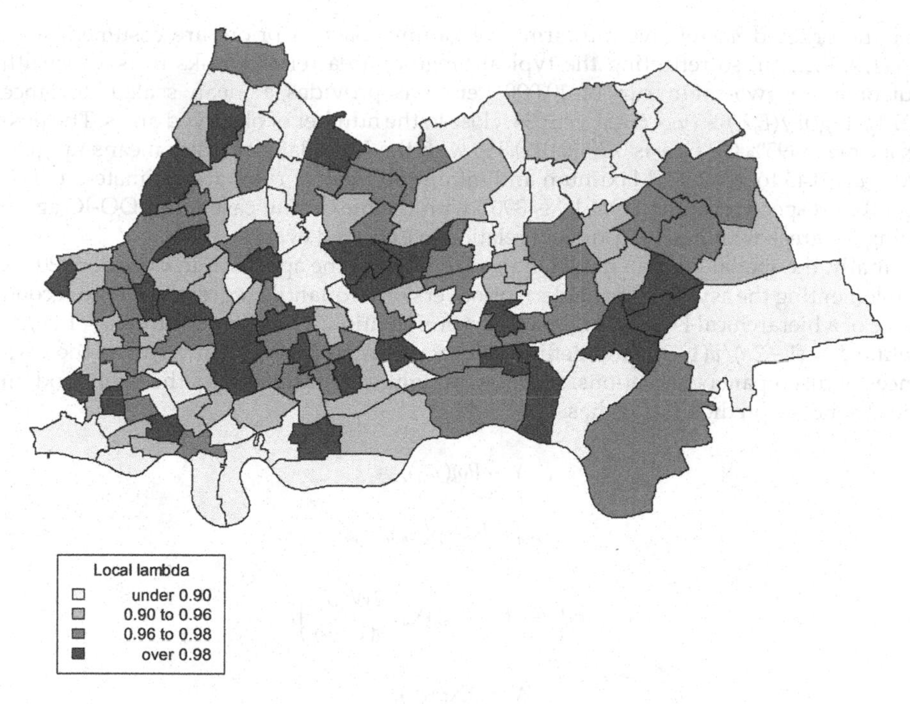

FIGURE 6.2
Local Leroux dependence parameters.

The first analysis uses the nimble package in R, and estimates the BYM model. This provides a LOO-IC of 3905, with maximum and minimum posterior mean relative risks of 1.40 and 0.82 . The maximum casewise LOO-IC are for areas (such as 263, 573, and 512) which have high y_i counts in relation to expected suicides. 10% of the total LOO-IC is due to the 5% worst fitting cases. Incidentally, the estimated proportion of variation due to spatial dependence is relatively low, namely 0.23 (95% CRI from 0.07 to 0.44).

This feature is reproduced in an estimation of the model incorporating a proper CAR spatial effect. This is implemented via a sparse precision matrix method in rstan, and draws on Joseph (2016). The resulting estimate for ρ in (6.7) is 0.28 (with 95% CRI from 0.19 to 0.69). The LOO-IC is 3902, with maximum and minimum posterior mean relative risks of 1.41 and 0.76. Mixed predictive exceedance checks are included, based on replicate samples of the random spatial effects, and obtained as

$$p_{i,\text{mix}} = Pr(y_{i,\text{rep}} > y_i) + 0.5Pr(y_{i,\text{rep}} = y_i).$$

These show over-prediction (high $p_{i,\text{mix}}$) in a relatively high proportion of cases, with high predicted y_i deaths in relation to actual deaths.

An alternative to the BYM and proper CAR priors is the Potts prior. This is applied with an exponential $E(1)$ prior on ω, and with an ordering constraint on the latent cluster means, so $\mu_1 \leq \mu_2 \leq \ldots \leq \mu_K$, where K is set at 10. Since there is evidence of unstructured heterogeneity, the scheme is modified to include unstructured area effects, namely

$$y_i \sim Po(E_i \mu_{S_i} \nu_i),$$

$$\nu_i = \exp(u_i),$$

$$u_i \sim N(0, \delta_u).$$

For the ordered μ_k, relatively informative gamma $Ga(a_k,5)$ priors are assumed, with $a=(1,2,3,\ldots,10)$, so reflecting the typical range of area relative risks for such health outcomes. A two-chain run of 10,000 iterations provides a mean scaled deviance $2\Sigma_i\{y_i\log(y_i/(E_i\lambda_i))-(y_i-E_i\lambda_i)\}$ of 1034, close to the number of observed areas. The posterior mean (95% CI) of ω is 0.30 (0.01,0.85), with the $K=10$ latent cluster means ranging from $\mu_1=0.43$ to $\mu_K=1.41$. Maximum and minimum relative risks are estimated as 1.32 and 0.65 respectively. The LOO-IC is 3907, with the maximum casewise LOO-IC again being for areas with high y_i counts in relation to expected events.

Finally, the spatial median model is an adaptation of the approach in Congdon (2017), implementing the asymmetric Laplace prior version of quantile regression at the second stage of a hierarchical Poisson log-normal representation. Thus for quantiles $a=1,\ldots,A$, define $\xi_a=(1-2\alpha)/\alpha(1-\alpha)$, and define scale factors $W_{ai}\sim Exp(\delta_a)$ which inflate the variances of discrepant observations, and downweight their influence on the likelihood. In the absence of predictors, one has

$$Y_i\sim Poi(\mu_{ai}),$$

$$\mu_{ai}=E_i\exp(\nu_{ai}),$$

$$\nu_{ai}\sim N\left(\beta_{0a}+s_{ai}+\xi_a W_{ai},\frac{2W_{ai}\delta_a}{\alpha(1-\alpha)}\right),$$

$$W_{ai}\sim \text{Exp}(\delta_a).$$

Here median regression ($\alpha=0.5$) only is considered, with a gamma Ga(1,0.001) prior on $\delta_{0.5}$. This model has a LOO-IC of 3895, improving on the Potts, BYM, and proper CAR priors. Poorly fitted areas cases are similar, whether identified by casewise LOO-IC, or by the residual type measure $(\nu_i-\beta_0-s_i)/(8W_i\delta)^{0.5}$.

Compared to the Potts prior, extreme elevated relative risks are identified under the spatial median model, the highest posterior mean relative risk $\rho_i=\exp(\beta_0+s_i)$ being 1.50 (though the second and third ranking posterior mean ρ_i are 1.41 and 1.31). The Potts prior is distinctive in its broader spread of estimated relative risk, including a longer tail of low estimated relative risk, with 211 of the 983 areas having posterior mean ρ_i under 0.9. Figure 6.3 compares posterior mean relative risks under the Potts and BYM priors, and Figure 6.4 compares the Potts and spatial median relative risks.

6.6 Models for Point Processes

A continuous spatial framework is appropriate when point observations are made. Nevertheless, a continuous framework is often applied to discrete area or lattice data (Berke, 2004; Kelsall and Wakefield, 2002; Yanli and Wall, 2004). Consider metric observations $y(g)=(y_1(g_1),\ldots y_n(g_n))$ at points $g=\{g_1,g_2,\ldots g_n\}$ in two-dimensional space G^2. To represent the spatially driven component in the variation of y, define a Gaussian spatial process, or Gaussian process prior, for $(s_1,\ldots s_n)=s(g)=(s(g_1),\ldots,s(g_n))$ with covariance matrix $\Sigma(d_{ij})=\sigma_s^2C(d_{ij})$, where the off-diagonal correlations depend on distances $d_{ij}=\|g_i-g_j\|$ between points g_i and g_j, and $C(0)=1$.

Such a process is ergodic if the off-diagonal elements in $\Sigma(d)$ tend to zero as $d\rightarrow\infty$ (so that covariance between values at two points vanishes for large enough distances), and

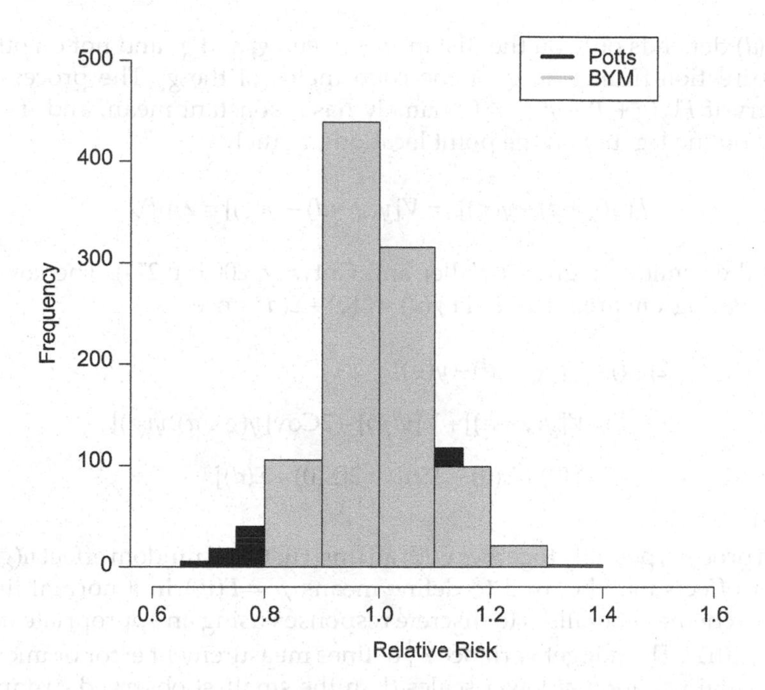

FIGURE 6.3
Posterior mean relative risks, Potts vs BYM.

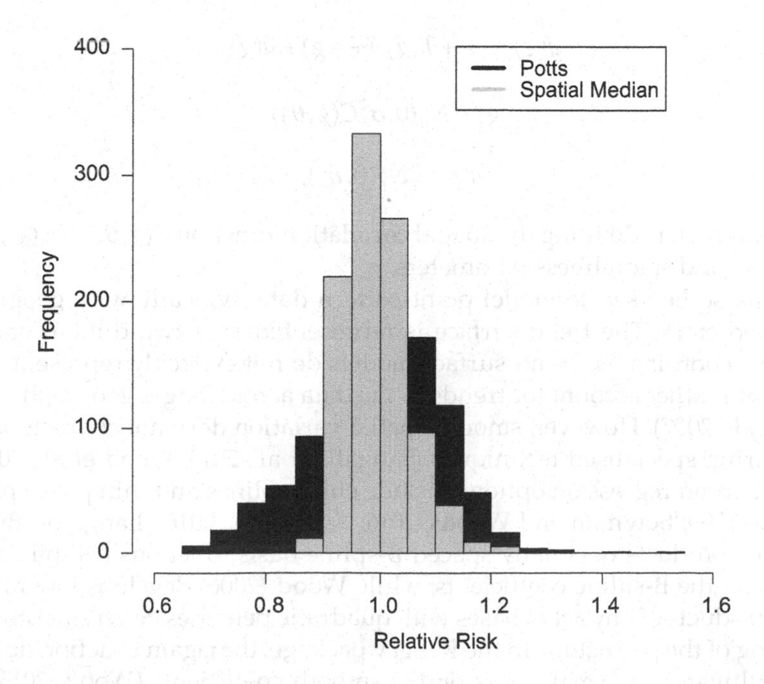

FIGURE 6.4
Posterior mean relative risks, Potts vs spatial median.

isotropic if $\Sigma(d)$ depends only on the distance between g_i and g_j, and not on other features such as the direction from g_i to g_j or the coordinates of the g_i. The process is intrinsically stationary if $E[y(g+d)-y(g)]=0$, namely has a constant mean, and if the variance depends only on the lag, not on the point locations, namely

$$E[y(g+d)-y(g)]^2 = V[y(g+d)-y(g)] = 2\gamma(d),$$

where $\gamma(d)$ is the semiovariogram (Waller and Gotway, 2004, p.274). The covariance $\Sigma(d)$ and the semiovariogram are related via $\gamma(d)=\Sigma(0)-\Sigma(d)$ since

$$\begin{aligned}2\gamma(d) &= V[y(g+d)-y(g)]\\ &= V[y(g+d)]+V[y(g)]-2\text{Cov}[y(g+d),y(g)]\\ &= \Sigma(0)+\Sigma(0)-2\Sigma(d) = 2[\Sigma(0)-\Sigma(d)]\end{aligned}$$

so that $\gamma(0)=0$.

A Gaussian process, possibly together with an unstructured random effect $u(g) \sim N(0,\sigma_u^2)$, and regressor effects may be used to define means $\mu_i = E(y_i)$ in a normal linear model. However, this scheme generalises to discrete responses using an appropriate link function (Eidsvik et al., 2012). The nugget variance σ_u^2 defines measurement error or micro scale spatial effects (spatial variation at lower scales than the smallest observed distance between sampled points). In Bayesian modelling, it is possible to take account of interplay between the nugget and the parameters of the spatial correlation function (Gramacy and Lee, 2008). Regressor effects might include a trend surface $T(g)$ defined by the coordinates of g_i (Diggle and Ribeiro, 2002, p.133), such as a quadratic polynomial with terms $(g_{1i}, g_{2i}, g_{1i}^2, g_{2i}^2, g_{1i}g_{2i})$. So, for y continuous, one might have

$$\begin{aligned}y(g) &= \beta_0 + T(g)\beta + s(g) + u(g), \qquad (6.8)\\ s(g) &\sim N_n(0,\sigma_s^2 C(g,\theta)),\\ u(g) &\sim N_n(0,\sigma_u^2),\end{aligned}$$

where θ are parameters defining the spatial correlation function $C(g,\theta)=[c_{ij}(g_i,g_j;\theta)]$, such as spatial decay and smoothness parameters.

Splines can also be used to model point pattern data, typically with geographic coordinates as predictors. The trend-surface is represented as a two-dimensional spline in the geographic coordinates. Trend-surface models do not explicitly represent local spatial dependence, but rather account for trends in the data across longer geographical distances (Dormann et al., 2007). However, smooth spatial variation does not characterise all applications, requiring specialised techniques (Sangalli et al., 2013; Wood et al., 2008). Widely applied spline trend regression options include cubic splines and thin plate splines (Mitas and Mitasova, 1999; Bowman and Woods, 2016; Yang et al., 2016). Lang and Brezler (2004) propose tensor products of equally spaced B-spline basis functions combined with symmetric priors on the B-spline coefficients, while Wood (2006) develops low rank smooths from tensor products of any set of bases with quadratic penalties. Such smooths are invariant to rescaling of the predictors. In the R mgcv package, the jagam function develops JAGS code with multivariate normal priors on the smooth coefficients (Wood, 2016). The prior precision matrix incorporates the smoothing parameters and smoothing penalty matrices.

To avoid a smoothing penalty not corresponding to a full rank precision matrix (and hence an improper prior), null space penalties, as in Marra and Wood (2012), are added to the usual penalties. The smooths are centred to improve identifiability.

6.6.1 Covariance Functions

Defining d_{ij} as a distance measure between points g_i and g_j, there are several common isotropic schemes with $C(d_{ij})$, and hence $\gamma(d_{ij})$, parameterised to reflect anticipated distance decay in the correlation between points (e.g. Grunwald, 2005). For example, the exponential distance model has

$$C(d_{ij}) = \exp(-\phi d_{ij}),$$

with range parameter $\phi > 0$, and larger values of ϕ leading to more pronounced distance decay. Note that different parameterisations of the exponential are used in different packages (e.g. in spBayes and spNNGP as opposed to gstat). The covariance function for $(s_1, \ldots, s_n)$ is then

$$\Sigma(d_{ij}) = \sigma_u^2 I(i = j) + \sigma_s^2 \exp(-\phi d_{ij}),$$

while the semivariogram is

$$\gamma(d_{ij}) = \sigma_u^2 + \sigma_s^2 [1 - \exp(\phi d_{ij})].$$

As d_{ij} tends to infinity, the semivariogram trends to an upper limit of $\sigma_u^2 + \sigma_s^2$, known as the sill. The powered exponential variant (Diggle and Ribeiro, 2007) has

$$C(d_{ij}) = \exp[-(\phi d_{ij})^{\kappa}],$$

for $\phi > 0$ and $0 < \kappa \leq 2$.

The spherical model (Zhang, 2002) has non-zero covariance only within a certain range δ, namely

$$C(d_{ij}) = 1 - \frac{3d_{ij}}{2\delta} + \frac{d_{ij}^3}{2\delta^3},$$

for $d < \delta$, whereas $C(d_{ij}) = 0$ for $d_{ij} \geq \delta$. Hence the spherical function has covariance

$$\Sigma(d_{ij}) = \sigma_u^2 I(i = j) + \sigma_s^2 \left[1 - \frac{3d_{ij}}{2\delta} + \frac{d_{ij}^3}{2\delta^3}\right] I(d_{ij} < \delta),$$

and semivariogram

$$\gamma(d_{ij}) = \sigma_u^2 + \sigma_s^2 \left[\frac{3d_{ij}}{2\delta} - \frac{d_{ij}^3}{2\delta^3}\right] \text{ for } d_{ij} < \delta,$$

$$\gamma(d_{ij}) = \sigma_u^2 + \sigma_s^2 \text{ for } d \geq \delta.$$

Finally, Matern covariances (Diggle et al., 2003) set

$$C(d_{ij}) = \frac{\sigma_s^2}{\Gamma(\nu)2^{\nu-1}}(\kappa d_{ij})^{\nu} K_{\nu}(\kappa d_{ij}),$$

where $K_\nu(u)$ is a modified Bessel function of order ν. The parameter ν controls the smoothness of the process, while κ is a scaling parameter. Together they define the range $r = (8\nu)^{0.5}/\kappa$ at which the covariance is diminished to low levels (close to 0.1). INLA parameterises the Matern in terms of a parameter $\alpha = \lambda + 1$, with $\alpha = 2$ as the default setting (Lindgren and Rue, 2015). Paciorek and Schervish (2006) use kernel convolution (Section 6.6) to develop nonstationary covariance functions, including a nonstationary version of the Matérn covariance. Implementation of this method in R is described in Risser and Calder (2017).

Prediction at new locations is a major aspect of geostatistical modelling. Suppose continuous observations $y = (y_1,\ldots,y_n) = (y_1(g_1),\ldots,y_n(g_n))$ are made at locations $g = (g_1,\ldots,g_n)$, and that predictions $y_0 = (y_{01},\ldots,y_{0k})$ are required at k new locations $g_0 = (g_{01},\ldots,g_{0k})$. These are based on the posterior predictive density

$$p(y_0 \mid y) = \int p(y_0, \xi \mid y) d\xi = \int p(y_0 \mid y, \xi) p(\xi \mid y) d\xi,$$

where ξ is the vector of parameters involved in the model for y, namely those defining its mean, and the covariance parameters for spatial and unstructured errors (Banerjee et al., 2014). For example, Diggle et al. (2003) consider a model $y(g) = \mu + s(g) + u(g)$ with $u(g) \sim N_n(0, \sigma_u^2)$, and spatial error process

$$s(g) \sim N_n(0, \sigma_s^2 C),$$

where prediction is required at a single new location g_0. With d_0 denoting a $n \times 1$ vector of distances between g_0 and $g = (g_1, \ldots, g_n)$, and with $Q = \sigma_u^2 I + \sigma_s^2 C$, one has

$$p(y_0 \mid \theta, y) = N(\mu + \sigma_s^2 d_0' Q^{-1}(y - \mu 1_n), \sigma_s^2 - \sigma_s^2 d_0' Q^{-1} \sigma_s^2 d_0)$$

For $k > 1$, univariate predictions may be obtained separately at each new site $g_{01}, g_{02}, \ldots, g_{0k}$, though multivariate predictions may be more precise.

6.6.2 Sparse and Low Rank Approaches

For large numbers n of points, the computational burden involved in operations using dense covariance matrices becomes prohibitive, and alternative strategies have been proposed. Under the stochastic partial differentiation (SPDE) approach (Lindgren et al., 2011), included in R-INLA, the continuous spatial domain $y(g) = \{y_1(g_1), \ldots y_n(g_n)\}$ is approximated by a discrete Gaussian Markov random field process. In particular, the Gaussian Markov random field (GMRF) of the stationary Matern family for $y(g)$ is obtained as

$$(\kappa^2 - \Delta)^{\alpha/2}(\tau y(g)) = W(g)$$

where Δ is the Laplacian, $W(g)$ is a white noise process, α and κ are as above, and τ controls the marginal variance σ_s^2.

The GMRF approximation involves a triangulation (with m nodes) of the spatial domain, and the density of the triangulation mesh determines how close the approximation is. However, increasing the mesh density also increases the computations involved. A projector matrix A of dimension $n \times m$, containing 0 or 1 entries, is used to link the original points to the mesh (Lindgren, 2012; Bakka et al., 2018). Unlike stationary covariance models, it is straightforward to allow nonstationarity in SPDE models.

Computational burden is also reduced by using a low-rank representation of the spatial field (e.g. Finley et al., 2009; Finley et al., 2015). This involves defining a set of knots $g^* = \{g_1^*, g_2^*, \ldots, g_r^*\}$ where $r \ll n$ is considerably less than the dimension of the actual data. Then denoting $s^* = \{s(g_1^*), s(g_2^*), \ldots, s(g_r^*)\}$ and distances between the knots as d^* one has

$$s^*(g^*) \sim N_r(0, \sigma_s^2 C^*(d^*, \theta)),$$

with predictions or interpolations $\tilde{s}(g)$ at generic locations g obtained as

$$\tilde{s}(g) = c(g;\theta)[C^*(d^*,\theta)]^{-1} s^*,$$

where $c(g;\theta)$ is an $r \times 1$ vector with ith element $[c(g, g_i^*;\theta)]$.

For a Gaussian outcome, and spatially reference predictors $X(g)$, a predictive process model is then defined as

$$y(g) = X(g)'\beta + \tilde{s}(g) + u(g).$$

For a non-Gaussian response, the predictive process is included in the link regression, such as for y binary with probability $\pi(g_i)$,

$$\text{logit}[\pi(g_i)] = X(g_i)'\beta + \tilde{s}(g_i).$$

Estimation of predictive process models for large n is further facilitated (Eidsvik et al., 2012) by using the latent approximation approach of Rue et al. (2009).

Under the nearest neighbour Gaussian process (NNGP) approach (Datta et al., 2016; Zhang et al., 2018), a sparse precision matrix of the joint density $p[s(g)]$ of the spatial process $s(g)$ is achieved by using neighbour sets $N(g_i)$. Following Vecchia (1988), the sets $N(g_i)$ can be specified as the m nearest neighbours of the point g_i. These sets are used to provide an approximate conditional specification of the joint density of the spatial process $p[s(R)]$ for a set of k reference locations R (that can be taken as the n observed locations). This approach is incorporated in the R package spNNG. The approximation to the joint density is provided by the conditional density representation

$$\tilde{p}(s[R]) = \prod_{i=1}^{k} p(s(g_i) \mid s(N(g_i)).$$

Different model formulations can be specified according to whether estimated spatial random effects are of interest, or simply regression and other hyperparameters, with the spatial effects then integrated out. These are denoted as the sequential and response options in the R package spNNGP. Thus, under the sequential model, and for hyperparameters $\xi = (\beta, \sigma_s^2, \sigma_u^2, \theta)$, the posterior density is

$$p(\xi) \times N(s(g) \mid 0, \tilde{C}) \times N(X(g)'\beta + s(g), \sigma_u^2),$$

where $\tilde{C}^{-1} = (I - A)^T D^{-1}(I - A)$ is the precision matrix for $s(g)$, A is a sparse lower triangular matrix with at least m non-zero elements in each row, and D is diagonal. The construction of these matrices is set out in Finley et al. (2017).

Example 6.5 COPD Prevalence

This example considers spatial covariance modelling for binomial disease prevalence data $y_i \sim \text{Bin}(N_i, \pi_i)$, specifically cases y of chronic obstructive pulmonary disease (COPD) in 2016–17 in outer NE London. Observed prevalence data y_i is available for 81 GP (general practitioner) practices, with predictions of prevalence required for $k=11$ GP practices, since their prevalence is not provided. GP practice locations (eastings, northings) are available for all 92 GP practices, based on their postcode. Locations are randomly jiggered to avoid colocation, as some practices are close to each other. As a predictor of prevalence, a deprivation score x_i is available for all 92 GP practices. Binomial population denominators N_i are age weighted and so adjust for higher COPD prevalence at older ages.

The first analysis uses rstan, with an exponential decay covariance for spatial effects η_i, namely

$$C_{ij}(d) = \exp[-\phi d],$$

with

$$\text{logit}(\pi_i) = \beta_0 + \beta_1 x_i + \eta_i,$$

and with the spatial effects covariance multivariate normal prior encompassing all 92 units in the analysis. A Cholesky decomposition is used to represent the multivariate normal covariance. The prevalence predictions for the 11 practices with missing prevalence data are obtained as generated quantities under an inverse logit transform.

A second analysis uses R2OpenBUGS and a powered exponential distance model for the spatial effects s_i, namely

$$C_{ij}(d) = \exp[-(\phi d)^{\kappa}],$$

with $\phi > 0$, $\kappa \in (0,2]$, and with univariate predictions $(s_{01}, \ldots, s_{0k})$ for the 11 new points. The coding in R2OpenBUGS is hierarchically centred (Thomas et al., 2014). A $Ga(1,0.001)$ prior adopted on $1/\sigma_s^2$.

The two models provide similar LOO-IC, respectively 691.4 and 691.7. The posterior mean (95% CRI) for ϕ under the simple exponential decay option are obtained as 0.92 (0.06, 1.25), with the posterior 95% interval for β_1 mostly positive, so that the deprivation score improves on the prediction of missing prevalence rates. The latter range from 1.9% to 2.3%, with a 0.95 correlation between the estimated missing prevalence rates between the two models.

Example 6.6 Recorded Earthquakes in Europe and Asia Minor

The data here are recorded earthquake locations across Europe (including Turkey and the mid-Atlantic ridge) as catalogued by the Seismic Hazard Harmonization in Europe (SHARE) project (Giardini et al., 2014). There are 29,542 records for earthquakes of magnitude 3.5 and higher during the period 1000–2007.

A first analysis uses the INLA spde representation of the Matern correlation function to model spatial dependence in the patterning of earthquake magnitudes. The spde method uses a triangulation of the spatial domain, with the mesh extended outside the region of interest to reduce boundary effects. The density of the mesh can be varied by changing max.edge and cutoff in the inla.mesh.2d command. Here we initially select a relatively coarse grid, setting $k=0.1$ and define the mesh using

```
mesh=inla.mesh.2d(coordinates,max.edge=c(1/k,2/k),cutoff=0.1/k).
```

There are no explanatory variables, so predictions are based only on the estimated spatial effects at the grid nodes.

With this relatively coarse grid, a correlation of 0.46 is obtained between actual and predicted magnitudes. Setting $k=1$ as opposed to $k=0.1$ produces a denser grid with around 24 times as many nodes (20,603 as against 866), and so is more computationally intensive. However, the correlation between actual and predicted magnitudes is increased to 0.54.

A second analysis is based on the spBayes package, and uses a 10% sample of the full data. The data involve repeated observations at the same locations which may cause numerical problems. Therefore, the actual locations are randomly jiggered to avoid repeat locations. A further 10% subsample of the coordinates (i.e. of 294 coordinates) is used to provide a set of knots. As an illustration of a particular covariance option, consider an exponential decay function, which using the notation in the package, assumes a covariance model

$$\sigma^2 \exp(-\delta d_{ij}) + \tau^2,$$

where σ^2 is the partial sill. To provide initial values for σ^2, τ^2 and the decay parameter δ, the variogram and fit.variogram options in gstat are used. This provides an estimated range of $\phi=3.2$, and hence an initial value for the decay parameter in spBayes of 0.31 (the spBayes parameterisation of the exponential uses a decay parameter $\delta=1/\phi$). Tuning values for the Metropolis sampler are chosen to produce an acceptance rate of around 30%. With an MCMC sample of 2,000 iterations and burn-in of 1,000, a correlation of 0.485 is obtained between actual and predicted magnitudes. Posterior means (and sd) for σ^2, τ^2 and δ are 0.13 (0.02), 0.31 (0.01) and 0.38 (0.07).

The gstat commands

```
v = variogram(Y~1, D)
fit.variogram(v, vgm(c("Exp", "Mat", "Sph","Gau")))
```

suggest a spherical model as better fitting, but a lower correlation (of 0.474) between actual and fitted values is obtained under this option. The GPD criterion of Gelfand and Ghosh (1998) also prefers the exponential model.

Finally, again using the full dataset, but with jiggering to avoid repeat locations, the nearest neighbour Gaussian Process approach is applied using spNNG. An exponential covariance and $m=10$ neighbours, are assumed. This provides a correlation between actual and predicted magnitudes of 0.51. Increasing the number of neighbours m from 10 to 15 makes no difference to the fit. Figure 6.5 shows the predicted magnitude surface. For $m=15$, the estimated posterior means (and sd) for σ^2, τ^2 and δ are 0.08 (0.05), 0.32 (0.03), and 0.33 (0.13).

6.7 Discrete Convolution Models

Assuming a stationary Gaussian process described through its mean and covariance structure may result in slow estimation when there are a large number of points and is relatively inflexible when stationarity and isotropy assumptions are violated. An alternative

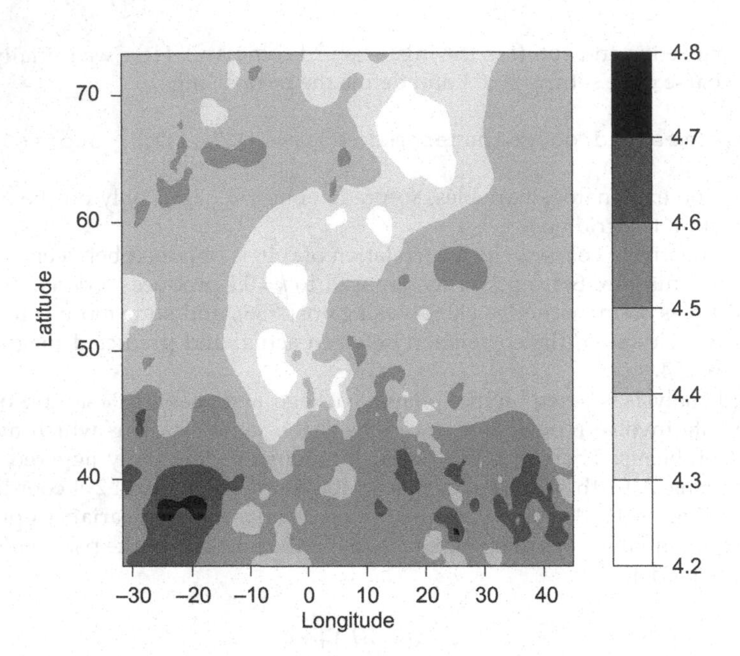

FIGURE 6.5
Magnitude predictions from NNGP.

representation, based on the Gaussian process, but one that adapts to spatial nonstationarity and anisotropy, is the process convolution approach (Higdon, 1998; Lee et al., 2005; Higdon, 2007; Liang and Lee, 2014). This involves convolving a continuous white noise process $w(g)$ with a symmetric smoothing kernel $K(g)$, with the spatial effect obtained as

$$s(g) = \int_G K(g-u)w(u)du,$$

where G is the region of interest. The spatial process might be combined with fixed effect regression impacts and with appropriate regression links for non-normal observations. For example, if $y(g)$ were binary, such as species presence or absence at site g (Gelfand et al., 2005a), then $y(g) \sim \text{Bern}(\pi(g))$ and

$$\text{logit}[\pi(g)] = \beta_0 + s(g).$$

where β_0 defines the average intensity.

In practice, the continuous underlying process can be approximated by a discretised process (e.g. one defined on a regular lattice over G) provided the discretisation is not too coarse relative to the smoothing kernel (Calder, 2003; Calder, 2007). So if there are $i = 1,\ldots,n$ observations at points $g_1, \ldots, g_n$ and grid locations $\{t_j, j = 1,\ldots,m\}$ with $t_j = (t_{1j}, t_{2j})$, over the region, one may define the discretised kernel smoother as

$$s(g_i) = \sum_{j=1}^{m} K(g_i - t_j)w_j,$$

where for large m, the w_j can be taken as a collection of random effects (Higdon, 2007, p.245). Lee et al. (2005) consider options for representing the kernel, possibly by a form with known variance (e.g. a standard normal), and consequent ways for modelling the w_j. Note that if both the K function and w series have unknown variances, then there is potential non-identifiability. Options for the w_j include exchangeable effects or low order random walks, with unknown precision τ_w. Assuming K is a normal kernel, by varying τ_w one can mimic the effect of the range parameter in a conventional Gaussian process model with a Gaussian variogram.

For example, Lee et al. (2005) consider $n=12$ observations y_i in G^1 at equally spaced locations g_i between 0 and 10. These are generated according to a Gaussian process $s(g)$ with mean 0 and covariance matrix

$$C(d_{ij}) = \exp(-d_{ij}^2/25),$$

where d_{ij} relates to distances between points g_i and g_j on the line. A white noise error u_i with standard deviation 0.2 is also used to define y_i, so that $y_i = s(g_i)+u_i$. They then fit a discrete convolution model to the y_i so generated, using a grid with $m=20$ points t_j equally spaced between −2 and 12. They assume the w_j follow a 1st order random walk, and assume the kernel is a normal density with standard deviation 0.6.

Best et al. (2000) consider a convolution model for health counts $y_i \sim Po(P_i\lambda_i)$ observed for areas rather than points, where P_i are populations and λ_i are latent rates. In this case, a rectangular grid is defined over m points in the region, and an additive (rather than log link) regression is used for modelling the latent rates. So, with a single predictor x_i taking positive values only, one has

$$\lambda_i = \beta_0 + \beta_1 x_i + \beta_2 \sum_{j=1}^{m} K(g_i - t_j) w_j,$$

where the w_j (and the β parameters) are gamma distributed and the kernel function K has a known variance. One can decompose the total risk parameter into three sources: one due to the background rate β_0, one reflecting the known predictor, and one the latent spatially configured risk over the region.

Semiparametric approaches to spatial modelling based on the stick-breaking prior can also be related to this theme (Reich and Fuentes, 2007). Thus there are kernel functions for each of m potential clusters, with the kernel centres $t_j = (t_{1j}, t_{2j})$ being unknowns, and the cluster allocation probabilities for sites or areas i at location $g_i = (g_{1i}, g_{2i})$ incorporating spatial information. While the cluster effects $w_j \sim N(0, 1/\tau_w)$ are unstructured, the cluster for area or point i is chosen using indicators

$$J_i \sim \text{Categorical}(p_{i1},..,p_{im}),$$

with the p_{ij} determined both via beta distributed $V_j \sim Be(c,d)$, and by cluster specific kernels K_{ij} constrained to lie in [0,1]. The realised spatial effect for area or point i is then w_{J_i}. Defining $R_{ij} = K_{ij}V_j$, one has

$$p_{i1} = R_{i1}$$

$$p_{ij} = R_{ij}(1-R_{i1})\ldots(1-R_{i,j-1}), \quad j = 2,\ldots,m-1$$

$$p_{im} = (1-R_{i1})\ldots(1-R_{i,m-1})$$

where (for example)

$$K_{ij} = \exp[-|g_i - t_j| / 2\gamma_j]$$

defines a normal kernel with bandwidth γ_j. Bandwidths can be taken equal across kernel functions or vary across kernel functions according to a positive prior (e.g. inverse gamma).

Example 6.7 Earthquake Magnitudes (Continued)

This example continues the analysis of the earthquake magnitude data. We consider a 10% sample of the original dataset, with $n = 2945$. A two way 10×10 grid cell subdivision of the region of interest is obtained using rasterisation. There are then $m = 100$ interior points $t_j = (t_{j1}, t_{j2})$ defining the grid.

Initially, a discrete kernel approach is applied via a single intercept linear regression

$$y_i \sim N(\mu_i, \sigma^2),$$

$$\mu_i = \beta_0 + \sum_j K_{ij} w_j,$$

with a bivariate exponential kernel (Clark et al., 1999)

$$K_{ij}(d_{ij}) = \frac{1}{2\pi\eta} \exp(-d_{ij} / \eta),$$

with distances $d_{ij} = [(g_{i1} - t_{j1})^2 + (g_{i2} - t_{j2})^2]^{0.5}$. The grid effects w_j are assumed iid random normal with zero mean, and with standard deviation σ_w.

Because of confounding between the grid effects and the kernel, for identifiability, it is assumed that $\eta = 1$, but that the w_j have an unknown variance, with σ_w assigned a $U(0,100)$ prior. Using jagsUI for estimation, this model provides a correlation between actual and predicted magnitudes of 0.32. Computation is slower if η is taken as an unknown, and σ_w is set to 1. Also, the fit is not improved.

However, a much-improved fit is obtained by a two-group mixture intercept, with preset probabilities on the two groups of 0.95 and 0.05 to facilitate identifiability. Thus

$$\mu_i = \beta_{0J_i} + \sum_j K_{ij} w_j,$$

$$J_i \sim \text{Categoric}(0.95, 0.05).$$

This increases the correlation between actual and predicted magnitudes to 0.76. The estimates (posterior means and sd) for the intercepts β_{01} and β_{02} are 4.40 (0.02) and 6.07 (0.05). Further improvements in fit might be obtained by taking additional groups in the discrete mixture intercept.

For this model, site-specific effects are obtained by comparing the μ_i to their overall average. Then 359 of the 2945 sites have a posterior probability over 95% that the effect is positive. Figure 6.6 maps out three significance categories, and in particular shows spatial clustering of sites with over 0.95 probability of elevated earthquake magnitudes.

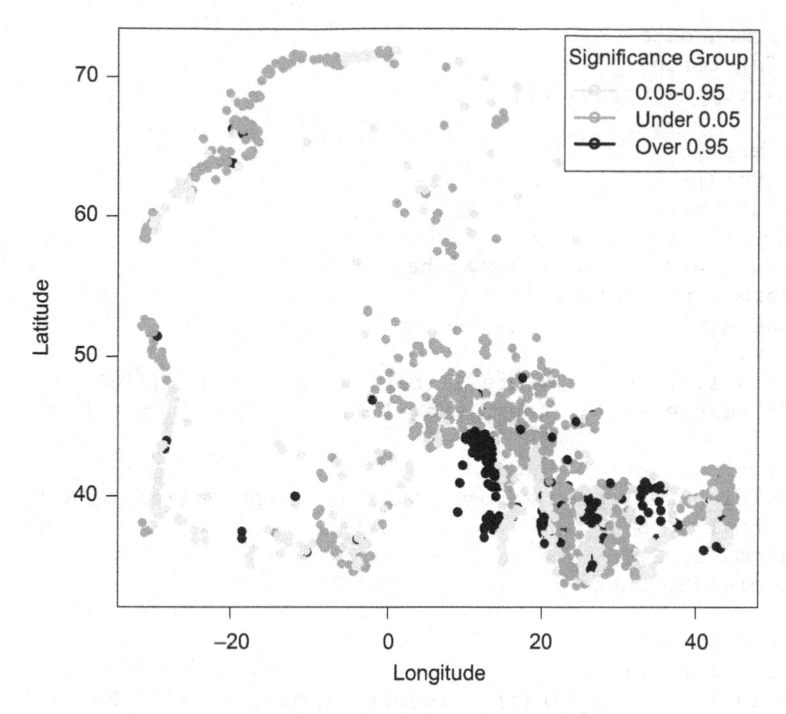

FIGURE 6.6
Significance of site effects.

6.8 Computational Notes

[1] With d[i] denoting the vector of neighbour numbers (the number of areas adjacent to area i), and W the interaction matrix, the Leroux et al. (1999) prior has the form

```
D=diag(d)
R=D-W
I <- diag(N)
# data inputs
D = list(n = N, # number of observations
y = y, # observed number of cases
T=T,
x=x,
R = R,
I=I)
model="
data {
int<lower = 1> n;
int<lower = 0> y[n];
real x[n];
int T[n];
matrix[n, n] R;
matrix[n, n] I;
}
```

```
transformed data{
vector[n] zeros;
zeros = rep_vector(0, n);
}
parameters {
real beta[2];
vector[n] phi;
real<lower = 0> tau;
real<lower = 0, upper = 1> alpha;}
transformed parameters {
real theta[n];
real eta[n];
for (i in 1:n) {eta[i]=beta[1]+beta[2]*x[i] + phi[i];
theta[i]=exp(eta[i])/(1+exp(eta[i]));}
}
model {
phi ~multi_normal_prec(zeros, tau * ((1-alpha)*I+alpha*R));
beta~normal(0, 5);
tau ~gamma(2, 2);
y ~binomial(T, theta);
}
generated quantities
{real log_lik[n];
for (i in 1:n) {log_lik[i]= binomial_lpmf(y[i]T[i],theta[i]);}
}
"
sm = stan_model(model_code=model)
fit = sampling(sm,data =D,iter = 2500,warmup=250,chains = 2,seed=
12345)
summary(fit,pars=c("beta","alpha"), probs=c(0.025,0.975))$summary
# Fit
loo(as.matrix(fit,pars="log_lik"))
```

References

Adin A, Lee D, Goicoa T, Ugarte M (2018) A two-stage approach to estimate spatial and spatio-temporal disease risks in the presence of local discontinuities and clusters. *Statistical Methods in Medical Research*, In press

Allard D, Beauchamp M, Bel L, Desassis N, Gabriel É, Geniaux G, Malherbe L, Martinetti D, Opitz T, Parent É, Romary T, Saby N (2017) Analyzing spatio-temporal data with R: Everything you always wanted to know – but were afraid to ask. *Journal de la Société Française de Statistique*, 158(3), 124–158.

Anselin L (2010) Thirty years of spatial econometrics. *Papers in Regional Science*, 89(1), 3–25.

Anselin L, Bera A (1998) Spatial dependence in linear regression models, with an introduction to spatial econometrics, pp 237–290, in *Handbook of Applied Economic Statistics*, eds A Ullah, D Giles. Marcel Dekker, New York.

Bakka H, Rue H, Fuglstad G, Riebler A, Bolin D, Krainski E, Simpson D, Lindgren F (2018) Spatial modelling with R-INLA: A review. arXiv preprint arXiv:1802.06350.

Banerjee S, Carlin B, Gelfand A (2014) *Hierarchical Modeling and Analysis for Spatial Data*. Chapman and Hall/CRC.

Bell B, Broemeling L (2000) A Bayesian analysis for spatial processes with application to disease mapping. *Statistics in Medicine*, 19, 957–974.

Berke O (2004) Exploratory disease mapping: kriging the spatial risk function from regional count data. *International Journal of Health Geographics*, 3, 18.

Bernardinelli L, Clayton D, Pascutto C, Montomoli C, Ghislandi M, Songini M (1995) Bayesian analysis of space–time variation in disease risk. *Statistics in Medicine*, 14(21–22), 2433–2443.

Besag J (1974) Spatial interaction and the statistical analysis of lattice systems. *Journal of the Royal Statistical Society: Series B*, 36, 192–236.

Besag J (1989) Towards Bayesian image analysis. *Journal of Applied Statistics*, 16, 395–407.

Besag J, Green P (1993) Spatial statistics and Bayesian computation. *Journal of the Royal Statistical Society: Series B*, 55, 25–37.

Besag J, Green P, Higdon D, Mengersen K (1995) Bayesian computation and stochastic systems. *Statistical Science*, 10, 3–66.

Besag J, Kooperberg C (1995) On conditional and intrinsic autoregressions. *Biometrika*, 82, 733–746.

Besag J, York J, Mollie A (1991) Bayesian image restoration, with two applications in spatial statistics. *Annals of the Institute of Statistical Mathematics*, 43, 1–21.

Best N, Arnold R, Thomas A, Waller L, Conlon E (2000) Bayesian models for spatially correlated disease and exposure data, p 131, in *Bayesian Statistics 6: Proceedings of the Sixth Valencia International Meeting*, Vol. 6. Oxford University Press.

Bhattacharjee A, Jensen-Butler C (2006) Estimation of the spatial weights matrix under structural constraints. *Regional Science and Urban Economics*, 43(4), 617–634.

Bivand R, Gomez-Rubio V, Rue H (2014) Approximate Bayesian inference for spatial econometrics models. *Spatial Statistics*, 9, 146–165.

Bivand R, Gomez-Rubio V, Rue H (2015). Spatial data analysis with R-INLA with some extensions. *Journal of Statistical Software*, 63(20), 1–31.

Bivand R, Pebesma E, Gómez-Rubio V, Pebesma E (2013). *Applied Spatial Data Analysis with R*, 2nd Edition. Springer, New York.

Blangiardo M, Cameletti M (2015) *Spatial and Spatio-temporal Bayesian models with R-INLA*. John Wiley & Sons.

Bowman V, Woods D (2016) Emulation of multivariate simulators using thin-plate splines with application to atmospheric dispersion. *SIAM/ASA Journal on Uncertainty Quantification*, 4(1), 1323–1344.

Brown P (2015) Model-based geostatistics the easy way. *Journal of Statistical Software*, 63(12), 1–24.

Brunsdon C (2018) Using rstan and spdep for Spatial Modelling. https://rstudio-pubs-static.s3.amazonaws.com/

Brunsdon C, Comber L (2015) *An Introduction to R for Spatial Analysis and Mapping*. Sage.

Calder K (2003) Exploring latent structure in spatial temporal processes using process convolutions. PhD Thesis, Duke University, Durham, NC. https://www2.stat.duke.edu/people/theses/CalderK.html

Calder K (2007) Dynamic factor process convolution models for multivariate space–time data with application to air quality assessment. *Environmental and Ecological Statistics*, 14, 229–247.

Clark J, Silman M, Kern R, Macklin E, HilleRisLambers J (1999) Seed dispersal near and far: Patterns across temperate and tropical forests. *Ecology*, 80, 1475–1494.

Clayton D, Kaldor J (1987) Empirical Bayes estimates of age-standardised relative risks for use in disease mapping. *Biometrics*, 43, 671–682.

Congdon P (2007) Mixtures of spatial and unstructured effects for spatially discontinuous health outcomes. *Computational Statistics and Data Analysis*, 51, 3197–3212.

Congdon P (2008) A spatially adaptive conditional autoregressive prior for area health data. *Statistical Methodology*, 5, 552–563.

Congdon P (2017) Quantile regression for overdispersed count data: a hierarchical method. *Journal of Statistical Distributions and Applications*, 4, 18.

Conlon E, Louis T (1999) Addressing multiple goals in evaluating region-specific risk using Bayesian methods, pp 31–47, in *Disease Mapping and Risk Assessment for Public Health*, eds A Lawson, A Biggeri, D Bohning, E Lesaffre, J Viel, R Bertollini. John Wiley, Chichester, UK.

Cressie N, Kapat P (2008) Some diagnostics for Markov random fields. *Journal of Computational and Graphical Statistics*, 17, 726–749.

Cressie N, Wikle CK (2011) *Statistics for Spatio-Temporal Data*. John Wiley & Sons, Inc., New York.

Datta A, Banerjee S, Finley A, Gelfand A (2016) Hierarchical nearest-neighbor Gaussian process models for large geostatistical datasets. *Journal of the American Statistical Association*, 111, 800–812.

De Oliveira V (2012) Bayesian analysis of conditional autoregressive models. *Annals of the Institute of Statistical Mathematics*, 64(1), 107–133.

Diggle P, Ribeiro P (2002) Bayesian inference in Gaussian model-based geostatistics. *Geographical and Environmental Modelling*, 6, 129–146.

Diggle P, Ribeiro P, Christensen O (2003) An introduction to model based geostatistics, pp 43–86, in *Spatial Statistics and Computational Methods*, ed Möller J. *Lecture Notes in Statistics*, Vol. 173. Springer.

Diggle P, Ribeiro P (2007) *Model-based Geostatistics*. Springer-Verlag, New York

Dormann, C., McPherson J, Araújo M, Bivand R, Bolliger J (2007) Methods to account for spatial autocorrelation in the analysis of species distributional data: A review. *Ecography*, 30(5), 609–628.

Duan J, Guindani M, Gelfand A (2007) Generalized spatial Dirichlet process models. *Biometrika*, 94, 809–825.

Eidsvik J, Finley A, Banerjee S, Håvard R (2012) Approximate Bayesian inference for large spatial datasets using predictive process models. *Computational Statistics & Data Analysis*, 56(6), 1362–1380.

Fernandez C, Green P (2002) Modelling spatially correlated data via mixtures: A Bayesian approach. *Journal of the Royal Statistical Society: Series B*, 64, 805–826.

Finley A, Sang H, Banerjee S, Gelfand A (2009) Improving the performance of predictive process modeling for large datasets. *Computational Statistics & Data Analysis*, 53(8), 2873–2884.

Finley A, Banerjee S, Gelfand A (2015) spBayes for large univariate and multivariate point-referenced spatio-temporal data models. *Journal of Statistical Software*, 63(13), 1–28.

Finley, A, Datta A, Cook B, Morton D, Andersen H, Banerjee S (2017) Applying Nearest Neighbor Gaussian Processes to Massive Spatial Data Sets: Forest Canopy Height Prediction Across Tanana Valley Alaska. https://arxiv.org/abs/1702.00434

Furrer R, Sain S (2010) spam: A sparse matrix R package with emphasis on MCMC methods for Gaussian Markov random fields. *Journal of Statistical Software*, 36(10), 1–25.

Gelfand A, Kottas A, MacEachern S (2005b) Bayesian nonparametric spatial modeling with Dirichlet process mixing. *Journal of the American Statistical Association*, 100(471), 1021–1035.

Gelfand A, Latimer A, Wu S, Silander J (2005a) Building statistical models to analyse species distributions, in *Hierarchical Modelling for the Environmental Sciences, Statistical Methods and Applications*, eds J Clark, A Gelfand. OUP.

Gelfand AE, Ghosh SK (1998) Model choice: A minimum posterior predictive loss approach. *Biometrika*, 85(1), 1–11.

Gerber F, Furrer R (2015) Pitfalls in the implementation of Bayesian hierarchical modeling of areal count data: An illustration using BYM and Leroux models. *Journal of Statistical Software, Code Snippets*, 63(1), 1–32. http://www.jstatsoft.org/v63/c01/

Giardini D, Woessner J, Danciu L (2014) Mapping Europe's seismic hazard. *EOS*, 95(29): 261–262.

Goméz-Rubio V, Bivand R (2018) R Package 'INLABMA', Bayesian Model Averaging with INLA. https://rdrr.io/rforge/INLABMA/

Goméz-Rubio V, Bivand R, Rue H (2018) Estimating spatial econometrics models with integrated nested laplace approximation. arXiv preprint arXiv:1703.01273.

Gotway C, Wolfinger R (2003) Spatial prediction of counts and rates. *Statistics in Medicine*, 22, 1415–1432.

Gramacy R, Lee H (2008) Gaussian processes and limiting linear models. *Computational Statistics & Data Analysis*, 53, 123–136.

Green P, Richardson S (2002) Hidden Markov models and disease mapping. *Journal of the American Statistical Association*, 97, 1055–1070.

Griffin J, Steel M (2006) Order-based dependent Dirichlet processes. *Journal of the American Statistical Association*, 101, 179–194.

Grunwald S (2005) *Environmental Soil-Landscape Modeling: Geographic Information Technologies and Pedometrics*. CRC Press.

Gschlößl S, Czado C (2006) Modelling count data with overdispersion and spatial effects. Technische Universität München, Statistical Papers. DOI: 10.1007/s00362-006-0031-6

Haran M, Hodges J, Carlin B (2003) Accelerating computation in Markov random field models for spatial data via structured MCMC. *Journal of Computational & Graphical Statistics*, 12, 249–264.

Hepple L (2003) Bayesian and maximum likelihood estimation of the linear model with spatial moving average disturbances. Working Papers Series, School of Geographical Sciences, University of Bristol.

Higdon D (1998) A process-convolution approach to modelling temperatures in the North Atlantic Ocean. *Environmental and Ecological Statistics*, 5, 173–190.

Higdon D (2007) A primer on space-time modelling from a Bayesian perspective, Chapter 6, in *Statistical Methods for Spatio-Temporal Systems*, eds B Finkelstadt, L Held, V Isham. CRC Press.

Hodges J, Carlin B, Fan Q (2003) On the precision of the conditionally autoregressive prior in spatial models. *Biometrics*, 59, 317–322.

Jiruše M, Machek J, Beneš V, Zeman P (2004) A Bayesian estimate of the risk of tick-borne diseases. *Applications of Mathematics*, 49, 389–404.

Joseph M (2016) Exact Sparse CAR Models in Stan. http://mc-stan.org/users/documentation/case-studies/mbjoseph-CARStan.html

Kelsall J, Wakefield J (2002) Modelling spatial variation in disease risk: A geostatistical approach. *Journal of the American Statistical Association*, 97, 692–770.

Knorr-Held L, Becker N (2000) Bayesian modelling of spatial heterogeneity in disease maps with application to German cancer mortality data. *Journal of the German Statistical Society*, 84, 121–140.

Knorr-Held L, Rasser G (2000) Bayesian detection of clusters and discontinuities in disease maps. *Biometrics*, 56, 13–21.

Lacombe D, LeSage J (2015) Using Bayesian posterior model probabilities to identify omitted variables in spatial regression models. *Papers in Regional Science*, 94(2), 365–383.

Lang S, Brezger A (2004) Bayesian P-splines. *Journal of Computational and Graphical Statistics*, 13(1), 183–212.

Lavine M (1999) Another look at conditionally Gaussian Markov random fields, in *Bayesian Statistics 6*, eds J Bernardo, J Berger, P Dawid, A Smith. Oxford University Press, Oxford, UK.

Lawson A (2008) *Bayesian Disease Mapping: Hierarchical Modeling in Spatial Epidemiology*. CRC Press.

Lawson A, Clark A (2002) Spatial mixture relative risk models applied to disease mapping. *Statistics in Medicine*, 21, 359–370.

Lee D (2011) A comparison of conditional autoregressive models used in Bayesian disease mapping. *Spatial and Spatio-Temporal Epidemiology*, 2(2), 79–89.

Lee D (2013) CARBayes: An R package for Bayesian spatial modeling with conditional autoregressive priors. *Journal of Statistical Software*, 55(13), 1–24.

Lee H, Higdon D, Calder C, Holloman C (2005) Efficient models for correlated data via convolutions of intrinsic processes. *Statistical Modelling*, 5, 53–74.

Leroux B, Lei X, Breslow N (1999) Estimation of disease rates in small areas: a new mixed model for spatial dependence, pp 135–178, in *Statistical Models in Epidemiology, the Environment and Clinical Trials*, eds M Halloran, D Berry. Springer-Verlag, New York.

LeSage J (1999) Spatial econometrics, in *The Web Book of Regional Science* (www.rri.wvu.edu/regscweb.htm), ed R W Jackson. Regional Research Institute, West Virginia University, Morgantown, WV.

Liang W, Lee H (2014) Sequential process convolution gaussian process models via particle learning. *Statistics and Its Interface*, 7(4), 465–475.

Lindgren F (2012) Continuous domain spatial models in R-INLA. The ISBA Bulletin, 19(4), 14–20.

Lindgren F, Rue H, Lindstrom J (2011) An explicit link between Gaussian fields and Gaussian Markov random fields: The stochastic partial differential equation approach. *Journal of the Royal Statistical Society: Series B*, 73(4), 423–498.
Lindgren F, Rue H (2015) Bayesian spatial modelling with R-INLA. *Journal of Statistical Software*, 63(19), 1–25
MacNab Y (2014) On identification in Bayesian disease mapping and ecological–spatial regression models. *Statistical Methods in Medical Research*, 23(2), 134–155.
MacNab Y, Kmetic A, Gustafson P, Shaps S (2006) An innovative application of Bayesian disease mapping methods to patient safety research. *Statistics in Medicine*, 25, 3960–3980.
Mardia K, Watkins A (1989) On multimodality of the likelihood in the spatial linear model. *Biometrika*, 76, 289–295.
Marra G, Wood S (2012) Coverage properties of confidence intervals for generalized additive model components. *Scandinavian Journal of Statistics*, 39(1), 53–74.
Mitas L, Mitasova H (1999) Spatial interpolation, pp 481–492, in *Geographical Information Systems: Principles, Techniques, Management and Applications*, eds P Longley, M Goodchild, D Maguire, D Rhind, 1st Edition. Wiley.
Moores M, Hargrave C, Deegan T, Poulsen M, Harden F, Mengersen K (2015) An external field prior for the hidden Potts model with application to cone-beam computed tomography. *Computational Statistics & Data Analysis*, 86, 27–41.
Morris M (2018) Spatial Models in Stan: Intrinsic Auto-Regressive Models for Areal Data. http://mc-stan.org/users/documentation/case-studies/icar_stan.html
Norton J, Niu X (2009) Intrinsically autoregressive spatiotemporal models with application to aggregated birth outcomes. *Journal of the American Statistical Association*, 104, 638–649.
Paciorek CJ, Schervish MJ (2006) Spatial modelling using a new class of nonstationary covariance functions. *Environmetrics*, 17(5), 483–506.
Pettitt A, Weir I, Hart A (2002) A conditional autoregressive Gaussian process for irregularly spaced multivariate data with application to modelling large sets of binary data. *Statistics and Computing*, 12, 353–367.
Reich BJ, Fuentes M (2007) A multivariate semiparametric Bayesian spatial modeling framework for hurricane surface wind fields. *The Annals of Applied Statistics*, 1(1), 249–264.
Ribeiro P, Diggle P (2018) Package 'geoR'. https://cran.r-project.org/web/packages/geoR/geoR.pdf
Richardson S, Guihenneuc C, Lasserre V (1992) Spatial linear models with autocorrelated error structure. *The Statistician*, 41, 539–557.
Richardson S, Monfort C (2000) Ecological correlation studies, in *Spatial Epidemiology Methods and Applications*, eds P Elliott, J Wakefield, N Best, D Briggs. Oxford University Press.
Richardson S, Thomson A, Best N, Elliott P (2004) Interpreting posterior relative risk estimates in disease-mapping studies. *Environmental Health Perspectives*, 112, 1016–1025.
Riebler A, Sørbye S, Simpson D, Rue H (2016) An intuitive Bayesian spatial model for disease mapping that accounts for scaling. *Statistical Methods in Medical Research*, 25, 1145–1165.
Riggan W, Manton K, Creason J, Woodbury M, Stallard E (1991) Assessment of spatial variation of risks in small populations. *Environmental Health Perspectives*, 96, 223–238.
Risser M, Calder C (2017) Local likelihood estimation for covariance functions with spatially-varying parameters: The convoSPAT package for R. *Journal of Statistical Software*, 81(14), 1–32.
Rodrigues A, Assuncao R (2008) Propriety of posterior in Bayesian space varying parameter models with normal data. *Statistics & Probability Letters*, 78, 2408–2411.
Rue H, Held L (2005) *Gaussian Markov Random Fields: Theory and Applications*. Chapman & Hall, London, UK.
Rue H, Martino S, Chopin, N (2009) Approximate Bayesian inference for latent Gaussian models by using integrated nested Laplace approximations. *Journal of the Royal Statistical Society: Series B*, 71(2), 319–392.
Rue H, Tjelmeland H (2002) Fitting Gaussian Markov random fields to Gaussian fields. *Scandinavian Journal of Statistics*, 29, 31–49.

Sangalli L, Ramsay J, Ramsay T (2013) Spatial spline regression models. *Journal of the Royal Statistical Society: Series B*, 75(4), 681–703.

Schabenberger O, Gotway C (2004) *Statistical Methods for Spatial Data Analysis*. Chapman & Hall/CRC.

Schrödle B, Held L (2011) A primer on disease mapping and ecological regression using INLA. *Computational Statistics*, 26(2), 241–258.

Sethuraman J (1994) A constructive definition of dirichlet priors. *Statistica Sinica*, 4, 639–665.

Seya H, Tsutsumi M, Yamagata Y (2012) Income convergence in Japan: A Bayesian spatial Durbin model approach. *Economic Modelling*, 29(1), 60–71.

Thomas A, Best N, Lunn D, Arnold R, Spiegelhalter D (2014) GeoBUGS User Manual. https://www.mrc-bsu.cam.ac.uk

Vecchia AV (1988) Estimation and model identification for continuous spatial processes. *Journal of the Royal Statistical Society: Series B (Methodological)*, 50(2), 297–312.

Wakefield J (2007) Disease mapping and spatial regression with count data. *Biostatistics*, 8, 158–183.

Wall M (2004) A close look at the spatial structure implied by the CAR and SAR models. *Journal of Statistical Planning and Inference*, 121, 311–324.

Waller L (2002) Hierarchical models for disease mapping, in *Encyclopedia of Environmetrics*, eds A El-Shaarawi, W Piegorsch. Wiley, Chichester, UK.

Waller L, Carlin B (2010) Disease mapping, Chapter 14, pp 217–243, in *Handbooks of Modern Statistical Methods*. ed G Fitzmaurice, Chapman & Hall/CRC.

Waller L, Gotway C (2004) *Applied Spatial Statistics for Public Health Data*. Wiley.

Webster R, Oliver M, Muir K, Mann J (1994) Kriging the local risk of a rare disease from a register of diagnoses. *Geographical Analysis*, 26, 168–185.

Wood S (2006) Low-rank scale-invariant tensor product smooths for generalized additive mixed models. *Biometrics*, 62(4), 1025–1036.

Wood S, Bravington M, Hedley S (2008) Soap film smoothing. *Journal of the Royal Statistical Society: Series B*, 70(5), 931–955.

Wood S N (2016) Just another Gibbs additive modeller: Interfacing JAGS and mgcv. arXiv preprint arXiv:1602.02539.

Yang C, Xu J, Li Y (2016) Bayesian geoadditive modelling of climate extremes with nonparametric spatially varying temporal effects. *International Journal of Climatology*, 36(12), 3975–3987.

Yanli Z, Wall M (2004) Investigating the use of the variogram for lattice data. *Journal of Computational and Graphical Statistics*, 13, 719–738.

Zhang H (2002) On estimation and prediction for spatial generalised linear mixed models. *Biometrics*, 58, 129–136.

Zhang L, Datta A, Banerjee S (2018) Practical Bayesian Modeling and Inference for Massive Spatial Datasets On Modest Computing Environments. arXiv preprint arXiv:1802.00495.

Zhu L, Gorman D, Horel S (2006) Hierarchical Bayesian spatial models for alcohol availability, drug hot spots and violent crime. *International Journal of Health Geographics*, 5, 54.

7

Regression Techniques Using Hierarchical Priors

7.1 Introduction

This chapter is concerned with the application of hierarchical prior schemes to regressions involving univariate responses, where the observations are non-nested, but may be spatially or temporally configured. Nested data applications are considered in Chapters 8 and 10. A range of Bayesian packages in R for regression, often in specialised applications, are detailed at https://cran.r-project.org/web/views/Bayesian.html, and include BayesLogit (Windle, 2016), BMA (Raftery et al., 2005), and BMS (Zeugner and Feldkircher, 2015). The treatment here is intended as providing a generic overview of regression applications involving hierarchical principles, and development of flexible data analysis through application-specific coding.

As a first illustration of regression modelling invoking hierarchical principles, much attention has focused on Bayesian methods for predictor selection, commonly using selection indicators or shrinkage priors, and these are discussed in Section 7.2. Many regression selection applications involve categorical predictors, including analysis of variance, and the particular issues raised are considered in Section 7.3. Hierarchical specfications also apply when latent responses or random effects are used to (a) improve data representation in overdispersed general linear models (Section 7.4), or (b) generate latent continuous responses (augmented data) underlying discrete observations (Section 7.5). Section 7.6 considers heterogeneity in regression relationships or variance parameters over exchangeable sample units. Heterogeneous regression effects and predictor selection are then considered for responses structured in time or space (Sections 7.7 and 7.8).

7.2 Predictor Selection

Regression model uncertainty most commonly focuses on which predictors to retain, though other aspects of regression specification may be considered also. Predictor selection methods generally also aim for improved predictive performance, through developing an encompassing model, or model simplification without adversely affecting predictive accuracy (Piironen and Vehtari, 2017a). Formal model choice is simplified for normal linear regression, as marginal likelihoods may be obtained analytically, but for a large number of predictors, comparison of the many possible models becomes infeasible.

One option for a more feasible analysis involves a form of discrete mixture: predictor selection indicators, possibly combined with particular priors on regression coefficients or variances, are introduced to enable additional inferences (e.g. marginal retention probabilities on predictors) (Rockova et al., 2012; Malsiner-Walli and Wagner, 2019). Alternatively, regularisation or shrinkage priors include a penalty (e.g. an L_1 norm) or some other mechanism to shrink unnecessary regression effects towards zero (e.g. Polson and Scott, 2010; Carvalho et al., 2009).

A leading motivation for predictor selection or effect regularisation is to estimate coefficients which alleviate for predictor collinearity. If not controlled for, collinearity may lead to low precision for regression coefficients, and coefficients with effect sizes or signs contrary to subject matter expectations (Winship and Western, 2016).

7.2.1 Predictor Selection

Predictor selection recognises model uncertainty, and a predictive target which acknowledges such uncertainty may be of interest. An example might be a mean treatment success rate conditional on predictors (Garcia-Donato and Martinez-Beneito, 2013). Using predictor selection, one is implicitly averaging over a set of plausible regression models, so providing an encompassing model with potential predictive advantages (Piironen and Vehtari, 2017a).

The discrete mixture approach involves binary selection indicators γ_j for predictors $j=1,\ldots,p$. These indicators may directly determine inclusion (e.g. Kuo and Mallick, 1998), or define prior regression coefficient variances consistent with inclusion or effective exclusion, as under stochastic search variable selection (SSVS) (George and McCullogh, 1993). Selection usually applies to all predictors except the intercept.

With normal priors on regression coefficients under univariate linear regression, and with response y_i and predictors X_i, one has under SSVS

$$y_i \sim N(\beta_0 + X_i\beta, \sigma^2),$$

$$\beta_j \mid \gamma_j = 1 \sim N(0, \tau_j^2),$$

$$\beta_j \mid \gamma_j = 0 \sim N(0, c_j\tau_j^2),$$

$$\gamma_j \mid \omega_j \sim \text{Bern}(\omega_j),$$

$$\omega_j \sim \pi(\omega_j).$$

The setting for τ_j^2 (at a known value) allows unrestricted search over the potential parameter space, while c_j is set suitably small, so that $\gamma_j=0$ corresponds to effective exclusion from the regression. The spike and slab prior (Kuo and Mallick, 1998), denoted SSP for short, specifies the inclusion and exclusion options as

$$\beta_j \mid \gamma_j = 1 \sim N(0, \tau_j^2),$$

$$\beta_j \mid \gamma_j = 0 \sim \delta_0(),$$

where $\delta_B()$ is a discrete measure concentrated at value B.

Whether SSVS or SSP priors are adopted, the setting for τ_j^2 affects the level of parsimony in the selection. Higher values of τ_j^2 favour more parsimonious models, with

Lee and Chen (2015) proposing a predictive mean square error criterion to select τ_j^2. Thus for predictions (replicate observations) $\tilde{y}_i$ and n cases, the MSE is $\Sigma_i(y_i - \tilde{y}_i)^2/n$.

The regression term becomes $\beta_0 + \gamma_1\beta_1 X_{i1} + \ldots + \gamma_p\beta_p X_{ip}$, and to assess the effect of the jth predictor, one would monitor the product $\xi_j = \gamma_j\beta_j$. Posterior marginal retention probabilities $Pr(\gamma_j = 1|y)$ are estimated by the proportion of MCMC iterations when $\gamma_j = 1$, while posterior model probabilities are based on the sampled frequency of different combinations of retained predictors.

It is often useful to have a measure of significance for individual predictors (analogous to classical t statistics for each predictor), and marginal Bayes factors for retention (Ghosh and Ghattas, 2015) can be obtained by comparing posterior retention odds $Pr(\gamma_j = 1|y)/Pr(\gamma_j = 0|y)$ with prior odds $\omega_j/(1-\omega_j) = Pr(\gamma_j = 1)/Pr(\gamma_j = 0)$. Interest generally lies with predictors having posterior retention probabilities exceeding their prior probability. For example, assuming $Pr(\gamma_j = 1) = 0.5$, Barbieri and Berger (2004) define the median probability model as that defined by predictors with posterior inclusion probabilities exceeding 0.5.

Settings for τ_j in the SSVS and SSP priors are facilitated by standardising predictors. Then reasonable priors are $\beta_j | \gamma_j = 1 \sim N(0, k^2)$, with $k \in (0.5,4)$ (Kuo and Mallick, 1998), or $k = 1$ (McElreath, 2016). Lee and Bing-Chen (2015) select $k = 10$ in a sparse group selection linear regression with standardised predictors and response. In logistic regression, very large k values are, in fact, informative (disproportionately weighting coefficient values consistent with fitted 0 or 1 probabilities), and the setting $k^2 = 3$ is used by Nott and Leng (2010).

Settings for ω_j correspond to prior beliefs about the potential importance of the predictor, with simplifying options being $\omega_j = \omega$, with ω either preset or an additional unknown, for example, beta distributed a priori. Taking ω as an unknown measure of model complexity is advantageous in applications with many predictors (Piironen and Vehtari, 2016a), with ωp amounting to a prior guess at model size (Bhattacharya et al., 2015). The indifference setting $\omega_j = 0.5$ implies 2^p equally probable models, and that about half the variables are to be retained a priori (Ishwaran and Rao, 2005; O'Hara and Sillanpaa, 2009).

In multivariate regression (e.g. for gene expression measures y_{ik}, $k = 1, \ldots, q$), the discrete selection approach may involve indicators γ_{jk} determining retention of the jth predictor in the regression for the kth outcome. For example, Jia and Xu (2007) propose

$$\beta_{jk} \sim (1-\gamma_{jk})N(0,c) + \gamma_{jk}N(0, \tau_k^2),$$

$$\gamma_{jk} \sim \text{Bern}(\omega_j),$$

where c is a fixed small constant, and a hierarchical prior is set on τ_k^2. Thus, Richardson et al. (2010) suggest a model with selection parameters ω_{jk} determined both by the outcome and predictor,

$$\omega_{jk} = \omega_k \rho_j,$$

where ρ_j captures the propensity for predictor j to influence several outcomes, and ω_k controls the complexity of the regression for outcome k.

A similar hierarchical indicators procedure is proposed by Chen et al. (2016) for sparse group selection, where covariates can be formed into substantively defined groups. The aim is to select the most important groups of predictors, and within those selected groups, select the more important predictors. Thus, retention for group j is determined by binary indicators $\rho_j \sim \text{Bern}(\omega_\rho)$, so that for predictors k within groups, the selection rule is

$$\gamma_{jk} \sim (1-\rho_j)\delta_0 + \rho_j \text{Bern}(\omega_\gamma).$$

Hence under an SSP prior

$$\beta_{jk} \sim (1-\gamma_{jk}\rho_j)\delta_0 + \gamma_{jk}\rho_j N(0, \tau_k^2).$$

7.2.2 Shrinkage Priors

Shrinkage priors seek a sparse representation of the regression coefficients without necessarily including a mechanism to actually formally exclude unnecessary predictors, with potential advantages in MCMC sampling (Bhattacharya et al., 2015; Makalic and Schmidt, 2016). For example, the Lasso prior specifies a heavy tailed double exponential or Laplace prior density for regression coefficients, where this density is defined as

$$\text{DE}(x \mid \mu, \lambda) = \frac{\lambda}{2}\exp\left(-\lambda|x-\mu|\right).$$

The prior $\beta_j \sim \text{DE}(0, \lambda)$ assigns higher weight to values near zero than the normal prior and favours shrinkage, with the scale parameter λ controlling the amount of shrinkage. Larger values of λ imply greater shrinkage with lower variance around the zero prior mean. This prior can be expressed in hierarchical terms (Kotz et al., 2001) as

$$\beta_j \sim N(0, \eta_j^2),$$

$$\eta_j^2 \sim E(\lambda^2/2).$$

In a normal linear regression with residual variance σ^2, the first stage of the prior should be expressed as $\beta_j \sim N(0, \sigma^2\eta_j^2)$, (Park and Casella, 2008). One may also allow the second stage parameters λ_j^2 to vary between coefficients e.g. following a gamma prior (Yi and Ma, 2012).

Shrinkage priors can be represented generically (Polson and Scott, 2010; Bhadra et al., 2016) as

$$\beta_j \sim N(0, \tau^2\eta_j^2), \tag{7.1}$$

with different possible choices of prior density for the η_j^2 (local shrinkage parameters) and τ^2 (the global shrinkage parameter).

The horseshoe prior specifies a half-Cauchy prior for the η_j, allowing considerable shrinkage for unnecessary coefficients (Carvalho et al., 2009; Polson and Scott, 2012). There is some debate about a suitable prior for τ^2 (Piironen and Vehtari, 2017b; Piironen and Vehtari, 2017c), with Carvalho et al. (2009) recommending a half-Cauchy prior also, namely $\tau \sim C^+(0,1)$.

This can be expressed in terms of a Beta(0.5,0.5) density for the shrinkage parameters

$$\kappa_j = 1/(1+\eta_j^2). \tag{7.2}$$

The estimated κ_j can be interpreted as "the amount of weight that the posterior mean for β_j places on 0" (Carvalho et al., 2009; Piironen and Vehtari, 2017c); so higher κ_j correspond to irrelevant predictors. Accordingly, Piironen and Vehtari (2017c) propose an effective number of coefficients measure

$$\sum_{j=1}^{p}(1-\kappa_j).$$

The horseshoe prior is a special case of the prior

$$\eta_j^2 \sim t_\nu^+(0,1),$$

namely a half-Student t-prior with ν degrees of freedom (Piironen and Vehtari, 2016). Piironen and Vehtari (2017c) mention using this prior (with small ν) to alleviate divergent transitions produced by the No U-Turn Sampler (NUTS) algorithm, but this implies a loss of sparsity.

Example 7.1 Diabetes Progression

This example illustrates spike-slab and shrinkage priors using the diabetes data considered by Ishwaran and Rao (2005), and included in the R spikeslab library (Ishwaran et al., 2010), and the R lars library. These data have a continuous measure of disease progression as a response for n = 442 patients, with p = 64 predictors. The latter consist of ten baseline measures (age, sex, body mass index (BMI), blood pressure, and five blood serum measurements; 45 pairwise interactions formed between baseline variables, and quadratic terms for the nine continuous baseline measurements.

Here predictors are arranged as in the lars library, and are standardised. Classical estimation shows four predictors with t statistics above 2, namely x_2 (sex), x_3 (BMI), x_4 (MAP, mean arterial pressure) and x_{20} (sex-age interaction). In the Bayesian analysis here, 95% credible intervals are considered on these predictors, and also on any other predictors shown as relevant.

The first analysis uses the spike-slab scheme of Ishwaran and Rao (2005, equation 4), whereby v_0 denotes a small constant,

$$\beta_j \sim N(0, \gamma_j \tau_j^2),$$

$$\gamma_j \sim (1-\omega)\delta_{v_0}() + \omega\delta_1(),$$

with gamma priors on $1/\tau_j^2$ and a beta or uniform prior on ω. A shrinkage mechanism operates whereby when $\gamma_j = 0$ the variance of β_j is very small. The parameter ω, if taken unknown, acts as a complexity parameter, controlling model size. Here $v_0 = 0.005$ and $\omega \sim U(0,1)$. It is assumed that $1/\tau_j^2 \sim \text{Exp}(1)$, centred around 1, since predictors are standardised.

The second half of a two-chain run of 5,000 iterations provides a posterior mean for ω of 0.84. The control for collinearity implicit in predictor selection reveals x_9 (LTG serum) as a significant predictor (Table 7.1), but with 95% credible intervals for β_2 and β_{20} straddling zero. The predictor x_7 (high-density lipoprotein or HDL cholesterol) has a 95% interval concentrated on negative values, albeit with an inclusion probability of 0.95. Retention probabilities are close to 1 for BMI, MAP, and LTG.

A second analysis uses a hierarchical version of the Lasso prior, namely

$$\beta_j \sim N(0, \sigma^2 \eta_j^2),$$

$$1/\sigma^2 \sim Ga(a,b),$$

$$\eta_j^2 \sim E(\lambda^2/2).$$

TABLE 7.1

Predictor Selection, Diabetes Progression, Spike-Slab Prior

		β_j			γ_j
Predictor	**Notation**	**Mean**	**2.5%**	**97.5%**	**Mean**
Sex	X_2	−4.09	−12.15	0.45	0.93
BMI	X_3	26.64	19.55	33.74	1.00
MAP	X_4	11.91	3.56	19.06	1.00
HDL	X_7	−5.84	−15.73	0.43	0.95
LTG	X_9	25.34	18.01	32.55	1.00
Age-Sex	X_{20}	4.05	−0.38	10.94	0.96

TABLE 7.2

Predictor Selection, Lasso Shrinkage Prior

		β_j		
Predictor	**Notation**	**Mean**	**2.5%**	**97.5%**
Sex	X_2	−7.88	−13.75	−2.09
BMI	X_3	23.14	15.86	30.22
MAP	X_4	13.52	7.37	19.80
LTG	X_9	23.39	15.74	31.18
Age-Sex	X_{20}	5.86	0.44	11.88

The residual precision $1/\sigma^2$ is assigned a Ga(1,0.001) prior, and λ is assigned a uniform $U(0.001,100)$ prior. Results are as in Table 7.2, with five predictors x_2, x_3, x_4, x_9, and x_{20} judged significant in terms of 95% credible intervals either entirely negative or positive. In the sense that regression including predictor selection is still a model for the data, fit statistics - penalised DIC (deviance information criterion), WAIC (widely applicable information criterion) and LOO-IC (leave-one-out information criterion - are very similar between the spike-slab model and Lasso models. For example, their respective LOO-IC are 4797 and 4794.

A third analysis uses a horseshoe prior, implemented in rstan using the scheme

$$\beta_j \sim N(0, \tau^2 \lambda_j^2),$$

$$1/\tau^2 \sim Ga(1, 0.001),$$

$$\lambda_j \sim C^+(0,1),$$

where $C^+(0,1)$ is a half-Cauchy density. A two-chain run of 2000 iterations provides posterior mean estimates for κ_j below 0.05 only for x_3.

One may also include a predictor selection mechanism when shrinkage priors are used for the coefficients (Yuan and Lin, 2005). Thus in a selection version of the Lasso prior

$$\beta_j \mid \gamma_j = 1 \sim N(0, \sigma^2 \eta_j^2),$$

$$\beta_j \mid \gamma_j = 0 \sim \delta_0(),$$

$$1/\sigma^2 \sim Ga(a,b),$$

$$\eta_j^2 \sim E(\lambda^2/2),$$

with ω and λ assigned Be(1,1) and $U(0.001,100)$ priors. This provides posterior inclusion probabilities above 0.95 for x_2, x_3, x_4, and x_9, though the median probability model also includes x_7, x_{20}, and x_{37}.

7.3 Categorical Predictors and the Analysis of Variance

Categorical predictors (with ordered or unordered categories) commonly occur in general regression settings, and in the more specific form of analysis of variance. They raise particular issues in terms of predictor selection: whether a categorical predictor should be retained or excluded in its entirety (group level selection), whether some categories are retained but others excluded (within group selection), and whether categories need to be distinguished within a particular categorical variable (Xu and Ghosh, 2015; Tutz and Gertheiss, 2016). The latter issue can be expressed alternatively as whether categories should be fused or merged.

For ordered predictors, a selection prior should enforce sparsity but also take account of the ordering. Thus consider an ordinal predictor x_j with Q_j levels. Taking the first category as reference with $\beta_{j1}=0$ for identifiability, a regularising prior can be expressed in terms of a prior on successive differences

$$\delta_{jq} = \beta_{j,q+1} - \beta_{jq}, \quad q = 1,\ldots Q_j - 1.$$

For p ordinal predictors, the Lasso prior may be taken on p sets of differences:

$$\delta_{jq} \sim N(0, \tau^2\eta_j^2), \quad j = 1,\ldots,p; q = 1,\ldots,Q_j - 1$$

$$\tau^2 \sim IG(a,b),$$

$$\eta_j^2 \sim E(\lambda^2/2).$$

Wagner and Pauger (2016) propose Normal shrinkage with

$$\delta_{jq} \sim N(0, \tau_j^2), \quad j = 1,\ldots,p; q = 1,\ldots,Q_j - 1$$

$$\tau_j^2 \sim IG(a,b).$$

These types of penalty apply shrinkage to groups of parameters representing the same categorical predictor, and also smooth over successive ordered categories. They can be combined with spike-slab discrete mixture priors where the spike either sets coefficients to zero, or scales the coefficient variances to very small positive values (i.e. effective exclusion).

For nominal predictors, the regularising prior can be applied to all possible contrasts between categories (Wagner and Pauger, 2016). Thus, with $\beta_{j1}=0$, consider contrasts

$$\delta_{jqr} = \beta_{jq} - \beta_{jr}, \; q = 1,\ldots,Q_j - 1; q > r$$

including the reference category. The prior is then

$$\delta_{jq} \sim N(0, R_j \tau^2 \eta_j^2),$$

with $R_j = (Q_j + 1)/2$ reflecting the number of categories.

For a nominal factor with several categories, the number of contrasts increases considerably, and one might instead assume a random prior over all categories (e.g. Albert, 1996; Gelman, 2005), as in

$$\beta_{jq} \sim N(0, \tau_j^2) \;\; q = 1,\ldots,Q_j$$

with identifiability possibly enforced by centring (Tingley, 2012). Xu and Ghosh (2015) consider the hierarchical group lasso representation

$$\beta_{jq} \sim N(0, \tau_j^2),$$

$$\tau_j^2 \sim Ga\left(\frac{Q_j + 1}{2}, \frac{\lambda^2}{2}\right).$$

This can be a combined scheme with a spike-slab discrete mixture with indicators γ_j, where the spike option $\gamma_j = 0$ sets the entire group j coefficient set $(\beta_{j1}, \beta_{j2}, \ldots, \beta_{jQ_j})$ to zero. This is consistent with a principle of group level sparsity. To allow for coefficient selection within groups one can use products of selection indicators $\gamma_j \gamma_{jq}$, with γ_j and γ_{jq} following separate Bernoulli densities (Xu and Ghosh, 2015, p.924).

7.3.1 Testing Variance Components

Bayesian methods may have benefit in analysis of variance applications beyond variable selection considerations. Conventional analysis of variance (ANOVA) estimation does not allow for factor combinations for which there are no observations (Tingley, 2012). Fixed effects ANOVA estimation also do not allow predictions of a future observation (Geinitz and Furrer, 2016). The analogue of F tests for effects of factor variables under classical estimation are provided by comparisons of variance estimates.

Thus, in a balanced normal one-way analysis of variance, one has a factor (indexed by *i*) with replicates (indexed by *j*)

$$y_{ij} = a + \beta_i + \varepsilon_{ij} \;\; i = 1,\ldots,n_I; j = 1,\ldots,n_J$$

where $\varepsilon_{ij} \sim N(0, \sigma_\varepsilon^2)$, and factor effects β_i are estimated either as fixed or random effects. Classical assessment of the hypothesis $\beta_1 = \ldots = \beta_{n_I} = 0$ uses F tests comparing mean squares due to the factor and the errors.

One possible perspective, broadening into multilevel applications (Geinitz et al., 2015), considers the parameters of this model as random effects (Gelman, 2005). For example, as a baseline representation for the one-way ANOVA,

$$y_{ij} \sim N(\beta_i, \sigma_\varepsilon^2),$$

$$\beta_i \sim N(a, \sigma_\beta^2),$$

with significant variation in the factor assessed by the comparison $Pr(\sigma_\beta > \sigma_\varepsilon \mid y)$, which can be estimated from MCMC sampling. An analogous comparison may be made using the marginal variance s_β, estimated from the sampled β_i during MCMC runs. The latter can be regarded as estimating the variance over the observed units, rather than some broader population of units.

Example 7.2 Horseshoe Crabs

As described in Brockmann (1996), horseshoe crabs arrive in pairs at particular beach sites for springtime spawning, but unattached ("satellite") males compete with attached males for fertilisations. Some couples are ignored, while some attract many satellites, and this is related to characteristics of females. The presence of satellites is described by a binary variable with value 1 for at least one satellite. Potential predictors are the female crab's weight, carapace width, colour, and spine condition. The latter two predictors are ordinal, with colour having values 1 = dark, 2 = medium dark, 3 = medium, and 4 = medium light, while spine condition has values 1 = both good, 2 = one worn or broken, and 3 = both worn or broken. Older crabs tend to have darker shells.

An analysis (using rjags) without predictor selection shows the medium and medium dark colour categories as significant in terms of a 95% credible interval entirely one side of zero (Table 7.3). A second model uses Lasso shrinkage priors, with four η_j^2 parameters: on weight and width, and on the collective differences applying to colour and spine condition. Thus, for the two continuous (and standardised) predictors

$$\beta_j \sim N(0, \eta_j^2), \ j = 1, 2,$$

while for the ordinal predictors

$$\beta_{j1} = 0,$$

$$\beta_{j,q+1} = \beta_{jq} + \delta_{jq} \quad j = 3, 4; q = 1, \ldots, Q_j - 1$$

$$\delta_{jq} \sim N(0, \eta_j^2),$$

with priors on hyperparameters

$$\eta_j^2 \sim E(\lambda^2/2) \quad j = 1, \ldots 4,$$

$$\lambda \sim U(0.001, 100).$$

This model produces an improved penalised deviance (Plummer, 2008), at 197 compared to 202 under the first model (though not a better Brier score). There is an enhanced role for the width predictor, with a shrinkage in the coefficient for weight (Table 7.3). This effect may reflect better control for impact of multicollinearity (the correlation between weight and width is 0.89). There is also an attenuation in the impacts of the colour categories, though there is still a 95% probability that the impact of medium colour is positive.

Posterior inference may be sensitive to the prior for λ, with large values producing overshrinkage (Xu and Ghosh, 2015). A sensitivity analysis assumes a prior $\lambda \sim E(1)$. This

TABLE 7.3

Horseshoe Crabs. Logistic Regression for Satellite Presence

Predictor (Category)	Parameter	Mean	2.5%	97.5%	$Pr(\beta_j>0)$
		Without Selection			
Width	β_1	0.58	−0.27	1.44	0.91
Weight	β_2	0.51	−0.32	1.38	0.89
Spine (category 2)	β_{32}	−0.11	−1.55	1.30	0.44
Spine (category 3)	β_{33}	0.44	−0.57	1.45	0.81
Color (category 2)	β_{42}	1.27	0.06	2.51	0.98
Color (category 3)	β_{43}	1.66	0.54	2.79	1.00
Color (category 4)	β_{44}	1.86	−0.05	3.90	0.97
		Lasso Prior, $\lambda \sim U(0.001,100)$			
		Mean	**2.5%**	**97.5%**	**$Pr(\beta_j>0)$**
Width	β_1	0.53	−0.06	1.18	0.95
Weight	β_2	0.43	−0.12	1.10	0.92
Spine (category 2)	β_{32}	0.01	−0.73	0.73	0.52
Spine (category 3)	β_{33}	0.15	−0.48	0.93	0.66
Color (category 2)	β_{42}	0.49	−0.16	1.54	0.90
Color (category 3)	β_{43}	0.79	−0.06	1.96	0.95
Color (category 4)	β_{44}	0.82	−0.23	2.40	0.91
Laplace scale	λ	3.4	0.9	8.4	
		Lasso Prior, $\lambda \sim E(1)$			
		Mean	**2.5%**	**97.5%**	**$Pr(\beta_j>0)$**
Width	β_1	0.54	−0.08	1.25	0.95
Weight	β_2	0.47	−0.17	1.16	0.92
Spine (category 2)	β_{32}	−0.01	−0.83	0.77	0.51
Spine (category 3)	β_{33}	0.18	−0.53	1.02	0.67
Color (category 2)	β_{42}	0.58	−0.14	1.60	0.93
Color (category 3)	β_{43}	0.93	0.01	2.06	0.98
Color (category 4)	β_{44}	0.97	−0.20	2.53	0.94
Laplace scale	λ	2.1	0.7	4.2	
		Lasso Prior and Selection			
		Mean	**2.50%**	**97.50%**	**$Pr(\beta_j>0)$**
Width	β_1	0.59	−0.03	1.34	0.90
Weight	β_2	0.42	−0.11	1.26	0.82
Spine (category 2)	β_{32}	0.04	−0.56	0.82	0.54
Spine (category 3)	β_{33}	0.17	−0.44	1.01	0.65
Color (category 2)	β_{42}	0.72	−0.09	2.00	0.88
Color (category 3)	β_{43}	0.93	−0.02	2.24	0.94
Color (category 4)	β_{44}	0.98	−0.20	2.63	0.92
Laplace scale	λ	1.6	0.4	3.6	
Selection probability	ω	0.64	0.15	0.98	

produces both a lower deviance and improved Brier score as compared to a model without selection. There is still considerable shrinkage in the colour coefficients.

A third model introduces selection indicators γ_j, such that when $\gamma_j=0$ the variances η_j^2 are scaled by a small constant ρ. This may be preset or assigned a prior centred on an informative value. Here $\rho=0.0001$, and the prior $\lambda \sim E(1)$ is retained. This model enables one to assess fusion probabilities, namely, that successive ordinal category coefficients are equated. Focusing on the colour coefficients, we find a probability of 0.33 (fusecol in the code) that $\beta_{42} = \beta_{43} = \beta_{44}$, amalgamating over iterations where β_{42} is retained or excluded.

Example 7.3 Rails Data, Analysis of Variance

This dataset contains measurements of times taken by an ultrasonic wave to travel the rail length; these data are included in the R package nlme (https://stat.ethz.ch/R-manual/R-devel/library/nlme/html/Rail.html). There are three replicates for six rails.

A first analysis assumes a normal random prior for the rails (the factor variable). This provides posterior means (sd) for α, σ_β, and s_β of 66.5 (10.3), 23.6 (7.5), and 24.7 (1.1). The probabilities $Pr(\sigma_\beta > \sigma_\varepsilon \mid y)$ and $Pr(s_\beta > \sigma_\varepsilon \mid y)$ are both 1.

Estimates of hyperparameters (variability over the rails and of the grand mean) may be affected by departures from the assumed normal hyperprior, and, in particular, by outliers. Thus Figure 7.1 shows the discrepant profiles of the 95% intervals for β_i. As one alternative, a student t scale mixture with known degrees of freedom ($\nu=4$) is considered, whereby

$$\beta_i \sim N(\alpha, \sigma_\beta^2/\delta_i),$$

$$\delta_i \sim Ga(2,2).$$

This analysis provides $\delta_2=0.66$ as the lowest estimated scale parameter. Downweighting the influence of rail 2 leads to a higher posterior means (sd) for α of 67.3 (11.1), while σ_β is reduced to 19.7 (7.5).

A third analysis assumes a double exponential prior for β_i,

$$\beta_i \sim DE(\alpha, \lambda/\sigma),$$

combined with an exponential $E(1)$ prior on λ. This provides a posterior mean (sd) for α of 68.0 (13.8) and for λ of 0.22 (0.10).

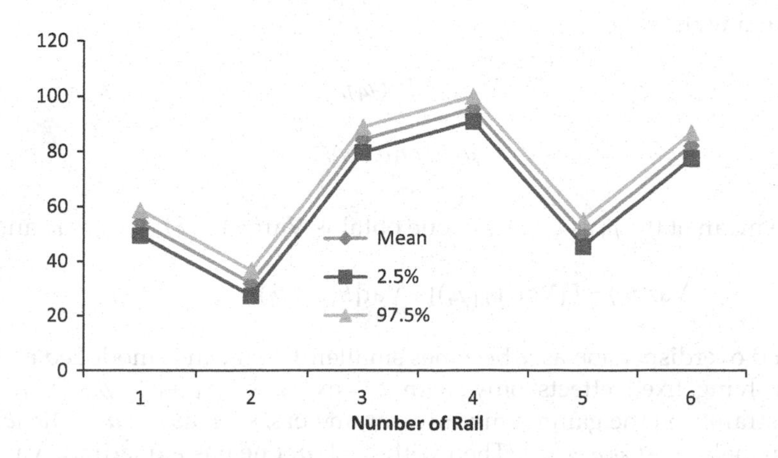

FIGURE 7.1
Profiles of β coefficients.

7.4 Regression for Overdispersed Data

Let $\{y_1,\ldots,y_n\}$ be observations from the exponential family density

$$p(y_i \mid \theta_i,\phi) = \exp\left[\frac{y_i\theta_i - b(\theta_i)}{a_i(\phi)} + c(y_i,\phi)\right]$$

with canonical parameter θ_i, dispersion function $a_i(\phi) = \phi/w_i$, and w_i known. Under the generalised linear model (GLM) framework, the mean of y is $\mu_i = E(y_i \mid \theta_i) = b'(\theta_i)$, predicted via a monotone link function $g(\mu_i) = X_i\beta$, and the variance is

$$\text{var}(y_i \mid \theta_i) = a_i(\phi)b''(\theta_i) = a_i(\phi)\,\text{var}(\mu_i).$$

The exponential family includes as special cases the normal, binomial, Poisson, multinomial, negative binomial, exponential, and gamma densities.

However, GLM regressions often show a residual variance larger than expected under the exponential family models, due to unknown omitted covariates, clustering in the original units, or inter-subject variations in propensity (Zhou et al., 2012). Particular types of response pattern (e.g. an excess proportion of zero counts as compared to the expected frequency) may also cause overdispersion (Garay et al., 2015; Musio et al., 2010) (Section 7.6).

Without correction for such extra-variability, regression parameter estimates may be biased, and their credible intervals will be too narrow, so that incorrect inferences about significance may be obtained. The solution involves regression with additional random effects to account for excess residual variation, and the focus in Monte Carlo Markov Chain (MCMC) is usually on the complete data likelihood, rather than the marginal model obtained by integrating over the random effects.

7.4.1 Overdispersed Poisson Regression

Thus, the Poisson regression model for count data assumes that the mean and variance are equal, but overdispersion, as compared to the Poisson assumption, is routinely encountered. As discussed in Chapter 4, the conjugate mixture model for count data is the Poisson-gamma with

$$y_i \sim Po(\mu_i),$$

$$\mu_i \sim Ga(a_i, \eta_i).$$

Denoting the mean of the μ_i as $\xi_i = a_i/\eta_i$, one obtains $\text{Var}(\mu_i) = a_i/\eta_i^2 = \xi_i^2/a_i$ and

$$\text{Var}(y_i) = E[\text{Var}(y_i \mid \mu_i)] + \text{Var}[E(y_i \mid \mu_i)] = \xi_i + \xi_i^2/a_i,$$

with increased overdispersion as α_i becomes smaller. The mean is modelled by regression, typically involving fixed effects only, with $\xi_i = \exp(\beta_0 + \beta_1 x_{1i} + \ldots + \beta_p x_{pi})$. Identification requires constraints on the gamma mixture parameters, such as $\alpha_i = \alpha$ in the $\{\xi_i, \alpha_i\}$ parameterisation, namely $\mu_i \sim Ga(\alpha, \alpha/\xi_i)$. Then with $\phi = 1/\alpha$, one has a quadratic variance function, with

$$\text{Var}(y_i) = E[\text{Var}(y_i \mid \mu_i)] + \text{Var}[E(y_i \mid \mu_i)] = \xi_i + \phi\xi_i^2.$$

Another possibility is to set $\mu_i = \xi_i\omega_i$ where $\omega_i \sim Ga(a, a)$ so that the frailties average 1, with variance $\phi = 1/\alpha$. Integrating out the ω_i leads to a marginal negative binomial (NB2) density for the y_i, namely

$$p(y_i \mid \beta, \alpha) = \frac{\Gamma(\alpha + y_i)}{\Gamma(\alpha)\Gamma(y_i + 1)} \left(\frac{\alpha}{\alpha + \xi_i} \right)^{\alpha} \left(\frac{\xi}{\alpha + \xi_i} \right)^{y_i}$$

Regarding the dispersion parameter, one may adopt a $Ga(a,b)$ prior for α, with $a = 1$, and with $b \sim Ga(1, 0.005)$ as an extra unknown (Fahrmeir and Osuna, 2006).

Setting

$$p_i = \left(\frac{\alpha}{\alpha + \xi_i}, \right)$$

the negative binomial (NB2) can also be denoted as NB(p_i, α). For example, Zhou et al. (2012) propose predictor impacts be represented via a logit regression for p_i, with the regression including an additional error term to partly represent heterogeneity.

More general negative binomial forms have been suggested, such as the NBk (Winkelmann and Zimmermann, 1995), with variance function

$$\text{Var}(y_i \mid X_i) = E[\text{Var}(y_i \mid \mu_i)] + \text{Var}[E(y_i \mid \mu_i)] = \xi_i + \phi\xi_i^{k+1}, \ (k \geq -1)$$

obtained with the gamma prior,

$$\mu_i \sim Ga\left(\frac{\xi_i^{1-k}}{\phi}, \frac{\xi_i^{-k}}{\phi} \right).$$

The values $k = 0$ and $k = 1$ lead to the NB1 and NB2 variance forms. The NB-P model (Greene, 2008) replaces α in the NB2 formulation by $\alpha\xi_i^{2-P}$, with $P = 2$ corresponding to the NB2 and $P = 1$ to the NB1.

Nonconjugate random mixture models are often adopted for count data (Kim et al., 2002), as in

$$\log(\mu_i) = X_i\beta + \log(\varepsilon_i),$$

with lognormal or log-t distributed errors ε_i. The prior

$$\varepsilon_i \sim LN\left(\frac{-\tau^2}{2}, \tau^2 \right),$$

ensures $E(\varepsilon_i) = 1$, while variance matching priors can be adopted for the Poisson log-normal and Poisson-gamma models (Millar, 2009). The nonconjugate approach is convenient when multivariate, multiple, or multilevel random effects are to be considered. An example of multiple effects is the convolution prior (Neyens et al., 2012) for area disease events y_i with expected totals P_i. Thus $y_i \sim Po(P_i\mu_i)$, where

$$\log(\mu_i) = X_i\beta + \varepsilon_i + s_i$$

and both random effects $\{\varepsilon_i, s_i\}$ may account for overdispersion, but the ε_i are unstructured (i.e. exchangeable with regard to area identifiers), while the s_i are spatially structured.

For count regressions only involving an unstructured error, one may specify

$$\mu_i = E(y_i \mid X_i, \varepsilon_i) = \exp(X_i\beta + \sigma\varepsilon_i),$$

with $\varepsilon_i \sim N(0,1)$. Denoting $v_i = \exp(X_i\beta)$, the conditional mean (Greene, 2008) is

$$E(y_i \mid X_i) = E_\varepsilon[E(y_i \mid X_i, \varepsilon_i)] = v_i \exp(\sigma^2/2),$$

and the conditional variance is

$$\begin{aligned}\text{Var}(y_i \mid X_i) &= E_\varepsilon[\text{Var}(y_i \mid X_i, \varepsilon_i)] + \text{Var}_\varepsilon[E(y_i \mid X_i, \varepsilon_i)] \\ &= v_i \exp(\sigma^2/2)\left\{1 + v_i \exp(\sigma^2/2)[\exp(\sigma^2) - 1]\right\}.\end{aligned}$$

Taking $\phi = e^{\sigma^2} - 1$

$$\text{Var}(y_i \mid X_i) = E(y_i \mid X_i, \varepsilon_i)\left[1 + \phi E(y_i \mid X_i, \varepsilon_i)\right],$$

showing that the variance has a quadratic form, as for the NB2 form of the negative binomial.

Example 7.4 School Attendance Data

This example considers Poisson overdispersion in terms of different negative binomial regression forms. The dataset concerns number of days absent for 314 high school juniors at two schools, with predictors including a standardised maths test score, gender (1 = Female, 0 = Male), and type of program enrolled on (general, academic, vocational). Overdispersion in the data is apparent with variance of 49.5 exceeding the mean of 6. A simple Poisson regression using glm in R has a scaled deviance of 1747.

By contrast, under a Poisson-gamma mixture (NB2) one has

$$y_i \sim Po(\xi_i \omega_i),$$

$$\omega_i \sim Ga(a, a),$$

$$\log(\xi_i) = \beta_1 + \beta_2 \text{Math}_i + \beta_3 \text{Female}_i + \beta_4 I(\text{Academic})_i + \beta_5 I(\text{Vocational})_i.$$

This can be estimated using a Poisson-gamma hierarchical likelihood or a negative binomial NB2 likelihood, with the former approach having the benefit of providing observation specific frailties. These models are estimated using rstan.

With a N(0,1) prior on β_j $(j = 2,\ldots 5)$ and $a \sim Ga(1, 0.01)$, the posterior mean scaled deviance under the Poisson-gamma model is 330.1, with posterior mean (95% CrI) for ϕ of 1.04 (0.84, 1.28), so the model accounts for Poisson overdispersion. The model also achieves a satisfactory representation of the data, since the mixed predictive strategy of Marshall and Spiegelhalter (2007) shows a relatively low number (18/314) of pupils with probabilities of overprediction over 0.95 or under 0.05. This model suggests the effect of gender, β_3, is marginally significant with 95% credible interval straddling zero.

A variant of the NB2 model (and of other overdispersion models) takes dispersion parameters as case-specific and determined by regression (Barreto-Souza and Simas,

2016). Thus, one has $NB(p_i, \alpha_i)$, where $\log(\alpha_i) = X_i\delta$. This model shows dispersion to be greater among general program students, though shows no benefit in terms of fit, with higher LOO-IC than the Poisson-gamma model.

The suitability of NB1 or NB2 forms of negative binomial regression may be assessed using the NBP (negative binomial P) model. To assess this, an exponential prior centred at 2 is used for P following the Greene (2008) approach. We obtain posterior mean (95% interval) for P of 1.23 (0.73,1.65). This excludes 2, and so does not favour the NB2 parameterisation. The LOO-IC is improved as compared to the NB2 model (Poisson-gamma version), namely 1540 vs 1559. Substantive inferences are affected in that the gender effect β_3 now has an entirely positive 95% credible interval, (0.02,0.45).

7.4.2 Overdispersed Binomial and Multinomial Regression

Binomial regression with excess variation may occur when responses are arranged in clusters and responses from the same cluster are correlated: examples occur in teratological studies (e.g. when the observation unit is a litter of animals, and litters differ in terms of unknown genetic factors). A conjugate approach to such overdispersion involves a beta-distributed success probability, leading to a beta-binomial regression model (Kahn and Raftery, 1996).

Thus with $y_i \sim \text{Bin}(n_i, p_i)$, one assumes

$$p_i \sim Be(\gamma\pi_i, (1-\pi_i)\gamma)$$

with mean π_i and variance

$$\pi_i(1-\pi_i)/(\gamma+1).$$

where $\gamma \geq 0$. Regression involves a logit or other link,

$$g(\pi_i) = X_i\beta.$$

Setting $\varphi = (\gamma+1)^{-1}$, the unconditional variance of a beta-binomial response is of the form

$$\text{Var}(y_i) = n_i p_i(1-p_i)[1+(n_i-1)\varphi].$$

Nonconjugate random mixture models are often adopted for binomial data, with normal or Student t errors in the regression link. The presence of an error term permits predictor selection using a g-prior approach (Gerlach et al., 2002; Kinney and Dunson, 2007) in mixed logistic models. For y binomial or binary with probabilities p_i, and n observations, one might specify

$$\text{logit}(p_i) = X_i\beta + \varepsilon_i,$$

$$\varepsilon_i \sim N(0, \sigma^2)$$

with

$$\beta \sim N(B, \sigma^2 g(X'X)^{-1}),$$

and g an unknown, with prior such as $g \sim IG(0.5, 0.5n)$ (Zellner and Siow, 1980; Perrakis et al., 2015). This prior can be combined with spike-slab binary selection indicators. For y binary, and data augmentation (Section 7.5), the g-prior can also be used.

For multinomial data (e.g. on voting patterns y_{ij} for parties j by constituency i), overdispersion may occur when choice probabilities vary between the N_i individuals in each observation unit, but clusters of individuals within each unit have similar probabilities. The individual-level factors associated with such clustering are not observed, so a random effect will proxy such unobserved factors; for example, voters with different education levels may differ in their voting preferences, but only the average education in each constituency is observed. The raw percentages y_{ij}/N_i are also likely to show erratic features, whereas hierarchical models for pooling strength provide frequency smoothing and model interdependencies between categories.

This form of data may be modelled as a product multinomial likelihood conditioning on known $N_i = y_{i+}$. With probabilities π_{ij} of choices $j = 1, \ldots J$, the sampling model is

$$y_{ij} \sim \text{Mult}(N_i, [\pi_{ij}, \ldots, \pi_{iJ}]) \quad i = 1, \ldots n.$$

The conjugate approach for such heterogeneity is the multinomial-Dirichlet mixture, where the Dirichlet is the multivariate generalisation of the beta density. However, the Dirichlet has a restricted covariance structure when there are dependencies between the response categories j within units i. For example, for n constituencies and J political parties, one may expect both negative and positive correlations between π_{ij} for different parties. Greater flexibility is provided by modelling heterogeneity within the regression link, as in random effects multiple logit models (Hensher and Greene, 2003), or via multinomial probit models.

Under the multiple logit form, define a $J-1$ dimensional random effect $\alpha_i = (\alpha_{i1}, \ldots \alpha_{i,J-1})$ representing subject or unit level intercepts; these might be exchangeable or correlated (if, say, the units were areas and behaviours were spatially clustered). Then with X_i excluding an intercept,

$$\pi_{ij} = \exp(\alpha_{ij} + X_i\beta_j) \Big/ \sum_{k=1}^{J} \exp(\alpha_{ik} + X_i\beta_k)$$

with $\alpha_{iJ} = \beta_J = 0$ for identification. For example, one may assume

$$(\alpha_{i1}, \ldots \alpha_{i,J-1}) \sim N_{J-1}(H_i, D),$$

where D is an unknown covariance matrix, $H_i = (H_{i1}, \ldots, H_{i,J-1})$, where $H_{ij} = A_j + X_i\beta_j$, and A_j is the intercept for choice j.

This multiple logit may also be fitted by Poisson regression using the fact that the multinomial is equivalent to a Poisson distribution conditional on a fixed total. This involves defining n fixed effect predictors a_i to ensure the unit totals N_i are maintained. Thus $y_{ij} \sim Po(\mu_{ij})$, with

$$\log(\mu_{ij}) = a_i + \alpha_{ij} + X_i\beta_j,$$

for $j = 1, \ldots, J$, where the a_i would typically be fixed effects assigned vague priors e.g. $a_i \sim N(0, 1000)$.

Example 7.5 Voting in Florida; Multinomial Overdispersion

This example considers normal random effects to model multinomial overdispersion via multiple logit links. The analysis relates to the 2000 US presidential election voting data y_{ij} for $i = 1,\ldots,67$ Florida polling districts and with N_i denoting total votes (Mebane and Sekhon, 2004). There are $J=5$ choices of candidate (Buchanan, Nader, Gore, Bush, other) and three predictors:

a) x_1, the proportion of each county's votes for different presidential candidates in 1996;
b) x_2, changes between 1996 and 2000 in party registration;
c) x_3, percentage of census population Cuban in district i.

Specification of x_1 and x_2 (predictors specific for area and candidate) follows Mebane and Sekhon (2004), but x_3 differs from their variable. Mebane and Sekhon (2004) find substantial overdispersion in these data.

The sampling model for the random effects multinomial is

$$y_{ij} \sim \text{Mult}(N_i, [\pi_{i1}, \ldots, \pi_{i5}]) \quad i = 1, \ldots n$$

$$\pi_{ij} = \phi_{ij} \Big/ \sum_{ij} \phi_{ij}$$

and to account for overdispersion, normal effects $\alpha_i = \{\alpha_{i1}, \ldots, \alpha_{i(J-1)}\}$ are included in multiple logit links. These have non-zero means A_j, namely the intercepts for the first four candidate choices. In a hierarchical parameterisation

$$\log(\phi_{ij}) = \alpha_{ij},$$

$$(\alpha_{i1}, \ldots, \alpha_{i,J-1}) \sim N_{J-1}(H_{i1,\ldots,i,J-1}; D),$$

$$H_{ij} = A_j + X_i \beta_j,$$

$$\log(\phi_{iJ}) = 0,$$

with $\{\beta_{j,k};\ j = 1,\ldots,4,\ k = 1,\ldots,3\}$, A_j assigned diffuse priors, and the precision matrix D^{-1} assigned a Wishart prior with identity scale matrix and $J-1=4$ degrees of freedom[2].

MCMC convergence is considerably assisted by the hierarchical parameterisation above and by centring predictors. Inferences are based on the second half of a two-chain run of 20,000 iterations. The β coefficients show 1996 voting to influence later voting, except for Nader, while change in party registration is important for all candidates, except Gore. The proportion of Cuban-Americans has a positive effect on voting for Gore and Bush. A posterior predictive check comparing chi-square values for replicate and observed data is satisfactory at around 0.49.

Posterior predictive checks are not satisfactory when a fixed effects only model is applied with

$$\log(\phi_{ij}) = A_j + X_i \beta_j, \quad j = 1, \ldots, J-1.$$

There is then zero posterior probability that $\chi^2(y_{\text{rep}}, \theta) > \chi^2(y, \theta)$. Standard deviations of predictor effects are also considerably understated if allowance is not made for excess variation, and, in fact, all coefficient effects are significant (95% CRI either entirely positive or negative) under this model.

7.5 Latent Scales for Binary and Categorical Data

Sampling and inference in Bayesian general linear models are complicated to the extent that conjugate priors are only available for normal regression (Holmes and Held, 2006). The auxiliary variable approach circumvents this by introducing latent continuous responses underlying binary or categorical observations, resulting in a specification (including priors) that effectively replicate normal regression. This provides simplified MCMC sampling and improved residual tests.

Consider first binary responses, and assume latent metric data y^* such that $y=1$ when $y^*>0$ and $y=0$ when $y^*\leq 0$ (Hooten and Hobbs, 2015; Albert and Chib, 1993). In economic choice applications (e.g. regarding economic participation or not), the latent scale y^* arises by comparing utilities U_{1i} and U_{0i} of options 1 and 0 with

$$U_{ji}=V_{ji}+\varepsilon_{ji}=X_i\beta_j^*+\varepsilon_{ji},$$

$$y_i^*=U_{1i}-U_{0i}.$$

Under the above scheme, one has

$$Pr(y_i=1)=Pr(y_i^*>0)=Pr(\varepsilon_{0i}-\varepsilon_{1i}<V_{1i}-V_{0i})=Pr(\varepsilon_{0i}-\varepsilon_{1i}<X_i\beta)$$

where $\beta=\beta_1^*-\beta_0^*$. Alternative forms for ε lead to different links: taking ε_{ji} to be normal with mean zero and variance σ^2 leads to a probit link with $Pr(y_i=1)=\Phi(X_i\beta/\sigma)$. It is apparent that β and σ cannot be separately identified, and the commonest identifying device takes $\sigma^2=1$.

A probit regression with binary responses y_i may therefore be obtained by truncated normal sampling for y_i^*, with the form of constraint determined by the observed y. Thus, if $y_i=1$, y_i^* is constrained to be positive, and sampled from a normal with mean $X_i\beta$ (including an intercept in p-dimensional X_i) and variance 1. If $y_i=0$, y_i^* is sampled from the same density, but constrained to be negative. With a normal prior on the coefficients $\beta\sim N_p(B_0,V_0)$, the full conditional distribution of β is also normal, namely

$$\beta|y^*\sim N(B,V)$$

$$B=V^{-1}(V_0^{-1}B_0+X'y^*)$$

$$V=(V_0^{-1}+X'X)^{-1}.$$

Improved MCMC mixing is obtained by updating y^* and β jointly (Holmes and Held, 2006), and justified by the factorisation,

$$p(\beta,y^*|y)=p(y^*|y)p(\beta|y^*)$$

where updating of β is as above, but y^* is updated from its marginal distribution integrated over β.

Heavier tailed links are obtained by sampling y_i^* directly from a Student t with ν degrees of freedom, or by using the scale mixture version of the Student t density (Chang et al., 2006).

This again involves constrained normal sampling but with gamma subject-specific precisions $\lambda_i \sim Ga(\nu/2, \nu/2)$, so that

$$y_i^* \sim N(X_i\beta, 1/\lambda_i)\, I(0,\infty), \quad \text{when } y_i = 1$$

$$y_i^* \sim N(X_i\beta, 1/\lambda_i)\, I(-\infty,0), \quad \text{when } y_i = 0.$$

Skew densities for ε have also been proposed, such as a skew-probit link (Bazan et al., 2010) with augmentation scheme

$$y_i^* = X_i\beta + \varepsilon_i,$$

$$\varepsilon_i = \sigma\left[-\varphi V_i - (1-\varphi^2)W_i\right],$$

where V_i is half normal $V_i \sim N^+(0,1)$, $W_i \sim N(0,1)$, $\varphi \sim U(-1,1)$, and $\sigma=1$ for identifiability. In hierarchical form, one has

$$y_i^* \sim N(X_i\beta - \varphi V_i, 1-\varphi^2).$$

Taking ε to be logistic, a logit regression is obtainable (e.g. Holmes and Held, 2006), by the augmentation scheme

$$y_i^* \sim \text{Logist}(X_i\beta, 1)\, I(0,\infty), \quad \text{when } y_i = 1$$

$$y_i^* \sim \text{Logist}(X_i\beta, 1)\, I(-\infty,0), \quad \text{when } y_i = 0$$

where $y \sim \text{Logist}(\mu, \tau)$, namely

$$p(y\,|\,\tau,\mu) = \tau \exp(\tau[y-\mu])/\{1+\exp(\tau[y-\mu])\}^2$$

with variance κ^2/τ^2, where $\kappa^2 = \pi^2/3$. A logit link can be approximated by Student t sampling when $\nu=8$, or equivalently by scale mixture normal sampling with $\lambda_i \sim Ga(\nu/2, \nu/2)$, combined with constrained sampling according to the observed y values. Specifically, a t_8 variable is approximately 0.634 times a logistic variable, so that

$$y_i^* \sim t_8\left(X_i\beta, \frac{1}{0.634^2}\right) I(0,\infty), \quad \text{when } y_i = 1$$

$$y_i^* \sim t_8\left(X_i\beta, \frac{1}{0.634^2}\right) I(-\infty,0), \quad \text{when } y_i = 0.$$

Equivalently with $\lambda_i \sim Ga(4,4)$,

$$y_i^* \sim N\left(X_i\beta, \frac{1}{\lambda_i(0.634)^2}\right) I(0,\infty), \quad \text{when } y_i = 1$$

$$y_i^* \sim N\left(X_i\beta, \frac{1}{\lambda_i(0.634)^2}\right) I(-\infty,0), \quad \text{when } y_i = 0.$$

A different approximation follows since a $t_{7.3}$ variable is approximately 0.647 times a logistic variable (Kinney and Dunson, 2007). So, with $\lambda_i \sim Ga(\nu/2, \nu/2)$, where $\nu = 7.3$, and $\tilde{\sigma}^2 = \pi^2(\nu-2)/3\nu$,

$$y_i^* \sim N(X_i\beta, \tilde{\sigma}^2/\lambda_i)\, I(0,\infty), \quad \text{when } y_i = 1$$

$$y_i^* \sim N(X_i\beta, \tilde{\sigma}^2/\lambda_i)\, I(-\infty,0), \quad \text{when } y_i = 0.$$

Logit models relate responses $y_i = 0$ or 1 to predictors X_i through proportional exponential functions of regressors,

$$Pr(y_i = k) \propto \exp\{X_i\beta\}$$

and a latent exponential variable version (Scott, 2011) of the logit link involves sampling $\{z_{0i}, z_{1i}\}$ from exponential densities $E(\lambda_{ji})$, with parameters $\lambda_{0i} = 1$ and $\lambda_{1i} = \exp(X_i\beta)$. If $y_i = \arg\min(z_{0i}, z_{1i})$, then $Pr(y_i = k \mid X_i) \propto \lambda_{ki}$ as under a logit regression. This principle extends to multiple logit regression by sampling $\{z_{0i}, z_{1i}, \ldots, z_{J-1,i}\}$.

Augmented data sampling for the logit model can also be achieved using a discrete mixture approximation of the type 1 extreme value error (Fruhwirth-Schnatter and Fruhwirth, 2010). With U_{0i} and U_{1i} as utilities of categories 0 and 1, and

$$U_{1i} = X_i\beta + \varepsilon_i$$

the binary logit is obtained when U_{0i} and ε_i follow type 1 extreme value distributions. Using the relation between the exponential and type 1 extreme distributions, and with $\nu_i = \exp(X_i\beta)$, one has

$$\exp(-U_{0i}) \sim E(1), \exp(-U_{1i}) \sim E(\nu_i).$$

with the minimum of these variables also exponential,

$$\min[\exp(-U_{0i}), \exp(-U_{1i})] \sim E(1+\nu_i).$$

When $y_i = 1$, one has $U_{1i} > U_{0i}$, or equivalently $\exp(-U_{1i}) < -\exp(-U_{0i})$, so that,

$$\exp(-U_{1i}) \sim E(1+\nu_i).$$

When $y_i = 0$, one has $U_{0i} > U_{1i}$, or equivalently $\exp(-U_{0i}) < -\exp(-U_{1i})$, so that

$$\exp(-U_{0i}) \sim E(1+\nu_i),$$

$$\exp(-U_{1i}) = \exp(-U_{0i}) + \delta_i,$$

where $\delta_i \sim E(\nu_i)$.

A useful diagnostic feature resulting from the augmented data approach is that the residuals $y_i^* - X_i\beta$ are nominally a random sample from the assumed cumulative distribution for ε (Johnson and Albert, 1999). So for the latent data probit, $\varepsilon_i = y_i^* - X_i\beta$ is approximately N(0,1) if the model is appropriate for case i, whereas if the posterior distribution of ε_i is significantly different from N(0,1) then the model conflicts with the observed y. So one might obtain the probability $Pr(|\varepsilon_i| > 2 \mid y)$ and compare to its prior value of 0.045. For the latent data logit, one may obtain $Pr(|\varepsilon_i| / \kappa > 2 \mid y)$.

Example 7.6 Low Birthweight

This example illustrates data augmentation and predictor selection for binary responses. The data concerns low birthweight from the R library glmulti. There are n = 189 observations with binary response 1 for birth weight under 2.5 kg, 0 otherwise. Potential predictors are mother's age (standardised), lwt (mother's weight at last menstrual period, standardised), race (1 = white, 2 = black, 3 = other), smoke (smoking status during pregnancy), ptl (number of previous premature labours), ht (history of hypertension), ui (presence of uterine irritability), and ftv (number of physician visits during the first trimester).

Collinearity in this application has been noted, though the interactions ui*smoke and ftv*age are mentioned as potentially significant by Calcagno and de Mazancourt (2010), giving p = 11 predictors. Probit and logit regressions with data augmentation, and initially without any predictor selection, are implemented using R2OpenBUGS. These show lwt, smoke, ht, ui, and ftv*age as having 95% credible intervals not straddling zero. However, the ethnic categories (black, other) and the interaction ui*smoke also have probabilities $Pr(\beta_j > 0 \mid y)$ under 0.95 or over 0.05.

As one approach to predictor selection, a horseshoe prior is applied as part of a probit regression. Table 7.4 shows the consequent shrinkage in coefficient estimates, together with a more clarified indication of significance, with only lwt, smoke, ui, ht, and ftv*age having probabilities $Pr(\beta_j > 0 \mid y)$ under 0.95 or over 0.05.

Using a g-prior method, combined with spike-slab selection, gives a more parsimonious result. An inverse gamma IG(0.5,0.5n) prior on the g parameter is adopted, implying a Cauchy prior on regression coefficients. With a prior retention probability of 0.5, only the ftv*age interaction has a posterior retention probability γ_j over 0.95. Adjusting the g-prior to include a ridge parameter (set to 1/p) as in Baragatti and Pommeret (2012) does not greatly affect the results.

A final analysis considers the skew probit model, but there is no evidence supporting skewed errors, with the parameter φ having posterior mean close to zero. Although there is a predominance of 1 responses in the data, using a skew regression also does not markedly affect the regression coefficients as compared to standard probit regression (cf. Pérez-Sánchez et al., 2018).

7.5.1 Augmentation for Ordinal Responses

Suppose that a categorical response y_i has J categories, with the observations measuring a latent response y^* according to the model

$$y_i = j \quad \text{if } a_{j-1} \le y_i^* < a_j.$$

The α_j are cutpoints dividing the values of y^* according to the observed y values (Bürkner and Vuorre, 2018). The regression in the latent data is then

$$y_i^* = X_i\beta_j + \varepsilon_{ji}$$

TABLE 7.4

Birthweight Data. Probit Regressions

Predictor	Name	Probit			Probit Horseshoe				Probit g-prior		
		β, Mean	β, St devn	$Pr(\beta_j>0)$	β, Mean	β, St devn	κ_j, Mean	$Pr(\beta_j>0)$	β, Mean	β, St devn	γ, Mean
x1	Age	0.26	0.16	0.948	0.14	0.14	0.77	0.843	0.06	0.12	0.38
x2	Lwt	−0.36	0.14	0.003	−0.26	0.14	0.72	0.023	−0.22	0.16	0.78
x3	Black	0.59	0.33	0.960	0.35	0.32	0.66	0.878	0.11	0.23	0.28
x4	Other	0.51	0.28	0.972	0.32	0.27	0.68	0.889	0.11	0.21	0.33
x5	Smoke	0.81	0.27	1.000	0.51	0.27	0.60	0.974	0.31	0.28	0.67
x6	Ptl	0.29	0.20	0.924	0.27	0.20	0.70	0.920	0.23	0.24	0.59
x7	Ht	1.20	0.44	0.997	0.84	0.44	0.51	0.980	0.83	0.55	0.84
x8	Ui	1.11	0.40	0.996	0.62	0.36	0.57	0.965	0.30	0.38	0.48
x9	Ftv	−0.09	0.13	0.260	−0.04	0.10	0.82	0.328	−0.01	0.05	0.20
x10	ui*smoke	−1.00	0.57	0.042	−0.30	0.47	0.67	0.271	−0.03	0.31	0.26
x11	ftv*age	−0.49	0.15	0.000	−0.38	0.14	0.64	0.002	−0.31	0.13	0.96

where ε_{ji} is usually either normally or logistically distributed (Albert and Chib, 2001). So $P(\varepsilon) = \Phi(\varepsilon)$, where Φ is the cumulative normal function, or $P(\varepsilon) = 1/(1+\exp(-\varepsilon))$, the cumulative logistic.

The corresponding model for cumulative probabilities is

$$Pr(y_i^* \le a_j) = Pr(X_i\beta_j + \varepsilon_{ji} \le a_j),$$
$$= Pr(\varepsilon_{ji} \le a_j - X_i\beta_j).$$

Thus

$$Pr(y_i^* \le a_j) = \Phi(a_j - X_i\beta_j),$$

or

$$Pr(y_i^* \le a_j) = 1/(1+\exp(-[a_j - X_i\beta_j])),$$

according to the assumed form for ε_{ji}. Let $\gamma_{ji} = Pr(y_i^* \le a_j)$, then

$$Pr(y_i = j) = Pr(a_{j-1} \le y_i^* < a_j) = \gamma_{ji} - \gamma_{j-1,i}.$$

The probability that $y_i = 1$, namely

$$Pr(y_i = 1) = Pr(a_0 \le y_i^* < a_1) = \gamma_{1i},$$

is obtained by setting $\alpha_0 = -\infty$, while the probability that $y_i = J$,

$$Pr(y_i = J) = Pr(a_{J-1} \le y_i^* < a_J) = 1 - \gamma_{J-1,i}$$

is obtained by setting $\alpha_J = \infty$.

Assuming X_i excludes an intercept, the remaining $J-1$ cut points $\{a_1, a_2 \ldots, a_{J-1}\}$ are unknowns subject to an order constrained prior $a_1 \le a_2 \ldots \le a_{J-1}$. By the reparameterisation

$$a_j = a_{j-1} + \exp(\Delta_j) \quad (J > j > 1),$$

$$a_1 = \Delta_1,$$

one may, however, specify unconstrained normal priors, such as $\Delta_j \sim N(0, V_\Delta)$ where V_Δ is preset or possibly itself unknown.

An equivalent specification of this model involves sets of $J-1$ binary variables for each subject, namely $z_{ji} = 1$ if $y_i \le j$, and $z_{ji} = 0$ otherwise. So if $J = 3$, and if $y_i = 1$, then $z_{1i} = 1$, $z_{2i} = 1$; if $y_i = 2$, then $z_{1i} = 0$, $z_{2i} = 1$. So, for ε normal,

$$Pr(y_i \le j) = Pr(y_i^* \le a_j) = Pr(z_{ji} = 1) = \Phi(a_j - X_i\beta_j).$$

Example 7.7 Delegation of Discretion in Trade Policy

This example involves direct and augmented data options for ordinal data, with ordered probit model analysing changes in discretion in trade policy delegated to the US President by Congress between 1890 and 1990 ($T=99$ observations) (Epstein and O'Halloran, 1996). The response has $J=3$ categories: 3 if discretion is increased between successive years, 2 if it stays the same, and 1 if it is reduced. Changes in discretion are related to $p=4$ predictors: changes in log GNP (x_1), changes in log unemployment rate (x_2), changes in the logged producer price index (x_3), and to a measure of changes in government disunity (x_4), where disunity in a particular year is measured by a trichotomy according as one or both Congress chambers are in the same political party as the President. So x_4 can take values $\{-2,-1,0,1,2\}$.

In a first analysis using rjags, a proportional odds model over responses j is assumed, namely $\beta_{jk} = \beta_k$ $(k=1,\ldots,p)$. Order constrained $N(0,1)$ priors are assumed on the unknown cut points $\{\alpha_1,\alpha_2\}$ (using the rjags sort option), and an MVN prior on $\{\beta_1,\ldots,\beta_4\}$, with mean zero and diagonal precision matrix B_0, with prior variances of 1000.

A two-chain run of 10,000 iterations (with the last 5,000 for inference) gives a non-significant posterior mean (95% CrI) for the impact β_1 of x_1, namely of 1.02 (−5.3,7.1). The 95% interval for the impacts of x_2 is inconclusive, though the posterior density is decidedly concentrated on negative values. The coefficients β_3 and β_4 have respective means (95% CrI) of −3.1 (−6.5,0.5) and −0.42 (−0.81,−0.04). The percentage of years where the actual response is accurately predicted by replicate responses is 62.6%.

Similar coefficients are obtained from an augmented data approach involving binary responses $z_{tj}=1$ if $y_t \leq j$, and $z_{tj}=0$ otherwise. This form of the model is coded in R2OpenBugs, initially with normality assumed. Plots of the resulting two sets of residuals suggests some unusual observations, and a Student t modification of the truncated sampling is adopted, with scale factors $\lambda_{tj} \sim Ga(2,2)$ (corresponding to a fixed 4 degrees of freedom in the Student t). This shows nine observations with posterior medians for λ_{tj} under 0.5, and downweighting their influence improves predictions: the percentage of years accurately predicted by replicate responses is raised to 64.5%.

7.6 Heteroscedasticity and Regression Heterogeneity

For data assumed conditionally normal, the canonical normal linear regression specifies

$$y_i = X_i\beta + \varepsilon_i,$$

where $\varepsilon_i \sim N(0,\sigma^2)$. Potential limitations of this specification reflect both the assumptions regarding errors and the same form of regression effect for all subjects. Assumptions of discrete regression (e.g. Poisson, binomial) are also vitiated by excess observations at particular outcomes (e.g. clumping at zero).

7.6.1 Nonconstant Error Variances

Thus, assuming homoscedastic normal errors in linear regression (or in overdispersed regressions such as the Poisson lognormal) may be restrictive due to the relatively thin tails of the normal, particularly when unusual observations are present. By contrast, scale mixtures of normals accommodate a wide variety of heavy-tailed distributions (Fonseca et al., 2008; Fernandez and Steele, 2000).

In the linear model $y_i = \eta_i + \sigma z_i$, with $z_i \sim N(0,1)$, a scale mixture is generated by assuming the residuals are distributed as

$$\varepsilon_i = z_i / \lambda_i^{0.5}$$

where the λ_i are independent positive random variables. The t_ν distribution results as a scale mixture of normal distributions by taking λ_i gamma with scale and shape $\nu/2$, with the Cauchy when $\nu = 1$ (Boris Choy and Chan, 2008). An alternative scale mixture of uniforms method can lead to both heavier and lighter tails than the normal (Qin et al., 2000).

Other approaches to heteroscedasticity include variance transformation and variance regression modelling (Cepeda and Gamerman, 2000; Chib and Greenberg, 2013). As an alternative to the canonical linear regression, one may consider a heteroscedastic model (Wang and Zhou, 2007)

$$y_i = \eta_i + \sigma_i z_i,$$

with $\eta_i = X_i\gamma$, $z_i \sim N(0,1)$, and σ_i or σ_i^2 taken as a function of η_i, such as

$$\sigma_i = \exp(\eta_i),$$

$$\sigma_i = \sigma|\eta_i|^\lambda,$$

or

$$\sigma_i = \sigma(1 + \lambda\eta_i^2)^{0.5}.$$

7.6.2 Varying Regression Effects via Discrete Mixtures

A potential limitation of the standard normal linear model, and of other generalised linear models, is the assumption of identical regression relationships for all cases. Alternatives include random coefficient models and discrete mixture regressions.

With regard to the former, residual heteroscedasticity may reflect varying predictor effects. Consider a linear model specified as $y_i = x_i\beta + w_i\gamma + \varepsilon_i^*$ when the true model is $y_i = x_i\beta + w_i(\gamma + v_i) + \varepsilon_i$ with $\text{var}(v_i) = \sigma_v^2$ and $\text{var}(\varepsilon_i) = \sigma_\varepsilon^2$. The error ε_i^* will then have non-constant variance $w_i^2\sigma_v^2 + \sigma_\varepsilon^2$. To address such issues, a random regression effects linear model specifies

$$y_i = a + \sum_{j=1}^{p} x_{ji}(\beta_j + v_{ji}) + \varepsilon_i,$$

where $\{v_{1i}, \ldots, v_{pi}\}$ are zero mean random effects. Random regression effects are often applied to structured data (e.g. time or spatially configured data).

Variation in regression effects is also approached using discrete regression mixtures, with form

$$p(y_i | X_i) = \sum_{k=1}^{K} \pi_k f_k(X_i, \beta_k, \phi_k),$$

where β_k are component specific regression effects, and ϕ_k are other parameters defining densities f_k. Such mixtures are useful for detecting subpopulations with different behaviours, while accounting for excess heterogeneity (e.g. overdispersion) related to varying regression relationships. Examples include normal regression mixtures

$$p(y_i \mid X_i) = \sum_{k=1}^{K} \pi_k N\left(\alpha_k + \sum_{j=1}^{p} x_{ji}\beta_{jk}, \sigma_k^2\right),$$

Poisson regression mixtures

$$p(y_i \mid X_i) = \sum_{k=1}^{K} \pi_k Po\left(\exp\left(\alpha_k + \sum_{j=1}^{p} x_{ji}\beta_{jk}\right)\right),$$

and logit regression mixtures for binary or binomial data

$$p(y_i \mid X_i) = \sum_{k=1}^{K} \pi_k \,\text{Bern}\left(\frac{\exp\left(\alpha_k + \sum_{j=1}^{p} x_{ji}\beta_{jk}\right)}{1+\exp\left(\alpha_k + \sum_{j=1}^{p} x_{ji}\beta_{jk}\right)}\right).$$

The probabilities π_k for the components may be predicted for each individual via regression (e.g. multinomial logit).

In Bayesian applications, MCMC sampling is facilitated by the introduction of latent allocation indicators $G_i \in (1,\ldots,K)$, with full conditionals based on multinomial probabilities

$$\pi_k f_k(X_i, \beta_k, \phi_k) \Big/ \sum_{k=1}^{K} \pi_k f_k(X_i, \beta_k, \phi_k).$$

Certain identification and estimation issues apply to discrete regression mixtures, and a variety of sampling and post-processing methods, and priors to gain or improve identifiability, have been proposed. Different component labels cannot be distinguished during MCMC sampling unless some identifiability constraint is imposed. Another issue involves small components (with low probabilities π_k), especially when combined with small samples, since at particular MCMC iterations, no cases may be allocated to a particular group, so that the associated parameters are not updated.

Sampling and estimation methods for discrete regression mixtures differ in whether they impose identifying constraints or allow switching between different numbers of components. For example, Viele and Tong (2002) apply identifying restrictions in linear regression mixtures. Ordering of variances may work better when the variances are well separated, whereas ordering of particular regression parameters works well when subpopulations are distinct in substantive terms. Such features might be established by preliminary classical estimation.

7.6.3 Other Applications of Discrete Mixtures

Discrete mixtures are also relevant for excess observations at particular points. For example, Poisson overdispersion may result from an excess number of zero counts. Under a zero-inflated Poisson (ZIP) model, or more generally zero modified model (ZMP) (Conceição et al., 2013), zero counts may be either true zeroes, or result from a stochastic mechanism, when the process is "active," but sometimes produces zero events. A distinction is similarly made between structural and random zeroes (Martin et al., 2005).

Denote the active stochastic mechanism as $f(y)$, and let $d_i = 1$ for true zeroes as against stochastic zeroes, obtained when $d_i = 0$. Setting $Pr(d_i = 1) = \omega$, one has for discrete $f(y)$,

$$Pr(y_i = 0) = Pr(d_i = 1) + f(y_i = 0 \mid d_i = 0)Pr(d_i = 0)$$

$$Pr(y_i = j) = f(y_i = j \mid d_i = 0)Pr(d_i = 0) \quad j = 1, 2, \ldots$$

Regressors X_i may be relevant both to the binary inflation mechanism, and to the parameters defining the density (Czado et al., 2007). A useful representation for programming the zero-inflated Poisson involves the data augmentation scheme (Ghosh et al., 2006):

$$d_i \sim \text{Bern}(\omega_i),$$

$$y_i \sim Po(\mu_i(1 - d_i)),$$

with analogous representations for other zero-inflated densities.

For a zero-inflated Poisson, with ω_i and μ_i defined by regression, one has

$$P(y_i = 0 \mid X_i) = \omega_i + (1 - \omega_i)e^{-\mu_i},$$

$$P(y_i = j \mid X_i) = (1 - \omega_i)e^{-\mu_i}\mu_i^{y_i} / y_i!, \quad j = 1, 2, \ldots$$

with variance then

$$\text{Var}(y_i \mid \omega_i, \mu_i) = (1 - \omega_i)[\mu_i + \omega_i \mu_i^2] > \mu_i(1 - \omega_i) = E(y \mid \omega_i, \mu_i).$$

So, the modelling of excess zeros implies overdispersion.

Discrete mixture approaches are also used for outlier accommodation and detection (Verdinelli and Wasserman, 1991). For example, in linear regression, one may have

$$p(y_i \mid X_i) = \sum_{k=1}^{2} \pi_k N\left(\alpha + \sum_{j=1}^{p} x_{ji}\beta_j, \sigma_k^2\right)$$

where $\sigma_2^2 \gg \sigma_1^2$, and π_2 is taken small (e.g. $\pi_2 = 0.05$). This provides variance-inflation for outliers. Mohr (2007) advocates a two-group model allowing for both clustered outliers (defined by similar predictor values), and for scattered outliers, generated by a variance inflation mechanism.

Example 7.8 Radioimmunoassay and Esterase

This example compares three heteroscedastic linear models and a varying regression model for n = 113 radioimmunoassay observations (y) in relation to a single predictor, namely esterase (x). Fit is based on sampling replicate data, with fit and penalty criteria derived as in Gelfand and Ghosh (1998).

So, with $z_i \sim N(0,1)$ and

$$y_i = \eta_i = \beta_0 + \beta_1 x_i + \sigma_i z_i,$$

the first model is a variance regression model with

$$\sigma_i^2 = \exp(\gamma_0 + \gamma_1 x_i).$$

This yields a significant γ_1, with mean (95% interval) of 0.068 (0.044,0.094), so indicating heteroscedasticity. The penalty criterion C_P (obtained by summing the posterior variances of replicates) is 1.41*E*+06, while the predictive fit criterion, the posterior mean of $\Sigma_i(y_i - y_{\text{rep},i})^2$, is $C_F = 2.635E{+}06$.

A variance power model in absolute predictor values, namely

$$\sigma_i = \sigma(1+|\eta_i|)^{\lambda}.$$

is then applied (Bonate, 2011). A $U(-2,2)$ prior is taken on λ, and a U(0,250) prior on σ, which includes the observed standard deviation of 213. The final 5000 iterations of a two-chain run of 15,000 iterations give an estimate for $\lambda = 0.61(0.44,0.82)$, and provide improved fit criteria $(C_P, C_F) = (1.21E+06, 2.42E{+}06)$.

A student t with ν degrees of freedom via normal scale mixing (centred on a single variance parameter σ^2) is then applied. A $U(0.01,1)$ prior is applied on the inverse of the degrees of freedom $1/\nu$. This shows 21 datapoints with posterior mean precision adjustment factors κ_i below 0.5, and ν estimated at 2.74. Although such estimates clearly show non-normality, the fit criteria deteriorate to $(C_P, C_F) = (1.66E{+}06, 2.87E{+}06)$.

Finally, a discrete mixture regression is applied, with group varying intercept, slope, and scale, namely

$$y_i = \alpha_{G_i} + \beta_{G_i} x_i + \sigma_{G_i} z_i,$$

$$G_i \sim \text{Mult}(1, \pi)$$

with π of dimension K assigned a Dirichlet prior, $\pi \sim \text{Dir}(1,1,\ldots,1)$. A prior constraint that regression slopes are increasing is applied, using the sort() option in rjags. As well as the predictive fit criterion C_F, choice between different values of K is based on the Bayesian Information Criterion (BIC) (Lee et al., 2016; Utazi et al., 2016).

There are considerable reductions in both measures in moving from $K=4$ to $K=5$, but increases in moving to $K=6$. For $K=5$, the posterior means for BIC and C_F are 621 and 1.10E+06. Hence, heteroscedasticity seems in this dataset to be linked in part to varying regression relationships between sample sub-groups.

Example 7.9 Predictor Selection in Discrete Mixture Regression, Baseball Salary Data

This example considers discrete mixture linear regression, combined with predictor selection within each component of the mixture (e.g. Khalili and Chen, 2007; Lee et al., 2016). The response is salaries for 337 major league baseball players (in units of 100,000 dollars) in the year 1992 with performance measures from the year 1991. Of the 16 available predictors, three are focused on here, and are in standardised form: x_1, batting average; x_2, on-base percentage; and x_3, number of runs. All these predictors are positively correlated with the response (with significant impacts when used as single predictors), but they are also intercorrelated. Using the flexmix package and comparing K=2, 3, and

4 components shows the lowest BIC for K = 3, and also shows a considerable differentiation in the residual standard deviations between the three components.

For a Bayesian analysis in rjags, initially without predictor selection and with K = 3, an identifying constraint based on ordering of the residual variances is adopted, with normal N(0,1000) priors on the regression coefficients β_{jk}. A Dirichlet prior assigns weights of 5 on each component π_k. The final 5,000 iterations of a two-chain run of 15,000 iterations leads to components distinguished firstly by salary level: the respective means on the response within components are 0.4, 1.3, and 10.8 (this is the node avg.sal in the code). The first component shows a significant effect of x_3, with posterior mean (95% CrI) of 0.60 (0.45, 0.74). The second component shows a relatively strong impact for x_1, namely 1.9 (−0.1,3.9), while the third component shows a pronounced impact for x_3, namely 9.0 (7.1, 10.9).

The second analysis uses Laplace priors on the regression coefficients, with Laplace parameters and selection rate parameters component specific

$$\beta_{jk} \mid \gamma_{jk} = 1 \sim N(0, \sigma_k^2 \eta_{jk}^2),$$

$$\beta_{jk} \mid \gamma_{jk} = 0 \sim \delta_0(),$$

$$\eta_{jk}^2 \sim E(\lambda_k^2 / 2),$$

$$\lambda_k \sim U(0.01, 100),$$

$$\gamma_{jk} \sim \text{Bern}(\omega_k).$$

The final 5,000 of a two-chain run of 30,000 iterations shows a high retention rate only for x_3 in the first and third component, with mean (95% CrI) for the realised coefficient $\xi_{jk} = \gamma_{jk}\beta_{jk}$ in the third component of 8.5 (6.7,10.1). The estimated λ_k for the second component is relatively high, reflecting the lack of significant predictor effects.

A final analysis uses Laplace priors again, but without a binary selection mechanism. The impact of x_3 in the third component is unaffected, whereas that in the first component is eliminated. Again, the estimated λ_k for the first two components are relatively high, in line with shrinkage in predictor effects.

Example 7.10 ZIP Regression for Pursuit Behaviours

This example involves zero-inflated regression to investigate impacts of education (x_1) and anxious attachment (x_2) on numbers of unwanted pursuit behaviour (UPB) incidents in couple separation contexts, with $n = 387$ cases (Loeys et al., 2012).

The response y_i is the UPB total, and the first model is a zero-inflated Poisson with a regression for the inflation mechanism. Thus

$$P(y_i = 0 \mid X_i) = \omega_i + (1 - \omega_i)e^{-\mu_i},$$

$$P(y_i = j \mid X_i) = (1 - \omega_i)e^{-\mu_i}\mu_i^{y_i} / y_i!, \quad j = 1,2,\ldots$$

$$\text{logit}(\mu_i) = \beta_0 + X_i\beta.$$

$$\text{logit}(\omega_i) = \gamma_0 + X_i\gamma.$$

Anxious attachment has a significant effect in both regressions, with β_2 and γ_2 having respective mean (95% CrI) coefficients of 0.13 (0.06,0.20) and −0.49 (−0.71,−0.27). The posterior mean marginalised likelihood (Millar, 2009) is −805.5.

As a mixed predictive check (Marshall and Spiegelhalter, 2007), replicate zero inflation indicators $d_i^* \sim \text{Bern}(\omega_i)$ are sampled, and replicate responses sampled from the corresponding shifted mean $y_i^* \sim Po(\mu_i(1-d_i^*))$. There is found to be only 1 case with probabilities of overprediction, $Pr(y_i^* > y_i)+0.5Pr(y_i^* = y_i)$, exceeding 0.95, but 28 cases with probabilities under 0.05, indicating underprediction of some larger counts. Hence, the ZIP model may not be representing the full extent of overdispersion.

A more general representation is obtained by using a zero-inflated negative binomial. This increases the posterior mean marginalised likelihood to –570 and the number of underpredicted cases is reduced to 15.

7.7 Time Series Regression: Correlated Errors and Time-Varying Regression Effects

Time series regression for generalised linear models raises distinct issues, such as serial correlation in regression residuals, and time-varying regression coefficients or dispersion (Jung et al., 2006; Kedem and Fokianos, 2002). There may also be dependence on earlier responses, observed or latent, and time-varying dependence on predictors (Kitagawa and Gersch, 1985; Nicholls and Quinn, 1982).

If autocorrelation in the regression errors is suspected or postulated, as opposed to dependence on past responses or latent data, one option is models with autoregressive and moving average random effects (Cox, 1981; Chiogna and Gaetan, 2002). For linear regression over time points $t=1,\ldots,T$

$$y_t = X_t\beta + \varepsilon_t,$$

an ARMA(p,q) error scheme specifies

$$\varepsilon_t - \rho_1\varepsilon_{t-1} - \rho_2\varepsilon_{t-2}\ldots - \rho_p\varepsilon_{t-p} = u_t - \theta_1 u_{t-1} - \theta_2 u_{t-2}\ldots - \theta_q u_{t-q},$$

where $u_t \sim N(0,\sigma^2)$ are white noise errors.

Widely applied options in practice for ARMA error dependence in time series models are the simple $AR(1)$ and $MA(1)$ schemes. The $AR(1)$ model with

$$\varepsilon_t = \rho\varepsilon_{t-1} + u_t,$$

where $u_t \sim N(0,\sigma^2)$ are iid and independent of ε_t, is an effective scheme for controlling for temporal error dependence if (as often) most correlation from previous errors is transmitted through the impact of ε_{t-1}. This assumption is widely used in longitudinal models (e.g. Chi and Reinsel, 1989). With $\sigma_\varepsilon^2 = \text{var}(\varepsilon_t)$, and assuming stationarity with $|\rho|<1$, $AR(1)$ error dependence implies

$$\text{var}(\varepsilon_t) = \rho^2\,\text{var}(\varepsilon_{t-1}) + \sigma^2 + 2\rho\,\text{cov}(\varepsilon_{t-1},u_t) = \rho^2\sigma_\varepsilon^2 + \sigma^2,$$

so that $\sigma_\varepsilon^2 = \sigma^2/(1-\rho^2)$, and the initial condition for the stationary case is

$$\varepsilon_1 \sim N\left(0,\frac{\sigma^2}{1-\rho^2}\right).$$

$AR(1)$ error dependence for non-metric responses is illustrated by the Poisson count outcomes case $y_t \sim Po(\mu_t)$ (Chan and Ledolter, 1995; Nelson and Leroux, 2006), with

$$\log(\mu_t) = X_t\beta + \varepsilon_t,$$

$$\varepsilon_t = \rho\varepsilon_{t-1} + u_t.$$

Bayesian analysis of $AR(1)$ errors for count data is exemplified by Oh and Lim (2001) and Jung et al. (2006), who also consider augmented data sampling for count responses, while Ibrahim and Chen (2000) set out sampling algorithms under a power prior approach (that assumes historic data with the same form of design are available).

The Durbin–Watson statistic for $AR(1)$ error dependence, namely

$$\text{DW} = \frac{\sum(\varepsilon_t - \varepsilon_{t-1})^2}{\sum \varepsilon_t^2} = 2 - 2\frac{\sum(\varepsilon_t - \varepsilon_{t-1})}{\sum \varepsilon_t^2 \sum \varepsilon_{t-1}^2} = 2 - 2\rho,$$

is often used to test temporal autocorrelation (when predictors exclude lagged responses), and in a Bayesian context can be applied in a posterior predictive check. For example, Spiegelhalter (1998, p.126) considers a Poisson time series for cancer cases y_{ijt} in ages $i = 1,\ldots I$, districts $j = 1,\ldots,J$, and years $t = 1,\ldots,T$, with μ_{ijt} being Poisson means. At each iteration, deviance residuals $d_{ijt} = -2\log\{p(y_{ijt} \mid \mu_{ijt})\}$ are obtained, and an average DW statistic derived for each age and district, namely

$$\text{DW}_{ij} = \frac{\sum_{t=2}^{T}(d_{ijt} - d_{ij,t-1})^2}{\sum_{t=1}^{T}(d_{ijt} - \overline{d}_{ij.})^2}.$$

A summary statistic for autocorrelation is then $\overline{\text{DW}} = \sum_i \sum_j \text{DW}_{ij}/IJ$, which can be obtained for both actual and replicate data.

The latent process driving autocorrelation may also be modelled using discrete mixture formulations. For example, one may define Markov Poisson regression in which for each observed count y_t, there corresponds an unobserved categorical variable $S_t \in (1,\ldots,K)$, representing the state by which y_t is generated (Wang and Puterman, 1999). The latent states are generated according to a stationary Markov chain with transition probabilities

$$Pr(S_t = k \mid S_{t-1} = j) = \pi_{jk} \quad \{j,k = 1,\ldots,K\}.$$

Conditional on $S_{t-1} = j$, the tth observation y_t is Poisson with mean $\mu_t = \exp(X_t\beta_k)$.

7.7.1 Time-Varying Regression Effects

Autocorrelated or heteroscedastic disturbances in time series regression may be caused by assuming predictor effects are constant when in fact they are time-varying. Consider a dynamic normal linear model for metric response

$$y_t = X_t\beta_t + \varepsilon_t,$$

with R predictors. A simple way to allow coefficient variation is simply to take

$$\beta_t = \beta_\mu + u_t$$

with u_t taken as iid random effects. However, in time series contexts, it is likely that deviations from the central coefficient effect β_μ will be correlated with nearby deviations in time.

A flexible framework for time-varying parameter effects is provided by the linear Gaussian state space model (Shumway, 2016), involving first order random walks in scalar or vector coefficients β_t

$$y_t = X_t\beta_t + \varepsilon_t, \quad \varepsilon_t \sim N_T(0, \Sigma_t)$$

$$\beta_t = G_t\beta_{t-1} + \omega_t. \quad \omega_t \sim N_R(0, V_t).$$

Often $G_t = I$, $\Sigma_t = \Sigma$, and $V_t = V$, but if there is stochastic volatility, the variances or log variances can also be brought into a random walk scheme.

Subject matter considerations are likely to govern the anticipated level of smoothness in the regression effects. For example, the $RW(2)$ scheme

$$\beta_t = 2\beta_{t-1} - \beta_{t-2} + \omega_t \quad \omega_t \sim N(0, V)$$

provides a more plausible smoothly changing evolution for changing regression effects (Beck, 1983). Dangl and Halling (2012) consider dynamic linear models for asset returns, and formal model choice between constant regression effects with $V=0$, and differing levels of variation in β_t, via a discrete prior over a set of covariance matrix discount factors.

Varying regression effects are important in particular applications of dynamic generalised linear models for discrete responses (Gamerman, 1998; Fruhwirth-Schnatter and Fruhwirth, 2007; Ferreira and Gamerman, 2000). Consider y from an exponential family density

$$p(y_t \mid \theta_t) \propto \frac{\exp(y_t\theta_t + b(\theta_t))}{\phi_t}$$

$$\mu_t = E(y_t \mid \theta_t) = b'(\theta_t)$$

where the predictors X_t may include past responses $\{y_{t-k}, y^*_{t-k}\}$, both observed and latent (Fahrmeir and Tutz, 2001, p.345). For example, y^*_t, the latent response (e.g. utility in economic applications) when y_t is binary, may depend on previous values of both y^*_t and y_t. The link for μ_t involves random regression parameters

$$g(\mu_t) = X_t\beta_t,$$

where the parameter vector evolves according to a linear Gaussian transition model,

$$\beta_t = G_t\beta_{t-1} + \omega_t,$$

with multivariate normal errors $\omega_t \sim N_R(0, V_t)$ independent of lagged responses, and of the initial condition $\beta_0 \sim N_R(B_0, V_0)$.

Models for binary time series with state-space priors on the coefficients have been mentioned in several studies. Thus, Fahrmeir and Tutz (2001) consider a binary dynamic logit model involving trend and varying effects of a predictor and lagged response,

$$\text{logit}(\pi_t) = \beta_{1t} + \beta_{2t}x_t + \beta_{3t}y_{t-1}$$

$$\beta_t \sim N_3(\beta_{t-1}, V),$$

while Gamerman (1998) consider nonstationary random walk priors in a marketing application with binomial data, where $\text{logit}(\pi_t) = \beta_{1t} + \beta_{2t}x_t$, and x_t is a measure of cumulative advertising expenditure.

Example 7.11 Epileptic Seizures

This example considers correlated error schemes for a count response, specifically data from a clinical trial into the effect of intravenous gamma-globulin on suppression of epileptic seizures (Wang et al., 1996). Daily seizure counts are recorded for a single patient for a period of 140 days, where the first 27 days are a baseline period without treatment, and the remaining 113 days are the treatment period. Predictors are x_{1t} = treatment, x_{2t} = days treated, and an interaction $x_{3t} = x_{1t}x_{2t}$ between days treated and treatment.

A simple Poisson regression is applied initially, using jagsUI, and a predictive p test based on the DW statistic applied. In fact, this does not appear to be significant, having a value of 0.76. However, the 95% credible interval for DW is entirely below 2, indicating positive error autocorrelation. Monte Carlo estimates of CPO statistics also indicate model failures (Figure 7.2), with the LPML standing at −591.

A stationary AR1 error model increases the LPML to −395, with ρ having a posterior mean (95% CrI) of 0.23 (−0.04, 0.50). The dependence structure may also be modelled using a latent Markov chain with K = 2 states (Wang and Puterman, 1999). Conditional on state $S_t = j$, the Poisson mean for the seizure count on day t is represented as

$$\mu_t = \exp(\beta_{0j} + \beta_{1j}x_{1t} + \beta_{2j}x_{12} + \beta_{3j}x_{3t}), \quad j = 1, 2.$$

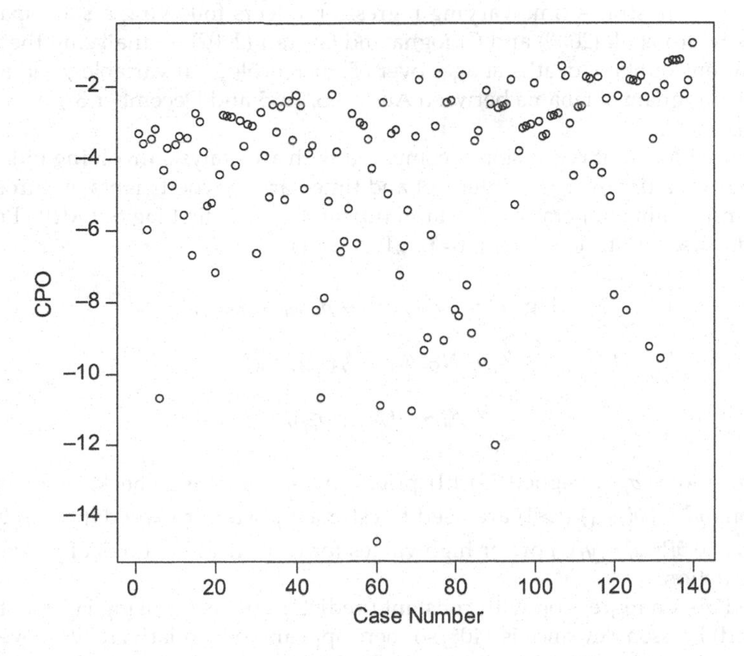

FIGURE 7.2
CPO estimates, seizures data.

TABLE 7.5

Seizures Data. Parameters of Markov Chain Model

	Mean	St Devn	2.5%	97.5%
β_{01}	−6.96	0.50	−7.94	−6.10
β_{11}	7.41	0.55	6.43	8.49
β_{21}	−0.26	0.06	−0.36	−0.12
β_{31}	−2.28	0.15	−2.58	−2.02
β_{02}	−0.23	0.44	−0.99	0.64
β_{12}	1.24	0.53	0.20	2.16
β_{22}	−0.38	0.13	−0.61	−0.11
β_{32}	−0.43	0.16	−0.72	−0.11
π_{11}	0.75	0.05	0.65	0.83
π_{21}	0.62	0.08	0.45	0.78
π_{12}	0.25	0.05	0.17	0.35
π_{22}	0.38	0.08	0.22	0.55

An identifiability (ordered parameter) constraint is applied to the intercepts, though classical estimation makes clear that the two regimes have markedly different treatment effects β_{1j}, and a constraint could be applied to them instead.

This option shows a further improved LPML of −348. State 1 has a much higher positive treatment effect β_{11} (Table 7.5), and a more negative interaction effect. If a subject is in that state on day t, the probability π_{11} of remaining there on the next day is 0.75, with probability $\pi_{12}=0.25$ of moving to state 2. If a subject currently occupies state 2, the respective probabilities are 0.62 and 0.38.

Example 7.12 Mortality and Environment

This example illustrates time-varying regression effects following a state space prior. It follows Smith et al. (2000) and Chiogna and Gaetan (2002) in analysing the relationship between counts of deaths at ages over 65, meteorological variables, and air pollution in Birmingham, Alabama between August 3, 1985 and December 31, 1988 ($T=1247$ observations).

Here a time constant regression is compared with an analysis involving independent RW1 priors on a time-varying intercept and time-varying coefficients on three predictors ($x_1=$minimum temperature, $x_2=$humidity and $x_3=$the first lag of PM10). Predictors are standardised. With $y_t \sim Po(\mu_t)$, t=1,…,T, one has

$$\log(\mu_t) = \beta_{0t} + \beta_{1t}x_{1t} + \beta_{2t}x_{2t} + \beta_{3t}x_{3t},$$

$$\beta_{0t} \sim N(2\beta_{0,t-1} - \beta_{0,t-2}, \sigma^2_{\beta 0}),$$

$$\beta_{jt} \sim N(\beta_{j,t-1}, \sigma^2_{\beta j}),$$

with $1/\sigma^2_a$ and $1/\sigma^2_{\beta j}$ assigned $Ga(1,1)$ priors. As a predictive check, one step ahead predictions $y^*_t \sim Po(\mu_{t-1})$ $(t>1)$ are used to estimate posterior exceedance probabilities $Pr(y^*_t > y_t)+0.5Pr(y^*_t = y_t)$. Low or high values for Q_t indicate failures of fit and/or forward prediction.

For the Poisson regression with constant predictor effects (fit using jagsUI), the average (scaled) Poisson deviance is 1406, so there appears to be relatively little overdispersion in relation to the 1247 observations. Of the three regression coefficients, only β_1 has a 95% posterior interval that excludes zero, namely −0.12 to −0.03. One step ahead predictive checks show 143 of 1246 values of Q_t (11.5%) exceeding 0.95 or under 0.05. The LOO-IC and LPML are respectively 7046 and −3524.

For the second model, with convergence readily obtained using rstan, a slightly improved fit is obtained. (Convergence is delayed if an RW2 prior is adopted in the intercept). Note that standardisation of covariates is important in this example to avoid numeric errors. One step ahead predictive checks now show 117 of 1246 values of Q_t (9.4%) exceeding 0.95 or under 0.05. The LOO-IC and LPML are respectively 7028 and −3514.

Figures 7.3 and 7.4 plot the time-varying coefficients β_{2t} and β_{3t}. Significant effects of PM10 (Figure 7.4) are limited to a central period, similar to the findings of Chiogna and Gaetan (2002).

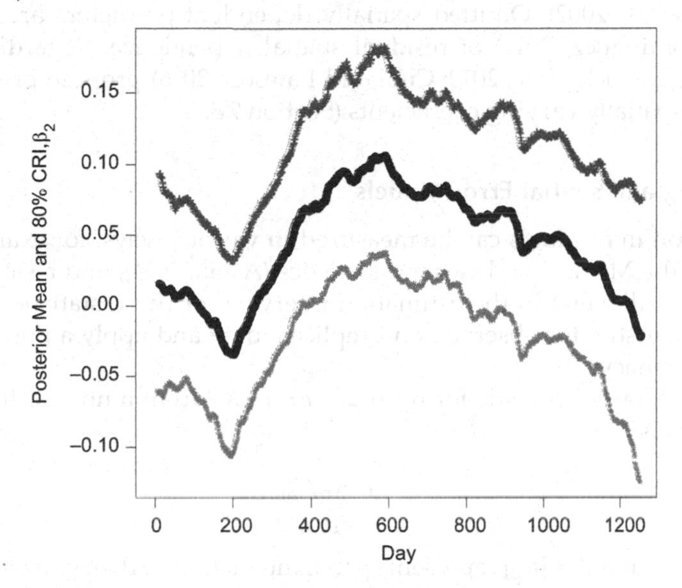

FIGURE 7.3
Varying beta coefficients, beta2.

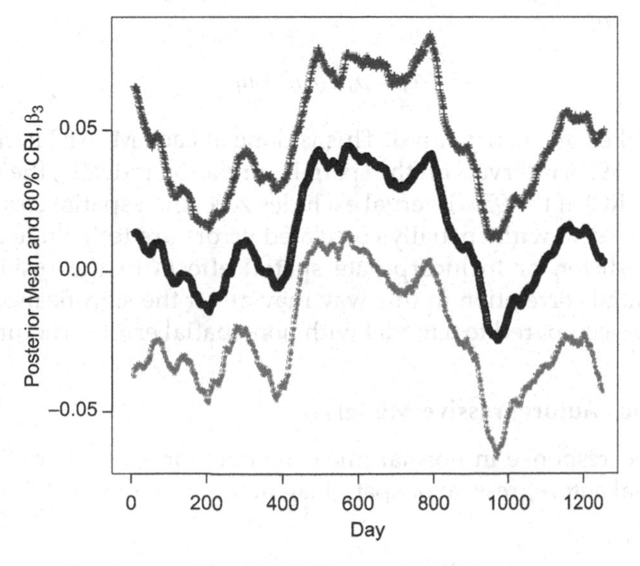

FIGURE 7.4
Varying beta coefficients, beta3.

7.8 Spatial Regression

In spatial data modelling, just as for time series regression, there may be correlated residuals, spatially varying predictor effects, and/or predictor collinearity. Correlated errors can bias regression parameter estimates and cause standard errors to be mis-stated (Boyd et al., 2005). Nonlinear predictor effects may contribute to residual spatial correlation (Dormann et al., 2007), as may incorrectly assuming homogenous regression coefficients when a nonstationary process (a varying regression coefficient approach) is appropriate (Fotheringham et al., 2002). Omitted spatially dependent predictors are another source (LeSage and Dominguez, 2012) of residual spatial dependence. Regarding collinearity, recent studies (e.g. Reich et al., 2010; Choi and Lawson, 2016) propose predictor selection combined with spatially varying coefficients (Section 7.8.2).

7.8.1 Spatial Lag and Spatial Error Models

Spatial correlation in residuals can be measured in various ways, for example, via modified versions of the Moran I and Geary c statistics (Arbia, 2014), and a satisfactory model will include the null value in the estimation intervals for these statistics. One may also calculate these statistics for observed and replicate data and apply a posterior predictive test of model adequacy.

For example, Moran's I statistic for residuals $e=y-X\beta$ from a normal linear regression involving n areas is

$$I = \frac{e'We/S_0}{e'e/n}$$

where $S_0 = \Sigma_i \Sigma_j w_{ij}$, and $W=[w_{ij}]$ represent spatial interactions. Alternatively, Congdon et al. (2007) apply a measure suggested by Fotheringham et al. (2002, p.106) obtained via linear regression of appropriately defined residuals e_i on the spatial lag $e_i^* = \sum_j w_{ij} e_j / \sum_j w_{ij}$. The regression is simply

$$e_i = \rho_0 + \rho_1 e_i^* + u_i,$$

where the u_i are taken as unstructured. This is done at each MCMC iteration to provide a posterior mean and 95% intervals on the spatial correlation index ρ_1, the spatial lag regression coefficient (SLRC). If the 95% interval excludes zero, then spatial correlation is present.

Standard ways to deal with spatially correlated errors are to include a spatially lagged response as a predictor, or to incorporate spatial effects in the residual specification. Correcting for spatial correlation in this way may affect the significance and direction of predictor effects, as compared to a model with non-spatial error structure (Kuhn, 2007).

7.8.2 Simultaneous Autoregressive Models

Including a lagged response in normal linear models for spatially defined observations provides the spatial autoregressive or spatial lag model (Darmofal, 2015) whereby:

$$y_i = \rho \sum_j c_{ij} y_j + X_i \beta + u_i,$$

where $-1 \leq \rho \leq 1$, $u_i \sim N(0, \sigma^2)$ are iid, and $c_{ij} = w_{ij} / \sum_j w_{ij}$ are row-standardised spatial interactions, with $\sum_j c_{ij} = 1$. This model has been proposed for binary responses, possibly by introducing augmented data, with a widely applied approach being known as the spatial probit (Holloway et al., 2002; Franzese and Hays, 2007). The spatial lag model can be estimated using a Bayesian approach in R-INLA, and in rstan (see Chapter 6), while the spatial probit spatial lag model can be estimated using the spatialprobit package (Wilhelm and de Matos, 2013).

So, for y_i binary, z_i is a latent metric variable, positive when $y=1$ and negative when $y=0$. Then the spatial lag model is

$$z_i = \rho \sum_j c_{ij} z_j + X_i \beta + u_i,$$

$$u_i \sim N(0, \sigma_u^2),$$

with $\sigma_u^2 = 1$ for identifiability. In econometric or voting applications, this might amount to expecting individuals located at similar points in space to exhibit similar choice behaviour (Smith and Lesage, 2004). In matrix terms

$$z = \rho C z + X\beta + u,$$

and solving for z gives

$$z = (I - \rho C)^{-1} X\beta + u^*,$$

where $u^* = (I - \rho C)^{-1} u$ are correlated disturbances with $u^* \sim N(0, \Omega)$, where $\Omega = (I - \rho C)^{-1} [(I - \rho C)^{-1}]'$.

An alternative solution to spatially correlated residuals – especially if there is no strong evidence for spatial lag effects – is to include spatial structure in the errors. A rationale is that effects of unknown predictors spill across adjacent areas, causing spatially correlated errors. Instead of normal linear regression with iid errors, the spatial error model specifies

$$y_i = X_i \beta + \varepsilon_i,$$

$$\varepsilon_i = \rho \sum_j c_{ij} \varepsilon_j + u_i,$$

with $u_i \sim N(0, \Sigma_u)$, and a maximum possible value of 1 for ρ, since the spatial weights are standardised. A lower prior limit for ρ of 0 may be assumed, since negative values are implausible.

Writing the equation for $\varepsilon = (\varepsilon_1, \ldots, \varepsilon_n)$ as $\varepsilon = (I - \rho C)^{-1} u$, the covariance matrix for ε is

$$(I - \rho C)^{-1} \Sigma_u (I - \rho C')^{-1},$$

and with $D = I - \rho C$, the joint prior for ε is obtained as

$$(\varepsilon_1, \ldots, \varepsilon_n) \sim N_n(0, D^{-1} \Sigma_u (D')^{-1}).$$

Assuming $\Sigma_u = \sigma^2 I$ the likelihood is

$$L(a,\rho,\sigma^2 \mid y) = \frac{1}{2\pi\sigma^n} |D'D|^{0.5} \exp\left[-\frac{1}{2\sigma^2}\left[(y - X\beta)'D'D(y - X\beta)\right]\right]$$

7.8.3 Conditional Autoregression

By contrast to SAR spatial models, conditional autoregressive error schemes (Besag, 1974) specify ε_i conditional on remaining effects $\varepsilon_{[i]}$. One option takes unstandardised spatial interactions with

$$E(\varepsilon_i \mid \varepsilon_{[i]}) = \lambda \sum_{j \neq i} w_{ij}\varepsilon_j,$$

$$\text{Var}(\varepsilon_i \mid \varepsilon_{[i]}) = \sigma^2,$$

with joint covariance $\sigma^2(I - \lambda W)^{-1}$. In this case (Bell and Broemeling, 2000), λ is constrained by the eigenvalues E_k of W, namely $\lambda \in [1/E_{\min}, 1/E_{\max}]$. Conditional variances may differ between subjects with $M = \text{diag}(\sigma_i^2)$ and the covariance is then $(I - \lambda W)^{-1}M$ (Lichstein et al., 2002).

If predictor effects are written $\eta_i = X_i\beta$, this formulation may be restated in terms of an own area regression effect, and a filtered effect of neighbouring regression residuals. Thus for y metric

$$y_i \sim N\left(\eta_i + \lambda \sum_{j \neq i} w_{ij}(y_j - \eta_j), \sigma^2\right).$$

In many spatial health applications y_i are Poisson counts, with means $\nu_i = E_i\rho_i$ where E_i are expected events, and ρ_i are unknown relative risks. One may then (Bell and Broemeling, 2000; Assunção and Krainski, 2009) assume $r_i = \log(\rho_i)$ are Normal with

$$r_i \sim N\left(\eta_i + \lambda \sum_{j \neq i} w_{ij}(r_j - \eta_j), \sigma^2\right).$$

The other conditional autoregressive option takes standardised spatial interactions, with conditional means and variances

$$E(\varepsilon_i \mid \varepsilon_{[i]}) = \kappa \sum_{j \neq i} c_{ij}\varepsilon_j,$$

$$\text{Var}(\varepsilon_i \mid \varepsilon_{[i]}) = \sigma^2 \Big/ \sum_{j \neq i} w_{ij}.$$

The joint covariance for the ε_i is then $\sigma^2(D - \kappa W)^{-1}$, where D is diagonal with $d_i = \Sigma_{j \neq i} w_{ij}$ (Sun et al., 1999, p.342). Equivalently, for binary w_{ij}, the diagonal terms of the precision matrix are τd_i where $\tau = 1/\sigma^2$ (Kruijer et al., 2007), while off-diagonal terms equal $-\tau\kappa$ when

i and j are neighbours and 0 otherwise. For $\kappa=1$, one obtains the CAR(1) prior of Besag et al. (1991) with joint covariance matrix no longer positive definite.

7.8.4 Spatially Varying Regression Effects: GWR and Bayesian SVC Models

As mentioned above, the assumption of constant parameter values over space may often be unrealistic, and allowing spatial variation in regression parameters may both improve fit and account for spatially correlated residuals (e.g. Leung et al., 2000; Osborne et al., 2007). A widely applied method is geographically weighted linear regression (GWR), which involves re-using the data n times, with the ith regression regarding area i as origin. With R predictors, coefficients $\beta_{1i},\ldots,\beta_{Ri}$ for the ith regression are derived using spatial interaction weights w_{ik}, which in concert with a precision parameter τ_i define area-specific precision parameters $\tau_i w_{ik}$. So for the ith regression (centred on area i), the response for area k is modelled as

$$y_k \sim N\left(\mu_{ik}, \frac{1}{\tau_i w_{ik}}\right) \quad k=1,\ldots,n$$

$$\mu_{ik} = \beta_0 + \beta_{1i}x_{1k} + \ldots + \beta_{Ri}x_{Rk}.$$

The corresponding weighted least squares estimator for $\beta_i = (\beta_{1i},\ldots,\beta_{Ri})$ is

$$\beta_i = (X'W_iX)^{-1}X'W_iy,$$

where W_i is an $n \times n$ diagonal matrix with entries w_{ik} (Assuncao, 2003).

Lesage (2004) notes that GWR estimates may suffer from weak identification as the effective number of observations used to produce estimates for some points in space may be small. This problem can be alleviated under a Bayesian approach by incorporating prior information. Lesage (2004) and Lesage and Kelley Pace (2009) reframe the GWR scheme to allow spatially nonconstant variance scaling parameters v_i, subject to an exchangeable chi-square prior density, $v_i \sim \chi^2(r)$, with r a hyperparameter. Lesage (2004) also redefines the w_{ik} as normalised distance-based weights (with $w_{ii}=0$). Then

$$W_iy = W_iX\beta_i + \varepsilon_i,$$

with smoothing of regression effects across space represented as

$$\beta_i = (w_{i1} \otimes I_R,\ldots,w_{in} \otimes I_R)\begin{pmatrix}\beta_1\\ \cdots\\ \cdots\\ \cdots\\ \beta_n\end{pmatrix} + u_i.$$

With $V_i = \text{diag}(v_1,\ldots,v_n)$, the error terms have priors

$$\varepsilon_i \sim N(0,\sigma^2 V_i),$$

$$u_i \sim N(0,\sigma^2\delta^2(X'W_i^2X)^{-1}),$$

with the specification on u_i being a form of Zellner g-prior, in which δ^2 governs adherence to the smoothing specification.

7.8.5 Bayesian Spatially Varying Coefficients

An alternative to the GWR approach is provided by spatially varying coefficient (SVC) models (Gelfand et al., 2003; Assuncao, 2002; Gamerman et al., 2003; Wheeler and Calder, 2006, 2007). For a continuous space perspective, let $Y(s)$ be the $n \times 1$ response for locations $s = (s_1, \ldots, s_n)$, and β be a $nR \times 1$ stacked vector of spatially varying regression coefficients. Then the normal SVC model is (Gelfand et al., 2003),

$$Y(s) \sim N(X(s)'\beta(s), \sigma^2 I)$$

where $X(s)$ is a $n \times nR$ block diagonal matrix of predictors.

The prior for β is

$$\beta(s) \sim N(1_{n\times 1} \otimes \mu_\beta, V_\beta)$$

where $\mu_\beta = (\mu_{\beta_1}, \ldots, \mu_{\beta_R})'$ contains mean regression effects, and V_β is the $nR \times nR$ covariance matrix defined as

$$V_\beta = C(\eta) \otimes \Lambda$$

where Λ is a $R \times R$ matrix containing covariances between regression coefficients at any particular location, and $C(\varsigma) = [c(s_i - s_j; \varsigma)]$ is a $n \times n$ correlation matrix representing spatial interaction between locations or areas, with denoting hyperparameters. For example, under exponential spatial interaction (Wheeler and Calder, 2007)

$$c(s_i - s_j; \varsigma) = \exp(-d_{ij}/\varsigma)$$

where ς is a positive parameter.

For discrete areas, distance-based kernel schemes for spatial interaction have a less substantive basis. For $i = 1, \ldots, n$ such areas, let $\beta_i = (\beta_{1i}, \ldots, \beta_{Ri})$ denote spatially varying regression effects in the linear predictor

$$\eta_i = \sum_{r=1}^{R} \beta_{ri} x_{ri},$$

of a general linear model with mean $\mu_i = E(y_i)$, and link $g(\mu_i) = \eta_i$. With $\beta = (\beta_1, \ldots, \beta_n)$, one possible spatially structured scheme is a pairwise difference prior (Assuncao, 2003)

$$p(\beta \mid \Phi) \propto |\Phi|^{n/2} \exp\left\{-0.5 \sum_i \sum_j w_{ij} (\beta_i - \beta_j)' \Phi (\beta_i - \beta_j)\right\},$$

with $R \times R$ precision matrix Φ, and spatial interactions w_{ij} usually binary ($w_{ij} = 1$ when areas i and j are adjacent, zero otherwise). When y_i is metric with $\mu_i = \eta_i$, with residual precision $\tau = 1/\sigma^2$, one may, following Gamerman et al. (2003), scale the covariance by τ, namely

$$p(\beta \mid \Phi, \tau) \propto \tau^{nR/2} |\Phi|^{n/2} \exp\left\{-0.5\, \tau \sum_i \sum_j w_{ij} (\beta_i - \beta_j)' \Phi (\beta_i - \beta_j)\right\}.$$

The covariance matrix for β in these specifications is respectively $K^{-1} \otimes \Phi^{-1}$ and $\sigma^2 K^{-1} \otimes \Phi^{-1}$, where K has elements

$$k_{ii} = w_{i+} = \sum_{j \neq i} w_{ij},$$

$$k_{ij} = -w_{ij} \quad i \neq j.$$

Hence these priors are improper because the elements in each row of K add to zero. Assuncao (2003, p.460) notes that propriety can be obtained by a constraint such as $\sum_i \beta_i = A$, where A is any preset R-vector. This consideration leads to a practical strategy representing β_i as $\beta_i = \mu_\beta + b_i$ where the b_i follow the pairwise difference prior, but are zero centred at each MCMC iteration, and the mean regression effect is $\mu_\beta = (\mu_{\beta_1}, \ldots, \mu_{\beta_R})$. This can be implemented using the car.normal or mvcar options in BUGS.

7.8.6 Bayesian Spatial Predictor Selection Models

For spatially configured datasets, the relevance of particular predictors, and whether their impact should vary spatially, may itself vary between areas or locations. Allowing for spatial heterogeneity may affect inferences regarding the importance of predictors. However, appropriate models are generally highly parameterised and careful specification of priors may be needed to achieve effective sampling, with potential problematic identifiability, sensitivity to priors, and mixing issues. As discussed above, either selection indicators or shrinkage priors may be invoked. Assuming a selector indicator approach, let $\beta_{ij}^{(r)}$ denote realised regression coefficients for area/location i and predictor j, as determined by priors on both coefficients and selection indicators.

Under a scheme proposed by Reich et al. (2010), assessing predictor relevance has two stages: (a) whether a predictor is relevant or irrelevant with a homogenous spatial effect (constant over locations), and conditional on a homogenous effect being relevant, then (b) assessing whether a spatially varying effect is justified. Let γ_j denote a binary selection indicator for stage (a) and δ_j denote an indicator for stage (b). Then the three possible relevant indicator pairings are $\{\gamma_j = 0, \delta_j = 0\}$, $\{\gamma_j = 1, \delta_j = 0\}$, and$\{\gamma_j = 1, \delta_j = 1\}$. So, one can define indicators $\gamma_{aj} = \gamma_j$ and $\gamma_{bj} = \gamma_j \delta_j$ as relevant to each stage, with two sets of suitably corresponding priors on regression coefficients. For example, for stage (a) a spike-slab prior could be set on coefficients μ_{β_j}, and for stage (b) an SVC prior on zero centred spatial effects b_{ij}, with realised coefficients $\beta_{ij}^{(r)} = \mu_{\beta_j} + b_{ij}$ when $\gamma_{bj} = 1$, and $\beta_{ij}^{(r)} = \mu_{\beta_j}$ when $\gamma_{bj} = 0$.

Under a scheme proposed by Choi and Lawson (2015), spatial structure is applied to the selection indicators, which are area and predictor specific. Thus, one option is to take a hierarchical prior on the regression coefficients, through the inclusion mechanism specifying spatial dependence. Thus

$$\beta_{ij}^{(r)} = \beta_{ij} \gamma_{ij},$$

$$\beta_{ij} \sim N(\mu_{\beta j}, 1/\tau_{\beta j})$$

$$\gamma_{ij} \sim \text{Bern}(\rho_{ij}),$$

$$\text{logit}(\rho_{ij}) = \omega_j + r_{ij},$$

where the r_{ij} are entirely spatially structured, as under a CAR(1) prior, or admit spatial structure, as under the Leroux et al. (1999) scheme.

Example 7.13 Very Low Birthweight, New York Counties

This example considers numbers of very low birthweight babies (under 1500 g) in 62 New York counties over 2008–12. Expected VLBW births, E_i, are obtained as total births in a county multiplied by the region-wide VLBW rate; so $\sum_i y_i = \sum_i E_i$. Predictors are x_1 = median household income; and x_2 = income inequality (ratio of household income at the 80th percentile to income at the 20th percentile). These predictors are standardised. A negative effect of x_1 and positive effect of x_2 would be expected.

A Poisson regression is fitted first, and shows significant impacts (negative and positive respectively) for the predictors. β_1 and β_2 have respective means (95% CrI) of −0.077 (−0.09,−0.065) and 0.037 (0.03,0.045). This model has a DIC of 589. The spatial lag regression coefficient (SLRC) discussed above has a 95% credible interval (−0.06,0.48) only just overlapping zero, indicating spatially correlated residuals.

One possible remedy is to include a spatially structured residual in the regression. Accordingly, a second model includes an additive CAR(1) spatial error s_i, and for identifiability omits the intercept, which is estimated as the mean of the s_i. The prior on the residual is specified from first principles, rather than using the car.normal function in BUGS. Thus with $y_i \sim Po(E_i\rho_i)$,

$$\log(\rho_i) = \beta_1 x_{1i} + \beta_2 x_{2i} + s_i,$$

with a Ga(0.5,0.0005) prior on the precision of s_i. This gives a considerably improved DIC of 464, with significant effects remaining for both predictors, including an enhanced effect of x_2. Thus, β_1 and β_2 have respective means (95%CrI) of −0.08 (−0.10,−0.02) and 0.06 (0.01,0.12). Such a change in the strength of predictor effects demonstrates the importance of correct error specification for inferences regarding risk factors in spatial data. Spatial correlation in residuals is removed as judged by an SLRC with 95% interval (−1.03,0.74).

An alternative possible solution to spatially correlated residuals is to consider spatial nonstationarity in predictor effects. Here

$$\log(\rho_i) = \beta_0 + \beta_{1i} x_{1i} + \beta_{2i} x_{2i},$$

where the β_{ki} follow independent CAR(1) priors. This model gives a DIC of 486, while posterior means (95% intervals) for μ_{β_1} and μ_{β_2} are obtained as −0.03 (−0.09,0.04) and 0.17 (0.09,0.25).

Adding a spatial residual to this model leads to

$$\log(\rho_i) = \beta_{1i} x_{1i} + \beta_{2i} x_{2i} + s_i,$$

where priors on spatial effects are specified from first principles. This improves the DIC to 472. This is a slight loss of fit compared to the spatial residual model, but acknowledging regression heterogeneity in regression effects may often be important on substantive grounds, and may impact on average regression effects over all areas. Posterior means (95% intervals) for μ_{β_1} and μ_{β_2} are obtained as −0.05 (−0.13,0.03) and 0.13 (0.05,0.22), so that recognising heterogeneity has much enhanced the inequality effect, as compared to the spatial residual model.

Wheeler and Tiefelsdorf (2005, p.169) mention implausibly signed effects when using a classical GWR approach. The Bayesian SVC approach reveals four counties with posterior probabilities $Pr(\beta_{1i} > 0 \mid y)$ over 0.25 (the maximum being 0.27), and no county with a posterior probability $Pr(\beta_{2i} < 0 \mid y)$ under 0.75 (the minimum being 0.76).

Example 7.14 EU Referendum Voting

This example considers voting data for 380 local authorities in England, Wales, and Scotland, namely the proportion voting for Britain's exit ("Brexit") from the European Union in the 2016 Referendum. The level of Brexit voting has been linked, inter alia, to age structure in different areas, proportions of adults with higher qualifications, proportions of residents born outside the UK, and urban status. Here the proportion of over 65s is used to measure age structure, and population density (1,000 persons per hectare) is used as a measure of urbanity.

An indication of collinearity distorting regression parameters is provided by spatial lag probit regression (via the spatialprobit package) applied to the outcome variable, $y=1$ for Brexit votes above the median proportion of 0.543, and $y=0$ otherwise. The spatial interaction matrix for this analysis is based on the five nearest neighbours to each of the 380 areas, with a sparse matrix representation adopted to assist computation.

A univariate regression on the proportion non-UK-born shows a significant negative effect on Brexit voting, but this coefficient changes direction (albeit is again significant) when all four predictors are included in the regression. By far the strongest predictor, with a negative effect on Brexit voting, is the area proportion with higher education with posterior mean (sd) of −20.8 (2.28). The population density covariate is not significant.

A spatial lag model is also fitted in R-INLA with response defined by the logit of the proportion voting Brexit. The predictors are higher education, over 65s, non-UK-born, and population density. This analysis provides a posterior mean (sd) on the higher education and age variables of −4.16 (0.57) and 2.18 (1.05). The other predictors have non-significant effects.

A conditional autoregressive approach is also applied using R2OpenBUGS, without a spatial lag on Brexit voting. This analysis is based on spatial adjacency of the local authorities, with binomial response based on Brexit voters y_i and total voters V_i. To ensure all areas have spatial adjacencies, links are introduced for island areas.

An initial analysis retains all predictors (which are standardised), and assumes a Leroux et al. (1999) prior (denoted LLB prior) for random effects in the logit regression. From a two-chain run of 10,000 iterations, the LLB spatial dependence parameter has mean (sd) of 0.92 (0.05) showing high spatial association between regression residuals. The regression coefficients show a negative effect of the higher education and population density predictors on Brexit voting, and a positive effect of over 65s, but with the impact of UK born predictor inconclusive.

Conventional predictor selection using an SSVS prior (George and McCullogh, 1993) is then applied. The prior precision on the regression coefficients is set at 1 under retention, and 1,000 under exclusion. From a two-chain run of 100,000 iterations, this model shows posterior retention probabilities of 0.95 for higher education, 1 for the over 65s predictor, 0.93 for population density, but below 0.5 (namely 0.10) for the non-UK-born predictor. Fit is essentially unchanged under the DIC criterion, namely 5,177 as compared to 5,173 in the analysis without selection.

A final analysis adopts the approach of Choi and Lawson (2015) with spatial selection indicators γ_{ij} (for area i, and predictor j) following CAR(1) priors. Thus

$$y_i \sim \text{Bin}(V_i, \pi_i),$$

$$\text{logit}(\pi_i) = \beta_0 + \sum_j X_{ij}\beta_{ij}^{(r)} + s_i,$$

$$\beta_{ij}^{(r)} = \beta_{ij}\gamma_{ij},$$

$$\beta_{ij} \sim N(\mu_j, 1/\tau_j)$$

$$y_{ij} \sim \text{Bern}(\rho_{ij}),$$

$$\text{logit}(\rho_{ij}) = \omega_j + r_{ij},$$

where s_i are Leroux et al. (1999) effects, whereas the r_{ij} are CAR(1). A two-chain run of 100,000 iterations does not attain convergence according to Brooks-Gelman-Rubin (BGR) statistics. Inferences at this stage show the highest retention probabilities (found by averaging γ_{ij} over areas) for the higher education and population density variables. These predictors both have negative effects, with posterior means (sd) of −0.33 (0.02) and −0.10 (0.04), as assessed from posterior mean $\beta_{ij}^{(r)}$ averaged over areas. The age 65+ and non-UK-born predictors have retention probabilities below 0.50.

7.9 Adjusting for Selection Bias and Estimating Causal Effects

The aim of much of social and health science research is to understand the effect of a treatment, intervention, or exposure on an outcome when only observational non-randomised data are available. Conventional regression may misrepresent treatment effects when there is selection bias (e.g. treated subjects differ from untreated subjects in terms of baseline health status or income). While many applications focus on the effects of a treatment or instructional program, the conception of treatment extends to demographic variables (Davis et al., 2017), and includes events such as dropping out of school (e.g. Vaughn et al., 2011), where the outcome of interest may be adult earnings or verbal ability, and the goal is to assess impacts of dropout after control for confounders. Since there may be intervening variables in the causal pathway, there is also often interest in decomposing the effect of a treatment or exposure into direct and mediated effects.

7.9.1 Propensity Score Adjustment

Propensity score (PS) methods are used in an attempt to reduce the potential bias in estimated effects (e.g. of a treatment or intervention) obtained from observational studies, when a treatment is not randomly assigned to subjects. They allow adjustment for multiple confounders without needing to specify a model for their association with the outcome. Estimated effects may be of a treatment or focus risk factor X (often binary), while C denotes remaining covariates, often considered as confounders. The propensity score is the estimated probability of treatment assignment conditional on confounders. The propensity score acts as a balancing score, such that conditional on the score, the distribution of confounders will be similar between treated and untreated subjects.

Let $Y(x)$ denote a possibly counterfactual outcome when the treatment or exposure has value $X = x$. So for a binary treatment, the outcome values are $Y(1)$ and $Y(0)$ of which only one is observed. Let $A \perp B \mid C$ denote that A is independent of B given C. Then a propensity score is sufficient to adjust for confounding provided

$$Y(0), Y(1) \perp X \mid C,$$

namely, treatment assignment is ignorable, given the confounders (Vansteelandt and Daniel, 2014).

Suppose a logit regression is used to predict $Pr(X_i = 1 | C_i, \gamma)$, so that the propensity score is $S_i = 1/[1 + \exp(-\gamma C_i)]$. It is potentially important to exclude insignificant confounder variables (Weitzen et al., 2004), so one may include Bayesian variable selection in the estimation of the propensity score, for example, using binary retention indicators δ_k,

$$\text{logit}(S_i) = \gamma_0 + \sum_k \delta_k \gamma_k C_{ik}. \tag{7.3}$$

Subsequent analysis options are then to apply the score in a subsequent regression to predict Y, stratify the sample according to propensity score (e.g. into quintiles or deciles), match on the propensity score, or use inverse weighting by the propensity score. Suppose the subsequent analysis involves regression. Regression to assess effects of exposure or treatment can then be (a) on X and S; (b) on X and groupings of S (e.g. a categorical variable based on a decile grouping of the S scores; or (c) on X, S and C.

A Bayesian approach may be specified using a joint likelihood (Zigler et al., 2013; McCandless et al., 2009). Suppose Y is binary with $p_i = Pr(Y_i = 1 | X_i, C_i)$, then the joint likelihood consists of (7.3) and an outcome model such as (Zigler and Dominici, 2014, section 2.3)

$$\text{logit}(p_i) = \beta_0 + \beta_1 X_i + \beta_2 S_i + \sum_k \delta_k \theta_k C_{ik} \tag{7.4}$$

where the selection indicators δ_k are common to both regressions. An average treatment or exposure effect

$$\Delta = E(Y = 1 | X = 1, C) - E(Y = 1 | X = 0, C)$$

may be calculated by comparing estimated responses for each subject at $X=1$ and $X=0$ in (7.4) (Davis et al., 2017).

The estimation of the propensity score S via a joint likelihood contrasts with a separate stage perspective whereby the propensity score is intended to approximate the design stage of a randomised study, without access to the outcome. In accord with a two-stage perspective, one may instead apply a quasi-Bayesian approach whereby feedback is cut between the PS model and the outcome model (McCandless et al., 2010; Zigler and Dominici, 2014, section 3.2).

For hierarchical data (e.g. subjects nested within institutions, or within areas) contributions of covariates to treatment assignment may vary across institutions. In terms of multilevel coefficients, this implies that a random slope analysis is needed to represent institutionally varying or area-varying effects of covariates. Such a cross-level interaction effect on the probability of receiving treatment means that each institution then has a different propensity equation. If C_{ij} denotes individual confounders and W_j denotes institutional confounders, then the ignorability assumption is now stated as

$$(Y(0), Y(1) \perp X | C, W)$$

The aim of the propensity score method is to ensure that within groups homogeneous on the propensity score, the distributions of the covariates are essentially the same for treated and untreated subjects (Austin, 2009). The achievement of covariate balance may be tested (Baser, 2006) e.g. by testing for significant differences in covariate distributions within propensity score strata.

Example 7.15 Patients Hospitalised for Suspected Myocardial Infarction

This example uses the data at http://web.hku.hk/~bcowling/examples/propensity.htm. The data consist of 400 subjects in a retrospective cohort study of men aged 40–70 admitted to hospital with suspected myocardial infarction, with a binary outcome Y for 30-day mortality. To be assessed is the impact of a newer clot-busting drug ($X=1$) versus a standard therapy ($X=0$) on the risk of mortality. Confounders are, respectively, age, an admission severity score (on a scale from 0 to 10, 10 being worst), and a risk factor score (on a scale from 0 to 5, 5 being worst).

Two models are compared using jagsUI. The first involves a logit propensity score model predicting S_i from the three confounders, namely

$$\text{logit}(S_i) = \gamma_0 + \sum_{k=1}^{3} \gamma_k C_{ik}.$$

and with an outcome model assuming no residual confounding, namely

$$\text{logit}(p_i) = \beta_0 + \beta_1 X_i + \beta_2 S_i.$$

This model has a satisfactory performance in reproducing the data based on posterior predictive tests using the Brier score.

The three confounders have significant positive effects in the propensity score regression, so that patients receiving the new drug have a distinctly adverse risk profile. Within quintiles of the propensity score, differences between confounder profiles are not significant (these are represented by match.C1, match.C2, and match.C3 in the code). In the outcome model, β_1 and β_2 have respective posterior means (sd) of −0.47 (0.28) and 2.83 (1.15). An estimate of the causal effect (of the new drug in reducing mortality) is based on evaluating outcomes $p_{1i} = \text{logit}^{-1}(\beta_0 + \beta_1 X_i + \beta_2 S_i)$ and $p_{0i} = \text{logit}^{-1}(\beta_0 + \beta_2 S_i)$ for all subjects. Then the causal effect (mort.X in the code) is estimated as the average of the differences $p_{1i} - p_{0i}$, which has a mean (95% CRI) of −0.065 (−0.142, 0.006). LOO-IC criteria are obtained separately for the propensity score and outcome models as 529 and 369.

A second model extends the propensity score model to include quadratic and interaction terms ($C_1^2, C_2^2, C_3^2, C_1C_2, C_1C_3, C_2C_3$), and also allows for confounder selection feedback, with a residual confounding effect in the outcome model. Additionally, rather than selection via SSVS or other spike-slab priors (Zigler and Dominici, 2014), horseshoe shrinkage priors are used, with a sharing of the shrinkage parameters between propensity and outcome models. Thus

$$\text{logit}(S_i) = \gamma_0 + \sum_{k=1}^{9} \gamma_k C_{ik},$$

$$\gamma_k \sim N(0, \tau_\gamma^2 \rho_k),$$

$$\kappa_k \sim Be(0.5, 0.5),$$

$$\rho_k = 1/\kappa_k - 1,$$

$$\text{logit}(p_i) = \beta_0 + \beta_1 X_i + \beta_2 S_i + \sum_{k=1}^{9} \eta_k C_{ik},$$

$$\eta_k \sim N(0, \tau_\eta^2 \rho_k).$$

LOO-IC criteria for the propensity score and outcome models are, in fact, now slightly higher at 533 and 372. Values of κ_k are above 0.5 (indicating redundant regressors) except for the main linear terms in C_1, C_2 and C_3. The average causal effect $\overline{p_{1i} - p_{0i}}$ is little changed, with mean (95% CRI) of −0.063 (−0.137, 0.009).

7.9.2 Establishing Causal Effects: Mediation and Marginal Models

One may often have preliminary knowledge about different predictors from a substantive perspective, as in epidemiology, whereby effects of a specific exposure on a disease are confounded by other predictors (McNamee, 2005). Among questions that occur are establishing the causal effect of an exposure after controlling for confounders, and establishing the direct and indirect effects of an exposure or treatment when there are mediators in the relationship between the exposure and outcome.

Consider the issue of estimating the direct effect of an exposure or treatment (X) on an outcome (Y) in the presence of mediator (M) and confounders (C). A standard strategy would be (i) to estimate a regression for Y conditional on X and C, and then (ii) add M into the model. The extent to which the coefficient of X changes between these two models is then interpreted as measuring how far the effect of X on Y is mediated by M. The coefficient on X in stage (i) is taken to represent the total effect of X on Y, while that from (ii) is taken to represent the direct effect of X on Y not mediated by M. This strategy assumes no interaction between X and M in their impact on Y. Additionally, estimation of direct effects by such a strategy is complicated if confounders are affected by the exposure, or certain confounders affect both the mediator and outcome. However, with suitable regression modelling of both outcome and mediator, even in the presence of the above complications, one may decompose the total causal effect (the average effect on the outcome of the exposure or treatment) into an indirect effect mediated through the mediator, and the remaining direct effect. These are represented specifically in the causal mediation literature (Pearl, 2014; Tchetgen and Vanderweele, 2014; Greenland, 2000) via two quantities: the natural indirect effect and the natural direct effect.

Let $E[Y\{X, M(X)\}]$ denote the expected value of Y at a stipulated value X, and at the corresponding value of the mediator $M(X)$. This expected value will be counterfactual if the observed X differs from the stipulated X. Causal effects may be assessed by considering counterfactual settings of exposures, which may correspond to a minimal risk value, such as $X^* = 0$ for a binary exposure. For a continuous exposure such as body mass index (BMI), the counterfactual level might be a minimum risk level such as $X^* = 21$.

7.9.3 Causal Path Sequences

Assume the model for the mediator M specifies dependence on the treatment X and confounders C, and the model for the outcome Y allows dependence on X, M, on interactions between X and M (denoted $X.M$), and on confounders. Symbolically

$$M \sim f_1(X, C)$$

$$Y \sim f_2(X, M, X.M, C).$$

Let $M(X^*)$ denote the prediction of the mediator M under the setting $X = X^*$. For example, suppose M is continuous, and a normal linear regression is specified with

$$M_i \sim N(\eta_i, \sigma_M^2),$$

where

$$\eta_i = \beta_1 + \beta_2 X_i + \beta_3 C_i.$$

Then expected values of $M(X)$ and $M(X^*)$ can be obtained as equal to the corresponding regression terms, namely as η_i and

$$\eta_i^* = \beta_1 + \beta_2 X_i^* + \beta_3 C_i,$$

respectively. Under a Bayesian perspective, they can also potentially be obtained as the respective predictions

$$M_{\text{new},i} \sim N(\eta_i, \sigma_M^2),$$

$$M_{\text{new},i}^* \sim N(\eta_i^*, \sigma_M^2).$$

The total effect is

$$E\left[Y\{X, M(X)\}\right] - E\left[Y\{X^*, M(X^*)\}\right],$$

while the natural direct effect (Lange et al., 2012) compares $E[Y\{X, M(X^*)\}]$ and $E[Y\{X^*, M(X^*)\}]$, namely

$$E\left[Y\{X, M(X^*)\}\right] - E\left[Y\{X^*, M(X^*)\}\right].$$

The natural indirect, or mediated, effect is the difference between the total effect and the natural direct effect, namely

$$E\left[Y\{X, M(X)\}\right] - E\left[Y\{X, M(X^*)\}\right].$$

An additional effect sometimes of interest (Naimi et al., 2014b; Vanderweele, 2013), namely the controlled direct effect CDE, the effect of exposure on outcome if the mediator is controlled uniformly at a particular value of M, say $M.c$. Then

$$\text{CDE} = E\left[Y\{X, M.c\}\right] - E\left[Y\{X^*, M.c\}\right].$$

Consider a structural model, as in Figure 7.5, with dependencies specified via normal linear regressions, with regression terms

$$E[M \mid X, C] = \beta_1 + \beta_2 X + \beta_3 C,$$

$$E[Y \mid X, M, C] = \theta_1 + \theta_2 X + \theta_3 M + \theta_4 X.M + \theta_5 C.$$

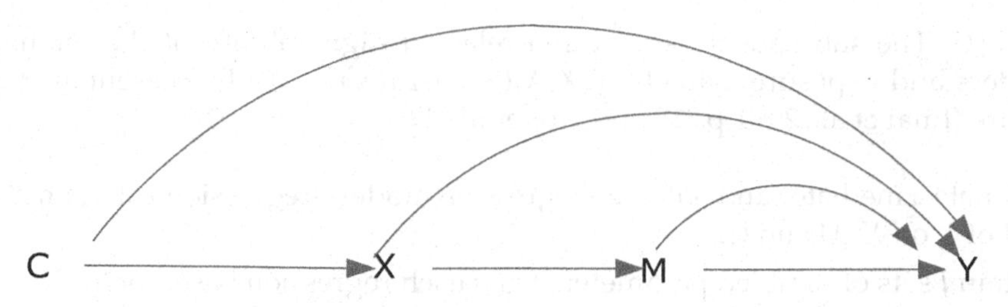

FIGURE 7.5
Causal path example.

Then the Natural Direct Effect (NDE) and the Natural Indirect Effect (NIE) can be obtained by effect substitution (substitution of regression means). So defining expected $M(X)$ and $M(X^*)$ from the corresponding regressions,

$$E[M(X)] = \beta_1 + \beta_2 X + \beta_3 C,$$

$$E[M(X^*)] = \beta_1 + \beta_2 X^* + \beta_3 C,$$

one has

$$E[Y\{X, M(X)\}] = \theta_1 + \theta_2 X + \theta_3 M(X) + \theta_4 X.M(X) + \theta_5 C,$$

$$E[Y\{X, M(X^*)\}] = \theta_1 + \theta_2 X + \theta_3 M(X^*) + \theta_4 X.M(X^*) + \theta_5 C,$$

$$E[Y\{X^*, M(X^*)\}] = \theta_1 + \theta_2 X^* + \theta_3 M(X^*) + \theta_4 X^*.M(X^*) + \theta_5 C,$$

The natural direct effect is then obtained as

$$\begin{aligned}\text{NDE} &= E[Y\{X, M(X^*)\}] - E[Y\{X^*, M(X^*)\}] \\ &= (X - X^*)(\theta_2 + \theta_4\beta_1 + \theta_4\beta_2 X^* + \theta_4\beta_3 C).\end{aligned}$$

Also, the mediated, or natural indirect, effect is

$$\text{NIE} = (X - X^*)(\theta_3\beta_2 + \theta_4\beta_2 X).$$

Underlying the effect decomposition in the above model are assumptions of conditional ignorability

$$Y\{X, M(X)\} \perp X \mid C$$

$$Y\{X, M(X)\} \perp M(X) \mid X, C.$$

These specify independence of exposure and outcome, given the confounders (Vanderweele, 2015), and of mediator and outcome, given confounders and exposure. Additional assumptions (VanderWeele and Vansteelandt, 2014) are $M(X) \perp X \mid C$ and $Y\{X, M(X)\} \perp M(X^*) \mid C$.

An alternative method for estimating causal effects is set out by Imai et al. (2010, p.312) based on assumptions of sequential ignorability. The initial assumptions relate to ignorability of the treatment (or exposure) given confounders, namely $Y\{X, M(X)\} \perp X \mid C$ and

$M(X) \perp X \mid C$. The subsequent assumption relates to ignorability of the mediator given confounders and exposure, namely $Y\{X, M(X)\} \perp M(X) \mid C, X$. The essential steps in the method are (Imai et al., 2010, p.317 and Appendix D)

i) Estimate a mediator and outcome regression models (regressions of M on X and C, and of Y on X, M and C);
ii) Obtain J sets of sampled parameters from each regression based on the estimated sampling distributions;
iii) For each of $j = 1, \ldots, J$ samples from stage ii, carry out K further simulations, namely (a) of mediator values $M \mid X,C$ for both $X=1$ and $X=0$, and (b) of two potential outcomes given the simulated mediator values, and then (c) compute the causal mediation effect for sample j by averaging over the K simulations.
iv) Compute summary statistics by averaging over all samples j.

Assuming a binary treatment, the two causal mediation effects are defined for treatment settings x = 0,1 as

$$\delta_i(x) = Y_i(x, M_i(1)) - Y_i(x, M_i(0)).$$

Similarly, direct treatment effects are defined as

$$\zeta_i(x) = Y_i(1, M_i(x)) - Y_i(0, M_i(x)).$$

Assuming that causal mediation and direct effects do not vary according to treatment status, so that $\delta_i(1) = \delta_i(0) = \delta_i$ and $\zeta_i(1) = \zeta_i(0) = \zeta_i$, one has that the total treatment effect $\tau_i = Y_i(1, M_i(1)) - Y_i(0, M_i(0))$ is the sum of the causal mediation and direct effects, $\tau_i = \delta_i + \zeta_i$.

The Imai et al. method may be characterised as a quasi-Bayesian Monte Carlo algorithm, and one may adapt the method using fully Bayesian principles, with substitution of appropriate replicates. Assuming the treatment X is binary, and based on sampled parameters at each MCMC iteration, one samples replicate mediator values $M_1 = M_{rep}(C, X = 1)$ and $M_0 = M_{rep}(C, X = 0)$ at different treatment levels. One then substitutes these (as mediator values) in the regression term for predicting Y, along with settings $X=1$ and $X=0$ on the treatment values. This provides predictions $Y_{rep}(1, M_1)$, $Y_{rep}(0, M_1)$, $Y_{rep}(1, M_0)$, and $Y_{rep}(0, M_0)$, with the total treatment effect $Y_{rep}(1, M_1) - Y_{rep}(0, M_0)$.

In real situations, the exposure X may influence one or more confounders C. Schematically,

$$C \sim f_1(X),$$

$$M \sim f_2(X, C),$$

$$Y \sim f_3(X, M, X.M, C).$$

Then more detailed calculations are obtained, since $C(X)$ and $C(X^*)$ will differ. To illustrate linear effect substitution, suppose C denotes a confounder influenced by X, and D denotes confounders independent of X. Then one will have an additional linear regression with expectation such as

$$E[C \mid X, D] = a_1 + a_2 X + a_3 D,$$

whereby $E[C(X)] = \alpha_1 + \alpha_2 X + \alpha_3 D$, and $E[C(X^*)] = \alpha_1 + \alpha_2 X^* + \alpha_3 D$. Then one has

$$E[M(X)] = \beta_1 + \beta_2 X + \beta_3 C(X) + \beta_4 D,$$

$$E[M(X^*)] = \beta_1 + \beta_2 X^* + \beta_3 C(X^*) + \beta_4 D,$$

and

$$E[Y\{X, M(X)\}] = \theta_1 + \theta_2 X + \theta_3 M(X) + \theta_4 X.M(X) + \theta_5 C(X) + \theta_6 D,$$

$$E[Y\{X, M(X^*)\}] = \theta_1 + \theta_2 X + \theta_3 M(X^*) + \theta_4 X.M(X^*) + \theta_5 C(X) + \theta_6 D,$$

$$E[Y\{X^*, M(X^*)\}] = \theta_1 + \theta_2 X^* + \theta_3 M(X^*) + \theta_4 X^*.M(X^*) + \theta_5 C(X^*) + \theta_6 D.$$

Example 7.16 Framing and Public Opinion

This example uses data from Tingley et al. (2014) in which randomised subjects are exposed to different media perspectives on immigration (the binary treatment, X) with the aim of assessing how this affects a political outcome measure (Y, also binary), namely whether or not a letter is sent regarding immigration policy to a Congress representative. Anxiety is posited as a continuous mediating variable, with the measure of anxiety (anx) influenced by the form of media framing. By virtue of the binary response, direct and indirect effects are obtained in terms of probability differences.

The mediator regression specifies dependence on X and four confounders C (age, gender, income, and educational category: less than high school, high school, some college, bachelor's degree or higher, with the first level as reference). These are taken not to be influenced by the exposure. The anxiety mediator has positive values only, and assuming a truncated normal density, the regression is coded in jags as

```
anx[i] ~dnorm(mu.anx[i],tau.anx) T(0,)
mu.anx[i]  <- a[1]+a[2]*treat[i]+a[3]*age.c[i]+a[4]*equals(edu[i],2)
+a[5]*equals(edu[i],3)+a[6]*equals(edu[i],4)+a[7]*gend[i]+a[8]
*income[i]
```

Predictions at values $X=1$ and $X=0$ are obtained as

```
anx.1[i] ~dnorm(mu.anx.1[i],tau.anx) T(0,)
anx.0[i] ~dnorm(mu.anx.0[i],tau.anx) T(0,)
```

with corresponding regression settings:

```
mu.anx.1[i] <- a[1]+a[2]+a[3]*age.c[i]+a[4]*equals(edu[i],2)
+a[5]*equals(edu[i],3)+a[6]*equals(edu[i],4)+a[7]*gend[i]+a[8]
*income[i]
mu.anx.0[i] <- a[1]+a[3]*age.c[i]+a[4]*equals(edu[i],2)
+a[5]*equals(edu[i],3)+a[6]*equals(edu[i],4)+a[7]*gend[i]+a[8]
*income[i]
```

The binary outcome has predictors X and C, but also involves the mediator. So assuming a probit regression one has

```
congmesg[i] ~dbern(p[i])
p[i] <- phi(b[1]+b[2]*treat[i]+b[3]*anx[i]+b[4]*age.c[i]+b[5]*equals
(edu[i],2)
+b[6]*equals(edu[i],3)+b[7]*equals(edu[i],4)+b[8]*gend[i]+b[9]*incom
e[i])
```

Four alternative predictions of the outcome are obtained at settings X=1 and X=0, crossed with mediator values set at the predictions $M_1 = M_{rep}(C, X=1)$ and $M_0 = M_{rep}(C, X=0)$ (anx.1[i] and anx.0[i] in the code). The direct causal effect is defined as

$$E\left[Y\{1, M(0)\}\right] - E\left[Y\{0, M(0)\}\right].$$

Two assumptions regarding the density for the positive anxiety score are made. Under a truncated normal assumption, a two-chain run using jagsUI provides means (95% CRI) for the average mediation, direct and total effects as 0.083 (0.011, 0.160), 0.012 (−0.113, 0.140) and 0.096 (−0.042, 0.238). These estimates are similar in location to, but less precise than, those contained in Tingley et al. (2014). In inference terms, the direct impact of the treatment is insignificant, and the impact of the treatment on the response is mainly due to its effect on the anxiety mediator. The LOO-IC for the mediator and outcome models are obtained as 2,099 and 298 respectively.

Assuming instead a lognormal density for anxiety, the respective means (95% CRI) become 0.090 (0.011, 0.181), 0.016 (−0.106, 0.142), and 0.106 (−0.034, 0.253). The LOO-IC for the mediator model is reduced to 2,092.

Example 7.17 Effect of Alcohol on SBP

This example uses a dataset for $n = 10{,}000$ subjects from Daniel et al. (2011), which illustrates a case where the exposure affects a confounder. The interest is in estimating direct and indirect effects of alcohol consumption (ALC, in units per day), which is the exposure, on the outcome, systolic blood pressure (SBP, measured in mmHg), while allowing for the mediating effect of a liver enzyme, GGT (gamma-glutamyl transpeptidase). GGT is the logarithm of the enzyme measured in grams per litre. The causal sequence is that alcohol intake affects levels of the liver enzyme, which in turn affect SBP, though there is also potentially a direct influence of alcohol on SBP.

Regarding confounders, body mass index (BMI) may affect both the mediator GGT and the outcome SBP, while socioeconomic status (SES, a trinomial category) may affect alcohol intake, BMI and SBP. Furthermore, alcohol intake may have a direct effect on BMI (i.e. the exposure potentially influences one of the confounders).

Data on the four continuous variables is subject to missingness. This is handled by assuming missingness at random, and by using a sequence of one-dimensional conditional distributions (Lipsitz and Ibrahim, 1996)

$$\text{ALC} \sim (\text{ALC} \mid \text{SES})$$

$$\text{BMI} \sim (\text{BMI} \mid \text{ALC}, \text{SES})$$

$$\text{GGT} \sim (\text{GGT} \mid \text{BMI}, \text{ALC})$$

$$\text{SBP} \sim (\text{SBP} \mid \text{ALC}, \text{GGT}, \text{ALC.GGT}, \text{BMI}, \text{SES}).$$

A different imputation strategy is adopted by Daniel et al. (2011) which may affect findings. Normal linear regressions are adopted for each outcome. In full, the regression term assumed for predicting the outcome SBP is

$$\mu_{SBP,i} = \theta_1 + \theta_2 \text{ALC} + \theta_3 \text{GGT} + \theta_4 \text{ALC.GGT} + \theta_5 \text{BMI}$$
$$+ \theta_6 I(\text{SES} = 2) + \theta_7 I(\text{SES} = 3).$$

One aim is to estimate the natural direct effect NDE, defined (in generic symbols) as the expected value of the difference $Y(X, M(X^*)) - Y(X^*, M(X^*))$. Accordingly, a counterfactual alcohol consumption level ALC* = 0 is defined, with corresponding predictions (obtained as Bayesian replicates)

$$\text{BMI}^* = \text{BMI}(\text{ALC}^*, \text{SES})$$

and

$$\text{GGT}^* = \text{GGT}(\text{ALC}^*, \text{BMI}^*).$$

These are bmi.star.new[i] and ggt.star.new[i] in the code. Then

$$\text{NDE} = E\left[\text{SBP}\left\{\text{ALC}, \text{GGT}^*, \text{BMI}^*, \text{SES}\right\}\right] - E\left[\text{SBP}\left\{\text{ALC}^*, \text{GGT}^*, \text{BMI}^*, \text{SES}\right\}\right]$$

with the first and second components defining NDE at subject level denoted NDE.a[i] and NDE.star[i] in the code.

The natural indirect effect NIE is defined generically as the expected value of the difference $Y(X, M(X)) - Y(X, M(X^*))$. In terms of the application, we have

$$\text{NIE} = E\left[\text{SBP}\left\{\text{ALC}, \text{GGT}(\text{BMI}, \text{ALC}), \text{BMI}(\text{ALC}, \text{SES}), \text{SES}\right\}\right]$$
$$- E\left[\text{SBP}\left\{\text{ALC}, \text{GGT}^*, \text{BMI}^*\right\}\right].$$

In the first component, GGT and BMI are replicates (ggt.new[i] and bmi.new[i] in the code). Including the prediction $\text{GGT}^* = \text{GGT}(\text{ALC}^*, \text{BMI}^*)$ in the second component allows for the fact that BMI (a confounder) depends on the exposure ALC, so that $\text{BMI}^* = \text{BMI}(\text{ALC}^*, \text{SES})$ and BMI(ALC,SES) differ.

The total causal effect TCE is the sum of NDE and NIE. Additionally, the controlled direct effect CDE may be obtained at the setting GGT.c = 3, namely

$$\text{CDE} = E\left[\text{SBP}\left\{\text{ALC}, \text{GGT}.c, \text{BMI}, \text{SES}\right\}\right]$$
$$- E\left[\text{SBP}\left\{\text{ALC}^*, \text{GGT}.c, \text{BMI}, \text{SES}\right\}\right].$$

Table 7.6 shows the posterior summary for these quantities and the regression parameters θ from a two-chain run using jagsUI. The estimated total causal effect (TCE) implies that the reduction of alcohol consumption to zero would reduce average SBP by 8.04 units (95% CRI from 7.71 to 8.40). A relatively small part of the reduction (with posterior mean 1.30 units) is mediated through GGT. It may be noted that the impact of alcohol on SBP is possibly nonlinear, with evidence of a U-shaped effect, and an extended model might allow nonlinearity (Jackson et al., 1985).

TABLE 7.6
Prediction of SBP, Posterior Parameter Summary

Parameter	Predictor	Mean	St devn	2.5%	97.5%
TCE		8.04	0.18	7.71	8.40
NDE		6.74	0.16	6.42	7.06
NIE		1.30	0.10	1.10	1.51
CDE		6.63	0.15	6.32	6.93
θ_1	Intercept	89.86	1.25	87.63	92.58
θ_2	ALC	5.94	0.19	5.58	6.33
θ_3	GGT	7.03	0.16	6.74	7.35
θ_4	GGT.ALC	−0.99	0.05	−1.10	−0.89
θ_5	BMI	0.51	0.05	0.41	0.59
θ_6	SES_2	−5.34	0.24	−5.80	−4.87
θ_7	SES_3	−10.32	0.31	−10.92	−9.71

7.9.4 Marginal Structural Models

Another strategy to estimate causal effects focuses on a marginal structural model (MSM) relating Y to exposure X, and possibly other selected risk factors V of interest, after adjusting for other confounders (Snowden et al., 2011; Joffe et al., 2004). Sometimes the marginal model will involve a regression on X and effect modifiers V (Robbins et al., 2000, p.556).

The inverse probability of treatment weighted (IPTW) approach involves a weighted likelihood for the MSM. Consider the case when the marginal structural model involves regression of Y on X only, adjusting for confounders C. The IPTW method involves first deriving weights

$$w_i = 1/Pr(X_i = x \mid C_i)$$

that $X=x$ given confounders C, following binary regression of X on C. The weights are

$$w_i = 1/P(X_i = 1 \mid C_i) = S_i$$

for subjects with observed $X=1$, and

$$w_i = 1/P(X_i = 0 \mid C_i) = 1/(1-S_i)$$

for subjects with observed $X=0$. Equivalently

$$\begin{aligned} w_i &= X_i/P(X_i = 1 \mid C_i) + (1-X_i)/P(X_i = 0 \mid C_i) \\ &= X_i/S_i + (1-X_i)/(1-S_i). \end{aligned}$$

Estimating the marginal structural model then involves a weighted likelihood (normal, logistic, etc.) of Y on X. Applying weights in this way creates an artificial population which tends to balance on covariates X used in deriving the weights (Naimi et al., 2014a). Doubly robust weights may also be defined that estimate causal effects if either the propensity score model or the outcome model is correctly specified. Davis et al. (2017) use Bayesian methods to estimate the parameters needed to define a propensity score in spatial applications, and substitute relevant posterior means to estimate IPTW weights, with the latter considered as frequentist.

Marginal structural models may also be estimated using a regression of Y on X (exposure) and C (confounders) to predict counterfactual outcomes for all subjects. This is in line with g-computation principles (Wang and Arah, 2015). Then $E(Y[X,C])$ denotes the prediction, possibly counterfactual, at the value X. So for X binary, and $X=1$ as exposed, the total causal effect is estimated as

$$\text{TCE} = E(Y[1,C]) - E(Y[0,C]).$$

Snowden et al. (2011) use an additional regression step, involving $2n$ outcomes (half being actual responses at observed X, half being counterfactual responses at the counterfactual X^*) and estimate the treatment effect by regression of the expanded outcome vector on corresponding X (or X^*) values. However, the TCE may also be estimated by averaging over predictions at appropriate settings of X and C (Example 7.18).

Example 7.18 Lung Function and Ozone Exposure

This example involves simulated data from Snowden et al. (2011), with Y being forced expiratory volume in 1 second (FEV1) measured in litres, X being ozone exposure (binary), and confounders C_1 (male=1, female=0), and C_2, controller medication use (1=effective use, 0=ineffective). The true model underlying the simulated Y-data involves terms in X, C_1, and an interaction between X and C_2.

In the absence of such knowledge about the data generation, a researcher might consider a full linear regression of Y on X, C_1, C_2, and the three possible interactions involving X, C_1, and C_2. Thus the regression term considered here is

$$\beta_1 + \beta_2 X + \beta_3 C_1 + \beta_4 C_2 + \beta_5 X.C_1 + \beta_6 X.C_2 + \beta_7 C_1.C_2$$

where $X.C_1$ denotes an interaction between X and C_1, etc. We estimate this model using jagsUI, and find the regression coefficient β_2 to have posterior mean (sd) of −0.486 (0.08). By contrast, the marginal causal effect has posterior mean (sd) of −0.337 (0.054). This is estimated by averaging the difference between the predictions

$$Y(1,C) = \beta_1 + \beta_2 + \beta_3 C_1 + \beta_4 C_2 + \beta_5 C_1 + \beta_6 C_2 + \beta_7 C_1.C_2$$

$$Y(0,C) = \beta_1 + \beta_3 C_1 + \beta_4 C_2 + \beta_7 C_1.C_2$$

So the conventional regression overstates the impact of ozone exposure in reducing FEV1. The estimates by Snowden et al. (2011, Table 3) are similar.

To alleviate impacts of insignificant predictors, one can include spike-slab predictor selection, whereby

$$\beta_j = \gamma_j J_j,$$

$$\gamma_j \sim N(0,10),$$

$$J_j \sim \text{Bern}(0,0.5).$$

This gives posterior probabilities of 1 for retaining β_2, and β_3, and 0.95 for retaining β_6, as expected in line with the data generation mechanism. Other coefficients have retention probabilities below 0.05. Including predictor selection affects estimates slightly: the regression coefficient β_2 now has a posterior mean (sd) of −0.481 (0.059), while the marginal causal effect has a posterior mean (sd) of −0.340 (0.054).

We also consider the propensity score approach of Section 7.9.1, regressing the probability S_i that $X_i=1$ on C_{1i}, C_{2i} and the interaction $C_{1i}C_{2i}$. One may then either simply regress the response Y_i on S_i and X_i, or also allow for residual confounding (Zigler and Dominici, 2014), namely,

$$Y \sim N(\beta_1 + \beta_2 X_i + \beta_3 S_i + \beta_4 C_{1i} + \beta_6 C_{2i} + \beta_7 C_{1i}.C_{2i}, \sigma^2).$$

This is carried out using a joint likelihood, though feedback between the Y-model and the X-model can be avoided using the BUGS "cut" function. Results for the coefficient β_2, and hence the ACE, are very similar whether or not residual confounding is allowed for, and also whether or not feedback is avoided. Allowing feedback, and without allowing residual confounding, the mean (sd) of the ACE is estimated as −0.352 (0.053).

To illustrate the IPTW approach, we again regress the probability that $X=1$ on C_1, C_2, and the interaction $C_1.C_2$. This provides probabilities $S_i = Pr(X_i = 1 \mid C_{1i}, C_{2i})$, and the weights

$$w_i = X_i / S_i + (1 - X_i)/(1 - S_i)$$

are then used in a weighted linear regression of Y on X with weights σ^2/w_i. To avoid feedback between the logit regression for X on $\{C_1,C_2\}$, and the marginal structural regression of Y on X, the "cut" function in BUGS is applied to the predicted probabilities S_i before they are inserted in the weights. From the second half of a 10,000 iteration sequence, the posterior mean (sd) for the marginal causal effect (MCE) is −0.36 (0.10). If feedback between the two regressions is allowed, convergence in the coefficients of the X-model is impeded, and the MCE has a value closer to null, around −0.27.

References

Albert J (1996) Bayesian selection of log-linear models. *Canadian Journal of Statistics*, 24, 327–347.

Albert JH, Chib S (1993) Bayesian analysis of binary and polychotomous response data. *Journal of the American statistical Association*, 88(422), 669–679.

Albert J, Chib S (2001) Sequential ordinal modeling with applications to survival data. *Biometrics*, 57(3), 829–836.

Arbia G (2014) *A Primer for Spatial Econometrics: With Applications in R*. Palgrave.

Assunçao RM (2003) Space varying coefficient models for small area data. *Environmetrics*, 14(5), 453–473.

Assunção R, Krainski E (2009) Neighborhood dependence in Bayesian spatial models. *Biometrical Journal*, 51(5), 851–869.

Austin P (2009) Balance diagnostics for comparing the distribution of baseline covariates between treatment groups in propensity-score matched samples. *Statistics in Medicine*, 28, 3083–3107.

Baragatti M, Pommeret D (2012) A study of variable selection using g-prior distribution with ridge parameter. *Computational Statistics and Data Analysis*, 56(6), 1920–1934.

Barbieri M, Berger J (2004) Optimal predictive model selection. *Annals of Statistics*, 32, 870–897.

Barreto-Souza W, Simas A (2016) General mixed Poisson regression models with varying dispersion. *Statistics and Computing*, 26, 1263–1280.

Baser O (2006) Too much ado about propensity score models? Comparing methods of propensity score matching. *Value in Health*, 9(6), 377–385.

Bazán J, Bolfarine H, Branco M (2010) A framework for skew-probit links in binary regression. *Communications in Statistics, Theory and Methods*, 39(4), 678–697.
Beck N (1983) Time-varying parameter regression models. *American Journal of Political Science*, 27, 557–600.
Bell S, Broemeling LD (2000) A Bayesian analysis for spatial processes with application to disease mapping. *Statistics in Medicine*, 19(7), 957–974.
Besag J (1974) Spatial interaction and the statistical analysis of lattice systems. *Journal of the Royal Statistical Society B*, 36, 192–225.
Besag J, York J, Mollié A (1991) Bayesian image restoration, with two applications in spatial statistics. *Annals of the Institute of Statistical Mathematics*, 43(1), 1–20.
Bhadra A, Datta J, Polson N, Willard B (2016) Default Bayesian analysis with global-local shrinkage priors. *Biometrika*, 103, 955–969.
Bhattacharya A, Pati D, Pillai N, Dunson D (2015) Dirichlet–Laplace priors for optimal shrinkage. *Journal of the American Statistical Association*, 110(512), 1479–1490.
Bonate P (2011) *Pharmacokinetic-Pharmacodynamic Modeling and Simulation*, 2nd Edition. Springer, New York.
Boris Choy S, Chan J (2008) Scale mixtures distributions in statistical modelling. *Australian & New Zealand Journal of Statistics*, 50(2), 135–146.
Boyd H, Flanders W, Addiss D, Waller L (2005) Residual spatial correlation between geographically referenced observations: A Bayesian hierarchical modeling approach. *Epidemiology*, 16, 532–541.
Brockmann H (1996) Satellite male groups in horseshoe crabs, Limulus polyphemus. *Ethology*, 102(1), 1–21.
Bürkner P, Vuorre M (2018, February 28) Ordinal Regression Models in Psychology: A Tutorial. https://doi.org/10.31234/osf.io/x8swp
Calcagno V, de Mazancourt C (2010) glmulti: An R package for easy automated model selection with (generalized) linear models. *Journal of Statistical Software*, 34(12), 1–29.
Carvalho C, Polson N, Scott J (2009) Handling Sparsity via the Horseshoe. *Proceedings of Machine Learning Research*, 5, 73–80.
Cepeda E, Gamerman D (2000) Bayesian modeling of variance heterogeneity in normal regression models. *Brazilian Journal of Probability and Statistics*, 14(2), 207–221.
Chan K, Ledolter J (1995) Monte Carlo EM estimation for time series models involving counts. *Journal of the American Statistical Association*, 90, 242–252.
Chang Y, Gianola D, Heringstad B, Klemetsdal G (2006) A comparison between multivariate Slash, Student's t and probit threshold models for analysis of clinical mastitis in first lactation cows. *Journal of Animal Breeding and Genetics*, 123, 290–300.
Chen R, Chu C, Yuan S, Wu Y (2016) Bayesian sparse group selection. *Journal of Computational and Graphical Statistics*, 25(3), 665–683.
Chi EM, Reinsel GC (1989) Models for longitudinal data with random effects and AR (1) errors. *Journal of the American Statistical Association*, 84(406), 452–459.
Chib S, Greenberg E (2013) On conditional variance estimation in nonparametric regression. *Statistics and Computing*, 23(2), 261–270.
Chiogna M, Gaetan C (2002) Dynamic generalized linear models with application to environmental epidemiology. *Journal of the Royal Statistical Society: Series C (Applied Statistics)*, 51(4), 453–468.
Choi J, Lawson A (2016, June 16) Bayesian spatially dependent variable selection for small area health modeling. *Statistical Methods in Medical Research*. pii: 0962280215627184.
Choi J, Lawson AB (2018) Bayesian spatially dependent variable selection for small area health modeling. *Statistical Methods in Medical Research*, 27(1), 234–249.
Conceição K, Andrade M, Louzada F (2013) Zero-modified Poisson model: Bayesian approach, influence diagnostics, and an application to a Brazilian leptospirosis notification data. *Biometrical Journal*, 55(5), 661–678.
Congdon P, Almog M, Curtis S, Ellerman R (2007) A spatial structural equation modelling framework for health count responses. *Statistics in Medicine*, 26(29), 5267–5284.
Cox D (1981) Statistical analysis of time series: Some recent developments. *Scandinavian Journal of Statistics*, 8, 93–115.

Czado C, Erhardt V, Min A, Wagner S (2007) Zero-inflated generalized Poisson models with regression effects on the mean, dispersion and zero-inflation level applied to patent outsourcing rates. *Statistical Modelling*, 7(2), 125–153.

Dangl T, Halling M (2012) Predictive regressions with time-varying coefficients. *Journal of Financial Economics*, 106(1), 157–181.

Daniel R, De Stavola B, Cousens S (2011) gformula: Estimating causal effects in the presence of time-varying confounding or mediation using the g-computation formula. *Stata Journal*, 11(4), 479–517.

Darmofal D (2015) *Spatial Analysis for the Social Sciences*. Cambridge University Press.

Davis M, Neelon B, Nietert P, Hunt K, Burgette L, Lawson A, Egede L (2017) Addressing geographic confounding through spatial propensity scores: A study of racial disparities in diabetes. *Statistical Methods in Medical Research*, 28(3), 734–748.

Dormann C, McPherson N, Araújo M et al. (2007) Methods to account for spatial autocorrelation in the analysis of species distributional data: A review. *Ecography*, 30(5), 609–628.

Epstein D, O'Halloran S (1996) The partisan paradox and the US tariff, 1877–1934. *International Organization*, 50(2), 301–324.

Fahrmeir L, Osuna E (2006) Structured additive regression for overdispersed and zero-inflated count data. *Applied Stochastic Models in Business and Industry*, 22(4), 351–369.

Fahrmeir L, Tutz G (2001) *Multivariate Statistical Modelling Based on Generalized Linear Models*, pp 69–137. Springer, New York.

Fernández C, Steel MF (1998) On Bayesian modeling of fat tails and skewness. *Journal of the American Statistical Association*, 93(441), 359–371.

Ferreira M, Gamerman D (2000) Dynamic generalized linear models, pp 57–72, in *Generalized Linear Models: A Bayesian Perspective*, eds D Dey, S Ghosh, B Mallick. Marcel Dekker, New York.

Fokianos K, Kedem B (2003) Regression theory for categorical time series. *Statistical Science*, 18(3), 357–376.

Fonseca TC, Ferreira MA, Migon HS (2008) Objective Bayesian analysis for the Student-t regression model. *Biometrika*, 95(2), 325–333.

Fotheringham A, Brunsdon C, Charlton M (2002) *Geographically Weighted Regression: The Analysis of Spatially Varying Relationships*. Wiley, Chichester, UK.

Franzese RJ, Hays JC (2007) Spatial econometric models of cross-sectional interdependence in political science panel and time-series-cross-section data. *Political Analysis*, 15(2), 140–164.

Fruhwirth-Schnatter S, Fruhwirth R (2007) Auxiliary mixture sampling with applications to logistic models. *Computational Statistics and Data Analysis*, 51, 3509–3528.

Frühwirth-SchnatterS, Frühwirth R (2010) Data augmentation and MCMC for binary and multinomial logit models, pp 111–132, in *Statistical Modelling and Regression Structures*, eds T Kneib, G Tutz. Physica-Verlag HD.

Gamerman D (1998) Markov chain Monte Carlo for dynamic generalised linear models. *Biometrika*, 85(1), 215–227.

Gamerman D, Moreira A, Rue H (2003) Space-varying regression models: Specifications and simulation. *Computational Statistics and Data Analysis*, 42, 513–533.

Garay A, Lachos V, Bolfarine H, Ortega E (2015) Bayesian estimation and case influence diagnostics for the zero-inflated negative binomial regression model. *Journal of Applied Statistics*, 42(6), 1148–1165.

Garcia-Donato G, Martinez-Beneito M (2013) On sampling strategies in Bayesian variable selection problems with large model spaces. *Journal of the American Statistical Association*, 108(501), 340–352.

Geinitz S, Furrer R (2016) Conjugate distributions in hierarchical Bayesian ANOVA for computational efficiency and assessments of both practical and statistical significance. arXiv:1303.3390.

Geinitz S, Furrer R, Sain S (2015) Bayesian multilevel analysis of variance for relative comparison across sources of global climate model variability. *International Journal of Climatology*, 35(3), 433–443.

Gelfand A, Kim H, Sirmans C, Banerjee S (2003) Spatial modelling with spatially varying coefficient models. *Journal of the American Statistical Association*, 98, 387–396.

Gelfand AE, Ghosh SK (1998) Model choice: A minimum posterior predictive loss approach. *Biometrika*, 85(1), 1–11.

Gelman A (2005) Analysis of variance—Why it is more important than ever. *The Annals of Statistics*, 33(1), 1–53.

George E, McCullogh R (1993) Variable selection via Gibbs sampling. *Journal of the American Statistical Association*, 85, 398–409.

Gerlach R, Bird R, Hall A (2002) Bayesian variable selection in logistic regression: Predicting company earnings direction. *Australian & New Zealand Journal of Statistics*, 44, 155–168.

Ghosh J, Ghattas A (2015) Bayesian variable selection under collinearity. *The American Statistician*, 69(3), 165–173.

Ghosh S, Mukhopadhyay P, Lu J-C (2006) Bayesian analysis of zero-inflated regression models. *Journal of Statistical Planning and Inference*, 136, 1360–1375.

Greene W (2008) Functional forms for the negative binomial model for count data. *Economics Letters*, 99(3), 585–590.

Greenland S (2000) Causal analysis in the health sciences. *Journal of the American Statistical Association*, 95, 286–289.

Hensher DA, Greene WH (2003) The mixed logit model: the state of practice. *Transportation*, 30(2), 133–176.

Holloway G, Shankar B, Rahmanb S (2002) Bayesian spatial probit estimation: A primer and an application to HYV rice adoption. *Agricultural Economics*, 27(3), 383–402.

Holmes C, Held L (2006) Bayesian auxiliary variable models for binary and multinomial regression. *Bayesian Analysis*, 1, 145–168.

Hooten M, Hobbs N (2015) A guide to Bayesian model selection for ecologists. *Ecological Monographs*, 85(1), 3–28.

Ibrahim JG, Chen MH (2000) Power prior distributions for regression models. *Statistical Science*, 15(1), 46–60.

Imai K, Keele L, Tingley D (2010) A general approach to causal mediation analysis. *Psychological Methods*, 15(4), 309.

Ishwaran H, Kogalur U, Rao J (2010) spikeslab: Prediction and variable selection using spike and slab regression. *R Journal*, 2(2), 68–73.

Ishwaran H, Rao J (2005) Spike and slab variable selection: Frequentist and Bayesian strategies. *Annals of Statistics*, 33, 730–773.

Jackson R, Stewart A, Beaglehole R, Scragg R (1985) Alcohol consumption and blood pressure. *American Journal of Epidemiology*, 122(6), 1037–1044.

Jia Z, Xu S (2007) Mapping quantitative trait loci for expression abundance. *Genetics*, 176, 611–623.

Joffe MM, Ten Have TR, Feldman HI, Kimmel SE (2004) Model selection, confounder control, and marginal structural models: Review and new applications. *The American Statistician*, 58(4), 272–279.

Johnson VE, Albert JH (1999) *Ordinal Data Modeling*. Springer-Verlag.

Jung R C, Kukuk M, Liesenfeld R (2006) Time series of count data: Modeling, estimation and diagnostics. *Computational Statistics & Data Analysis*, 51, 2350–2364.

Kahn M, Raftery A (1996) Discharge rates of Medicare stroke patients to skilled nursing facilities: Bayesian logistic regression with unobserved heterogeneity. *Journal of the American Statistical Association*, 91, 29–41.

Khalili A, Chen J (2007) Variables selection in finite mixture of regression models. *Journal of the American Statistical Association*, 102, 1025–1038.

Kim H, Sun D,Tsutakawa R K (2002) Lognormal vs. gamma: Extra variations. *Biometrical Journal*, 44(3), 305–323.

Kinney S, Dunson D (2007) Fixed and random effects selection in linear and logistic models. *Biometrics*, 63, 690–698.

Kitagawa G, Gersch W (1985) A smoothness priors time-varying AR coefficient modeling of nonstationary covariance time series. *IEEE Transactions on Automatic Control*, 30, 48–56.

Kotz S, Kozubowski T, Podgórski K (2001) *The Laplace Distribution and Generalizations: A Revisit with Applications to Communications, Economics, Engineering, and Finance*. Springer.

Kruijer W, Stein A, Schaafsma W, Heijting S (2007) Analyzing spatial count data, with an application to weed counts. *Environmental and Ecological Statistics*, 14, 399–410.

Kuhn I (2007) Incorporating spatial autocorrelation may invert observed patterns. *Diversity and Distributions*, 13, 66–69.

Kuo L, Mallick B (1998) Variable selection for regression models. *Sankhya B*, 60, 65–81.

Lange T, Vansteelandt S, Bekaert M (2012) A simple unified approach for estimating natural direct and indirect effects. *American Journal of Epidemiology*, 176(3), 190–195.

Lee K, Chen R, Wu Y (2016) Bayesian variable selection for finite mixture model of linear regressions. *Computational Statistics & Data Analysis*, 95, 1–16.

Lee K-J, Chen R-B (2015) BSGS: Bayesian sparse group selection. *The R Journal*, 7(2), 122–133.

Leroux B, Lei X, Breslow N (1999) Estimation of disease rates in small areas: A new mixed model for spatial dependence, pp 135–178, in *Statistical Models in Epidemiology, the Environment and Clinical Trials*, eds M Halloran, D Berry. Springer-Verlag, New York.

LeSage J (2004) A family of geographically weighted regression models, Chapter 11, pp 241–264, in *Advances in Spatial Econometrics: Methodology, Tools and Applications*, eds L Anselin, R Florax, S Rey. Springer, New York.

LeSage J, Dominguez M (2012) The importance of modeling spatial spillovers in public choice analysis. *Public Choice*, 150(3–4), 525–545.

LeSage J, Kelley Pace R (2009) *Introduction to Spatial Econometrics*. CRC Press/Taylor & Francis.

Leung Y, Mei C-L, Zhang W-X (2000) Statistical tests for spatial nonstationarity based on the geographically weighted regression model. *Environment and Planning A*, 32(1), 9–32.

Lichstein J, Simons T., Shriner S, Franzreb, K (2002) Spatial autocorrelation and autoregressive models in ecology. *Ecological Monographs*, 72, 445–463.

Lipsitz SR, Ibrahim JG (1996) A conditional model for incomplete covariates in parametric regression models. *Biometrika*, 83(4), 916–922.

Loeys T, Moerkerke B, De Smet O, Buysse A (2012) The analysis of zero-inflated count data: Beyond zero-inflated Poisson regression. *British Journal of Mathematical and Statistical Psychology*, 65(1), 163–180.

Makalic E, Schmidt D (2016) High-dimensional Bayesian regularised regression with the BayesReg package. arXiv preprint arXiv:1611.06649.

Malsiner-Walli G, Wagner H (2019) Comparing spike and slab priors for Bayesian variable selection. arXiv preprint arXiv:1812.07259.

Marshall E, Spiegelhalter D (2007) Identifying outliers in Bayesian hierarchical models: A simulation-based approach. *Bayesian Analysis*, 2, 409–444.

Martin T, Wintle B, Rhodes J, Kuhnert P, Field S, Low-Choy S, Tyre A, Possingham H (2005) Zero tolerance ecology: improving ecological inference by modelling the source of zero observations. *Ecology Letters*, 8(11), 1235–1246.

McCandless L, Douglas I, Evans S, Smeeth L (2010) Cutting feedback in Bayesian regression adjustment for the propensity score. *International Journal of Biostatistics*, 6(2), 16.

McCandless L, Gustafson P, Austin P (2009) Bayesian propensity score analysis for observational data. *Statistics in Medicine*, 28, 94–112.

McElreath R (2016) *Statistical Rethinking: A Bayesian Course with Examples in R and Stan*, Vol. 122. CRC Press.

McNamee R (2005) Regression modelling and other methods to control confounding. *Occupational and Environmental Medicine*, 62(7), 500–506, 472.

Mebane W, Sekhon J (2004) Robust estimation and outlier detection for overdispersed multinomial models of count data. *American Journal of Political Science*, 48(2): 391–410.

Millar R (2009) Comparison of hierarchical Bayesian models for overdispersed count data using DIC and Bayes' factors. *Biometrics*, 65, 962–969.

Mohr D (2007) Bayesian identification of clustered outliers in multiple regression. *Computational Statistics & Data Analysis*, 51, 3955–3967.
Musio M, Sauleau E, Buemi A (2010) Bayesian semi-parametric ZIP models with space-time interactions: An application to cancer registry data. *Mathematical Medicine and Biology*, 27(2), 181–194.
Naimi A, Kaufman J, MacLehose R (2014b) Mediation misgivings: Ambiguous clinical and public health interpretations of natural direct and indirect effects. *International Journal of Epidemiology*, 43(5):1656–1661.
Naimi A, Moodie E, Auger N, Kaufman J (2014a) Constructing inverse probability weights for continuous exposures: A comparison of methods. *Epidemiology*, 25, 292–299.
Nelson K, Leroux B (2006) Statistical models for autocorrelated count data. *Statistics in Medicine*, 25, 1413–1430.
Neyens T, Faes C, Molenberghs G (2012) A generalized Poisson-gamma model for spatially-overdispersed data. *Spatial and Spatio-temporal Epidemiology*, 3(3), 185–194.
Nicholls D, Quinn B (1982) *Random Coefficient Autoregressive Models: An Introduction*. Springer-Verlag, New York.
Nott D, Leng C (2010) Bayesian projection approaches to variable selection in generalized linear models. *Computational Statistics & Data Analysis*, 54(12), 3227–3241.
Oh MS, Lim YB (2001) Bayesian analysis of time series Poisson data. *Journal of Applied Statistics*, 28(2), 259–271.
O'Hara R, Sillanpaa M (2009) A review of Bayesian variable selection methods: What, how and which. *Bayesian Analysis*, 4(1), 85–118.
Osborne P, Foody G, Suárez-Seoane S (2007) Non-stationarity and local approaches to modelling the distributions of wildlife. *Diversity and Distributions*, 13, 313–323.
Park T, Casella G (2008) The Bayesian Lasso. *Journal of the American Statistical Association*, 103, 681–686.
Pearl J (2014) Interpretation and identification of causal mediation. *Psychological Methods*, 19(4), 459–481.
Pérez-Sánchez J, Salmerón-Gómez R, Ocaña-Peinado F (2018) A Bayesian asymmetric logistic model of factors underlying team success in top-level basketball in Spain. *Statistica Neerlandica*, 73(1), 22–43.
Perrakis K, Fouskakis D, Ntzoufras I (2015) Bayesian variable selection for generalized linear models using the power-conditional-expected-posterior prior, pp 59–73, in *Bayesian Statistics from Methods to Models and Applications: Research from BAYSM 2014*, eds S Fruhwirth-Schnatter, A Bitto, G Kastner, A Posekany, Vol. 126. Springer, New York.
Piironen J, Vehtari A (2016) Projection predictive variable selection using Stan+R. Proceedings of the 2016 IEEE 26th International Workshop on Machine Learning for Signal Processing. http://arxiv.org/abs/1508.02502
Piironen J, Vehtari A (2017a) Comparison of Bayesian predictive methods for model selection. *Statistics and Computing*, 27, 711–735.
Piironen J, Vehtari A (2017b) On the hyperprior choice for the global shrinkage parameter in the horseshoe prior, in *Proceedings of the 20th International Conference on Artificial Intelligence and Statistics (AISTATS), PMLR*, 54, 905–913.
Piironen J, Vehtari A (2017c) Sparsity information and regularization in the horseshoe and other shrinkage priors. *Electronic Journal of Statistics*, 11(2), 5018–5051.
Plummer M (2008) Penalized loss functions for Bayesian model comparison. *Biostatistics*, 9(3), 523–539.
Polson N, Scott J (2010) Shrink globally, act locally: Sparse Bayesian regularization and prediction, pp 501–553, in Bayesian Statistics 9, eds J Bernardo, M Bayarri, J Berger, A Dawid, D Heckerman, A Smith, M West. Oxford University Press, New York.
Polson N, Scott J (2012) On the half-Cauchy prior for a global scale parameter. *Bayesian Analysis*, 7(4), 887–902.
Qin ZS, Damien P, Walker S (2003) Scale mixture models with applications to Bayesian inference, pp 394–395, in AIP Conference Proceedings, Vol. 690, No. 1. AIP.

Raftery A, Painter I, Volinsky C (2005) BMA: An R package for Bayesian model averaging. *R News*, 5(2), 2–8.

Reich B, Fuentes M, Herring A, et al. (2010) Bayesian variable selection for multivariate spatially-varying coefficient regression. *Biometrics*, 66, 772–782.

Richardson S, Bottolo L, Rosenthal J (2010) Bayesian models for sparse regression analysis of high dimensional data. *Bayesian Statistics*, 9, 539–569.

Robins J, Hernan M, Brumback B (2000) Marginal structural models and causal inference in epidemiology. *Epidemiology* 11(5), 550–560.

Rockova V, Lesaffre E, Luime J, Löwenberg B (2012) Hierarchical Bayesian formulations for selecting variables in regression models. *Statistics in Medicine*, 31(11–12), 1221–1237.

Scott S (2011) Data augmentation, frequentist estimation, and the Bayesian analysis of multinomial logit models. *Statistical Papers*, 52(1), 87–109.

Shumway R (2016) State space models, Chapter 6, in *Time Series Analysis and Its Applications*, eds R Shumway, D Stoffer. Springer, New York.

Smith RL, Davis JM, Sacks J, Speckman P, Styer P (2000) Regression models for air pollution and daily mortality: Analysis of data from Birmingham, Alabama. *Environmetrics: The official journal of the International Environmetrics Society*, 11(6), 719–743.

Smith T, LeSage J (2004) A Bayesian probit model with spatial dependencies, pp 127–160, in *Pace Advances in Econometrics: Vol 18: Spatial and Spatiotemporal Econometrics*, eds J LeSage, R Kelley. Elsevier Science.

Snowden JM, Rose S, Mortimer KM (2011) Implementation of G-computation on a simulated data set: Demonstration of a causal inference technique. *American Journal of Epidemiology*, 173(7), 731–738.

Spiegelhalter D (1998) Bayesian graphical modelling: A case-study in monitoring health outcomes. *Applied Statistics*, 47, 115–133.

Sun D, Tsutakawa RK, Speckman PL (1999) Posterior distribution of hierarchical models using CAR (1) distributions. *Biometrika*, 86(2), 341–350.

Tchetgen E, Vanderweele T (2014) Identification of natural direct effects when a confounder of the mediator is directly affected by exposure. *Epidemiology*, 25(2), 282–291.

Tingley D, Yamamoto T, Hirose K, Keele L, Imai K (2014) Mediation: R package for causal mediation analysis. *Journal of Statistical Software*, 59. https://www.jstatsoft.org/article/view/v059i05

Tingley M (2012) A Bayesian ANOVA scheme for calculating climate anomalies, with applications to the instrumental temperature record. *Journal of Climate*, 25(2), 777–791.

Tutz G, Gertheiss J (2016) Regularized regression for categorical data. *Statistical Modelling*, 16(3), 161–200.

Utazi C, Sahu S, Atkinson P, Tejedorc N, Tatem A J (2016) A probabilistic predictive Bayesian approach for determining the representativeness of health and demographic surveillance networks. *Spatial Statistics*, 17, 161–178.

VanderWeele T (2013) Policy-relevant proportions for direct effects. *Epidemiology*, 24(1), 175–176.

VanderWeele T (2015) *Explanation in Causal Inference: Methods for Mediation and Interaction*. OUP.

VanderWeele T, Vansteelandt S (2014) Mediation analysis with multiple mediators. *Epidemiologic Methods*, 2(1), 95–115.

Vansteelandt S, Daniel R (2014) On regression adjustment for the propensity score. *Statistics in Medicine*, 33, 4053–4072.

Vaughn M, Beaver K, Wexler J, DeLisi M, Roberts G (2011) The effect of school dropout on verbal ability in adulthood: A propensity score matching approach. *Journal of Youth and Adolescence*, 40(2), 197–206.

Verdinelli I, Wasserman L (1991) Bayesian analysis of outlier problems using the Gibbs sampler. *Statistics and Computing*, 1(2), 105–117.

Viele K, Tong B (2002) Modeling with mixtures of linear regressions. *Statistics and Computing*, 12(4), 315–330.

Wagner H, Pauger D (2016) Discussion: Bayesian regularization and effect smoothing for categorical predictors. *Statistical Modelling*, 16(3), 220–227.

Wang A, Arah O (2015) G-computation demonstration in causal mediation analysis. *European Journal of Epidemiology*, 30(10), 1119–1127.
Wang L, Zhou XH (2007) Assessing the adequacy of variance function in heteroscedastic regression models. *Biometrics*, 63(4), 1218–1225.
Wang P, Puterman M (1999) Markov Poisson regression models for discrete time series, part 1: Methodology. *Journal of Applied Statistics*, 26, 855–869.
Wang P, Puterman M, Cockburn I, Le N (1996) Mixed poisson regression models with covariate dependent rates. *Biometrics*, 52, 381–400.
Weitzen S, Lapane K, Toledano A Y, Hume A L, Mor V (2004) Principles for modeling propensity scores in medical research: A systematic literature review. *Pharmacoepidemiology and Drug Safety*, 13(12), 841–853.
Wheeler D, Calder C (2006) Bayesian spatially varying coefficient models in the presence of collinearity. ASA Section on Bayesian Statistical Science, Proceedings of the Joint Statistical Meetings, Seattle, WA, August 6–10, 2006.
Wheeler D, Calder C (2007) An assessment of coefficient accuracy in linear regression models with spatially varying coefficients. *Journal of Geographical Systems*, 9, 145–166.
Wheeler D, Tiefelsdorf M (2005) Multicollinearity and correlation among local regression coefficients in geographically weighted regression. *Journal of Geographical Systems*, 7, 161–187.
Wilhelm S, de Matos M (2013) Estimating spatial probit models in R. *The R Journal*, 5(1), 130–143.
Windle J (2016) BayesLogit. https://www.rdocumentation.org/packages/BayesLogit/versions/0.6
Winkelmann R, Zimmermann K F (1995) Recent developments in count data modelling: Theory and application. *Journal of Economic Surveys*, 9(1), 1–24.
Winship C, Western B (2016) Multicollinearity and model misspecification. *Sociological Science*, 3, 627–649.
Xu X, Ghosh M (2015) Bayesian variable selection and estimation for group lasso. *Bayesian Analysis*, 10(4), 909–936.
Yi N, Ma S (2012) Hierarchical shrinkage priors and model fitting for high-dimensional generalized linear models. *Statistical Applications in Genetics and Molecular Biology*, 11(6). DOI: https://doi.org/10.1515/1544-6115.1803.
Yuan M, Lin Y (2005) Efficient empirical Bayes variable selection and estimation in linear models. *Journal of the American Statistical Association*, 100, 1215–1224.
Zellner A, Siow A (1980) Posterior odds ratios for selected regression hypotheses, pp 585–603, in *Bayesian Statistics: Proceedings of the First International Meeting Held in Valencia*, eds Bernardo J, DeGroot M, Lindley D, Smith A. University of Valencia Press.
Zeugner S, Feldkircher M (2015) Bayesian model averaging employing fixed and flexible priors: The BMS package for R. *Journal of Statistical Software*, 68(4), 1–37.
Zhou M, Li L, Dunson D, Carin L (2012) Lognormal and gamma mixed negative binomial regression. *Proceedings of the 29th International Conference on Machine Learning*, 2012, 1343–1350.
Zigler C, Dominici F (2014) Uncertainty in propensity score estimation: Bayesian methods for variable selection and model-averaged causal effects. *Journal of the American Statistical Association*, 109(505), 95–107.
Zigler C, Watts K, Yeh R, Wang Y, Coull B, Dominici F (2013) Model feedback in Bayesian propensity score estimation. *Biometrics*, 69(1), 263–273.

8

Bayesian Multilevel Models

8.1 Introduction

The rationale for applying multilevel models to hierarchical data is well-established (Snijders and Bosker, 1999; Skrondal and Rabe-Hesketh, 2004). When lower level units are nested within one or more higher level strata, conventional single-level regression analysis is not appropriate, since observations are no longer independent: pupils in the same schools, or households in the same communities, tend to be more similar to one another than pupils in different schools or households in different communities. Such dependency means standard errors are downwardly biased if the nesting is ignored, and spurious inferences regarding predictor or treatment effects may be made (Hox, 2002; Aarts et al., 2015; Bliese and Hanges, 2004).

In multilevel analysis, predictors may be defined at any level and the interest focuses on adjusting predictor effects for the simultaneous operation of contextual and individual variability in the outcome. This may be important in health applications, for example, if impacts of individual-level risk factors vary by geographic context (Congdon and Lloyd, 2010). Another major goal is variance partitioning (Goldstein et al., 2002; Gelman and Pardoe, 2006); for example, what proportion of area variations in crime rates is due to characteristics of those areas (what is sometimes termed "contextual variation"), and how much is due to the characteristics of the individuals who live in these areas (termed "compositional variation") (Subramanian et al., 2003).

One may also be interested in estimates for geographic areas or institutions that include both individual and area information; for example, the multilevel model for county radon estimates discussed by Gelman (2006). Gelman (2006) notes that compared to estimates involving no pooling or complete pooling, inferences from multilevel models are more reasonable. Complete pooling leads to identical estimates for all units, while a no-pooling model (no borrowing strength) overfits the data, giving implausibly high or low estimates for particular units and low precisions for such estimates.

As well as predictor effects at any level, a multilevel model is likely to involve random effects defined over the clusters at higher level(s), and possibly correlation between different cluster effects. As in Chapter 4, one seeks to pool strength in inferences about clusters when the number of observations for each cluster might be quite small. While exchangeable cluster effects dominate the multilevel literature, there may well be instances where random cluster effects are better regarded as non-exchangeable, as recognised in the general design general linear mixed model of Zhao et al. (2006). For example, it is possible that the significance level of cluster effects is overstated in area multilevel applications that disregard spatial dependence between clusters (Chaix et al., 2005; Dong et al., 2016).

Application of multilevel models from a Bayesian perspective exemplifies many of the issues referred to in earlier chapters; these include sensitivity to priors and setting priors to ensure identifiability and satisfactory mixing (Draper, 2006). Devices such as hierarchical centring may reduce correlation in the joint posterior and increase MCMC effective sample sizes (Givens and Hoeting, 2012; Browne, 2004). The Bayesian approach has benefits in ensuring that uncertainty in variance components is fully reflected in posterior inferences (e.g. regarding cluster effects), an important issue when the number of clusters is small and the likelihood function of level 2 variance parameters may be asymmmetric (Seltzer et al., 1996, 2002).

Improved software for Bayesian multilevel analysis is exemplified by the rstan based brms package (Buerkner, 2017), with an overview provided by Mai and Zhang (2018). The remaining sections of the chapter consider the normal linear multilevel model (Section 8.2), general linear and conjugate models for multilevel discrete data (Section 8.3), crossed factor and multiple member random effect models (Section 8.4), and robust multilevel models (Section 8.5).

8.2 The Normal Linear Mixed Model for Hierarchical Data

A multilevel model typically assumes observations to be independent conditional on fixed regression and random effects defined at one or more levels in the data hierarchy. The prototype two-level model for a continuous response y_{ij} with repetitions $j=1,\dots n_i$ (e.g. pupils, patients, households) in clusters $i=1,\dots m$ (e.g. schools, hospitals, communities) tackles a similar scenario to that considered in Chapter 4, but assumes individual observations to be available, rather than cluster averages. Consider observation level attribute vectors x_{ij} of dimension p, and z_{ij} of dimension q, typically a subvector of x_{ij} with $q \le p$ (Chen and Dunson, 2003).

Then a widely used form of the normal linear mixed model for nested data (e.g. Snijders and Berkhof, 2002) specifies

$$y_{ij} = x_{ij}\beta + z_{ij}b_i + u_{ij}, \tag{8.1}$$

with b_i and u_{ij} denoting random cluster effects and observation level random effects respectively. The intercept $x_{1ij}=1$ with parameter β_1 is included in x_{ij}. With $N=\Sigma_{i=1}^{m} n_i$ total observations, the nested form of the model is

$$y = X\beta + Zb + u,$$

where y is $N \times 1$, $X \equiv \begin{bmatrix} X_1 \\ \dots \\ X_m \end{bmatrix}$ is $N \times p$, with $X_i = (x_{i1}, \dots x_{in_i})'$ of dimension $n_i \times p$, and where the $N \times mq$ matrix Z is block diagonal with m diagonal blocks $Z_i = (z_{i1}, \dots z_{in_i})'$ of dimension $n_i \times q$ (Gamerman, 1997, p.61; Zhao et al., 2006, p.3). Here β is a $(p \times 1)$ vector of population parameters and $b_i = (b_{1i}, \dots, b_{qi})'$ is a $q \times 1$ vector of zero mean cluster specific deviations around those population parameters, with b_i assumed random.

While random effects models offer a way to borrow strength (e.g. when level 2 cluster sizes n_i are relatively small), fixed effect models, especially for varying intercepts are, however, advocated in longitudinal applications, especially in econometrics. Fixed effects for parameter collections are sometimes used in cross-sectional multilevel applications (Snijders and Berkhof, 2002). The choice between the two depends on the purpose of the statistical inference and how far the level 2 units can be regarded as a sample from a policy-relevant population (Draper, 1995). If the sampled clusters are representative of (exchangeable with) a wider population, then a random coefficient model is, in principle, appropriate (Hsiao, 1996). If statistical inference is confined to the particular unique set of level 2 units included in a data set, then a fixed effects model may be more appropriate.

The conjugate linear normal model with random cluster effects assumes multivariate normality for these effects and for the observation level errors. Assuming the z_{ij} are a subvector of x_{ij}, the cluster effects have zero mean, so that

$$(b_{1i},\ldots,b_{qi})' \sim N_q(0,\Sigma_b).$$

The total impact of x_{rij} is then obtained by cumulating over fixed and random components as $\beta_r + b_{ri}$.

Assume the unstructured level 1 errors $u_i = (u_{i1},\ldots u_{in_i})'$ have prior $u_i \sim N_{n_i}(0,H_i)$ where H_i represents the within-cluster dispersion matrix. The stacked form of the linear mixed model at cluster level, namely $y_i = X_i\beta + Z_ib_i + u_i$, may then be expressed in joint likelihood form as

$$\begin{bmatrix} y_i \\ b_i \end{bmatrix} \sim N_{n_i+q}\left(\begin{bmatrix} X_i\beta \\ 0 \end{bmatrix}, \begin{bmatrix} Z_i\Sigma_bZ_i' + H_i & Z_i\Sigma_b \\ \Sigma_bZ_i' & \Sigma_b \end{bmatrix}\right),$$

or in marginal form as

$$y_i \sim N_{n_i}(X_i\beta, Z_i\Sigma_bZ_i' + H_i).$$

The level 1 errors are typically assumed to be independent, given cluster effects and regression terms, often with $H_i = \sigma^2 I$ for all clusters.

The conjugate model then takes inverse gamma and inverse Wishart priors for σ^2 and Σ_b respectively (or gamma and Wishart priors on σ^{-2} and Σ_b^{-1}), and common practice is to adopt just proper priors e.g. $\sigma^2 \sim IG(\varepsilon,\varepsilon)$ where ε is small. Recent research shows that such priors can lead to effectively improper posteriors and also that inferences are sensitive to the choice of hyperparameters (Natarajan and McCulloch, 1998). Alternatives for the level 1 variance include uniform or half t priors on σ (Gelman, 2006b), while hierarchical models for Σ_b are considered by Daniels and Kass (1999) and Daniels and Zhao (2003). A separation strategy using the LKJ (Lewandowski, Kurowicka and Joe) prior is another option (McElreath, 2016).

Following Gamerman (1997), one may sometimes also include random predictor effects at observation level

$$y_{ij} = x_{ij}\beta + z_{ij}b_i + w_{ij}u_{ij},$$

which is one way of specifying what is known as complex level 1 variation or heteroscedasticity related to level 1 attributes (Browne et al., 2002). This means that variances depend on subject level predictors (when subjects j are nested in clusters i) or in panel data applications that variances are changing over time (when times t are nested in subjects i). For categorical w_{ij}, one may equivalently specify complex variation in terms of category-specific variances. Thus Goldstein (2005) considers school exam data y_{ij} (pupils j nested in schools i), with a single predictor gender x_{ij} (=1 for boy, 0 for girl). Then level 1 heteroscedasticity can be represented as

$$y_{ij} = \beta_1 + \beta_2 x_{ij} + x_{ij} u_{1ij} + (1 - x_{ij}) u_{0ij}$$

where $u_{0ij} \sim N(0, \sigma_0^2)$ is the prior for girl observation level errors, and $u_{1ij} \sim N(0, \sigma_1^2)$ is the prior for boy observation level errors. Equivalently, setting $w_{ij} = x_{ij}$,

$$y_{ij} \sim N(\beta_1 + \beta_2 x_{ij}, \sigma^2_{w_{ij}}).$$

It can be seen that random variation over clusters or at level 1 in specification (8.1) raises questions of empirical identification (see Chapter 1), as the fixed regression effects are confounded with the mean of the associated cluster random effect. Suppose $x_{ij} = (x_{fij}, x_{hij})$ and $\beta = (\beta_f, \beta_h)$, where x_{fij} of dimension p–q contains predictors where no variation in clusters is posited, while x_{hij} contains predictors (usually including the constant term) which have a randomly varying effect over clusters.

Under hierarchical centring of the cluster effects, which has been argued to improve MCMC convergence (Gelfand et al., 1995), varying cluster effects γ_{ri} are centred on β_r so that the rth varying predictor effect is $\gamma_{ri} = \beta_{hr} + b_{ri}$ in cluster i. The parameterisation $(\beta, b_i) = ([\beta_f, \beta_h], b_i)$ with zero mean b_i, is replaced by the parameterisation (β_f, γ_i) where $\gamma_i = \beta_h + b_i$. Then

$$y_{ij} = x_{ij}\beta + z_{ij}\gamma_i + u_{ij} \quad (= x_{fij}\beta_f + x_{hij}\gamma_i + u_{ij}) \tag{8.2}$$

$$(\gamma_{1i}, \dots, \gamma_{qi}) \sim N_q(\beta_h, \Sigma_\gamma),$$

where the vectors z_{ij} and x_{ij} are now distinct, with x_{ij} now containing only x_{fij}, while $z_{ij} = x_{hij}$.

8.2.1 The Lindley–Smith Model Format

An alternative fully hierarchical presentation of the normal linear multilevel model (e.g. Seltzer, 1993; Candel and Winkens, 2003) is based on the scheme of Lindley and Smith (1972). It is assumed that all the effects of level 1 predictor (e.g. pupil characteristics in a two-level educational attainment application) vary randomly over clusters, with their variability explained by cluster predictors $W_i = (w_{1i}, \dots, w_{ri})'$ (e.g. school-level attributes).

The two-level scheme is

$$y_i = Z_i \beta_i + u_i, \quad i = 1, \dots, m \tag{8.3}$$

$$\beta_i = \kappa W_i + b_i$$

where $y_i = (y_{i1}, \dots, y_{in_i})'$ is $n_i \times 1$, κ is $q \times r$, Z_i is $n_i \times q$, β_i is a $q \times 1$ vector of random cluster regression parameters, and the errors $u_i = (u_{i1}, \dots, u_{in_i})'$ have prior $u_{ij} \sim N(0, \sigma^2)$. The level 2 regression for β_i involves a fixed effect parameter matrix κ, and errors $b_i = (b_{1i}, \dots, b_{qi})'$ with mean zero and precision matrix T_b. Substituting the second equation in (8.3) into the first yields the model

$$y_i = Z_i \kappa W_i + Z_i b_i + u_i.$$

To constrain the effect of one or more level 1 predictors to have an identical effect across all clusters, the model may be reformulated as the mixed model (8.2) above.

In (8.3), one may assume flat (uniform) priors for κ, and gamma and Wishart priors for σ^{-2} and T_b, namely $1/\sigma^2 \sim Ga(a_u, b_u)$, $T_b \sim W(S_e, \nu_e)$. Also define $r_{ij} = y_{ij} - Z_{ij}\beta_i$, $\hat{\beta}_i = (Z_i'Z_i)^{-1}Z_i y_i$, $\tilde{V}_i = (\sigma^{-2}Z_i'Z_i + T_b)^{-1}$, $V_i = \sigma^2 Z_i'Z_i$, $\Lambda_i = (V_i^{-1} + T_b)V_i^{-1}$, $U_i = (\beta_i - \kappa W_i)$ and $G = [\Sigma W_i' T_b W_i]^{-1}$. Then the full conditionals for Gibbs sampling are

$$1/\sigma^2 \sim Ga\left(0.5(a_u + m), 0.5\left(b_u + \sum_{i=1}^{m}\sum_{j=1}^{n_i} r_{ij}^2\right)\right)$$

$$\beta_i \sim N_q\left(\Lambda_i \hat{\beta}_i + (I - \Lambda_i)\kappa W_i, \tilde{V}_i\right)$$

$$T_b \sim W\left(S_e + \sum_{i=1}^{m} U_i U_i', m + \nu_e\right)$$

$$\kappa \sim N_r\left(G\sum_{i=1}^{m} W_i T_b \beta_i, G\right).$$

Example 8.1 Maths Achievement

Hierarchical centring, random predictor effects, and level 1 complex variation are illustrated in an educational example from Kreft and de Leeuw (1998) concerning 519 pupils in 23 schools (the "clusters"). Analysis uses the brms, rstan, and R2OpenBUGS packages. The response is for Math achievement, with predictors being homework hours per week and gender (F=1, M=0). An initial model (model 1) assumes school-varying intercepts only, while an extension (model 2) assumes varying intercepts and slopes on homework, with the effect of gender not varying by cluster.

Model 2 is

$$y_{ij} = x_{ij}\beta_f + z_{ij}\gamma_i + u_{ij},$$
$$u_{ij} \sim N(0, \sigma^2)$$
$$(\gamma_{1i}, \gamma_{2i}) \sim N_2(\beta_h, \Sigma_\gamma),$$

where $x_{ij} = (\text{gend})$ excludes an intercept, and $z_{ij} = (1, \text{homework})$, with β_{h1} providing the regression intercept.

The brms package is applied to assess gain in fit, using WAIC (widely applicable information criterion) and LOO-IC (leave-one-out information criterion), through adding the extra source of cluster variability. The command form

```
BRMS2=brm(y~1+homework+gend+(1+homework|sch), data = D,
family="gaussian", chains=2)
```

ensures that mean random effects in model 2 are zero. The default setting for the LKJ prior for the random effects correlation matrix is adopted, with shape parameter 1 (Buerkner, 2017, p.4). Sensitivity may be assessed, for example, by specifying set_prior("lkj(2)", class="cor") or set_prior("lkj(0.5)", class="cor").

There is a substantial gain in fit in adding homework random slopes according to both WAIC and LOO information criteria, which in this example have very similar values. The WAIC falls from 3712.7 to 3578, and the LOOIC from 3712.9 to 3579.6.

R2OpenBUGS codes for these models include exceedance checks $Pr(y_{ij,\text{rep}} > y_{ij}|y)$ based on the mixed predictive method (Marshall and Spiegelhalter, 2007; Green et al., 2009). Exceedance checks are also included at cluster level, obtained by checking school averaged replicates of $y_{ij,\text{rep}}$ against school averages on the response. In the R2OpenBUGS code for the second model, a Wishart prior with identity scale matrix and 2 degrees of freedom is assumed for the cluster precision matrix Σ_γ^{-1}, and a $Ga(1,0.001)$ prior for the observation level precision σ^{-2}. Predictors are centred, but not standardised (as in BRMS).

WAIC measures are very similar between the LKJ and Wishart approaches to model 2, at just under 3580. However, there is sensitivity to priors in covariance estimates: the LKJ prior identifies a negative correlation of −0.78 between school intercepts and slopes, whereas the Wishart prior method estimates a positive correlation of around 0.35. In fact, the 23 observed school-level averages on achievement and homework also show a positive correlation of 0.40. Random intercepts under the Wishart model 2 have a correlation of 0.90 with observed average school achievement levels, as against a corresponding correlation of 0.48 under the LKJ prior. Sensitivity may be partly related to small cluster sizes (e.g. schools 2 and 3 have under 10 pupils).

Cross-validatory checks at school level (testmx.sch in the R2OpenBUGS code) under model 2 show a 96% probability of overprediction for school 17, and a 6% probability of overprediction for school 2. This may indicate the need to adjust for school-level predictors, or to adopt a cluster effects scheme that is more robust to outlier schools. However, this is an improved performance over model 1 which shows three schools with mixed exceedance probabilities under 0.05 or over 0.95 (8, 17 and 18).

Individual pupils with extreme cross-validatory checks differ according to cluster random effects approach. Both models have under 10% of cases with mixed cross-validatory probabilities either exceeding 0.95 or under 0.05 (cvtail[1] and cvtail[2] in the code). For the random intercepts model, the lowest (highest) exceedance probabilities are for subjects 51 and 88 respectively, subject 51 having zero homework hours but a relatively high achievement of 67, while subject 88 has 5 homework hours but achievement of 33.

The random intercepts and slopes model shows widely discrepant homework effects between schools (under both LKJ and Wishart priors). Hence, outlier pupils may be identified if they are discrepant with the cluster sub-model defined by school-specific intercepts and slopes.

A third analysis illustrates the economy of coding possible with rstan and assumes σ^2 differing by gender (complex level 1 variation). An LKJ prior is assumed on the intercepts-slopes correlation. The posterior mean residual standard deviation is found to be slightly lower for females as compared to males (7.05 vs. 7.64), but the LOO-IC is unchanged (in fact slightly increased) at 3581.

8.3 Discrete Responses: GLMM, Conjugate, and Augmented Data Models

While conjugate multilevel structures can be developed for discrete responses such as counts or proportions (see Section 8.3.1), a more flexible approach is based on the general linear mixed model (GLMM), which extends the linear normal formulation to discrete outcomes. Thus, consider univariate observations y_{ij}, with repetitions j nested in clusters i, that, conditional on cluster effects b_i, follow an exponential family density

$$f(y_{ij} \mid b_i) \propto \exp\left\{\left[\frac{y_{ij}\theta_{ij} - d(\theta_{ij})}{\phi_{ij}} + c(y_{ij}, \phi_{ij})\right]\right\},$$

where θ_{ij} is the canonical parameter and ϕ_{ij} is usually a known scale parameter. Additionally, $E(y_{ij} \mid \theta_{ij}) = d'(\theta_{ij})$ and $\mathrm{Var}(y_{ij} \mid \theta_{ij}, \phi_{ij}) = d''(\theta_{ij})\phi_{ij}$. For example, under the Poisson, $d(u) = \exp(u)$, and for binomial data, $d(u) = \log(1 + e^u)$. Taking the regression terms as $\eta_{ij} = g(\theta_{ij})$ where g is a link function, the observation level model (including a level 2 regression on cluster attributes) is

$$\eta_{ij} = x_{ij}\beta + z_{ij}b_i,$$

$$b_i = \kappa W_i + e,$$

where β and b_i are of dimension p and q respectively.

It is also common to include a residual term $u_i = (u_{i1}, \ldots, u_{in_i})$ to account for overdispersion, so that

$$\eta_{ij} = x_{ij}\beta + z_{ij}b_i + u_{ij}. \tag{8.4}$$

Assume priors $\beta \sim N_p(a, R)$, $b_i \sim N_q(0, \Sigma_b)$ and $u_{ij} \sim N_r(0, \Sigma_u)$, with inverse Wishart priors $\Sigma_b \sim IW(\nu_b, S_b)$ and $\Sigma_u \sim IW(\nu_u, S_u)$. Then the full posterior conditional for each b_i vector is

$$p(b_i \mid b_{[i]}, \beta, u, \Sigma_b, \Sigma_u) \propto \exp\left\{-0.5 b_i' \Sigma_b^{-1} b_i + \sum_{j=1}^{n_i} \frac{y_{ij}\theta_{ij} - d(\theta_{ij})}{\phi_{ij}}\right\},$$

while the full conditional for each u_{ij} vector is

$$p(u_{ij} \mid u_{[ij]}, b, \beta, u, \Sigma_b, \Sigma_u) \propto \exp\left\{-0.5 u_{ij}' \sum_u^{-1} u_{ij} + \frac{y_{ij}\theta_{ij} - d(\theta_{ij})}{\phi_{ij}}\right\}.$$

Additionally, the covariance matrices have inverse Wishart full conditionals, namely

$$\Sigma_b \sim IW\left(\nu_b + m, S_b + \sum_{i=1}^{m} b_i b_i'\right),$$

$$\Sigma_u \sim IW\left(\nu_u + \sum_{i=1}^{m} n_i, S_u + \sum_{i,j} u_{ij} u_{ij}'\right).$$

The GLMM approach includes multilevel multinomial observations in a choice setting (e.g. brand, political party),

$$(d_{ij1}, \ldots, d_{ijK}) \sim \mathrm{Mult}(1, [p_{ij1}, \ldots, p_{ijK}]),$$

with probability π_{ijk} that option k is chosen by subject j in cluster i, namely that $y_{ij}=k$ (or $d_{ijk}=1$) where options are unordered. A particular choice ($k \in 1,\ldots,K$) made by subject j in cluster i results from comparing the latent utilities of all options $(\eta_{ij1},\ldots,\eta_{ijK})$, with

$$p_{ijk} = Pr(y_{ij} = k) = Pr(\eta_{ijk} > \eta_{ijm}),\ \ m \neq k$$

where the η_{ijk} include systematic effects and random errors ε_{ijk}. Suppose the errors follow a Gumbel (extreme value type I) density, namely $P(\varepsilon) = \exp(-\varepsilon - \exp(-\varepsilon))$, then since differences between Gumbel errors follow a standard logistic distribution, the choice probabilities reduce to the multinomial logit (Hedeker, 2003, p.1439).

Predictors in the systematic term may be defined at option-subject, or at option level, but consider subject level predictors x_{ij} and z_{ij} (e.g. voter age) of respective dimensions p and q, that may vary according to cluster i. Then with the final category as a reference, fixed effect parameters and random effects are specific to choices k, with $K-1$ sets of random effects b_{ih} each of dimension q,

$$Pr(y_{ij} = k) = \frac{\exp(\alpha_k + x_{ij}\beta_k + z_{ij}b_{ik})}{1+\sum_{h=1}^{K-1} \exp(\alpha_h + x_{ij}\beta_h + z_{ij}b_{ih})} \qquad k = 1,\ldots,K-1$$

$$Pr(y_{ij} = K) = \frac{1}{1+\sum_{h=1}^{K-1} \exp(\alpha_h + x_{ij}\beta_h + z_{ij}b_{ih})}.$$

The $b_i = (b_{i1},\ldots b_{i,K-1})$ are multivariate zero mean effects, typically assumed multivariate normal.

8.3.1 Augmented Data Multilevel Models

Another option for multilevel binary and multinomial responses is to introduce augmented metric data y^*_{ij} with sampling constrained according to the observed y_{ij}, and apply the linear mixed model to y^*_{ij}. The data augmentation density depends on the assumed link. Thus, a logit link for two-level binary data implies truncated standard logistic sampling to generate the augmented data, namely

$$y^*_{ij} \sim \text{Logistic}(\eta_{ij},1)\ I(A_{ij},B_{ij}),$$

where $A_{ij} = -\infty$ or 0, and $B_{ijk}=0$ or ∞, according as $y_{ij}=0$ or 1. As mentioned in Chapter 7, data augmentation leads to simpler MCMC sampling and improved residual tests. In a multilevel setting, it may further assist in assessing variance partitioning. Consider a regression with a random level 2 intercept

$$\eta_{ij} = x_{ij}\beta + b_i,$$

where $b_i \sim N(0,\sigma_b^2)$. Since the variance of the standard logistic is $\pi^2/3$, the intraclass correlation at level 2 may be obtained as $\sigma_b^2/(\sigma_b^2 + \pi^2/3)$, and monitored over MCMC iterations. Moreover, if the composite fixed effect term $x_{ij}\beta$ is monitored and its posterior variance $\tilde{\sigma}_F^2$ obtained, one may obtain a proportion of variance explained by covariates as $\tilde{\sigma}_F^2/[\tilde{\sigma}_F^2 + \tilde{\sigma}_b^2 + \pi^2/3]$, where $\tilde{\sigma}_b^2$ is the posterior mean of σ_b^2.

For multilevel ordinal outcomes with K levels, the observations

$$(d_{ij1},\ldots,d_{ijK}) \sim \text{Mult}(1,[p_{ij1},\ldots,p_{ijK}]),$$

provide information about an underlying metric variable y^*_{ij} defined by cutpoints such that

$$y_{ij} = k, \quad (\text{i.e. } d_{ijk} = 1)$$

if

$$\kappa_{k-1} < y^*_{ij} \le \kappa_k,$$

for $k = 1,\ldots,K$, where $\kappa_0 = -\infty$ and $\kappa_K = \infty$. The corresponding augmented data regression is

$$y^*_{ij} = x_{ij}\beta + z_{ij}b_i + \varepsilon_{ij},$$

where ε is normal or logistic. If x_{ij} excludes an intercept, there are $K-1$ unknown cutpoints $(\kappa_1,\ldots,\kappa_{K-1})$, with $y_{ij}=1$ if $y^*_{ij} \le \kappa_1$, $y_{ij}=2$ if $\kappa_1 < y^*_{ij} \le \kappa_2$, etc., and $y_{ij}=K$ if $y^*_{ij} > \kappa_{K-1}$.

A standard logistic density for ε_{ij} with mean 0, variance $\pi^2/3$, and distribution function $F(\varepsilon) = \exp(\varepsilon)/(1+\exp(\varepsilon))$ leads to a logit link for the cumulative probabilities

$$\gamma_{ijk} = \sum_{m=1}^{k} p_{ijm} = Pr(y^*_{ij} \le \kappa_k) = Pr(y_{ij} \le k), \quad k = 1,\ldots,K-1$$

with $p_{ijK} = 1 - \Sigma_{m=1}^{K-1} p_{ijm}$. Taking $\varepsilon_{ij} \sim N(0,1)$ corresponds to a probit link for γ_{ijk}. For ε logistic, the hierarchical regression is expressed as follows

$$\begin{aligned}\gamma_{ijk} &= Pr(x_{ij}\beta + z_{ij}b_i + \varepsilon_{ij} \le \kappa_k),\\ &= Pr(\varepsilon_{ij} \le \kappa_k - x_{ij}\beta - z_{ij}b_i),\\ &= \frac{\exp(\kappa_k - x_{ij}\beta - z_{ij}b_i)}{1+\exp(\kappa_k - x_{ij}\beta - z_{ij}b_i)},\end{aligned}$$

that is,

$$\text{logit}(\gamma_{ijk}) = \kappa_k - x_{ij}\beta - z_{ij}b_i.$$

8.3.2 Conjugate Cluster Effects

An alternative to generalised linear mixed models for count and binomial data is provided by conjugate random effects at different levels. Daniels and Gatsonis (1999) also consider hierarchical conjugate priors one-parameter exponential family densities (e.g. Poisson, binomial). For binomial data $y_{ij} \sim \text{Bin}(T_{ij}, p_{ij})$, p_{ij} are taken to be beta distributed with means π_{ij}, and cluster-specific scale parameters δ_i.

With a logit link to predictors, and level 2 regression involving cluster level predictors, W_i one has

$$p_{ij} \sim \text{Beta}(\pi_{ij}\delta_i,(1-\pi_{ij})\delta_i),$$
$$\text{logit}(\pi_{ij}) = x_{ij}\beta + z_{ij}b_i,$$
$$b_i = \kappa W_i + e_i.$$

To provide robustness (e.g. to outlier clusters), the e_i may be taken as Student t distributed (see Section 8.5). The prior on δ_i has the form

$$p(\delta_i) \propto \frac{\delta_i}{(h_i+\delta_i)^2},$$

where $h_i = \min_{j\in 1,\ldots,n_i}(T_{ij})$. For Poisson data, one has $y_{ij} \sim Po(o_{ij}\theta_{ij})$, where o_{ij} is an offset for the expected response, and $\theta_{ij} \sim Ga(\mu_{ij}\delta_i,\delta_i)$. The regression model then involves a log link for the μ_{ij},

$$\log(\mu_{ij}) = x_{ij}\beta + z_{ij}b_i.$$

More specialised models apply for particular data structures. For example, Van Duijn and Jansen (1995) suggest a model for repeated counts (e.g. tests $j=1,\ldots,n_i$ within students $i=1,\ldots,m$) with Poisson means

$$\mu_{ij} = \nu_i\delta_{ij},$$

and gamma distributed student ability effects $\nu_i \sim Ga(a_1,a_2)$, where a_1 and a_2 are additional parameters, and the δ_{ij} represent subject specific difficulty parameters for tests j, with identifiability constraint $\Sigma_j\delta_{ij}=1$, and prior

$$(\delta_{i1},\ldots\delta_{iJ}) \sim \text{Dir}(\xi_1,\ldots,\xi_J),$$

where the ξ_j are also unknowns. If the subjects fall into known (or possibly unknown) groups $k=1,\ldots K$ with allocation indicators $S_i \in (1,\ldots,K)$, then a more general model specifies $(\nu_i \mid S_i = k) \sim Ga(a_{1k},a_{2k})$.

A conjugate structure for stratified area health counts is considered by Dean and MacNab (2001). Thus for micro areas $j=1,\ldots,n_i$ nested within larger areas $i=1,\ldots,m$, let μ be an average event rate across all m areas, and T_{ij} be populations at risk. Assume first cluster level overdispersion represented by effects ρ_i, so that $y_{ij} \sim Po(\mu T_{ij}\rho_i)$, where ρ_i have mean 1, and let the mean and variance of y_{i+} be $T_{i+}\mu$ and $T_{i+}\mu(1+\sigma_\rho^2)$. Under gamma mixing

$$\rho_i \sim Ga\left(\frac{T_{i+}\mu}{\sigma_\rho^2},\frac{T_{i+}\mu}{\sigma_\rho^2}\right),$$

with variance $\sigma_\rho^2/(T_{i+}\mu)$. The interpretation is that ρ_i represents the average relative risk over the T_{i+} individuals in area i.

Example 8.2 Well-Being and Hours Worked

This example illustrates multilevel binary and augmented data methods applied to data from the study of Bliese and Halverson (1996), which are included in the R package multilevel (Bliese, 2016a). There are N=7382 subjects nested in m=99 US Army companies, with the study investigating group level influences on reported well-being. This is an example of a macro-micro multilevel situation, where a response measured at a lower level is predicted by variables measured both at that level and a higher level (Croon and van Veldhoven, 2007).

A binary response analysis sets $y_{ij}=1$ for well-being score above 3.5, and $y_{ij}=0$ otherwise. Then Bliese (2016b) specifies random intercept variation in well-being at group level, together with subject level impacts. These apply to group average hours worked, and subject level hours worked respectively.

This model is investigated using a Bernoulli likelihood, and logit regression in brms and rstan, and an augmented data logistic regression using R2OpenBUGS. Under the latter approach, a logit link involves the sampling mechanism

$$y_{ij}^{*} \sim \text{Logistic}(\eta_{ij},1)\, I(A_{ij},B_{ij}),$$

where $A_{ij}=-\infty$ or 0 (and $B_{ij}=0$ or ∞) according as $y_{ij}=0$ or 1. The equivalent code is

```
for (h in 1:N) {z[h] ~ dlogis(eta[h],1) %_% I(A[h],B[h])
A[h] <- -100*equals(wb[h],0)
B[h] <- 100*equals(wb[h],1)}
```

where wb[h] denotes the observed binary outcome. The R2OpenBUGS analysis centres the group intercepts around the impact of group average hours worked.

Both methods of estimation report a stronger impact of group average hours worked than individual hours worked on individual well-being, with respective posterior means (sd) of −0.27 (0.05) and −0.10 (0.02) (respectively beta[2] and beta[1] in the R2OpenBUGS code).

Subject level mixed predictive checks (Marshall and Spiegelhalter, 2007) are based on sampling replicate cluster intercepts, and these predictive checks are aggregated to group level (testmx.grp in the R2OpenBUGS code). These show well-being in some groups to be much better explained than in others, with average predictive success varying from 0.50 (group 68) to 0.84 (group 57). Both the brms logistic regression and the augmented data logistic regression show around 22% of subjects with predictive concordance below 0.50 (the model does not improve on guesswork for such subjects).

Example 8.3 Ordinal Three Level Model

Vermunt (2013) and others have considered data from the Television, School and Family Smoking Prevention and Cessation Project (Flay et al., 1987). Schools included in the project were randomised to one of four categories, defined by the presence or absence of a TV intervention (TV), and by the presence or absence of a social-resistance classroom curriculum (CC). One outcome measure was a tobacco and health knowledge (THK) score, here represented using $H=4$ ordinal categories, with predictors being a pre-intervention THK score, the binary intervention variables TV and CC, and a CC by TV interaction.

The analysis is three level, with schools $i=1,\ldots,m$ at level 3 (with $m=28$), and classrooms j within schools at level 2. With responses $y_{ijk} \in (1,\ldots,H)$, where $H=4$, for subjects $k=1,\ldots,n_{ij}$ in each class, consider a logit model for the cumulative probabilities

$$\gamma_{ijkh} = Pr(y_{ijk} \leq h),$$

with two sets of higher level errors, namely level 3 random errors u_{3i} with variance σ_3^2, and level 2 class errors u_{2ij} with variance σ_2^2 (pertaining to effects of classrooms within schools). Then

$$\text{logit}(\gamma_{ijkh}) = \kappa_h - \mu_{ijk}, \quad h = 1,\ldots,H-1,$$

$$\mu_{ijk} = x_{ijk}\beta + u_{3i} + u_{2ij}$$

with N(0,1000) priors on fixed effects and U(0,1000) priors on the random effect standard deviations. Predictors are centred.

A two-chain run of 2,500 iterations with 500 burn-in gives posterior means for the cutpoints $(\kappa_1, \kappa_2, \kappa_3)$ of –1.39, –0.11, and 1.10, with a significant coefficient of 0.83 on the curriculum intervention and significant influence also of pre-intervention score, but no significant effects for TV or the interaction term. The posterior means for σ_2 and σ_3 are 0.41 and 0.30 (for classrooms and schools respectively) with densities bounded away from zero. Similar estimates are obtained using brms and rstan.

By contrast, a maximum likelihood analysis using numerical quadrature reported by Rabe-Hesketh et al. (2004) finds an insignificant school variance, and Vermunt (2013) also finds a model with class effects only to be the best fitting.

The analysis was also carried out using the latent data approach, which may be useful for obtaining intraclass correlations or for model checking. This produces larger estimates of σ_2 and σ_3, namely 0.59 and 0.36, but similar fixed predictor and cutpoint estimates. The worst fit (using pointwise WAIC) is for subjects 952 and 190, who have respectively high (low) THK scores, despite low (high) pre-intervention THK scores and absence (presence) of the curriculum intervention.

8.4 Crossed and Multiple Membership Random Effects

Crossed random effects at level 2 and above occur when classifications in a model are not completely nested. For a two-level example, let i denote the main level 2 nesting classification, and h_{ij} denote a crossed nesting. Raudenbush (1993), Browne et al. (2001, Section 3.2) and Snijders and Bosker (1999) mention educational examples, namely pupils classified by primary school i, and by secondary school h_{ij}, or pupils classified by school i and neighbourhood h_{ij}. In the latter situation, a school can draw pupils from multiple neighbourhoods, and residents in a neighbourhood can choose between multiple schools for their children. By extension, if pupils are classified by primary school, secondary school and neighbourhood, then two crossed nestings (denoted say as h_{1ij} and h_{2ij}) will be involved. An important issue is the relationship between the crossed classifications since they may well not be independent, and introducing extra crossed factors will typically reduce the variance explained by the main level 2 nesting.

A straightforward extension of the normal linear mixed model to account for a single extra crossed factor is to add varying intercepts and slopes according to that extra factor. Assuming h_{ij} varies between 1 and H, then adapting the (8.1) format for z_{ij} of dimension q,

$$y_{ij} = x_{ij}\beta + z_{ij}(b_i + c_{h_{ij}}) + u_{ij},$$

where

$$(b_{1i},\ldots,b_{qi}) \sim N_q(0,\Sigma_b), \quad i = 1,\ldots,m,$$

$$(c_{h1},\ldots c_{hq}) \sim N_q(0,\Sigma_c), \quad h = 1,\ldots,H.$$

Alternatively, variation over the extra crossed factor may be applied to a different predictor than those subject to random variation over the main level 2 classification. Often the additional random effects would be confined to intercept variation over the extra crossed factor, so that with $q=1$ and $z_{i1}=1$ also, one has

$$y_{ij} = x_{ij}\beta + b_i + c_{h_{ij}} + u_{ij}.$$

In these situations, the random effects are confounded and empirical identification may be impeded. Selection between random effects may well be needed (Browne et al., 2001).

Another possible source of variation in crossed models is defined in cells formed by cross-classification of two or more higher level factors. For example, N patients living in a particular administrative health district may be classified into subpopulations s based on intersections of their primary care general practitioner $i_1 = 1,\ldots,m_1$, and small area of residence $i_2 = 1,\ldots,m_2$ (Congdon and Best, 2000). Often there may be no subjects in certain combinations of higher level factors. So define total non-empty cells as S_n, equal to or less than the total $S=m_1m_2$ of all possible combinations, with different values $s = 1,\ldots S_n$ defined by cross-hatched factor identifiers $[i_1,i_2]$. Let $r = 1,\ldots,N$ denote a single string subject level identifier. Subjects will be classified by subpopulation $s_r \in \{1,\ldots S_n\}$, by higher level factor 1 classification indicator h_{1r} (general practitioner), higher level factor 2 classification indicator h_{2r} (small area of residence), and so on. Random intercept variation in a metric response over the two factors and the cells then takes the form

$$y_r = x_r\beta + z_r(a_{1,h_{1r}} + a_{2,h_{1r}} + \eta_{s_r}) + u_r, \quad (8.5)$$

where a_{1i_1}, a_{2,i_2}, η_s and u_r are random effects.

Multiple membership schemes define a generic weighting scheme applicable to cross-classified data (Browne et al., 2001), and may be illustrated by the case where subjects at level 1 may belong to more than one level 2 unit. Supposing a pupil's entire primary school career is of interest, there may then be moves between schools. Multiple affiliations then need to be taken account of in terms of school impacts on attainment. Another example is analysis of neighbourhood health effects to take account of changes in residence (Subramanian, 2004). Suppose there are m level 2 units (clusters such as schools) that are included in the analysis, and that subjects $j = 1,\ldots,J$ (not taken to be nested within schools) have K_j level 2 affiliations with weights $\{w_{j1}, w_{j2},\ldots w_{jK_j}\}$ where $\Sigma_{k=1}^{K_j} w_{jk} = 1$. The weights would in many situations be taken as known (e.g. based on the number of terms spent by a pupil in different schools). Then for pupil level predictors z_j of dimension q not varying over affiliations, the normal linear mixed model becomes

$$y_{ij} = x_{ij}\beta + z_{ij}\sum_{k=1}^{K_j} w_{jk}b_k + u_{ij},$$

where $(b_{1i},\ldots,b_{qi}) \sim N_q(0,\Sigma_b)$, $i = 1,\ldots,m$. If the pupil predictors vary over affiliations, then

$$y_{ij} = x_{ij}\beta + \sum_{k=1}^{K_j} w_{jk}z_{jk}b_k + u_{ij}.$$

Multiple member schemes extend to data frames which are structured spatially or temporally rather than nested. A particular kind of multiple member prior can be applied to spatially configured count responses y_i subject to random intercept variation. Thus let $y_i \sim Po(o_i\mu_i)$ where o_i are expected events, and where the μ_i measure the Poisson intensity relative to expected levels (in spatial health applications the μ_i are termed relative risks). Then the impact of K_i neighbouring areas can be represented by random effects b_k while own area effects are represented by effects u_i in a model

$$\log(\mu_i) = x_i\beta + \sum_{k=1}^{K_i} w_{ik} b_k + u_i,$$

where the w_{ik} are row standardised with $\Sigma_{k=1}^{K_i} w_{ik} = 1$, obtained from spatial interactions $C = c_{ik}$. These might be based on binary spatial interactions c_{ik} ($c_{ik}=1$ if areas i and k are contiguous, $c_{ik}=0$ otherwise), or based on distances d_{ik} between area centres, such as $c_{ik} = \exp(-\eta d_{ik})$ where η is positive; then $w_{ik} = c_{ik} / \Sigma_{k=1}^{K_i} c_{ik}$.

Example 8.4 Neighbourhood Effects on Educational Attainment

Raudenbush and Bryk (2002) consider attainment data from a Scottish education authority with $R=2310$ pupils classified by crossed factors (neighbourhood and secondary school). Some neighbourhoods send pupils to multiple schools, while schools generally draw pupils from several neighbourhoods. They consider effects of a "place variable," namely, neighbourhood social deprivation, on educational attainment, after controlling for the impacts of pupil attributes (pupil aptitude and family background). The data involves $m_1=524$ neighbourhoods, and $m_2=17$ schools, with child-specific predictors being gender (1=M, 0=F), and a verbal reasoning quotient (VRQ) and a reading test score (RTS), both obtained when the child was at primary school. Parent-specific predictors are father's status, FSTAT, and three binary indicators (whether father educated beyond age 15, whether the mother educated beyond 15, and whether father unemployed).

Two models are considered. In the first, there are random effects for both neighbourhoods and schools, with varying neighbourhood effects linked to social deprivation. Let h_{1r} and h_{2r} denote neighbourhood and school indices for pupils $r=1,\ldots,R$ in the model

$$y_r = x_r\beta + \alpha_{1,h_{1r}} + \alpha_{2,h_{2r}} + u_r,$$

$$\alpha_{1i_1} \sim N(\gamma_2 \text{Dep}_{i_1}, \sigma_1^2), \quad i_1 = 1,\ldots,m_1$$

$$\alpha_{2i_2} \sim N(\gamma_1, \sigma_2^2), \quad i_2 = 1,\ldots,m_2$$

$$u_r \sim N(0, \sigma_3^2),$$

with x_r excluding a constant term, and all predictors centred. Centring the random effects around the intercept γ_1, and neighbourhood deprivation effect γ_2 improves convergence.

In the R2OpenBUGS analysis, gamma priors are adopted on neighbourhood, school, and pupil random effect precisions. The model is also estimated using the brms library in R, but with half-t priors on standard deviation parameters.

Model 1 results from R2OpenBUGS show residual pupil standard variation (σ_3 has posterior mean 0.67) as more substantial than either school or neighbourhood variation (σ_1 and σ_2 have means 0.09 and 0.08). A negative deprivation impact γ_2 on attainment (with mean −0.156 and 95% interval from −0.202 to −0.106) operates via (is mediated by) the neighbourhood effects.

A second model allows the deprivation effect to vary by school – expressing potentially varying effectiveness on schools h_{2r} in countering catchment area effects (also known as contextual value-added effects). There are now four random variances, with

$$y_r = x_r\beta + \alpha_{1,h_{1r}} + \alpha_{2,h_{2r}} + \delta_{h_{2r}}\text{Dep}_{h_{1r}} + u_r,$$

$$\alpha_{1i_1} \sim N(0,\sigma_1^2),$$

$$\alpha_{2i_2} \sim N(\gamma_1,\sigma_2^2),$$

$$\delta_{i_2} \sim N(\gamma_2,\sigma_3^2),$$

$$u_r \sim N(0,\sigma_4^2).$$

In R2OpenBUGS, convergence of sampling the first three random effect components was improved by partitioning, using a Dirichlet density applied to a total precision parameter. This model is also estimated by brms.

In fact, this model has a slightly worse WAIC than the first model, increasing from 4792 to 4797; and the performance with regard to the mixed predictive check (cvtail in the code) is also not improved. The school deprivation effects δ_{i_2} on attainment from R2OpenBUGS (delta[] in the code) have a mean γ_2 of −0.17, and vary from −0.22 for school 11 to −0.10 for school 9. Their standard deviation σ_3 is 0.08. The brms estimates show less variability between schools in the deprivation effect.

8.5 Robust Multilevel Models

Under normality assumptions regarding errors at different levels, extreme data points can influence estimates of fixed effect and variance component parameters, and reduce the precision of estimates (i.e. widening the width of credible intervals). Sensitivity of the level 2 fixed effect estimates κ to alternative assumptions regarding the prior on level 2 effects b_i in (8.3) is the focus of Seltzer (1993). Estimates of level 2 random cluster effects may also be sensitive to normality assumptions (Seltzer et al., 1996, p.137). Multilevel logistic regression in particular may be sensitive to multicollinearity and small cluster sizes (Shieh and Fouladi, 2003; Moineddin et al., 2007).

For a two-level model, outliers may occur both in level 2 cluster effects and in level 1 within-cluster errors (Pinheiro et al., 2001; Langford and Lewis, 1998), and the two sources may be confounded. For example, a discordant school effect might be due to a systematic effect across all pupils, or because a few pupils in the school are responsible for the discrepancy. Seltzer et al. (2002) investigate how level-1 outliers affect estimation of fixed effect regression parameters and inferences regarding level 2 cluster effects (e.g. treatment contrasts for individual clusters) in two-level models for continuous outcomes.

A robust alternative to the normal linear mixed model based on the multivariate t density is proposed by Pinheiro et al. (2001), and shown to outperform normality assumptions when outliers are present in multilevel data. Daniels and Gatsonis (1999, p.31) assume multivariate t random effects at level 2 by default in a generalised linear mixed model, while Seltzer et al. (1996) present Gibbs sampling steps for the linear mixed model case where a multivariate t with a single degrees of freedom parameter is assumed for level 2 random effects. Seltzer et al. (2002) adopt Student t priors at both levels and apply a U(0, 1) prior to sample from a discrete grid of values on the degrees of freedom parameter. Thus for an equally spaced grid of potential values {2.1, 2.2, 2.3, …, 49.9} with equal prior probabilities, the cumulative probability $Pr(\nu=2.1)+Pr(\nu=2.2)+\ldots$ is calculated for each point, and the $U(0, 1)$ draw determines which is sampled.

Following the Pinheiro et al. (2001) scheme, assume a gamma-normal hierarchical representation with scale mixture parameters $s_i \sim Ga(0.5\nu, 0.5\nu)$, and also that $e_i \sim N_q(0, I)$. Then for continuous responses $y_i = (y_{i1}, \ldots, y_{in_i})'$, a level 2 assumption of t distributed random effects $b_i = (b_{1i}, \ldots, b_{qi})'$ with dispersion Σ_b leads to

$$y_i = X_i\beta + Z_ib_i + u_i, \quad i = 1, \ldots, m$$

$$b_i = \kappa W_i + \Sigma_b^{0.5} e_i / \sqrt{s_i}$$

For outlier clusters with low s_i the overall dispersion Σ_b / s_i^2 is inflated, but the fixed effect κ will be less distorted than under normal level 2 errors.

The degrees of freedom parameters ν_i of the level 2 multivariate t prior may be taken to vary between clusters, namely

$$\begin{bmatrix} y_i \\ b_i \end{bmatrix} \sim t_{n_i+q}\left(\begin{matrix} X_i\beta \\ 0 \end{matrix}, \begin{bmatrix} Z_i\Sigma_bZ_i' + \Lambda_i & Z_i\Sigma_b \\ \Sigma_bZ_i' & \Sigma_b \end{bmatrix}, \nu_i\right).$$

or under a gamma-normal hierarchical representation,

$$\begin{bmatrix} y_i \\ b_i \end{bmatrix} \sim N_{n_i+q}\left(\begin{matrix} X_i\beta \\ 0 \end{matrix}, \frac{1}{s_i}\begin{bmatrix} Z_i\Sigma_bZ_i' + \Lambda_i & Z_i\Sigma_b \\ \Sigma_bZ_i' & \Sigma_b \end{bmatrix}\right).$$

where $s_i \sim Ga(0.5\nu_i, 0.5\nu_i)$. The s_i can then be used for identifying cluster outliers. An alternative to assuming cluster specific degrees of freedom is to take $\nu_i = \nu_{g_i}$, according to a known or possibly unknown grouping variable $g_i \in (1, \ldots, G)$ applicable to clusters, for example, type of school in an educational application.

Discrete mixtures of random effects are also possible for outlier accommodation, modelling non-normality or other asymmetry in random effects. Latent mixtures of regression effects may also be present: Muthén and Asparouhov (2009) show how latent regression classes may be misrepresented as random cluster variation. To detect outlier random effects, Daniels and Gatsonis (1999, p.36) adapt the approach of Albert and Chib (1997) in their models for hierarchical conjugate priors for discrete data.

For nested binomial data $y_{ij} \sim \text{Bin}(n_{ij}, p_{ij})$, a mechanism to detect level 1 outliers may be specified with p_{ij} drawn for a two-group mixture of beta densities, both with means π_{ij}. For the main group, the dispersion parameters are δ_i, while for the outlier group they are deflated as δ_i/K where $K \gg 1$. Then

$$p_{ij} \sim (1-\lambda)\text{Beta}(\pi_{ij}\delta_i, (1-\pi_{ij})\delta_i) + \lambda\text{Beta}\left(\pi_{ij}\frac{\delta_i}{K}, (1-\pi_{ij})\frac{\delta_i}{K}\right).$$

If the outlier probability λ is preset to a low value (e.g. $\lambda = 0.05$), then K might be taken as an extra parameter. Weiss et al. (1999) suggest a similarly motivated prior for mixtures of normal random effects at levels 1 and 2 in (8.1) and (8.3), namely

$$b_i \sim (1-\lambda_b)N_q(0, \Sigma_b) + \lambda_b N_q(0, K_b\Sigma_b),$$

$$u_{ij} \sim (1-\lambda_u)N(0, \sigma_u^2) + \lambda_u N(0, K_u\sigma_u^2),$$

An alternative mixture prior to reduce the impact of parametric assumptions is the mixture of Dirichlet process approach (Kleinman and Ibrahim, 1998; Guha, 2008). Thus, a conventional first stage likelihood

$$y_i \sim N(X_i\beta + Z_i b_i, \sigma^2),$$

may be combined with a semiparametric approach for $b_i = (b_{1i}, \ldots b_{qi})'$, typically with a multivariate normal base G_0 as in

$$b_i \sim G,$$
$$G \sim DP(a, G_0),$$
$$G_0 = N_q(0, D),$$
$$D^{-1} \sim \text{Wishart}(d_0, R_0),$$

Gibbs sampling for D^{-1} is modified for clustering among the sampled b_i (Kleinman and Ibrahim, 1998, p.94).

Example 8.5 Police Stops and Ethnicity

This example considers representation of cluster effects in multilevel Poisson regression analysis of 900 counts y_{ij} of "stop and frisk" over a 15-month period in 1998–99 (Gelman and Hill, 2006). For each of m = 75 New York police precincts, counts are disaggregated both by ethnic group (1 = black, 2 = hispanic, 3 = white), and crime type (1 = violent, 2 = weapons, 3 = property, 4 = drug), so that there are $j = 1, \ldots, n_i$ observations, with $n_i = 12$, for each precinct i. These 12 categories are called classes here. An offset, o_{ij}, is provided by arrests according to precinct, ethnicity, and type in 1997 (multiplied by 15/12); in fact $o_{ij} + 1$ is used instead, since some 1997 arrest counts are zero.

Here, an initial analysis (using jagsUI) has normal random errors at both precinct and precinct-class level (model 1), the latter introduced to account for overdispersion, while the former measure overall crime levels in a precinct. Cluster (precinct) effects are taken as normal. So $y_{ij} \sim Po(\mu_{ij}[o_{ij} + 1])$ with

$$\log(\mu_{ij}) = \beta_0 + \beta_{eth_{ij}} + b_i + u_{ij},$$

$$u_{ij} \sim N(0, \sigma_u^2), b_i \sim N(0, \sigma_b^2).$$

The ethnicity fixed effect $(\beta_1, \beta_2, \beta_3)$ has black ethnicity as reference. In practice, the u_{ij} are centred around the regression term $\beta_0 + \beta_{eth_{ij}}$ to improve convergence. $U(0,100)$ priors are adopted for the random standard deviations.

Using replicate random effects $u_{ij,\text{rep}}$ and $b_{i,\text{rep}}$, and the resulting replicate data $y_{ij,\text{rep}}$ sampled from the model, predictive checks involve the mixed predictive exceedance criterion $Pr(y_{ij,\text{rep}} > y_{ij} \mid y)$ (Green et al., 2009). The observation level log posterior predictive densities (LPPDs) associated with the WAIC are also obtained. The significance of individual precinct effects b_i is assessed using the probabilities $Pr(b_i > 0 \mid y)$.

The scaled deviance (DV in the code) is estimated as 925, so overdispersion is accounted for. The Hispanic and white ethnic coefficients (β_2 and β_3) have 95% intervals (−0.12,0.22), and (−0.59,−0.23), so whites have lower chances of being subject to "stop and frisk." Specifically, they have a 33% lower relative risk, namely 100(1 − exp(−0.4)) where −0.4 is the posterior mean of β_3. Despite the presence of precinct-cell error terms (which might reduce the need for separate precinct effects), a relatively high number (25 out of 75)

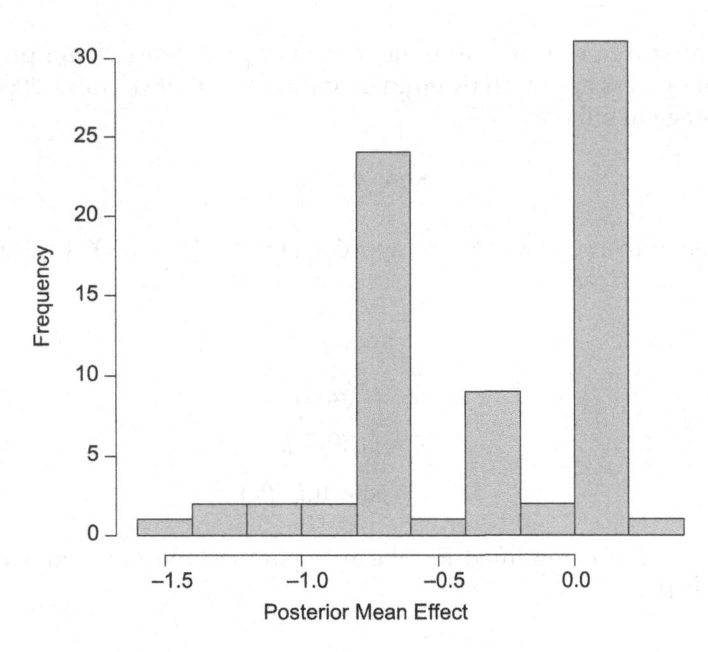

FIGURE 8.1
Precinct random effects, truncated Dirichlet prior.

of the precinct effects b_i are significant in the sense that the probabilities $Pr(b_i > 0 \mid y)$ exceed 0.95 or are under 0.05.

Around 8.7% of the mixed predictive exceedance checks are in the extreme tails (under 0.05 or over 0.95), so the model is reproducing the data effectively. The lowest LPPD values and extreme exceedance probabilities are for subjects with very high stop counts, and for subjects with zero stop counts, despite relatively large offsets.

Of interest in terms of the robustness of the model assumptions are the characteristics of the posterior estimates of b_i and u_{ij}. The proportion of extreme values for the precinct effects may cast doubt on a normality assumption, and as an alternative, a truncated Dirichlet process prior is adopted for these effects (model 2). A fixed Dirichlet concentration parameter is assumed, namely $\alpha = 1$, to aid convergence. The base density involves normal random effects over a maximum of 20 clusters.

A slight reduction in WAIC (from 6897 to 6891) is obtained. Around 9% of the mixed predictive exceedance checks are in the extreme tails (under 0.05 or over 0.95), similar to model 1. Similar results regarding the fixed effects and ethnic differences in risk of stop and frisk are also estimated for this model. However, a histogram of the posterior mean b_i suggests non-normality (Figure 8.1), shown, for example, by a bimodal pattern, and five precincts (2,26,28,51,70) with unusually low b_i.

Example 8.6 Prenatal Care

This example illustrates the possible sensitivity of both fixed effects and variance component estimates to multicollinearity. It involves a study of the adoption of modern prenatal care (binary response) among Guatemalan women, based on 2,449 births (in a 5-year period before the survey) to 1,558 mothers living in one of $m = 161$ communities. The predictor variables are at all levels, namely communities i at level 3, mothers ij at level 2, and pregnancy episodes ijk at level 1, and denoted X_i, X_{ij} and X_{ijk} respectively. The predictors include at level 3: proportion of community population indigenous and distance to the nearest clinic; at level 2: mother's ethnicity, mother's education, husband's education, husband's occupation, and presence of a modern toilet; and at level 1: (existing) child's age, mother's age, and birth order.

Then with $y_{ijk} \sim \text{Bern}(p_{ijk})$, a binary multilevel model specifies normal random intercept variation at levels 2 and 3, namely according to both mother and community. Using a non-centred parameterisation and logit link, one has

$$logit(p_{ijk}) = \alpha + b_{i3} + b_{ij2} + X_i\beta_3 + X_{ij}\beta_2 + X_{ijk}\beta_1,$$

where $b_{i3} \sim N(0,\sigma_3^2)$ are community effects, and $b_{ij2} \sim N(0,\sigma_2^2)$ are mother level effects. Alternatively, community effects and mother effects could be centred at $X_i\beta_3$ and $X_{ij}\beta_2$ respectively. Instability across estimation methods in this dataset is noted by Rodriguez and Goldman (2001) and Guo and Zhao (2000).

This instability may be related partly to small cluster sizes at both levels as well as the binary form of outcome. Here we illustrate the potential impacts of (fixed effect) predictor collinearity, comparing a diffuse normal prior on predictors with a horseshoe prior. Under the horseshoe prior, the student-t prior of the local shrinkage parameters

TABLE 8.1

Modern Pregnancy Advice

	Diffuse Normal Prior on Fixed Regression Effects			Horseshoe Prior on Fixed Regression Effects		
Fixed effects	Mean	2.50%	97.50%	Mean	2.50%	97.50%
Intercept	5.3	0.3	11.1	3.0	0.5	5.9
Pregnancy Level						
Child aged 3–4 years	−1.38	−2.16	−0.66	−0.89	−1.51	−0.29
Mother aged > 25 years	1.28	0.00	2.65	0.60	−0.16	1.57
Birth order 2–3	−1.05	−2.18	0.03	−0.35	−1.16	0.18
Birth order 4–6	−0.54	−2.06	0.98	−0.01	−0.88	0.76
Birth order > 7	−1.29	−3.45	0.73	−0.32	−1.75	0.65
Mother Level						
Indigenous, no Spanish	−7.85	−13.20	−3.59	−5.30	−8.95	−1.70
Indigenous Spanish	−4.21	−7.68	−1.18	−2.60	−5.18	−0.10
Mother's education primary	2.66	0.85	4.79	1.59	0.18	3.03
Mother's education secondary	5.73	1.40	10.95	3.51	0.00	7.22
Husband's education primary	1.14	−0.84	3.22	0.57	−0.45	2.18
Husband's education secondary	4.89	1.04	9.21	3.35	0.19	6.64
Husband's education missing	0.07	−3.03	3.11	−0.05	−1.38	1.25
Husband professional etc.	−0.54	−5.53	4.36	0.65	−0.87	2.70
Husband agric. self-employed	−2.73	−7.09	1.46	−0.57	−2.40	0.60
Husband agric. employee	−3.82	−8.49	0.44	−1.33	−3.52	0.16
Husband skilled service	−1.17	−5.46	3.17	0.25	−1.03	1.83
Modern toilet in households	2.80	0.08	5.87	1.82	−0.02	4.00
Television not watched daily	2.18	−1.81	6.37	0.59	−0.74	3.03
Television watched daily	2.14	−0.38	4.93	1.00	−0.32	3.08
Community Level						
Proportion indigenous, 1981	−6.60	−11.84	−1.98	−4.95	−8.96	−1.08
Distance to nearest clinic	−0.07	−0.14	−0.02	−0.06	−0.10	−0.02
Random effect variances						
Family	10.5	7.7	14.2	7.3	5.7	9.3
Community	5.6	3.8	7.8	4.0	2.8	5.3

(see Equation 7.1) has 1 degree of freedom (Piironen and Vehtari, 2016), while the global parameter has a Cauchy prior with scale 1. Using rstan, convergence is achieved in two chain runs of 2000 iterations.

Table 8.1 shows that posterior mean random intercept variances at both family and community level are reduced by about 30% under the horseshoe prior. Fixed regression effects show considerable shrinkage, but significant predictor effects (on 8 of the 21 predictors) are maintained (as assessed by 95% credible intervals either entirely negative or positive). The indicators κ_j (see Equation 7.2) show that the "indigenous, no Spanish" (mother) and "proportion indigenous" (community) predictors have the highest relevance, with posterior mean κ_j around 0.12 for both (kappa[6] and kappa[20] in the code), and posterior median κ_j around 0.06. This strategy improves fit: the LOO-IC falls from 2653 to 1765 on adopting shrinkage priors.

References

Aarts E, Dolan C, Verhage M, van der Sluis S (2015) Multilevel analysis quantifies variation in the experimental effect while optimizing power and preventing false positives. *BMC Neuroscience*, 16, 94.

Albert J, Chib S (1997) Bayesian tests and model diagnostics in conditionally independent hierarchical models. *Journal of the American Statistical Association*, 92(439), 916–925.

Bliese P (2016a) Package 'Multilevel' Manual. https://cran.r-project.org/web/packages/multilevel/

Bliese P (2016b) Multilevel Modeling in R: A Brief Introduction to R, the multilevel Package and the nlme Package. Darla Moore School of Business, University of South Carolina.

Bliese P, Halverson R (1996) Individual and nomothetic models of job stress: An examination of work hours, cohesion, and well-being. *Journal of Applied Social Psychology*, 26(13), 1171–1189.

Bliese P, Hanges P (2004) Being both too liberal and too conservative: The perils of treating grouped data as though they were independent. *Organizational Research Methods*, 7(4), 400–417.

Browne W (2004) An illustration of the use of reparameterisation methods for improving MCMC efficiency in crossed random effect models. *Multilevel Modelling Newsletter*, 16, 13–25.

Browne W, Draper D, Goldstein H, Rasbash J (2002) Bayesian and likelihood methods for fitting multilevel models with complex level-1 variation. *Computational Statistics and Data Analysis*, 39, 203–225.

Browne W, Goldstein H, Rasbash J (2001) Multiple membership multiple classification (MMMC) models. *Statistical Modelling*, 1, 103–124.

Buerkner P (2017) brms: An R package for Bayesian multilevel models using Stan. *Journal of Statistical Software*, 80(1), 1–28.

Candel J, Winkens B (2003) Performance of empirical Bayes estimators of level-2 random parameters in multilevel analysis: A Monte Carlo study for longitudinal designs. *Journal of Educational and Behavioral Statistics*, 28, 169–194.

Chaix B, Merlo J, Chauvin P (2005) Comparison of a spatial approach with the multilevel approach for investigating place effects on health: The example of healthcare utilisation in France. *Journal of Epidemiology and Community Health*, 59, 517–526.

Chen Z, Dunson D (2003) Random effects selection in linear mixed models. *Biometrics*, 59, 762–769.

Congdon P, Best N (2000) Small area variation in hospital admission rates: Adjusting for referral and provider variation. *Journal of the Royal Statistical Society: Series C*, 49(2), 207–226.

Congdon P, Lloyd P (2010) Estimating small area diabetes prevalence in the US using the behavioral risk factor surveillance system. *Journal of Data Science*, 8(2), 235–252.

Croon M, van Veldhoven M (2007) Predicting group-level outcome variables from variables measured at the individual level: A latent variable multilevel model. *Psychological Methods*, 12(1), 45–57.

Daniels M, Gatsonis C (1999) Hierarchical generalized linear models in the in the analysis of variations in health care utilization. *Journal of the American Statistical Association*, 94, 29–42.
Daniels M, Kass R (1999) Nonconjugate Bayesian estimation of covariance matrices and its use in hierarchical models. *Journal of the American Statistical Association*, 94, 1254–1263.
Daniels M, Zhao Y (2003) Modelling the random effects covariance matrix in longitudinal data. *Statistics in Medicine*, 22, 1631–1647.
Dean C, MacNab Y (2001) Modeling of rates over a hierarchical health administrative structure. *Canadian Journal of Statistics*, 29, 405–419.
Dong G, Ma J, Harris R, Pryce G (2016) Spatial random slope multilevel modeling using multivariate conditional autoregressive models: A case study of subjective travel satisfaction in Beijing. *Annals of the American Association of Geographers*, 106(1), 19–35.
Draper D (1995) Inference and hierarchical modeling in the social sciences. *Journal of Educational and Behavioral Statistics*, 20, 115–147.
Draper D (2006) Bayesian multilevel analysis and MCMC, Chapter 2, in *Handbook of Quantitative Multilevel Analysis*, eds J de Leeuw, E Meijer. Springer, New York.
Flay B, Hansen W, Johnson C, Collins L, Dent C, Dwyer K, Grossman L, Hockstein G, Rauch J, Sobol J, Sobel D, Sussman S, Ulene A (1987) Implementation effectiveness trial of a social influences smoking prevention program using schools and television. *Health Education Research*, 2, 385–400.
Gamerman D (1997) Sampling from the posterior distribution in generalized linear mixed models. *Statistics and Computing*, 7, 57–68.
Gelfand A, Sahu S, Carlin BP (1995) Efficient parameterisations for normal linear mixed models. *Biometrika*, 82, 479–488.
Gelman A (2006) Multilevel (hierarchical) modeling: What it can and can't do. *Technometrics*, 48, 432–435.
Gelman A, Hill J (2006) *Data Analysis Using Regression and Multilevel/Hierarchical Models*. Cambridge University Press.
Gelman A, Pardoe I (2006) Bayesian measures of explained variance and pooling in multilevel (hierarchical) models. *Technometrics*, 48(2), 241–251.
Givens G, Hoeting J (2012) *Computational Statistics*, 2nd Edition. John Wiley.
Goldstein H (2005) Heteroscedasticity and complex variation, pp 790–795, in *Encyclopedia of Statistics in Behavioral Science*, Vol. 2, eds B Everrit, D Howell. Wiley, New York.
Goldstein H, Browne W, Rasbash J (2002) Partitioning variation in multilevel models. *Understanding Statistics*, 1, 223–232.
Green MJ, Medley GF, Browne WJ (2009) Use of posterior predictive assessments to evaluate model fit in multilevel logistic regression. *Veterinary Research*, 40(4), 1–10.
Guha S (2008) Posterior simulation in the generalized linear mixed model with semiparametric random effects. *Journal of Computational and Graphical Statistics*, 17, 410–425.
Guo G, Zhao H (2000) Multilevel modeling for binary data. *Annual Review of Sociology*, 26, 441–462.
Hedeker D (2003) A mixed-effects multinomial logistic regression model. *Statistics in Medicine*, 22, 1433–1446.
Hox J (2002) *Multilevel Analysis: Techniques and Applications*. Lawrence Erlbaum Associates, Mahwah, NJ.
Hsiao C (1996) Random coefficient models, pp 77–99, in *The Econometrics of Panel Data*, eds L Matyas, P Sevestre. Kluwer, Dordrecht, Netherlands.
Kreft I, de Leeuw J (1998) *Introducing Multilevel Modeling*. Sage, Thousand Oaks, CA.
Kleinman K, Ibrahim J (1998) A semi-parametric Bayesian approach to generalized linear mixed models. *Statistics in Medicine*, 17, 2579–2596.
Langford I, Lewis T (1998) Outliers in multilevel data. *Journal of the Royal Statistical Society: Series A*, 161, 121–160.
Lindley DV, Smith AF (1972) Bayes estimates for the linear model. *Journal of the Royal Statistical Society: Series B (Methodological)*, 34(1), 1–18.
Mai Y, Zhang Z (2018) Software Packages for Bayesian Multilevel Modeling. *Structural Equation Modeling*, 25(4), 650–658.

Marshall E, Spiegelhalter D (2007) Identifying outliers in Bayesian hierarchical models: A simulation-based approach. *Bayesian Analysis*, 2(2), 409–444.

McElreath R (2016) *Statistical Rethinking: A Bayesian Course with Examples in R and Stan*. Chapman & Hall/CRC.

Moineddin R, Matheson F, Glazier R (2007) A simulation study of sample size for multilevel logistic regression models. *BMC Medical Research Methodology*, 7, 34.

Muthén B, Asparouhov T (2009) Multilevel regression mixture analysis. *Journal of the Royal Statistical Society. Series A*, 172, 639–657.

Natarajan R, McCulloch C (1998) Gibbs sampling with diffuse proper priors: A valid approach to data-driven inference? *Journal of Computational and Graphical Statistics*, 7, 267–277.

Piironen J, Vehtari A (2016) Projection predictive variable selection using Stan+ R. arXiv preprint arXiv:1508.02502.

Pinheiro J, Liu C, Wu Y (2001) Efficient algorithms for robust estimation in linear mixed-effects models using the multivariate t distribution. *Journal of Computational and Graphical Statistics*, 10, 249–276.

Rabe-Hesketh S, Skrondal A, Pickles A (2004) GLLAMM Manual. U.C. Berkeley Division of Biostatistics Working Paper 160.

Raudenbush S (1993) A crossed random effects model for unbalanced data with applications in cross-sectional and longitudinal research. *Journal of Educational Statistics*, 18, 321–349.

Raudenbush S, Bryk A (2002) *Hierarchical Linear Models: Applications and Data Analysis Methods*. Sage.

Rodriguez G, Goldman N (2001) Improved estimation procedures for multilevel models with binary response: A case study. *Journal of the Royal Statistical Society. Series A*, 164, 339–355.

Seltzer M (1993) Sensitivity analysis for fixed effects in the hierarchical model: A Gibbs sampling approach. *Journal of Educational Statistics*, 18, 207–235.

Seltzer M, Wong W, Bryk A (1996) Bayesian inference in applications of hierarchical models: Issues and methods. *Journal of Educational and Behavioral Statistics*, 21, 131–167.

Seltzer M, Novak J, Choi K, Lim N (2002) Sensitivity analysis for hierarchical models employing *t* level-1 assumptions. *Journal of Educational and Behavioral Statistics*, 27, 181–222.

Shieh Y, Fouladi R (2003) The effect of multicollinearity on multilevel modeling parameter estimates and standard errors. *Educational and Psychological Measurement*, 63(6), 951–985.

Skrondal A, Rabe-Hesketh S (2004) *Generalized Latent Variable Modelling: Multilevel, Longitudinal and Structural Equation Models*. Chapman & Hall/CRC, Boca Raton, FL.

Snijders T, Berkhof J (2002) Diagnostic checks for multilevel models, in *Handbook of Quantitative Multilevel Analysis*, eds J de Leeuw, I Kreft. Kluwer, Boston/Dordrecht/London.

Snijders T, Bosker R (1999) *Multilevel Analysis. An Introduction to Basic and Advanced Multilevel Modelling*. Sage, London, UK.

Subramanian S (2004) The relevance of multilevel statistical methods for identifying causal neighborhood effects. *Social Science & Medicine*, 58, 1961–1967.

Subramanian S, Jones K, Duncan C (2003) Multilevel methods for public health research, pp 65–111, in *Neighborhoods and Health*, eds I Kawachi, L Berkman. Oxford University Press, New York.

van Duijn M, Jansen M (1995) Modelling repeated count data: some extensions of the Rasch Poisson counts model. *Journal of Educational and Behavioral Statistics*, 20, 241–258.

Vermunt J (2013) Categorical response data, Chapter 16, pp 289–297, in *The SAGE Handbook of Multilevel Modeling*, eds M Scott, J Simonoff, B Marx. Sage.

Weiss R, Cho M, Yanuzzi M (1999) On Bayesian calculations for mixture priors and likelihoods. *Statistics in Medicine*, 18, 1555–1570.

Zhao Y, Staudenmayer J, Coull B, Wand M (2006) General design Bayesian generalized linear mixed models. *Statistical Science*, 21, 35–51.

9

Factor Analysis, Structural Equation Models, and Multivariate Priors

9.1 Introduction

A range of multivariate techniques are available both for modelling multivariate collections of metric, binary, or count data, and for modelling multivariate random effects or regression residuals. These include data reduction (reduced dimension) methods such as factor and principal component analysis (e.g. Hayashi and Arav, 2006; Lopes and West, 2004), structural equation modelling (Schumacker and Lomax, 2016), discriminant analysis (e.g. Brown et al., 1999; Rigby, 1997), and data mining, as well as direct (full dimension) modelling of the joint density of the observations or regression residuals (e.g. Chib and Winkelmann, 2001; Martinez-Beneito, 2013). Structured multivariate effects in the analysis of spatial or time configured data raise additional issues, such as representing inter-variable correlation within units as well as non-exchangeability between units (Song et al., 2005). Bayesian applications of factor analysis and structural equation modelling have grown considerably in recent years; for overviews, see Palomo et al. (2007), Merkle and Wang (2016), Kaplan and Depaoli (2012), Lee (2007), Stromeyer et al. (2015), and Levy and Mislevy (2016).

The rationale for introducing latent variables lies in parsimonious representation of the covariance structure of multivariate data, while also revealing underlying clustering of, or associations between, the variables, ideally with substantive interpretability. The latent variables are typically unobservable constructs (e.g. authoritarianism, population morbidity, or a common trend over time) that can only be imperfectly measured by observed indicators. The latent variables may be continuous, as in factor analysis (Fokoue, 2004; Lopes and West, 2004), or categorical, as in latent class analysis (Berkhof et al., 2003). The original variables might themselves also be discrete or continuous. For example, item response models typically involve multiple binary observed items and a single latent continuous ability score (Bazan et al., 2006; Luo and Jiao, 2017; Albert and Ghosh, 2000). Bayesian latent variable packages in R include blavaan (Merkle and Rosseel, 2017), brms (Byrnes, 2017), BayesFM (Piatek, 2017), bfa (Murray, 2016), and BayesLCA (White, 2017). Preliminary analysis using classical estimation is often useful in problem definition, for example, using the lavaan or openMX packages (Boker et al., 2011).

The extraction of information from multivariate observed indicators to derive a smaller set of latent variables defines a measurement model, as in confirmatory and explanatory factor analysis (Bartholomew, 1987; Skrondal and Rabe-Hesketh, 2007). The subsequent use of the latent constructs in describing causal relationships or associations leads into structural equation modelling (Lee, 2007). Both types of model have been developed, especially in areas such as psychology, marketing, educational testing, and sociology, where it is not possible to measure underlying constructs directly. Newer areas of development include environmental modelling (Malaeb et al., 2000; Nikolov et al., 2007), biomass models (Arhonditsis et al., 2006a), and time series and spatial data analysis using common factor approaches.

The observed variables in a measurement model are variously known as "items" (e.g. in psychometric tests), as "indicators," or as "manifest variables." Canonical assumptions are that (a) conditional on the constructs, the observed indicators are independent, in which case the constructs explain the observed correlations between the indicators, and (b) that the construct scores are independent over subjects. As Bollen (2002) points out, the local independence property in (a) is not an intrinsic feature of structural equation models, while spatial and time series factor and structural equation models (Hogan and Tchernis, 2004; Congdon et al., 2007) exemplify how construct scores may be dependent over space or time.

This chapter presents a selective review of multivariate techniques, namely

a) factor modelling via continuous latent constructs, as applied in normal linear and general linear model contexts (Sections 9.2, 9.3, and 9.4);
b) models for multivariate discrete area (lattice) data, including spatial factor models (Sections 9.6 and 9.7); and
c) models for multivariate time series (Section 9.8), with a focus on dynamic linear and general linear models.

A Bayesian approach is arguably of benefit in such multivariate applications. Many classical applications of factor and structural equation methods assume multivariate normality of the indicators, with estimation based on minimising a discrepancy between the observed and predicted covariance matrix – under multivariate normality, the covariance matrix is sufficient for describing the correlations between observed indicators (Sanchez et al., 2005). Considerations of robustness to outliers and other departures from normality, and the ease with which parameter restrictions may be imposed and predictions made for new cases, may point to a Bayes approach which retains the full observation set as input (Lee, 2007) (see Section 9.5). The fully Bayes method has further potential advantages in allowing flexible prior specification (e.g. hierarchical priors on loadings as against fixed effects priors), and for describing the densities of the parameters of structural equation models without making asymptotic approximations (Aitkin and Aitkin, 2005). Simpler fitting of models involving interactions between latent factors is also a feature (Merkle and Wang, 2016).

9.2 Normal Linear Structural Equation and Factor Models

Following Joreskog (1973) and classical presentations (e.g. Bollen, 1989; Hoyle, 1995), Bayesian treatments of normal linear or general linear structural equation models are now substantially represented (e.g. Nikolov et al., 2007; Song et al., 2006; Palomo et al., 2007; Kaplan and Depaoli, 2012). Consider observed multivariate metric indicators y and x, and continuous endogenous and exogenous construct vectors, denoted F and H respectively. For subjects $i = 1,\ldots,n$, the measurement model components of a normal linear SEM are

$$y_i = a_y + \Lambda_y F_i + u_i,$$

$$x_i = a_x + \Lambda_x H_i + e_i,$$

where $y_i = (y_{1i},\ldots,y_{P_y i})'$ is a $P_y \times 1$ vector of indicators describing or measuring an endogenous construct vector $F_i = (F_{1i},\ldots,F_{Q_y i})'$ of dimension Q_y less than or equal to P_y; and x_i is a $P_x \times 1$ vector of indicators measuring an exogenous construct vector H_i of dimension $Q_x \leq P_x$. The individual factor variables F_{qi} may be independent of each other or intercorrelated, and similarly for the H_{ri}. The matrices Λ_y and Λ_x are of dimension $P_y \times Q_y$ and $P_x \times Q_x$ and contain loading parameters describing how observed indicators are related to the latent constructs. The $\{F,H\}$ are sometimes known as common factors while the errors $\{u,e\}$ are sometimes called unique factors (Skrondal and Rabe-Hesketh, 2007). The errors u_i and e_i are assumed independent of the common factors and are typically assumed to have diagonal covariance matrices (Merkle and Rosseel, 2017).

A structural model may describe (a) interrelations between the F_{qi} (namely reciprocal flows between endogenous variables such as social authoritarianism and religiosity) and (b) effects of exogenous constructs H_{ri} on the endogenous ones (e.g. effects of socioeconomic status on authoritarianism or religiosity). These effects are represented by the equation system

$$F_i = BF_i + CH_i + w_i,$$

$$w_i \sim N(0,\Phi),$$

where an intercept is typically not identified, and B is a $Q_y \times Q_y$ matrix with zero diagonal elements and off-diagonal parameters describing relations between endogenous constructs. The matrix C is $Q_y \times Q_x$ with parameters describing the impact of exogenous on endogenous constructs. The structural model may also contain further observed variables as responses or predictors.

Many multivariate reduction applications involve just a measurement model (i.e. a simple factor analysis), and so distinction between different types of observed indicator and factor is not needed. Then a normal linear factor model is

$$y_i = a + \Lambda F_i + u_i, \tag{9.1}$$

where $y_i = (y_{1i},\ldots,y_{Pi})'$ is $P \times 1$, $F_i = (F_{1i},\ldots,F_{Qi})'$ is of dimension $Q < P$, and $(u_{1i},\ldots,u_{Pi})' \sim N_P(0,\Sigma)$. Either interrelated factors may be posited with

$$F_i = BF_i + w_i,$$

or independent factors

$$F_i \sim N_Q(0,\Phi)$$

assumed, with diagonal Φ (Mavridis and Ntzoufras, 2014). Identifying assumptions on Λ and Φ are considered below. Under a local independence assumption, the residuals $(u_{1i},\ldots,u_{Pi})'$ are typically taken to be independent over cases i and variables, so that $\Sigma = \text{diag}(\sigma_1^2,\sigma_2^2,\ldots,\sigma_P^2)I$. This assumption can equivalently be stated as that the outcome variables are conditionally independent, given the latent variables (Skrondal and Rabe-Hesketh, 2007).

It may be noted that path analysis models, a special case of SEM, may be estimated straightforwardly using brms (Byrnes, 2017) [1]. Whereas SEM models in general may include latent variables, path analysis models assume observed variables measured without error. Only postulated structural relationships between observed variables are included in the model. This approach is often used when particular variables are thought to mediate relationships between others.

9.2.1 Forms of Model

If all loadings λ_{pq} in the $P \times Q$ matrix Λ are free parameters (apart from those subject to identification constraints, as discussed below), this structure is known as an exploratory factor analysis (EFA), and typically assumes independent factors, with $\Phi = I$ (Merkle and Wang, 2016). By contrast, in a confirmatory factor analysis (CFA) or measurement model, many of the loadings take preset values (usually zero) on the basis of substantive theory, and correlations between factors may be assumed. A particular form of confirmatory model is known as simple structure, such that each observed variable y_{pi} loads on only one of the constructs F_{qi}. For example, Fleishman and Lawrence (2003) apply a simple structure model to ordinal items from the SF12 questionnaire, assuming that each item reflects either a physical or mental health construct.

A multiple indicator-multiple cause (MIMIC) model extends confirmatory models by incorporating the effects of exogenous observed variables on latent factors (Joreskog and Goldberger, 1975; Tekwe et al., 2014). MIMIC models for normal outcomes consist of (a) measurement equations

$$y_i = a + \Lambda F_i + \varphi X_i + u_i,$$

relating multiple indicator variables y_i to latent constructs F_i, and possibly also to known influences X_i, and (b) structural equations. In the latter, the latent variables F_i are related, both to one another and to observed exogenous variables Z_i, which are viewed as causal influences on the factors, namely

$$F_i = BF_i + CZ_i + w_i,$$

where Z_i excludes a constant term, and the coefficient matrix B allows reciprocal effects between latent factors. A MIMIC model with a single latent construct, as applied, for instance, in analyses of the size of underground economies (Wang et al., 2006), would typically take the form

$$y_{pi} = a_p + \lambda_p F_i + \varphi_p X_i + u_{pi},$$

$$F_i = \gamma Z_i + w_i.$$

As noted by Breusch (2005), the correlation structure in a MIMIC model may need substantive support, as it typically assumes that (i) the indicators y are conditionally independent of the causes Z, given the latent construct(s) F, and (ii) that the indicators $y_1,\ldots,y_P$ are mutually independent given F. This amounts to saying that all connections that indicator variables y have with the causal variables Z, and with one another, are transmitted through the latent variable(s).

9.2.2 Model Definition

Bayesian analysis in the normal linear factor model has recently focused on model definition questions. These include selection of important factor-indicator loadings (analogous to predictor selection), covariance specification, and uncertainty in the number of factors. Predictor selection methods such as SSVS (George and McCulloch, 1993) can be adapted to selection of important loadings using binary indicators γ_{jk} for observed item j and latent factor k. These indicators provide information about which items are associated with particular factors, and which items are relevant or irrelevant to the overall latent structure. For a preset number of factors Q, this leads to confirmatory analysis, but subject to uncertainty (Lu et al., 2016). Thus, analogous to SSVS, one has for $\gamma_{jk}=0$,

$$\lambda_{jk} \sim N(0, \varphi_j^2),$$

where φ_j^2 is set very small so as to shrink λ_{jk} towards zero, whereas for $\gamma_{jk}=1$,

$$\lambda_{jk} \sim N(0, c_j^2),$$

where c_j is chosen large (e.g. $c_j=10$ or $c_j=100$) to enable effective search for non-zero λ_{jk} values. Alternatively a spike and slab prior may be used, with $\lambda_{jk}=0$ when $\gamma_{jk}=0$.

Such procedures can be extended with binary indicators δ_k that allow retention or exclusion of factors (Mavridis and Ntzoufras, 2014). This leads to item and factor selection in an exploratory factor analysis in which

$$\lambda_{jk} \mid \gamma_{jk}, \delta_k \sim (1-\gamma_{jk}\delta_k)N(0, \varphi_{jk}^2) + \gamma_{jk}\delta_k N(0, c_{jk}^2).$$

This involves a hierarchical prior on selection indicators, whereby

$$\delta_k \sim \text{Bern}(\pi_\delta),$$

$$\gamma_{jk} \mid \delta_k \sim \text{Bern}(\pi_\gamma \delta_k).$$

As opposed to the selection of loadings, sparsity-inducing priors may be applied (Feng et al., 2017; Bhattacharya and Dunson, 2011). Bhattacharya and Dunson propose shrinkage parameters τ_k for the kth column of Λ, combined with a hierarchical Student t_ν degrees of freedom, specified so that sparsity is encouraged for higher k. Thus

$$\lambda_{jk} \sim N(0, \phi_{jk}^{-1}\tau_k^{-1}),$$

$$\phi_{jk} \sim Ga\left(\frac{\nu}{2}, \frac{\nu}{2}\right),$$

$$\tau_k = \prod_{l=1}^{k} \delta_l,$$

$$\delta_1 \sim Ga(a_1,1),$$
$$\{\delta_2,\ldots,\delta_k\} \sim Ga(a_2,1),$$

where $a_2 > 1$, so that the precisions τ_k are necessarily increasing. Fokoue (2004) proposes to seek relatively simple structure (a Bayesian version of varimax rotation) by taking the precisions for each loading as unknown gamma variables, namely

$$\lambda_{jk} \sim N(0,\tau_{jk}^{-1})$$
$$\tau_{jk} \sim Ga(a,b).$$

In related work, Muthén and Asparouhov (2012) propose a modified constraint form of confirmatory analysis, labelled as a Bayesian SEM, and included in the Mplus package. Under this approach, the main loadings (those consistent with simple structure) have a prior variance large enough to represent non-zero effects. However, instead of constraining other (cross) loadings to zero, they are assigned informative priors with very low variance (e.g. 0.01), so are approximate rather than exact zeros. If certain cross-loadings are found to be significant (95% credibility interval excluding zero) despite these priors, such that an item loads on more than one construct, then this suggests simple structure no longer holds. The model may be re-estimated with those cross-loadings assigned a less informative prior (Smith et al., 2017).

Default covariance specifications, such as diagonal Σ and Φ in (9.1), may be restrictive in certain applications. The package blavaan (Merkle and Rosseel, 2017) uses a form of parameter expansion, involving phantom latent variables, to facilitate the estimation of non-diagonal covariance matrices.

Choice between models involving different numbers of factors may be tackled using parameter expansion, combined with a Bayes factor approximation (Ghosh and Dunson, 2008), by RJMCMC (reversible jump Markov chain Monte Carlo) methods (Lopes and West, 2004), or by marginal likelihood approximation using path sampling (Lee, 2007). The latter approach may be extended to full structural equation models (SEMs) (Lee and Song, 2008). The parameter expansion method may also improve MCMC performance (Ghosh and Dunson, 2009; Merkle and Wang, 2016), and involves a reference model with standardised factors, and a lower triangular structure for Λ (see Section 9.3), including the diagonals constraint $\lambda_{qq} > 0$.

Thus, the reference model is

$$y_i = a + \Lambda F_i + u_i, \quad F_i \sim N_Q(0,R),$$

where R allows correlations between factors, but has diagonal 1. The expanded model is

$$y_i = a + \Lambda^* F_i^* + u_i, \quad F_i^* \sim N_Q(0,\Psi)$$

where Ψ is unconstrained, the loadings Λ^* are not subject to the diagonals constraint $\lambda_{qq}^* > 0$, but Λ^* is still lower triangular. $Q(Q-1)$ parameters in Λ^* are set to zero when R is non-diagonal (Merkle and Wang, 2016). Priors on parameters in the expanded model induce priors on (Λ,F,R) in the reference model, via

$$\lambda_{pq} = S(\lambda_{qq}^*)\lambda_{pq}^*\psi_q^{0.5}, \tag{9.2}$$

$$F_{qi} = S(\lambda^*_{qq})F^*_{qi} / \psi_q^{0.5},$$

$$R_{qr} = S(\lambda^*_{qq})S(\lambda^*_{rr})\Psi_{qr} / (\Psi_{qq}\Psi_{rr}),$$

where a sign function, $S(x) = -1$ if $x < 0$ and $S(x) = 1$ if $x \geq 0$, is used to ensure a positive diagonals constraint in Λ.

9.2.3 Marginal and Complete Data Likelihoods, and MCMC Sampling

From (9.1), the conditional likelihood of the normal linear factor model is $p(y_i \mid F_i, a, \Lambda, \Phi, \Sigma) = N(a + \Lambda F_i, \Sigma)$, with conditional covariance matrix $V(y_i \mid F_i, \Sigma) = \Sigma$, and hence $\{\text{cov}(y_{ji}, y_{mi}) = 0, m \neq j\}$ if Σ is diagonal. The marginal likelihood obtained by integrating out the factor scores in the normal linear factor model (Lee and Shi, 2000, p.724; Fokoue, 2004) is $p(y_i \mid a, \Lambda, \Phi, \Sigma) = N(a, \Lambda\Phi\Lambda' + \Sigma)$. The joint likelihood of y_i and F_i, obtained by multiplying the marginal density of F, $F_i \sim N_Q(0, \Phi)$, and the conditional density of y_i given F_i, is

$$\begin{bmatrix} y_i \\ F_i \end{bmatrix} \sim N_{P+Q}\left(\begin{bmatrix} \alpha \\ 0 \end{bmatrix}, \begin{bmatrix} \Lambda\Phi\Lambda' + \Sigma & \Lambda\Phi \\ \Phi\Lambda' & \Phi \end{bmatrix}\right).$$

When the factors are standardised (Bartholomew et al., 2002, p.150; Lopes and West, 2004, p.44), the marginal variance of y_p is accordingly $\lambda^2_{p1} + \ldots \lambda^2_{pQ} + \sigma^2_p$ and the marginal covariance of y_p and y_m is $\lambda_{p1}\lambda_{m1} + \lambda_{p2}\lambda_{m2} \ldots + \lambda_{pQ}\lambda_{mQ}$. The contribution $\lambda^2_{p1} + \ldots \lambda^2_{pQ}$ of the common factors to explaining the marginal variability in the y_p is known as the "communality," while that part due to the residual error σ^2_p is called the "unique variance" or "uniqueness."

The marginal likelihood structure for cov(y) as $\Lambda\Phi\Lambda' + \Sigma$ does not lead to any simple form for the posterior distributions of the unknowns, though it can be used in RJMCMC approaches to estimation and factor model selection (Lopes and West, 2004). In Gibbs sampling estimation of linear Bayesian factor and SEM models, it is simplest to approach estimation of the parameters $(F, a, \Lambda, \Phi, \Sigma)$ indirectly through the conditional likelihood or complete data model (Aitkin and Aitkin, 2005; Fokoue, 2004), with the F scores regarded as missing data rather than integrated out (Lee and Shi, 2000). Setting $\theta = (a, \Lambda, \Phi, \Sigma)$, the posterior density is then

$$p(\theta, F \mid y) \propto L(\theta \mid y, F)p(\theta).$$

While MCMC sampling is typically used with the conditional likelihood, the marginal covariance $\Lambda\Phi\Lambda' + \Sigma$ may be useful in posterior checking of model assumptions (e.g. conditional independence between the y variables given the factor scores). For example, Lee and Shi (2000) suggest a posterior check using $D(y, \theta) = \sum_i y_i'(\Lambda\Phi\Lambda' + \Sigma)^{-1} y_i$. Following Gelman et al. (1996), replicate data $y_{\text{rep},i}$ are sampled from the predictive distribution $p(y_{\text{rep}} \mid y, \theta)$ and $D(y,\theta)$ compared to $D(y_{\text{rep}},\theta)$.

From a set of MCMC samples, one seeks the marginal posterior density $p(\theta \mid y)$ of the hyperparameters, and the predictive distribution $p(F \mid y)$ of the factor scores. Estimation at iteration $t+1$ proceeds by switching between (a) sampling $\theta^{(t+1)}$ from the posterior conditional $p(\theta \mid y, F^{(t)})$ for θ conditional on y and sampled F scores, and (b) updating $F^{(t+1)}$ from the conditional density $p(F \mid y, \theta^{(t+1)})$. The latter corresponds to the imputation step in data augmentation (Tanner, 1996).

A range of inference issues may occur, subject to identifiability being fully considered (Section 9.3). The patterns of significant loadings and subject factor scores raise questions of substantive theory, depending on the application area. As noted by Aitkin and Aitkin (2005), one can assess the significance of parameter or factor score contrasts on the basis of the MCMC sample, such as pairwise difference or ratio comparisons of scores on the kth factor for subjects i_1 and i_2, $F_{i_1k} - F_{i_2k}$ and F_{i_1k}/F_{i_2k}. Compared to classical analysis, the posterior means and variances of the factor scores (and of factor contrasts) are routinely obtained.

To illustrate MCMC complete data-sampling, assume Σ is diagonal in the conjugate normal model (9.1) with priors $\sigma_{pp}^{-1} \sim Ga(\alpha_{0p}, \beta_{0p})$, that the precision matrix for F has a Wishart prior $\Phi^{-1} \sim W(R_0, \rho_0)$, and that the prior for Λ follows the form proposed by Press and Shigemasu (1989). Specifically, with Λ_p as the pth row of Λ,

$$\Lambda_p \sim N_Q(\Lambda_{0p}, \sigma_{pp} H_{0p}),$$

where the $Q \times Q$ matrix H_{0p} is positive definite. Often, simple assumptions such as $H_{0p} = I_Q$ are made (Lee and Shi, 2000, p.729). Letting y_p' be the pth row of y, and denoting $\Omega_p = (H_{0p}^{-1} + F'F)^{-1}$, and $\eta_p = \Omega_p(H_{0p}^{-1}\Lambda_{0p} + Fy_p)$, the posterior conditional for the unique variances is (Lee and Shi, 2000, p.725)

$$\sigma_{pp}^{-1} \sim Ga(\alpha_{0p} + n/2, \beta_{0p} + 0.5[y_p' y_p - \eta_p' \Omega_p^{-1} \eta_p + \Lambda_{0p}' H_{0p}^{-1} \Lambda_{0p}]).$$

The conditional for Λ_p is a Q-variate normal with mean η_p and covariance $\sigma_{pp}\Omega_p$, and the conditional for Φ^{-1} is Wishart with scale matrix $FF' + R_0$ and degrees of freedom $n + \rho_0$. Finally, the conditional $p(F_i \mid y, \theta)$ for the factor scores for subject i is a Q-variate normal with mean $[\Phi^{-1} + \Lambda'\Sigma^{-1}\Lambda]^{-1}\Lambda'\Sigma^{-1}y_i$ and covariance $[\Phi^{-1} + \Lambda'\Sigma^{-1}\Lambda]^{-1}$.

9.3 Identifiability and Priors on Loadings

Under the model (9.1), the marginal covariance of y is $V = \Lambda\Phi\Lambda' + \Sigma$. It can be seen that the contribution $\Lambda\Phi\Lambda'$ of the factor scores to explaining variation in the y may be achieved by an infinite number of pairs (Λ,Φ), and constraints must be imposed to ensure a unique location and scale for the factor scores (Wedel et al., 2003, pp.358–359). One way of providing factor score identifiability (the scaling constraint) is to define the factors to be in standardised form, with zero means and variances of 1 (Mezzetti and Billari, 2005). Under the alternative anchoring constraint (Skrondal and Rabe-Hesketh, 2004), one among the set of loadings $\{\lambda_{pq}, p = 1,\ldots,P\}$ on each construct is preset for identification. The factors are still required to have zero means (providing unique location), but may have unknown variances.

For the measurement model to be identifiable, the number of unknown parameters in $\theta = (\Sigma,\Phi,\Lambda)$ must be less than the number, $P(P+1)/2$, of distinct elements in the residual variance-covariance matrix V of y. For example, in the standardised factor case, and with $\Phi = I$ excluding correlations, one has

$$V = \Lambda\Lambda' + \Sigma,$$

with $PQ+P$ parameters on the right-hand side under a local independence assumption (Σ taken as diagonal). For $P(P+1)/2 \geq PQ+P$ to apply requires that $P \geq 2Q+1$ (Geweke and Zhou, 1996).

In confirmatory models, certain elements of Λ are generally preset to zero, alleviating requirements such that Σ be diagonal or that Φ exclude covariances/correlations. However, in exploratory factor analysis (EFA) with multiple factors ($Q>1$), additional identifying constraints must be set to avoid rotation invariance. Otherwise, there is no unique solution because any orthogonal transformation of Λ leaves the likelihood unchanged (Everitt, 1984, p.16). Thus for $F^* = H'F$ and $\Lambda^* = \Lambda H$, where $HH' = I$,

$$y = 1a + \Lambda F + u = 1a + (\Lambda H)(H'F) + u = 1a + \Lambda^* F^* + u$$

where $\text{cov}(F^*) = H'\text{cov}(F)H = \text{cov}(F)$. The exception is the simple structure case (each observed variable loading on only one factor) when rotational identifiability is not an issue (Wedel et al., 2003, p.358; Liu et al., 2005, p.550).

In other cases, EFA identification may be achieved by fixing enough λ_{pq} to ensure a unique solution; thus in the case $Q=2$, setting any $\lambda_{p2} = 0$ would be sufficient. Provided the variables are ordered in such a way as to ensure substantive justification, a widely adopted option is to assume Λ to be lower triangular, as in Geweke and Zhou (1996), Ghosh and Dunson (2009), Zhou et al. (2014), and Mavridis and Ntzoufras (2014), namely

$$\Lambda = \begin{bmatrix} \lambda_{11} & 0 & 0 & \cdots & 0 & 0 \\ \lambda_{21} & \lambda_{22} & 0 & \cdots & 0 & 0 \\ \lambda_{31} & \lambda_{32} & \lambda_{33} & \cdots & 0 & 0 \\ \vdots & \vdots & \vdots & \ddots & \vdots & \vdots \\ \lambda_{Q-1,1} & \lambda_{Q-1,2} & \lambda_{Q-1,3} & \cdots & \lambda_{Q-1,Q-1} & 0 \\ \lambda_{Q1} & \lambda_{Q2} & \lambda_{Q3} & \cdots & \lambda_{Q,Q-1} & \lambda_{QQ} \\ \vdots & \vdots & \vdots & \ddots & \vdots & \vdots \\ \lambda_{P1} & \lambda_{P2} & \lambda_{P3} & \cdots & \lambda_{P,Q-1} & \lambda_{PQ} \end{bmatrix}.$$

The required structural zeros can be chosen according to prior knowledge, perhaps requiring rearrangement of the indicators. A possible drawback with this constraint is order dependence (Bhattacharya and Dunson, 2011), whereby the choice of the first Q responses becomes an important model feature. Conti et al. (2014) avoid assuming a lower triangular Λ by including identifying criteria into prior densities for model parameters. This leads to an EFA in which indicators are uniquely allocated to only one factor, but where neither the number of factors nor the structure of the loading matrix are specified a priori. This approach is applied in the R package BayesFM (Piatek, 2017).

To avoid potential labelling issues, a lower triangular Λ can be combined with the diagonals constraint

$$\lambda_{qq} > 0.$$

If the λ_{qq} are unknowns under a standardised factor scale with $\Phi = I$, one might take

$$\lambda_{qq} \sim N(0, \delta_{qq})I(0,), \tag{9.3}$$

or some other positive prior (e.g. lognormal). Otherwise, without such a constraint, and since $\Lambda F = (-\Lambda)(-F)$, loadings on (and hence scores for) a particular factor may flip over during MCMC iterations (Geweke and Zhou, 1996, p.566). In fact, this may happen even if a necessarily positive prior, as in (9.3), is adopted. The effectiveness of the qth indicator, in acting as a "factor founder" (Aßmann et al., 2016) or "anchor item," and hence guiding the remaining loadings on the qth factor, may be influenced in substantive applications by the ordering of indicators (see Example 9.2). This may be so in applications with a large number of indicators and/or relatively modest correlations.

To completely avoid possible label-switching, a positivity constraint may be applied to all loadings (Ghosh and Dunson, 2009; Sahu, 2002). A positivity constraint on all difficulty loadings is in fact standard in item response theory (IRT) (Section 9.4.2) (Natesan et al., 2016; Luo and Jiao, 2017). Setting one loading for each construct to be fixed (usually at 1.0) under an anchoring constraint, also usually ensures remaining loadings conform to a consistent interpretation and direction of the factor (Levy and Mislevy, 2016).

9.3.1 An Illustration of Identifiability Issues

To exemplify identifiability constraints, consider a spatial example involving English local authorities, and suppose six observed indicators $\{y_1,\ldots,y_6\}$ are taken to measure two latent area constructs F_1 and F_2, deprivation and fragmentation. Thus, several studies have shown that area material deprivation (i.e. meaning economic hardship represented by observed variables such as high unemployment, and low car and home ownership) tends to be associated with higher psychiatric morbidity and suicide mortality (Gunnell et al., 1995). So also does social fragmentation, meaning relatively weak community ties associated with observed indices such as one person households, high population turnover and many adults outside married relationships (Evans et al., 2004). Indicators $\{y_1, y_2, y_3\}$ of deprivation are provided by square roots (a normalising transform) of the UK Census rates of renting from social (public sector) landlords, and of unemployment among the economically active, together with the square root of the rate of households claiming income support. Indicators $\{y_4, y_5, y_6\}$ of social fragmentation are provided by square roots of census rates of one person households, migration in the precensal year, and people over 15 not married.

A confirmatory factor model (with simple structure) is assumed with $\{y_1,.,y_3\}$ loading only on a deprivation score F_1, and with $\{y_4,\ldots,y_6\}$ loading only on a fragmentation score F_2. Let D_{pi} be the denominator (e.g. total population) used to define the transformed census index y_{pi}. Then the measurement model has the form

$$y_{1i} = \alpha_1 + \lambda_{11}F_{1i} + u_{1i}$$

$$y_{2i} = \alpha_2 + \lambda_{21}F_{1i} + u_{2i}$$

$$y_{3i} = \alpha_3 + \lambda_{31}F_{1i} + u_{3i}$$

$$y_{4i} = \alpha_4 + \lambda_{42}F_{2i} + u_{4i}$$

$$y_{5i} = \alpha_5 + \lambda_{52}F_{2i} + u_{5i}$$

$$y_{6i} = \alpha_6 + \lambda_{62}F_{2i} + u_{6i}$$

where the u_{ji} are mutually uncorrelated, with $u_{pi} \sim N(0, \tau_p / D_{pi})$ (Hogan and Tchernis, 2004, p.316).

Since F_1 and F_2 have arbitrary location and scale, one way of providing identifiability (the variance scaling or standardisation constraint) is to define them to be in standard form with zero means and variances of 1 (while still possibly allowing a non-zero correlation between the two factors, which is possible under this confirmatory model). Under the alternative anchoring constraint (Skrondal and Rabe-Hesketh, 2007), one loading on each construct is preset for identification, for example, $\lambda_{11} = \lambda_{42} = 1$. The F_{qi} may be assumed independent of one another, although correlation over areas i may still be incorporated via two separate univariate CAR (conditional autoregressive) priors (Besag et al., 1991). Alternatively, correlation both between factors and over areas may be assumed, so that $\{F_{1i}, F_{2i}\}$ follow a bivariate CAR prior (see Section 9.6). Under an anchoring constraint, the within area factor covariance matrix would then contain three unknowns $\{\phi_{11}, \phi_{22}, \rho\}$

$$\Phi = \begin{pmatrix} \phi_{11} & \rho\sqrt{\phi_{11}\phi_{22}} \\ \rho\sqrt{\phi_{11}\phi_{22}} & \phi_{22} \end{pmatrix},$$

whereas under a standardisation constraint, the diagonal elements in Φ are set to 1, and only ρ would be unknown.

Adopting an anchoring constraint has utility in helping to prevent "relabelling" of the construct scores F_{qi} during MCMC sampling. Since the indicators $\{y_1, \ldots, y_3\}$ in this example are positive measures of material deprivation, setting $\lambda_{11} = 1$ is consistent with the construct F_{1i} being a positive deprivation measure. If, however, one adopted the standardised factor assumption with $\phi_{pp} = 1$ and all the λ_{pq} free, it would be necessary, in order to prevent label switching, to set a prior on one or possibly more loadings constraining positivity, for example,

$$\lambda_{p1} \sim N(1,1)I(0,),$$

$$\lambda_{p2} \sim N(1,1)I(0,),$$

for one or more p.

Example 9.1 Wechsler Intelligence Scale

This example involves a dataset used by Tabachnick and Fidell (2006) to illustrate confirmatory factor analysis, and consisting of subtest scores on the second version of the Wechsler Intelligence Test for Children (WISC-R). The 175 subjects are school-aged children diagnosed as learning-disabled. The Wechsler Intelligence Scale for Children-Revised (WISC-R) is a general test of intelligence, defined as "the global capacity of the individual to act purposefully, to think rationally, and to deal effectively with his environment." Considering intelligence as an aggregate of mental aptitudes, the WISC-R data here consists of 11 tests divided into two groups, verbal and performance. The six verbal tests (Information, Comprehension, Arithmetic, Similarities, Vocabulary, Digit Span) use language-based items, whereas the five performance tests (Picture Completion, Picture Arrangement, Block Design, Object Assembly, Coding) are visual-motor in character. In all analyses below, the observations are standardised. Item scores range from 0 to 20 and are considered continuous.

Two models are compared using rstan, with factor score covariation modelled using the cholesky_factor_corr function. The first is an exact confirmatory factor analysis, with factor 1 corresponding to the verbal tests, and factor 2 to the performance items. So exact zeros are used to define loadings $(\lambda_{71}, \lambda_{81}, \lambda_{91}, \lambda_{10,1}, \lambda_{11,1})$ of the performance items on factor 1, and of the verbal item loadings $(\lambda_{12}, \lambda_{22}, \lambda_{32}, \lambda_{42}, \lambda_{52}, \lambda_{62}$ on factor 2. Thus, with exact zero loadings not shown, one has

$$y_{1i} = \alpha_1 + \lambda_{11}F_{1i} + u_{1i}$$

$$y_{2i} = \alpha_2 + \lambda_{21}F_{1i} + u_{2i}$$

$$y_{3i} = \alpha_3 + \lambda_{31}F_{1i} + u_{3i}$$

$$y_{4i} = \alpha_4 + \lambda_{41}F_{1i} + u_{4i}$$

$$y_{5i} = \alpha_5 + \lambda_{51}F_{1i} + u_{5i}$$

$$y_{6i} = \alpha_6 + \lambda_{61}F_{1i} + u_{6i}$$

$$y_{7i} = \alpha_7 + \lambda_{72}F_{2i} + u_{7i}$$

$$y_{8i} = \alpha_8 + \lambda_{82}F_{2i} + u_{8i}$$

$$y_{9i} = \alpha_9 + \lambda_{92}F_{2i} + u_{9i}$$

$$y_{10,i} = \alpha_{10} + \lambda_{10,2}F_{2i} + u_{10i}$$

$$y_{11,i} = \alpha_{11} + \lambda_{11,2}F_{2i} + u_{11i}$$

with uncorrelated normally distributed u_{pi}. In the second model, exact zeros are replaced by approximate zeros specified using informative normal priors with a small variance of 0.01, so that 95% of the prior variation is between 0.2 and 0.2 (Muthen and Asparouhov, 2012).

This model has a lower WAIC (widely applicable information criterion), namely 4828 compared to 4841, than the exact zero CFA. Table 9.1 compares the two sets of estimated loadings. Estimated main loadings under the second model are similar to those under the exact zero CFA, and both show that indicator 11 (Coding) is essentially unrelated to the second factor (the loading $\lambda_{11,2}$). The second model also suggests a significant cross-loading (λ_{22}) of indicator 2 (Comprehension) on the performance factor, with 95% posterior interval (0.07,0.34).

A third analysis via rjags uses binary factor-indicator selection indicators (Mavridis and Ntzoufras, 2014), while also retaining the approximate zero prior formulation. Spike-slab priors are adopted on the selection indicators. Thus cross-loadings $(\lambda_{71}, \lambda_{81}, \lambda_{91}, \lambda_{10,1}, \lambda_{11,1})$ and $(\lambda_{12}, \lambda_{22}, \lambda_{32}, \lambda_{42}, \lambda_{52}, \lambda_{62})$ are assigned informative N(0,0.01) priors, while for identifiability the main loadings are assigned N(0,1) priors constrained to positive values.

One objective of this analysis is to detect indicators not relevant to the postulated confirmatory analysis scheme. The analysis with rstan indicated that indicator 11 may not be relevant. Selection indicators γ_{pq} therefore have Bernoulli probabilities π_p that are indicator specific. For indicator p, one has

$$y_{pi} = \alpha_p + \gamma_{p1}\lambda_{p1}F_{1i} + \gamma_{p2}\lambda_{p2}F_{2i} + u_{pi},$$

$$\gamma_{pq} \sim \text{Bernoulli}(\pi_p),$$

TABLE 9.1

Posterior Summary. Exact Zero vs Approximate Zero Confirmatory Factor Analysis

	Analysis 1 (Cross Loadings Preset to Zero)			
	Mean	**St Devn**	**2.5%**	**97.5%**
λ_{11}	0.77	0.07	0.63	0.91
λ_{21}	0.70	0.07	0.56	0.85
λ_{31}	0.57	0.08	0.42	0.73
λ_{41}	0.71	0.07	0.57	0.86
λ_{51}	0.78	0.07	0.64	0.93
λ_{61}	0.39	0.08	0.24	0.55
λ_{72}	0.59	0.09	0.42	0.76
λ_{82}	0.47	0.09	0.3	0.65
λ_{92}	0.69	0.08	0.53	0.85
$\lambda_{10,2}$	0.57	0.08	0.41	0.74
$\lambda_{11,2}$	0.11	0.07	0.01	0.26
Factor Correlation	0.58	0.08	0.41	0.72
	Analysis 2 (Informative Prior on Cross Loadings)			
Main loadings	**Mean**	**St Devn**	**2.5%**	**97.5%**
λ_{11}	0.78	0.07	0.65	0.92
λ_{21}	0.59	0.08	0.44	0.74
λ_{31}	0.55	0.08	0.39	0.71
λ_{41}	0.64	0.08	0.49	0.79
λ_{51}	0.76	0.08	0.61	0.91
λ_{61}	0.39	0.08	0.23	0.55
λ_{72}	0.55	0.10	0.36	0.73
λ_{82}	0.43	0.09	0.25	0.61
λ_{92}	0.65	0.09	0.47	0.83
$\lambda_{10,2}$	0.56	0.09	0.39	0.75
$\lambda_{11,2}$	0.09	0.06	0	0.24
Cross Loadings	**Mean**	**St Devn**	**2.5%**	**97.5%**
Loadings of Performance Items on Verbal Factor				
λ_{71}	0.10	0.06	0.01	0.24
λ_{81}	0.08	0.05	0.01	0.19
λ_{91}	0.08	0.05	0	0.20
$\lambda_{10,1}$	0.05	0.04	0	0.15
$\lambda_{11,1}$	0.06	0.04	0	0.17
Loadings of Verbal Items on Performance Factor				
λ_{12}	0.04	0.03	0	0.13
λ_{22}	0.20	0.07	0.07	0.34
λ_{32}	0.06	0.05	0	0.17
λ_{42}	0.14	0.06	0.02	0.27
λ_{52}	0.06	0.05	0	0.17
λ_{62}	0.05	0.04	0	0.14
Factor Correlation	0.35	0.11	0.12	0.57

TABLE 9.2

Posterior Summary. Predictor Selection Combined with Approximate Zero CFA

Main loadings	Mean	St Devn	2.5%	97.5%	Mean Selection Probability (γ_{jk})
λ_{11}	1.15	0.29	0.64	1.80	1
λ_{21}	0.84	0.23	0.45	1.37	1
λ_{31}	0.81	0.22	0.44	1.29	1
λ_{41}	0.92	0.24	0.49	1.47	1
λ_{51}	1.11	0.28	0.62	1.72	1
λ_{61}	0.59	0.19	0.28	1.01	1
λ_{72}	0.65	0.22	0.29	1.14	1
λ_{82}	0.50	0.19	0.19	0.93	1
λ_{92}	0.74	0.27	0.31	1.35	1
$\lambda_{10,2}$	0.65	0.23	0.28	1.17	1
$\lambda_{11,2}$	0.02	0.06	0.00	0.21	0.14
Cross Loadings	**Mean**	**St Devn**	**2.5%**	**97.5%**	
Loadings of performance items on verbal factor					
λ_{71}	0.03	0.07	−0.11	0.20	0.67
λ_{81}	0.01	0.07	−0.13	0.17	0.63
λ_{91}	0.01	0.07	−0.14	0.16	0.64
$\lambda_{10,1}$	−0.03	0.07	−0.19	0.11	0.65
$\lambda_{11,1}$	0.01	0.05	−0.09	0.15	0.36
Loadings of verbal items on performance factor					
λ_{12}	−0.06	0.07	−0.22	0.06	0.71
λ_{22}	0.15	0.08	0.00	0.30	0.93
λ_{32}	−0.01	0.06	−0.15	0.12	0.60
λ_{42}	0.08	0.08	−0.04	0.24	0.80
λ_{52}	−0.01	0.06	−0.16	0.10	0.61
λ_{62}	−0.04	0.07	−0.20	0.08	0.67
Factor Correlation	0.54	0.10	0.33	0.71	

$$\pi_p \sim \text{Beta}(1,1),$$

with the covariance matrix of the factor scores taken as unknown.

Table 9.2 shows the posterior summaries for the realised loadings $\gamma_{pq}\lambda_{pq}$ and means for the selection probabilities, $Pr(\gamma_{pq}=1)$. It can be seen that there is support for a significant loading of indicator 2 (Comprehension) on the performance factor, with $\gamma_{22}=0.93$, and so a marginal Bayes factor of 14 for $Pr(\gamma_{22}=1)$. Also $\gamma_{11,2}=0.14$, suggesting the indicator 11 (Coding) is essentially unrelated to the second factor.

Example 9.2 Job Applicant Data

This example illustrates the application of the parameter expansion method, as in (9.2), combined with hierarchical priors on loadings. The data are from Kendall (1975) and correspond to 48 applicants for a position in firm who have been judged on 15 variables, treated as continuous ($y_1,\ldots,y_{15}$): form of letter of application; appearance; academic ability; likeability; self-confidence; lucidity; honesty; salesmanship; experience; drive; ambition; grasp; potential; keenness to join; and suitability.

The third and fourth columns of Table 9.3 show the results of a maximum likelihood factor analysis in Stata with two factors; the first two factors account for 59% of the

TABLE 9.3
Estimated Factor Loadings, Two Factors. Applicant Data

		Maximum Likelihood		Bayesian EFA, Posterior Summary					
				Factor 1			Factor 2		
	Variable	Factor1	Factor2	Mean	2.5%	97.5%	Mean	2.5%	97.5%
y_1	Form of Letter	0.37	0.56	0.62	0.33	0.93	0.00	−0.17	0.19
y_2	Appearance	0.53	0.03	0.16	−0.21	0.54	0.44	0.09	0.79
y_3	Academic ability	0.12	0.19	0.21	−0.14	0.58	−0.01	−0.38	0.35
y_4	Likeability	0.51	0.08	0.20	−0.17	0.60	0.38	−0.01	0.74
y_5	Self confidence	0.84	−0.39	−0.22	−0.67	0.36	0.98	0.67	1.36
y_6	Lucidity	0.88	−0.18	0.02	−0.40	0.57	0.88	0.52	1.23
y_7	Honesty	0.36	−0.29	−0.18	−0.57	0.22	0.46	0.11	0.83
y_8	Salesmanship	0.91	−0.05	0.16	−0.26	0.68	0.83	0.42	1.17
y_9	Experience	0.32	0.73	0.81	0.48	1.21	−0.16	−0.68	0.33
y_{10}	Drive	0.86	0.10	0.30	−0.09	0.77	0.68	0.25	1.03
y_{11}	Ambition	0.89	−0.14	0.05	−0.37	0.59	0.87	0.51	1.22
y_{12}	Grasp	0.90	0.00	0.21	−0.20	0.72	0.79	0.37	1.13
y_{13}	Potential	0.89	0.09	0.31	−0.09	0.79	0.71	0.27	1.06
y_{14}	Keeness to join	0.63	0.06	0.21	−0.16	0.61	0.50	0.12	0.85
y_{15}	Suitability	0.61	0.67	0.83	0.49	1.22	0.11	−0.43	0.58

variance in the original indicators. It can be seen that the second factor emphasises form of letter (y_1), experience (y_9), and suitability (y_{15}), while the first factor is more generic, with loadings over 0.8 on seven of the 15 indicators. Similar results are obtained using the R package lavaan.

With intercorrelation between the two factors allowed under an EFA approach, there are $Q^2=4$ restrictions needed when loadings are treated as fixed effects parameters (Merkle and Wang, 2016). Under fixed effects priors, and counting of parameter restrictions under a degrees of freedom approach, this can be achieved by (a) assuming standardised factors in the reference model; (b) setting $\lambda^*_{12}=\lambda_{12}=0$ (the loading of indicator 1 on factor 2) for a lower triangular structure; and (c) setting an additional loading on the first factor to zero (e.g. $\lambda^*_{p1}=\lambda_{p1}=0$ for some p). However, setting one of the loadings λ^*_{p1} on the first factor to be zero is potentially arbitrary, and may affect substantive findings.

We avoid this, and formal parameter counting, by (a) adopting an approximate zero prior on λ^*_{12}, namely $\lambda^*_{12} \sim N(0,0.01)$, instead of an exact zero, and by (b) adopting hierarchical priors, rather than fixed effects priors, on the remaining λ^*_{p2}, and on the λ^*_{p1}. Thus

$$\lambda^*_{p1} \sim N(0,\omega_1)$$

$$\lambda^*_{p2} \sim N(0,\omega_2) \ (p>1),$$

where ω_1 and ω_2 are unknown variances. Because of the approximate zero restriction on λ^*_{12}, one might anticipate that factor 1 in the EFA would be the most relevant to explaining indicator y_1 (form of application letter).

A two-chain run of 20,000 iterations in jagsUI shows estimated loadings under Bayesian and MLE estimation as in Table 9.3. The estimated loadings λ_{p1} on factor 1 are highest for y_1, y_9, and y_{15}, with other loadings having credible intervals straddling zero. By contrast, the second factor is a more generic factor, similar to factor 1 in the MLE estimation, with highest loadings on y_5, y_6, y_8, y_{10}, y_{11}, y_{12}, and y_{13}. The 90% highest posterior density interval for the factor correlation (rho in the jags code) is (0.04,0.75). Both types of analysis (Bayesian and maximum likelihood) show low loadings of academic ability (y_3) on the first two factors, and one might adopt a variable selection approach to confirm its relevance, or add an additional factor.

A maximum likelihood analysis with three factors identifies a third factor, loading highly on likeability (y_4), honesty (y_7), and keenness to join (y_{14}). A Bayesian analysis with the implicit constraint $\lambda_{qq}>0$ identifies a similar factor (as factor 3) if the originally labelled indicators y_3 and y_4 are reordered, so that likeability becomes y_3. The Bayesian analysis then identifies a factor with high positive loadings on likeability and honesty (y_3 and y_7 in the revised sorting), with respective 90% hpd intervals (0.15,1.34) and (−0.04,1.21). This factor is not detected if the original ordering of indicators is retained.

9.4 Multivariate Exponential Family Outcomes and Generalised Linear Factor Models

The normal linear factor and structural equation models considered above extend straightforwardly to generalised linear factor and SEM models for non-normal data from the exponential family density: namely binomial, Poisson, and multinomial or ordinal data. Consider multivariate observations $y_i=(y_{1i},\dots,y_{Pi})'$, that conditional on factor scores $F_i=(F_{1i},\dots,F_{Qi})'$ follow an exponential family density, namely

$$p(y_{pi}\mid F_i)\propto \exp\left\{\frac{y_{pi}\theta_{pi}-b(\theta_{pi})}{\phi_{pi}}+c(y_{pi},\phi_{pi})\right\}$$

where θ_{pi} is the canonical parameter, with the ϕ_{pi} typically taken as known scale parameters. Denoting regression terms as $\eta_{pi}=g(\theta_{pi})$ where g is a link function, and $\eta_i=(\eta_{1i},\dots,\eta_{Pi})'$, intercept $a=(a_1,\dots,a_P)'$ and $P\times Q$ loading matrix Λ, the regression term without extra-variation is

$$\eta_i=a+\Lambda F_i,$$

while allowing extra-variation

$$\eta_i=a+\Lambda F_i+u_i,$$

with $u_i=(u_{1i},\dots,u_{Pi})'$, where the u_{pi} are independent of each other under conditional independence. The errors u (if present) and factor scores F are also independent.

Normality of errors and factors is often assumed with $(u_{1i},\dots,u_{Pi})'\sim N_P(0,\Sigma)$, where Σ is diagonal, and $F_i\sim N_Q(0,\Phi)$, where Φ may be non-diagonal according to the form of model (e.g. exploratory or confirmatory) assumed. Compared to the normal data-normal factor model, the marginal densities of y are no longer simply derived, but involve integration over F, namely

$$p(y_i\mid\theta,\psi)=\int\prod_{p=1}^{P}p(y_{pi}\mid F_i,\theta)p(F_i\mid\psi)dF_i,$$

where ψ are hyperparameters defining the density of F_i. The usual conditional independence assumptions are made. For example, for a P-variate categorical response (K_p categories for the pth response), the conditional probability that subject i with factor scores $F_i = (F_{1i}, \ldots, F_{Qi})'$ exhibits a particular set of responses is the product of separate categorical likelihoods

$$Pr(y_{1i} = k_1, y_{2i} = k_2, \ldots, y_{Pi} = k_P \mid F_i) = Pr(y_{1i} = k_1 \mid F_i) Pr(y_{2i} = k_2 \mid F_i) \ldots Pr(y_{Pi} = k_P \mid F_i).$$

For factor reduction of binary, multinomial, or ordinal data, there may be benefit (e.g. in simplified MCMC sampling algorithms) in considering latent variables posited to underlie the observed discrete responses. The missing data then consists not only of factor scores but of the latent scale data y^*_{pi} that underlie the observed data y_{pi}. Thus for y_{pi} binary, and $y_{pi} = 1$ if $y^*_{pi} > 0$ and $y_{pi} = 0$ otherwise, one might take $y^*_i = (y^*_{1i}, \ldots y^*_{Pi})$ to be normal or logistic, with the diagonal terms in the unique covariance matrix Σ set (usually to 1) for identifiability. For instance, a normal model taking the underlying responses to be conditionally independent given the factors, would be

$$y^*_{pi} \sim N(\alpha_p + \Lambda_p F_i, 1) I(A_{pi}, B_{pi}),$$

where Λ_p is the pth row of Λ, and the truncation ranges are determined by the observed y_{pi}.

9.4.1 Multivariate Count Data

Factor models with $Q < P$ may be more parsimonious than full dimension error models for multivariate exponential family data. However, multivariate reduction may not always be preferred in terms of fit, so parsimony may sometimes be at the expense of predictions that reproduce the data satisfactorily. Chib and Winkelmann (2001) illustrate how multivariate count data may not always be suitable for reduction using latent factors. In their full-dimension model $y_{pi} \sim Po(\mu_{pi})$, with outcome-specific predictors $x_{pi} = (x_{1pi}, x_{2pi}, \ldots, x_{Rpi})'$ and

$$\mu_{pi} = \exp(\beta_p x_{pi} + u_{pi}),$$

$$(u_{1i}, \ldots, u_{Pi}) \sim N_P(0, D),$$

where D is an unrestricted covariance matrix; see also Inouye et al. (2017) on Poisson mixture formulations, and Rodrigues-Motta et al. (2013) for the case where y_{pi} may follow different count densities. The y_{pi} are conditionally independent given the correlated errors $u_i = (u_{1i}, \ldots, u_{Pi})'$. Defining $v_{pi} = \exp(u_{pi})$, one has equivalently $y_{pi} \sim Po(\lambda_{pi} v_{pi})$ with

$$\lambda_{pi} = \exp(\beta_p x_{pi})$$

$$(v_{1i}, \ldots, v_{Pi}) \sim LN_P(\mu_v, \Sigma_v).$$

That is, the v_{pi} are multivariate lognormal with mean vector $\mu_v = \exp(0.5\text{diag}(D))$, and covariance $\Sigma_v = \text{diag}(\mu_v)[\exp(D) - 11']\text{diag}(\mu_v)$.

Other ways to generate correlated count data include the overlapping sums technique (Madsen and Dalthorp, 2007). Thus consider independent Poisson variables Z_{12}, Z_1 and Z_2 with means θ_{12}, θ_1 and θ_2; then $y_1 = Z_1 + Z_{12}$ and $y_2 = Z_2 + Z_{12}$ are correlated with marginal means $\theta_1+\theta_{12}$ and $\theta_2+\theta_{12}$ and covariance θ_{12}. The mean and covariance of the corresponding joint Poisson density for three variables is provided by Karlis and Meligkotsidou (2005, p.257).

Factor models for count data typically may include both normal factor scores and residuals u_{pi}, taken as uncorrelated if the usual conditional independence assumption is made. Thus

$$y_{pi} \sim Po(\mu_{pi}),$$

$$\mu_{pi} = \exp(\beta_p x_{pi} + \Lambda_p F_i + u_{pi}).$$

where Λ_p is $1\times Q$, $F_i = (F_{1i},\ldots,F_{Qi})'$ and under a standardised factor constraint $F_i \sim N_Q(0,R_F)$ where R_F is a correlation matrix, with possibly unknown off-diagonal terms subject to identifiability. Alternatively, Wedel et al. (2003) consider gamma distributed factors in an identity link model, as well as normal F scores combined with a log link. Gamma factors would have mean 1 to avoid location invariance, and taking their cumulative impact to be multiplicative, one could have

$$\mu_{pi} = \exp(\beta_p x_{pi} + u_{pi})(F_{1i}^{\omega p1} F_{2i}^{\omega p2} \ldots F_{Qi}^{\omega pQ})$$

with comparable identification restrictions on the loadings $\Omega_p = (\omega_{p1},\ldots,\omega_{pQ})$ to those in the normal linear factor model (see also Dunson and Herring, 2005). The constraints would differ according to whether the variance of the F scores were unknown, as in

$$F_{qi} \sim Ga(\varphi_q,\varphi_q),$$

with φ_q to be estimated, or whether the variance of F is preset, as in $F_{qi} \sim Ga(1,1)$.

An alternative to outcome-specific residuals u_{pi} in the above models is a common residual factor, especially when the F_i are derived as part of a broader structural model involving further observed indicators. For example, consider count observations y_{pi} $(p=1,\ldots,P)$ on clinical outcomes for a set of hospitals, while also available are metric measures x_{ri} $(r=1,\ldots,R)$ of resource inputs, efficiency, etc. The latter variables are relevant to defining a multivariate latent "care quality" construct F_i in a MIMIC framework that assists in explaining the clinical outcomes, but this construct may not explain all the covariation among (or overdispersion in) the y variables and correlated residuals and/or common residual factors are needed. The structural equations for the metric data might take the form (for standardised x)

$$F_{qi} = \beta_q x_i + w_{qi}.$$

while the errors in the Poisson likelihood measurement equations for $y_{pi} \sim Po(\mu_{pi})$ are correlated over outcomes under a common factor model

$$\log(\mu_{pi}) = \alpha_p + \Lambda_p F_i + \kappa_p u_i.$$

Assuming u_i is univariate, one of the loadings κ_p is preset if the variance σ_u^2 of the common residual scores u_i is unknown. Spatial applications of common factors are exemplified by Wang and Wall (2003) and Nethery et al. (2015).

9.4.2 Multivariate Binary Data and Item Response Models

As for counts, models for P-variate binary outcomes $y_{pi} \sim \text{Bern}(\pi_{pi})$ may retain the observed binary data likelihood and represent joint or residual correlations by additive full-dimension multivariate effects u_{pi}, for example,

$$\text{logit}(\pi_{pi}) = \beta_p x_{pi} + u_{pi},$$

$$(u_{1i}, \ldots, u_{Pi}) \sim N_P(0, D),$$

where D is an unrestricted covariance matrix. By contrast, multivariate probit or logit models may also follow from an augmented data perspective in which unobserved metric variables $y_i^* = (y_{1i}^*, y_{2i}^*, \ldots, y_{Pi}^*)$ result in the observed binary vector (Chen and Dey, 1998; Chen and Dey, 2000).

Thus, Bayesian estimation of the multivariate probit typically involves augmenting the data with the latent normal variables obtained by truncated multivariate normal sampling (Chib and Greenberg, 1998; Talhouk et al., 2012). Denoting $\eta_{pi} = \beta_p x_{pi}$,

$$(y_{1i}^*, y_{2i}^*, \ldots, y_{Pi}^*) \sim N_P(\eta_{pi}, \Sigma) I(A_i, B_i), \tag{9.4}$$

with the observations generated according to

$$y_{pi} = I\{y_{pi}^* > 0\}.$$

The lower and upper sampling limits in the vectors $A_i = (A_{1i}, \ldots, A_{Pi})$ and $B_i = (B_{1i}, \ldots, B_{Pi})$ depend on the observations: sampling of the constituent y_{pi}^* is confined to values above zero when $y_{pi} = 1$, and to zero or negative values when $y_{pi} = 0$. Scale mixtures of multivariate normal densities for the y_{pi}^* are also possible, and equivalent to a multivariate Student t for $(y_{1i}^*, y_{2i}^*, \ldots, y_{Pi}^*)$, which for particular degrees of freedom approximates a multivariate logit link (Chen and Dey, 1998). A multivariate logit regression may be achieved directly with suitable mixing strategies (Chen and Dey, 2000; O'Brien and Dunson, 2004).

The covariance matrix Σ in (9.4) is not identified, and when the predictor effects vary by response, only the correlation matrix can be identified (Rossi et al., 2005). The identification criteria for the multivariate probit differ from those of the multinomial probit where identification is obtained by setting one of the diagonal variance elements σ_{pp} (e.g. the first) to 1 (McCulloch et al., 2000). It is possible to sample the correlation matrix R directly (Barnard et al., 2000; Chib and Greenberg, 1998). Talhouk et al. (2012) use a parameter expansion method to sample R, and the LKJ (Lewandowski, Kurowicka and Joe) prior may be used (see Example 9.6). One may also (Edwards and Allenby, 2003; McCulloch and Rossi, 1994) sample the Σ matrix or its inverse from an unrestricted prior, and then scale both the fixed effects and the covariance matrix to their identified forms, namely

$$\beta_p^* = \beta_p / \sqrt{\sigma_{pp}},$$

and the correlation matrix

$$R = \Delta \Sigma \Delta,$$

where $\Delta = \text{diag}(\sqrt{\sigma_{pp}})$.

Factor models for multiple binary data most typically have the general linear mixed form

$$y_{pi} \sim \text{Bern}(\pi_{pi}),$$

$$g(\pi_{pi}) = \beta_p x_{pi} + \Lambda_p F_i,$$

where g is the link, and $F_i = (F_{1i},\ldots,F_{Qi})'$. As for the normal linear factor model, a common assumption for the density of F is normal with a known scale. If, additionally, factors are independent, then $F_{qi} \sim N(0,1)$, $q = 1,\ldots,Q$. If instead the assumption $F_{qi} \sim \text{Logist}(0,1)$ is made, with loadings κ_{pq} in

$$\eta_{pi} = \alpha_p + \kappa_{p1}F_{1i} + \ldots + \kappa_{pQ}F_{Qi},$$

then $\kappa_{pq} \approx (\sqrt{3}/\pi)\lambda_{pq}$, since the variance of a standard logistic is $\pi^2/3$ (Bartholomew, 1987).

A widely applied method in educational and psychometric evaluation (Albert, 1992; Rupp et al., 2004; Fox and Glas, 2005) is based on item response theory (IRT for short). Typically, the observation vector $y_i = (y_{1i},\ldots,y_{Pi})$ consists of binary items measuring ability, with 1 denoting a correct answer and 0 an incorrect answer, and a model seeks a single latent ability factor score F_i. Factor score identifiability is generally obtained by assuming $F_i \sim N(0,1)$. Under conditional independence, the joint success probability given F_i is

$$Pr(y_{1i} = 1, y_{2i} = 1,\ldots, y_{pi} = 1 \mid F_i) = Pr(y_{1i} = 1 \mid F_i)Pr(y_{2i} = 1 \mid F_i)\ldots Pr(y_{Pi} = 1 \mid F_i).$$

With a Bernoulli likelihood, $y_{pi} \sim \text{Bern}(\pi_{pi})$, and link g, one has a factor model

$$g(\pi_{pi}) = \eta_{pi} = \alpha_p + \lambda_p F_i. \tag{9.5}$$

IRT rests on relatively strong assumptions, namely that a unidimensional factor is appropriate, conditional independence of the items given the latent factor, and a monotonic relationship between latent ability and performance on the items (Arima, 2015).

The intercepts α_p can be interpreted as measures of difficulty of item p, with more negative α_p implying greater difficulty under the parameterisation in (9.5), while λ_p measures an item's power to discriminate ability between subjects. A now frequent practice assigns positive (e.g. lognormal, gamma, truncated normal) priors to the discrimination parameters λ_p, and draws these parameters from a hierarchical density with common variance (Curtis, 2010; Luo and Jiao, 2017). A hierarchical prior may also be assumed for the difficulty parameters. Using hierarchical priors may improve convergence. An alternative is to adopt fixed effects priors (Sahu, 2002), as illustrated in the Stan Case Studies. This model may also be parameterised as

$$g(\pi_{pi}) = \lambda_p(F_i - \alpha_p),$$

so that α_p increases with difficulty. These are called two-parameter logistic or probit IRT models. The three-parameter model includes a guessing (or threshold) parameter c_p, whereby

$$\pi_{pi} = c_p + (1 - c_p)g^{-1}[\lambda_p(F_i - \alpha_p)].$$

An IRT information function $I_p(F)$ measures how precisely an item measures the latent ability scale. For example, easy items may provide little information about higher ability subjects, while difficult items will provide little information regarding lower ability subjects. Assuming a two-parameter logistic, the item information function can be obtained as

$$I_p(F) = \lambda_p^2 \pi_p(F)(1 - \pi_p(F)),$$

where $\pi_p(F) = \text{logit}^{-1}[\lambda_p(F - a_p)]$, and can be displayed graphically with F is taken over the range of ability scores. The total information function is the sum of the item-specific functions.

Soares et al. (2009) and Fox and Glas (2005) describe Bayesian IRT models allowing differential item functioning (DIF), for example, when one or more items are not appropriate for measuring ability because the knowledge needed for a correct answer is culturally specific. Thus, let $x_i = 0$ for a reference population and $x_i = 1$ for a focal group (e.g. disadvantaged or minority group) (Magis et al., 2015). Then DIF is indicated if the extended model

$$Pr(y_{pi} = 1 \mid F_i) = g(\eta_{pi}),$$

$$\eta_{pi} = a_p + \lambda_p F_i + x_i(\gamma_p + \delta_p F_i),$$

has better fit than the standard model without group differentiation (Choi et al., 2011).

9.4.3 Latent Scale IRT Models

As an alternative to binary likelihood modelling in IRT and binary SEM applications, the latent scale method may be applied with the appropriate underlying density defined by the link g (Albert, 2015). Thus, for a probit link with $g^{-1} = \Phi$, the latent metric scale y^* is normal, such that $y_{pi} = 1$ corresponds to the imputation scheme

$$y_{pi}^* \sim N(\eta_{pi}, 1) \;\; I(0,)$$

and $y_{pi} = 0$ corresponds to $y_{pi}^* \sim N(\eta_{pi}, 1)I(,0)$. For a logit link $g^{-1}(u) = L(u) = e^u/(1 + e^u)$ and sampling of y^* is from a standard logistic. Sahu (2002) considers an extra data imputation to provide three-parameter IRT models. The three-parameter probit IRT model specifies

$$\pi_{pi} = c_p + (1 - c_p)\Phi(a_p + \lambda_p F_i)$$

while the three-parameter logistic IRT is

$$\pi_{pi} = c_p + (1 - c_p)L(a_p + \lambda_p F_i).$$

Lee and Song (2003) adopt a latent scale approach to a structural equation model for multiple binary observations. Their model specifies

$$y_i^* = a + \Lambda F_i + u_i,$$

where the latent constructs F_i are partitioned into endogenous and exogenous vector components $F_i = (F_{1i}, F_{2i})$ of dimension Q_1 and Q_2 respectively, with structural model

$$F_{1i} = BF_{1i} + CF_{2i} + w_i.$$

For identification $u_i \sim N_P(0,I)$, while $F_{1i} \sim N_{Q_1}(0,\Phi_1), F_{2i} \sim N_{Q_2}(0,\Phi_2), w_i \sim N_{Q_1}(0,\Sigma_w)$ and each row of Λ follows a separate normal prior. The observed binary data y is augmented with latent data $\{y^*,F\}$ to provide complete data $\{y,y^*,F\}$. Setting $\theta = (\alpha,\Lambda,\Phi_1,\Phi_2,\Sigma_\delta,B,\Gamma)$, the updating sequence involves sampling from conditionals $p(\theta^{(t+1)} \mid F^{(t)},y^{*(t)}), p(F^{(t+1)} \mid \theta^{(t+1)},y^{*(t)})$, and $p(y^{*(t+1)} \mid F^{(t+1)},\theta^{(t+1)})$.

Dunson and Herring (2005) consider instead the case where the underlying y^*_{pi} (e.g. tumour counts) are Poisson, or overdispersed Poisson, and the observations y_{pi} (e.g. whether tumours are present) are binary. Thus

$$y^*_{pi} \sim Po(\exp(x_{pi}\beta)\Lambda_p\xi_i)\ I(A_{pi},B_{pi})$$

where $\xi_i = (\xi_{1i},\ldots,\xi_{Qi})$ are gamma distributed latent constructs, and the loadings Λ_p are also gamma distributed. The sampling limits are $(A_{pi} = 0, B_{pi} = 0)$ when $y_{pi}=0$, and $(A_{pi}=1, B_{pi}=\infty)$ when $y_{pi}=1$.

9.4.4 Categorical Data

For unordered polytomous indicators y_{pi} with M_p categories $(p = 1,\ldots,P)$, intercept and loading parameters are typically specific to the category of each item, with one category (e.g. the final one) as reference. Assume a multiple logit link, with multinomial parameter $\pi_{pi} = (\pi_{pi1},\ldots\pi_{piM_p})$ for subject i and indicator p. Then while factors are common across categories, loadings are specific to indicator p and category h of that indicator,

$$y_{pi} \sim \text{Categoric}(\pi_{pi1},\pi_{pi2},\ldots\pi_{piM_p})$$

$$\pi_{pih} = \varphi_{pih} \Big/ \sum_{m=1}^{M_p} \varphi_{pim} \quad h = 1,\ldots M_p$$

$$\log(\varphi_{pih}) = \alpha_{ph} + \lambda_{ph1}F_{1i} + \ldots + \lambda_{phQ}F_{Qi} \quad h = 1,\ldots,M_{p-1}$$

$$\varphi_{piM_p} = 1$$

with the usual constraints on Λ and/or F to avoid scale and rotational invariance.

Factor models for multiple ordinal items $y_{pi} \in (1,\ldots,K_p)$ refer to locations on an underlying continuous scales z_{pi}. Thus $y_{pi}=j$ when $\alpha_{p,j-1} \le z_{pi} < \alpha_{pj}$, where α_{pj} are cutpoints on the underlying scale. Define binary indicators $d_{pij}=1$ if $y_{pi}=j$, and $d_{pij}=0$ otherwise, and denote $d_{pi} = (d_{pi1},\ldots,d_{piK_p})$. With $\pi_{pi} = (\pi_{pi1},\ldots,\pi_{piK_p})$, η_{pi} denoting a regression term potentially including latent factors, and $z_{pi} = \eta_{pi} + \varepsilon_{pi}$, where errors ε_{pi} have cdf $P(\varepsilon)$, one has

$$d_{pi} \sim \text{Mult}(1,\pi_{pi}),$$

$$\pi_{pij} = Pr(y_{pi} = j) = Pr(a_{p,j-1} \leq z_{pi} < a_{pj}),$$
$$= Pr(a_{p,j-1} \leq \eta_{pi} + \varepsilon_{pi} < a_{pj}),$$
$$= P(a_{pj} - \eta_{pi}) - P(a_{p,j-1} - \eta_{pi})$$
$$= \gamma_{pij} - \gamma_{pi,j-1},$$

where

$$\gamma_{pij} = Pr(y_{pi} \leq j) = P(a_{pj} - \eta_{pi}), \quad j = 1, \ldots K_p - 1$$

are cumulative probabilities over ordered categories, $\gamma_{pij} = p_{pi1} + \ldots + p_{pij}$.
If ε follows a logistic density, then

$$\text{logit}(\gamma_{pij}) = a_{pj} - \eta_{pi},$$

where taking η_{pi} as uniform across response categories j defines the proportional odds assumption. For example, assuming a univariate latent factor F_i, and other predictors X_i, one has

$$\text{logit}(\gamma_{pij}) = a_{pj} - \lambda_p F_i - \beta_p X_i.$$

One application is in the graded response model for ordinal outcome IRT (Luo and Jiao, 2017). Assuming X_i excludes an intercept, the $K_p - 1$ thresholds $\{a_{p1}, a_{p2} \ldots, a_{K_p-1}\}$ are unknowns subject to the order constraint $a_{p1} \leq a_{p2} \ldots \leq a_{K_p-1}$. An augmented data approach may also be used for latent variable analysis with ordinal responses (Lee and Tang, 2006; Poon and Wang, 2012).

Example 9.3 Greek Crime Totals

This example considers count data and compares a full dimension covariance structure to a common factor model. The data relate to $P=4$ counts of crimes (rapes, arsons, manslaughter, smuggling of antiquities) in n=49 Greek prefectures, as used in Karlis and Meligkotsidou (2005).* The counts are assumed to be Poisson with offset being prefecture populations, pop_i (in millions). Predictors are unemployment rate (x_1), a binary indicator (x_2) for whether the prefecture is on the Greek border, GDP per capita in euros (x_3), and a binary indicator (x_4) for whether the prefecture has at least one large city (over 150,000 inhabitants). GDP and unemployment are centred.

Event rates are low in relation to populations at risk, so a reasonable sampling model takes $y_{pi} \sim Po(pop_i \rho_{pi})$ with ρ_{pi} then being crime rates per million. The full dimension covariance model specifies

$$\rho_{pi} = \exp(a_p + X_i \beta_p + u_{pi}), \; p = 1, \ldots, P$$

$$(u_{1i}, \ldots, u_{Pi}) \sim N_P(0, D).$$

* Data kindly provided by Dimitris Karlis.

with a Wishart prior on the precision matrix

$$D^{-1} \sim W(PI, P).$$

With jagsUI for estimation, the mean scaled deviance is obtained as 209, comparing closely to the number of observations, namely 196, so that Poisson extra-variation is accounted for. Most predictor effects are insignificant: the only significant effects are of unemployment on the rape crime rate (with β_{11} having mean 0.064), and of GDP on manslaughter rates. Most correlations $r_{jm} = \text{corr}(u_{ji}, u_{mi})$ in the regression residuals have credible intervals straddling zero, though r_{13} has a posterior mean 0.46 with 95% CRI (−0.03,0.81). Adequate model performance is shown by the fact that only 9 of the 196 observations have mixed predictive p values under 0.05 or over 0.95 (Marshall and Spiegelhalter, 2007).

To illustrate a factor analytic approach to these data, the four predictors are taken to be causes of a single underlying crime construct, F_i, in a MIMIC analysis. The Poisson regressions form a measurement model in which crime levels are indicators of F_i. A further common factor u_i is included in the model for the crime types to account for residual variation in the crime data. So

$$\rho_{pi} = \exp(\alpha_p + \lambda_p F_i + \kappa_p u_i),$$

$$F_i = b_1 x_{1i} + b_2 x_{2i} + b_3 x_{3i} + b_4 x_{4i} + w_i,$$

$$w_i \sim N(0, 1/\tau_w),$$

$$u_i \sim N(0, 1/\tau_u).$$

Anchoring constraints are used to define the scale of the factor scores F_i and u_i. So $\lambda_1 = 1$ and $\kappa_2 = 1$, with the latter setting corresponding to a belief that arson is relatively distinct from the other variables in its pattern.

Inferences are based on a 15,000 iterations run with two chains using jagsUI. The posterior means (sd's) of the unknown λ_p coefficients ($p = 2,3,4$) are respectively −0.94 (0.95), 0.69 (0.38), 0.82 (0.54). These loadings tend to confirm F as a positive crime construct with positive loadings on all crime variables except for arson. The posterior mean F scores range from −0.96 to 1.22, with high F scores in prefectures with above average violent crime (such as prefecture 13), or where high violent crime is combined with smuggling. By contrast, low F scores occur in prefectures with little crime (prefecture 40), or in areas where arson is unduly elevated (e.g. prefecture 48). The x_i are relatively weak predictors of F_i though the GDP coefficient (b_3) has a mainly positive 95% interval (−0.03, 0.28).

The average scaled deviance of this model (278) indicates some residual over-dispersion. The estimated parameter total is lower at 129 (compared to 170 for model 1), though the DIC is higher at 756 (compared to 730 for model 1). Model checks are, however, adequate: 14 of the 196 observations have mixed predictive p values under 0.05 or over 0.95.

The fact that this particular data reduction method did not yield a better fit may be taken to illustrate caveats to discrete data factor reduction, as also illustrated by Chib and Winkelmann (2001). They undertake a Poisson regression analysis of six health use outcomes, and conclude that "a flexible model with a full set of correlated latent effects is needed to adequately describe the correlation structure [in the regression residuals]."

Example 9.4 Attitudes to Science

This example considers a unifactorial model for data on four ordinal indicators of attitudes to science, a subsample from the International Social Survey Program (ISSP) 1993

(Greenacre and Blasius, 2006). The indicators are Lickert scales with five levels, with wording as follows: y_1, we believe too often in science, and not enough in feelings and faith; y_2, overall, modern science does more harm than good; y_3, any change humans cause in nature, no matter how scientific, is likely to make things worse; and y_4, modern science will solve our environmental problems. Responses range from 1 = strongly agree to 5 = strongly disagree. Except for the fourth question, agreement suggests a negative attitude toward science, while disagreement (higher ordinal ranks) suggests a positive attitude.

A logit regression for ordinal responses $y_{pi} \in (1,\ldots,K)$ $(p=1,\ldots,P; k=1,\ldots K)$, where $P=4$ and $K=5$, assumes underlying continuous variables z_{pi} such that $y_{pi}=k$ when $a_{p,k-1} \leq z_{pi} < a_{pk}$. Define binary indicators $d_{pik}=1$ if $y_{pi}=k$, and $d_{pik}=0$ otherwise, and denote $d_{pi}=(d_{pi1},\ldots,d_{piK_p})$. So with $\pi_{pi}=(\pi_{p1i},\ldots,\pi_{pKi})$, and assuming Q latent factors, the sampling model is $d_{pi} \sim \text{Mult}(1,\pi_{pi})$, where

$$\pi_{pki} = Pr(a_{p,k-1} \leq z_{pi} < a_{pk}),$$

$$= P(a_{pk} - \lambda_{p1}F_{1i} \ldots - \lambda_{pQ}F_{Qi}) - P(a_{p-1,k} - \lambda_{p1}F_{1i} \ldots - \lambda_{pQ}F_{Qi})$$

$$= \gamma_{pki} - \gamma_{p-1,ki}$$

where $\gamma_{pki} = Pr(y_{pi} \leq k)$, $p=1,\ldots K-1$, are cumulative probabilities.

Here a single factor is assumed so that $p_{pki} = P(a_{pk} - \lambda_p F_i) - P(a_{p-1,k} - \lambda_p F_i)$. To ensure consistent labelling, we set $\lambda_1 = 1$ (i.e. an anchoring constraint), so the factor will most likely measure positive attitudes to science. We also assume a MIMIC model, whereby factor scores are assumed to be influenced by three observed predictors: x_1 = sex (M = 0, F = 1), x_2 = age and x_3 = education. The latter two covariates are categorical with 6 levels, but taken to be continuous for simplicity. The categories are for age: 16–24, 25–34, 35–44, 45–54, 55–64, 65 and older, and for education: primary incomplete, primary completed, secondary incomplete, secondary completed, tertiary incomplete, tertiary completed. So

$$F_i \sim N(\beta_1 x_{1i} + \beta_2 x_{2i} + \beta_3 x_{3i}, \sigma_F^2),$$

where σ_F^2 is an unknown by virtue of the anchoring constraint.

Table 9.4 shows the estimated parameters obtained using jagsUI. It can be seen that positive attitudes to science are more likely among males, and among younger, more highly educated, subjects. Also apparent is the irrelevance of the fourth item to the latent scale. Outlier diagnostics can be taken at the subject-indicator level, here using the leave one out criterion. Thus elevated LOO-IC values occur for subjects such as 784, an older, less educated subject (65+, primary completed), with indicator profile for (y_1, y_2, y_3, y_4) of (2,1,5,2), so that the third indicator value is unusual in terms both of covariate profile and the other indicators. We can aggregate the LOO-IC criteria within subjects, and this shows subject 781 with the most extreme criterion. This subject is also older and less educated subject, with an indicator profile for (y_1, y_2, y_3, y_4) of (4,1,5,3). The second indicator value is unusual, and the otherwise favourable science attitudes are unusual given the covariate profile.

Example 9.5 LSAT Data Item Response Model

This example compares item analysis (IRT) using two and three parameter logit models. The application involves Law School Admission Test (LSAT) data from the R ltm package,with $n=1000$ subjects and $P=5$ items. The two-parameter model for subjects i and item p is $y_{pi} \sim \text{Bern}(\pi_{pi})$, with π_{pi} specified as

$$\pi_{pi} = \text{logit}^{-1}[\lambda_p(\theta_i - \alpha_p)],$$

TABLE 9.4

Parameter Estimates. Scientific Attitudes

Parameter	Mean	St Devn	2.5%	97.5%
β_1	−0.346	0.112	−0.566	−0.133
β_2	−0.086	0.034	−0.153	−0.018
β_3	0.219	0.043	0.128	0.305
α_{11}	−2.561	0.146	−2.869	−2.295
α_{12}	−0.183	0.108	−0.416	0.018
α_{13}	1.157	0.119	0.924	1.387
α_{14}	3.363	0.198	3	3.751
α_{21}	−3.839	0.282	−4.457	−3.354
α_{22}	−1.703	0.18	−2.087	−1.386
α_{23}	−0.141	0.139	−0.432	0.117
α_{24}	2.306	0.184	1.957	2.672
α_{31}	−2.432	0.176	−2.779	−2.087
α_{32}	−0.011	0.124	−0.259	0.223
α_{33}	1.455	0.137	1.187	1.734
α_{34}	3.549	0.224	3.138	4.024
α_{41}	−2.617	0.129	−2.875	−2.375
α_{42}	−0.689	0.072	−0.834	−0.547
α_{43}	0.273	0.069	0.142	0.403
α_{44}	1.568	0.09	1.406	1.757
λ_2	1.45	0.199	1.141	1.882
λ_3	1.24	0.141	0.984	1.536
λ_4	−0.019	0.061	−0.135	0.103
$\tau_\Phi=1/\sigma^2_\Phi$	0.642	0.103	0.474	0.881

where $\theta_i \sim N(0,1)$ are ability scores, and the items are all positive measures of ability. The λ_p are assigned a hierarchical $LN(0,\sigma_\lambda^2)$ prior. The model is checked by assessing whether mixed predictive replicates $y_{\text{new},pi}$ sampled from the model (Marshall and Spiegelhalter, 2007) are concordant with actual values y_{pi}, though one may also compare actual and predicted totals falling into particular item response patterns (Sahu, 2002).

The three-parameter logit model includes guessing parameters γ_p (also called threshold parameters), whereby

$$\pi_{pi} = \gamma_p + (1-\gamma_p)\text{logit}^{-1}[\lambda_p(\theta_i - \alpha_p)],$$

The $\gamma_p = \text{logit}^{-1}(\xi_p)$ are obtained as inverse logits of ξ_p, which are assigned a hierarchical normal prior.

As an example of IRT outputs, we obtain the item-specific information functions and the test information function (or total information function), using the formulas in Baker (2001). The test information function indicates where the set of items provides most information about students with varying ability.

The analysis is implemented using rstan, with the rstan analysis having a substantial advantage in early convergence. Parameters from the 2PL and 3PL models are shown in Table 9.5. Measures of global fit show little gain in adopting a 3PL model instead of the 2PL, with the latter in fact having a lower LOO-IC (4910 vs 4914), but a higher WAIC (4903 vs 4897).

In terms of discrimination, item 3 shows maximum difficulty (highest α_p) under both models. Posterior mean rates of predictive concordance for the five items are (0.87, 0.69,

TABLE 9.5

LSAT Data, Parameter Summary, 2PL vs 3PL IRT

	Two Parameter Logistic				
	Mean	**St Devn**	**2.5%**	**50%**	**97.5%**
λ_1	0.90	0.20	0.54	0.89	1.33
λ_2	0.76	0.16	0.46	0.76	1.07
λ_3	0.86	0.18	0.54	0.85	1.24
λ_4	0.76	0.17	0.46	0.76	1.10
λ_5	0.78	0.17	0.44	0.78	1.11
α_1	−3.26	0.64	−4.81	−3.14	−2.31
α_2	−1.36	0.28	−2.03	−1.31	−0.95
α_3	−0.30	0.11	−0.52	−0.29	−0.11
α_4	−1.79	0.37	−2.71	−1.72	−1.26
α_5	−2.83	0.64	−4.43	−2.69	−2.03
	Three Parameter Logistic (Hierarchical Normal on Inv_Logit(gamma)				
	Mean	**St Devn**	**2.5%**	**50%**	**97.5%**
λ_1	1.15	0.56	0.65	1.02	2.33
λ_2	1.15	0.99	0.67	1.01	2.15
λ_3	1.09	0.37	0.68	1.01	1.93
λ_4	1.06	0.34	0.65	1.00	1.76
λ_5	1.09	0.48	0.65	1.01	1.93
α_1	−0.68	0.84	−2.56	−0.49	0.46
α_2	−0.12	0.41	−1.01	−0.10	0.69
α_3	0.18	0.32	−0.28	0.10	0.94
α_4	−0.26	0.47	−1.33	−0.20	0.58
α_5	−0.51	0.69	−2.09	−0.34	0.54
γ_1	0.74	0.15	0.29	0.80	0.88
γ_2	0.37	0.12	0.08	0.39	0.56
γ_3	0.16	0.10	0.01	0.14	0.38
γ_4	0.45	0.13	0.11	0.48	0.63
γ_5	0.64	0.15	0.21	0.69	0.8
	Three Parameter Logistic (Hierarchical Beta prior on gamma)				
	Mean	**St Devn**	**2.5%**	**50%**	**97.5%**
λ_1	1.11	0.40	0.68	1.02	2.02
λ_2	1.05	0.33	0.66	1.00	1.67
λ_3	1.05	0.29	0.69	1.00	1.71
λ_4	1.03	0.27	0.65	1.00	1.69
λ_5	1.07	0.39	0.63	1.00	1.91
α_1	−0.88	1.01	−3.02	−0.58	0.45
α_2	−0.23	0.46	−1.19	−0.16	0.60
α_3	0.11	0.29	−0.34	0.05	0.80
α_4	−0.36	0.54	−1.52	−0.26	0.60
α_5	−0.60	0.79	−2.39	−0.38	0.52
γ_1	0.69	0.23	0.04	0.79	0.88
γ_2	0.34	0.14	0.01	0.37	0.54
γ_3	0.14	0.09	0.00	0.13	0.34
γ_4	0.42	0.15	0.03	0.46	0.63
γ_5	0.61	0.19	0.05	0.69	0.80

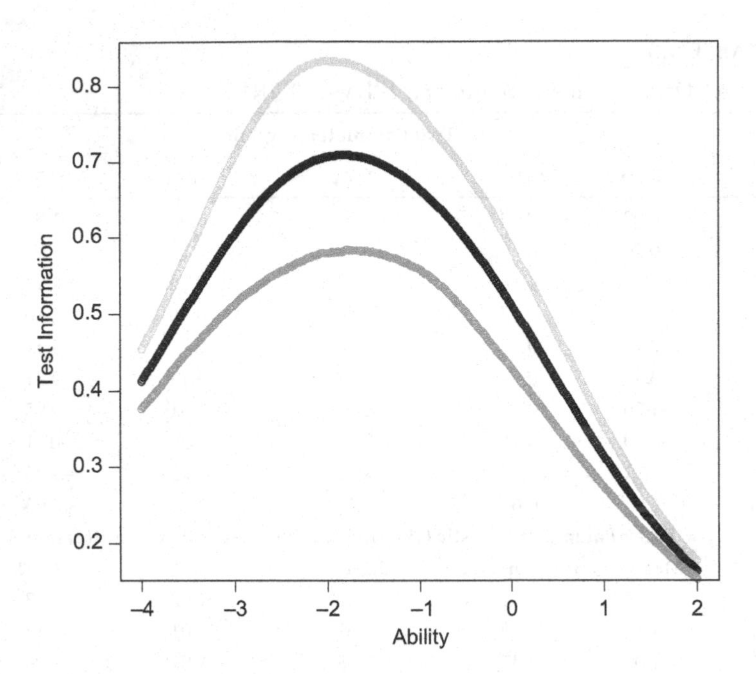

FIGURE 9.1
Test information plot (mean and 60% CRI).

0.57, 0.74, 0.82) under the 2PL model, suggesting that the third item is the least well explained by, and possibly less relevant to, the latent structure model. The test information plot (Figure 9.1) for the 2PL model peaks at around –2, indicating that this set of items best identifies learners with an ability less than average.

Estimates (and fit) for the 3PL may be sensitive to the prior adopted for γ_p. For example, a hierarchical Beta(a_1,b_1) prior for the γ_p, with a_1 and b_1 assigned Exponential(1) priors, provides differing estimates of the difficulty parameters, and a lower LOO-IC of 4910.

Example 9.6 Latent Regression vs Differential Item Functioning

The data in this example are from Thissen et al. (1993) and concern student spelling performance (correct/incorrect) using four words: infidelity, panoramic, succumb, and girder. The sample includes 284 male and 374 female undergraduate students.

One option, considered in Zheng and Rabe-Hesketh (2007) is latent regression, with the factor scores depending on gender. Generically

$$y_{pi} \sim \text{Bern}(\pi_{pi}),$$

$$\theta_i \sim N(\delta X_i, 1),$$

$$\pi_{pi} = \text{logit}^{-1}[\lambda_p(\theta_i - \alpha_p)],$$

where X_i consists of centred covariates (an intercept not being identifiable). The λ_p and α_p parameters are assigned hierarchical normal priors, with λ_p constrained to positive values. Table 9.6 shows the 2PL parameter estimates for this model (obtained using jagsUI), with a significant effect δ of male gender in improving spelling ability. One feature to note is the more informative nature (compared to Example 9.5) of the total information function. Figure 9.2 shows that this provides a higher information level centred at average ability. The LOO-IC is 3033.

TABLE 9.6
Spelling Data, Parameter Estimates Compared

	2PL Latent Regression			
	Mean	**Sd**	**2.50%**	**97.50%**
α_1	−1.66	0.26	−2.26	−1.24
α_2	−0.54	0.11	−0.77	−0.35
α_3	0.90	0.14	0.66	1.21
α_4	−0.12	0.08	−0.28	0.03
λ_1	0.97	0.18	0.64	1.36
λ_2	1.26	0.24	0.84	1.78
λ_3	1.22	0.23	0.83	1.73
λ_4	1.47	0.32	0.98	2.26
Predictive Concordance Item 1	0.67	0.02	0.63	0.70
Predictive Concordance Item 2	0.53	0.02	0.49	0.57
Predictive Concordance Item 3	0.58	0.02	0.54	0.62
Predictive Concordance Item 4	0.51	0.02	0.47	0.54
Δ	0.23	0.12	0.01	0.46
	Differential Item Functioning			
	Mean	**Sd**	**2.50%**	**97.50%**
$\alpha_{1,1}$	−1.50	0.29	−2.22	−1.08
$\alpha_{2,1}$	−1.39	0.32	−2.14	−0.91
$\alpha_{1,2}$	−0.53	0.15	−0.87	−0.27
$\alpha_{2,2}$	−0.55	0.15	−0.87	−0.29
$\alpha_{1,3}$	1.15	0.26	0.73	1.75
$\alpha_{2,3}$	0.68	0.14	0.44	1.00
$\alpha_{1,4}$	0.18	0.12	−0.03	0.42
$\alpha_{2,4}$	−0.52	0.15	−0.86	−0.26
$\lambda_{1,1}$	1.35	0.35	0.75	2.16
$\lambda_{2,1}$	0.99	0.25	0.57	1.54
$\lambda_{1,2}$	1.18	0.31	0.67	1.87
$\lambda_{2,2}$	1.45	0.37	0.85	2.30
$\lambda_{1,3}$	0.94	0.24	0.56	1.47
$\lambda_{2,3}$	1.88	0.58	1.05	3.29
$\lambda_{1,4}$	1.31	0.39	0.74	2.31
$\lambda_{2,4}$	1.39	0.39	0.80	2.30
Predictive Concordance Item 1	0.67	0.02	0.63	0.70
Predictive Concordance Item 2	0.53	0.02	0.49	0.57
Predictive Concordance Item 3	0.58	0.02	0.54	0.62
Predictive Concordance Item 4	0.52	0.02	0.48	0.55
	Differential Item Functioning (Reduced Model)			
	Mean	**Sd**	**2.50%**	**97.50%**
$\alpha_{1,1}$	−1.49	0.15	−1.80	−1.21
$\alpha_{2,1}$	−1.15	0.15	−1.45	−0.86
$\alpha_{1,2}$	−0.47	0.11	−0.69	−0.25
$\alpha_{2,2}$	−0.55	0.13	−0.81	−0.31
$\alpha_{1,3}$	0.90	0.12	0.66	1.15
$\alpha_{2,3}$	0.79	0.14	0.53	1.07

(Continued)

TABLE 9.6 (CONTINUED)

Spelling Data, Parameter Estimates Compared

	Differential Item Functioning (Reduced Model)			
	Mean	**Sd**	**2.50%**	**97.50%**
$\alpha_{1,4}$	0.17	0.11	−0.04	0.38
$\alpha_{2,4}$	−0.49	0.13	−0.75	−0.24
Λ	1.27	0.10	1.08	1.45
Predictive Concordance Item 1	0.67	0.02	0.63	0.70
Predictive Concordance Item 2	0.53	0.02	0.49	0.57
Predictive Concordance Item 3	0.58	0.02	0.54	0.61
Predictive Concordance Item 4	0.52	0.02	0.48	0.55

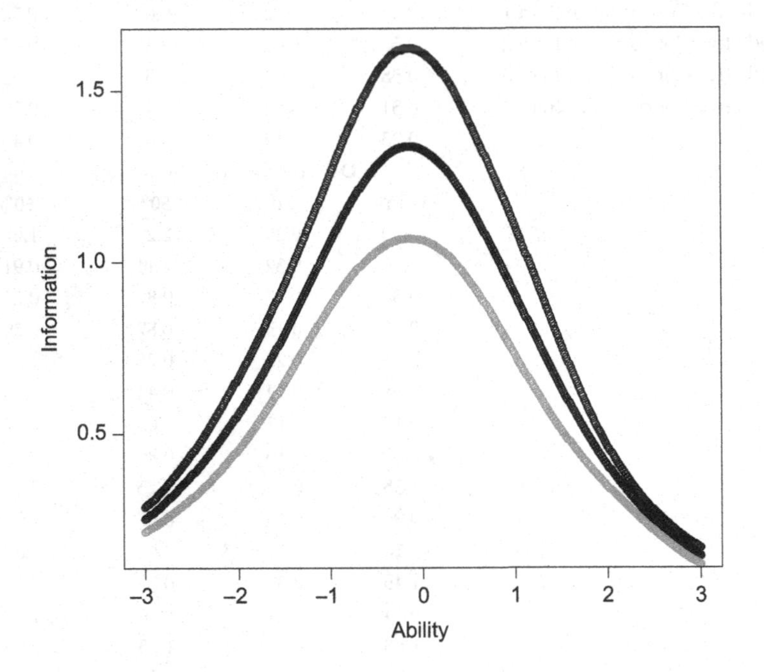

FIGURE 9.2
Test information function, mean, and 80% CRI.

Fit is improved (with LOO-IC reduced to 3015) using a DIF model with difficulty and discrimination parameters varying by group. Thus

$$y_{pi} \sim \text{Bern}(\pi_{pi}),$$

$$\theta_i \sim N(0,1)$$

$$\pi_{pi} = \text{logit}^{-1}[\lambda_{g_i p}(\theta_i - \alpha_{g_i,p})],$$

where gender g_i is coded as ($F=1, M=2$), and where the λ_{gp} and α_{gp} parameters are again assigned hierarchical normal priors. There is some evidence favouring differential functioning. For example, Table 9.6 shows higher difficulty for females on items 3 and 4, and higher discrimination for males on items 2 and 3.

A simplified DIF approach is discussed by Magis et al. (2015) involving a single discrimination parameter (homogenous across groups and items), and α_{gp} parameters subject to (classical) Lasso penalisation. One possible Bayesian option is a Lasso shrinkage prior

$$\alpha_{gp} \sim N(0, \sigma_a^2 \eta_{gp}^2),$$

$$\eta_{gp}^2 \sim \text{Exponential}\left(\frac{\rho^2}{2}\right),$$

$$\rho \sim U(0.001, 1000),$$

$$1/\sigma_a^2 \sim \text{Exponential}(1),$$

with other parameterisations possible. This option provides a further reduction in LOO-IC to 3007, with the discrimination parameter more precisely estimated than under the full DIF method. Another option, assuming a half-Cauchy prior with 1 d.f. (a horseshoe prior) on the η_{gp}, leads to a LOO-IC of 3010, with shrinkage to zero most marked for α_{14}. Posterior mean percentage predictive concordances for the four items are (66.8,53.0,57.8,51.5).

Since there appears to be an association between gender and spelling ability in these data, this can be explored more fully using a multivariate probit, as implemented in rstan. The code involves a Cholesky decomposition of the correlation matrix $R = LL'$, combined with an LKJ(1) prior on L. Estimation shows that the effect of male gender β_{2p} is significantly positive for the last of the four words, and significantly negative for the first (see Table 9.7). The intercepts show item 3 as the most difficult. The highest element in the residual correlation matrix is 0.38, between items 2 and 4.

TABLE 9.7

Multivariate Probit, Spelling Data, Posterior Summary

	Mean	St devn	2.5%	5%	95%	97.5%
β_{11}	0.89	0.07	0.75	0.77	1.01	1.03
β_{12}	0.29	0.06	0.17	0.19	0.40	0.42
β_{13}	−0.54	0.07	−0.67	−0.65	−0.42	−0.41
β_{14}	−0.10	0.07	−0.24	−0.21	0.00	0.02
β_{21}	−0.20	0.11	−0.41	−0.38	−0.02	0.01
β_{22}	0.07	0.10	−0.13	−0.10	0.22	0.26
β_{23}	0.06	0.10	−0.15	−0.11	0.23	0.26
β_{24}	0.42	0.10	0.24	0.27	0.59	0.62
R_{12}	0.30	0.06	0.17	0.19	0.40	0.42
R_{13}	0.28	0.07	0.14	0.16	0.39	0.41
R_{14}	0.33	0.06	0.20	0.23	0.43	0.45
R_{23}	0.37	0.06	0.25	0.27	0.47	0.48
R_{24}	0.38	0.06	0.26	0.28	0.47	0.49
R_{34}	0.36	0.06	0.24	0.26	0.46	0.48

9.5 Robust Density Assumptions in Factor Models

To improve estimability of factor models with data containing unusual observation, heavy tailed or skew densities may be considered (Yuan et al., 2004; Lai and Zhang, 2017). Consider the normal linear factor reduction model (9.1), namely $y_i = a + \Lambda F_i + u_i$. Instead of conventional normality assumptions for residuals $u_i = (u_{1i}, \ldots, u_{Pi})'$ or factor scores $F_i = (F_{1i}, F_{2i}, \ldots, F_{Qi})'$, one might use options that are robust to measurement or construct outliers. For example, a Student t model with ν_{1p} degrees of freedom for the measurement model regressions for y_{pi} is obtainable via scale mixing, with

$$y_{pi} \sim N(a_p + \Lambda_p F_i, \sigma_p^2 / \zeta_{pi}),$$

$$\zeta_{pi} \sim Ga(0.5\nu_{1p}, 0.5\nu_{1p}).$$

To identify possible observation outliers, one may monitor the lowest weights ζ_{pi}. Assume also standardised and uncorrelated factor scores, but following a Student t rather than normal density. Then the corresponding heavy tailed construct score model for identifying construct outliers is

$$F_{qi} \sim N(0, 1/\omega_{qi})$$

$$\omega_{qi} \sim \text{Gamma}(0.5\nu_{2q}, 0.5\nu_{2q}).$$

Skewness in outcomes or factor scores may also be present. Following Azzalini (1985), let f and g be symmetric probability density functions, with G being the cumulative distribution function associated with g. Then for location parameter μ and scale parameter σ, the density

$$\frac{2}{\sigma} f\left(\frac{x-\mu}{\sigma}\right) G\left(\kappa \frac{x-\mu}{\sigma}\right)$$

is a skew pdf for any κ. If $f = \phi$ and $G = \Phi$ (respectively the normal pdc and cdf), one obtains the skew-normal distribution. Positive (negative) values of κ indicate positive (negative) skewness, while $\kappa = 0$ provides the normal density. Bazan et al. (2006) consider the application of the skew-normal density in item analysis. For binary items $p = 1, \ldots, P$, and with

$$\delta_p = \frac{\kappa_p}{\sqrt{(1+\kappa_p^2)}},$$

they define a skew probit IRT model involving a common factor F_i and item-specific effects V_{pi} to allow for skew errors. So

$$y_{pi}^* \sim N(a_p + \lambda_p F_i + \delta_p V_{pi}, 1 - \delta_p^2) \quad I(A_{pi}, B_{pi})$$

$$F_i \sim N(0,1),$$

$$V_{pi} \sim HN(0,1),$$

with sampling limits $\{A_{pi}, B_{pi}\}$ defined according to the observed binary responses. This parameterisation necessitates priors for δ_p in the interval [−1,1].

Example 9.7 Greek Crimes by Prefecture: Non-Parametric Prior for Random Effects

The analysis of the Greek crime data in Example 9.3 assumed normally distributed errors u_{pi} in the log-link model for the crime rates ρ_{pi}. As noted by Knorr-Held and Rasser (2000) a fully parametric specification of the random effects distribution may result in oversmoothing, and mask local discontinuities, especially when the true distribution is characterised by a finite number of locations. Here a truncated Dirichlet process prior (DPP) is adopted to model the density of the residuals u_{pi}, with potential values $\{u^*_{pk}, p=1,\ldots,P\}$ from K clusters centred on the multivariate normal $G_0 = N_P(0,D)$, where $P=4$. D^{-1} has a Wishart prior with identity scale matrix and P degrees of freedom.

Thus the infinite DPP representation is approximated by one truncated at $K \le n$ components, with appropriate values u_{pi} for prefecture i chosen according to an allocation indicator $S_i \in (1,\ldots K)$. The probabilities π_k of allocation to clusters $\{1,\ldots,K\}$ are determined by $K-1$ beta distributed random variables $V_k \sim \text{Beta}(1,\kappa)$, with unknown concentration parameter κ, and $V_K=1$ to ensure the random weights π_k sum to 1 (Ishwaran and James, 2001; Sethuraman, 1994). Then $\pi_1 = V_1$ and

$$\pi_k = (1-V_1)(1-V_2)\ldots(1-V_{k-1})V_k \quad k>1.$$

Following Ishwaran and Zarepour (2000, p.377), the gamma prior for κ, namely $\kappa \sim Ga(\nu_1,\nu_2)$ has relatively large ν_1 and ν_2, with ν_2 set larger than ν_1. Such a setting discourages small and large values for κ. Here $\nu_2=4$ and $\nu_1=2$. The maximum possible clusters is set at $K=20$.

Estimation using jagsUI shows early convergence and replicates Example 9.3 in showing mostly non-significant predictor effects. There are, however, significant positive effects of unemployment on rape, and of urban centre on manslaughter, and a significant negative effect of GDP levels on manslaughter. The posterior mean for κ is 1.26, with the average number of non-empty clusters K^* being 8.26.

Extreme residuals, and departures from normality, are associated with poorly fitted cases (with extreme response values and high pointwise LOO-IC). For example, Figure 9.3 plots out positively skewed mean residuals for u_{3i} (manslaughter), with the most extreme positive residual for the elevated observation $y_{3,14}$.

Example 9.8 Maths Aptitude; Skew Probit

This example uses binary item data from Tanner (1996) concerning maths aptitude; there are $n=39$ students and $P=6$ items. An augmented data probit regression is applied, and an extension to a skew probit link is adopted, following the approach of Bazan et al. (2006).

So latent metric data y^*_{pi} underlying the observed binary response are sampled according to

$$y^*_{pi} \sim N(\alpha_p + \lambda_p F_i + \delta_p V_{pi}, 1-\delta_p^2).$$

with $F_i \sim N(0,1)$ and the λ_p all being unknowns. A $U(-1,1)$ prior is adopted on the δ_p parameters, with the prior on discrimination parameters

$$\lambda_p \sim N(1,0.5)\ I(0,),$$

providing an identifying constraint (Sahu, 2002).

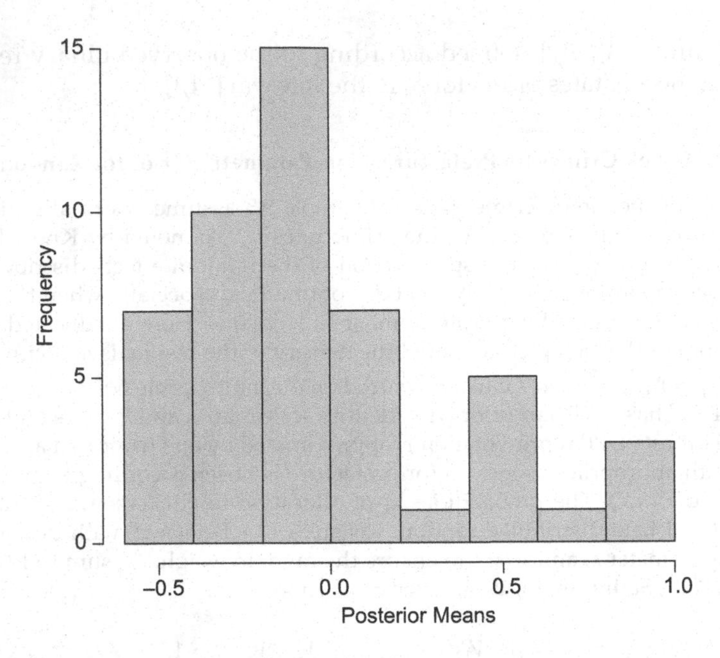

FIGURE 9.3
Histogram of residuals, manslaughter.

A two-chain run of 10,000 iterations (with convergence at under 1,000) shows none of the δ_p (and hence κ_p) parameters to be significantly positive or negative. Despite the apparent absence of skew, this model has a lower LOO-IC than the symmetric probit, 297 as against 322. Posterior mean percentages of predictive concordance for the six items are also higher under the skew probit, namely (57.5,59.1,65.9,66.5,70.5,66.3).

Example 9.9 Student t Factor Model

This example involves simulated continuous data (n=200 observations) derived using a multivariate t_5 density for the factor scores. The simulated data are then re-analysed using different assumptions of multivariate normal factors and multivariate student factors. The simulation is focused on explanatory factor analysis with $P=6$ indicators and $Q=2$ factors, and with the same assumptions on priors for loadings as in Example 9.2. The code for the simulation takes account of inferences being potentially influenced by order dependence in the indicators, and the loading of the second factor on the second indicator is set to ensure the second indicator is an effective factor founder. The code is as follows, assuming 5 degrees of freedom in the Student t, and requires the library mvtnorm:

```
n = 200 # number of observations
p = 6 # number of indicators
q = 2 # number of factors
# Loading matrix
Lambda = matrix(c(1.1,-0.1,
-0.2,1.2,
0.8,0,
-0.9,0.6,
0.3,0.7,
-0.8,0.9),
```

```
nrow=p, ncol=q, byrow=T)
DF=5
S=matrix(c(1,0,0,1), nrow=q, ncol=q, byrow=T)
mean=c(0, 0)
F = rmvt(n, sigma=S*(DF-2)/DF, df=DF) + mean # MVT Factor Scores
e <- rmvnorm(n, rep(0, p), diag(p)) # N(0,1) errors
y <- F %*% t(Lambda) + e # indicator matrix
y <- scale(y)
```

Table 9.8 shows the estimated loadings obtained under maximum likelihood (via the lavaan package), using Bayesian analysis with MVN factors (via jagsUI), and using Bayesian estimation with MVT factors, obtained using scale mixing (via the rube package). The prior for the unknown degrees of freedom ν follows Juárez and Steel (2010).

The first factor from the maximum likelihood analysis is reverse signed, but otherwise the loadings are similar between the alternative estimation methods. The respective LOO-IC values for the MVN and MVT factor models are 1577 and 1557. All estimated loadings show shrinkage from the generating loadings.

The MVT analysis provides a posterior mean (95% CRI) of 12.4 (4.0,34.7) for ν, as compared to the generating value $\nu=5$. The posterior median for ν of 10.1 is a better estimator of the generating value. Around 10% of the observations have scale adjustments under 0.8, and two observations (103, 152) have scale adjustments with 95% credible intervals entirely below 1.

9.6 Multivariate Spatial Priors for Discrete Area Frameworks

Consider multivariate spatial responses $(y_{1i},\ldots y_{Pi})'$ of dimension P from an exponential family density observed over n discrete areas (e.g. administrative regions). Conditional on random spatial effects $s_i=(s_{1i},\ldots,s_{Pi})'$ of the same dimension, and predictors $x_i=(x_{1i},\ldots,x_{Ri})'$, one then has

$$p(y_{pi}\mid s_{pi},x_i)\propto\exp\left\{\frac{y_{pi}\theta_{pi}-b(\theta_{pi})}{\phi_{pi}}+c(y_{pi},\phi_{pi})\right\}$$

where θ_{pi} is the canonical parameter, and ϕ_{pi} a known scale. Denoting regression terms as $\eta_{pi}=g(\theta_{pi})$ with link g, the s_{pi} are included to measure spatially configured but unmeasured predictors. So, one has at a minimum the representation

$$\eta_{pi}=\alpha_p+\beta_p x_i+s_{pi},$$

where the spatial effects for area i, $s_i=(s_{1i},\ldots,s_{Pi})'$, follow a multivariate spatial prior. For certain definitions of spatial effects, it may be appropriate to also include unstructured (i.e. exchangeable over areas) multivariate effects, in line with a multivariate form of the Besag et al. (1991) convolution prior. Thus, the full dimension analogue to the convolution prior is

$$\eta_{pi}=\alpha_p+\beta_p x_i+s_{pi}+u_{pi},\tag{9.6}$$

where the u_{pi} also follow a multivariate prior. Other possibilities, following Chapter 6, include regression effects β_{pi} that vary spatially as well as over response variables.

TABLE 9.8

EFA of Simulated Data, Estimated Loadings

	Maximum Likelihood		Bayesian Estimation MVN Factors (Mean, 95% CRI)						Bayesian Estimation MVT Factors (Mean, 95% CRI)					
			Factor 1			Factor 2			Factor 1			Factor 2		
Indictor	**Factor1**	**Factor2**	**Mean**	**2.5%**	**97.5%**	**Mean**	**2.5%**	**97.5%**	**Mean**	**2.5%**	**97.5%**	**Mean**	**2.5%**	**97.5%**
y_1	−0.71	0.00	0.69	0.50	0.88	0.01	−0.19	0.25	0.61	0.44	0.79	0.00	−0.18	0.17
y_2	0.30	0.78	−0.11	−0.96	0.90	0.84	0.48	1.41	−0.13	−0.79	0.66	0.71	0.43	1.23
y_3	−0.75	0.00	0.75	0.46	1.06	0.08	−0.25	0.49	0.65	0.44	0.91	0.05	−0.21	0.35
y_4	0.68	0.19	−0.63	−0.88	−0.28	0.14	−0.20	0.48	−0.56	−0.78	−0.28	0.14	−0.10	0.41
y_5	0.45	0.00	0.15	−0.42	0.92	0.52	0.28	1.06	0.12	−0.37	0.72	0.46	0.26	0.85
y_6	0.59	0.43	−0.47	−0.94	0.16	0.45	0.10	0.90	−0.44	−0.84	0.09	0.40	0.15	0.75

Conditions for a valid multivariate spatial prior, specifically a multivariate Gaussian Markov random field (MGMRF), are discussed by MacNab (2018), Rue and Held (2005, section 2.2), and Banerjee et al. (2004, section 9.4). Thus denote the nP length vector over all areas as $s=(s_1,\ldots,s_n)'$, and denote the mean vector, possibly including regression effects, as $\mu_i=(\mu_{1i},\ldots,\mu_{Pi})$, with $\mu=(\mu_1,\ldots,\mu_n)'$. Also denote the matrix describing observed spatial interactions in the region by $W=[w_{ij}]$, with $w_{ij}=w_{ji}$, and set

$$D=\text{Diag}(d_1,\ldots,d_n),$$

where $d_i=\sum_{j\neq i} w_{ij}$. If $w_{ij}=1$ when areas i and j are contiguous, and zero otherwise (binary adjacency), then d_i is the number of neighbours for area i. The neighbourhood for area i is often denoted ∂_i, and if area j is a neighbour of area i, then the neighbour relation (under binary interaction) is denoted $j\sim i$.

The joint density for a normal MGMRF for P spatial effects and with $nP\times nP$ precision matrix Q may be expressed

$$p(s\mid Q)=\left(\frac{1}{2\pi}\right)^{nP/2}|Q|^{0.5}\exp\left[(s-\mu)'Q(s-\mu)\right]$$

$$=\left(\frac{1}{2\pi}\right)^{nP/2}|Q|^{0.5}\sum_{ij}\exp\left[(s_i-\mu_i)'Q_{ij}(s_j-\mu_j)\right].$$

Q is block diagonal with $P\times P$ sub-matrix elements Q_{ij} that are non-zero (zero) if area j is (is not) a neighbour of area i. Retaining the possibility of a regression model in the means μ_i (Rue and Held, 2005), the corresponding full conditional density is

$$s_i\mid s_{[i]}\sim N\left(\mu_i-Q_{ii}^{-1}\sum_{j\neq i}Q_{ij}(s_j-\mu_j),\frac{1}{Q_{ii}}\right),$$

with conditional precision matrices

$$\text{Prec}(s_i\mid s_{[i]})=Q_{ii}=\Delta_i.$$

Equivalently define $P\times P$ matrices $B_{ij}=-Q_{ij}/Q_{ii}$, with $B_{ii}=0$, and $\Delta_i=Q_{ii}$. Then

$$E(s_i\mid s_{[i]})=\mu_i+\sum_{j\neq i}B_{ij}(s_j-\mu_j) \tag{9.7}$$

$$\text{Prec}(s_i\mid s_{[i]})=\Delta_i.$$

Most commonly the μ_i are set to zero.

Under the parameterisation in (9.7), the joint density has mean μ and precision matrix $Q=\Delta(I-B)$, where Δ is block diagonal with blocks Δ_i, and the $nP\times nP$ matrix B is block diagonal with (i,j)th block B_{ij}. The requirements for a valid joint density to exist (for example, if specification is starting from a prior involving the full conditionals) are that (Sain and Cressie, 2007; Rue and Held, 2005, p.31)

$$\Delta_i B_{ij} = \Delta_j B_{ji}.$$

For example, setting $B_{ij} = [w_{ij}/d_i]I_{P\times P}$, and $\Delta_i = d_i\zeta$ (where ζ is a $P\times P$ precision matrix) will ensure a valid joint density.

A number of multivariate priors which incorporate spatial dependence between areas have been proposed. The generalisation of the intrinsic univariate CAR to a multivariate setting is denoted as the multivariate CAR or MCAR prior (Mardia, 1988; Jin et al., 2005, equation 6; Song et al., 2006, p.254). This takes the vector of multivariate area effects $s = (s_{11}, s_{12}, \ldots, s_{1n}; \ldots; s_{P1}, s_{P2}, \ldots, s_{Pn})$ as multivariate normal with mean consisting of a vector of zeros of length nP, and with $nP\times nP$ precision matrix, $Q = (D - aW)\otimes\zeta$, namely

$$p(s \mid \zeta, a) = \left(\frac{1}{2\pi}\right)^{nP/2} |D - aW|^{P/2} |\zeta|^{n/2} \exp\left[-\frac{1}{2}s'Qs\right], \tag{9.8}$$

where $a \in (0,1)$ is a propriety parameter. The $P\times P$ positive definite symmetric matrix ζ^{-1} describes covariation between the outcomes, and $D - \alpha W$ is the precision matrix for the spatial effects. The latter matrix can also be written as $D(I - \alpha B)$ where $B = D^{-1}W$. Let the effects be arranged by variable rather than subject, so that $S_1 = (s_{11}, s_{12}, \ldots, s_{1n})'$, $S_2 = (s_{21}, s_{22}, \ldots, s_{2n})'$, etc., then for $P = 2$, the joint prior is

$$\begin{pmatrix} S_1 \\ S_2 \end{pmatrix} \sim N\left(\begin{pmatrix} 0 \\ 0 \end{pmatrix}, \begin{bmatrix} \zeta_{11}(D-\alpha W) & \zeta_{12}(D-\alpha W) \\ \zeta_{12}(D-\alpha W) & \zeta_{22}(D-\alpha W) \end{bmatrix}^{-1}\right),$$

where each submatrix $\zeta_{pq}(D - aW)$ is of dimension $n\times n$.

The conditional prior under (9.8) for $s_i = (s_{1i}, \ldots, s_{Pi})$ given the remaining effects $s_{[i]} = (s_1, \ldots s_{i-1}, s_{i+1}, \ldots s_n)$ is multivariate normal with means $E(s_i \mid s_{[i]}) = (M_{1i}, \ldots, M_{Pi})$, where

$$E(s_{pi} \mid s_{[i]}) = M_{pi} = a\sum_{j\neq i} w_{ij}s_{pj} \Big/ \sum_{j\neq i} w_{ij}$$

and with precisions

$$\text{Prec}(s_i \mid s_{[i]}) = d_i\zeta.$$

If the w_{ij} are set to 1 for neighbouring areas and to 0 otherwise, then the $M_{pi} = a\sum_{j\in\partial_i} s_{pj}/d_i$ are locality averages (times α) of the spatial effect for the pth response. Setting $\alpha = 1$ provides the multivariate version of the intrinsic CAR prior of Besag et al. (1991); such intrinsic GMRFs (for spatial and non-spatial priors) are considered by Rue and Held (2005, Chapter 3).

MacNab (2007) discusses a multivariate extension of the Leroux et al. (1999) prior, which allows the data to determine the appropriate mix between spatial or exchangeable dependence. This may be achieved with a single set of random effects r_{pi} rather than the two sets $\{s_{pi}, u_{pi}\}$ present in the multivariate extension (9.6) of the convolution prior. Thus with $r_i = (r_{1i}, \ldots, r_{Pi})'$, parameter $\kappa \in (0,1)$, and spatial interactions $W = [w_{ij}]$

$$E(r_i \mid r_{[i]}) = [M_{1i},\dots M_{Pi}] = \kappa \sum_{j \neq i} w_{ij} I_P r_j \Bigg/ \left[1-\kappa+\kappa\sum_{j\neq i} w_{ij}\right]$$

$$\text{Prec}(r_i \mid r_{[i]}) = \left[1-\kappa+\kappa\sum_{j\neq i} w_{ij}\right]\zeta$$

where, as above, ζ is the within area covariance of dimension $P \times P$. Thus

$$B_{ij} = \left[\frac{\kappa w_{ij}}{\left[1-\kappa+\kappa\sum_{j\neq i} w_{ij}\right]}\right] I_{P\times P}$$

$$\Delta_i = \left[1-\kappa+\kappa\sum_{j\neq i} w_{ij}\right]\zeta$$

and $\Delta_i B_{ij} = \Delta_j B_{ji}$ holds. When the w_{ij} are binary adjacency indicators, with d_i the number of neighbours of area i, the conditional expectations become

$$E(r_{pi} \mid r_{[i]}) = M_{pi} = \frac{\kappa \sum_{j\in\partial_i} r_{pj}}{[1-\kappa+\kappa d_i]}.$$

Define

$$H = \text{diag}(1-\kappa+\kappa\sum_{j\neq 1} w_{1j},\dots,1-\kappa+\kappa\sum_{j\neq n} w_{nj}) = (1-\kappa)I_n + \kappa D.$$

Then the joint density is multivariate normal with mean vector 0 and $np \times np$ precision matrix $(H-\kappa W)\otimes\zeta$.

Jin et al. (2005) propose a generalised MCAR (GMCAR) model whereby the joint distribution for a multivariate spatial effect is obtained by specifying a sequence of conditional and marginal models. Let effects be arranged by variable rather than subject. Then for a bivariate spatial effect with $P(S_1,S_2) = P(S_1 \mid S_2)P(S_2)$, where $S_1 = (s_{11}, s_{12},\dots,s_{1n})'$ and $S_2 = (s_{21}, s_{22},\dots,s_{2n})'$, one has

$$\begin{pmatrix} S_1 \\ S_2 \end{pmatrix} \sim N\left(\begin{pmatrix} 0 \\ 0 \end{pmatrix}, \begin{bmatrix} \Sigma_{11} & \Sigma_{12} \\ \Sigma_{12} & \Sigma_{22} \end{bmatrix}\right),$$

where $E(S_1 \mid S_2) = \Sigma_{12}\Sigma_{22}^{-1}S_2$, and $\text{var}(S_1 \mid S_2) = \Sigma_{11.2} = \Sigma_{11} - \Sigma_{12}\Sigma_{22}^{-1}\Sigma_{12}'$. Hence with $G = \Sigma_{12}\Sigma_{22}^{-1}$, one has equivalently

$$\begin{pmatrix} S_1 \\ S_2 \end{pmatrix} \sim N\left(\begin{pmatrix} 0 \\ 0 \end{pmatrix}, \begin{bmatrix} \Sigma_{11.2} + G\Sigma_{22}G' & G\Sigma_{22} \\ (G\Sigma_{22})' & \Sigma_{22} \end{bmatrix}\right).$$

To specify the joint distribution of S_1 and S_2, it is therefore necessary to specify the matrices $\Sigma_{11.2}$, Σ_{22}, and G.

Taking $\Sigma_{11.2}^{-1} = \tau_1[D - a_1 W]$, $\Sigma_{22}^{-1} = \tau_2[D - a_2 W]$ and $G = \gamma_0 I + \gamma_1 W$, the marginal joint prior for the second set of effects is then

$$S_2 \sim N(0, \tau_2^{-1}[D - a_2 W]^{-1}),$$

and the conditional prior for the first set of effects is

$$S_1 \mid S_2 \sim N(GS_2, \tau_1^{-1}[D - a_1 W]^{-1}).$$

As above the $0 < a_p < 1$ are propriety parameters, and the γ_0 parameter links different variable-same area effects, namely regresses s_{1i} on s_{2i}, while γ_1 links s_{1i} with other variable-other area effects $\{s_{2j}, j \neq i\}$. This approach is possibly more suitable for small P, as $P!$ conditional density sequences are possible, and may give different inferences or fits – though Jin et al. (2005, p.957) demonstrate how initial regression analysis may lead one to prefer one sequence to another.

The linear co-regionalisation model of Jin et al. (2007) avoids dependence on any particular ordering. Assuming binary adjacency, the most general option in Jin et al. (2007), namely Case 3 (dependent and non-identical latent processes), specifies a conditional mean

$$E(s_{pi} \mid s_{p,k\neq i}, s_{q\neq p,i}, s_{q\neq p,k\neq i}) = \alpha_{pp} \sum_{k\sim i} s_{pk}/d_i + \sum_{q\neq p}\left(\alpha_{pq} \sum_{k\sim i} s_{qk}/d_i\right),$$

where α_{pp} is the spatial autocorrelation measure for the pth outcome, and α_{pq} is a crossspatial correlation between S_p and S_q. The joint distribution (Martinez-Beneito, 2013, p.4) may be represented

$$s \sim N_{nP}\left\{0, \begin{pmatrix} (D-\alpha_{11}W)\zeta_{11} & \dots & (D-\alpha_{1P}W)\zeta_{1P} \\ \dots & \dots & \dots \\ (D-\alpha_{1P}W)\zeta_{1P} & \dots & (D-\alpha_{PP}W)\zeta_{PP} \end{pmatrix}\right\}$$

with $\vartheta = \zeta_{pq}$ denoting the within area between disease precision matrix.

Martinez-Beneito (2013) represents the joint prior for the spatial error s of length nP in the generic form $s \sim N_{nP}(0, \Sigma_b \otimes \Sigma_w)$ where Σ_b and Σ_w represent between and within disease covariance matrices. Denoting $\tilde{\Sigma}_b$ and $\tilde{\Sigma}_w$ as lower triangular matrices such that $\Sigma_b = \tilde{\Sigma}_b \tilde{\Sigma}_b^T$ and $\Sigma_w = \tilde{\Sigma}_w \tilde{\Sigma}_w^T$, one has that $s = \tilde{\Sigma}_w \varepsilon \tilde{\Sigma}_b^T$ with ε of dimension $n \times P$, consisting of independent N(0,1) variates. Representing $\phi = \tilde{\Sigma}_w \varepsilon$ as a matrix with P columns containing a set of particular spatial distributions (e.g. P independent ICAR densities), then interdependence is induced via the product form $s = \text{vec}(\phi \tilde{\Sigma}_b^T)$, which has covariance $\Sigma_b \otimes (D-W)^{-1}$. If ϕ consists of independent ICAR(α_p) densities, then Case 2 of Jin et al. (2007) is obtained. More flexibility is obtained by representing Σ_b as $\Sigma_b = \tilde{\Sigma}_b C C^T \tilde{\Sigma}_b^T$ where C is any square orthogonal matrix, which enables reproduction of Case 3 of Jin et al. (2007).

9.7 Spatial Factor Models

When high correlations are evident in ζ^{-1}, common spatial factor models may be more parsimonious (Tzala and Best, 2007; Congdon et al., 2007; Liu et al., 2005: Gielen et al., 2017). These can extend to full structural equation models (e.g. Arhonditsis et al., 2006b). Standard presentations of the normal linear and generalised linear factor models assume factor scores are independent over subjects, though in fact they might be spatially or temporally structured. So for P outcomes, the factor scores F of dimension $Q<P$ may be correlated over both variables and areas. Then for Poisson or binomial responses with mean $\mu_{pi} = g^{-1}(\eta_{pi})$ for the pth dependent variable and area i, one might have a regression term

$$\eta_{pi} = a_p + \beta_p x_i + \Lambda_p F_i,$$

where the vector Λ_p is of dimension Q, and the factor score variables $F_i = (F_{1i},\ldots,F_{Qi})'$ are spatially dependent over areas i, as well as mutually intercorrelated. For example, a MCAR prior would specify the joint pairwise difference density for the factor scores

$$p(F \mid \Sigma_F) \propto |\Sigma_F|^{-n/2} \exp\left[-0.5 \sum_{i,j} w_{ij}(F_i - F_j)'\Sigma_F^{-1}(F_i - F_j)\right].$$

As in other factor models, constraints are required to deal with label switching and location, scale, and rotational indeterminacy. Constraining one or more loadings to be positive is one strategy for avoiding label switching (Mar-Dell'Olmo et al., 2011). In the multivariate CAR model for $F_i = (F_{1i},\ldots,F_{Qi})'$, the location is fixed in practice by centring each of the Q sets of spatial factor scores at each MCMC iteration. Scale may be determined by fixing the Q variances of the F_{qi} scores at 1, or by fixing one of the loadings $(\lambda_{1q},\ldots,\lambda_{Pq})$ linking the P manifest indicators to the qth factor. Additional loadings would need to be fixed to avoid rotational indeterminacy, typically $\lambda_{pq}=0$ for $q>p$. For example, if $Q=2$, and the variances of the F scores are free parameters, then the two loadings λ_{qq} may be set to 1 to define the scale, while rotational invariance is avoided by setting $\lambda_{12}=0$.

Example 9.10 Chronic Disease Prevalence

This example contrasts covariance models for multivariate spatial outcomes with a common spatial factor approach. It considers prevalence totals for common chronic diseases for 56 wards (small political areas) in three London boroughs (Barking and Dagenham, Havering, and Redbridge). The $P=3$ outcomes are diagnosed counts y_{pi} of diabetes, hypertension, and chronic kidney disease (CKD) in 2016. Offsets are expected prevalence totals E_{pi}, based on region-wide age-specific rates.

The first model applied is the multivariate generalisation of the Leroux et al. (1999) conditional autoregressive prior under a Poisson likelihood, with binary adjacencies w_{ij}, and d_i the total number of areas adjacent to area i. The regression involves a constant term and spatially configured effects $(s_{1i},\ldots,s_{Pi})$ of the same dimension as the response vector

$$y_{pi} \sim Po(E_{pi}e^{\eta_{pi}}),$$

$$\eta_{pi} = a_p + s_{pi}.$$

The conditional mean of s_{pi} is

$$E(s_{pi} \mid s_{[i]}) = S_{pi} = \frac{\kappa \sum_{k \in \partial_i} s_{pk}}{[1-\kappa+\kappa d_i]},$$

where κ is between 0 and 1. A Wishart prior for the conditional precision matrix Ψ, with prior mean covariance I, is assumed.

This model is estimated using both CARBayes and R2OpenBUGS option. The first option, using the MVS.Carleroux command (Kavanagh et al., 2016), provides an estimate of 0.92 for κ and DIC of 1748.

Under the second option, early convergence is attained in a two-chain run of 10,000 iterations, with a LOO-IC of 1811 and DIC of 1749. Setting $\Phi = \Psi^{-1}$, posterior mean correlations $r_{jk} = \Phi_{jk}/(\Phi_{jj}\Phi_{kk})^{0.5}$ are highest (0.335) between hypertension and CKD. The spatial parameter κ is estimated at 0.97. Six of the $3 \times 56 = 168$ observations have mixed predictive p-tests under 0.05 or over 0.95 (Marshall and Spiegelhalter, 2007), with under-prediction most apparent for CKD in ward 16, and over-prediction most apparent for diabetes in ward 56.

A second analysis uses the Martinez-Beneito (2013) implementation of Case 3 of Jin et al. (2007), which allows for distinct spatial parameters for each outcome, and also for between disease dependence within areas. The CAR(γ_p) prior, implemented via the proper.car function within BUGS, is used to model between area spatial dependence within outcomes, where γ_p represents spatial dependence for outcome p. The precision parameters (tau[j] in the proper.car function in the code) are set to 1 for identifiability. Binary adjacency is assumed with conditional spatial variances proportional to $1/d_i$, where d_i is the number of areas adjacent to area i. The parameter γ_p has a value between bounds given by the inverse of the minimum and maximum eigenvalues of the matrix $M^{-0.5}W_sM^{0.5}$, where W_s is the row standardised adjacency matrix and $M = \text{diag}(1/d_i)$.

A two-chain run of 100,000 iterations gives significant correlations (a) between CKD and hypertension, with mean (95% CRI) of 0.57 (0.30, 0.72), and (b) between hypertension and diabetes, namely 0.51 (0.03,0.71). The γ parameters are similar between the outcomes, with respective posterior means 0.937, 0.933, and 0.943. The overall LOO-IC is estimated at 1828.

The third model combines a common spatial factor and unstructured outcome-specific random effects u_{pi}, whereby

$$\eta_{pi} = \alpha_p + \lambda_p F_i + u_{pi}$$

where F_i follows a univariate Leroux et al. (1999) prior. Thus for $\kappa \in (0,1)$, precision parameter τ_F, and with $F_{[i]} = (F_1, \ldots, F_{i-1}, F_{i+1}, \ldots, F_n)$, and binary spatial interactions, the conditional mean and precision for ward i are

$$E(F_i \mid F_{[i]}) = \frac{\kappa \sum_{j \in \partial_i} F_j}{[1-\kappa+\kappa d_i]},$$

and

$$\text{Prec}(F_i \mid F_{[i]}) = [1-\kappa+\kappa d_i]\,\tau_F.$$

With $\tau_F = \sigma_F^{-2}$ taken as an unknown, with prior $\sigma_F \sim U(0,1000)$, one of the loadings λ_j must be fixed for identification, and accordingly $\lambda_1 = 1$. This model has a LOO-IC of 1823, with κ estimated as 0.89.

9.8 Multivariate Time Series

Multivariate time series can occur in several ways. One example is where the same measurement process (e.g. repeated environmental readings) is carried out at several locations and where high correlation between the series is expected. Another situation occurs with financial data, such as exchange rates or stock returns, where high correlations raise questions such as whether there are feedbacks between different series, or whether common factors (e.g. market risk) affect all series. Multivariate time series for count data are also an increasing focus (Aktekin et al., 2017; Chapados, 2014), with applications in political science (Brandt and Freeman, 2005) and ecology (Wang et al., 2012). Overviews of Bayesian methods for multivariate time series include Koop and Korobilis (2010) and Sims and Zha (1998).

Classical multivariate time series analysis includes extending the ARMA model to vector responses (Tiao and Tsay, 1989; Reinsel, 2003). For observation vector $y_t = (y_{1t},\ldots y_{Pt})'$, the vector ARMA($R$,$S$) model has the form

$$y_t = \mu + \Phi_1 y_{t-1} + \ldots + \Phi_R y_{t-R} + u_t - \Theta_1 u_{t-1} \ldots - \Theta_S u_{t-S},$$

where the coefficient matrices are all of order $P \times P$, and u_t denotes P-variate white noise, with $E(u_t) = 0$, and

$$E(u_t u'_{t-k}) = 0 \quad k \neq 0;$$

$$E(u_t u'_{t-k}) = \Sigma \quad k = 0.$$

For the vector autoregressive or VAR model obtained on omitting moving average terms, stationarity requires that the roots of the characteristic equation

$$\det(I - \Phi_1 z + \ldots + \Phi_r z^R) = 0$$

lie outside the unit circle.

Bayesian analysis of VAR models are extensive, and include treatments of cointegration* (Koop et al., 2006; Kleibergen and Paap, 2002), model selection and averaging (Andersson and Karlsson, 2007), variable selection (Karlsson, 2015), and informative and restricted priors (Litterman, 1986; Sims and Zha, 1998; Brandt and Freeman, 2005). Relevant packages in R include MTS (Tsay, 2014), BMR (Bayesian Macroeconometrics in R) and MSBVAR (https://cran.revolutionanalytics.com/web/packages/MSBVAR/index.html).

9.8.1 Multivariate Dynamic Linear Models

The structural model approach is widely applied in Bayesian time series studies (e.g. Petris et al., 2009; Commandeur and Koopman, 2007; West and Harrison, 1997; Durbin and

* Classical approaches using autoregressive moving average models may rest on assumptions of stationarity, following transformation or differencing: a time series is integrated of order d, or $I(d)$, when differencing to order d is needed for stationarity. Such series are cointegrated if some linear combination of the series has a lower order of integration than the individual series (Phillips and Durlauf, 1986), for example, when two series y_t and x_t are both $I(1)$, but there is a parameter α such that $u_t = y_t - \alpha x_t$ is stationary (integrated of order zero).

Koopman, 2012) and focuses on underlying components of multiple series without requiring initial differencing. The multivariate normal dynamic linear model specifies

$$y_t = F_t\theta_t + e_t, \quad e_t \sim N(0,V_t), \quad t = 1,\ldots,T$$

$$\theta_{t+1} = G_t\theta_t + H_t u_t, \quad u_t \sim N(0,W_t),$$

where y_t is a $P \times 1$ observation vector, and θ_t is a $Q \times 1$ latent state vector following a Markov process. The disturbance vectors e_t and u_t are assumed normally distributed, and uncorrelated with each other and over time. The initialising prior for the state vector is typically assumed to be a normal fixed effect with mean m_1 and covariance matrix C_1, $\theta_1 \sim N(m_1,C_1)$. The system matrices F_t, G_t, V_t, W_t and H_t may be assumed to be known, in which case simple updating, forecasting and filtering densities can be derived – see West and Harrison (1997, p.582). In more realistic settings where the covariances V_t and W_t are unknown, time-invariant assumptions such as $V_t = \Sigma_e$ and $W_t = \Sigma_u$ are one possible parameterisation. A simple case occurs (Koopman and Durbin, 2000) when V_t is diagonal, the assumption being that the observations are independent conditional on the latent states.

Common model forms include the local level (LL) model with measurement and transition equations

$$y_t = \theta_t + e_t, \quad e_t \sim N(0,\Sigma_e), \quad t = 1,\ldots,T$$

$$\theta_{t+1} = \theta_t + u_t, \quad u_t \sim N(0,\Sigma_u),$$

where y_t is a $P \times 1$ metric observation, θ_t also has dimension P, and Σ_e and Σ_u are of dimension $P \times P$. A local linear trend (LLT) includes a trend in the underlying level, as in

$$y_t = \theta_t + e_t, \quad e_t \sim N(0,\Sigma_e), \quad t = 1,\ldots,T$$

$$\theta_{t+1} = \theta_t + \delta_t + u_t, \quad u_t \sim N(0,\Sigma_u),$$

$$\delta_{t+1} = \delta_t + w_t, \quad w_t \sim N(0,\Sigma_w).$$

For example, Proietti (2007) applies a multivariate local level model to measuring core inflation, while Moauro and Savio (2005) apply a LLT approach to temporal disaggregation of multiple economic series. Multivariate signal models may be applied to measure latent risk, as in the accident rate and credit card use examples of Bijleveld et al. (2005). This approach involves time series or panel data on exposure totals (x_t or x_{it}), outcomes (y_t or y_{it}), and what may be generically termed "losses" (z_t or z_{it}). A simple bivariate case with x_t = vehicle registrations and y_t = motor accidents would lead to a model

$$\log(x_t) = \theta_t^{(E)} + e_t^{(x)}$$

$$\log(y_t) = \theta_t^{(E)} + \theta_t^{(R)} + e_t^{(y)}$$

where the components of $\theta_t = (\theta_t^{(E)}, \theta_t^{(R)})$ represent underlying log exposure and log risk, which evolve according to a bivariate local linear trend

$$\theta_{t+1} = \theta_t + \delta_t + u_t, \quad u_t \sim N(0, \Sigma_u),$$

$$\delta_{t+1} = \delta_t + w_t, \quad w_t \sim N(0, \Sigma_w).$$

A simplifying "homogenous" model (Harvey, 1989, Chapter 8) for the covariance matrices is obtained for the LL model by setting

$$\Sigma_u = q\Sigma_e$$

where q is an unknown signal-to-noise ratio, and for the LLT model by setting

$$\Sigma_u = q_1\Sigma_e$$

$$\Sigma_w = q_2\Sigma_e.$$

Generalisations to include trend, seasonal, and cyclical effects can be made in which each sort of effect is independent of the other and each follows its own multivariate evolution prior (Durbin and Koopman, 2001, p.44). These assumptions lead to what is termed a seemingly unrelated time series equations or SUTSE model (Harvey and Shephard, 1993; Harvey and Koopman, 1997), since the individual series are connected only via the correlated disturbances in the measurement and transition equations. More complex matrix normal priors (West and Harrison, 1997, p.597) result from assuming interdependence between different types of parameter.

A model with level, seasonal, and cyclical effects for multivariate $y_t = (y_{1t}, \ldots, y_{Pt})$ would specify

$$y_t = \theta_t + \gamma_t + \psi_t + e_t, \quad t = 1, \ldots, T$$

$$e_t \sim N(0, \Sigma_e),$$

$$\theta_{t+1} = \theta_t + u_t,$$

$$u_t \sim N(0, \Sigma_u),$$

where the seasonal components for the pth variable (with s seasons) evolve according to

$$\gamma_{pt} = \gamma_{p,t-1} + \gamma_{p,t-2} \ldots + \gamma_{p,t-s+1} + \omega_{pt},$$

with

$$(\omega_{1t}, \omega_{2t}, \ldots, \omega_{Pt}) \sim N(0, \Sigma_\omega).$$

Following Harvey and Koopman (1997), the cyclical effects ψ_t may be assumed "similar," namely to have the same damping factor ρ and frequency $0 \le \lambda \le \pi$ across variables. The period is then $2\pi/\lambda$ with the full prior being

$$\psi_t = (\psi_{1t}, \psi_{2t}, \ldots, \psi_{Pt}) \sim N(m_\psi, \Sigma_\eta),$$

with additional shadow period effects

$$\psi_t^* = (\psi_{1t}^*, \psi_{2t}^*, \ldots, \psi_{Pt}^*) \sim N(m_{\psi^*}, \Sigma_{\eta^*}),$$

where means m_{ψ_p} and $m_{\psi_p^*}$ for the pth variable are obtained according to

$$\begin{bmatrix} \psi_{pt} \\ \psi_{pt}^* \end{bmatrix} = \rho \begin{bmatrix} \cos(\lambda) & \sin(\lambda) \\ -\sin(\lambda) & \cos(\lambda) \end{bmatrix} \begin{bmatrix} \psi_{p,t-1} \\ \psi_{p,t-1}^* \end{bmatrix} + \begin{bmatrix} \eta_{pt} \\ \eta_{pt}^* \end{bmatrix}.$$

It may be noted that multivariate DLMs occur in the analysis of univariate data, for example for categorical and ordinal outcomes. Thus Cargnoni et al. (1997) propose a model for time series of a multinomial outcome with M categories, and denominators n_t. One has

$$(y_{1t}, y_{2t}, \ldots y_{Mt}) \sim \text{Mult}(n_t, [\pi_{1t}, \pi_{2t}, \ldots, \pi_{Mt}])$$

$$\pi_{mt} = \exp(\eta_{mt}) \Big/ \sum_{h=1}^{M} \exp(\eta_{ht})$$

$$\eta_{mt} = \alpha_{mt} + \beta_m x_t, \quad m = 1, \ldots M-1$$

$$\eta_{Mt} = 0$$

where the time-varying category intercepts $\alpha_t = (\alpha_{1t}, \ldots, \alpha_{Mt})$ follow a multivariate normal random walk prior

$$\alpha_t \sim N_{M-1}(\alpha_{t-1}, \Sigma_\alpha).$$

Example 9.11 Minks and Muskrats: Multivariate Dynamic Linear Model

Harvey and Koopman (1997) consider a bivariate series, namely numbers of skins of minks and muskrats traded annually (logarithms of the annual sales) for $T=64$ years (1848–1911) by the Hudson Bay Company. There is a prey-predator relationship between the $P=2$ species (minks are the main predators of muskrats) leading to inter-linked cycles. A model is fitted including trends and similar cycles, so that

$$y_t = \theta_t + \psi_t + e_t, \quad t = 1, \ldots, T$$

$$e_t \sim N(0, \Sigma_e),$$

$$\theta_{t+1} = \theta_t + \beta_t + u_t,$$

$$u_t \sim N(0, \Sigma_u),$$

$$\beta_{t+1} = \beta_t + w_t,$$

$$w_t \sim N(0, \Sigma_w),$$

$$\psi_t = (\psi_{1t}, \psi_{2t}) \sim N(m_\psi, \Sigma_\eta)$$

$$\psi_t^* = (\psi_{1t}^*, \psi_{2t}^*) \sim N(m_{\psi^*}, \Sigma_{\eta^*})$$

where the cyclical effects for the two species have the same damping factor $\rho \sim U(0,1)$ and frequency λ, and the non-diagonal covariance matrices are of order $P \times P$.

Since the series contains 64 points, an informative assumption is made that the period is between 4.2 and 21, namely that $\lambda \sim U(0.3,1.5)$. Taking a simple uniform prior on λ between 0 and π is associated with implausibly low λ. Covariances are linked using the homogeneity assumption, namely $\Sigma_u = q_u \Sigma_e$, $\Sigma_w = q_w \Sigma_e$, $\Sigma_\eta = q_\eta \Sigma_e$, and $\Sigma_{\eta^*} = q_{\eta^*} \Sigma_e$, with the signal to noise ratios $\{q_u, q_w, q_\eta, q_{\eta^*}\}$ all assumed to follow Exponential(1) priors. For Σ_e^{-1}, a Wishart prior assumes 5 degrees of freedom and a prior covariance matrix based on the observed covariance.

Inferences are from the final 75,000 of a two-chain run of 100,000 iterations, using R2OpenBUGS. One finds the cycles to have a mean period of 9.9 years, with 95% CRI (9.3,10.9). Figures 9.4 and 9.5 show modelled trends in the mink and muskrat series (theta.var[1,] and theta.var[2,] in the code) together with the original data. The posterior means for q_u, q_w, q_η, and q_{η^*} are (0.093, 0.004,0.058, 0.16). The LOO-IC is 11.1, with point-wise LOO-IC identifying the discordant observation in 1908, when muskrat sales were unduly low.

The interlinking of the two series (and its predator-prey nature) also shows in a VAR(1) model with

$$y_{1t} = \gamma_1 + \alpha_{11} y_{1,t-1} + \alpha_{12} y_{2,t-1} + u_{1t}, \quad t = 2,\ldots,T$$

$$y_{2t} = \gamma_2 + \alpha_{21} y_{1,t-1} + \alpha_{22} y_{2,t-1} + u_{2t},$$

$$u_t \sim N(0, \Sigma_u),$$

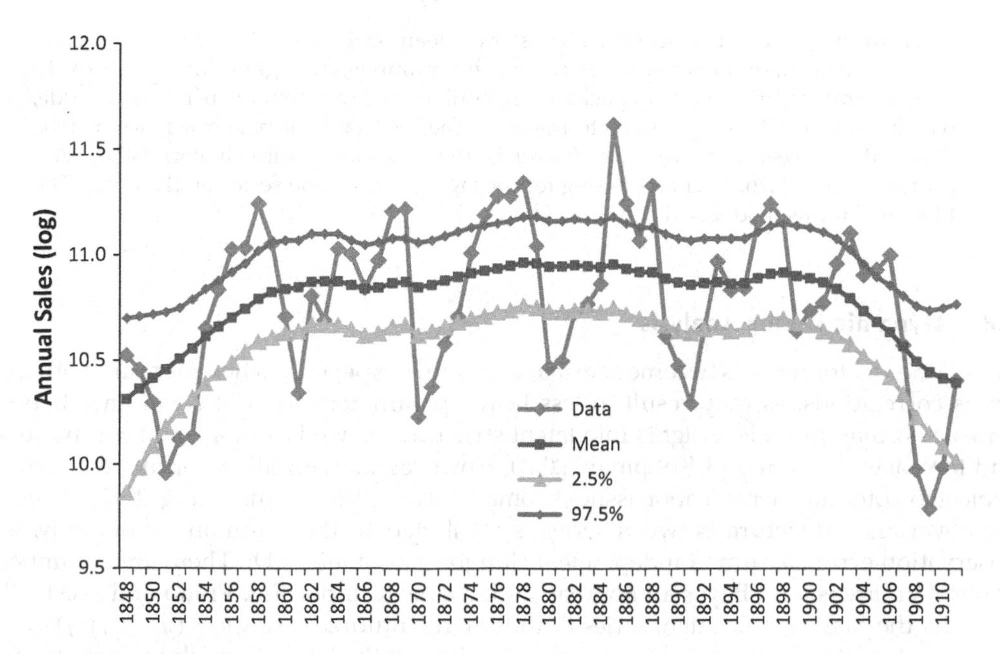

FIGURE 9.4
Annual mink sales, 1848–1911 (logarithm).

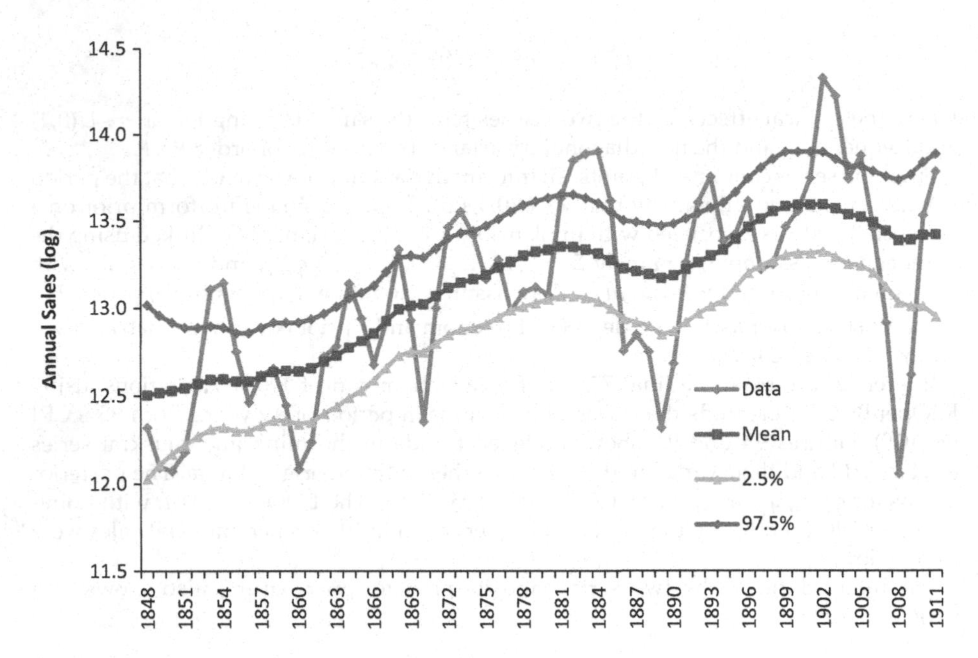

FIGURE 9.5
Annual muskrat sales, 1848–1911 (logarithm).

with y_{11} and y_{21} taken as known, and where Σ_u is non-diagonal. The estimated α coefficient matrix from a two-chain run of 25,000 iterations is

$$\begin{pmatrix} 0.61 & 0.21 \\ -0.49 & 0.91 \end{pmatrix}$$

where the negative α_{21} coefficient, with posterior mean (95% interval) of −0.49 (−0.69,−0.12), shows muskrat numbers are lower when mink number are higher. Maximum likelihood estimates from the vars package are similar, as are estimates using rstan code, which uses the Cholesky parameterisation of the bivariate normal covariance matrix. Residuals between the two series (u.corr in the code) are positively correlated, with posterior mean 0.26, after accounting for the lag 1 effect of one series on the other. The LOO-IC for this model is 41.

9.8.2 Dynamic Factor Analysis

Time series factor models become sensible for large P, especially when there are high inter-series correlations, as they result in less heavy parameterisation of covariance between series, and may provide insights into latent structure, as well as more efficient inferences and forecasts (Durbin and Koopman, 2001). However, as with all factor models, they are subject to potential identification issues (Aßmann et al., 2016; Bai and Wang, 2015). Typically, the covariance structure between series is attributed to the common factors only, with observation errors assumed independent (Jungbacker et al., 2009). There are a number of application areas, and Bayesian approaches have been important. Prado and West (1997) consider the case a single latent series F_t underlying multiple series $y_t = (y_{1t}, \ldots, y_{Pt})$ of EEG readings, and discuss TVAR autoregressive models for the latent F_t involving time-varying

AR1 coefficients which follow random walk priors. Thus, first order random walk priors in $r=1,\ldots,R$ autoregressive parameters ϕ_{rt} leads to

$$y_{pt} = a_p + \lambda_p F_t + e_{pt},$$

$$e_{pt} \sim N(0, \sigma_p^2),$$

$$F_t \sim N\left(\sum_{r=1}^{R} \phi_{rt} F_{t-r}, \sigma_F^2\right),$$

$$\phi_{rt} \sim N(\phi_{r,t-1}, \sigma_\phi^2),$$

with a preset λ_p (anchoring constraint) if σ_F^2 is an unknown. Autoregressive dependence in the residuals e_{pt} may also be considered (Jackson et al., 2016; Kaufmann and Schumacher, 2013), or lagged effects of the latent factor(s) in the model for y_{pt} (Aßmann et al., 2016). Another application occurs in econometric modelling of asset returns, where the number of assets may exceed the length of the time series and factor models for returns are a clear option (Zivot and Wang, 2006). Factor models may also be a component of multivariate volatility models – see Section 9.8.3.

A relatively simple approach for reducing a P dimensional vector y_t to a Q dimensional vector F_t involves a dynamic linear model for the factor score vector. Thus, a local linear factor or factor trend model would propose a measurement model linking indicators and factors

$$y_t = a + \Lambda F_t + e_t, \quad e_t \sim N(0, \Sigma_e), \quad t = 1,\ldots,T$$

with the transition equation specifying a random walk in the factors, namely

$$F_{t+1} = F_t + u_t, \quad u_t \sim N(0, \Sigma_u),$$

where F_t is a Q dimensional latent construct, with $Q<P$, and Λ is of dimension $P \times Q$. If the series e_t and F_t are uncorrelated, then the marginal mean and covariance of y_t are α and $\Lambda\Sigma_u\Lambda + \Sigma_e$ respectively. To avoid location invariance in the F scores, devices such as centring at each iteration or setting initial factor scores to known values can be used (see Example 9.12).

The loadings matrix Λ and/or the factor score covariance matrix Σ_u are parameterised to ensure identification and avoid various forms of invariance. If all elements in Σ_e are taken as unknown (i.e off-diagonal as well as diagonal terms) and Σ_u is also unknown, then Harvey and Koopman (1997) mention the loadings matrix formulation

$$\Lambda = \begin{pmatrix} I_Q \\ \Lambda^* \end{pmatrix}$$

with Λ^* of dimension $(P-Q) \times Q$ containing unknown loadings. If Σ_e is diagonal (only residual variances assumed unknown), and Σ_u is also diagonal, but contains unknown factor variances, then one may set $\lambda_{pp}=1$ and $\lambda_{pq}=0$ for $q>p$. This is the anchoring constraint of Skrondal and Rabe-Hesketh (2004), with the latter constraint used to avoid rotation invariance (Geweke and Zhou, 1996, pp.565–566).

If Σ_e contains just residual variances, and Σ_u is diagonal with known factor variances (typically of 1), then constraints on the λ_{pq} to ensure scale identification are not needed, but the rotational constraint $\lambda_{pq}=0$ for $q>p$ still applies. However, Geweke and Zhou (1996) suggest $\lambda_{pp}>0$ as an identification device in this case, to ensure a unique labelling of factors.

9.8.3 Multivariate Stochastic Volatility

Many multivariate series (e.g. share prices, exchange rates, asset returns) may be subject to volatility clustering, with the clustering often correlated over different series (Kastner et al., 2017; Chib et al., 2006). For example, Yu and Meyer (2006) mention that financial decision-making needs to take correlations into account when market volatilities move together across multiple assets. With initial transformation or differencing of series, the main focus of stochastic volatility modelling may be on modelling the changing covariances. In other applications, it may be necessary to model time variation both in autoregressive (or VAR) coefficients and the covariance matrix of error terms (Primiceri, 2005; Krueger, 2016). Alternatively, one may investigate stochastic volatility in tandem with a multivariate dynamic factor model.

With regard to the latter option, a general model may be stated for untransformed data $y_t=(y_{1t},y_{2t},\ldots,y_{Pt})'$ with factor scores vector $F_t=(F_{1t},F_{2t},\ldots,F_{Qt})'$. Thus following Zhou et al. (2014)

$$y_t=\alpha_t+\Lambda_t F_t+e_t,$$

$$e_t\sim N(0,\Sigma_t),$$

$$F_t=\Gamma F_{t-1}+w_t,$$

$$w_t\sim N(0,\Phi_t),$$

where e_t and w_t are respectively residuals and factor innovations, α_t is an intercept, Λ_t of dimension $P\times Q$ denotes time-varying (autoregressive) loadings, $\Sigma_t=\text{diag}(\sigma^2_{1t},\ldots,\sigma^2_{Pt})$ contains time-varying residual variances, $\Gamma=\text{diag}(\gamma_1,\ldots,\gamma_Q)$ governs AR(1) dependence in the factor scores F_t, and Φ_t is a diagonal factor volatility matrix.

To exemplify autoregressive dependence in loadings, consider a univariate factor model ($Q=1$). Then with AR(1) dependence in the loadings and Σ_t diagonal one has

$$y_{pt}=\alpha_{pt}+\lambda_{pt}F_t+e_{pt},$$

$$e_{pt}\sim N(0,\sigma^2_p),$$

$$F_t=\gamma F_{t-1}+w_t,$$

$$w_t\sim N(0,\sigma^2_w),$$

$$\lambda_{pt}=\mu_p+\rho_p(\lambda_{p,t-1}-\mu_p)+\eta_{pt},$$

$$\eta_{pt}\sim N(0,\xi_p).$$

For multivariate factors, sparsity-inducing priors on the coefficients λ_{pqt} may be indicated, with Zhou et al. (2014) proposing a threshold mechanism.

Simplifications to such a scheme are often the focus, involving decompositions of the residual variance. Applications typically involve metric series $y_t = (y_{1t}, y_{2t}, \ldots, y_{Pt})'$ either mean centred, or in transformed form (e.g. logs of share prices compared between successive time points), with effectively zero means. A latent factor will not necessarily be involved. Thus, for centred or appropriately transformed prices or returns y_{pt}, one possible model (Asai et al., 2006) for a response $y_t = (y_{1t}, y_{2t}, \ldots, y_{Pt})'$ is

$$y_t = H_t e_t,$$

$$H_t = \text{diag}(\exp(h_{1t}/2),)\exp(h_{2t}/2), \ldots, \exp(h_{Pt}/2)),$$

where e_t is a vector of independent standard normal variates, and $h_t = (h_{1t}, \ldots, h_{Pt})'$ is a vector of unobserved log variances (or volatilities), evolving according to stationary autoregressive schemes,

$$h_{pt} = \mu_p + \phi_p(h_{p,t-1} - \mu_p) + u_{pt}, \quad t > 2,$$

$$u_{pt} \sim N(0, \tau_p),$$

$$h_{p1} \sim N(\mu_p, \tau_p/(1 - \phi_p^2))$$

with persistence parameters $\phi_p \in (-1,1)$ (Kastner, 2016). The autoregression in h_t can be extended to VAR or VARMA form with heavier parameterisation (Asai et al., 2006).

The errors in the price series and in the volatilities may be stated in multivariate normal or multivariate t form, with the MVN assumption expressed

$$\begin{pmatrix} e_t \\ u_t \end{pmatrix} \sim N\left[\begin{pmatrix} 0 \\ 0 \end{pmatrix}, \begin{pmatrix} R_e & 0 \\ 0 & \Sigma_u \end{pmatrix}\right].$$

where R_e is a positive definite correlation matrix with a diagonal of ones, and Σ_u is a $P \times P$ covariance matrix for volatility shocks. Taking R_e to be non-diagonal means shocks in prices may be correlated, while taking Σ_u to be non-diagonal allows volatility shocks to be correlated (Yu and Meyer, 2006, 365–366). Thus, in a bivariate example, taking

$$\phi = \begin{pmatrix} \phi_{11} & \phi_{12} \\ \phi_{21} & \phi_{22} \end{pmatrix}$$

to be non-diagonal in a VAR(2) regression for (h_{1t}, h_{2t}) amounts to allowing bilateral Granger causality in volatility between the two series. Taking R_e to evolve through time according to

$$R_{et} = \begin{pmatrix} 1 & \rho_t \\ \rho_t & 1 \end{pmatrix}$$

means that not only log volatilities h_t, but also correlations between the observed series are time-varying. Specifically, with

$$\rho_t = \frac{\exp(g_t)-1}{\exp(g_t)+1}$$

an additional autoregression can be set, with

$$g_{t+1} = \mu_g + \phi_g(g_t - \mu_g) + v_t.$$

Factor analytic models may also include correlated volatility (Pitt and Shephard, 1999; Chan et al., 2006; Zhou et al., 2014). As an example, for two series $\{y_{pt}, p = 1,2\}$ and a univariate factor F_t, one might have

$$y_{1t} = \lambda_1 F_t + e_{1t}$$

$$y_{2t} = \lambda_2 F_t + e_{2t}$$

with evolving variances for F_t and the e_{pt}. The stochastic variance prior for the residuals e_{pt} may include autoregressive dependence, since a factor structure may be sufficient to account for the non-diagonal elements of the residual variance matrix of the outcomes, but not sufficient to explain all the marginal persistence in volatility (Pitt and Shephard, 1999, p.551). Thus, one might have

$$F_t \sim N(0, e^{h_{1t}}),$$

$$e_{1t} \sim N(0, e^{h_{2t}}),$$

$$e_{2t} \sim N(0, e^{h_{3t}}),$$

with first-order autoregressive dependence in the log variances h_{pt}

$$h_{pt} = \phi_p h_{p,t-1} + u_{pt} \quad t = 2, \ldots, T$$

with possibly unknown initial conditions h_{p1}. For identification, one may set one or other of the λ_p parameters to 1 (an anchoring constraint). Alternatively, a standardised factor constraint might be implemented by setting the scale of the factors at one time point, for instance by taking $F_1 \sim N(0,1)$, that is $h_{11}=0$.

Adaptivity in the modelling of stochastic variances can be combined with factor reduction. Chib et al. (2006) propose a multivariate stochastic volatility factor model that permits both series-specific jumps at each time, and Student-t innovations with unknown degrees of freedom. For bivariate data and a univariate factor, this model has the form

$$y_{1t} = \lambda_1 F_t + \delta_{1t} q_{1t} + \varepsilon_{1t}$$

$$y_{2t} = \lambda_2 F_t + \delta_{2t} q_{2t} + \varepsilon_{2t}$$

where $q_{pt}=1$ with probability π_p, and the ε_{pt} follow independent Student t densities with unknown degrees of freedom ν_p. In hierarchical form

$$\varepsilon_{pt} = e_{pt}/\gamma_{pt}^{0.5},$$

$$\gamma_{pt} \sim Ga\left(\frac{v_p}{2}, \frac{v_p}{2}\right),$$

$$[e_{1t}, e_{2t}] \sim N_2(0, V_t).$$

where V_t is diagonal with elements $\exp(h_{pt})$, with evolution scheme

$$h_{p,t+1} - \mu_p = \phi_p(h_{pt} - \mu_p) + \sigma_p u_{pt},$$

$$u_{pt} \sim N(0,1).$$

The variables $\zeta_{pt} = \log(1 + \delta_{pt})$ are assumed to be $N(-0.5\zeta_p^2, \zeta_p^2)$ where ξ_p are additional unknowns. The more general form for $y_t = (y_{1t}, \ldots, y_{Pt})'$ and $F_t = (F_{1t}, \ldots F_{Qt})'$, $Q \leq P$ is

$$y_t = BF_t + \Delta_t q_t + \varepsilon_t$$

with identification constraints $\lambda_{pp}=1$ and $\lambda_{pq}=0$ for $q > p$. These constraints set a scale and prevent rotation invariance. The covariance matrix for F_t is diagonal with evolution scheme as for the log diagonal elements of V_t.

Example 9.12 Common Factor Model for Flour Prices

Tiao and Tsay (1989) analyse a trivariate series formed by the logarithms of indices of monthly flour prices in Buffalo, Minneapolis and Kansas City between August 1972 and November 1980 ($T=100$). The data are plotted in Figure 9.6 which shows that the series are closely related, and a common factor model is indicated. The variance of the factor scores F_t is taken as unknown, so a loading constraint is needed. The factor scores

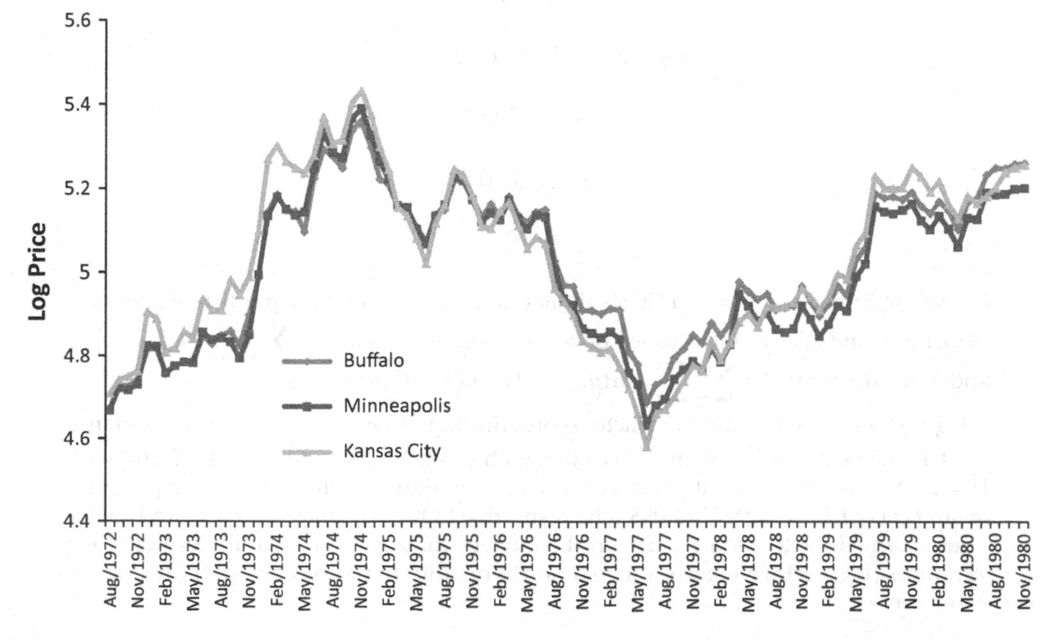

FIGURE 9.6
Flour prices in three cities.

are assumed to follow a random walk with $F_1 = 0$ to identify the level of the scores. The observation residuals after accounting for the common factor are assumed multivariate normal, with a Wishart prior on precision matrix with 3 degrees of freedom and a diagonal scale matrix. The elements of the Wishart scale matrix are based on the observed variances V_p of the three series, leading to a data-based prior.

Thus, with $P = 3$, and an anchoring constraint on the loadings,

$$y_{pt} = \alpha_p + \lambda_p F_t + u_{pt}$$

$$(u_{1t},\ldots,u_{Pt}) \sim N_P(0,\Sigma_u),$$

$$\Sigma_u^{-1} \sim W(PS,P),$$

$$S = \text{diag}(V_1,V_2,..,V_P),$$

$$\lambda_1 = 1;$$

$$\lambda_k \sim N(1,1),\ k = 2,\ldots,P,$$

$$F_t \sim N(F_{t-1},\sigma_F^2)\ t = 2,\ldots,T,$$

$$F_1 = 0,$$

$$\sigma_F \sim U(0,10).$$

An alternative model for the factor scores adopts a locally adaptive prior, allowing for changing variance through time (Lang et al., 2002). Thus

$$F_t \sim N(F_{t-1},\exp(\eta_t))\quad t = 2,\ldots,T,$$

$$\eta_t \sim N(\eta_{t-1},\tau_\eta^{-1})\quad t = 2,\ldots,T,$$

$$\tau_\eta \sim Ga(1,0.001),$$

$$\eta_1 \sim N(0,1),$$

$$F_1 = 0.$$

Following Migon and Moreira (2004), fit may be assessed using the predictive approach of Gelfand and Ghosh (1998), based on a goodness of fit term $G = \sum_p \sum_t (y_{\text{rep},pt} - y_{pt})^2$ and a penalty term $H = \sum_p \sum_t \text{var}(y_{\text{rep},pt})$. The LOO-IC is also used.

Figure 9.7 shows the estimated factor scores through time under the constant variance model. The posterior mean for σ_F^2 is 0.0018, with posterior mean for G of 1.417 and with H = 1.108. The non-constant variance model has similar fit criteria, namely a posterior mean for G of 1.397 with H=1.098. The respective LOO-IC values are −1027 and −1028. There seems little to choose between the models, though the plot of the evolving log variances (Figure 9.8) suggests a reduction in volatility in the second half of the observation period.

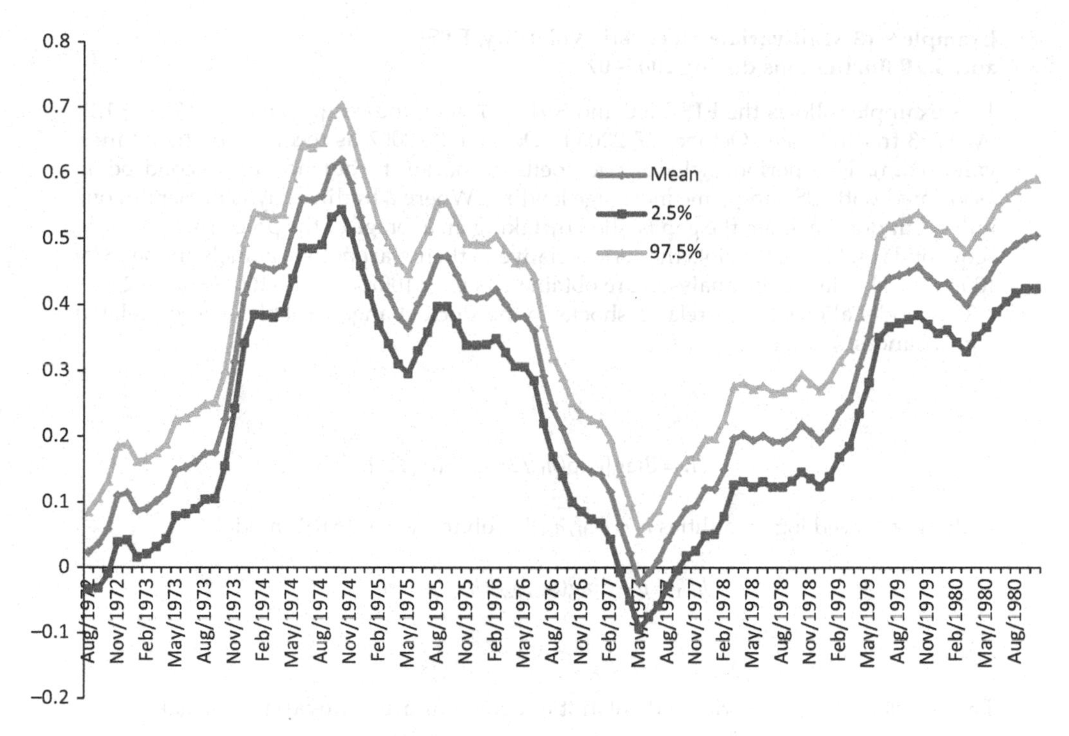

FIGURE 9.7
Factor scores, Model 1, Example 9.12.

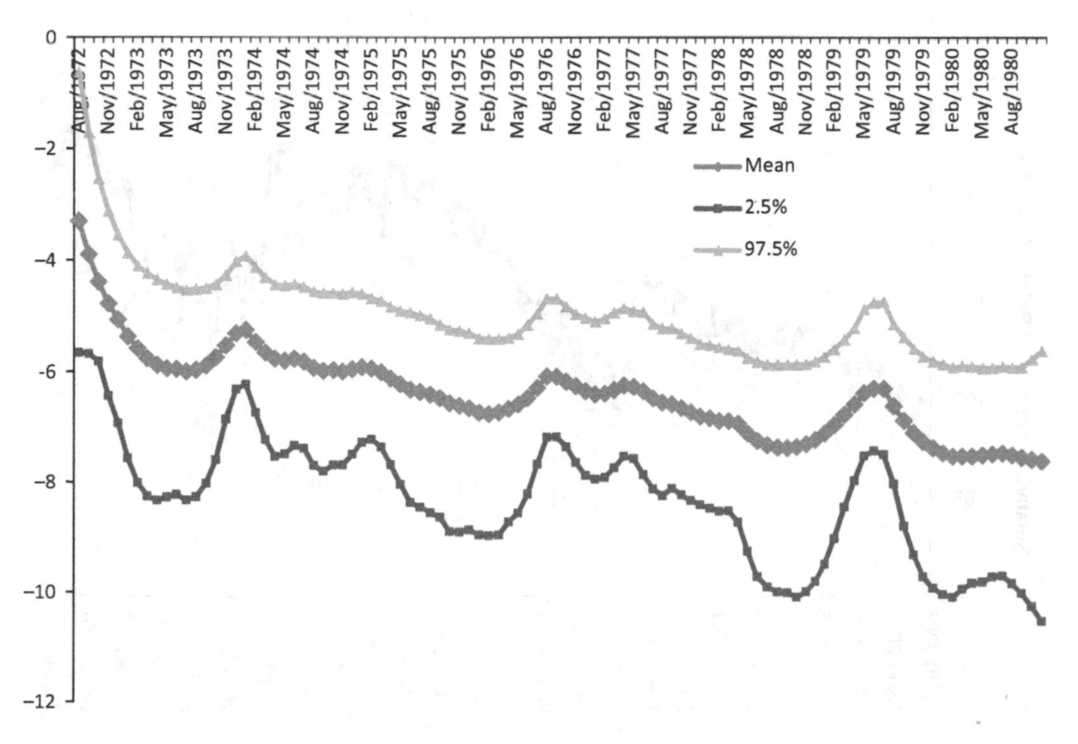

FIGURE 9.8
Log variances of factor scores, Example 9.12.

Example 9.13 Multivariate Stochastic Volatility, FTSE, and S&P fluctuations during 2006–07

This example follows the FTSE 100 and S&P 500 stock indices $\{r_{pt}, t = 0,\ldots,252; p = 1,2\}$ over 253 trading days (October 27, 2006 to October 19, 2007, as recorded by uk.finance.yahoo.com). This period includes two spells of market turbulence, the second being associated with US sub-prime mortgage lending. Where a trading day is present in one index, but not the other, the gap is filled by taking an average of the preceding and subsequent days. Figure 9.9 plots the series relative to their start points, namely in the form $100r_{pt}/r_{p1}$. The data to be analysed are obtained as $y_{pt} = 100r_{p,t}/r_{p,t-1} - 100$, $t=1, \ldots, 252$.

The model allows for correlated shocks in the stock change variables and correlated log variances, so that

$$y_t = H_t e_t,$$

$$H_t = \text{diag}[\exp(h_{1t}/2),)\exp(h_{2t}/2)],$$

with unobserved log volatilities $h_t = (h_{1t}, h_{2t})'$ evolving via a VAR(1) model

$$h_{t+1} = \mu + \text{diag}(\phi_{11}, \phi_{22})(h_t - \mu) + u_t,$$

$$h_1 = \mu + u_1.$$

The errors in the price series and volatilities equations are multivariate normal

$$\begin{pmatrix} e_t \\ u_t \end{pmatrix} \sim N\left[\begin{pmatrix} 0 \\ 0 \end{pmatrix}, \begin{pmatrix} R_e & 0 \\ 0 & \Sigma_u \end{pmatrix}\right].$$

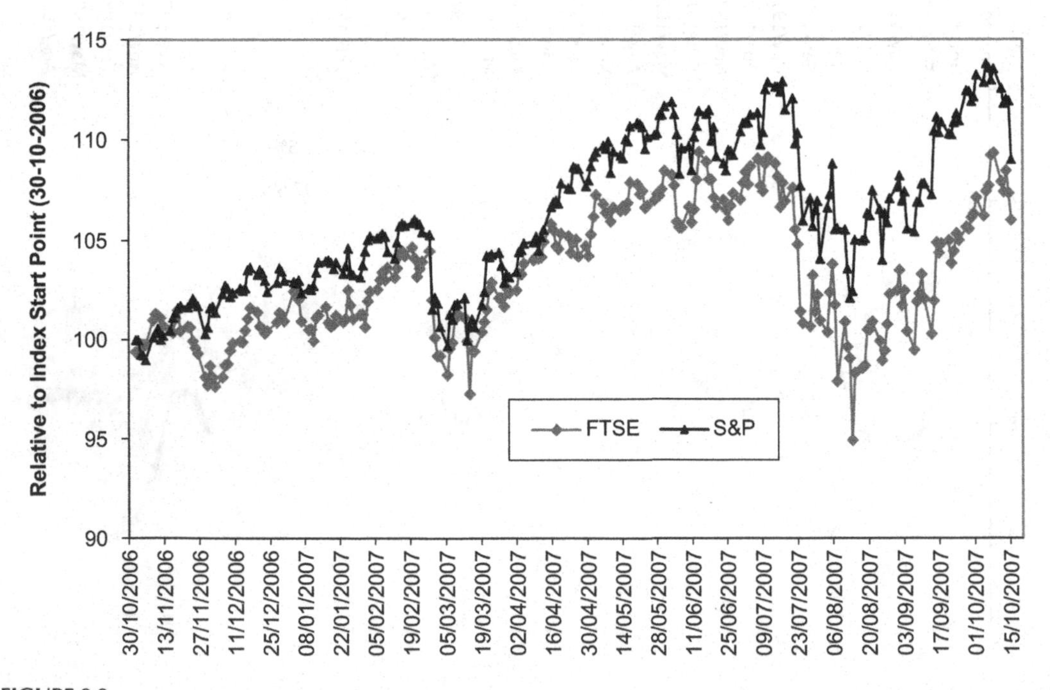

FIGURE 9.9
US and GB share indices, October 30, 2006 to October 19, 2007.

where R_e is a non-diagonal correlation matrix, and Σ_u is also non-diagonal, allowing volatility shocks to be correlated (Yu and Meyer, 2006).

The jags code samples h^*_{1t} and h^*_{2t} as standard normal, and applies the standard deviations $(\sigma_{u1}, \sigma_{u2})$, and correlation ρ_u, of the u_{1t} and u_{2t} series to the standard normal log volatilities. Thus

$$h_{1t} = \sigma_{u1} h^*_{1t},$$

$$h_{2t} = \sigma_{u2}\rho_u h^*_{1t} + \sigma_{u2}(1-\rho_u^2)^{0.5} h^*_{2t}.$$

This raises an identification issue for the correlation ρ_u, since $\rho_u h^*_{1t} = (-\rho_u)(-h^*_{1t})$, which is resolved by assuming ρ_u to be U(0,1) rather than U(−1,1). The correlation ρ_e in R_e is taken as $U(-1,1)$. The stationary autocorrelation parameters are obtained as $\phi_{pp} = 2\phi^*_{pp} - 1$, where $\phi^*_{pp} \sim$ Beta(19,1). The diagonal terms in Σ_u^{-1} are taken to be Exponential(1).

Estimation using jagsUI provides early convergence, with posterior means for ρ_e and ρ_u of 0.54 and 0.87, and with the autoregressive coefficients in the AR1 log volatility equations having means 0.88 and 0.74. Figure 9.10 plots the resulting log volatility series (posterior means of h_{1t} and h_{2t}) with the two periods of market turbulence apparent. The pointwise LOO-ICs detect aberrant observations for $t=85$ (comparing 27/02/2007 and 26/02/2007) when there was a sharp fall in the S&P index, and $t=206$ (comparing 16/08/2007 and 15/08/2007) when there was a sharp fall in the FTSE100.

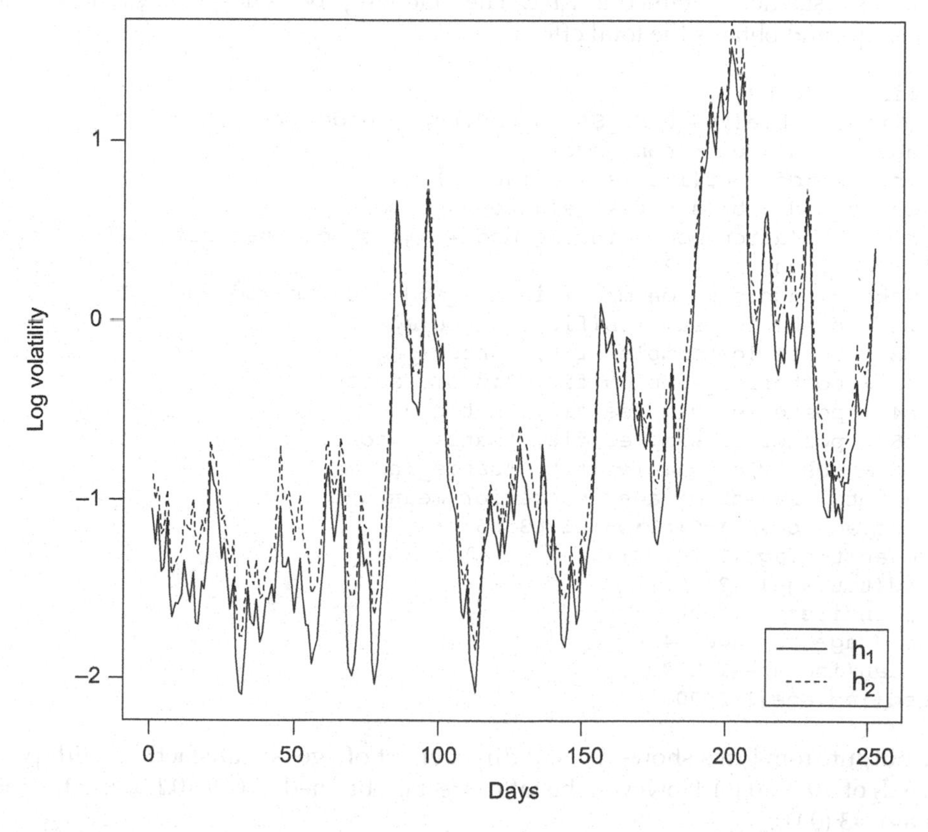

FIGURE 9.10
Log volatility plot.

9.9 Computational Notes

1. The application of brms to path models can be illustrated using data from an analysis for job satisfaction (Bryman and Cramer, 2005) and considered in Congdon (2001, p.98). Thus, job survey data on age, income, job satisfaction, and job autonomy is available for 68 workers. A path model is proposed with y_1 (age) influencing the three remaining variables, namely y_2=autonomy, y_3=income, and y_4=satisfaction. Autonomy is postulated to influence both income and satisfaction, while all three variables – age, income, and autonomy – affect satisfaction. All variables are standardised. Hence, the following regressions are involved

$$y_2 = b_1 y_1 + e_2,$$

$$y_3 = b_2 y_1 + b_3 y_2 + e_3,$$

$$y_4 = b_4 y_1 + b_5 y_2 + b_6 y_3 + e_4.$$

We may be interested in calculating the total effect of age on satisfaction, which involves the direct effect b_4, a path from age to income to satisfaction, calculated as $b_2 b_6$, a path from age to autonomy to satisfaction, obtained as $b_1 b_5$, and a path from age to autonomy to income to satisfaction, obtained as $b_1 b_3 b_6$ The following code encapsulates the anticipated relationships and obtains the total effect:

```
library(brms)
D=read.table("DS_BRMS_SATIS_CH9.txt",header=T)
auton_mod = bf(auton ~age)
income_mod = bf(income ~age+auton)
satis_mod = bf(satis ~age+auton+income)
fit= brm(auton_mod + income_mod + satis_mod+ set_rescor(FALSE), data
   = D, chains = 2)
make_stancode(auton_mod + income_mod + satis_mod, data=D)
b1 = posterior_samples(fit, "auton_age")
b2 = posterior_samples(fit, "income_age")
b3 = posterior_samples(fit, "income_auton")
b4 = posterior_samples(fit, "satis_age")
b5 = posterior_samples(fit, "satis_auton")
b6 = posterior_samples(fit, "satis_income")
# total effect of age, posterior mean and sd
tot.age=b4+b2*b6+b1*b5+b1*b3*b6
mean(tot.age[1:2000,])
sd(tot.age[1:2000,])
# indirect effect
ind.age=tot.age-b4
mean(ind.age[1:2000,])
sd(ind.age[1:2000,])
```

The output from brms shows a small direct effect of age on satisfaction, with posterior mean (sd) of −0.06 (0.10). However, the total effect is obtained as 0.36 (0.12), and the indirect effect as 0.43 (0.11).

References

Aßmann C, Boysen-Hogrefe J, Pape M (2016) Bayesian analysis of static and dynamic factor models: An ex-post approach towards the rotation problem. *Journal of Econometrics*, 192(1), 190–206.

Aitkin M, Aitkin I (2005) Bayesian inference for factor scores, in *Contemporary Psychometrics*, eds A Maydeu-Olivares, J McArdle. Lawrence Erlbaum Associates.

Aktekin T, Polson N, Soyer R (2017) Sequential Bayesian analysis of multivariate count data. *Bayesian Analysis*, 13(2), 385–409.

Albert J (1992) Bayesian estimation of normal ogive response curves using Gibbs sampling. *Journal of Educational Statistics*, 17, 251–269.

Albert J (2015) Introduction to Bayesian item response modelling. *International Journal of Quantitative Research in Education*, 2(3–4), 178–193.

Albert J, Ghosh M (2000) Item response modeling, pp 173–193, in *Generalized Linear Models: A Bayesian Perspective*, eds D Dey, S Ghosh, B Mallick. Addison–Wesley, New York.

Andersson M, Karlsson S (2007) Bayesian forecast combination for VAR models. Working paper, Öebro University.

Anselin L, Hudak S (1992) Spatial econometrics in practice: A review of software options. *Regional Science & Urban Economics*, 22, 509–536.

Arhonditsis G, Paerl H, Valdes-Weaver L, Stow C, Steinberg J, Reckhow K (2006a) Application of Bayesian structural equation modeling for examining phytoplankton dynamics in the Neuse River Estuary. *Estuarine, Coastal & Shelf Science*, 72, 63–80.

Arhonditsis G, Stow C, Steinberg L, Kenney M, Lathrop R, McBride S, Reckhow K (2006b) Exploring ecological patterns with structural equation modeling and Bayesian analysis. *Ecological Modelling*, 192(3–4), 385–409.

Arima S (2015) Item selection via Bayesian IRT models. *Statistics in Medicine*, 34(3), 487–503.

Asai M, McAleer M, Yu J (2006) Multivariate stochastic volatility: A review. *Econometric Reviews*, 25, 145–175.

Azzalini A (1985) A class of distributions which includes the normal ones. *Scandinavian Journal of Statistics*, 12, 171–178.

Bai J, Wang P (2015) Identification and Bayesian estimation of dynamic factor models. *Journal of Business & Economic Statistics*, 33(2), 221–240.

Baker F (2001) *The Basics of Item Response Theory*, 2nd Edition. ERIC Clearinghouse on Assessment and Evaluation.

Banerjee S, Carlin BP, Gelfand AE (2004) *Hierarchical Modeling and Analysis for Spatial Data*. Chapman and Hall/CRC, Boca Raton, FL.

Barnard J, McCulloch R, Meng X-L (2000) Modeling covariance matrices in terms of standard deviations and correlations, with applications to shrinkage. *Statistica Sinica*, 10, 1281–1312.

Bartholomew D (1987) *Latent Variable Models and Factor Analysis*. Charles Griffin, London, UK.

Bartholomew D, Steele F, Moustaki I, Galbraith J (2002) *The Analysis and Interpretation of Multivariate Data for Social Scientists*. CRC Press.

Bazan J, Branco M, Bolfarine H (2006) A skew item response model. *Bayesian Analysis*, 1, 861–892.

Berkhof J, van Mechelen I, Gelman A (2003) A Bayesian approach to the selection and testing of mixture models. *Statistica Sinica*, 13, 423–442.

Besag J, York J, Mollie A (1991) Bayesian image restoration, with two applications in spatial statistics. *Annals of the Institute of Statistical Mathematics*, 43, 1–59.

Bhattacharya A, Dunson D (2011) Sparse Bayesian infinite factor models. *Biometrika*, 98(2), 291–306.

Bijleveld F, Commandeur J, Gould P, Koopman, S (2005) Model-based measurement of latent risk in time series with applications. Tinbergen Institute Discussion Paper No. 05-118/4. Available at SSRN: http://ssrn.com/abstract=873466

Boker S, Neale M, Maes H, Wilde M, Spiegel M, Brick T, Spies J, Estabrook R, Kenny S, Bates T, Mehta P (2011) OpenMx: An open source extended structural equation modeling framework. *Psychometrika*, 76(2), 306–317.

Bollen K (1989) *Structural Equations with Latent Variables*. Wiley, New York.

Bollen KA (2002) Latent variables in psychology and the social sciences. *Annual Review of Psychology*, 53(1), 605–634.

Brandt P, Freeman J (2005) Advances in Bayesian time series modeling and the study of politics: Theory testing, forecasting, and policy analysis. *Political Analysis*, 14(1), 1–36.

Breusch T (2005) Estimating the Underground Economy using MIMIC Models. Working Paper, National University of Australia, Canberra.

Brown P, Fearn T, Haque M (1999) Discrimination with many variables. *Journal of the American Statistical Association*, 94, 1320–1329.

Bryman A, Cramer D (2005) *Quantitative Data Analysis with SPSS 12 and 13: A Guide for Social Scientists*. Routledge.

Byrnes J (2017) Bayesian SEM with BRMS. http://rpubs.com/jebyrnes/343408

Cargnoni C, Müller P, West M (1997) Bayesian forecasting of multinomial time series through conditionally Gaussian dynamic models. *Journal of the American Statistical Association*, 92(438), 640–647.

Chan D, Kohn R, Kirby C (2006) Multivariate stochastic volatility models with correlated errors. *Econometric Reviews*, 25, 245–274.

Chapados N (2014) Effective Bayesian modeling of groups of related count time series. Proceedings of the 31st International Conference on Machine Learning, Beijing, China. JMLR: W&CP volume 32.

Chen M-H, Dey D (1998) Bayesian modeling of correlated binary responses via scale mixture of multivariate normal link functions. *Sankhya*, 60A, 322–343.

Chen M-H, Dey D (2000) Bayesian analysis for correlated ordinal data models, in *Generalized Linear Models: A Bayesian Perspective*, eds D Dey, S Ghosh, B Mallick. Marcel Dekker, New York.

Chib S, Greenberg E (1998) Analysis of multivariate probit models. *Biometrika*, 85, 347–361.

Chib S, Nardari F, Shephard N (2006) Analysis of high dimensional multivariate stochastic volatility models. *Journal of Econometrics*, 134, 341–371.

Chib S, Winkelmann R (2001) Markov chain Monte Carlo analysis of correlated count data. *Journal of Business & Economic Statistics*, 19, 428–435.

Choi S, Gibbons L, Crane P (2011) Lordif: An R package for detecting differential item functioning using iterative hybrid ordinal logistic regression/item response theory and Monte Carlo simulations. *Journal of Statistical Software*, 39(8), 1.

Commandeur J, Koopman S (2007) *An Introduction to State Space Time Series Analysis*. Oxford University Press.

Congdon P (2001) *Bayesian Statistical Modelling*. Wiley.

Congdon P, Almog M, Curtis S, Ellerman R (2007) A spatial structural equation modelling framework for health count responses. *Statistics in Medicine*, 26(29), 5267–5284.

Conti G, Fruhwirth-Schnatter S, Heckman J, Piatek R (2014) Bayesian exploratory factor analysis. *Journal of Econometrics*, 183(1), 31–57.

Curtis S (2010) BUGS code for item response theory. *Journal of Statistical Software*, 36(1), 1–34.

Dunson D, Herring A (2005) Bayesian latent variable models for mixed discrete outcomes. *Biostatistics*, 6, 11–25.

Durbin J, Koopman S (2001) *Time Series Analysis by State Space Methods*, 1st Edition. OUP.

Durbin J, Koopman S (2012) *Time Series Analysis by State Space Methods*. Oxford University Press.

Edwards Y, Allenby G (2003) Multivariate analysis of multiple response data. *Journal of Marketing Research*, 40, 321–334.

Evans J, Middleton N, Gunnell D (2004) Social fragmentation, severe mental illness and suicide. *Social Psychiatry and Psychiatric Epidemiology*, 39, 165–170.

Everitt BS (1984) *Introduction to Latent Variable Models*. Chapman and Hall, London, UK.

Feng X, Wu H, Song X (2017) Bayesian adaptive Lasso for ordinal regression with latent variables. *Sociological Methods & Research*, 46(4), 926–953.

Fleishman J, Lawrence W (2003) Demographic variation in SF–12 scores: True differences or differential item functioning? *Medical Care*, 41, 75–86.

Fokoue E (2004) Stochastic determination of the intrinsic structure in Bayesian factor analysis. SAMSI Technical Report #2004-17. http://www.samsi.info/reports/index.shtml
Fox J, Glas C (2005) Bayesian modification indices for IRT models. *Statistica Neerlandica*, 59, 95–106.
Gelfand A, Ghosh S (1998) Model choice: A minimum posterior predictive loss approach. *Biometrika*, 85, 1–11.
Gelman A, Meng X-L, Stern H (1996) Posterior predictive assessment of model fitness. *Statistica Sinica*, 6, 733–807.
George E, McCulloch R (1993) Variable selection via Gibbs sampling. *Journal of the American Statistical Association*, 88, 881–889.
Geweke J, Zhou G (1996) Measuring the pricing error of the arbitrage pricing theory. *Review of Financial Studies*, 9, 557–587.
Ghosh J, Dunson D (2008) Bayesian model selection in factor analytic models, pp 151–163, in *Random Effect and Latent Variable Model Selection*, ed D Dunson. Springer.
Ghosh J, Dunson D (2009) Default prior distributions and efficient posterior computation in Bayesian factor analysis. *Journal of Computational and Graphical Statistics*, 18(2), 306–320.
Gielen E, Riutort-Mayol G, Palencia-Jiménez J, Cantarino I (2017) An urban sprawl index based on multivariate and Bayesian factor analysis with application at the municipality level in Valencia. *Environment and Planning B*, 45(5), 888–914.
Greenacre M, Blasius J (eds) (2006) *Multiple Correspondence Analysis and Related Methods*. CRC Press.
Gunnell D, Peters T, Kammerling R, Brooks J (1995) Relation between parasuicide, suicide, psychiatric admissions and socio-economic deprivation. *British Medical Journal*, 311, 226–230.
Harvey A (1989) *Forecasting, Structural Time Series Models and the Kalman Filter*. Cambridge University Press, Cambridge, UK.
Harvey A, Koopman S (1997) Multivariate structural time series models, pp 269–298, in *Systematic Dynamics in Economic and Financial Models*, eds C Heij, H Schumacher, B Hanzon, C Praagman. Wiley, Chichester, UK.
Harvey A, Shephard N (1993) Structural time series models, in *Handbook of Statistics*, Vol. 11, eds G S Maddala et al. Elsevier Science Publishers, Barking, UK.
Hayashi K, Arav M (2006) Bayesian factor analysis when only a sample covariance matrix is available. *Educational and Psychological Measurement*, 66, 272–284.
Hogan J, Tchernis R (2004) Bayesian factor analysis for spatially correlated data, with application to summarizing area-level material deprivation from census data. *Journal of the American Statistical Association*, 99, 314–324.
Hoyle R (ed) (1995) *Structural Equation Modeling: Concepts, Issues, and Applications*. Sage.
Inouye D, Yang E, Allen G, Ravikumar P (2017) A review of multivariate distributions for count data derived from the Poisson distribution. *Wiley Interdisciplinary Reviews: Computational Statistics*, 9(3), e1398.
Ishwaran H, James L (2001) Gibbs sampling methods for stick-breaking priors. *Journal of the American Statistical Association*, 96, 161–173.
Ishwaran H, Zarepour M (2000) Markov chain Monte Carlo in approximate Dirichlet and beta two-parameter process hierarchical models. *Biometrika*, 87, 371–390.
Jackson L, Kose M, Otrok C, Owyang M (2016) Specification and estimation of Bayesian dynamic factor models: A Monte Carlo analysis with an application to global house price comovement, in *Advances in Econometrics*, Vol. 35, eds S Koopman, E Hillebrand. Emerald Publishing.
Jin X, Banerjee S, Carlin BP (2007) Order-free co-regionalized areal data models with application to multiple-disease mapping. *Journal of the Royal Statistical Society: Series B (Statistical Methodology)*, 69(5), 817–838.
Jin X, Carlin B, Banerjee S (2005) Generalized hierarchical multivariate CAR models for areal data. *Biometrics*, 61, 950–961.
Jöreskog KG (1973) Analysis of covariance structures, pp 263–285, in *Multivariate Analysis–III*. ed P Krishnaiah, Academic Press.
Joreskog K, Goldberger A (1975) Estimation of a model with multiple indicators and multiple causes of a single latent variable. *Journal of the American Statistical Association*, 70, 631–639.

Juárez M, Steel M (2010) Model-based clustering of non-Gaussian panel data based on skew-t distributions. *Journal of Business & Economic Statistics*, 28, 52–66.

Jungbacker B, Koopman S, van der Wel M (2009) Dynamic factor models with smooth loadings for analyzing the term structure of interest rates. Tinbergen Institute Discussion Paper, TI 2009-041/4.

Kaplan D, Depaoli S (2012) Bayesian structural equation modeling, pp 650–673, in *Handbook of Structural Equation Modeling*, eds R Hoyle. Guilford Publications Inc.

Karlis D, Meligkotsidou L (2005) Multivariate Poisson regression with covariance structure. *Statistics and Computing*, 15, 255–265.

Karlsson S (2015) Forecasting with Bayesian vector autoregression. *Handbook of Economic Forecasting*, 2B, 791–897.

Kastner G (2016) Dealing with stochastic volatility in time series using the R package stochvol. *Journal of Statistical Software*, 69(5), 1–30.

Kastner G, Frühwirth-Schnatter S, Lopes H (2017) Efficient Bayesian inference for multivariate factor stochastic volatility models. *Journal of Computational and Graphical Statistics*, 26(4), 905–917.

Kaufmann S, Schumacher C (2013) Bayesian estimation of sparse dynamic factor models with order-independent identification (No. 13.04). Working Paper, Study Center Gerzensee.

Kavanagh L, Lee D, Pryce G (2016) Is poverty decentralising? Quantifying uncertainty in the decentralisation of urban poverty. *Annals of the American Association of Geographers*, 106(6), 1286–1298.

Kendall M (1975) *Multivariate Analysis*. Charles Griffin & Co., London, UK.

Kleibergen F, Paap R (2002) Priors, posteriors and Bayes factors for a Bayesian analysis of cointegration. *Journal of Econometrics*, 111(2), 223–249.

Knorr-Held L, Rasser G (2000) Bayesian detection of clusters and discontinuities in disease maps. *Biometrics*, 56, 13–21.

Koop G, Korobilis D (2010) Bayesian multivariate time series methods for empirical macroeconomics. *Foundations and Trends in Econometrics*, 3(4), 267–358.

Koop G, Strachan R, van Dijk H, Villani M (2006) Bayesian approaches to cointegration, in *The Palgrave Handbook of Theoretical Econometrics*, eds K Patterson, T Mill. MacMillan.

Koopman S, Durbin J (2000) Fast filtering and smoothing for multivariate state space models. *The Journal of Time Series Analysis*, 21, 281–296.

Krueger F (2016) bvarsv: Bayesian Analysis of a Vector Autoregressive Model with Stochastic Volatility and Time-Varying Parameters. https://sites.google.com/site/fk83research/code

Lai M, Zhang J (2017) Evaluating fit indices for multivariate t-based structural equation modeling with data contamination. *Frontiers in Psychology*, 8, 1286.

Lang S, Fronk E, Fahrmeir L (2002) Function estimation with locally adaptive dynamic models. *Computational Statistics*, 17, 479–500.

Lee S-Y (2007) *Structural Equation Modelling: A Bayesian Approach*. Wiley.

Lee S-Y, Shi J (2000) Joint Bayesian analysis of factor score and structural parameters in the factor analysis models. *Annals of the Institute of Statistical Mathematics*, 52, 722–736.

Lee S-Y, Song X-Y (2003) Bayesian model selection for mixtures of structural equation models with an unknown number of components. *British Journal of Mathematical and Statistical Psychology*, 56, 145–165.

Lee S-Y, Song X-Y (2008) Bayesian model comparison of structural equation models, pp 121–149, in *Random Effect and Latent Variable Model Selection*, ed D Dunson. Springer.

Lee S-Y, Tang N (2006) Bayesian analysis of structural equation models with mixed exponential family and ordered categorical data. *British Journal of Mathematical and Statistical Psychology*, 59, 151–172.

Leroux B, Lei X, Breslow N (1999) Estimation of disease rates in small areas: A new mixed model for spatial dependence, pp 135–178, in *Statistical Models in Epidemiology, the Environment and Clinical Trials*, eds M Halloran, D Berry. Springer-Verlag, New York.

Levy R, Mislevy R (2016) *Bayesian Psychometric Modeling*. CRC Press.

Litterman R (1986) Forecasting with Bayesian vector autoregressions – Five years of experience. *Journal of Business & Economic Statistics*, 4, 25–38.

Liu X, Wall M, Hodges J (2005) Generalized spatial structural equation modeling. *Biostatistics*, 6, 539–557.
Lopes H, West M (2004) Bayesian model assessment in factor analysis. *Statistica Sinica*, 14, 41–67.
Lu Z-H, Chow S, Loken E (2016) Bayesian factor analysis as a variable-selection problem: Alternative priors and consequences. *Multivariate Behavioral Research*, 51(4), 519–539.
Luo Y, Jiao H (2017) Using the Stan program for Bayesian item response theory. *Educational and Psychological Measurement*, 77, 1–25.
MacNab Y (2007) Mapping disability-adjusted life years: A Bayesian hierarchical model framework for burden of disease and injury assessment. *Statistics in Medicine*, 26(26), 4746–4769.
MacNab Y (2018) Some recent work on multivariate Gaussian Markov random fields. *Test*, 27(3), 1–45.
Madsen L, Dalthorp D (2007) Simulating correlated count data. *Environmental and Ecological Statistics*, 14, 129–148.
Magis D, Tuerlinckx F, De Boeck P (2015) Detection of differential item functioning using the lasso approach. *Journal of Educational and Behavioral Statistics*, 40(2), 111–135.
Malaeb Z, Summers K, Pugesek B (2000) Using structural equation modeling to investigate relationships among ecological variables. *Environmental and Ecological Statistics*, 7, 93–111.
Mardia K (1988) Multi-dimensional multivariate Gaussian Markov random fields with application to image processing. *Journal of Multivariate Analysis*, 24, 265–284.
Mar-Dell'Olmo M, Martnez-Beneito M, Borrell C, Zurriaga O, Nolasco A, Domínguez-Berjón M (2011) Bayesian factor analysis to calculate a deprivation index and its uncertainty. *Epidemiology*, 22(3), 356–364.
Marshall EC, Spiegelhalter DJ (2007) Identifying outliers in Bayesian hierarchical models: A simulation-based approach. *Bayesian Analysis*, 2(2), 409–444.
Martinez-Beneito MA (2013) A general modelling framework for multivariate disease mapping. *Biometrika*, 100(3), 539–553.
Mavridis D, Ntzoufras I (2014) Stochastic search item selection for factor analytic models. *British Journal of Mathematical and Statistical Psychology*, 67(2), 284–303.
McCulloch R, Rossi P (1994) An exact likelihood analysis of the multinomial probit model. *Journal of Econometrics*, 64, 207–240.
McCulloch R, Polson N, Rossi P (2000) A Bayesian analysis of the multinomial probit model with fully identified parameters. *Journal of Econometrics*, 99, 173–193.
Merkle E, Rosseel Y (2017) blavaan: Bayesian structural equation models via parameter expansion. *Journal of Statistical Software*, 85(4), 1–30.
Merkle E, Wang T (2016) Bayesian latent variable models for the analysis of experimental psychology data. *Psychonomic Bulletin & Review*, 25(1), 256–270.
Mezzetti M, Billari F (2005) Bayesian correlated factor analysis of socio-demographic indicators. *Statistical Methods and Applications*, 14(2), 223–241.
Migon H, Moreira A (2004) Core inflation: Robust common trend model forecasting. *Brazilian Review of Econometrics*, 24, 1–19.
Moauro P, Savio G (2005) Temporal disaggregation using multivariate structural time series models. *Journal of Econometrics*, 8, 214–234.
Murray J (2016) R Package 'bfa', Bayesian Factor Analysis. https://cran.r-project.org/web/packages/bfa/bfa.pdf
Muthén B, Asparouhov T (2012) Bayesian structural equation modeling: A more flexible representation of substantive theory. *Psychological Methods*, 17(3), 313–335.
Natesan P, Nandakumar R, Minka T, Rubright J (2016) Bayesian prior choice in IRT estimation using MCMC and variational Bayes. *Frontiers in Psychology*, 7, 1422.
Nethery R, Warren J, Herring A, Moore K, Evenson K, Diez-Roux A (2015) A common spatial factor analysis model for measured neighborhood-level characteristics: The multi-ethnic study of atherosclerosis. *Health & Place*, 36, 35–46.
Nikolov M, Coull B, Catalano P (2007) An informative Bayesian structural equation model to assess source-specific health effects of air pollution. *Biostatistics*, 8, 609–624.

O'Brien S, Dunson D (2004) Bayesian multivariate logistic regression. *Biometrics*, 60, 739–746.
Palomo J, Dunson D, Bollen K (2007) Bayesian structural equation modeling, in *Handbook of Latent Variable and Related Models*, ed S-Y Lee. Elsevier.
Petris G, Petrone S, Campagnoli P (2009) *Dynamic Linear Models with R*. Springer.
Piatek R (2017) R Package 'BayesFM', Bayesian Inference for Factor Modeling. https://cran.r-project.org/web/packages/BayesFM/BayesFM.pdf
Pitt M, Shephard N (1999) Time varying covariances: A factor stochastic volatility approach, pp 547–570, in *Bayesian Statistics 6*, eds J Bernardo, J Berger, A Dawid, A Smith. Oxford University Press.
Poon W, Wang H (2012) Latent variable models with ordinal categorical covariates. *Statistics and Computing*, 22(5), 1135–1154.
Prado R, West M (1997) Exploratory modelling of multiple non-stationary time series: Latent process structure and decompositions, in *Modelling Longitudinal and Spatially Correlated Data*, ed T Gregoire. Springer-Verlag.
Press S, Shigemasu K (1989) Bayesian inference in factor analysis, pp 271–287, in *Contributions to Probability and Statistics*. eds L Gleser, M Perlman, S Press, A Sampson. Springer, New York.
Primiceri G E (2005) Time varying structural vector autoregressions and monetary policy. *The Review of Economic Studies*, 72(3), 821–852.
Proietti T (2007) Measuring core inflation by multivariate structural time series models, in *Optimisation, Econometric and Financial Analysis*, eds E J Kontoghiorghes, C Gatu. *Advances in Computational Management Science*, Vol. 9. Springer, Berlin/Heidelberg, Germany.
Reinsel G (2003) *Elements of Multivariate Time Series Analysis*. Springer Science & Business Media.
Rigby R (1997) Bayesian discrimination between two multivariate normal populations with equal covariance matrices. *Journal of the American Statistical Association*, 92, 1151–1154.
Rodrigues-Motta M, Pinheiro H, Martins E, Araujo M, dos Reis S (2013) Multivariate models for correlated count data. *Journal of Applied Statistics*, 40(7), 1586–1596.
Rossi P, Allenby G, McCulloch R (2005) *Bayesian Statistics and Marketing*. Wiley.
Rue H, Held L (2005) *Gaussian Markov Random Fields: Theory and Applications*. Chapman and Hall/CRC.
Rupp A, Dey D, Zumbo B (2004) To Bayes or not to Bayes, from whether to when: Applications of Bayesian methodology to modeling. *Structural Equation Modeling*, 11, 424–451.
Sahu S (2002) Bayesian estimation and model choice in item response models. *Journal of Statistical Computation and Simulation*, 72, 217–232.
Sain S, Cressie N (2007) A spatial model for multivariate lattice data. *Journal of Econometrics*, 140, 226–259.
Sanchez B, Butdz-Jorgensen E, Ryan L, Hu H (2005) Structural equation models: A review with applications to environmental epidemiology. *Journal of the American Statistical Association*, 100, 1443–1455.
Schumacker R, Lomax R (2016) *A Beginner's Guide to Structural Equation Modeling*, 4th Edition. Routledge.
Sethuraman J (1994) A constructive definition of Dirichlet priors. *Statistica Sinica*, 4, 639–665.
Sims C, Zha T (1998) Bayesian methods for dynamic multivariate models. *International Economic Review*, 39, 949–968.
Skrondal A, Rabe-Hesketh S (2004) Generalized latent variable modeling: Multilevel, longitudinal and structural equation models. Chapman & Hall/CRC, Boca Raton, FL.
Skrondal A, Rabe-Hesketh S (2007) Latent variable modelling: A survey. *Scandinavian Journal of Statistics*, 34, 712–745.
Smith D, Harvey P, Lawn S, Harris M, Battersby M (2017) Measuring chronic condition self-management in an Australian community: Factor structure of the revised Partners in Health (PIH) scale. *Quality of Life Research*, 26(1), 149–159.
Soares T, Gonçalves F, Gamerman D (2009) An integrated Bayesian model for DIF analysis. *Journal of Educational and Behavioral Statistics*, 34(3), 348–377.
Song J, Ghosh M, Miaou S, Mallick B (2005) Bayesian multivariate spatial models for roadway traffic crash mapping. *Journal of Multivariate Analysis*, 97, 246–273.

Song X-Y, Lee S-Y, Ng M, So W-Y, Chan J (2006) Bayesian analysis of structural equation models with multinomial variables and an application to type 2 diabetic nephropathy. *Statistics in Medicine*, 26, 2348–2369.

Stromeyer W, Miller J, Sriramachandramurthy R, DeMartino R (2015) The prowess and pitfalls of Bayesian structural equation modeling: Important considerations for management research. *Journal of Management*, 41(2), 491–520.

Tabachnick B, Fidell L (2006) *Using Multivariate Statistics*, 5th Edition. Allyn & Bacon, Inc., Needham Heights, MA.

Talhouk A, Doucet A, Murphy K (2012) Efficient Bayesian inference for multivariate probit models with sparse inverse correlation matrices. *Journal of Computational and Graphical Statistics*, 21(3), 739–757.

Tanner M (1996) *Tools for Statistical Inference: Methods for the Exploration of Postrior Distributions and Likelihood Functions*, 3rd Edition. Springer-Verlag, New York.

Tekwe C, Carter R, Cullings H, Carroll R (2014) Multiple indicators, multiple causes measurement error models. *Statistics in Medicine*, 33(25), 4469–4481.

Thissen D, Steinberg L, Wainer H (1993) Detection of differential item functioning using the parameters of item response models, pp 67–113, in *Differential Item Functioning: Theory and Practice*, eds P W Holland, H Wainer. Lawrence Erlbaum Associates, Hillsdale, NJ.

Tiao G, Tsay R (1989) Model specification in multivariate time series. *Journal of the Royal Statistical Society: Series B*, 51, 157–213.

Tsay R (2014) *Multivariate Time Series Analysis with R and Financial Applications*. Wiley, Hoboken, NJ

Tzala E, Best N (2007) Bayesian latent variable modelling of multivariate spatio-temporal variation in cancer mortality. *Statistical Methods in Medical Research*, 2007 Sep 13 (epub).

Wang D, Lin J-Y, Yu T (2006) A MIMIC approach to modeling the underground economy in Taiwan. *Physica A*, 371, 536–542.

Wang F, Wall M (2003) Generalized common spatial factor model. *Biostatistics*, 4(4), 569–582.

Wang Y, Neuman U, Wright S, Warton D (2012) mvabund: an R package for model-based analysis of multivariate abundance data. *Methods in Ecology and Evolution*, 3, 471–473.

Wedel M, Bockenholt U, Kamakura W (2003) Factor models for multivariate count data. *Journal of Multivariate Analysis*, 87, 356–369.

West M, Harrison J (1997) *Bayesian Forecasting and Dynamic Models*, 2nd Edition. Springer Verlag.

White A, Murphy T (2014) BayesLCA: An R Package for Bayesian Latent Class Analysis. *Journal of Statistical Software*, 61(13). https://www.jstatsoft.org/article/view/v061i13

Yu J, Meyer A (2006) Multivariate stochastic volatility models: Bayesian estimation and model comparison. *Econometric Reviews*, 25, 361–384.

Yuan K-H, Bentler P, Chan W (2004) Structural equation modeling with heavy tailed distributions. *Psychometrika*, 69, 421–436.

Zheng X, Rabe-Hesketh S (2007) Estimating parameters of dichotomous and ordinal item response models with gllamm. *Stata Journal*, 7(3), 313–333.

Zhou X, Nakajima J, West M (2014) Bayesian forecasting and portfolio decisions using dynamic dependent sparse factor models. *International Journal of Forecasting*, 30(4), 963–980.

Zivot E, Wang J (2006) *Modeling Financial Time Series with S-PLUS*. Springer, Berlin, Germany.

10

Hierarchical Models for Longitudinal Data

10.1 Introduction

Longitudinal data sets occur when continuous or discrete observations y_{it} on a set of subjects, or units $i = 1,\ldots,n$, are repeated over a number of measuring occasions $t = 1,\ldots,T_i$ possibly differing between subjects (with $N = \Sigma_i T_i$ total observations). There are many contexts for such data to occur, with variation in type of unit, study design, and data form. For instance, in economic and marketing applications (Keane, 2015; Rossi et al., 2005), the unit is typically an individual consumer, household, or firm, whereas in actuarial applications (Antonio and Beirlant, 2007), the units may consist of groups of policyholders (risk classes) with responses being insurance claim counts. Longitudinal studies often feature an intervention or treatment comparison, with intervention studies (Thiese, 2014) including both observational studies and controlled clinical trials with random treatment assignment. In balanced studies, repeat measurements on all subjects are contemporaneous, whereas measurement at different times for different units leads to unbalanced longitudinal data (Daniels and Hogan, 2008), with unit specific times $\{a_{it}, t = 1,\ldots,T_i\}$ at which events are recorded. Furthermore, measuring occasions may be over more than one time scale. An example is disease incidence by calendar time and age at onset or death, leading to a further implicit cohort scale defined by the difference between age and time (Schmid and Held, 2007; Lagazio et al., 2003); see Section 10.7.

There are also a variety of approaches to analysing longitudinal data, such as random effects (conditional) models on the one hand, and marginal or population-averaged approaches on the other (Heagerty and Zeger, 2000; Lee and Nelder, 2004), with conditional model and marginal model parameters not necessarily having the same interpretation (Verbeke et al., 2010). The focus of this chapter is on conditionally specified hierarchical and random effect models, and on MCMC estimation via conditional likelihood with random effects as part of the parameter set (e.g. Daniels and Hogan, 2008; Chib and Carlin, 1999); see Section 10.2.

Longitudinal data offer major advantages over cross-sectional designs in the analysis of causal interrelationships between variables, including developmental and growth processes and clinical studies, and before-after studies (Menard, 2002; Chen et al., 2016). The accumulation of information over both times and subjects increases the power of statistical methods to identify treatment effects or values-added (Lockwood et al., 2003), and permits the estimation of parameters (e.g. permanent random effects or "frailties" for subjects i) that are not identifiable from cross-sectional analysis or from repeated cross-sections on different subjects.

On the other hand, analysis of longitudinal data may be problematic if the longitudinal sequences are subject to missing observations; see Section 10.8. Missingness may involve intermittently missing values of responses or predictors, or total loss to observation after a certain point. The latter is variously known as attrition (Schafer and Graham, 2002) or dropout (Hogan et al., 2004). A particular question relevant to estimation of the main structural model is whether such permanent exit is random (independent of the response that would otherwise have been observed) or informatively related to the missing response.

Bayesian estimation via repeated sampling from posterior densities facilitates hierarchical modelling of longitudinal data, whether of permanent subject effects, correlated or iid observation level errors, time-varying regressor effects, or common factors in multivariate longitudinal data. As noted by Davidian and Giltinan (2003), random effects are treated as parameters in Bayesian MCMC estimation, and not integrated out, as may be done in frequentist approaches. The integrated or marginal likelihood estimation approach may become infeasible in complex varying coefficient models (Tutz and Kauermann, 2003), and different parameter estimates may be obtained according to the maximisation methods used (Molenberghs and Verbeke, 2004). Bayesian modelling perspectives are also important in the application of latent metric augmentation to categorical longitudinal outcomes (binary, multinomial) (Chib and Carlin, 1999), and for dealing with missing data, especially attrition of subjects (Little and Rubin, 2002; Ibrahim and Molenberghs, 2009).

10.2 General Linear Mixed Models for Longitudinal Data

General linear mixed models (GLMMs) provide a framework for modelling longitudinal data, and may be characterised by distributional and structural assumptions. Conditional on predictors and random effects, it is assumed that data y_{it} for subjects i and times t are distributed according to the exponential family,

$$p(y_{it} \mid \theta_{it}, \phi) = \exp\left(\frac{y_{it}\theta_{it} - a(\theta_{it})}{\phi} + C(y_{it}, \phi)\right)$$

where θ_{it} denotes the natural parameter (Tsai and Hsiao, 2008; Fong et al., 2010; Natarajan and Kass, 2000). The structural assumption governs the forms assumed for the conditional means $E(y_{it}) = \mu_{it} = a'(\theta_{it})$, with regression link $g(\mu_{it}) = \eta_{it}$, and for the variances $V_{it} = \phi \text{Var}(\mu_{it})$. This involves questions such as whether the conditional mean is linear or nonlinear in predictors and random effects, and at what levels random effects are present.

As in Chapter 9, random effects may be used at different levels: enduring differences between subjects may be represented by time-invariant random effects, while representing excess dispersion may require observation level random effects. Fixed effects are also used to represent subject level heterogeneity, especially in econometrics (Frees, 2004). This is equivalent to generating dummy variables for each subject, and works best for relatively few subjects and more time periods, as there is no pooling strength in fixed effects models and the parameter count increases with the number of subjects n. In this chapter, the focus is on random subject iid effects (exchangeable over units), though if the units are spatially configured (say), then a structured prior for the permanent effects can be used.

Conjugate priors may be suitable for handling random variation at subject or observation level for exponential family responses, especially if random variation does not involve predictors (Lee and Nelder, 2000). However, more flexible models, possibly involving subject specific regression effects, as well as varying intercepts, involve a vector of subject-level effects $b_i = (b_{1i}, \ldots, b_{Qi})'$ in a general linear mixed model format. These are typically taken to be normal with dispersion matrix D, with elements $\{\sigma_{bij}, i, j \in 1,\dots,Q\}$, and with mean $B = (B_1,\dots,B_Q)$, with elements zero or non-zero depending on how the predictors are defined (see Section 10.2.1). With link g, the structural assumption specifies

$$g[E(y_{it} \mid b_i.u_{it})] = \eta_{it} = X_{it}\beta + Z_{it}b_i + u_{it}, \tag{10.1}$$

where $Z_{it} = (z_{1it},\dots,z_{Qit})$ is $1 \times Q$. In a typical analysis, b_i and $u_i = (u_{i1},\dots,u_{iT_i})'$ are assumed independent, and both also taken to be normal, at least initially. Residual autocorrelation may necessitate a correlation structure in the observation errors (e.g. Franco and Bell, 2015). In some applications (e.g. Poisson data without overdispersion), the observation level errors u_{it} may not be present.

A particular widely applied GLMM is the normal linear mixed model, with

$$y_{it} = X_{it}\beta + Z_{it}b_i + u_{it},$$

with $u_i = (u_{i1},\dots,u_{iT_i})'$ iid normal with mean zero, $T_i \times T_i$ covariance matrix $\Sigma_i = \sigma^2 I$, and conditional expectations

$$E(y_{it} \mid b_i) = \eta_{it} = X_{it}\beta + Z_{it}b_i.$$

The normal linear mixed model may be achieved with latent data y^*_{it} underlying observed data, either binary or categorical (e.g. Chib, 2008; Chib and Jeliazkov, 2006). Thus for binary y_{it}

$$y^*_{it} \mid b_i \sim N(X_{it}\beta + Z_{it}b_i, 1) \quad I(0,) \text{ if } y_{it} = 1$$

$$y^*_{it} \mid b_i \sim N(X_{it}\beta + Z_{it}b_i, 1) \quad I(,0) \text{ if } y_{it} = 0,$$

with iid residuals under conditional independence having known variance $\sigma^2 = 1$, but with D still unknown.

Stacked over times, the conditional mean in (10.1) is expressed as

$$\eta_i = X_i\beta + Z_i b_i + u_i, \tag{10.2}$$

where η_i is $T_i \times 1$, X_i is $T_i \times P$, and Z_i is $T_i \times Q$, while the normal linear mixed model is

$$y_i = X_i\beta + Z_i b_i + u_i.$$

For the normal linear mixed model, the marginal model (with b_i integrated out) is obtainable analytically as

$$y_i \sim N(X_{it}\beta, Z_i D Z_i' + \sigma^2 I),$$

which is a feature not present for the broader class of general linear mixed models (Molenberghs and Verbeke, 2006).

Given the fixed regression effects, and subject permanent effects $b = (b_1,\dots,b_n)$, repeated observations on the same subject are conditionally independent (Kleinman and Ibrahim, 1998; Tutz and Kauermann, 2003), and the conditional likelihood factors as

$$p(y \mid b, \beta, \sigma^2) = \prod_{i=1}^{n} p(y_i \mid b_i, \beta, \sigma^2),$$

where $p(y_i \mid b_i,\beta,\sigma^2)=\prod_{t=1}^{T_i} p(y_{it} \mid b_i,\beta,\sigma^2)$. Similarly, the joint density $p(y,b \mid \beta,D,B,\sigma^2)=p(y \mid b,\beta,\sigma^2)p(b \mid B,D)$ factors into subject-specific elements

$$\left(\prod_{t=1}^{T_i} p(y_{it} \mid b_i,\beta,\sigma^2)\right)p(b_i \mid B,D).$$

If model checking in fact reveals the conditional independence assumption does not provide an adequate fit, then the model requires elaboration. For example, if checking shows regression errors correlated through time, then the iid assumption for u_{it} may have to be reconsidered, or lagged effects in the response included in predictor sets X_{it} or Z_{it} (Frees, 2004, p.279); see Sections 10.3 and 10.5.

10.2.1 Centred or Non-Centred Priors

The parameterisation adopted in the prior for b_i depends on whether X_{it} and Z_{it} are specified to be overlapping or distinct, and also on MCMC convergence considerations. In many presentations of the GLMM, Z_{it} are assumed to be a subset of X_{it}, in which case the b_i are typically taken to be zero mean random effects, usually following a standard density (e.g. multivariate normal). If the X_{it} and Z_{it} are non-overlapping, then the b_{qi} may be taken to have non-zero means equal to the central (fixed effect) regression parameters B_q for z_{qit} (Chib and Carlin, 1999, p.20), namely

$$g[E(y_{it} \mid b_i.u_{it})]=X_{it}\beta+Z_{it}b_i+u_{it},$$
$$b_i \sim N(B,D).$$

Such hierarchical centring may assist precise identification and MCMC convergence. If no predictors have fixed effect coefficients, one has what is sometimes termed a random coefficient regression, namely

$$g[E(y_{it} \mid b_i.u_{it})]=Z_{it}b_i+u_{it},$$

where b_{qi} have non-zero means B_q (e.g. Daniels and Hogan, 2008).

Papaspiliopoulos et al. (2003) compare MCMC convergence for centred, non-centred, and partially non-centred hierarchical model parameterisations, and mention that hierarchical centring may be less effective when the latent effects b_i are relatively weakly identified. Consider the normal linear mixed model in the form

$$y_i=X_i\beta+Z_ib_i+(\sigma^2 I_{T_i})e_i,$$
$$b_i=B+D^{0.5}v_i,$$

where e_i and v_i, of dimension T_i and Q respectively, are standard normal variables.

Then the non-centred parameterisation (NCP) and partially non-centred parameterisations (PNCP) are respectively

$$\tilde{b}_i=b_i-B,$$

and

$$\tilde{b}_i^w = b_i - W_i B,$$

where $W_i = U_i D^{-1}, U_i = (1/\sigma^2) Z_i Z_i + D^{-1}$, and

$$\tilde{b}_i \sim N(0, D),$$
$$\tilde{b}_i^w \sim N(B - W_i B, D).$$

The proportion of B subtracted from b_i under the PNCP form (that has favourable MCMC convergence properties) is observation specific. The longitudinal model under the NCP and PNCP parameterisations become

$$y_i = X_i \beta + Z_i B + Z_i \tilde{b}_i + (\sigma^2 I_{T_i}) e_i,$$
$$y_i = X_i \beta + Z_i W_i B + Z_i \tilde{b}_i^w + (\sigma^2 I_{T_i}) e_i.$$

The NCP form has potential use in random effects selection (see Section 10.2.3).

10.2.2 Priors on Unit Level Random Effects

The most commonly adopted prior for the random subject or cluster effects $b_i = (b_{1i}, \ldots, b_{Qi})$ is an iid multivariate normal

$$(b_{1i}, \ldots, b_{Qi}) \sim N_Q(B, D), \tag{10.3}$$

where the means $B = (B_1, \ldots, B_Q)$ are either zeroes or unknown fixed effects, and $D = [d_{rs}] = [\sigma_{brs}]$ represents covariation within subjects between the rth and sth random effects b_{ri} and b_{si}. If the Z_{it} are a subset of the X_{it}, then the means B_q will be zero. For robustness against non-normality or outliers, other forms of mixture, including scale mixtures of normals, or discrete mixtures of random effects, may be assumed for subject effects (section 10.6). For spatially configured units, a prior for $(b_{1i}, \ldots, b_{Qi})$ including correlation over areas is likely to be relevant. For doubly nested data (e.g. observations y_{ijt} within subjects i within clusters j), the second stage parameters are likely to be cluster specific and possibly also randomly varying, as in

$$(b_{1ij}, \ldots, b_{Qij}) \sim N_Q(B_j, D_j),$$
$$(B_{1j}, \ldots, B_{Qj}) \sim N_Q(M_B, C_B).$$

In many applications, the Z_{it} will be of relatively small dimension, confined to the intercept or simple time functions. For example, if $Q = 1$ and $Z_{it} = 1$, one has the normal linear form

$$y_{it} = b_i + X_{it} \beta + u_{it}, \tag{10.4}$$

where b_i represent permanent subject effects, namely enduring differences between subjects due to unmeasured attributes. If X_{it} excludes (or includes) an intercept, then the b_i will be normal with mean B (or zero) and variance D.

In growth curve applications, the Z_{it} typically include transforms of time or age, and the mean level for an individual changes with time or age (e.g. linearly or quadratically) with growth rates specific to each subject. For example, under a linear growth model with $Q=2$, each subject has their own linear growth rate (Weiss, 2005)

$$y_{it} = b_{1i} + b_{2i}t + X_{it}\beta + u_{it},$$

where D_{12} measures the correlation between intercepts and slopes. Assuming X_{it} omits an intercept and linear time term, one may take

$$(b_{1i}, b_{2i}) \sim N([B_1, B_2], D).$$

In particular, under the random intercept and slope (RIAS) model

$$y_{it} = b_{1i} + b_{2i}t + u_{it},$$

so that an individual's response will differ from his/her mean level at a particular time or age by a random term u_{it}. Another option is to replace known time functions by an unknown time-varying function, δ_t, as in

$$y_{it} = b_{1i} + b_{2i}\delta_t + X_{it}\beta + u_{it},$$

with δ_t subject to identifying constraints (e.g. $\delta_1=0$, and $\delta_T=1$), or with the variance of b_{2i} preset (Zhang et al., 2007; Zhang, 2016).

To illustrate MCMC sampling, consider the random coefficient normal linear model, namely

$$y_{it} = Z_{it}b_i + u_{it},$$

$$u_{it} \sim N(0, \sigma^2),$$

$$(b_{1i}, b_{2i}, \ldots, b_{Qi}) \sim N_Q([B_1, B_2, \ldots, B_Q], D).$$

Let $\tau = 1/\sigma^2$ and assume, following Wakefield et al. (1994), that $\tau \sim Ga(\nu_0/2, \tau_0\nu_0/2)$. Also assume a multivariate normal prior for the second-stage population means B, and a Wishart prior for D^{-1}, namely

$$B \sim N(B_0, C),$$

$$D^{-1} \sim W([\rho R]^{-1}, \rho).$$

Setting

$$N = \sum_{i=1}^{n} T_i,$$

$$E_i^{-1} = \tau Z_i' Z_i + D^{-1},$$

$$V^{-1} = nD^{-1} + C^{-1},$$

$$\bar{b} = \sum_{i=1}^{n} b_i / n,$$

then Gibbs sampling involves the posterior conditionals

$$b_i \sim N(E_i[\tau Z_i'y + D^{-1}B], E_i), \quad i = 1,\ldots,n$$

$$B \sim N(V[nD^{-1}\bar{b} + C^{-1}B_0], V)$$

$$D^{-1} \sim W\left(\left[\sum_{i=1}^{n}(b_i - B)(b_i - B)' + \rho R\right]^{-1}, n + \rho\right)$$

$$\tau \sim Ga\left(\frac{v_0 + N}{2}, \frac{1}{2}\left[\sum_{i=1}^{n}(y_i - Z_i b_i)'(y_i - Z_i b_i) + v_0\tau_0\right]\right).$$

When predictors are available that might explain heterogeneity between subjects (e.g. treatment allocations), regression priors may be used as means for unit random effects b_{qi} (Chib, 2008). Thus, consider a model with varying intercepts b_{1i}, varying linear and quadratic growth effects $\{b_{2i}, b_{3i}\}$, and observations at differentially spaced time points $\{a_{i1}, a_{i2}, \ldots, a_{iT_i}\}$ (Muthen et al., 2002). So

$$y_{it} = b_{1i} + b_{2i}a_{it} + b_{3i}a_{it}^2 + u_{it},$$

where random growth coefficients are related to an intervention variable Tr_i according to

$$\begin{aligned} b_{1i} &= B_1 + e_{1i} \\ b_{2i} &= B_2 + \delta_2 Tr_i + e_{2i} \\ b_{3i} &= B_3 + \delta_3 Tr_i + e_{3i} \end{aligned} \tag{10.5}$$

with $(e_{1i}, \ldots, e_{Qi}) \sim N_Q(0, D)$. Treatment is randomised so the baseline effects b_{1i} are taken to be independent of the intervention Tr_i.

10.2.3 Priors for Random Covariance Matrix and Random Effect Selection

Inferences may be sensitive to the form of prior adopted for the dispersion matrix D, and the amount of information it contains. Improper or overly diffuse priors on D or other variance hyperparameters may be associated with actual or effectively improper posterior densities. For example, the Jeffreys rule prior, namely

$$p(D) \propto \det(D)^{-(Q+1)/2}$$

may lead to an improper joint posterior for D and β under certain conditions (Natarajan and Kass, 2000). The conjugate model for $Q > 1$ random effects involves a Wishart prior for D^{-1}, $D^{-1} \sim \text{Wish}(A, \nu)$, or

$$p(D^{-1} | A, \nu) \propto |A|^{\nu/2} |D^{-1}|^{0.5(\nu - Q - 1)} \exp(-0.5tr(AD^{-1})),$$

where $E(D^{-1}) = \nu A^{-1}$ and $E(D) = A/(\nu - Q - 1)$.

Setting the elements in A may be difficult in the absence of substantive information. Greater flexibility, including random effects selection, may be gained with matrix decomposition alternatives to the Wishart (e.g. Alvarez et al., 2016; Kinney and Dunson, 2007; Frühwirth-Schnatter and Tüchler, 2008; Tutz and Kauermann, 2003; Hedeker, 2003), or with adaptations of the uniform shrinkage prior (Natarajan and Kass, 2000; Tsai and Hsiao, 2008). One can follow a separation strategy (Barnard et al., 2000), decomposing the dispersion matrix into a product of a correlation matrix R and diagonal matrices $\Delta = \text{diag}(\sigma_{b1},\ldots,\sigma_{bQ})$, namely

$$D = \Delta R \Delta.$$

Barnard et al. (2000) construct the correlation matrix R from an inverse Wishart distribution, but a more versatile approach is provided by the $LKJ(\nu)$ prior (Lewandowski et al., 2009), available in rstan. Thus

$$R \sim LKJ(\nu),$$

where, as ν increases, large correlations are less plausible and the prior concentrates around the unit correlation matrix. At $\nu=1$, the $LKJ(\nu)$ correlation distribution reduces to the identity distribution over correlation matrices, so that all correlations are equally plausible. A setting such as $LKJ(1.5)$ might be taken as applicable in many situations, where extreme correlations of −1 or +1 are downweighted slightly, but relatively high correlations are not to be ruled out. Any suitable prior (inverse gamma, lognormal, uniform, half Cauchy) may be used for the standard deviations σ_{bj}.

Cholesky decomposition methods also provide flexibility. Consider the Cholesky decomposition $D=CC'$ where C is a lower triangular matrix, with $D_{pq} = \sum_{r=1}^{Q} c_{pr}c_{qr}$ and variances obtained as

$$D_{qq} = \sum_{r=1}^{Q} c_{qr}^2.$$

Then if Z_{it} is a subset of X_{it}, (10.1) may be expressed

$$\eta_{it} = X_{it}\beta + Z_{it}C\zeta_i + u_{it}$$

where $(\zeta_{1i},\ldots,\zeta_{Qi}) \sim N_Q(0,I)$. For example, with $Q=2$,

$$(z_{1it}, z_{2it})\begin{pmatrix} c_{11} & 0 \\ c_{21} & c_{22} \end{pmatrix}\begin{pmatrix} \zeta_{1i} \\ \zeta_{2i} \end{pmatrix} = \zeta_{1i}(z_{1it}c_{11} + z_{2it}c_{21}) + \zeta_{2i}z_{2it}c_{22},$$

Instead of a Wishart prior on D^{-1}, priors are then adopted for each element of C. To ensure D is positive definite, the diagonal terms c_{11} and c_{22} need to be assigned positive priors, while the prior c_{21} is unconstrained. For $Q=3$, one has

$$(z_{1it}, z_{2it}, z_{3it})\begin{pmatrix} c_{11} & 0 & 0 \\ c_{21} & c_{22} & 0 \\ c_{31} & c_{32} & c_{33} \end{pmatrix}\begin{pmatrix} \zeta_{1i} \\ \zeta_{2i} \\ \zeta_{3i} \end{pmatrix} = \zeta_{1i}(z_{1it}c_{11} + z_{2it}c_{21} + z_{3it}c_{31})$$

$$+ \zeta_{2i}(z_{2it}c_{22} + z_{3it}c_{32}) + \zeta_{3i}z_{3it}c_{33},$$

with three positive unknowns c_{qq} and three unconstrained lower diagonal unknowns.

An alternative Cholesky decomposition (Cai and Dunson, 2006; Chen and Dunson, 2003) has

$$D = \Lambda\Omega\Omega'\Lambda,$$

where $\Lambda = \text{diag}(\lambda_1, \ldots \lambda_Q)$, and Ω is lower triangular,

$$\Omega = \begin{pmatrix} 1 & 0 & \ldots & 0 \\ \omega_{21} & 1 & \ldots & 0 \\ \ldots & \ldots & \ddots & 0 \\ \omega_{Q1} & \omega_{Q2} & \ldots & 1 \end{pmatrix}.$$

Hence C in $D = CC'$ can be written

$$C = \begin{pmatrix} \lambda_1 & 0 & \ldots & 0 \\ \omega_{21}\lambda_2 & \lambda_2 & \ldots & 0 \\ \ldots & \ldots & \ddots & 0 \\ \omega_{Q1}\lambda_Q & \omega_{Q2}\lambda_Q & \ldots & \lambda_Q \end{pmatrix}.$$

Positive priors (e.g. lognormal, gamma) are taken for λ_q, while normal $N(0,V_\Omega)$ priors may be assumed for unconstrained elements of Ω, with $V_\Omega = 0.5$ providing relatively diffuse priors on correlations between the b_{qi}. Retention of terms in Λ is determined by binary indicators $\gamma_{qq} \sim \text{Bern}(\pi_{qq})$, where π_{qq} may be preset or unknown. Retention of the unknown terms in Ω is determined both by binary indicators $\gamma_{qr} \sim \text{Bern}(\pi_{qr})$, and also by whether λ_q and λ_r are retained; if either of $\{\lambda_q, \lambda_r\}$ is omitted, then ω_{qr} necessarily is.

If Z_{it} is not a subset of X_{it}, one may consider the non-centred parameterisation (Frühwirth-Schnatter and Tüchler, 2008)

$$\eta_{it} = X_{it}\beta + Z_{it}B + Z_{it}C\zeta_i + u_{it}$$

where $(\zeta_{1i}, \ldots, \zeta_{Qi}) \sim N_Q(0,I)$. As above, diagonal terms c_{qq} need to be assigned positive priors, while priors for c_{qr} $(q > r)$ are unconstrained. Selection of which c_{qq} and c_{qr} terms to retain may be based on binary indicators $\{\gamma_{qq} \sim \text{Bern}(\pi_{qq}), \pi_{qq} \sim Be(a_{qq}, b_{qq})\}$, $\{\gamma_{qr} \sim \text{Bern}(\pi_{qr}), \pi_{qr} \sim Be(a_{qr}, b_{qr}), q > r\}$ where $a_{qq} = b_{qq} = a_{qr} = b_{qr} = 1$ is a default option. In effect, the model involves composite terms,

$$G_{qq} = c_{qq}\gamma_{qq}, \tag{10.6}$$

$$G_{qr} = c_{qr}\gamma_{qr}, \quad q > r$$

so that for $Q = 2$

$$\eta_{it} = X_{it}\beta + \zeta_{1i}(z_{1it}G_{11} + z_{2it}G_{21}) + \zeta_{2i}z_{2it}G_{22} + u_{it}.$$

The posterior estimate for D would be based on MCMC monitoring of $D_{qr} = \Sigma_{s=1}^{Q} G_{qs}G_{rs}$.

A regression approach to covariance estimation for longitudinal data is proposed by Pourahmadi (1999, 2000) and Pourahmadi and Daniels (2002). For the essence of the method, consider normal metric data y_{it} for subjects $i=1,\ldots,n$ with individual specific covariance matrices Σ_i of dimension $T\times T$. The model $y_{it} \sim N(\mu_{it},\Sigma_i)$ may be re-expressed (for the purposes of decomposing Σ_i) as an antedependence model

$$y_{it}-\mu_{it}=\sum_{j=1}^{t-1}\phi_{itj}(y_{ij}-\mu_{ij})+u_{it},$$

where the errors $u_{it} \sim N(0,h_{it})$ are uncorrelated. Denote

$$H_i=\text{diag}(h_{i1},\ldots,h_{iT}),$$

together with the lower triangular matrix,

$$F_i=\begin{bmatrix} 1 & & & & \\ -\phi_{i21} & 1 & & & \\ -\phi_{i31} & -\phi_{i32} & 1 & & \\ \cdots & \cdots & \cdots & \cdots & \\ -\phi_{iT1} & -\phi_{iT2} & \cdots & -\phi_{iT,T-1} & 1 \end{bmatrix}.$$

One then has the decomposition

$$\text{Var}(u_i)=H_i=F_i\Sigma_iF_i'.$$

The parameters ϕ_{itj} and h_{it} may be referred to respectively as the generalised autoregressive parameters and the innovation variances of Σ_i (Pourahmadi and Daniels, 2002).

A parsimonious covariance model, especially for large T, may then be achieved by using predictors z_{it} and w_{itj} in the regressions

$$\log(h_{it})=z_{it}\gamma,$$

$$\phi_{itj}=w_{itj}\lambda.$$

Often one might take $\Sigma_i=\Sigma$, in which case

$$y_{it}-\mu_{it}=\sum\phi_{tj}(y_{ij}-\mu_{ij})+u_{it},$$

with $u_{it} \sim N(0,h_t)$, $H=\text{diag}(h_1,\ldots,h_T)$, and

$$F=\begin{bmatrix} 1 & & & & \\ -\phi_{21} & 1 & & & \\ -\phi_{31} & -\phi_{32} & 1 & & \\ \cdots & \cdots & \cdots & \cdots & \\ -\phi_{T1} & -\phi_{T2} & \cdots & -\phi_{T,T-1} & 1 \end{bmatrix},$$

with $\text{Var}(u_i) = H = F\Sigma F'$. The covariates used for covariance model become $\{z_t, w_{tj}\}$, where the w_{tj} might simply be powers in $(t-j)$ as illustrated by Cepeda and Gamerman (2004), and the z_t are simply powers of t. A possible drawback to using polynomial functions of time is the multicollinearity that may be encountered, and Bayesian regression selection may then be applied. One may also consider autoregressive or random walk priors in h_t and modelling ϕ_{tj} as a collection of iid random effects under a shrinkage prior strategy (Daniels and Pourahmadi, 2002, p.558).

10.2.4 Priors for Multiple Sources of Error Variation

Estimation of variance components and convergence of MCMC samplers for longitudinal data may also be sensitive to the assumed prior interlinkages (or not) between multiple sources of random variation. Consider a random intercept model

$$y_{it} = b_i + X_{it}\beta + u_{it},$$

with unknown variances $D = \text{var}(b_i)$ and $\sigma^2 = \text{var}(u_{it})$. The conjugate approach with the advantage of simple posterior conditionals involves separate gamma priors on D^{-1} and $\tau = \sigma^{-2}$. These could be informative (e.g. downweighted results from a maximum likelihood fit), but are often taken to be diffuse with small scale and shape parameters, leading to potentially delayed convergence of Gibbs sampling methods since sampling is from an almost improper posterior (Natarajan and McCulloch, 1998). These problems may increase if an autocorrelated error term is added to the white noise error as in

$$y_{it} = b_i + X_{it}\beta + u_{it} + \varepsilon_{it},$$

$$\varepsilon_{it} = \rho\varepsilon_{i,t-1} + v_{it},$$

where $v_{it} \sim N(0, \sigma_v^2)$, and there are three variances.

An alternative is to allow for potential interdependence between variance components via adaptations of the uniform shrinkage prior (Natarajan and Kass, 2000). The uniform shrinkage principle extends to beta priors on the relative shares for two variances and to Dirichlet priors on the relative shares for three or more variance components. So, one might set a prior on one or other of D or σ^2, but then specify a variance partitioning rule such as $\kappa = D/(D+\sigma^2) \sim U(0,1)$ to obtain the other. A related strategy might take $\kappa = D/(D+\sigma^2) \sim Be(a_\kappa, b_\kappa)$, where a_κ and b_κ are preset or hyperparameters. Lee and Hwang (2000) use uniform shrinkage priors in a multilevel longitudinal context when repetitions $t = 1,\ldots,T_{ij}$ are for subjects i nested within clusters j, and where there is an autocorrelated error ε_{ijt} as well as a white noise error u_{ijt}. An extension to multivariate b_i of the uniform shrinkage prior proposed by Natarajan and Kass (2000) takes the form

$$p(D) \propto \det\left[I_Q + \left\{ \frac{1}{n} \sum_{i=1}^{n} Z_i' W_i Z_i \right\} D \right]$$

where W_i is diagonal of dimension T_i with elements $1/V_{it}[\partial\eta_{it}/\partial\mu_{it}]^2)$ where $V_{it} = \phi\text{Var}(\mu_{it})$ and $g(\mu_{it}) = \eta_{it}$.

Example 10.1 Growth Model Simulation

This example illustrates sensitivity to prior specifications and also the application of random effects selection. Thus observations y_{it} $(i=1,\ldots,500; t=1,\ldots,5)$ are generated according to a prototypical growth model, namely:

$$y_{it} = b_{1i} + b_{2i}t + b_{3i}x_{it} + u_{it},$$

$$x_{it} \sim U(-1,1),$$

$$(b_{1i}, b_{2i}, b_{3i}) \sim N_3(B,D),$$

$$u_{it} \sim N(0,1/\tau),$$

$$\tau = 0.5,$$

$$D^{-1} = \begin{pmatrix} 1 & 0 & 0 \\ 0 & 100 & 0 \\ 0 & 0 & 10000 \end{pmatrix},$$

$$B = (5,0.5,0.5).$$

So, the known covariance structure involves uncorrelated random effects b_{qi} with respective standard deviations {1,0.1,0.01}.

The parameters are first re-estimated (in R2OpenBUGS) under a conjugate Wishart prior for the precision matrix $D^{-1} \sim W(I,3)$, and with $\tau \sim Ga(1,0.001)$ and $(B_q \sim N(0,100), q=1,\ldots,3)$. The last 4,000 of a 5,000 iteration two-chain run provides estimated means (sd) for $\sigma_{bq} = D_{qq}^{0.5}$ of 1.13 (0.047), 0.22 (0.022), and 0.35 (0.06). The standard deviations σ_{b2} and σ_{b3} are therefore overestimated as compared to the simulation parameters 0.1 and 0.01. The correlation $r(b_1,b_2)$ is estimated as 0.23 with 95% interval (0.02,0.45). A more diffuse prior on D^{-1} (e.g. a scale matrix with diagonal terms 0.1) provides lower posterior means for $(\sigma_{b2},\sigma_{b3})$, namely 0.14 and 0.16, but an increased posterior mean of 0.38 for $r(b_1,b_2)$. Possible limitations of the Wishart prior are mentioned in the literature (Alvarez et al., 2016).

The posterior means and standard deviations on σ_{bq} under the $W(I,3)$ prior are used to set gamma priors on the Cholesky terms c_{qq} in a selection model (10.6) (Fruhwirth-Schnatter and Tuchler, 2008), but with precision downweighted 100 times, namely $Ga(5.7,5.1)$, $Ga(1.1,4.7)$, and $Ga(0.35,0.95)$. Heavier downweighting (e.g. a thousandfold) is avoided, as it may lead to over-diffuse priors c_{qq}. Inferences from such selection may be sensitive to priors on the Cholesky elements, and a full analysis would consider several choices of prior. As in Section 10.2.3, with composite terms $G_{qr} = c_{qr}\gamma_{qr}$ and $Z_{it} = (1,t,x_{it})$, the linear predictor is

$$\begin{aligned} y_{it} &= Z_{it}B + Z_{it}G\zeta_i + u_{it} \\ &= B_1 + B_2 t + B_3 x_{it} + \zeta_{1i}(z_{1it}G_{11} + z_{2it}G_{21} + z_{3it}G_{31}) \\ &\quad + \zeta_{2i}(z_{2it}G_{22} + z_{3it}G_{32}) + \zeta_{3i}z_{3it}G_{33} + u_{it}, \end{aligned}$$

with selection indicators, $\gamma_{qr} \sim \text{Bern}(\pi_{qr})$, and with π_{qr} unknown and assigned uniform priors. The off-diagonal Cholesky terms c_{qr} $(q>r)$ are assigned $N(0,1)$ priors.

The last 4,000 of a 5,000 iteration two-chain run (in R2OpenBUGS) with $D=CC'$, provides estimated means (medians) for $\sigma_{bq} = D_{qq}^{0.5}$ of 1.12 (1.12), 0.033 (0), and 0.049 (0), with the densities for σ_{b2} and σ_{b3} both having spikes at zero (consistent with zero random variation). The retention probabilities γ_{22} and γ_{33} are respectively 0.30 and 0.38. It is not possible to monitor the correlation matrix, but the covariance D_{12} has posterior mean 0.02, with γ_{21} estimated at 0.27. The selection approach (with the priors adopted) provides a more accurate estimate of the original σ_{b3}, and of the correlation structure, but also essentially eliminates the random variability in the slopes on time.

Random effect selection is also undertaken with $D=\Lambda\Omega\Omega'\Lambda$, namely $C=\Lambda\Omega$. Priors on λ_q are the same as for c_{qq} under the decomposition $D=CC'$, while $N(0,0.5)$ priors are used for the elements of Ω, and Bernoulli priors with preset probability 0.5 for γ_{qr}. The last 4,000 of a 5,000 iteration two-chain run provides posterior means (medians) for $\sigma_{bq}=D_{qq}^{0.5}$ of 1.12 (1.12), 0.040 (0), and 0.021 (0). The densities for σ_{b2} and σ_{b3} again both have spikes at zero. So, a more accurate estimate of the original σ_{b3} is obtained than under a Wishart prior, but random variability in the slopes on time (as summarised in σ_{b2}) is understated.

Either Cholesky decomposition approach can also be applied without selection (in effect setting $\gamma_{qr}=1$). For example, with Ga(1,1) priors on λ_q and N(0,0.5) priors on the ω_{qr}, the second method gives posterior means for σ_{b2} and σ_{b3} of 0.09 and 0.11, with σ_{b3} less inflated (as compared to the true value) than under a Wishart prior.

Also considered is the rstan option whereby an $LKJ(\nu)$ prior is applied to the lower Cholesky factor of the correlation matrix between (b_{1i},b_{2i},b_{3i}) (Vaidyanathan, 2016; Baldwin, 2014). Half Cauchy priors with scale 5 are assumed for σ_{bq}. With a shape parameter $\nu=1$ for the $LKJ(\nu)$ prior, a two-chain run of 5,000 iterations provides posterior means for σ_{b2} and σ_{b3} of 0.09 and 0.15. So again the estimated σ_{b3} is less inflated (as compared to the true value) than under a Wishart prior. However, the estimated correlation matrix shows $r(b_1,b_2)$ to be positive with mean 0.54 and 95% limits (0.04,0.91). Setting $\nu=1.5$ provides posterior means for σ_{b2} and σ_{b3} of 0.08 and 0.10, with $r(b_1,b_2)$ having mean 0.48 and 95% limits (−0.09,0.87).

The Cholesky factor correlation matrix approach is also applied using a Dirichlet partition to a total random variance parameter σ_T^2. This recognises the interdependence of the sources of random variation. Thus, one has $\sigma_b^2=\phi_b\sigma_T^2$, where $(\phi_1,\phi_2,\phi_3)\sim \mathrm{Dir}(w_1,w_2,w_3)$, with the w vector itself Dirichlet distributed with prior weights 1. The total variance σ_T^2 is taken as half Cauchy. This provides posterior means for σ_{b2} and σ_{b3} of 0.08 and 0.10, with $r(b_1,b_2)$ estimated at 0.49.

Finally, a direct separation strategy is applied (McElreath, 2015, p.393). Thus with $D=\Delta R\Delta$, the correlation matrix R is assigned an $LKJ(1.5)$ prior, and σ_{bj} (the diagonal elements in Δ) taken as lognormal with variance 1. This gives posterior means for σ_{b2} and σ_{b3} of 0.09 and 0.13, with $r(b_1,b_2)$ estimated at 0.55.

One can say in conclusion that some of the options considered provide better performance in certain regards, but that no approach satisfactorily reproduces all aspects of the known covariance structure. It may be that a more extended longitudinal simulation (e.g. with T = 10), would be less sensitive, as there is more information on each unit.

Example 10.2 Joint Regression for Mean and Covariance

This example follows Cepeda and Gamerman (2004) in applying the method of Pourahmadi (1999) to $T=24$ monthly height readings y_{it} for $n=6$ students. The antedependence re-expression of the model $y_{it}\sim N(\mu_t,\Sigma)$ is

$$y_{it}-\mu_t=\sum \phi_{tj}(y_{ij}-\mu_j)+u_{it},$$

with $u_{it}\sim N(0,h_t)$. Taking $H=\mathrm{diag}(h_1,\ldots,h_T)$, and

$$F=\begin{bmatrix} 1 & & & & \\ -\phi_{21} & 1 & & & \\ -\phi_{31} & -\phi_{31} & 1 & & \\ \cdots & \cdots & \cdots & \cdots & \\ -\phi_{T1} & -\phi_{T1} & \cdots & -\phi_{T,T-1} & 1 \end{bmatrix},$$

provides the covariance decomposition $\text{Var}(u_i) = H = F\Sigma F'$. The covariates $\{w_{tj}, z_t\}$ used for the covariance regression model are powers of $t-j$ for w_{tj}, and powers of t for z_t. Then the model takes

$$\mu_t = \beta_1 + \beta_2 t + \beta_3 t^2 + \beta_4 t^3,$$

$$\log(h_t) = \gamma_1 + \gamma_2 t + \gamma_3 t^2,$$

$$\phi_{tj} = \lambda_1 + \lambda_2 (t-j) + \lambda_3 (t-j)^2,$$

with the model for ϕ_{tj} here extending only to a quadratic term in $(t-j)$ rather than a quartic as in Cepeda and Gamerman (2004).

The last 9,000 iterations from a two-chain run of 10,000 iterations in R2OpenBUGS provide posterior mean (sd) estimates for the parameters as follows: $\beta_1=94.2(0.39)$, $\beta_2=0.82(0.11)$, $\beta_3=-0.021(0.011)$, $\beta_4=4.0E-4(3.4E-4)$, $\gamma_1=0.36(0.37)$, $\gamma_2=-0.189(0.071)$, $\gamma_3=0.0085(0.0029)$, $\lambda_1=0.34(0.39)$, $\lambda_2=-0.053(0.011)$, and $\lambda_3=-0.00181(5.3E-4)$. Predictions from the model reproduce the observations satisfactorily, with 11 of the 144 data points having predictive exceedance probabilities $Pr(y_{\text{new},it} > y_{it} \mid y)$ under 0.05 or over 0.95.

Despite providing an insight into the temporal aspects of covariation, this model has worse fit measures than a standard approach, with a Wishart prior on trivariate normal random intercepts and slopes in a quadratic growth model. The LOO-IC for the latter is 532, compared to over 2,380 for the joint regression model.

Finally, a quadratic growth curve model is combined with AR1 dependence in the errors, using the appropriate form of error covariance matrix represented by a function (see section 10.3.1). This is implemented in rstan. We find a significant AR1 parameter with posterior mean (sd) 0.93 (0.03). This model can also be coded from first principles. The LOO-IC is reduced to 359.

10.3 Temporal Correlation and Autocorrelated Residuals

Correlation between regression errors at different times is obtained as a by-product of other random effect schemes, not only from explicit time series priors. In particular, the random intercept model in (10.4) illustrates how subject random effects induce temporal correlation. It is important to control for such heterogeneity to avoid spurious "state dependence," namely dependence of the current outcome or probability on past outcomes (Chib and Jeliazkov, 2006). Thus, for metric data, suppose

$$y_{it} = b_i + X_{it}\beta + u_{it},$$

where X_{it} includes an intercept, $b_i \sim N(0,D)$, and $u_{it} \sim N(0,\sigma^2)$. Assuming u_{it} and b_i are independent, the correlation between $\omega_{it} = u_{it} + b_i$ and $\omega_{is} = u_{is} + b_i$ at periods t and s is

$$\kappa = \text{cov}(\omega_{it}, \omega_{is})/\text{Var}(\omega_{it}) = D/(D+\sigma^2),$$

sometimes called the intraclass correlation. The random intercept model leads to the "compound symmetry" form for the intra-subject covariance matrix Σ_i (Weiss, 2005, pp.246–250), with diagonal terms $\Sigma_{itt} = \sigma^2$, and off-diagonal terms $\Sigma_{ist} = \sigma^2\kappa$, $s \neq t$. Equivalently

$$\Sigma_i = \sigma^2[(1-\kappa)I + \kappa J],$$

where J is an $T_i \times T_i$ matrix of ones.

A factor analytic form of the random intercept model (Weiss, 2005, p.269) includes period specific scale parameters D_t,

$$y_{it} = D_t^{0.5} b_i + X_{it}\beta + u_{it},$$

with $b_i \sim N(0,1)$ having a known variance to provide identification, so that

$$\kappa_t = D_t / (D_t + \sigma^2),$$

and the correlation between $\omega_{it} = D_t^{0.5} b_i + u_{it}$ at times t and s is $(\kappa_t \kappa_s)^{0.5}$. The corresponding RIAS model has loadings $D_{1t}^{0.5}$ and $D_{2t}^{0.5}$ and standard normal random effects b_{1i} and b_{2i}, so that

$$y_{it} = D_{1t}^{0.5} b_{1i} + D_{2t}^{0.5} b_{2i} t + X_{it}\beta + u_{it}.$$

For discrete data, the temporal correlation under random intercept and RIAS models may be confined to positive values only. Thus, for Poisson counts y_{it}, with $\log[E(y_{it} \mid b_i)] = \log(\mu_{it}) = X_{it}\beta + b_i$, and $\gamma_{it} = \exp(X_{it}\beta)$, one has under conditional independence that

$$\begin{aligned} \text{cov}(y_{it}, y_{is}) &= E[\text{cov}(y_{it}, y_{is} \mid b_i)] + \text{cov}[E(y_{it} \mid b_i), E(y_{is} \mid b_i)] \\ &= \text{cov}([e^{b_i}\gamma_{it}], [e^{b_i}\gamma_{is}]) = \gamma_{it}\gamma_{is}\,\text{var}(e^{b_i}), \end{aligned}$$

while

$$\begin{aligned} \text{var}(y_{it}) &= E[\text{var}(y_{it} \mid b_i)] + \text{var}[E(y_{it} \mid b_i)] \\ &= E[e^{b_i}\gamma_{it}] + \text{var}[e^{b_i}\gamma_{it}] = \mu_{it} + \gamma_{it}^2\,\text{var}(e^{b_i}), \end{aligned}$$

with correlation then necessarily positive.

10.3.1 Explicit Temporal Schemes for Errors

When the residuals from (10.1)–(10.2) show temporal correlation, autocorrelated residuals may be used instead of, or in addition to, the white noise errors u_{it}. Let these take the generic form

$$(\varepsilon_{i1}, \ldots, \varepsilon_{iT_i}) \sim N(0, \Sigma_i),$$

where Σ_i is a unit level covariance matrix of dimension $T_i \times T_i$. Commonly adopted schemes for such residuals include low order random walks (e.g. first order or *RW*1 priors), or low order stationary schemes (typically *AR*1 or *MA*1). For example, Xu et al. (2007), Oh and Lim (2001), and Ibrahim et al. (2000) adopt stationary *AR*1 errors in models for longitudinal count data, with $y_{it} \sim Po(\mu_{it})$,

$$\log(\mu_{it}) = X_{it}\beta + \varepsilon_{it},$$

$$\varepsilon_{it} = \rho\varepsilon_{i,t-1} + \upsilon_{it},$$

where $\upsilon_{it} \sim N(0,\sigma_\upsilon^2)$ are iid, and $|\rho|<1$. For metric data, a stationary AR1 error scheme with

$$y_{it} = X_{it}\beta + \varepsilon_{it},$$

$$\varepsilon_{it} = \rho\varepsilon_{i,t-1} + \upsilon_{it},$$

where $|\rho|<1$, and $\upsilon_{it} \sim N(0,\sigma_\upsilon^2)$, leads to error covariance matrix Σ_i with elements

$$\Sigma_{ist} = \text{var}(\varepsilon_{it})\rho^{|s-t|} = \frac{\sigma_\upsilon^2}{1-\rho^2}\rho^{|s-t|},$$

with

$$\Sigma_i = \frac{\sigma_\upsilon^2}{1-\rho^2}\begin{pmatrix} 1 & \rho & \rho^2 & \dots & \rho^{T_i-1} \\ \rho & 1 & \rho & \rho^2 & \dots \\ \dots & \dots & \dots & \dots & \dots \\ \dots & \rho^2 & \rho & 1 & \rho \\ \rho^{T_i-1} & \dots & \rho^2 & \rho & 1 \end{pmatrix}.$$

Assuming homogenous parameters across subjects so that $\Sigma_i = \Sigma$, and that subjects are independent, the full population covariance matrix is

$$\Phi = \frac{\sigma_\upsilon^2}{1-\rho^2} I_n \otimes \Sigma,$$

where I_n is an identity matrix of order n. With $e_{it} = y_{it} - X_{it}\beta$, the marginal likelihood for parameters $\chi = (\rho, \sigma_\upsilon^2, \beta)$ is then of the form $L(\chi|y) = \text{const} - 0.5\log|\Phi| + e'\Phi^{-1}e$.

A stationary first-order moving average or *MA*1 scheme, namely

$$y_{it} = X_{it}\beta + u_{it} + \theta u_{i,t-1},$$

and with $|\theta|<1$, leads to a particular form of a Toeplitz covariance matrix (Weiss, 2005, p.267). Thus set

$$\varphi^2 = \text{var}(u_{it} + \theta u_{i,t-1}) = \sigma^2(1+\theta^2),$$

$$\gamma = \theta/(1+\theta^2),$$

then

$$\Sigma_i = \varphi^2\begin{pmatrix} 1 & \gamma & 0 & 0 & \dots \\ \gamma & 1 & \gamma & 0 & \dots \\ \dots & \dots & \dots & \dots & \dots \\ \dots & 0 & \gamma & 1 & \gamma \\ \dots & \dots & 0 & \gamma & 1 \end{pmatrix}$$

$$= \sigma^2\begin{pmatrix} 1+\theta^2 & \theta & 0 & 0 & \dots \\ \theta & 1+\theta^2 & \theta & 0 & \dots \\ \dots & \dots & \dots & \dots & \dots \\ \dots & 0 & \theta & 1+\theta^2 & \theta \\ \dots & \dots & 0 & \theta & 1+\theta^2 \end{pmatrix}.$$

Stationary or random walk models for errors can be extended in various ways. Thus, for unequally spaced data at points $\{a_{i1}, a_{i2}, \ldots, a_{iT}\}$, the $AR1$ model becomes

$$\varepsilon_{it} = \rho^{|a_{it}-a_{i,t-1}|}\varepsilon_{i,t-1} + \upsilon_{it},$$

with covariance between errors given by

$$\text{cov}(\varepsilon_{it}, \varepsilon_{is}) = \rho^{|a_{it}-a_{is}|}\sigma_\upsilon^2.$$

Another option when T_i is relatively large are subject varying autocorrelation parameters, possibly independently distributed $\rho_i \sim U(-1,1)$ (Ryu et al., 2007) or hierarchically specified; see Example 10.3.

The use of autocorrelated or random walk effects raises issues about how to specify the initial conditions (initial random effects) such as ε_{i1} under an $AR1$ or $RW1$ prior on ε_{it}, and $\{\varepsilon_{i1}, \varepsilon_{i2}\}$ under an $AR2$ or $RW2$ prior. For stationary autoregressive errors, such as the $AR1$ prior

$$\varepsilon_{it} = \rho\varepsilon_{i,t-1} + \upsilon_{it},$$

the variances of ε_{it} and υ_{it} are analytically linked, so that the initial conditions are necessarily specified as part of the prior. So, for stationary $AR1$ dependence in ε_{it} and equally spaced data, one has

$$\varepsilon_{i1} = \upsilon_{i1}/(1-\rho^2)^{0.5},$$

and

$$\text{var}(\varepsilon_{i1}) = \sigma^2\upsilon/(1-\rho^2),$$

and the joint distribution of the ε_{it} is obtained (Xu et al., 2007) as

$$p(\varepsilon_{i1})\prod_{t=2} p(\varepsilon_{it} \mid \varepsilon_{i,t-1})$$

where

$$p(\varepsilon_{it} \mid \varepsilon_{i,t-1}) = \frac{1}{\sigma_\upsilon(2\pi)^{0.5}}\exp(-0.5[\varepsilon_{it} - \rho\varepsilon_{i,t-1}]^2/\sigma_\upsilon^2).$$

In non-stationary and random walk models with $|\rho| \geq 1$, initial conditions are usually specified by diffuse fixed effect priors, though Chib and Jeliazkov (2006) interlink the variance of the initial conditions with that of the main sequence of effects to provide a proper joint prior on $\{\varepsilon_{i1}, \ldots \varepsilon_{iT_i}\}$. One may also link initial conditions ε_{i1} and subject heterogeneity, as in

$$b_i \sim N(\psi\varepsilon_{i1}, \sigma_b^2),$$

where ψ can be positive or negative (Chamberlain and Hirano, 1999). This amounts to assuming a bivariate density for b_i and ε_{i1}.

Example 10.3 Capital Asset Pricing Model

This example considers residual autocorrelation and associated model checking in an application of the capital asset pricing model, considering links between the performance of a particular security and market performance in general (Frees, 2004). The particular application is to $n=90$ insurance firms observed over $T=60$ months (January 1995 to December 1999). The response y_{it} is the security return for firm i in excess of the risk-free rate, and the predictor x_t is the market return in excess of the risk-free rate.

To allow for varying impacts of x_t on y_{it}, a baseline model (model 1) is the RIAS specification

$$y_{it} = b_{1i} + b_{2i}x_t + u_{it},$$
$$u_{it} \sim N(0,1/\tau_u).$$

The coefficients b_{2i} measure how far the return of security i is attributable to market factors. A bivariate normal prior is assumed for$\{b_{1i}, b_{2i}\}$, with mean (B_1,B_2), and covariance D. A Wishart $W(I,2)$ prior for D^{-1} is assumed, with the prior mean for the covariance matrix D then being the identity matrix.

The last 4,000 of a 5,000 iteration two-chain run in R2OpenBUGS provide a significant effect for x_t with B_2 having posterior mean (95% credible interval) of 0.72 (0.63,0.81). The mixed predictive procedure (Marshall and Spiegelhalter) shows a satisfactory fit, around 8% of the 5,400 observations to have predictive exceedance probabilities $Pr(y_{\text{rep.mix},it} > y_{it} \mid y)$ over 0.95 or under 0.05. However, to assess whether first-order autoregressive dependence might be present, define realised residuals $e_{it} = y_{it} - b_{1i} - b_{2i}x_t$. Then a firm-specific measure of AR1 error dependence is

$$\tilde{\rho}_{1i} = \sum_{t=2}^{T} e_{it}e_{i,t-1} \Big/ \sum_{t=1}^{T} e_{it}^2.$$

Thus 58 of the 90 firms have probabilities below 0.05 that $\tilde{\rho}_{1i} > 0$, with the sample-wide AR1 dependence parameter (the mean of the $\tilde{\rho}_{1i}$) estimated at −0.097 with 95% CRI (−0.103,−0.091).

Another evaluation involves a posterior predictive check based on an average of Durbin-Watson (DW) statistics taken over all 90 firms. Thus at each iteration r, a DW statistic is derived for each firm, namely

$$DW_i^{(r)} = \sum_{t=2}^{T} (e_{it}^{(r)} - e_{i,t-1}^{(r)})^2 \Big/ \sum_{t=1}^{T} (e_{it}^{(r)})^2.$$

A summary statistic for autocorrelation is then the average over firms $\overline{DW}^{(r)} = \sum_i DW_i^{(r)}/n$, which is obtained for actual data $\overline{DW}_{\text{obs}}^{(r)}$, and for replicate data $\overline{DW}_{\text{new}}^{(r)}$ (Gelman et al., 1996). The resulting posterior probability $Pr(\overline{DW}_{\text{obs}} \geq \overline{DW}_{\text{new}} \mid y)$ is 1, indicating inadequate fit.

Accordingly, a revised model (model 2) includes a stationary $AR1$ error, so that

$$y_{it} = b_{1i} + b_{2i}x_t + \varepsilon_{it},$$

$$\varepsilon_{it} = \rho\varepsilon_{i,t-1} + v_{it},$$

and a stationary prior, $\rho \sim U(-1,1)$. A 5,000 iteration two-chain run (with the last 4,000 for inference) gives a significant ρ estimate, with posterior mean (sd) −0.088 (0.014).

The LOO-IC for this model is 39,696, compared to 39,733 for model 1. However, checking based on firm-specific $\tilde{\rho}_{1i}$ shows 39 firms with probability under 0.05 that $\tilde{\rho}_{1i} > 0$, and 30 firms with probability over 0.95 that $\tilde{\rho}_{1i} > 0$.

An extension to unit-specific AR1 parameters is therefore adopted. Thus

$$y_{it} = b_{1i} + b_{2i}x_t + \varepsilon_{it},$$

$$\varepsilon_{it} = \rho_i \varepsilon_{i,t-1} + v_{it},$$

with the prior on the firm-specific ρ_i specified indirectly in a hierarchical prior:

$$\delta_i \sim \text{Beta}(a_\delta, b_\delta),$$

$$\rho_i = 2\delta_i - 1,$$

$$a_\delta \sim E(1),$$

$$b_\delta \sim E(1).$$

Priors are as above on (b_{1i}, b_{2i}), and D. Estimation of this model shows checks that $\tilde{\rho}_{1i} > 0$ are now considerably less concentrated in the tails, with only one firm now having a probability under 0.05 that $\tilde{\rho}_{1i} > 0$, and no firms with probability over 0.95 that $\tilde{\rho}_{1i} > 0$. However, possibly illustrating that improved model checks are not necessarily associated with improved overall fit, the LOO-IC rises to 39737, as the complexity index (p_loo) rises to 174.

10.4 Longitudinal Categorical Choice Data

Repeated categorical data involving ordered or unordered options or choices $k = 1,\ldots,K$ by subjects $i = 1,\ldots,n$ for repetitions $t = 1,\ldots,T_i$ are often found in brand choice, labour market, political science, or clinical applications (Rossi et al., 2005; Pettitt et al., 2006; Terzi and Cengiz, 2013). These may be expressed via binary indicators $d_{ikt} = 1$ if category or choice k applies ($d_{ikt} = 0$ for remaining categories), or by categorical responses $y_{it} \in (1,\ldots,K)$. Clinical and pharmaceutical applications commonly involve ordinal rating scales (e.g. Zayeri et al., 2005; Qiu et al. 2002; Agresti and Natarajan, 2001). Particular issues raised by such data include the possibility that permanent subject effects vary between choices (or more generally between categories), and that predictor effects may vary over one or more of choices, as well as over subjects or times. If lagged effects of the dependent variable are included (Section 10.5), these may include both own category and cross-category lags, leading to categorical transition models (Fokianos and Kedem, 2003).

Chintagunta et al. (2001) considers repeated brand choice data and allows subject heterogeneity in relation to attributes of the choices (e.g. variable consumer responsiveness to brand prices), as well as randomly varying subject-choice intercepts b_{ik}. A Bayesian perspective, including optimal MCMC sampling schemes, on consumer heterogeneity in multinomial longitudinal data for purchase choices is provided by Rossi et al. (2005, Chapter 5). For identifiability, choice or category specific parameters must be set to a fixed value (usually zero) in a reference category. For example, the probability that a consumer chooses brand k in period t might be modelled using a multinomial logit (MNL) regression, with choice K as a reference,

$$\pi_{ikt} = Pr(y_{it} = k) = \phi_{ikt} \Big/ \sum_{k=1}^{K} \phi_{ikt},$$

$$\log(\phi_{ikt}) = \beta_{0k} + b_{ik} + P_{kt}\beta_k + A_{kt}\gamma, \quad k = 1,\ldots,K-1$$

$$\log(\phi_{iKt}) = A_{Kt}\gamma,$$

where β_{0k} are intercept terms, P_{kt} and A_{kt} are brand-time specific characteristics (e.g. price and advertising spend) varying in whether associated regression parameters are choice specific, and b_{ik} are random consumer-brand taste effects. These are typically taken as multivariate normal of dimension $K-1$, with $b_{iK}=0$ for identifiability (Malchow-Moller and Svarer, 2003).

Consumer variation in response to prices or attributes would involve making the β_k and γ coefficients specific to each consumer, and defining hyperparameters for the densities of β_{ki} and γ_i. For P_{kt} of dimension R, Rossi et al. (2005, p.136) propose a conjugate normal hierarchical prior structure for $\beta_i = (\beta_{1i},\ldots,\beta_{Ri})$, with mean $Z_i\Delta$, where Z_i are consumer attributes, and with variance V_β of dimension $\dim(\beta_i) = (R-1)R$. V_β is assigned an inverse Wishart prior having with expectation I and $\dim(\beta_i)+3$ degrees of freedom. They demonstrate the improved MCMC convergence for β_i obtained by using a random walk Metropolis with increments that have covariance $s^2(H_i + (V_\beta^{(r)})^{-1})^{-1}$, where H_i is the Hessian of a composite likelihood based on multiplying the MNL subject specific likelihood by the pooled (all subject) likelihood raised to power $\rho_i = T_i/cN$, and $c>1$ and $s = 2.93/\text{sqrt}[\dim(\beta_i)]$ are tuning constants.

Consider categorical longitudinal data with subject level predictors only, namely X_{it} and Z_{it} of dimension P and Q, and category-specific fixed regression effects, namely

$$\log(\phi_{ikt}) = \beta_{0k} + X_{it}\beta_k + Z_{it}b_{ik},$$

$$\log(\phi_{iKt}) = 0,$$

where $Z_{it}b_{ik} = z_{1it}b_{ik1} + z_{2it}b_{ik2} + \ldots + z_{Qit}b_{ikQ}$. Assuming the X_{it} and Z_{it} are non-overlapping, one may adopt Q independent sets of subject-category effects each of dimension $K-1$, one for each predictor z_{qit},

$$(b_{i1q},\ldots b_{i,K-1,q}) \sim N_{K-1}(B_q, D_q).$$

Alternatively, the covariance matrix of the random effects may of dimension $(K-1)Q$ with the b_{ik} correlated over both categories and predictors. In the case where Z_{it} is a subset of X_{it}, the b_{ikq} are zero mean random effects, and covariance matrices may be choice specific D_k of dimension Q, so that

$$(b_{ik1},\ldots,b_{ikQ}) \sim N_Q(0, D_k).$$

This permits a latent variable interpretation based on a Cholesky decomposition of D_k and standardised random effects ζ_{ik}, namely

$$\log(\phi_{ikt}) = \beta_{0k} + X_{it}\beta_k + Z_{it}C_k\zeta_{ik},$$

where $C_kC_k' = D_k$.

Examples of repeated ordinal observations are provided by labour market perception data (Spiess, 2006), changing attitudes to divorce (Berrington et al., 2005), and repeated ordinal scores in horticultural research (Parsons et al., 2006). Suppose responses y_{it} have K ordered categories, with corresponding latent responses y^*_{it} specified by thresholds, possibly time-varying κ_{kt}, or subject varying κ_{ik}. For time-varying thresholds

$$y_{it} = k \quad \text{if} \quad \kappa_{k-1,t} < y^*_{it} \le \kappa_{kt},$$

with predictor effects also possibly varying over (at least one of) categories, subjects or times. For example, Spiess (2006) considers predictor effects varying over times, as in

$$y^*_{it} = X_{it}\beta_t + \varepsilon_{ikt},$$

where $P(\varepsilon_{ikt})$ is usually a normal or logistic distribution. These distributions are very similar though the logistic places more probability in the tails (Hedeker, 2003). So

$$Pr(y^*_{it} \le \kappa_{kt}) = Pr(X_{it}\beta_t + \varepsilon_{ikt} \le \kappa_{kt}) = Pr(\varepsilon_{ikt} \le \kappa_{kt} - X_{it}\beta_t).$$

Depending on the form for $P(\varepsilon_{ikt})$, one has

$$Pr(y^*_{it} \le \kappa_{kt}) = \Phi(\kappa_{kt} - X_{it}\beta_t),$$

or

$$Pr(y^*_{it} \le \kappa_{kt}) = 1/(1 + \exp(-[\kappa_{kt} - X_{it}\beta_t])).$$

Let $\gamma_{ikt} = Pr(y^*_{it} \le \kappa_{kt})$, then

$$Pr(y_{it} = k) = Pr(\kappa_{k-1,t} < y^*_{it} \le \kappa_{kt}) = \gamma_{ikt} - \gamma_{i,k-1,t}.$$

An equivalent specification of this model involves sets of $K-1$ binary variables for each subject-time pairing, namely $d_{ikt}=1$ if $y_{it} \le k$, and $d_{ikt}=0$ otherwise. Then for ε logistic,

$$Pr(y_{it} \le k)) = Pr(d_{ikt} = 1) = 1/(1 + \exp(-\kappa_{kt} + X_{it}\beta_t).$$

Example 10.4 Yoghurt Purchases

Data on yoghurt brand choice from Chen and Kuo (2001) exemplify random effects to represent household heterogeneity in consumer behaviour, specifically longitudinal analysis of unordered choices data. The yoghurt choice data relate to repeated purchases by $i = 1,\ldots,n$ households ($n=100$) between $K=4$ brands, with widely varying numbers of repetitions T_i for each household (between 4 and 185). The total of observations is $N = \sum_{i=1}^{n} T_i = 2412$. Known influences on brand choice are brand and time specific, namely features A_{kt} (=1 if the brand k was subject to an advertising feature at the time t of purchase, =0 otherwise), and shelf price P_{kt}.

A baseline fixed effects model (model 1) has the form

$$\pi_{ikt} = Pr(d_{ikt} = 1) = \phi_{ikt} \Big/ \sum_{k=1}^{K} \phi_{ikt},$$

$$\log(\phi_{ikt}) = \beta_{0k} + A_{kt}\gamma_1 + P_{kt}\gamma_2, \quad k = 1,\ldots,K-1$$

$$\log(\phi_{iKt}) = A_{Kt}\gamma_1 + P_{Jt}\gamma_2,$$

As mentioned by Chen and Kuo (2001), observations from the same household are usually correlated in brand choice applications, and not accounting for such dependence may produce biased estimates. A random intercepts model (model 2) accordingly allows for heterogeneity at household-choice level, though retaining homogenous impacts for brand attributes. This has the form

$$\log(\phi_{ikt}) = A_{kt}\gamma_1 + b_{ik} + P_{kt}\gamma_2, \quad k = 1,\ldots,K-1$$

$$\log(\phi_{iKt}) = A_{Kt}\gamma_1 + P_{Kt}\gamma_2,$$

$$(b_{i1},\ldots,b_{i,K-1}) \sim N(B,D),$$

where the vector B denotes the average category intercepts $(\beta_{01},\ldots,\beta_{0,K-1})$. A Wishart prior for the precision matrix, $D^{-1} \sim W(I,3)$, is assumed.

Inferences (from jagsUI) show significant fixed effects, (γ_1,γ_2), under model 1, for both feature and price, with posterior means (sd) of 0.49 (0.12) and −36.7 (2.3). This model has a LOO-IC of 5,324, whereas the trivariate normal random intercept model has a LOO-IC of 2,181. Estimates of the correlation matrix under model 2 show brand 1 and 3 choices to be positively correlated, with $r(b_{i1},b_{i3})=0.44$. The impacts for feature, and to a lesser degree, price, are enhanced, though with reduced precision. Thus namely posterior means (sd) of (γ_1,γ_2) are now 0.86 (0.18) and −44.8 (3.8). While model 2 yields a pronounced gain in fit, it has not controlled for consumer variation in price or advertising responsiveness, which would involve making the γ_1 and γ_2 coefficients household specific.

Example 10.5 NIMH Schizophrenic Collaborative Study: Ordinal Symptom Score

This example illustrates model checks for longitudinal ordinal outcomes, and involves a study evaluating four drug treatments to alleviate symptoms in schizophrenia subjects: chloropromazine, fluphenazine, thioridazine, and a placebo (Hedeker and Gibbons, 2006). Similar effects were obtained for the three anti-psychotic drugs, and so here the treatment is reduced to a binary comparison of any drug vs the placebo. Symptom severity scores y_{it} are observed for n = 324 subjects on three occasions after the first reading (at week 0), which is coincident with treatment commencing, namely at weeks 1, 3, and 6. The score is ordinal with $K = 7$ levels, namely 1 = normal, 2 = borderline, 3 = mildly ill, 4 = moderately ill, 5 = markedly ill, 6 = severely ill, and 7 = extremely ill. Random baseline intercepts are assumed, together with random slopes on a time variable Z_t, obtained as the square root of weeks. Fixed effect predictors X_{it} are baseline treatment status, a treatment by time interaction, and the patient's sex.

The responses are ordinal subject-time pairs y_{it}, with corresponding binary indicators $d_{itj} = 1$ if $y_{it} = j$, $d_{itj} = 0$ otherwise. Then with $d_{it} = (d_{it1},\ldots,d_{itK})$, and P denoting the logistic distribution function, one has

$$d_{it} \sim \text{Mult}(1,\mathbf{p}_{it}),$$

$$\mathbf{p}_{it} = (p_{it1},\ldots p_{itK}),$$

$$
\begin{aligned}
p_{itj} &= Pr(y_{it} = j), \\
&= P(\kappa_j - \mu_{it}) - P(\kappa_{j-1} - \mu_{it}) \\
&= \gamma_{itj} - \gamma_{it,j-1},
\end{aligned}
$$

where

$$\gamma_{itj} = Pr(y_{it} \leq j) = P(\kappa_j - \mu_{it}), \quad j = 1,\ldots K-1$$

are cumulative probabilities over ranked categories, $\gamma_{itj} = p_{it1} + \ldots + p_{itj}$.

The regression term is a random intercepts and slopes specification,

$$\mu_{it} = b_{1i} + b_{2i}Z_{it} + \beta_1 Tr_i + \beta_2 Z_t Tr_i + \beta_3 \text{Gend}_i,$$

$$(b_{1i}, b_{2i}) \sim N_2(B, D),$$

and $B = (B_1, B_2)$ contains an overall intercept and time slope. Since the overall intercept is an unknown, identification of the $K-1=6$ thresholds requires setting κ_1 to zero. The remaining five threshold parameters are subject to monotonicity constraints: $\kappa_k = \kappa_{k-1} + \delta_k$, where $\delta_k \sim Ga(1,1)$.

The model is first fitted in rjags, with inferences from a two-chain run of 10,000 iterations. The coefficient β_1 (a measure of differences in symptom level between treatment options at baseline) is not significant, but there is a steeper decline in ill-health for treated subjects. Thus, the coefficient β_2 has a posterior mean (95%CRI) of −0.69 (−1.05,−0.32). Posterior means for σ_{b1} and σ_{b2} are 1.71 and 0.86 respectively, with $r(b_1,b_2)$ estimated at −0.47, showing steeper decline effects for higher initial symptom levels. A posterior predictive check based on comparing total likelihoods for actual and replicate data gave probability 0.54, indicating a satisfactory model. Diagnostic tests such as Q-Q plots and Jarque–Bera tests support normality of the permanent effects (b_{1i}, b_{2i}).

To assess sensitivity to alternative priors regarding random effect covariance, the above model is also fitted in rstan using an $LKJ(1.5)$ prior applied to the lower Cholesky factor of the correlation matrix between b_{1i} and b_{2i}. The code for this model involves six threshold parameters, with B_1 set at 0. From a run of 2,000 iterations, estimates for β_1 and β_2 are little changed, while posterior means for $(\sigma_{b1}, \sigma_{b2})$ are 1.80 and 0.92 respectively, with $r(b_1,b_2)$ estimated at −0.49.

10.5 Observation Driven Autocorrelation: Dynamic Longitudinal Models

Differences in behaviour or event proneness between individuals (e.g. in econometric or health applications) may operate through an autoregression in the observations, latent or observed. Longitudinal models including lagged observations are often termed "dynamic longitudinal models," whereas static longitudinal models do not include lagged response values (e.g. Nerlove, 2002; Liu et al., 2017). A canonical dynamic model for metric data involves lagged values of the dependent variable with the overall error combining a time-invariant individual effect and observation level random noise (Bond, 2002).

Thus, with a first order lag in the response, one has

$$y_{it} = \phi y_{i,t-1} + X_{it}\beta + b_i + u_{it}, \quad t = 2,\ldots,T$$

where the $u_{it} \sim N(0,\sigma^2)$ are independent of each other, and under standard assumptions are also uncorrelated with the initial observations y_{i1} and with permanent subject effects b_i. If X_{it} contains a constant term, then the b_i have mean zero, and $b_i \sim N(0,D)$. Allowing for subject level variation in a Q length vector of predictors Z_{it}, as well as for first-order lagged response, leads to

$$y_{it} = \phi y_{i,t-1} + X_{it}\beta + Z_{it}b_i + u_{it}.$$

Assuming a stationary process with $|\phi| < 1$, one possible model for y_{i1} is

$$y_{i1} = \frac{Z_{it}b_i}{1-\phi} + \frac{X_{it}\beta}{1-\phi} + u_{i1},$$

with $u_{i1} \sim N(0, \sigma^2/(1-\phi^2))$. A simplifying approach, more feasible for large T, is to condition on the first observation in a model involving a first-order lag in y, so that y_1 is non-stochastic (Bauwens et al. 1999, p.135). Geweke and Keane (2000) and Lancaster (2002) consider Bayesian approaches to the dynamic linear longitudinal model, in which the model for period 1 is not necessarily linked to those for subsequent periods in a way consistent with stationarity.

Maximum likelihood analysis of dynamic longitudinal models is subject to an initial conditions problem if in fact there is correlation between the permanent subject effects b_i and the initial observations (Hsiao, 1986). In case of such correlation, possible options are a joint random prior (e.g. bivariate normal) involving b_i and u_{i1} (Dorsett, 1999), or a prior for b_i that is conditional on y_{i1}, such as (Wooldridge, 2005; Hirano, 2002)

$$b_i \mid y_{i1} \sim N(\varphi y_{i1}, \sigma_1^2).$$

Dynamic linear models may be extended in several ways, to include ARMA(p,q) error schemes, effects of time functions, or random variation over subjects or times in the impacts of lagged predictors. For example, a dynamic model for earnings (e.g. Galler, 2001) might include $AR1$ autocorrelated errors as in

$$y_{it} = b_i + \phi y_{i,t-1} + X_{it}\beta + W_i\gamma + \varepsilon_{it},$$

$$\varepsilon_{it} = \rho\varepsilon_{i,t-1} + v_{it},$$

where W_i are fixed human capital attributes, or in RIAS form,

$$y_{it} = b_{1i} + b_{2i}t + \phi y_{i,t-1} + X_{it}\beta + W_i\gamma + \varepsilon_{it},$$

where the random effects b_{1i} and b_{2i} allow subject specific variation in wage level and wage growth. Taking the time function to be an unknown function of t, δ_t, lead to autoregressive latent trait models (Bollen and Curran, 2004). Allowing for time-varying coefficients on lagged responses $y_{i,t-1}$, as well as random subject intercepts and growth rates, one might then have

$$y_{it} = b_{1i} + b_{2i}\delta_t + X_{it}\beta + \phi_t y_{i,t-1} + u_{it},$$

with δ_t subject to identifying constraints, such as $\delta_1 = 1$.

10.5.1 Dynamic Models for Discrete Data

For discrete data, a range of dynamic longitudinal approaches have been proposed, varying according to form of response (e.g. count or binary) and initial conditions prior. These include using a conditional prior method relating b_i and y_{i1} (Wooldridge, 2005), or specifying an initial period model without subject effects or a lagged response effect (Pettitt et al., 2006).

For counts y_{it} taken as Poisson, $y_{it} \sim Po(\mu_{it})$, problems with taking a linear impact of the first lag outcome $y_{i,t-1}$, as in

$$\mu_{it} = \exp(X_{it}\beta + \phi y_{i,t-1} + b_i), \quad t = 2,\ldots,T$$

are mentioned by Fahrmeir and Tutz (2001). This option for modelling lag response impacts defines the Markov property scheme studied by Fotouhi (2007), under which the initial observation is modelled as

$$\mu_{i1} = \exp(X_{i1}\tilde{\beta} + c_i),$$

where X_{i1} includes any relevant predictors for the first period, and the subject effects b_i and c_i follow a bivariate normal with correlation ρ.

Alternatively, the impact of a lagged count response may be modelled by a log or other transform $g(y)$, with extra preset or unknown parameters in case the lagged y is zero. Thus if $g(y) = \log(y + c)$, where $c = 1$ (say), one has

$$\mu_{it} = \exp(X_{it}\beta + \phi g(y_{i,t-1}) + b_i), \quad t = 2,\ldots,T$$

where one might assume

$$b_i \mid y_{i1} \sim N(\varphi y_{i1}, \sigma_1^2).$$

By contrast, applying the conditional linear autoregressive process to longitudinal data (Grunwald et al., 2000) leads to means

$$\mu_{it} = \phi y_{i,t-1} + \exp(X_{it}\beta + b_i).$$

while the full autoregressive conditional Poisson specification (Jung et al., 2006) specifies

$$\mu_{it} = \phi y_{i,t-1} + \eta\mu_{i,t-1} + \exp(X_{it}\beta + b_i).$$

In contrast to count regression, regression for binary responses $y_{it} \sim \text{Bern}(\pi_{it})$ may straightforwardly include lags in observed outcomes $y_{i,t-s}$ leading to Markov Chain models (Kedem and Fokianos, 2005). First order Markov dependence, as in

$$\text{logit}(\pi_{it}) = \alpha_0 + \alpha_1 y_{i,t-1} + X_{it}\beta + b_i,$$

may be extended to higher order Markov dependence,

$$\text{logit}(\pi_{it}) = \alpha_0 + \sum_{s=1}^{L} \alpha_k y_{i,t-s} + X_{it}\beta + b_i,$$

with L preset or determined by selection (Erkanli et al., 2001). Alternatively, fixed predictor effects β, and parameters for random effects b_i, may vary according to the previous value s of the binary response; so $\{\beta_s, D_s\}$ are specific to previous response $y_{i,t-1} = s$ (Islam and Chowdhury, 2006).

Such alternatives extend in principle to multinomial outcomes $y_{it} \in (1,\ldots,K)$, or equivalently $d_{ikt}=1$ if category k applies (or is chosen), and $d_{itk}=0$ otherwise. So

$$(d_{it1},\ldots d_{itK}) \sim \text{Mult}(n_{it}, [\pi_{it1},\ldots,\pi_{itK}]),$$

where $n_{it}=1$. Use of lags is complicated by the possible influence of cross-category lags as well as own-category lags. Pettitt et al. (2006) consider a Bayesian hierarchical multinomial model for changes in employment status (a trichotomy), with one period lags in status as predictors. Thus, with employment status 1 as the reference (and so $\phi_{i1t}=1$), one has for $t>1$

$$\pi_{ikt} = Pr(y_{it} = k) = \phi_{ikt} \Big/ \sum_{k=1}^{K} \phi_{ikt},$$

$$log(\phi_{ikt}) = b_{ik} + \beta_k X_{it} + \gamma_{k1} I(y_{i,t-1} = 2) + \gamma_{k2} I(y_{i,t-1} = 3), \quad k = 2,\ldots,K$$

where b_{ik} are category specific random effects. For the initial period, a static multinomial logit model can be adopted, without lag effects or b_{ik}, and with distinct regression effects, namely

$$\log(\phi_{ik1}) = \delta_k X_{i1} \quad k = 2,\ldots,K.$$

This follows from a linear approximation to the reduced form obtained when lagged response variables are replaced by their specifications under the dynamic model for periods preceding $t=1$.

Dynamic modelling approaches may also be applied using latent metric responses, associated with binary or ordinal observations. Suppose observations y_{it} are binary such that the latent continuous response $y^*_{i,t} > 0$ if and only if $y_{it}=1$, and $y^*_{i,t} \le 0$ if $y_{it}=0$. Then one might specify

$$y^*_{i,t} = X_{it}\beta + \phi_1 y_{i,t-1} + \phi_2 y^*_{i,t-1} + u_{it},$$

with $u_{it} \sim N(0,1)$, and lag one dependence on both previous events and latent utilities. If there is serial correlation (e.g. AR1 dependence) in the errors, then $\varepsilon_{it} = \rho_1 \varepsilon_{i,t-1} + \upsilon_{it}$, with $\upsilon_{it} \sim N(0,1)$. In this way, one may avoid spurious state dependence in which previous responses proxy unobserved variation.

Example 10.6 National Longitudinal Study of Youth: Lagged Earnings Model

This example considers unbalanced data and the modelling of initial conditions and autocorrelation in such data. It involves earnings data from the US National Longitudinal Survey relating to young women aged between 14 and 26 in 1968, and either already in the labour market in 1968, or entering the labour market during the period 1968–88. In this period, there were fifteen measuring occasions, namely each year during 1968–88 except 1974, 1976, 1979, 1981, 1984, and 1986. There are 4,711 subjects varying considerably in their observed histories; many subjects are subject to attrition or intermittent observation. The analysis here is based on a 10% sample of the n = 4164 subjects who have at least two measurements on yearly log earnings, where earnings figures for each subject are divided by calendar year averages to correct for inflation. In this way, the earnings profile of a subject observed over 1968–1975 (say) can be compared with that for a subject observed over 1978–1985. An alternative might be to have fixed or random effects for each calendar year to model population trends in average income.

Although not all years were subject to survey updates, the analysis here takes a subject's entire observation span (obtained by comparing initial and last observation year) to define that subject's total times T_i. Any intervening years without observations are treated as missing data, whether this is due to intermittent missingness or the absence of an NLS update in particular years. Thus the first subject is observed on twelve occasions (in the studies in 1970, 1971, 1972, 1973, 1975, 1977, 1978, 1980, 1983, 1985, 1987, and 1988), but that subject's total times T_i is set at 19, with the intervening years without observations (e.g. 1974, 1976, etc.) treated as missing data. Missingness is taken to be at random, not depending on the possibly missing response value.

With y_{it} denoting (inflation corrected) log earnings, the initial regression model includes subject effects b_i, and fixed binary attributes $\{W_{1i}, W_{2i}, W_{3i}\}$, with W_{1i} for college graduate (=1, 0 otherwise), W_{2i} for white ethnicity (=1, 0 for other ethnicities), and an interaction $W_{3i} = W_{1i}W_{2i}$. So, for $i = 1,\ldots,n$,

$$y_{it} = b_i + W_i\gamma + u_{it}, \quad t = 1,\ldots,T_i$$

where $b_i \sim N(\beta_1, D)$, and $u_{it} \sim N(0,\sigma^2)$. Uniform $U(0,10)$ priors are assumed for σ and $\sigma_b = D^{0.5}$, and $N(0,1000)$ priors for fixed effects $\{\beta_1, \gamma_1, \gamma_2, \gamma_3\}$.

Estimation using jagsUI give posterior means (sd) for γ_1 and γ_2 of 0.32 (0.05) and 0.036 (0.022), showing significantly higher earnings for college graduates, and a positive but not significant white ethnicity effect. The effect of the interaction term is significantly negative, with mean (sd) of −0.12 (0.06), suggesting a greater positive impact of college education on earnings for non-white subjects. The posterior mean for the standard deviation of the b_i is 0.18, so that a subject for whom b_i is one standard deviation above the average would have earnings about 20%, namely 100exp(0.18), above average, given observed personal characteristics W_i. Taking $\hat{u}_{it} = y_{it} - \beta_1 - b_i - W_i\gamma$, there is evidence of autocorrelated errors, with the 95% interval for the statistic $r_u = \sum_{i=1}^{n}\sum_{t=2}^{T_i} \hat{u}_{it}\hat{u}_{i,t-1} \Big/ \sum_{i=1}^{n}\sum_{t=1}^{T_i} \hat{u}_{it}^2$ being (0.06,0.11).

To improve fit, a second dynamic model is non-stationary, in that there is a distinct model for the first period for each subject (Geweke and Keane, 2000), and a one period lag effect ϕ of earnings, with this effect not constrained to stationarity. Random subject effects are also included in the model for periods $t = 2,\ldots,T_i$ so that

$$y_{it} = b_i + W_i\gamma + \phi y_{i,t-1} + u_{it}, \quad t = 2,\ldots,T_i$$

$$y_{i1} = W_i\gamma_1 + u_{i1},$$

with an $N(0,1)$ prior on ϕ, and with $u_{it} \sim N(0,\sigma^2)$ and $u_{i1} \sim N(0,\sigma_1^2)$ taken independently. The 95% interval for ϕ is obtained as (0.57, 0.64), along with considerably reduced autocorrelation, with 95% interval for r_u now from −0.048 to −0.003. Fit is improved, with WAIC now lower at −1530, compared to −387 under the non-dynamic model. The γ coefficients are reduced in absolute size, but the college effect γ_1 remains significant, with 95% interval (0.08,0.17). The posterior mean for σ_b is also reduced, to 0.072. There is scope for further model development, as the probability that r_u is positive is low (under 0.02), and this might involve subject specific lag parameters, random slopes on time, or autocorrelated errors.

Example 10.7 Epileptic Seizure Data: Lagged Count Model

This example illustrates model checks for a dynamic count regression using the epileptic seizure data from Thall and Vail (1990). An anti-epileptic drug treatment (progabide) was applied for some of the n = 59 patients, with others receiving a placebo. A pre-treatment eight-week baseline seizure count was also obtained, and may be treated either as exogenous, or as an endogenous initial condition (Fotouhi, 2007). Here the baseline count is included in the outcome profile, so that $T=5$ with the baseline seizure count denoted y_{i1}. The analysis here follows Lindsey (1993) in including a lagged response as one predictor. The predictors X_{it} for $t \geq 2$ are age, treatment, treatment by time interaction, and lagged seizure count, while the predictor set X_{i1} consists of age at baseline, and treatment (to measure any differential baseline morbidity between treatment groups). A trivariate normal model correlates the random intercept and slopes (b_{i1}, b_{i2}) for periods $t \geq 2$ with the random intercepts c_i in the model for y_{i1}.

So, model 1 takes

$$y_{it} \sim Po(\mu_{it}), \quad t = 1, \ldots, 5$$

with the means modelled as

$$\mu_{it} = \exp(X_{it}\beta + \phi y_{i,t-1} + b_{i1} + b_{i2}t), \quad t = 2, \ldots, T$$

$$\mu_{i1} = \exp(X_{i1}\tilde{\beta} + c_i),$$

$$(b_{i1}, b_{i2}, c_i) \sim N_3(0, D),$$

where $D^{-1} \sim W(I, 3)$, and fixed effects are assigned N(0,10) priors. Estimation using jagsUI shows neither the main treatment effect or the treatment by time effect to be significant, while the 95% interval for the coefficient ϕ on lagged seizure counts is (−0.009,−0.002). The correlation between b_{i1} and c_i is 0.77. This model leaves excess dispersion: the mean scaled deviance of 440 (Fit[1] in the code) exceeds the number, $5 \times 59 = 295$, of observations, an issue returned to in Example 10.10. Predictive discrepancy shows in a posterior predictive check involving the deviance, with zero probability that the deviance involving replicate data exceeds the deviance for the actual data. On the other hand, mixed predictive checks (Marshall and Spiegelhalter, 2007), denoted exc.mx[i,t] in the code, do not show an excess of tail value probabilities: 10 under 0.05 and 10 over 0.95.

A second analysis replicates model 1 except in taking (b_{i1}, b_{i2}, c_i) as multivariate skew student t, to account for possibly heavy tailed or skew random effects. Thus

$$\log(\mu_{it}) = X_{it}\beta + \phi y_{i,t-1} + b_{i1} + \delta_2 W_{2i} + (b_{i2} + \delta_3 W_{3i})t, \quad t = 2, \ldots, T$$

$$\log(\mu_{i1}) = X_{i1}\tilde{\beta} + c_i + \delta_1 W_{1i},$$

$$(b_{i1}, b_{i2}, c_i) \sim N_3(0, D/\xi_i),$$

$$\xi_i \sim Ga\left(\frac{\nu}{2}, \frac{\nu}{2}\right),$$

where the W_{ji} are independently half normal $N^{+}(0,1)$, and the skew parameters have $\delta_k \sim N(0,10)$ priors. The degrees of freedom ν has a set value, $\nu = 4$, providing a robust setting (Gelman et al., 2014), as estimation of ν may be sensitive to priors adopted. The skew parameters have 95% credible intervals (−0.04,0.82), (−1.01,0.28), and (−0.16, 0.13). The lowest scale factors (xi[i] in the code) are for subjects 49, 18, 15, and 8, namely $\xi_{49} = 0.28$, $\xi_{18} = 0.37$, $\xi_{15} = 0.44$, and $\xi_8 = 0.45$ (cf. Fotouhi, 2007). This extension reduces the LOO-IC slightly, from 1,764 to 1,762.

10.6 Robust Longitudinal Models: Heteroscedasticity, Generalised Error Densities, and Discrete Mixtures

Preceding sections consider the normal linear mixed model for continuous longitudinal outcomes $y_i = (y_{i1},\ldots,y_{iT_i})'$ assuming normal errors at both levels, and constant variances (or dispersion matrices) across subjects and observations. Thus, assuming Z_{it} is a subset of X_{it}, one has

$$y_{it} = X_{it}\beta + Z_{it}b_i + u_{it},$$
$$(b_{1i}, b_{2i},\ldots,b_{Qi}) \sim N_Q(0,D),$$

and with residuals

$$(u_{i1},\ldots,u_{iT_i}) \sim N(0,\sigma^2 I_{T_i}).$$

The general linear mixed model for y possibly being a discrete response may not include observation level residuals, and for overlapping X_{it} and Z_{it}, then takes the form

$$g[E(y_{it} \mid b_i)] = X_{it}\beta + Z_{it}b_i,$$

with $(b_{1i}, b_{2i},\ldots,b_{Qi}) \sim N_Q(0,D)$, where, again, normality of errors and constant dispersion D are default assumptions.

Violation of standard assumptions regarding the forms of error density, or of homoscedasticity, are likely to affect inferences. Among principles that may provide a robust approach to departures from such standard assumptions is that of embedding the model in a more general framework (Zhang et al., 2014; Ma et al., 2004; Rice, 2005), with conventional assumptions (e.g. normality and homoscedasticity of errors) as special cases of a broader model.

Following Chapter 8, assumptions of homoscedasticity at level 1 (repeated observations within subjects) or at level 2 (heterogeneity between subjects) may be modified to allow more general variance functions varying over subjects, times, or both, including dependence of the variance on subject or observation level attributes. For example, heteroscedasticity may exist in the permanent random effects component of longitudinal models, which may be modelled by variance regression in a positive function. For varying intercepts b_i as in (10.4), one might relate subject specific variances D_i to predictor values averaged over time, $\overline{X}_i$, as in

$$D_i = \alpha^2(1+\varphi\overline{X}_i)^2,$$

where terms in the scalar or vector φ are positive. Heteroscedasticity may be considered at observation level, so that for $y_{it} = X_{it}\beta + Z_{it}b_i + u_{it}$ one might take

$$u_{it} \sim N(0,\sigma^2_{it}),$$
$$\sigma^2_{it} = \alpha^2(1+\varphi X_{it})^2.$$

Wakefield et al. (1994) in a nonlinear pharmacokinetic longitudinal analysis with positive structural effects η_{it} specify a Bayesian heteroscedastic model at observation level.

Thus $y_{it} \sim N(\eta_{it}, \eta_{it}^{\omega} / \tau)$, where ω is an unknown power and τ is an overall precision parameter, and $\omega = 0$ corresponds to homoscedasticity.

Similarly, more general error densities allowing for skewness, heavy tails, or other non-normal features may be adopted, with the standard assumptions embedded within them. Alternatives to assuming multivariate normal subject effects may include heavy tailed Student t heterogeneity (Zhang et al., 2014; Chib, 2008; Lin and Lee, 2006), skew normal and skew-t densities, and skew-elliptical densities (Ma et al., 2004). Thus, the normal linear mixed model can be embedded within a wider class of scale mixture normal densities, with the subject or observation level scale parameters measuring outlier status (Wakefield et al., 1994; Chib, 2008). Thus, the model of (10.2), with normal cluster effects b_i and normal residuals u_{it}, is a special case of a scale mixture model with

$$y_{it} = X_{it}\beta + Z_{it}b_i + u_{it},$$

$$u_{it} \sim N\left(0, \frac{1}{\lambda_i}\Sigma_i\right),$$

$$b_i \sim N\left(B, \frac{1}{\xi_i}D\right),$$

$$\lambda_i \sim G_\lambda,$$

$$\xi_i \sim G_\xi.$$

A widely applied option takes the densities $\{G_\lambda, G_\xi\}$ to be gamma with equal scale and shape, $\nu_\lambda/2$ and $\nu_\xi/2$ respectively, leading to multivariate t densities with $\{\nu_\lambda, \nu_\xi\}$ degrees of freedom. This provides resistance to atypical data at both observation and cluster levels.

For possibly skew residual or subject effects, skew-normal or skew-t densities may be adopted. Ghosh et al. (2007) consider bivariate skew-normal errors at both subject and observation level in a linear longitudinal model for metric responses, while Jara et al. (2008) allow both subject random effects and observation level errors to follow a multivariate skew-t distribution. Thus, for a linear mixed model for y of dimension T_i

$$y_i = X_i\beta + Z_ib_i + u_i, \tag{10.7}$$

suppose y_i follows the multivariate skew-t density (Sahu et al., 2003). Then

$$y_i \mid \beta, b_i, \sigma^2, R_i, \Delta_i \sim ST_\nu(X_i\beta + Z_ib_i, \sigma^2 R_i, \Delta_i),$$

where ν is the degrees of freedom, R_i is a $T_i \times T_i$ matrix, and $\Delta_i = \text{diag}(\delta_{1i}, \ldots, \delta_{T_i,i})$ contains skewness parameters relevant to the observation level residuals that may in principle be specific to individuals and times. The density of the entire observation set $y = (y_1, \ldots, y_n)$, conditional on collections of b_i, R_i and Δ_i, is (Jara et al., 2008)

$$p(y \mid \beta, b, \sigma^2, R, \Delta) \propto \prod_{i=1}^{n} 2^{T_i} t_{T_i,\nu}(y_i \mid X_i\beta + Z_ib_i, \sigma^2 R_i + \Delta_i^2)$$

$$\times \int_0^\infty t_{T_i,\nu}(w_i \mid \mu_w, \Sigma_w) dw_i,$$

where $t_{m,\nu}(x \mid \mu_x, \Sigma_x)$ denotes a multivariate t density of dimension m. When R_i reduces to an identity matrix I_{T_i}, and the subject-time skewness parameters δ_{it} to a global parameter δ, namely $\delta_{it} = \delta$, the conditional expectation and variances for each subject are

$$E(y_i \mid \beta, b_i, \sigma^2, \delta) = X_i\beta + Z_ib_i + (\nu/\pi)^{0.5}\frac{\Gamma(\nu-1)/2]}{\Gamma(\nu/2)}\delta 1_{T_i},$$

$$\text{Var}(y_i \mid \beta, b_i, \sigma^2, \delta) = \frac{\nu}{\nu-2}(\sigma^2 + \delta^2)I_{T_i} + (\nu/\pi)\left[\frac{\Gamma(\nu-1)/2]}{\Gamma(\nu/2)}\right]^2 \delta^2 I_{T_i}.$$

Under the reductions $R_i = I_{T_i}$, $\delta_{it} = \delta$, the conditional density may be described by a mixture of normal distributions by conditioning on positive variables $w_i = (w_{1i}, \ldots, w_{T_i i})$ obtained by truncated sampling from a multivariate normal with identity covariance matrix of dimension T_i and subject-specific scalings $\lambda_i \sim Ga(\nu/2, \nu/2)$, so that

$$y_i \mid \beta, b_i, \sigma^2, w_i, \lambda_i, \delta) \sim N_{T_i}\left(X_i\beta + Z_ib_i + \delta w_i, \frac{\sigma^2}{\lambda_i}I\right).$$

$$w_i \sim N_{T_i}\left(0, \frac{1}{\lambda_i}I\right) \quad I(0,).$$

In the (usual) case when $X_i\beta + Z_ib_i$ contains an intercept, then for identifiability, the elements in the vector w_i may be centred (subsequent to truncated sampling) (Jara et al., 2008). Thus, at each iteration, the average of the w_{it} can be obtained, and then the centred variables $W_{it} = w_{it} - \bar{w}_i$, so that

$$y_{it} \sim N\left(X_{it}\beta + Z_{it}b_i + \delta W_{it}, \frac{\sigma^2}{\lambda_i}\right).$$

Additionally, in the model (10.7), the permanent random effects b_i may also be taken as skew multivariate t. Assuming the Z predictors are a subset of the X predictors, one then has

$$b_i \mid D, \Gamma_i \sim ST_{\nu_b}(0, D, \Gamma_i),$$

where D is $Q \times Q$, ν_b is the degrees of freedom, and $\Gamma_i = \text{diag}(\gamma_{1i}, \ldots, \gamma_{Qi})$ contains skewness parameters relevant to the permanent effects. Assuming common skew parameters $\Gamma_i = \Gamma = \text{diag}(\gamma_1, \ldots, \gamma_Q)$, and conditional on a Q vector of positive variables, $h_i = (h_{1i}, \ldots, h_{Qi})$, with

$$h_i \sim N_Q\left(0, \frac{1}{\xi_i}I\right) \quad I(h_i > 0), \tag{10.8}$$

$$\xi_i \sim Ga\left(\frac{\nu_b}{2}, \frac{\nu_b}{2}\right),$$

the random effects are mixtures of normals, namely

$$b_i \sim N_Q\left(\Gamma h_i, \frac{1}{\xi_i}D\right).$$

For improved identification, the h_i can be centred around their means (at each MCMC iteration), namely $H_{qi} = h_{qi} - \overline{h_{q.}}$, so that

$$b_i \sim N_Q\left(\Gamma H_i, \frac{1}{\xi_i} D\right).$$

10.6.1 Robust Longitudinal Data Models: Discrete Mixture Models

Another way of reducing the impact of arbitrarily selecting a particular parametric form for random variation in b_i and/or u_{it} is by using discrete mixtures of random effects priors (e.g. Weiss et al., 1999). A discrete mixture prior may be more flexible in dealing with unusual cases, skewness, and multiple modes. The possibly conflicting criteria required in the case of a prior on b_i are considered by Muller and Rosner (1997): namely, that the prior should be flexible to allow for heterogeneity in the population, though on the other hand, unusual cases should not have an undue predictive influence.

An often suitable approach would involve two group normal mixture priors with the groups typically conceived of as a main group and outlier group (Weiss et al., 1999, p.1563). Such schemes may apply both for random intercepts b_i

$$b_i \sim \pi_b N_Q(0, D) + (1 - \pi_b) N_Q(0, \varphi_b^2 D),$$

and for iid observation level u_{it}

$$u_{it} \sim \pi_u N(0, \sigma^2) + (1 - \pi_u) N(0, \varphi_u^2 \sigma^2),$$

where the factors $\{\varphi_b > 1, \varphi_u > 1\}$ are used for variance inflation for the outlier group. The prior probabilities of being in the main population are set high (e.g. $\pi_b = \pi_u = 0.95$), and variance inflation factors are typically large e.g. $\varphi_b = \varphi_u = 5$ or 10. Provided one or other of the parameter sets $\{\pi_b, \pi_u\}$ or $\{\varphi_b, \varphi_u\}$ is assumed known (i.e. is assigned preset values), the other set may be taken as unknowns.

Another option is "switching" or shift priors whereby one group has zero effects, but a minority group has non-zero effects. These may be used for iid errors introduced to reflect overdispersion in count or binomial data. For example, for $y_{it} \sim Po(\mu_{it})$, one may have

$$\log(\mu_{it}) = X_{it}\beta + Z_{it} b_i + \sigma k_{it} u_{it}$$

where σ is a scale factor, $k_{it} \sim \text{Bern}(\pi_u)$, $u_{it} \sim N(0,1)$, such that observation level effects are zero when $k_{it} = 0$. One may preset π_u low, say $\pi_u = 0.05$. For a longitudinal series with level c_t subject to possible shifts, and X_{it} not containing an intercept, one may similarly propose that

$$y_{it} = c_t + X_{it}\beta + Z_{it} b_i + \sigma u_{it},$$

$$c_t = c_{t-1} + k_t \sigma w_t,$$

where $u_{it} \sim N(0,1)$, $w_t \sim N(0,1)$, and $k_t \sim \text{Bern}(\pi_c)$ with π_c low.

A different emphasis, as in latent growth curve models (Depaoli and Boyajian, 2014; Galatzer-Levy, 2015; Galatzer-Levy and Bonanno, 2012), is when there is a substantive rationale for assuming subject level effects b_{qi} follow a discrete prior at subject level. The hyperparameters governing the subject effects $\{b_{1i}, b_{2i}, \ldots, b_{Qi}\}$ then become specific for the latent category. Thus, in a growth curve model for modelling changes in aggression ratings, Muthen et al. (2002) assume that a small number of latent trajectories characterise growth in aggression. For subject i, let the latent category be denoted $k_i \in (1, \ldots, K)$. Then conditional on $k_i = k$, (10.5) would become

$$b_{1i} = B_{1k} + e_{1i},$$

$$b_{2i} = B_{2k} + \delta_{2k} Tr_i + e_{2i},$$

$$b_{3i} = B_{3k} + \delta_{3k} Tr_i + e_{3i},$$

where $(e_{1i}, e_{2i}, e_{3i}) \sim N_3(0, D_k)$. Observation level dispersion parameters may also differ according to latent group.

Flexible discrete mixture models are also obtained under Dirichlet process and related semiparametric priors (Dunson, 2009), as considered for repeated binary data by Quintana et al. (2008), for longitudinal count data by Kleinman and Ibrahim (1998), and for multiple membership longitudinal models by Savitsky and Paddock (2014) and Paddock and Savitsky (2013). Averaging over different number of mixture components K is possible under discrete parametric mixture models using the RJMCMC (reversible jump MCMC) algorithm – see Ho and Hu (2008) for an application to the linear mixed model. In the non-parametric mixture approach, the number of clusters is an outcome of other parameters such as the Dirichlet process mass parameter κ. Under the truncated Dirichlet process (Ohlssen et al., 2007), one may set a maximum K_m possible clusters, with the realised number at each iteration being $K \leq K_m$. The posterior density of K will indicate whether the assumed maximum K_m is sufficient.

Hirano (2000, 2002) discusses non-parametric alternatives regarding white noise observation errors u_{it} in longitudinal data, while Kleinman and Ibrahim (1998) and Muller and Rosner (1997) consider mixed Dirichlet process (MDP) modelling of Q dimensional unit level effects b_i. Under the MDP option, one has b_i following a density G which is itself unknown, centred on a specified base density G_0 with precision κ. For example, with a base density $G_0 = N_Q(B, D)$, one has

$$g[E(y_{it} \mid b_i)] = \eta_{it} = X_{it}\beta + Z_{it}b_i + u_{it},$$

$$u_{it} \sim N(0, \sigma^2),$$

$$b_i \sim G,$$

$$G \sim DP(\kappa G_0),$$

$$G_0 = N_q(B, D),$$

where priors on β, B, D, σ^2 are typically as considered above. This is the conjugate MDP prior for the normal linear mixed model which tends to the conventional hierarchical prior as $\kappa \rightarrow \infty$.

The model considered by Hirano (2002) is also conjugate, and based on a dynamic model

$$y_{it} = b_i + \rho y_{i,t-1} + u_{it},$$

where the b_i are zero mean effects that are modelled parametrically, and $u_{it} = y_{it} - b_i + \rho y_{i,t-1}$ may have non-zero means. One has for $\theta_{it} = \{\mu_{it}, \sigma_{it}^2\}$

$$u_{it} \sim N(\mu_{it}, \sigma_{it}^2),$$
$$\theta_{it} \sim G,$$
$$G \sim DP(\kappa G_0),$$

where G_0 specifies

$$G_0(\mu, \sigma^2): \frac{1}{\sigma^2} \sim \frac{\chi^2(s)}{sL}; \mu \sim N(m, b\sigma^2).$$

where s, L, m and b are preset. As discussed in Chapter 3, κ may be preset or taken as an unknown. Thus, Kleinman and Ibrahim (1998, p.2592) consider defaults such as κ=1.5 and κ=100, while Hirano (2002) takes $\kappa \sim Ga(2, 0.5)$.

Example 10.8 A Pharmacokinetic Application

To exemplify heteroscedastic longitudinal analysis, this example considers pharmacokinetic longitudinal data. The dataset consists of plasma concentrations y_{it} of the drug Cadralazine in n=10 cardiac failure patients at various times $t = 1, \ldots, T_i$ (in hours) following administration of a single dosage of G=30 mg. A one-compartment nonlinear model for these data (Bauer et al., 2007; Bonate, 2008) with mean concentration η_{it} at time t can be expressed as

$$\eta_{it} = (G/\alpha_i)\exp(-\beta_i t/\alpha_i),$$

where $\alpha_i > 0$ and $\beta_i > 0$ are respectively the volume of distribution and clearance parameters for each subject. A hierarchical model is proposed with the second stage consisting of a multivariate normal or multivariate Student t for the transformed subject effects $(b_{1i}, b_{2i}) = \{\log(\alpha_i), \log(\beta_i)\}$.

For the first stage density, one option is a log-normal since y is positive, or a truncated normal, with y_{it} constrained to be positive. Under the latter, a heteroscedastic power model, with a single precision parameter τ, leads to a variance η_{it}^{ω}/τ, and the first stage model is

$$y_{it} \sim N(\eta_{it}, \eta_{it}^{\omega}/\tau), \quad I(0,).$$

Note that zero y values are replaced by 0.001 to avoid conflict with this density assumption. Another option for the first stage model involves a normal scale mixture, namely

$$y_{it} \sim N(\eta_{it}, \eta_{it}^{\omega}/[\lambda_i \tau]) \quad I(0,),$$
$$\lambda_i \sim Ga(0.5\nu, 0.5\nu).$$

Here these options are compared under the priors $\tau \sim Ga(1, 0.001)$ and $\nu \sim Ga(2, 0.1)$. A uniform $U(0,5)$ prior is assumed for ω, as in Wakefield et al. (1994).

At the second stage, a bivariate normal for (b_{1i}, b_{2i}) is assumed with

$$(b_{1i}, b_{2i}) \sim N_2([B_1, B_2], D),$$
$$B \sim N(B_0, C),$$
$$D^{-1} \sim W([\rho R]^{-1}, \rho),$$

with ρ, R, B, and C as in Wakefield et al. (1994).

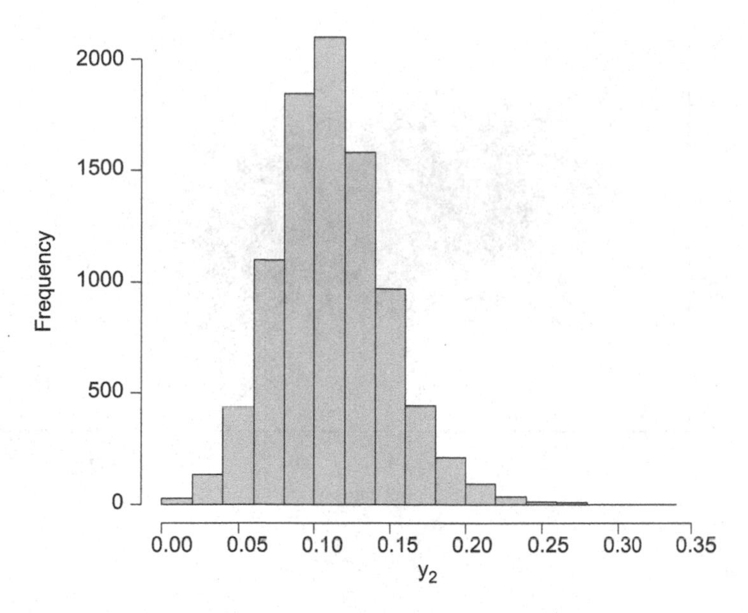

FIGURE 10.1
Predictive distribution of y_2.

Inferences are based on runs of 10,000 iterations using the rube package. The scale mixture model with variances $\eta_{it}^{\omega}/[\lambda_i\tau]$ has a lower DIC, namely −183.5, as compared to −180 for the model with variances η_{it}^{ω}/τ. The posterior mean ν under the scale mixture is 16, with the lowest scale parameter for subject 2, $\lambda_2 = 0.66$. The power ω is estimated at 0.86 under the better performing model, whereas a log-normal model would imply $\omega \simeq 2$.

Out-of-sample predictions of concentrations are made for a duration of 32 hours. For subject 2, whose plasma concentrations remain relatively high compared to other subjects, the mean prediction is 0.11. Figure 10.1 shows the predictive distribution. Inferences on the population distribution of concentration parameters are important, for example, the half-life (period of time required for a drug concentration to be reduced by one-half), which for patient i is $\alpha_i \log(2)/\beta_i$. The median half-life and clearance are obtained, and Figure 10.2 shows the corresponding bivariate posterior plot.

Example 10.9 Skewed Cholesterol Data

This example relates to longitudinal data on cholesterol levels collected during the Framingham heart study for $n = 200$ randomly selected subjects, as considered by Zhang and Davidian (2001) and Ma et al. (2004). Relevant subject attributes are sex (1 = M, 0 = F), and age at baseline. Several studies have re-considered the linear mixed model used by those authors, namely

$$\begin{aligned} y_{it} &= X_{it}\beta + Z_{it}b_i + u_{it}, \\ &= \beta_1 + \beta_2 \text{Sex}_i + \beta_3 \text{Age}_i + \beta_4 a_{it} + b_{1i} + b_{2i}a_{it} + u_{it}, \end{aligned} \tag{10.9}$$

where y_{it} is cholesterol level divided by 100, and a_{it} = (years − 5)/10, derived using years from baseline. Total periods T_i differ between subjects, varying from 1 to 6.

Here two models are considered to reflect positive skew apparent from plots of the outcome. One may, for example, consider multivariate skew normal or multivariate

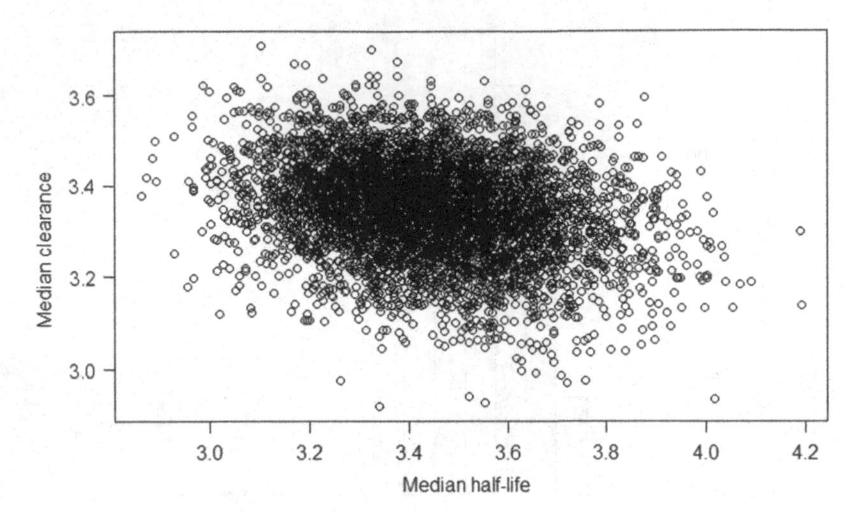

FIGURE 10.2
Bivariate posterior.

skew in the random effects (b_{1i}, b_{2i}). For skew bivariate normal random effects, one has (as model 1)

$$b_i \mid D, \Gamma \sim SN(0, D, \Gamma),$$

where D is 2×2, and $\Gamma = \text{diag}(\gamma_1, \gamma_2)$. Equivalently, conditional on the positive standard normal effects

$$h_i \sim N_Q(0, I) \quad I(0,),$$

(with $Q = 2$), the random intercepts and slopes in (10.8) are obtained as

$$b_i \sim N_Q(\Gamma h_i, D).$$

An alternative perspective (model 2) is provided by allowing changing skew through time. This involves a T_i vector of period-specific skewness parameters $\delta_i = (\delta_1, \ldots, \delta_{T_i})$, that is $\delta_{it} = \delta_t$, in a multivariate skew normal scheme for the observation level errors. Hence

$$u_{it} \mid \sigma^2, w_{it}, \delta_t \sim N_{T_i}(\delta_t w_{it}, \sigma^2),$$

$$w_{it} \sim N(0,1) \quad I(0,).$$

While centred positive variables h_i and w_{it} may be preferred for identification, this slows MCMC analysis considerably and uncentred effects are used for illustration.

Estimates for model 1 using jagsUI show significant skewness in subject intercepts b_{1i}, but not in the time slopes, with the respective γ parameters having 95% intervals (0.38,0.60) and (−0.26,0.23). The LOO-IC is obtained as −20.5 under both models. Under model 2, the δ_t parameters all have credible intervals straddling zero, but earlier ones are biased to positive values.

Example 10.10 Robust Random Effects for Epilepsy Data

This example considers forms of robust modelling for the seizure data discussed in Example 10.7. For example, Yau and Kuk (2002) consider sensitivity of fixed effects parameter estimates to specification of random effects at both subject and observation level. Following their analysis, we condition on the initial observation, treated as a baseline measure of severity; the five predictors (X_1 to X_5) are then: log of baseline seizure, treatment, treatment interaction with baseline, log of patient age, and a binary variable equal to 1 for the final visit. As Example 10.7 shows, including only subject level random effects alone does not eliminate excess dispersion.

Instead consider random variation at both subject and observation levels, with hierarchically centred random effect priors (cf. Roberts and Sahu, 2001). So $y_{it} \sim Po(\mu_{it})$, with

$$\log(\mu_{it}) = u_{it} + X_{it}\beta,$$

$$u_{it} \sim N(b_i, \sigma_u^2),$$

$$b_i \sim N(a, D),$$

where α is the regression intercept, and a U(0,100) prior is assumed for $D^{0.5}$. Additionally, a uniform shrinkage prior (Natarajan and Kass, 2000) is adopted in relation to the other variance component $\text{var}(u_{it}) = \sigma_u^2$, with

$$\phi = \frac{D}{D + \sigma_u^2} \sim U(0,1).$$

Estimation using jagsUI gives posterior means for σ_u^2 and D of 0.13 and 0.27. The posterior mean of the scaled deviance is now 271, and a posterior predictive check is satisfactory, providing a probability of 0.26 that the deviance involving replicate data exceeds the deviance for the actual data. The LOO-IC is 1200.

One may also model subject intercept heterogeneity using a discrete mixture of intercepts, so avoiding parametric assumptions about such heterogeneity. Thus

$$\log(\mu_{it}) = u_{it} + x_{it}\beta,$$

$$u_{it} \sim N(a_{k_i}, \sigma_u^2),$$

where the latent categorical allocation $k_i \in (1, \ldots, K)$ is multinomial with probabilities $(\pi_1, \ldots, \pi_K)$ following a diffuse Dirichlet prior. The intercepts $a_1, \ldots, a_K$ are subject to an order constraint. For illustrative purposes, this approach is applied with $K=2$, providing a LOO-IC of 1224. The intercepts are estimated with posterior means (sd) 1.57 (0.15) and 2.48 (0.19).

The third model adopts a selection mechanism for observation level effects u_{it}, adapting to a scenario where many patients exhibit a stable differential over the visits (modelled by a level 2 effect b_i), with only a subset of patients exhibiting erratic trajectories that require a random effect for each visit. Thus, binary indicators $\delta_i \sim \text{Bern}(\pi_\delta)$ are introduced for each subject in a model where

$$\log(\mu_{it}) = a + x_{it}\beta + b_i + \delta_i u_{it},$$

with a U(0,100) prior for $D^{0.5}$, and $\phi = D/(D + \sigma_u^2) \sim U(0,1)$. π_δ may be an unknown or preset. Here a value $\pi_\delta = 0.10$ is adopted, so that the posterior values $Pr(\delta_i = 1 \mid y)$ can provide clear contrasts to the prior values $Pr(\delta_i = 1) = \pi_\delta$. From a two-chain run of

10,000 iterations, it emerges that seven patients have sufficiently high posterior odds $Pr(\delta_i = 1 | y)/(1 - Pr(\delta_i = 1 | y))$ to provide marginal Bayes factors exceeding 3, namely 10, 11, 16, 25, 39, 53, and 56. This model gives a LOO-IC of 1,215.

A fourth analysis reverts to random subject effects only, but allows for possible non-normality via a mixed Dirichlet process. The random effects are bivariate, with non-zero means, one for the intercept B_1 (α in above models) and one for a linear slope B_2 on visit. Thus

$$\log(\mu_{it}) = X_i\beta + Z_{it}b_i,$$

where $X_i = (\text{Base}, Tr, \text{Base} * Tr, \text{Age})$, with predictor variables defined as in Kleinman and Ibrahim (1998), and with $Z_{it} = (1, \text{Visit})$, where the visit times are centred weeks/10. The patient random effects have prior

$$(b_{1i}, b_{2i}) \sim G,$$

$$G \sim DP(\kappa G_0),$$

$$G_0 = N_Q(B, D),$$

$$\kappa \sim Ga(2, 4),$$

$$(B_1, B_2) \sim N_2(0, 1000I),$$

$$D^{-1} \sim W(R, \rho),$$

$$R = \text{diag}(20),$$

$$\rho = 10,$$

so that $E(D^{-1}) = \text{diag}(0.5)$, as in Kleinman and Ibrahim (1998). The maximum number of possible clusters is set at $K_m = 20$.

Conditional on the particular choice made for the prior on κ, one obtains a mean scaled deviance of 417, still leaving excess variability. The posterior density for the realised number of clusters has 0.025 and 0.975 percentiles at 6 and 13, with mean 8.8, while κ has posterior mean 1.32. Histograms of the mean $\{b_{1i}, b_{2i}\}$ with superimposed normal curves show excess kurtosis (i.e. peaked densities) (Figures 10.3 and 10.4).

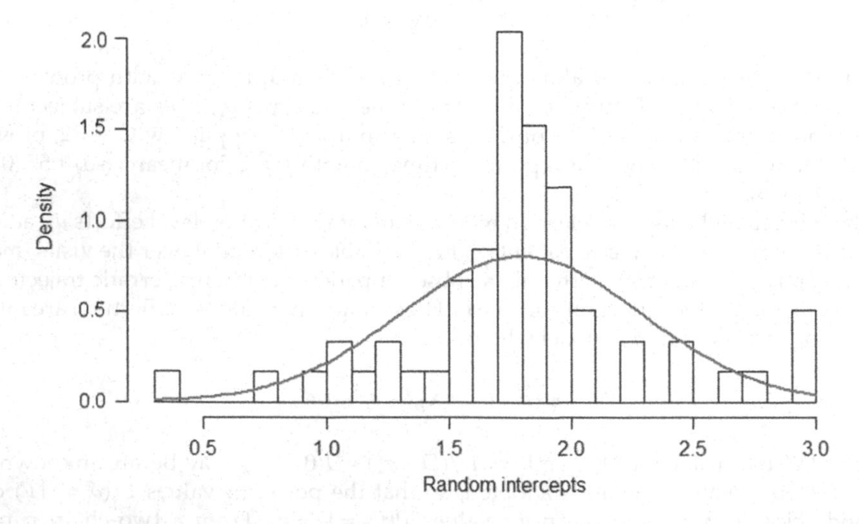

FIGURE 10.3
Histogram with normal curve, varying intercepts.

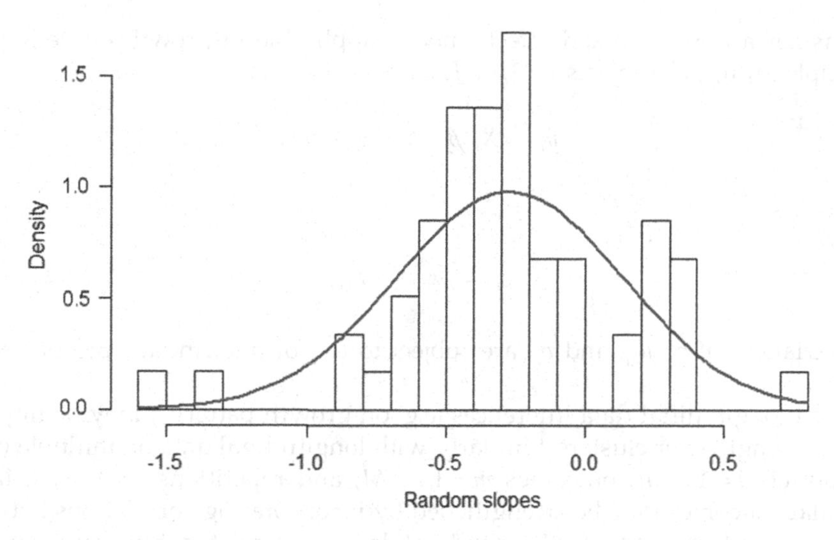

FIGURE 10.4
Histogram with normal curve, varying slopes.

10.7 Multilevel, Multivariate, and Multiple Time Scale Longitudinal Data

Applications involving longitudinal data often involve contextual nesting of subjects (Steele, 2008; Lockwood et al., 2003), multivariate responses (Verbeke et al, 2014; Curran et al., 2010), or multiple time scales. Consider data y_{ijt} for repetitions $t = 1,\ldots,T_{ij}$ for subjects $i = 1,\ldots,n_j$ nested within clusters $j = 1,\ldots,J$. The general linear mixed model assumes that conditional on predictors and random effects, the data are distributed independently according to the exponential family,

$$p(y_{ijt} \mid \theta_{ijt},\phi) = \exp\left(\frac{y_{ijt}\theta_{ijt} - a(\theta_{ijt})}{\phi} + C(y_{ijt},\phi)\right),$$

with conditional means $E(y_{ijt}) = \mu_{ijt} = a'(\theta_{ijt})$, and link $g(\mu_{ijt}) = \eta_{ijt}$ to regression terms η_{ijt}. The structural model may specify permanent random effects $\{d_j,b_{ij}\}$ for both clusters and subjects within clusters. Fixed effect regression parameters may now be cluster specific, namely

$$g[E(y_{ijt} \mid \beta_j,c_i,b_{ij})] = X_{ijt}\beta_j + W_{ijt}d_j + Z_{ijt}b_{ij}.$$

Taking β_j as cluster fixed effects is appropriate when the categorisation $j = 1,\ldots,J$ refers to a small number of treatment groups or demographic categories, as in the data from Oman et al. (1999) where four groups are formed by crossing treatment by gender – see Example 10.11.

Where predictor effects vary randomly over time, one may now include time and cluster-specific effects c_{jt}, so that

$$g[E(y_{ijt} \mid \beta_j,c_i,b_{ij})] = X_{ijt}\beta_j + W_{ijt}d_j + H_{ijt}c_{jt} + Z_{ijt}b_{ij}.$$

Autocorrelated errors may also be required to model temporal dependencies, so that the unexplained variance may be due to a number of sources. For example, Lee and Hwang

(2000) consider a normal mixed effects model, applicable in growth curve applications with multiple groups of subjects $j = 1,\ldots,J$, with

$$y_{ijt} = X_{ijt}\beta_j + b_{ij} + \varepsilon_{ijt} + u_{ijt},$$

where

$$\varepsilon_{ijt} = \rho\varepsilon_{ij,t-1} + v_{ijt},$$

and the variances of b_{ij}, u_{ijt}, and v_{ijt} are subject to uniform shrinkage priors (see Section 10.2.3).

For nested longitudinal data inferences (e.g. on growth patterns) may be improved by borrowing strength over clusters. Similarly, with longitudinal data on multiple outcomes, y_{mit} for subjects $i = 1,\ldots,n$, outcomes $m = 1,\ldots,M$, and repetitions $t = 1,\ldots,T$, inferences on particular outcomes may be strengthened by incorporating correlations between outcomes. An example might be for longitudinal data on correlated, but relatively rare, spatially configured health events, such as cancer types. Multiple outcome longitudinal data are common in clinical and educational applications, and the effectiveness of interventions may be judged in terms of multiple (usually) correlated outcomes rather than by a single criterion (Dunson, 2007). In environmental applications, multiple outcomes with related aetiology are likely to be correlated (e.g. Liu and Hedeker, 2006; Jorgensen et al., 1999).

With metric or discrete data y_{mit} for multiple outcomes, the general linear mixed model with time homogenous, time-varying, and subject varying predictor effects becomes

$$g[E(y_{mit} \mid \beta_m, b_{mi}, c_{mt})] = \eta_{mit} = X_{it}\beta_m + Z_{it}b_{mi} + H_{it}c_{mt},$$

where X_{it}, Z_{it}, and H_{it} are of length P, Q, and R. For example, consider multivariate repeated binary responses,

$$y_{mit} \sim \text{Bern}(\pi_{mit}),$$

and a prediction model based on outcome-subject and outcome-time effects (Agresti, 1997), namely

$$\text{logit}(\pi_{mit}) = \beta_m + b_{mi} + c_{mt}.$$

Given multivariate random outcome-subject effects b_{mi}, and fixed effects c_{mt} subject to an identifying corner constraint such as $c_{m1}=0$ or $c_{mT}=0$, the y_{mit} are assumed conditionally independent.

The corresponding normal linear mixed model for multivariate metric responses is

$$y_{mit} \mid \beta_m, b_{mi}, c_{mt}, \sigma_m^2 = X_{it}\beta_m + Z_{it}b_{mi} + H_{it}c_{mt} + u_{mit},$$

where the residuals u_{mit} are typically iid normal with variances σ_m^2 specific to outcome m. The M sets of permanent effect priors b_{mi}, each of dimension Q, may be correlated between predictors q within outcomes m, or between outcomes m within predictors q, or most generally over both predictors q and outcomes m. The same applies to the outcome-time effects c_{mt}, which may be random, and incorporate short range temporal dependence. For example, time-varying intercepts $c_t = (c_{1t}, \ldots, c_{Mt})$ in the case $R=1$ (and $H_{it}=1$) could follow autoregressive or random walk priors correlated over outcomes, as in

$$(c_{1t}, \ldots c_{Mt}) \sim N_M(c_{t-1}, \Sigma_c).$$

Suppose T_{mi} responses are observed on outcome m for subject i, with $S_i = \sum_m T_{mi}$. In vector form, the multivariate normal longitudinal model is then

$$Y_i = X_i\beta + Z_i b_i + H_i c_t + u_i,$$

where Y_i is of length S_i (Beckett et al., 2004).

Conjugate structures (e.g. Poisson-gamma, beta-binomial) may also be used instead of the GLMM approach for discrete multivariate longitudinal outcomes. For example, over-dispersed count data y_{mit} may be assumed Poisson with

$$y_{mit} \sim Po(\mu_{mit}\theta_{im}\xi_{mit}),$$

where $\mu_{mit} = \exp(X_{mit}\beta_m)$, and

$$\theta_{im} \sim Ga(a_m, a_m),$$

represent subject-outcome permanent random effects. The ξ_{mit} represent observation level effects that are iid, or autoregressive, as in

$$\xi_{mit} \sim Ga(b_m\xi_{mi,t-1}, b_m)$$

with variance parameters b_m (Jorgensen et al., 1999).

10.7.1 Latent Trait Longitudinal Models

As the number of responses M increases, the full dimensional approach becomes cumbersome, and factor analytic or latent trait approaches may pool information just as effectively and more parsimoniously (e.g. Roy and Lin, 2000; Jorgensen et al., 1999; Dunson, 2003, 2006, 2007). Longitudinal data on multiple outcomes raise the possibility of shared random effects across outcomes, instead of outcome-specific effects. For example, in spatio-temporal health applications, it is common to have correlated count responses y_{mit} such as different types of cancer or psychiatric illness for areas i. Observed risk factors for such outcomes may be limited or incomplete. Common unobserved area-time risks may be summarised in effects r_{it}, with loadings λ_m linking the common factor scores to each outcome. These may be taken as iid (Tzala and Best, 2008), or assumed to be spatially and/or temporally correlated. For identifiability, one may either set var(r_{it}) = 1, in which case the loadings are free parameters, or set one of the loadings to a fixed value, such as $\lambda_1=1$, in which case var(r_{it}) is an unknown. Time and area common effects may be combined with common area effects b_i with loadings γ_m, and common time effects c_t with loadings κ_m, and the same type of identifying rules. Then $y_{mit} \sim Po(\mu_{mit})$, with fixed regression effects that might vary over outcomes, as in

$$\log(\mu_{mit}) = X_{it}\beta_m + \lambda_m r_{it} + \gamma_m b_i + \kappa_m c_t.$$

Additionally, iid effects u_{mit} may be included to represent remaining overdispersion.

In item analysis and psychometric longitudinal applications, a measurement model might involve both constant and time-varying common factors. Thus for N items or tests carried out on S occasions, responses may be determined by the interaction of $M<N$ item specific factors b_{mi} and by $T<S$ time-specific factors r_{it} (Marsh and Grayson, 1994). The impact of these is governed by time and outcome-specific loadings γ_t and λ_m respectively, so that

$$y_{mit} = a_m + \gamma_t b_{mi} + \lambda_m r_{it} + u_{mit}.$$

Structural equation models for longitudinal data typically involve both response indicators y_{mit} of dimension P_y which measure latent outcomes η_{qit} of dimension $Q_y<P_y$, and exogenous predictors x_{kit} of dimension P_x which measure latent causal influences ξ_{qit} of dimension $Q_x<P_x$ (Dunson, 2007). For example, for $Q_y=Q_x=1$, ξ_{it} might be a time-varying stress severity scale related to short-term stressors $\{x_{kit}, k=1,\ldots P_x\}$, and η_{it} might be a time-varying latent depression scale related to mood scale measures $\{y_{mit}, m=1,\ldots P_y\}$. Then the measurement model is

$$y_{mit} = a_{1m} + \lambda_{1m}\eta_{it} + u_{1mit}, \quad m = 1,\ldots,P_y$$

$$x_{kit} = a_{2k} + \lambda_{2m}\xi_{it} + u_{2kit}, \quad k = 1,\ldots,P_x$$

while the structural model might include a linear effect, possibly time-varying, of ξ_{it} on η_{it}.

A simple common factor model may be applied when there are alternative measuring scales, typically a gold standard measure, and one or more measures of the same quantity, but less expensive to obtain. Consider a situation where bivariate data $\{y_{1ijt}, y_{2ijt}\}$ are obtained for subjects i within clusters j, where y_{1ijt} denotes repetitions on the standard measure, and y_{2ijt} denotes repetitions on the proxy measure. The goal is to assess the reliability of the proxy measure. One may postulate a shared permanent effect b_{ij} between the two outcomes, as well as a unique permanent effect c_{ij} for the proxy measure. In the absence of intercepts for the y_1 model, one has

$$y_{1ijt} = b_{ij} + u_{1ijt},$$

$$y_{2ijt} = a_j + \lambda_j b_{ij} + c_{ij} + u_{2ijt},$$

where the b_{ij} have non-zero cluster means B_j, but the c_{ij} are zero mean effects, namely:

$$b_{ij} \sim N(B_j, D_{1j}),$$

$$c_{ij} \sim N(0, D_{2j}).$$

The residuals are distributed as $u_{mijt} \sim N(0, 1/\tau_{mj})$. The hypothesis that $\{a_j = 0, \lambda_j = 1\}$ corresponds to y_1 and y_2 being identically calibrated in group j (Oman et al., 1999, p.43), that is, they both measure the same quantity on the same scale (see Example 10.11).

10.7.2 Multiple Scale Longitudinal Data

Aggregate health and demographic event data are often available as totals y_{ixt} for multiple time scales, for example, by age group $x=1,\ldots,X$, as well as by period $t=1,\ldots,T$, and possibly also by area or actuarial risk group $i = 1,\ldots,n$ (Chernyavskiy et al., 2019). A further cohort dimension $c = 1,\ldots,C$ is implicit in biological age-time data via the relation $c = t - x + X$, and there have been extensive developments in Bayesian age-period-cohort (APC) and area APC models (AAPC) models (Lagazio et al., 2003; Schmid and Held, 2004; Bray, 2002; Baker and Bray, 2005), which draw on developments in space-time models involving spatial and temporal autocorrelation (Quick et al., 2018; Rushworth et al., 2014; Donald et al., 2015). For rare event totals y_{ixt} in relation to large populations N_{ixt}, and assuming $y_{ixt} \sim Po(N_{ixt}\mu_{ixt})$, a baseline age-period (AP) model might assume independence of age and period dimensions, with

$$\mu_{ixt} = \exp(\eta_{ix})\exp(\theta_{it}),$$

or equivalently

$$\log(\mu_{ixt}) = \kappa + \eta_{ix} + \theta_{it},$$

where structured (e.g. random walk or autoregressive) priors might be adopted for age-area effects η_{ix} and area-time effects θ_{it}, and the intercept κ is identified according to possible constraints on the random effects.

Thus Clayton and Schifflers (1987) consider data of the form y_{xt} (i.e. without further stratification), with means μ_{xt} where

$$\log(\mu_{xt}) = \eta_x + \theta_t,$$

with both sets of effects assumed to be random, though fixed effects may be used when X or T is small. In the absence of an overall intercept in this model, one or other series (say η_x) sets the level, and identifiability may be gained by centring the remaining series θ_t at zero (possibly repeatedly at each MCMC iteration), or by setting one parameter in the remaining series to a fixed value e.g. $\theta_1 = 0$. If the model includes an overall intercept κ, then centring both sets of effects, namely $\sum_x \eta_x = \sum_t \theta_t = 0$, provides a way of ensuring identifiability. An APC model including a mean and structured age, period and cohort effects is

$$\log(\mu_{xt}) = \kappa + \eta_x + \theta_t + \gamma_c,$$

and identifiability requires either that the three sets of effects be centred, or that edge constraints such as $\eta_1 = \theta_1 = \gamma_1 = 0$ are used to avoid confounding of the three series. Additionally the relation $c = X - x + t$ means an extra constraint is needed for full identification, for example, by taking $\gamma_1 = \gamma_2 = 0$ (Clayton and Schifflers, 1987).

The convolution prior of Besag et al. (1991) may be generalised by adopting structured and iid effects for each time scale, as well as for areas (Knorr-Held, 2000). Hence an APC model would then become

$$\log(\mu_{xt}) = \kappa + \eta_x + \theta_t + \gamma_c + u_{1x} + u_{2t} + u_{3c},$$

where u_{1x}, u_{2t} and u_{3c} are iid zero mean random effects, while $\{\eta_x, \theta_t, \gamma_c\}$ follow structured (i.e. random walk or other autoregressive) form. For area-age-period data, $y_{ixt} \sim Po(N_{xit}\mu_{xit})$, this approach leads to

$$\log(\mu_{ixt}) = \kappa + \eta_x + \theta_t + \gamma_c + s_i + u_{1x} + u_{2t} + u_{3c} + u_{4i},$$

where s_i follows a structured spatial autoregressive prior, but the u_{4i} are iid zero mean random effects.

In the preceding models, the dimensions are independent and multiplicative in the risk scale (additive in the log risk scale). In practice, interactions between one or more of the different time scales, or between the time scales and the units (e.g. areas or actuarial risk groups), are likely. Interactions ψ_{xc} between age and cohort are relevant if the age slope is changing between cohorts (e.g. cancer deaths at younger ages are less common in recent cohorts), while in mortality forecasting, age-time interactions ψ_{xt} are of interest, since different age groups may be subject to different mortality improvements (Pedroza, 2006; Lee and Carter, 1992). In area APC models, area-cohort and area-time interactions might be relevant (Lagazio et al., 2003), while in area life table models (Congdon, 2006), age-area interactions may be investigated, since deprived areas may have relatively high "premature" mortality (sometimes defined by death before age 75).

In area-time (spatio-temporal) models, one may extend the RIAS principle, and assume area-specific random variation for both the level and a time covariate. This amounts to taking the interaction ψ_{it} as a linear trend model, with neighbouring areas having similar trend parameters, as in Bernardinelli et al. (1995). Thus with $y_{it} \sim Po(N_{it}\mu_{it})$,

$$\log(\mu_{it}) = \kappa + \omega_{1i} + \omega_{2i}(t - \bar{t}),$$

where ω_{1i} and ω_{2i} are spatially correlated over areas. One may further adopt a bivariate spatial (e.g. bivariate CAR) prior for $\{\omega_{1i}, \omega_{2i}\}$, allowing level and trend parameters to be correlated. Additionally, a convolution form may be adopted both for level and trend, so that

$$\log(\mu_{it}) = \kappa + \omega_{1i} + u_{1i} + (\omega_{2i} + u_{2i})(t - \bar{t}),$$

where u_{1i} and u_{2i} are iid random effects. Equivalently, letting $c_{ji} = \omega_{ji} + u_{ji}$ one has

$$\log(\mu_{it}) = \kappa + c_{1i} + c_{2i}(t - \bar{t}).$$

A variation is to introduce an overall nonlinear trend via parameters δ_t, along with time specific spatial and iid effects $\{\omega_{it}, u_{it}\}$, and stationary AR1 dependence in the total lagged spatial effect $c_{it} = \omega_{it} + u_{it}$ (Martinez-Beneito et al., 2008). Thus for $t > 2$,

$$\log(\mu_{it}) = \kappa + \delta_t + c_{it} + \rho c_{i,t-1},$$

with $\rho \in (-1, 1)$, while for $t = 1$,

$$\log(\mu_{i1}) = \kappa + \delta_1 + \frac{c_{i1}}{(1-\rho^2)^{0.5}}.$$

This is equivalent to assuming $\log(\mu_{it}) = \kappa + \delta_t + \rho^{t-1}(1-\rho^2)^{-0.5} c_{i1} + \sum_{k=2}^{t} \rho^{t-k} c_{ik}$, where the last term is zero when $t = 1$.

In area-age-time models for mortality counts $y_{ixt} \sim Po(\mu_{ixt})$, area-age-time interactions ψ_{ixt} may be parsimoniously modelled by separate linear time trends for each age and area, namely

$$\psi_{ixt} = (\omega_{1x} + \omega_{2i})(t - \bar{t}),$$

as in Sun et al. (2000), where the random coefficients ω_{1x} and ω_{2i} may be structured over ages and areas respectively. Sun et al. (2000) actually assume a spatial CAR(ρ) prior with mean zero for the ω_{2i} (section 6.3.3), but take the ω_{1x} to be unrelated fixed effects. The full model of Sun et al. (2000) also includes iid age-area-time effects, u_{ixt}, so that

$$\log(\mu_{ixt}) = \kappa + s_i + \eta_x + (\omega_{1x} + \omega_{2i})(t - \bar{t}) + u_{ixt}.$$

Alternatively, the time function in ψ_{ixt} may be unknown, as in

$$\psi_{ixt} = (\omega_{1x} + \omega_{2i})\delta_t,$$

where positive loadings ω_{1x} and ω_{2i} specify which ages are most sensitive to trend effects δ_t. For identification, the δ_t are centred at zero or have a corner constraint such as $\delta_1=0$, and the loadings ω_{1x} and ω_{2i} may be centred at 1, constrained to sum to 1, or have a minimum of 1. So, for declining mortality, represented by δ_t following (say) a 1st order random walk, larger ω_{1x} and ω_{2i} indicate which age groups and areas contribute most to the mortality decline. Lee and Carter (1992) apply the age-time product model $\psi_{xt} = \omega_x\delta_t$ in mortality forecasting, with identification obtained by ensuring δ_t sum to zero, and that the ω_x sum to 1.

Interaction priors may also be based on a Kronecker product of the structure matrices for the relevant dimensions (Knorr-Held, 2000; Clayton, 1996), where a structure matrix is a constituent part of the precision (inverse covariance) matrix. For example, if the structure matrix of separate area and age effects are denoted K_s and K_x, then $K_{sx} = K_s \otimes K_x$ defines the structure matrix for the joint prior for ψ_{ix}, and conditional priors on ψ_{ix} can be obtained from K_{sx}. Thus an *RW*1 prior in age has a structure matrix with off-diagonal elements $K_{x[ab]} = -1$ if ages a and b are adjacent, and $K_{x[ab]} = 0$ otherwise. Diagonal elements are 1 if $a=b=1$ or $a=b=X$, and equal 2 for other diagonal terms. An *RW*2 prior for age has structure matrix

$$K_x = \begin{bmatrix} 1 & -2 & 1 & & & & & & \\ -2 & 5 & -4 & & & & & & \\ 1 & -4 & 6 & -4 & 1 & & & & \\ & 1 & -4 & 6 & -4 & 1 & & & \\ & & . & . & . & . & & & \\ & & & 1 & -4 & 6 & -4 & 1 & \\ & & & & 1 & -4 & 6 & -4 & 1 \\ & & & & & 1 & -4 & 5 & -2 \\ & & & & & & 1 & -2 & 1 \end{bmatrix}.$$

Similarly, the CAR(1) prior for spatially structured errors $s = (s_1,\ldots,s_n)$ based on adjacency of areas is multivariate normal with precision matrix $\tau_s K_s$, where τ_s is an overall precision parameter, and off-diagonal terms $K_{s[ij]} = -1$ if areas i and j are neighbours, and $K_{s[ij]} = 0$ for non-adjacent areas. The diagonal terms in K_s are L_i where L_i is the cardinality of area i (its total number of neighbours). Then an area-age interaction effect ψ_{ix} formed by crossing an *RW*1 age prior with a *CAR*(1) spatial effect has joint precision

$$\frac{1}{\sigma_\psi^2} K_s \otimes K_x,$$

and full prior conditionals with variances σ_ψ^2/L_i when $x=1$ or $x=X$, and $\sigma_\psi^2/(2L_i)$ otherwise. With ∂_i denoting the neighbourhood of area i, the prior conditional means Ψ_{ix} for ψ_{ix} are

$$\Psi_{i1} = \psi_{i2} + \sum_{j\in\partial_i} \psi_{j1}/L_i - \sum_{j\in\partial_i} \psi_{j2}/L_i,$$

$$\Psi_{ix} = 0.5(\psi_{i,x-1} + \psi_{i,x+1}) + \sum_{j\in\partial_i} \psi_{jx}/L_i - \sum_{j\in\partial_i} (\psi_{j,x+1} + \psi_{j,x-1})/(2L_i), \quad 1 < x < X$$

$$\Psi_{iX} = \psi_{i,X-1} + \sum_{j\in\partial_i} \psi_{jX}/L_i - \sum_{j\in\partial_i} \psi_{j,X-1}/L_i.$$

For identification, the ψ_{ix} should be doubly centred at each iteration (over areas for a given age x, and over ages for a given area i).

Example 10.11 Alternative Measures of Creatinine Clearance

Oman et al. (1999) compare a standard measure of creatinine clearance (MCC) with a proxy measure ECC for 113 patients with 437 clinic visits. MCC is obtained as the ratio of the amount of creatinine (CR24) excreted in the urine over 24 hours, divided by creatinine concentration (SERUMCR) and by the number of minutes in the period, namely

$$\text{MCC} = \text{CR24}/(\text{SERUMCR} \times 60 \times 24).$$

ECC is obtained from patient age and weight WT as

$$\text{ECC} = (140 - \text{Age}) * \text{WT}/(\text{SERUMCR} \times 60 \times 24),$$

with a further scaling by 0.85 for women only. There are four patient groups formed by crossing gender with whether third-space body fluids were present on at least one visit. The $J=4$ groups are then (1 = female, no fluids; 2 = female, fluids; 3 = male, no fluids; 4 = male, fluids), with group sizes $n=(51,12,41,9)$ and total visits within groups $N=(211,42,148,36)$.

The repeated responses for patients $i=1,\ldots,n_j$ within groups $j=1,\ldots,J$ are $y_{1ijt} = \log(\text{MCC}_{ijt})$ and $y_{2ijt} = \log(\text{ECC}_{ijt})$. The model involves a patient-group common factor b_{ij}, and a unique factor c_{ij}, for each outcome-cluster pair, namely

$$y_{1ijt} = b_{ij} + u_{1ijt},$$

$$y_{2ijt} = \lambda_j b_{ij} + c_{ij} + u_{2ijt},$$

$$b_{ij} \sim N(B_j, D_{1j}),$$

$$c_{ij} \sim N(0, D_{2j}),$$

$$u_{mijt} \sim N(0, 1/\tau_{mj}).$$

The intercepts for y_2 are represented by the product of loadings and B-coefficients. Identification issues are lessened by the fact that period 1 defines the direction of the b_{ij} effects. Gamma priors with index and shape parameters of unity are assumed for the precisions $\{1/D_{mj}, \tau_{mj}\}$, and N(0,1000) priors for the fixed effects $\{\lambda_j, B_j\}$.

Estimation via jagsUI provides posterior means (95% CRI) for λ_j of 1.03 (1.01,1.06), 1.06 (0.98,1.14), 1.02 (0.99,1.04), and 1.05 (0.96,1.15). The representation adopted avoids including weakly identified separate intercepts for y_2, and the results support identical calibration except for the first cluster, where there is a high probability that the λ coefficient is positive. This probability is obtained from monitoring the node step.lambda[1:4] in the rjags code. These conclusions are unaffected by adopting a robust student t (with preset d.f. = 4) option for the b_{ij} effects.

Example 10.12 Mortality Change, with Area and Age Dimensions

This example considers age-area interactions in mortality level and trend, and how these can be modelled parsimoniously. Consider deaths and population data $\{d_{ixt}, P_{ixt}\}$ for areas $i = 1,\ldots,n$, ages $x = 1,\ldots,X$, times $t = 1,\ldots,T$. One may assume binomial sampling,

$$d_{ixt} \sim \text{Bin}(P_{ixt}, \mu_{ixt}),$$

with a logit link for the mortality rates μ_{ixt}. The application here involves annual deaths to male white non-Hispanics over the period 1999–2014 (T = 16 years), in n = 51 US states (including District of Columbia). Deaths d_{ixt} and SEER population data P_{ixt} are for X = 13 age bands (<1, 1–4, 5–9, 10–14, 15–19, 20–24, 25–34, 35–44, ..., 75–84, 85+). Recent research (Squires and Blumenthal, 2016; Case and Deaton, 2015) reports an unexpected rise in death rates among middle-aged, white Americans between 1999 and 2014; see also www.commonwealthfund.org/publications/issue-briefs/2016/jan/mortality-trends-among-middle-aged-whites.

Thus, one might adopt a linear trend model (model 1) with independent age and area impacts (η_x and r_i) on the mortality level, and parallel effects (ρ_{1x} and ρ_{2i}) on the trend also. This leads to

$$\text{logit}(\mu_{ixt}) = \kappa + \eta_x + r_i + (\rho_{1x} + \rho_{2i})(t - \bar{t}) + u_{ixt},$$

where the intercept κ is assigned a normal N(0,1000) prior, the area effects r_i follow the Leroux et al. (1999) prior allowing for spatial dependence, and age effects η_x follow a normal first order random walk. The associated conditional precisions (τ_r, τ_η) are assigned gamma Ga(1,0.01) priors. The ρ_{1x} and ρ_{2i} linear trend coefficients are taken to be iid normal random effects with zero means, and precisions $\tau_{\rho 1}$ and $\tau_{\rho 2}$ that are assigned gamma Ga(1,0.01) priors. Model 1 allows for miscellaneous departures (e.g. age-area interactions in level and trend) from a linear trend by adding iid Normal errors $u_{ixt} \sim N(0,1/\tau_u)$ for each observation, where $\tau_u \sim Ga(1,0.01)$.

Hierarchical centring is applied with $u_{ixt} \sim N(\eta_x + r_i + (\rho_{1x} + \rho_{2i})(t - \bar{t}), 1/\tau_u)$, with the spatial effects additionally centred around the overall intercept κ. For the continental states (i = 1,...,49), the prior is then

$$r_i \sim N\left(\kappa + \lambda \sum_{j \in L_i} \frac{(r_j - \kappa)}{1 - \lambda + \lambda D_i}, \frac{1}{\tau_r[1 - \lambda + \lambda D_i]}\right),$$

where $\lambda \in [0,1]$ measures spatial dependence, L_i denotes the locality of area i (i.e. the set of states adjacent to state i), and there are D_i states in that locality. For the remaining two states (Alaska, Hawaii) without neighbours, $D_i = \lambda = 0$, one specifies

$$r_i \sim N(\kappa, 1/\tau_r).$$

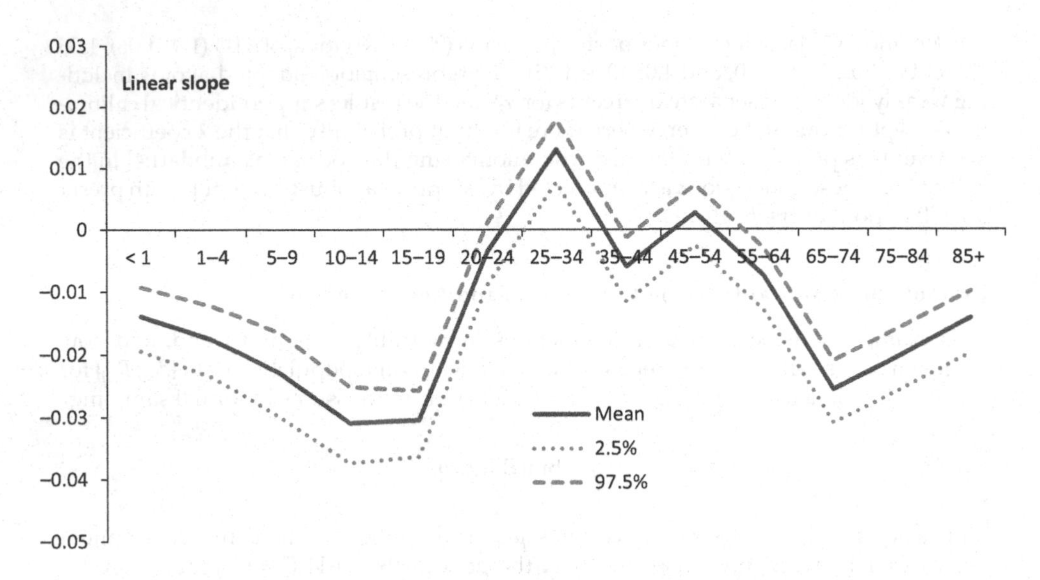

FIGURE 10.5
Linear trend slopes for mortality by age band, white non-Hispanic males, 1999–2014.

By contrast, model 2 allows explicit age-area interactions in trend. Thus

$$\text{logit}(\mu_{ixt}) = \kappa + \eta_x + r_i + \rho_{ix}(t - \bar{t}) + u_{ixt},$$

where priors on κ, r_i, η_x and (τ_r, τ_η) are as for Model 1.

Inferences for both models (using rube) are based on two-chain runs of 2,000 iterations, with thinning to retain every other sample. The DIC in fact prefers the less heavily parameterised model 1, namely 90,665 as against 90,807, so supporting independent age and area effects on the time evolution of US white non-Hispanic male mortality. Figure 10.5, based on Model 1, shows the most significant adverse trend (in terms of positive ρ_{1x}, namely mortality increase) to be for age groups 7 and 9 (namely 25–34 and 45–54).

10.8 Missing Data in Longitudinal Models

Attrition and intermittently missing data are frequently found in longitudinal data, and disregarding the process underlying such missingness may lead to biased and inefficient estimates of parameters in the outcome model, though different mechanisms may apply for attrition as opposed to intermittent missingness (Ma et al., 2005). In particular, missingness may be non-ignorable, meaning that the probability of a missing observation or of permanent drop-out is associated with the value or values of the variable that would otherwise have been observed (Troxel et al., 1998). Thus, in clinical trials, patients may drop out because of adverse treatment effects, or because they don't feel the treatment is of benefit, leading to biased estimates of treatment effects unless the missing data mechanism is allowed for.

Missingness generates an additional form of binary (or sometimes categorical) data R, depending on whether responses Y and/or predictor variables X are missing. Li et al. (2007)

obtain a categorical (trinomial) missing data indicator by distinguishing between intermittent and permanent missingness. Similarly, if missing data is entirely due to attrition, it may be summarised in a single multinomial indicator $R_i = j$ if an individual drops out between the $(j-1)$th and jth measurement (Hedeker and Gibbons, 2006, p.290; Fitzmaurice et al., 2004, section 14.4). In pattern mixture models for attrition, the dropout pattern may be summarised in various ways, most simply via a binary variable contrasting completers against dropouts, regardless of when the dropout occurred (Hedeker and Gibbons, 1997). Finally, for longitudinal datasets with continuous measurement at differing observation times a_{it}, one may record actual dropout times U_i (Hogan et al., 2004), and use these in the model for the observed Y.

However, initially consider binary indicators $R_{it} = 1$ when a response variable Y_{it} is missing, whether intermittently or permanently, and $R_{it} = 0$ for when the response is observed. Further, let $Y = (Y_{\text{obs}}, Y_{\text{mis}})$ denote the observed and unobserved response data. The totality (R,Y) is sometimes known as the complete or full data (Ibrahim et al., 1999; Daniels and Hogan, 2008, p.89).

How one deals with missing data depends on the generating mechanism assumed. Two broad missing data schemes (the selection approach and the pattern mixture approach) involve a different conditioning for the joint density $P(Y, R \mid \theta_Y, \theta_R)$ of the responses and the missingness indicators. The pattern mixture model (Little, 1993) starts with a model for the missing data $P(R \mid \theta_R)$, and models Y conditional on R, namely $P(Y \mid R, \theta_Y)$. When dropout times are discrete, the model for $P(R \mid \theta_R)$ is often not specified (Hogan et al., 2004), or when missingness is expressed in various dropout patterns, $P(R \mid \theta_R)$ may be specified simply by the relevant multinomial probabilities of different dropout options (Curran et al., 2002, p.13). By contrast, the selection model (Diggle and Kenward, 1994) starts with the data likelihood $P(Y \mid \theta_Y)$, and models missingness conditional on the responses, $P(R \mid Y, \theta_R)$, so that $P(Y, R \mid \theta_Y, \theta_R) = P(R \mid Y, \theta_R) P(Y \mid \theta_Y)$.

A classification of missingness mechanisms is set out by Little and Rubin (2002), and framed in terms of the selection approach, though is applicable also to pattern mixture analysis. They distinguish between

a) missingness completely at random (abbreviated as MCAR), when the probability $Pr(R = 1 \mid Y)$ of a missing response is independent of both observed and missing data $Y = (Y_{\text{obs}}, Y_{\text{mis}})$, namely $P(R \mid Y) = P(R)$;

b) missingness at random (MAR), when missingness is independent of the unobserved data Y_{mis}, but may depend on observed data Y_{obs}, such as when the chance that $R_{it} = 1$ depends on preceding non-missing observations $y_{i,t-s}$; in this case one has the simplification $P(R \mid Y) = P(R \mid Y_{\text{obs}})$;

c) missingness not at random (MNAR), when the probabilities of missingness depend on unobserved missing responses, namely $P(R \mid Y) = P(R \mid Y_{\text{obs}}, Y_{\text{mis}})$. Since the data are partly missing, and R now depends on the complete outcome data $(Y_{\text{obs}}, Y_{\text{mis}})$, the selection model factors the joint distribution into a complete outcome model, and a missing-data mechanism given the partially unobserved complete outcomes (Troxel et al., 2004).

An additional distinction is made between ignorable and non-ignorable missingness. Assume a MAR mechanism, and that the missing-data model is independent of the response data parameters θ_Y. Then the missing-data process is ignorable in the sense that a model for missingness is not needed in order to make valid inferences from the main

Y-likelihood (Rubin, 1976; Fichman and Cummings, 2003). However, for non-ignorable missingness, both the R-likelihood and Y-likelihood must be modelled.

As an illustration, drop-out at time t is classed as being at random if $Pr(R_t = 1 | Y) = Pr(R_t = 1 | Y_1,\ldots Y_{t-1})$, namely when the missingness probability is related to lagged observed responses. However, if the probability of missingness at time t is related also to the current outcome Y_t, possibly missing, so that $Pr(R_t = 1 | Y) = Pr(R_t = 1 | Y_1,\ldots,Y_t)$, then missingness is non-random or informative (Diggle and Kenward, 1994). In practice, informative missingness is assessed empirically, and would require a significant effect of (possibly missing) Y_{it} on $\pi_{it} = Pr(R_{it} = 1)$ in a binary regression also involving other influences on missingness, with the regression taken over subjects i and repetitions $t = 1,\ldots,T_i$. For dropouts, one takes $T_i = T_{i,obs} + 1$ where $T_{i,obs}$ is the last interval where data on subject i was obtained (Roy and Lin, 2002). Since MNAR missingness can never be excluded as a generating mechanism, a sensitivity analysis under different mechanisms may be considered (Kenward, 1998). This means estimating the model under a "range of assumptions about the non-ignorability parameters and assessing the impact of these parameters on key inferences" (Ma et al., 2005).

A common set of predictors X_{it} may be relevant to modelling both the data Y_{it}, and missingness indicators R_{it}, or different predictors W_{it} may be used in the R model. King (2001) accordingly presents a statement of the MCAR-MAR-MNAR alternatives as above, but replacing Y by $D = (Y,X)$, namely predictor and outcome data combined, and where $D = (D_{obs}, D_{mis})$ denotes the subdivision of the data according to observation status. For example, the MCAR assumption then requires $P(R|D) = P(R)$, while missingness at random requires $P(R | D) = P(R | D_{obs})$.

An alternative less stringent definition of MCAR missingness is used by Little (1995), in which missingness is independent of Y, whether observed or not, but may depend on fully observed covariates X (Curran et al., 2002, p.12; Daniels and Hogan, 2008, p.92). Such covariates might for instance include time, as missingness rates often increase at later stages of longitudinals (Hedeker and Gibbons, 2006, p.281). So, given X_{obs}, R is independent of both Y_{obs} and Y_{mis}, leading to what is sometimes termed "covariate dependent MCAR missingness."

10.8.1 Forms of Missingness Regression (Selection Approach)

A logit or probit regression is the most common approach to predicting $\pi_{it} = Pr(R_{it} = 1)$, and to assessing ignorability and MCAR assumptions. For example, π_{it} might at a minimum be a function of immediately preceding and current Y values, namely (Curran et al., 2002; Mazumdar et al., 2007)

$$\text{logit}(\pi_{it}) = \gamma_1 + \gamma_2 y_{it} + \gamma_3 y_{i,t-1},$$

with a significant γ_2 indicating non-ignorable missingness. Refinements, especially in problems with intermittently missing data, include transition probability approaches (Li et al., 2007) with the model for

$$\pi_{i01t} = Pr(R_{it} = 1 | R_{i,t-1} = 0),$$

having distinct parameters from that for

$$\pi_{i11t} = Pr(R_{it} = 1 | R_{i,t-1} = 1).$$

If missingness is restricted to dropout only (i.e. there is no intermittent missingness), then one may use a logit or clog-log link for the probability that $R_i = j \mid R_i \geq j$, where $R_i = j$ if a subject drops out between the $(j-1)$th and jth measurement.

Choice of additional predictors in the missingness model is an area of potential sensitivity in terms of whether the coefficient on the current Y value is found to be significant. Hedeker and Gibbons (2006) use logit or clog-log link models to assess whether a covariate-dependent MCAR assumption applies for a given data set. They relate $Pr(R_i = j \mid R_i \geq j)$ to observed covariates X_{obs} such as time and treatment, as well as to the history $h(y_{it})$ of observed Y values, and to interactions between X_{obs} and $h(y)$. For example, $h(y_{ij})$ might be the average of all y_{it} between periods 1 and j. Then to test for covariate-dependent MCAR, one might use a logit regression for $Pr(R_i = j \mid R_i \geq j)$ that includes main effects $\{t, Tr, h(y)\}$, as well as interactions between $h(y)$ and t, between $h(y)$ and Tr, and between $h(y)$, Tr and t jointly.

The missingness model may have a role not only as part of a likelihood analysis allowing non-random missing data, or testing for different types of missingness, but as a method for imputing missing data. Thus a "propensity score" analysis may be based on categorising the regression terms η_{it} in

$$\text{logit}(\pi_{it}) = \eta_{it},$$

into quantile groups (e.g. quartiles) (Rosenbaum and Rubin, 1983). Among subjects located within particular quantiles of η_{it}, some subjects will exit but some remain. Sampling of the missing y_{it} for exiting subjects may be based on sampling with replacement from the known y_{it} values of stayers in the same quantile – this is sometimes called the "approximate Bayesian bootstrap method" (Rubin and Schenker, 1986; Lavori et al., 1995). In multiple imputation, this imputation process would be repeated several times to provide multiple filled-in datasets.

10.8.2 Common Factor Models

Latent variables may be introduced to explain both the Y and R data. Thus, a latent data perspective on the selection model might consider bivariate data (Y,Z) where Y_{it} is observed if the latent data Z_{it} is positive (Copas and Li, 1997). Furthermore, let X_{it} be predictor data potentially relevant to explaining both Y and Z and define bivariate standard normal errors $(\varepsilon_{1i}, \varepsilon_{2i})$ with correlation ρ. Assume a linear regression for Y with

$$y_{it} = X_{it}\beta + \sigma_1\varepsilon_{1i},$$

and a missingness model

$$Z_{it} = X_{it}\gamma + \varepsilon_{2i}.$$

Then if $\rho \neq 0$, the missing data are informative or non-ignorable, whereas $\rho = 0$ corresponds to missingness at random.

A similar principle involves low dimension random effects F, also known as common factors, that are shared between outcome and missingness models; similar shared frailty models are used for models with outcome-dependent follow-up (Ryu et al., 2007). As often in factor models, the outcome data and missingness patterns may be viewed as conditionally independent, given the common factors (Song and Belin, 2004; Albert et al., 2002;

Roy and Lin, 2002; Ten Have et al., 1998). Equivalently, it is assumed that "all information about the missing data in the observed response is accounted for through the shared random effects" (Albert and Follmann, 2007). In fact, Li et al. (2007) and Yang and Shoptaw (2005) distinguish such models as an alternative to selection and pattern mixture methods, since under conditional independence one may represent the (R,Y,F) joint density as

$$\begin{aligned} P(R_i,Y_i,F_i \mid \theta_R,\theta_Y,\theta_F) &= P(R_i,Y_i \mid \theta_Y,\theta_R,F_i)P(F_i \mid \theta_F) \\ &= P(R_i \mid \theta_R,F_i)P(Y_{\text{obs},i},Y_{\text{mis},i} \mid \theta_Y,F_i)P(F_i \mid \theta_F). \end{aligned}$$

Integrating out the F_i, one has

$$P(R_i,Y_i \mid \theta_Y,\theta_R) = \int P(R_i \mid \theta_R,F_i)P(Y_{\text{obs},i},Y_{\text{mis},i} \mid \theta_Y,F_i)P(F_i \mid \theta_F)dF_i.$$

Other assumptions are possible, as under the "conditional linear model" (Daniels and Hogan, 2008, p.112), with the conditioning sequence

$$P(R_i,Y_i,F_i \mid \theta_R,\theta_Y,\theta_F) = P(Y_i \mid \theta_Y,F_i,R_i)P(F_i \mid \theta_F,R_i)P(R_i \mid \theta_R).$$

One form of common effect that may be used to model informative missingness is based on shared heterogeneity (e.g. Li et al., 2007; Chib, 2008, p.507). An example is a general linear mixed outcome model with permanent subject random effects $b_i = (b_{1i},\ldots,b_{Qi})$

$$g[E(y_{it} \mid b_i)] = X_{it}\beta + Z_{it}b_i,$$

where the missingness model for $Pr(R_{it}=1)$ also conditions on the b_i, and possibly on separate predictors W_{it}, and on the history of responses $H_{it} = \{y_{i1},\ldots,y_{it}\}$. Consider the case $Q=1$ with $z_{it}=1$, and suppose predictors W_{it} are relevant to dropout (e.g. baseline health status in a clinical trial). Then a common factor model adapted to predicting the missingness probability $\pi_{it} = Pr(R_{it}=1 \mid W_{it},H_{it})$ might take the form

$$g[E(y_{it} \mid b_i)] = X_{it}\beta + b_i \tag{10.10}$$

$$\text{logit}(\pi_{it}) = W_{it}\gamma + \lambda b_i + y_{it}\delta_1 + y_{i,t-1}\delta_2,$$

where b_i are zero mean random effects, and the predictors $\{X_{it},W_{it}\}$ both include an intercept. For example, Li et al. (2007) consider Poisson data with $y_{it} \sim Po(\lambda_{it})$,

$$\log(\lambda_{it}) = X_{it}\beta + b_i,$$

and with binary indicators for missingness, and a lagged outcome scheme adapted to counts, one would obtain

$$\text{logit}(\pi_{it}) = W_{it}\gamma + \lambda b_i + \log(y_{it}+1)\delta_1 + \log(y_{i,t-1}+1)\delta_2.$$

In fact, the model of Li et al. (2007) distinguishes between intermittently missing data and permanent attrition via a multinomial rather than binary regression, and uses a transition probability missingness model.

A model with shared latent effects exemplified by (10.9) imposes possibly restrictive assumptions on the correlations among repeated responses for a given subject. Conditional on the time-invariant shared effects b_i, observations on a subject are uncorrelated (Albert and Follmann, 2007). An alternative is a shared autoregressive process, as in

$$g[E(y_{it} \mid F_{it})] = X_{it}\beta + F_{it},$$

$$F_{it} = \rho F_{i,t-1} + u_{it},$$

$$\text{logit}(\pi_{it}) = W_{it}\gamma + \lambda F_{it} + Y_{it}\delta_1 + Y_{i,t-1}\delta_2,$$

where the u_{it} are white noise, and $\rho \in (-1,1)$.

For multivariate responses $\{y_{mit}, m = 1,\ldots,M\}$, one might propose common factors to model both correlation between the observed responses, and the probabilities of missing response, especially attrition affecting all outcomes (Lin et al., 2004). Thus, consider a single time-varying factor F_{it}, and loadings $\{\lambda_m, \kappa\}$ in the Y and R likelihoods, and let H_{it} denote a subset of the history of the observed X and Y variables up to time t. Then for outcomes $m = 1,\ldots,M$, one might have

$$g[E(y_{mit} \mid X_{it}, F_{it})] = X_{it}\beta + \lambda_m F_{it}$$

while the drop out probability $R_{it} \sim \text{Bern}(\pi_{it})$ is modelled as

$$\text{logit}(\pi_{it}) = W_{it}\gamma + \varphi H_{i,t-1} + \kappa F_{it}$$

for $t = 1,\ldots,T_i$, where for dropouts $T_i = T_{i,\text{obs}} + 1$ and $T_{i,\text{obs}}$ is the last interval where data was observed. Furthermore, the factor scores may depend on known predictors $\{U_{it}, Z_{it}\}$ and zero mean random permanent effects b_i, as in

$$F_{it} = U_{it}\eta + Z_{it}b_i + \upsilon_{it},$$

with $\upsilon_{it} \sim N(0,1)$ if all loadings κ and λ_m are unknowns, and with U_{it} omitting an intercept for identifiability (Roy and Lin, 2002, p.42). The missingness model is non-ignorable by virtue of dependence of π_{it} on F_{it}, which represents possibly missing y_{mit} (Roy and Lin, 2002, p.43).

10.8.3 Missing Predictor Data

Often longitudinal data will have missingness on covariates as well as on the response, so that binary or categorical indicators R_X are defined according as covariates have missing values or not. With $R = (R_Y, R_X)$, the joint density under a selection approach has the form

$$p(Y, X, R_Y, R_X \mid \eta, \beta, \theta) = p_R(R_X, R_Y \mid Y, X, \eta)p_Y(Y \mid X, \beta)p_X(X \mid \theta),$$

where p_X now models the likelihood of the predictors. If R_Y is conditional on all the components of R_X, one has

$$p(Y, X, R_Y, R_X \mid \eta, \beta, \theta) = p(R_Y \mid R_X, Y, X, \eta_Y)p(R_X \mid Y, X, \eta_X) \times p_Y(Y \mid X, \beta)p_X(X \mid \theta).$$

Alternatively, R_Y may be modelled jointly with the R_X, though complexity increases as the number of predictors subject to missingness rises, giving rise to different possible conditional sequences for R_Y and the components of R_X.

Suppose a subset of q predictors have missing values, with $R_{ji}=1$ if X_{ji} is missing, and $R_{ji}=0$ otherwise. If Y is fully observed, a selection approach specifies

$$p(Y,X,R_X \mid \eta,\beta,\theta) = p(R_X \mid Y,X,\eta)p_Y(Y \mid X,\beta)p_X(X \mid \theta),$$

where $p(R_X)$ is a multinomial with 2^q cells. To define p_X, one needs to specify the joint distribution of $X_{i,\text{mis}} = \{X_{1i},\ldots,X_{qi}\}$. Suppose the incompletely observed covariates $X_{\text{mis}} = (X_1,\ldots,X_q)$ are both categorical $X_{\text{mis},D} = \{X_1,\ldots,X_r\}$ and continuous $X_{\text{mis},C} = \{X_{r+1},\ldots,X_q\}$, with fully observed covariates denoted $X_{\text{obs}} = \{X_{q+1},\ldots,X_p\}$. Ibrahim et al. (1999) proposed the joint density of X_{mis} be specified as a series of conditional distributions, namely

$$p(X_1,\ldots,X_q \mid \theta\} = p_q(X_q \mid X_{q-1},\ldots,X_1,\theta_q,X_{\text{obs}}\}\ldots p_2(X_2 \mid X_1,\theta_2,X_{\text{obs}})p_1(X_1 \mid \theta_1,X_{\text{obs}})$$

though there may be sensitivity as to which of the $q!$ conditioning sequences is adopted. The completely observed predictors may be used in predicting the missing covariates. For continuous predictors, the form of density (e.g. gamma, normal) can be adapted to whether only positive values are observed. Ibrahim et al. (1999, p.180) suggested one-dimensional or joint distributions for the continuous predictors in the lower stages (p_1, p_2 etc.), with the higher stages being models for categorical predictors that are based on the imputed continuous covariates (e.g. logistic regression models).

Another general scheme for specifying the joint density of X_{mis} adopts a different strategy by first representing the joint density of categorical predictors. This is the general location model

$$\begin{aligned} p(X_{\text{mis}} \mid \theta, X_{\text{obs}}) &= p(X_{\text{mis},C}, X_{\text{mis},D} \mid \theta_C, \theta_D, X_{\text{obs}}) \\ &= p(X_{\text{mis},C} \mid X_{\text{mis},D}, \theta_C, X_{\text{obs}})p(X_{\text{mis},D} \mid \theta_D, X_{\text{obs}}), \end{aligned}$$

typically involving a multivariate normal or multivariate Student t distribution for the continuous predictors, conditional on a given combination of values of the categorical covariates. For example, means and covariances for the MVN model could be specific to each combination of the categorical predictors. The first stage of the joint density for predicting missing categorical covariates $p(X_{\text{mis},D} \mid \theta_D, X_{\text{obs}})$ would be a multinomial distribution, or possibly log-linear regression, over discrete outcomes, missing and observed.

Possible approaches for modelling the covariate missingness indicators $p(R_{ji} \mid Y_i, X_i, \eta)$ under a selection approach include a joint log-linear model with $X_i = (X_{i,\text{mis}}, X_{i,\text{obs}})$ as predictors, or equivalently a multinomial model with all possible classifications of non-response as categories (Schafer, 1997, chapter 9). For example, if $X_{i,\text{mis}}$ contains two variables subject to missingness, then there are four possible combinations of values of R_{1i} and R_{2i} for each subject. The joint density of missingness indicators can be expressed (Ibrahim et al., 1999) as a series of conditional distributions, namely

$$\begin{aligned} p(R_{1i},\ldots,R_{qi} \mid \eta, X_i, Y_i\} &= p(R_{qi} \mid R_{q-1,i},\ldots R_{1i}, \eta_q, X_i, Y_i) \\ &\quad \ldots p(R_{2i} \mid R_{1i}, \eta_2, X_i, Y_i)p(R_{1i} \mid \eta_1, X_i, Y_i), \end{aligned}$$

which in practice implies a series of binary regressions. For assessing non-randomness in covariate missingness, one allows $Pr(R_{2i}=1|R_{1i},\eta_2,X_i,Y_i)$ to depend on predictors X_i that may be subject to missing values, as well as on earlier R_{ji} in the conditional sequence.

In practice, a multivariate density for a set of continuous variables might be represented indirectly by a series of regressions, and missing values for binary or categorical data items modelled or imputed via regressions on other predictors – see Austin and Escobar (2005) for an illustration of such methods. Such procedures are related to multiple imputation procedures for covariates, and possibly responses also (Schafer, 1997; Allison, 2000). Consider the case where Z and X are predictors, with Z subject to missingness. Schafer (1997) proposes random regression imputation by initially regressing Z on X and Y, but using only cases with Z observed, and from this regression forms point estimates $\hat{Z}$ for cases with missing data. Let $\hat{\sigma}$ be the square root of the mean square error from the observed data regression, then for subjects with missing Z, one obtains imputations $\tilde{Z}=\hat{Z}+\hat{\sigma}U$ where U is a draw from a standard normal. For cases with observed Z, one sets $\tilde{Z}=Z$. One then carries out a filled-in data regression of Y on X and $\tilde{Z}$ for all subjects. The Z-imputation and filled-in data regressions may be repeated M times. Such a procedure is, however, not proper in the sense of Rubin (1987).

10.8.4 Pattern Mixture Models

Pattern mixture models may have a benefit in avoiding intricate modelling of the missingness indicators (Daniels et al., 2015). For regular longitudinal data (repeat measures at fixed intervals for all subjects) subject to missingness only through attrition, a pattern mixture analysis (Little, 1995) might simply involve differentiating regression effects in the Y-model according to discrete drop out times $U_i \in (2,\ldots,T-1)$, as well as completers with $U_i=T$. Thus "the missing-data patterns can be used as grouping variables in the analysis" (Hedeker and Gibbons, 1997). If there are h_m subjects belonging to M different missingness patterns, with associated proportions $\phi_m = h_m/n$, then the "marginal" or composite parameter (e.g. the regression impact of a predictor x_p) is obtained as a weighted average of the pattern specific parameters β_{pm}, namely $\beta_p = \Sigma_{m=1}^{M}\phi_m\beta_{pm}$ (Curran et al., 2002). A Bayesian analysis might involve repeated multinomial sampling of the ϕ_m at each MCMC iteration, and monitoring the composite parameters $\beta_p^{(r)} = \Sigma_{m=1}^{M}\phi_m^{(r)}\beta_{pm}^{(r)}$. Often the preliminary model for missingness $P(R|\theta_R)$ would be confined to such multinomial sampling.

For example, in a clinical application, separate intercepts, growth coefficients and treatment effects would be estimated according to dropout category. The variance or covariance parameters for random effects may also be differentiated. In an initial analysis, dropout category might just be binary, differentiating between completers and dropouts, regardless of the interval when the dropout occurred. Thus, set $G_i=1$ for dropouts and $G_i=2$ for completers, and consider a regression model for fixed interval (balanced) longitudinal data y_{it} with intercept, time, treatment effect (*Tr*), and time-treatment interaction (e.g. Hedeker and Gibbons, 1997; Mazumdar et al., 2007). Then a grouped linear regression with varying intercepts could take the form

$$y_{it} = \beta_{1,G_i} + \beta_{2,G_i}t + \beta_{3,G_i}Tr_i + \beta_{4,G_i}(t.Tr_i) + b_{1i} + b_{2i}t + e_{it},$$

$$b_i \sim N(0, D_{G_i}),$$

$$e_{it} \sim N(0, \sigma^2_{G_i}).$$

Curran et al. (2002, p.13) allow for an additional autocorrelated error ε_{it} with pattern specific covariance matrix R_{G_i}.

The conditional linear model (Paddock, 2007; Hogan et al., 2004) is a version of the pattern mixture model that may be applied to continuously recorded longitudinal data (rather than fixed interval longitudinal data). The impact of missingness on Y involves functions $\beta_j(U_i)$ of possibly continuous dropout times U_i though this reduces to a grouping approach for fixed intervals; that is, the $\beta_j(U_i)$ become step functions (Hogan et al., 2004, p.856). At their most simple, such functions are linear in U, but polynomial functions or non-parametric models (e.g. splines) can be used. In the preceding example, one might have

$$y_{it} = \beta_1(U_i) + \beta_2(U_i)t + \beta_3(U_i)Tr_i + \beta_4(U_i)(t.Tr_i) + b_{1i} + b_{2i}t + e_{it},$$
$$b_i \sim N(0, D),$$
$$e_{it} \sim N(0, \sigma^2),$$
$$\beta_j(U_i) = \alpha_{j0} + \alpha_{j1}U_i \quad j = 1, \ldots, 4,$$

and a test for missingness at random is whether the α_{j1} are zero. Paddock (2007) applies a Bayesian regression selection approach to coefficients in models involving quadratic effects of U_i.

Example 10.13 Cocaine Use and Desipramine

This example compares some of the missing data techniques described above, including common factor and pattern mixture approaches, for data subject to dropout and intermittent missing data. The data are from a 12-week trial of the antidepressant desipramine in cocaine-dependent patients with depressive comorbidity, with 106 patients, 52 in the treatment arm, and the remainder given a placebo (Ma et al., 2005). The responses y_{it} are average dollars per day spent on cocaine use. Only 47 patients completed the full 12 weeks of observation. Let $T_i^* = 12$ for completers, while for dropouts let T_i^* denote the week subsequent to the last week $T_i^* - 1$ when an observation is obtained. So $T_i^* = 7$ if a subject is observed (with $R_{it} = 1$) for the first six weeks, but is missing ($R_{it} = 0$) for all the last six weeks.

A plot of the average responses for the two arms (including the baseline) shows that the treatment group begins with a higher average baseline spending level, and reduces its cocaine spending more. The y-model involves predictors $X = \{1, Tr, t, Tr.t, \text{Base}\}$ where Base = baseline cocaine spending. So

$$y_{it} = \beta_1 + \beta_2 Tr_i + \beta_3(t - \bar{t}) + \beta_4 Tr_i(t - \bar{t}) + \beta_5 \text{Base}_i + u_{it},$$
$$u_{it} \sim N(0, 1/\tau_u).$$

Assessment of desipramine efficacy in the outcome model focuses especially on the coefficient for treatment-time interaction, $Tr \times (t - \bar{t})$. $N(0,100)$ priors are assumed on the first three predictors, but for numeric stability, an informative $N(0,0.1)$ prior is assumed for the impact of baseline spend (as large predictor values are observed). Assessing whether missingness is informative or not is initially based on a selection approach, with $R_{it} \sim \text{Bern}(\pi_{it}), t = 1, \ldots, T_i^*$, and

$$\text{logit}(\pi_{it}) = \gamma_1 + \gamma_2 y_{it} + \gamma_3 y_{i,t-1} + \gamma_4 Tr_i + \gamma_5 Tr_i(t - \bar{t}) + \gamma_6 \text{Base}_i,$$

using both y_{it} and $y_{i,t-1}$ as predictors (Mazumdar et al., 2007).

Estimates from iterations 1,001–10,000 of a two-chain run with rjags show no effect on missingness probabilities π_{it} of the possibly unobserved current outcome y_{it} (gamma[2] in the code) with 95% interval {−0.008, 0.012}. The treatment effect β_2 in the outcome model is not significant, but the treatment-time interaction parameter β_4 has a predominantly negative density, albeit with an inconclusive 95% credible interval {−16.2,0.7}. The WAIC is 17048 for the y-model, and 544 for the R-model.

An alternative model involves a common factor F_i (multiple indicator, multiple cause) that depends on Base_i = baseline spend (standardised). The prior mean for F_i specifies a regression with intercept omitted for identifiability; the prior variance of the factor scores is set at 1. The missingness model now involves a lagged response and the common factor, while the Y likelihood no longer involves baseline spending. Thus

$$F_i \sim N(\eta \times \text{Base}_i, 1),$$

$$y_{it} = \beta_1 + \beta_2 Tr_i + \beta_3(t - \bar{t}) + \beta_4 Tr_i(t - \bar{t}) + \lambda F_i + u_{it},$$

$$R_{it} \sim \text{Bern}(\pi_{it}), \quad t = 1, \ldots, T_i^*$$

$$\text{logit}(\pi_{it}) = \gamma_1 + \gamma_2 y_{i,t-1} + \kappa F_i,$$

where a $LN(0,1)$ prior is adopted for κ, and the prior on λ is $N(1,1)$ and constrained to positive values. Estimates show the common factor is a positive function of baseline spend with η having 95% CRI (6.1,15.9). Its impact on π_{it} is positive, with κ having 95% CRI (0.01,0.05). So, the chance of a missing value ($R_{it}=0$) tends to diminish with the score on the common factor. The WAIC for the y-data under this model is broadly similar (17055) to that for the earlier one.

Finally, a pattern mixture analysis is applied, distinguishing simply between noncompleter ($G_i=1$) and completer groups ($G_i=2$). The assumed model is

$$y_{it} = \beta_{1,G_i} + \beta_{2,G_i} Tr_i + \beta_{3,G_i} t + \beta_{4,G_i} Tr_i(t - \bar{t})$$

$$+ \beta_{5,G_i} \text{Base}_i + b_i + e_{it},$$

$$b_i \sim N(0, D_{G_i}),$$

$$e_{it} \sim N(0, \sigma^2).$$

The group-specific precisions $1/D_j$ $(j=1,\ldots,2)$ are assumed to follow independent gamma priors with shape 1 and scale 1.

Posterior estimates (from iterations 1,001–14,000 of a two-chain run) show a significantly positive baseline effect, β_{51}, for dropouts, with 95% interval (0.17,0.76), whereas completers do not have a significant baseline effect. Time-treatment interactions are similar between the two groups. The pooled estimates for β_2 and β_4 (pooling over dropout patterns) show an insignificant main treatment effect, but the interaction parameter β_4 has a 95% credible interval (−8.2,−1.4).

Example 10.14 Shared Effect Missingness Model for IMPS (Inpatient Multidimensional Psychiatric Scale) Data

Hedeker and Gibbons (2006, pp.297–302) consider data relating to psychiatric morbidity; in particular, item 79 of the IMPS scale is a positive measure of morbidity with values ranging from 0 (normal) to 7 (extremely ill). The analysis here follows Hedeker and Gibbons (2006) in treating the outcomes as metric, and adopts a shared effects model. The data involve n=437 patients with up to $T_i=5$ repeat measurements not necessarily at the same times (in weeks) after the baseline at 0 weeks; most patients have four measurements. Treatment is coded as Drug=1 for patients receiving any of the drugs

Chlorpromazine, Fluphenazine, and Thioridazine; and Drug = 0 otherwise. Follow up is terminated after six weeks, with most patients only measured at weeks $a_{i1}=0, a_{i2}=1, a_{i3}=3$ and $a_{i4}=6$. Completers are those terminating at six weeks, with all sequences ending in earlier weeks considered as dropouts. So, in a similar way to that used for discrete time hazards in Chapter 11 (Section 11.5), one may define event indicators $\{w_{ij}=0, j=1,\ldots,T_i\}$ for subjects whose last week is 6, and

$$w_{ij}=0, j=1,\ldots,T_i-1;$$
$$w_{iT_i}=1,$$

for subjects whose last observation is before six weeks.

Let $\text{Drug}_i=1$ for the treatment group subjects, with $\text{Drug}_i=0$ otherwise. Also let $S_{it}=a_{it}^{0.5}$ be the square root of the number of weeks at which the tth observation of patient i is obtained. The model for the morbidity outcome then has the form

$$y_{it}=\beta_1+\beta_2\text{Drug}_i+\beta_3 S_{it}+\beta_4 S_{it}\text{Drug}_i+b_{1i}+b_{2i}S_{it}+u_{it},$$

with priors

$$(b_{1i},b_{2i})\sim N([0,0],D),$$
$$D^{-1}\sim \text{Wish}(I,2),$$
$$u_{it}\sim N(0,1/\tau_u),$$
$$\tau_u\sim Ga(1,0.001).$$

The missing data model is a complementary log-log regression, sharing the random intercept b_{1i} and with an interaction between treatment and the shared effect. Thus

$$\log(-\log[1-Pr(w_{ij}=1\mid T_i\geq j)])=\gamma_{1j}+\gamma_2\text{Drug}_i+\alpha_1 b_{1i}+\alpha_2 b_{1i}\text{Drug}_i.$$

with $\{\gamma_{1j}\sim N(0,1000), j=1,\ldots,\max(T_i)\}, \gamma_2\sim N(0,1000)$, and $\{\alpha_k\sim N(1,1), k=1,\ldots,2\}$. Non-ignorable missingness corresponds to any of the α_k coefficients being distinct from zero (Hedeker and Gibbons, 2006, p.298).

A two-chain run of 5,000 iterations using rjags shows early convergence, and posterior means on β coefficients (from the last 4000 iterations) similar to those reported by Hedeker and Gibbons (2006). In particular, β_4 has posterior mean (sd) of −0.65 (0.08) consistent with a greater reduction in morbidity for the treatment group. Dropout is lower for treated patients, with γ_2 having mean (sd) of −0.65 (0.23). Both the α coefficients have 95% credible intervals excluding zero: α_1 has mean and 95% interval 0.84 (0.17,1.57) indicating that among untreated patients, those more ill are likely to drop out, while the sum $(\alpha_1+\alpha_2)$ has mean (95%CrI) −0.50 (−1.08,0.03), showing that for those being treated, the more ill are in fact less likely to drop out.

References

Agresti A (1997) A model for repeated measurements of a multivariate binary response. *Journal of the American Statistical Association*, 92, 315–321.

Agresti A, Natarajan R (2001) Modeling clustered ordered categorical data: A survey. *International Statistical Review*, 69, 345–371.

Albert P, Follmann D (2007) Random effects and latent processes approaches for analyzing binary longitudinal data with missingness: A comparison of approaches using opiate clinical trial data. *Statistical Methods in Medical Research*, 16, 417–439.
Albert PS, Follmann DA, Wang SA, Suh EB (2002) A latent autoregressive model for longitudinalbinary data subject to informativemissingness. *Biometrics*, 58, 631–642.
Allison P (2000) Multiple imputation for missing data: A cautionary tale. *Sociological Methods and Research*, 28, 301–309.
Alvarez I, Niemi J, Simpson M (2016) Bayesian inference for a covariance matrix. arXiv:1408.4050v2
Antonio K, Beirlant J (2007) Actuarial statistics with generalized linear mixed models. *Insurance: Mathematics and Economics*, 40, 58–76.
Austin P, Escobar M (2005) Bayesian modeling of missing data in clinical research. *Computational Statistics and Data Analysis*, 49, 821–836.
Baker A, Bray I (2005) Bayesian projections: What are the effects of excluding data from younger age groups? *American Journal of Epidemiology*, 162, 798–805.
Baldwin S (2014) Visualizing the LKJ Correlation Distribution. https://www.psychstatistics.com/2014/12/27/d-lkj-priors/
Barnard J, McCulloch R, Meng XL (2000) Modeling covariance matrices in terms of standarddeviations and correlations, with applications to shrinkage. *Statistica Sinica*, 10, 1281–1311.
Bauer R, Guzy S, Ng C (2007) A survey of population analysis methods and software for complex pharmacokinetic and pharmacodynamic models with examples. *The AAPS Journal*, 9(1), E60–E83.
Bauwens L, Lubrano M, Richard J-F (1999) *Bayesian Inference in Dynamic Econometric Models*. Oxford University Press, Oxford, UK.
Beckett L, Tancredi D, Wilson R (2004) Multivariate longitudinal models for complex change processes. *Statistics in Medicine*, 23, 231–239.
Bernardinelli L, Clayton D, Pascutto C, Montomoli C, Ghislandi M, Songini M (1995) Bayesian analysis of space-time variation in disease risk. *Statistics in Medicine*, 14, 2433–2443.
Berrington A, Hu Y, Ramirez-Ducoing K, Smith P (2005) Multilevel modelling of repeated ordinal measures: An application to attitude towards divorce. Southampton Statistical Sciences Research Institute Applications and Policy Working Paper M05/10 and ESRC Research Method Programme Working Paper No. 26.
Besag J, York J, Mollié A (1991) Bayesian image restoration, with two applications in spatial statistics. *Annals of the Institute of Statistical Mathematics*, 43(1), 1–20.
Bollen K, Curran P (2004) Autoregressive latent trajectory (ALT) models: A synthesis of two traditions. *Sociological Methods and Research*, 32, 336–383.
Bonate P (2008) *Pharmacokinetic-Pharmacodynamic Modeling and Simulation*, 2nd Edition. Springer, New York.
Bond S (2002) Dynamic panel data models: A guide to microdata methods and practice. *Portuguese Economic Journal*, 1, 141–162.
Bray I (2002) Application of Markov chain Monte Carlo methods to projecting cancer incidence and mortality. *Journal of the Royal Statistical Society: Series C (Applied Statistics)*, 51(2), 151–164.
Case A, Deaton A (2015) Rising morbidity and mortality in midlife among white non-hispanic Americans in the 21st century. *Proceedings of the National Academy of Sciences of the United States of America*, 112(49), 15078–15083.
Cepeda E, Gamerman D (2004) Bayesian modeling of joint regressions for the mean and covariance matrix. *Biometrical Journal*, 46, 430–440.
Chamberlain G, Hirano K (1999) Predictive distributions based on longitudinal earnings data. *Annales d'Economie et de Statistique*, 55–56, 211–242.
Chen Z, Kuo L (2001) A note on the estimation of the multinomial logit model with random effects. *The American Statistician*, 55, 89–95.
Chen Z, Rus H, Sen A (2016) Border effects before and after 9/11: Longitudinal data evidence across industries. *World Economics*. DOI:10.1111/twec.12413.
Chernyavskiy P, Little M, Rosenberg P (2019) A unified approach for assessing heterogeneity in age–period–cohort model parameters using random effects. *Statistical Methods in Medical Research*, 28(1). https://journals.sagepub.com/doi/abs/10.1177/0962280217713033

Chib S (2008) Panel data modeling and inference: A Bayesian primer, pp 479–515, in *The Econometrics of longitudinal Data*, 3rd Edition, eds L Matyas, P Sevestre. Springer-Verlag, Berlin, Germany.
Chib S, Carlin B (1999) On MCMC sampling in hierarchical longitudinal models. *Statistics and Computing*, 9, 17–26.
Chib S, Jeliazkov I (2006) Inference in semiparametric dynamic models for binary longitudinal data. *Journal of the American Statistical Association*, 101, 685–700.
Chintagunta P, Kyriazidou E, Perktold J (2001) Panel data analysis of household brand choices. *Journal of Economics*, 103, 111–153.
Clayton D (1996) Generalized linear mixed models, pp 275–301, in *Markov Chain MonteCarlo in Practice*, eds WR Gilks, S Richardson, DJ Spiegelhalter. Chapman & Hall, London, UK.
Clayton D, Schifflers E (1987) Models for temporal variation in cancer rates. II: Age-period-cohort models. *Statistics in Medicine*, 6, 467–810.
Congdon P (2006) A model for geographical variation in health and total life expectancy. *Demographic Research*, 14, 157–178.
Copas JB, Li HG (1997) Inference for non-random samples. *Journal of the Royal Statistical Society: Series B (Statistical Methodology)*, 59(1), 55–95.
Curran D, Molenberghs G, Aaronson N, Fossa S, Sylvester R (2002) Analyzing longitudinal continuous quality of life data with dropout. *Statistical Methods in Medical Research*, 11, 5–23.
Curran P, Obeidat K, Losardo D (2010) Twelve frequently asked questions about growth curve modeling. *Journal of Cognition and Development*, 11(2), 121–136.
Daniels M, Normand S (2006) Longitudinal profiling of health care units based on continuous and discrete patient outcomes. *Biostatistics*, 7, 1–15.
Daniels M J, Jackson D, Feng W, White I (2015) Pattern mixture models for the analysis of repeated attempt designs. *Biometrics*, 71(4), 1160–1167.
Davidian M, Giltinan D (2003) Nonlinear models for repeated measures data: An overview and update. *Journal of Agricultural, Biological, and Environmental Statistics*, 8, 387–419.
Depaoli S, Boyajian J (2014) Linear and nonlinear growth models: Describing a Bayesian perspective. *Journal of Consulting and Clinical Psychology*, 82(5), 784–802.
Diggle P, Kenward M (1994) Informative dropout in longitudinal data analysis. *Journal of the Royal Statistical Society: Series C*, 43, 49–94.
Donald M, Mengersen K, Young R (2015) A four dimensional spatio-temporal analysis of an agricultural dataset. *PLOS ONE*, 10(10), e0141120.
Dorsett R (1999) An econometric analysis of smoking prevalence among lone mothers. *Journal of Health Economics*, 18, 429–441.
Dunson D (2003) Dynamic latent trait models for multidimensional longitudinal data. *Journal of the American Statistical Association*, 98, 555–563.
Dunson D (2006) Bayesian dynamic modeling of latent trait distributions. *Biostatistics*, 7, 551–568.
Dunson D (2007) Bayesian methods for latent trait modeling of longitudinal data. *Statistical Methods in Medical Research*, 16, 399–415.
Dunson D (2009) Bayesian nonparametric hierarchical modeling. *Biometrical Journal*, 51(2), 273–284.
Erkanli A, Soyer R, Angold A (2001) Bayesian analyses of longitudinal binary data using Markov regression models of unknown order. *Statistics in Medicine*, 20, 755–770.
Fahrmeir L, Tutz G (2001) *Multivariate Statistical Modelling Based on Generalized Linear Models*, pp 69–137. Springer, New York.
Fichman M, Cummings J (2003) Multiple imputation for missing data: Making the most of what you know. *Organizational Research Methods*, 6, 282–308.
Fitzmaurice G, Laird N, Ware J (2004) *Applied Longitudinal Analysis*. Wiley.
Fokianos K, Kedem B (2003) Regression theory for categorical time series. *Statistical Science*, 18, 357–376.
Fong Y, Rue H, Wakefield J (2010) Bayesian inference for generalized linear mixed models. *Biostatistics*, 11(3), 397–412.
Fotouhi A (2007) The initial conditions problem in longitudinal count process: A simulation study. *Simulation Modelling Practice and Theory*, 15, 589–604.

Franco C, Bell W (2015) Borrowing information over time in binomial/logit normal models for small area estimation. *Statistics in Transition*, 16(4), 563–584.
Frees E (2004) *Longitudinal and Panel Data*. Cambridge University Press, Cambridge, UK.
Frühwirth-Schnatter S, Tüchler R (2008) Bayesian parsimonious covariance estimation for hierarchical linear mixed models. *Statistics and Computing*, 18, 1–13.
Galatzer-Levy I (2015) Applications of Latent Growth Mixture Modeling and allied methods to posttraumatic stress response data. *European Journal of Psychotraumatology*, 6, 27515.
Galatzer-Levy I, Bonanno G (2012) Beyond normality in the study of bereavement: Heterogeneity in depression outcomes following loss in older adults. *Social Science & Medicine*, 74(12), 1987–1994.
Galler H (2001) On the dynamics of individual wage rates – Heterogeneity and stationarity of wage rates of West German Men, pp 269–293, in *Econometric Studies. A Festschrift in Honour of Joachim Frohn*, eds R Friedmann, L Knüppel, H Lütkepohl. LIT, Münster, Germany.
Gelman A, Carlin J, Stern H, Dunson D, Vehtari A, Rubin D (2014) *Bayesian Data Analysis*. CRC, Boca Raton, FL.
Gelman A, Meng XL, Stern H (1996) Posterior predictive assessment of model fitness via realized discrepancies. *Statistica Sinica*, 733–760.
Geweke J, Keane M (2000) An empirical analysis of earnings dynamics among men in the PSID: 1968–1989. *Journal of Econometrics*, 96, 293–356.
Ghosh P, Branco M, Chakraborty H (2007) Bivariate random effect model using skew-normal distribution with application to HIV-RNA. *Statistics in Medicine*, 26, 1255–1267.
Grunwald GK, Hyndman RJ, Tedesco LM, Tweedie RL (2000) Non-Gaussian conditional linear AR(1) models. *Australian and New Zealand Journal of Statistics*, 42, 479–495.
Heagerty P, Zeger S (2000) Marginalized multilevel models and likelihood inference. *Statistical Science*, 15, 1–26.
Hedeker D (2003) A mixed-effects multinomial logistic regression model. *Statistics in Medicine*, 22, 1433–1446.
Hedeker D, Gibbons R (2006) *Longitudinal Data Analysis*. Wiley, New York.
Hedeker D, Gibbons R (1997) Application of random-effects pattern-mixture models for missing data in longitudinal studies. *Psychological Methods*, 2, 64–78.
Hirano K (2000) A semiparametric model for labor earnings dynamics, in *Practical Nonparametric and Semiparametric Bayesian Statistics*, eds D Dey, P Mueller, D Sinha. Springer-Verlag, New York.
Hirano K (2002) Semiparametric Bayesian inference in autoregressive panel data models. *Econometrica*, 70, 781–799.
Ho R, Hu I (2008) Flexible modelling of random effects in linear mixed models – A Bayesian approach. *Computational Statistics and Data Analysis*, 52, 1347–1361.
Hogan J, Lin X, Herman B (2004) Mixtures of varying coefficient models for longitudinal data with discrete or continuous non-ignorable dropout. *Biometrics*, 60, 854–864.
Hsiao C (2014) *Analysis of Panel Data* (No. 54). Cambridge University Press.
Ibrahim J, Chen M-H, Ryan L (2000) Bayesian variable selection for time series count data. *Statistica Sinica*, 10, 971–987.
Ibrahim J, Lipsitz S, Chen M-H (1999) Missing covariates in generalized linear models when the missing data mechanism is non-ignorable. *Journal of the Royal Statistical Society, Series B*, 61, 173–190.
Ibrahim J, Molenberghs G (2009) Missing data methods in longitudinal studies: A review. *Test*, 18(1), 1–43.
Islam M, Chowdhury R (2006) A higher order Markov model for analyzing covariate dependence. *Applied Mathematical Modelling*, 30, 477–488.
Jara A, Quintana F, San Martin E (2008) Linear effects mixed models with skew-elliptical distributions: A Bayesian approach. *Computational Statistics & Data Analysis*, 52, 5033–5045.
Jorgensen B, Lundbye-Christensen S, Song P, Sun L (1999) A state space model for multivariate longitudinal count data. *Biometrika*, 86, 169–181.
Jung R, Kukuk M, Liesenfeld R (2006) Time series of count data: Modeling, estimation and diagnostics. *Computational Statistics and Data Analysis*, 51, 2350–2364.

Keane M (2015) Longitudinal data discrete choice models of consumer demand, in *The Oxford Handbook of longitudinal Data*, ed B Baltagi. OUP.

Kedem B, Fokianos K (2005) Regression models for binary time series, pp 185–199, in *Modeling Uncertainty*, eds M Dror, P L'Ecuyer, F Szidarovszky. Springer.

Kenward M (1998) Selection models for repeated measurements with non-random dropout: An illustration of sensitivity. *Statistics in Medicine*, 17, 2723–2732.

King G (2001) Analyzing incomplete political science data: An alternative algorithm for multiple imputation. *American Political Science Review*, 95, 49–69.

Kinney S, Dunson D (2007) Fixed and random effects selection in linear and logistic models. *Biometrics*, 63, 690–698.

Kleinman K, Ibrahim J (1998) A semi-parametric Bayesian approach to generalized linear mixed models. *Statistics in Medicine*, 17, 2579–2596.

Knorr-Held L (2000) Bayesian modelling of inseparable space-time variation in disease risk. *Statistics in Medicine*, 19, 2555–2567.

Lagazio C, Biggeri A, Dreassi E (2003) Age-period-cohort models and disease mapping. *Environmetrics*, 14, 475–490.

Lancaster T (2002) Orthogonal parameters and panel data. *The Review of Economic Studies*, 69, 647–666.

Lavori P, Dawson R, Shera D (1995) A multiple imputation strategy for clinical trials with truncation of patient data. *Statistics in Medicine*, 14, 1913–1925.

Lee J, Hwang R (2000) On estimation and prediction for temporally correlated longitudinal data. *Journal of Statistical Planning and Inference*, 87, 87–104.

Lee RD and Carter LR (1992) Modeling and forecasting U.S. mortality. *Journal of the American Statistical Association*, 87(419), 659–671.

Lee Y, Nelder J (2000) Two ways of modelling overdispersion in non-normal data. *Applied Statistics*, 49, 591–598.

Lee Y, Nelder J (2004) Conditional and marginal models: another view. *Statistical Science*, 19, 219–238.

Leroux B, Lei X, Breslow N (1999) Estimation of disease rates in small areas: A new mixed model for spatial dependence, pp 135–178, in *Statistical Models in Epidemiology, the Environment and Clinical Trials*, eds M Halloran, D Berry. Springer-Verlag, New York.

Lewandowski D, Kurowicka D, Joe H (2009) Generating random correlation matrices based on vines and extended onion method. *Journal of Multivariate Analysis*, 100, 1989–2001.

Li J, Yang X, Wu Y, Shoptaw S (2007) A random-effects Markov transition model for Poisson-distributed repeated measures with non-ignorable missing values. *Statistics in Medicine*, 26, 2519–2532.

Lin H, McCulloch C, Rosenheck R (2004) Latent pattern mixture models for informative intermittent missing data in longitudinal studies. *Biometrics*, 60, 295–305.

Lin T, Lee J (2006) A robust approach to *t* linear mixed models applied to multiple sclerosis data. *Statistics in Medicine*, 25, 1397–1412.

Lindsey J (1993) *Models for Repeated Measurements*. Oxford University Press, New York.

Little RJA (1993) Pattern-mixture models for multi-variate incomplete data. *Journal of the American Statistical Association*, 88, 125–133.

Little R (1995) Modeling the drop-out mechanism in repeated-measures studies. *Journal of the American Statistical Association*, 90, 1112–1121.

Little R, Rubin D (2002) *Statistical Analysis with Missing Data*, 2nd Edition. Wiley-Interscience, Hoboken, NJ.

Liu F, Zhang P, Erkan I, Small D S (2017) Bayesian inference for random coefficient dynamic panel data models. *Journal of Applied Statistics*, 44(9), 1543–1559.

Liu L, Hedeker D (2006) A mixed-effects regression model for longitudinal multivariate ordinal data. *Biometrics*, 62, 261–268.

Lockwood J, Doran H, McCaffrey D (December 2003) Using R for estimating longitudinal student achievement models. *R Newsletter*, 3(3), 17–23.

Ma Y, Genton M, Davidian M (2004) Linear mixed effects models with semiparametric generalized skew elliptical random effects, pp 339–358, in *Skew-Elliptical Distributions and their Applications: A Journey Beyond Normality*, ed M Genton. Chapman and Hall/CRC, Boca Raton, FL.

Ma G, Troxel A, Heitjan D (2005) An index of local sensitivity to nonignorable drop-out in longitudinal modelling. *Statistics in Medicine*, 24, 2129–2150.
Malchow-Moller N, Svarer M (2003) Estimation of the multinomial logit model with random effects. *Applied Economics Letters*, 10, 389–392.
Marsh H, Grayson D (1994) Longitudinal confirmatory factor analysis: Common, time-specific, item-specific, and residual-error components of variance. *Structural Equation Modeling*, 1, 116–145.
Marshall EC, Spiegelhalter DJ (2007) Identifying outliers in Bayesian hierarchical models: A simulation-based approach. *Bayesian Analysis*, 2(2), 409–444.
Martinez-Beneito M, Lopez-Quilez A, Botella-Rocamora P (2008) An autoregressive approach to spatio-temporal disease mapping. *Statistics in Medicine*, 27, 2874–2889.
Mazumdar S, Tang G, Houck P, Dew M, Begley A, Scott J, Mulsant B, Reynolds C (2007) Statistical analysis of longitudinal psychiatric data with dropouts. *Journal of Psychiatric Research*, 41, 1032–1041.
McElreath R (2015) *Statistical Rethinking: A Bayesian Course with Examples in R and Stan*. CRC Press.
Menard S (2002) *Longitudinal Research*, 2nd Edition. Sage, London, UK.
Molenberghs G, Verbeke G (2004) An introduction to (generalized) (non)linear mixed models, pp 111–153, in *Explanatory Item Response Models: A Generalized Linear and Nonlinear Approach*, ed P de Boeck. Springer, New York.
Molenberghs G, Verbeke G (2006) *Models for Discrete Longitudinal Data*. Springer, New York.
Müller P, Rosner G (1997) A Bayesian population model with hierarchical mixture priors applied to blood count data. *Journal of the American Statistical Association*, 92, 1279–1292.
Muthén B, Brown C, Masyn K, Jo B, Khoo S, Yang C, Wang C, Kellam S, Carlin J, Liao J (2002) General growth mixture modeling for randomized preventive interventions. *Biostatistics*, 3, 459–475.
Natarajan R, Kass R (2000) Reference Bayesian methods for generalized linear mixed models. *Journal of the American Statistical Association*, 95, 227–237.
Natarajan R, McCulloch C (1998) Gibbs sampling with diffuse proper priors: A valid approach to data-driven inference? *Journal of Computational and Graphical Statistics*, 7, 267–277.
Nerlove M (2002) The history of panel data econometrics, 1861–1997, Chapter 1, in *Essays in Longitudinal Data Econometrics*. ed M Nerlove. Cambridge University Press.
Oh MS, Lim YB (2001) Bayesian analysis of time series Poisson data. *Journal of Applied Statistics*, 28(2), 259–271.
Ohlssen D, Sharples L, Spiegelhalter D (2007) Flexible random-effects models using Bayesian semi-parametric models: Applications to institutional comparisons. *Statistics in Medicine*, 26, 2088–2112.
Oman S, Meir N, Halm N (1999) Comparing two measures of creatinine clearance: An application of errors-in-variables and bootstrap techniques. *Applied Statistics*, 48, 39–52.
Paddock S (2007) Bayesian variable selection for longitudinal substance abuse treatment data subject to informative censoring. *Journal of the Royal Statistical Society: Series C*, 56, 293–311.
Paddock S, Savitsky T (2013) Bayesian hierarchical semiparametric modelling of longitudinal post-treatment outcomes from open enrolment therapy groups. *Journal of the Royal Statistical Society: Series A*, 176(3), 795–808.
Papaspiliopoulos O, Roberts G, Skold M (2003) Non-centered parameterisations for hierarchical models and data augmentation, pp 307–326, in *Bayesian Statistics 7*, eds J Bernardo, M Bayarri, J Berger, A Dawid, D Heckerman, A Smith, M West. OUP.
Parsons N, Edmondson R, Gilmour S (2006) A generalized estimating equation method for fitting autocorrelated ordinal score data with an application in horticultural research. *Applied Statistics*, 55, 507–524.
Pedroza C (2006) A Bayesian forecasting model: predicting U.S. male mortality. *Biostatistics*, 7, 530–550.
Pettitt A, Tran T, Haynes M, Hay J (2006) A Bayesian hierarchical model for categorical longitudinal data from a social survey of immigrants. *Journal of the Royal Statistical Society: Series A*, 127, 97–114.
Pourahmadi M (1999) Joint mean-covariance models with applications to longitudinal data: Unconstrained parameterisation. *Biometrika*, 86, 677–690.

Pourahmadi M (2000) Maximum likelihood estimation of generalized linear models for multivariate normal covariance matrix. *Biometrika*, 87, 425–435.

Pourahmadi M, Daniels M (2002) Dynamic conditionally linear mixed models. *Biometrics*, 58, 225–231.

Qiu Z, Song P, Tan M (2002) Bayesian hierarchical models for multi-level repeated ordinal data using WinBUGS. *Journal of Biopharmaceutical Statistics*, 12, 121–135.

Quick H, Waller L A, Casper M (2018) A multivariate space–time model for analysing county level heart disease death rates by race and sex. *Journal of the Royal Statistical Society: Series C (Applied Statistics)*, 67(1), 291–304.

Quintana F, Müller P, Rosner G (2008) A semiparametric Bayesian model for repeated binary measurements. *The Journal of the Royal Statistical Society: Series C (Applied Statistics)*, 57(4):419–431.

Rice K (2005) Bayesian measures of goodness of fit, in *Encyclopedia of Biostatistics*, eds P Armitage, T Colton. John Wiley, Chichester, UK.

Roberts GO, Sahu SK (2001) Approximate predetermined convergence properties of the Gibbs sampler. *Journal of Computational and Graphical Statistics*, 10(2), 216–229.

Rosenbaum P, Rubin D (1983) The central role of the propensity score in observational studies for causal effects. *Biometrika*, 70, 41–55.

Rossi P, Allenby G, McCulloch R (2005) *Bayesian Statistics and Marketing*. Wiley.

Roy J, Lin X (2000) Latent variable models for longitudinal data with multiple continuous outcomes. *Biometrics*, 56, 1047–1054.

Roy J, Lin X (2002) Analysis of multivariate longitudinal outcomes with non-ignorable dropouts and missing covariates: Changes in methadone treatment practices. *Journal of the American Statistical Association*, 97, 40–52.

Rubin DB (1976) Inference and missing data. *Biomelrika*, 63, 581–592.

Rubin D, Schenker N (1986) Multiple imputation for interval estimation from simple random samples with ignorable nonresponse. *Journal of the American Statistical Association*, 81, 366–374.

Rushworth A, Lee D, Mitchell R (2014) A spatio-temporal model for estimating the long-term effects of air pollution on respiratory hospital admissions in Greater London. *Spatial and Spatio-Temporal Epidemiology*, 10, 29–38.

Ryu D, Sinha D, Mallick B, Lipsitz S, Lipshultz S (2007) Longitudinal Studies with outcome-dependent follow-up: Models and Bayesian regression. *Journal of the American Statistical Association*, 102, 952–961.

Sahu S, Dey D, Branco M (2003) A new class of multivariateskew distributions with applications to bayesian regression models. *The Canadian Journal of Statistics*, 31, 129–150.

Savitsky T, Paddock S (2014) Bayesian semi-and non-parametric models for longitudinal data with multiple membership effects in R. *Journal of Statistical Software*, 57(3), 1–35.

Schafer J (1997) Imputation of missing covariates under a multivariate linear mixed model. Technical report, Dept. of Statistics, The Pennsylvania State University.

Schafer J, Graham J (2002) Missing data: Our view of the state of the art. *Psychological Methods*, 7, 147–177.

Schmid V, Held L (2004) Bayesian extrapolation of space–time trends in cancer registry data. *Biometrics*, 60(4), 1034–1042.

Schmid V, Held L (2007) Bayesian age-period-cohort modeling and prediction – BAMP. *Journal of Statistical Software*, 21(8). http://www.jstatsoft.org/

Song J, Belin TR (2004) Imputation for incomplete high-dimensional multivariate normal data using a common factor model. *Statistics in Medicine*, 23(18), 2827–2843.

Spiess M (2006) Estimation of a two-equation panel model with mixed continuous and ordered categorical outcomes and missing data. Jour Roy Stat Soc C 55: 525–538.

Squires D, Blumenthal D (2016) Mortality trends among working-age whites: The untold story. *Issue Brief (Commonwealth Fund)*, 3, 1–11.

Steele F (2008) Multilevel models for longitudinal data. *Journal of the Royal Statistical Society: Series A*, 171(1), 5–19.

Sun D, Tsutakawa R, Kim H, He Z (2000) Spatio-temporal interaction with disease mapping. *Statistics in Medicine*, 19, 2015–2035.

Ten Have T, Kunselman A, Pulkstenis E, Landis R (1998) Mixed effects logistic regression models for longitudinal binary response data with informative drop-out. *Biometrics*, 54, 367–383.

Terzi E, Cengiz M (2013) Bayesian hierarchical modeling for categorical longitudinal data from sedation measurements. *Computational and Mathematical Methods in Medicine*, 2013, 579214.

Thall P, Vail S (1990) Some covariance models for longitudinal count data with overdispersion. *Biometrics*, 46, 657–671.

Thiese M S (2014) Observational and interventional study design types; an overview. *Biochemia Medica*, 24(2), 199–210.

Troxel A, Harrington D, Lipsitz S (1998) Analysis of longitudinal data with non-ignorable non-monotone missing values. *Applied Statistics*, 47, 425–438.

Troxel A, Ma G, Heitjan D (2004) An index of local sensitivity to nonignorability. *Statistica Sinica*, 14, 1221–1237.

Tsai M-Y, Hsiao C (2008) Computation of reference Bayesian inference for variance components in longitudinal studies. *Computational Statistics*, 23(4), 587–604.

Tutz G, Kauermann G (2003) Generalized linear random effects models with varying coefficients. *Computational Statistics and Data Analysis*, 43, 13–28.

Tzala E, Best N (2008) Bayesian latent variable modelling of multivariate spatio-temporal variation in cancer mortality. *Statistical Methods in Medical Research*, 17, 97–118.

Vaidyanathan R (2016) Using a LKJ Prior in Stan. http://stla.github.io/stlapblog/posts/StanLKJprior.html

Verbeke G, Fieuws S, Molenberghs G, Davidian M (2014) The analysis of multivariate longitudinal data: A review. *Statistical Methods in Medical Research*, 23(1), 42–59.

Verbeke G, Molenberghs G, Rizopoulos D (2010) Random effects models for longitudinal data, Chapter 2, pp 37–96, in *Longitudinal Research with Latent Variables*, eds K van Montfort, J Oud, A Satorra. Springer.

Wakefield J, Smith A, Racine-Poon A, Gelfand A (1994) Bayesian analysis of linear and non-linear population models using the Gibbs sampler. *Journal of the Royal Statistical Society: Series C (Applied Statistics)*, 43, 201–221.

Weiss R (2005) *Modelling Longitudinal Data*. Springer, New York.

Weiss R, Cho M, Yanuzzi M (1999) On Bayesian calculations for mixture priors and likelihoods. *Statistics in Medicine*, 18, 1555–1570.

Wooldridge J (2005) Simple solutions to the initial conditions problem in dynamic, nonlinear panel data models with unobserved heterogeneity. *Journal of Applied Econometrics*, 20, 39–54.

Xu S, Jones R, Grunwald G (2007) Analysis of longitudinal count data with serial correlation. *Biometrical Journal*, 49, 416–428.

Yang X, Shoptaw S (2005) Assessing missing data assumptions in longitudinal studies: An example using a smoking cessation trial. *Drug and Alcohol Dependence*, 77, 213–225.

Yau KK, Kuk AY (2002) Robust estimation in generalized linear mixed models. *Journal of the Royal Statistical Society: Series B (Statistical Methodology)*, 64(1), 101–117.

Zayeri F, Kazemnejad A, Khanafshar N, Nayeri F (2005) Modeling repeated ordinal responses using a family of power transformations: Application to neonatal hypothermia data. *BMC Medical Research Methodology*, 5, 29.

Zhang D, Davidian M (2001) Linear mixed models with flexible distributions of random effects for longitudinal data. *Biometrics*, 57(3), 795–802.

Zhang Z (2016) Modeling error distributions of growth curve models through Bayesian methods. *Behavioral Research*, 48, 427–444.

Zhang Z, Hamagami F, Wang L, Grimm K, Nesselroade J (2007) Bayesian analysis of longitudinal data using growth curve models. *International Journal of Behavioral Development*, 31(4), 374–383.

Zhang Z, Keke L, Zhenqiu L, Xin T (2014) Bayesian inference and application of robust growth curve models using student's *t* distribution. *Structural Equation Modeling*, 20, 47–78.

11

Survival and Event History Models

11.1 Introduction

In many applications in the health and social sciences, the response of interest is duration to a certain event, such as age at first maternity, survival time after diagnosis, or times spent in different jobs or places of residence. In clinical applications, the interest is typically in representing and comparing the distribution of times to an event among different patient groups (e.g treatment vs control groups) (Brard et al., 2017), whereas in social science applications, the interest may focus on the impacts of demographic or socioeconomic attributes on human behaviours.

Typically, durations or event times are not observed for all subjects, either because not all subjects are followed up, or because for some events the event may never occur (e.g. age at first marriage). So some times are missing or censored, and the missingness mechanism is generally assumed to be at random. The most common form is right-censoring, when the event has not occurred by the end of the observation period; the unknown failure time exceeds the subject's survival time c when observation ceased. A failure time is left censored at c if its unobserved actual value is less than c (e.g. a population census may record limiting illness status by current age, but not the age when it commenced). A failure time is interval censored if it is known only that it lies in the interval (c_1,c_2).

Distributions of durations or survival times are equivalently described by hazard rates, also known as failure rates, exit rates, or forces of mortality according to the application. The modelling of the hazard rate through time may be undertaken parametrically. Alternatively, one may adopt semiparametric methods, such as assuming piecewise constancy in the rates within sub-intervals of the observation span (Ibrahim et al., 2001). Pooling strength through correlated priors is then relevant, as rates in successive intervals tend to be similar. Imposing smoothness conditions on the baseline hazard also provides stable estimators when observations are sparse at particular durations (Omori, 2003).

Variations in failure rates between subjects or other units may be explained to a large degree by observed covariates, the impact of which may also vary over intervals or time. Selection of covariates may be relevant in particular applications (Lee et al., 2015). However, unobserved random variations between subjects are present in many applications and may be modelled by introducing subject level frailty (see section 11.4). Additionally, duration times may be hierarchically stratified (e.g. patient survival by hospital or by area of residence) (e.g. Austin, 2017). Durations or survival times may also be differentiated by types of possible exit, as in competing risk analysis (see Section 11.7). One may also consider multivariate survival outcomes, as in multiple component failure (Damien and Muller, 1998) or in familial survival studies (Viswanathan and Manatunga, 2001). In such situations, shared frailty models may account for correlated unobserved variation over different strata or causes of exit.

Survival analysis options in R are summarised at https://cran.r-project.org/web/views/Survival.html, with a review provided by Crippa (2018). Recently developed R survival packages using Bayesian computing include biostan (https://github.com/jburos/biostan), BayesMixSurv (https://cran.r-project.org/web/packages/BayesMixSurv/index.html), dynsurv (https://cran.r-project.org/web/packages/dynsurv/dynsurv.pdf), MRH (Hagar et al., 2017), bamlss (Umlauf et al., 2017), R2BayesX, spBayesSurv (Zhou et al., 2017, 2018) for spatially nested data, CFC (https://cran.r-project.org/web/packages/CFC/CFC.pdf), and icensBKL (Bogaerts et al., 2017). The survivalstan package (https://jburos.github.io/survivalstan/index.html) is for Python, but with transferable rstan codes.

11.2 Survival Analysis in Continuous Time

Let T denote a survival time. The distribution function of T, providing the probability of exit before time $T=t$, is then

$$F(t) = Pr(T \leq t),$$

while the probability of surviving beyond t is $S(t) = 1 - F(t) = Pr(T > t)$. Note that one has $S(\infty) = 0$, except for applications with a cure fraction (Lambert, 2007). So, the density of T can be expressed as

$$f(t) = \frac{dF(t)}{dt} = -\frac{dS(t)}{dt}.$$

The chance of an event occurring in a short interval $(t, t+dt)$, given survival to t, is

$$Pr(t < T \leq t+dt \mid T > t) = \frac{Pr(t < T \leq t+dt)}{Pr(T > t)} = \frac{F(t+dt) - F(t)}{S(t)}.$$

The hazard function $h(t)$ is the instantaneous event rate, obtained as $dt \to 0$ in the ratio of the preceding probability to the length of the interval dt. That is

$$h(t) = \lim_{dt \to 0} \frac{F(t+dt) - F(t)}{dt} \frac{1}{S(t)} = \lim_{dt \to 0} \frac{S(t) - S(t+dt)}{dt} \frac{1}{S(t)} = \frac{f(t)}{S(t)}.$$

Since $-f(t)$ is the derivative of $S(t)$, one obtains that $h(t) = -S'(t)/S(t)$, and so

$$h(t) = \frac{-d \log S(t)}{dt}. \tag{11.1}$$

On integrating both sides in (11.1), one obtains the cumulative hazard rate

$$H(t) = \int_0^t h(u)du = \int_0^t \left[\frac{-d \log S(u)}{du}\right] du = -\int_0^{-\log S(t)} d \log S(u) = -\log S(t),$$

and so

$$S(t) = \exp[-H(t)] = \exp\left[-\int_0^t h(u)du\right].$$

The hazard function is the central focus for modelling variations in survival. Assume predictors Z_i are available (excluding a constant). Their impact is most simply modelled using a proportional hazards form (e.g. Kiefer, 1988; Li, 2007)

$$h(t \mid Z) = h_0(t)\exp(Z_i\beta),$$

where $h_0(t)$ is known as the baseline hazard, and the regression impact is constant across time. Letting $\eta_i = Z_i\beta$, the associated survivor function is

$$\begin{aligned} S(t \mid Z_i) &= \exp\left[-\int_0^t h(u \mid Z_i)du\right] \\ &= \exp\left[-H_0(t)e^{\eta_i}\right] \\ &= \left[S_0(t)\right]^{\exp(\eta_i)} \\ &= \exp\left\{-\exp\left[\eta_i + \log H_0(t)\right]\right\}, \end{aligned}$$

where $H_0(t)$ is the integrated baseline hazard. The proportional hazard assumption is often restrictive, though Yin and Ibrahim (2006) show the proportional hazard model (PHM) may be nested in a broader class of transformation hazard models, with parameter $0 \le \gamma \le 1$ and

$$h(t \mid Z) = \left[h_0(t)^{\gamma} + \exp(Z_i\beta)\right]^{1/\gamma}$$

which reduces to the proportional model when $\gamma = 0$ and to an additive model when $\gamma = 1$.

Consider an absorbing (non-repeatable) type of exit, and let $d_i = 1$ for an observed exit and $d_i = 0$ for a censored time. Assuming censoring is non-informative, the likelihood contribution for subject i is

$$f(t_i \mid Z_i) = h(t_i \mid Z_i)S(t_i \mid Z_i)$$

if $d_i = 1$, and $S(t_i \mid Z_i)$ if $d_i = 0$. The likelihood contribution may therefore be expressed in equivalent form as

$$h(t_i \mid Z_i)^{d_i} S_i(t_i \mid Z_i) = f(t_i \mid Z_i)^{d_i} S_i(t_i \mid Z_i)^{1-d_i}.$$

For a PHM, the likelihood contribution also may be written (Aitkin and Clayton, 1980; Orbe and Nunez-Anton, 2006) as

$$\left[h_0(t_i)\exp\{Z_i\beta - H_0(t_i)\exp(Z_i\beta)\}\right]^{d_i}\left[\exp\{-H_0(t_i)\exp(Z_i\beta)\}\right]^{1-d_i}$$

$$= \left(\mu_i^{d_i} e^{-\mu_i}\right)\left(\frac{h_0(t_i)}{H_0(t_i)}\right)^{d_i} \tag{11.2}$$

where

$$\mu_i = H_0(t_i)\exp(Z_i\beta) \tag{11.3}$$

and the second bracketed term in (11.2) depends only on the baseline hazard and is independent of β. The first term in (11.2) is the kernel of a Poisson likelihood for the event status indicators $d_i \sim Po(\mu_i)$. From (11.3), the corresponding log-linear model is

$$\log(\mu_i \mid t_i, Z_i) = \log(H_0(t_i) + Z_i\beta \tag{11.4}$$

where $\log(H_0(t_i))$ is an offset using the observed time, whether censored or uncensored.

11.2.1 Counting Process Functions

For repeated events in continuous time, especially with successive durations not necessarily independent, it may be advantageous to use additional functions. The count of failures $N(t)$ occurring over $(0,t]$ for a given individual or component system defines a counting process satisfying $N(s) \leq N(t)$ for $s<t$. For a non-repeatable event, the counting process may still be useful (e.g. in modelling time-varying impacts of predictors), and one may denote $N(t) = I(T \leq t)$, namely by an indicator of whether the event has occurred by t. The counting process formulation extends straightforwardly to time-dependent covariates and frailty (Chen et al., 2014).

For event types considered over sufficiently small intervals, the counting process increments $dN(t) = N(t) - N(t-)$ are either 1 or 0, where $N(t-)$ denotes $\lim_{\delta \downarrow 0} N(t-\delta)$ (Manda et al., 2005). Let $A(t-)$ denote the antecedent history of the event sequence up to, but not including, t. Then conditional on $A(t-)$, the probability that $dN(t)=1$ can be written in terms of an intensity process $\lambda(t)$, namely,

$$Pr\{N(t+\delta) - N(t-) = 1 \mid A(t-)\} \simeq \lambda(t)\delta.$$

Equivalently

$$Pr\{dN(t) = 1 \mid A(t-)\} \simeq d\Lambda(t),$$

where $\Lambda(t) = \int_0^t \lambda(u)du$ is the integrated intensity, with $\Lambda(t) = E(N(t))$.

The intensity is equal to the hazard while the subject or system is still under observation, that is, still at risk, but is zero when the event has happened (when the event is non-repeatable), or when a sequence of (repeatable) events has finished. An example of the latter might be when a repairable system subject to repeated breakdowns is finally decommissioned – see Watson et al. (2002) for a counting process analysis of failure times of water pipes. Let $Y(t) = I(T \geq t)$ denote the at-risk indicator, then

$$\lambda(t) = Y(t)h(t).$$

This representation of the intensity function generalises to include predictors and random effects (or frailties). So for proportional hazards, effects of predictor Z_i would be included via

$$\lambda(t_i \mid Z_i) = Y(t_i)h_0(t_i)\exp(Z_i\beta).$$

One may then compare observed and predicted counts via the Martingale residual at t, defined as

$$M_i(t) = N(t_i) - \Lambda_0(t_i \mid Z_i) = N(t_i) - \int_0^{t_i} Y_i(u)\exp(Z_i\beta)dH_0(u).$$

The total residual $M_i = M_i(\infty)$ for a subject with observation time t_i is obtainable for a non-repeatable event, and event indicator d_i, as

$$M_i = d_i - \Lambda_0(t_i \mid Z_i).$$

Deviance residuals r_i are obtained as

$$r_i = \text{sgn}(M_i)\sqrt{2\left[M_i - N_i(\infty)\log\left(\frac{N_i(\infty) - M_i}{N_i(\infty)}\right)\right]}$$

11.2.2 Parametric Hazards

The hazard rate $h(t)$ is called "duration dependent" if its value changes over t. Under negative duration dependence (often observed in occupational or residential careers), $h(t)$ decreases with time. In practice, plots of survivor proportions are often jagged with respect to time, and semiparametric or non-parametric methods for representing the hazard function reflect this. However, parametric lifetime models are also often applied to test whether certain basic features of duration dependence are supported by the data; see http://rstudio-pubs-static.s3.amazonaws.com/5560_c24449c468224fd4af9f3c512a24e07d.html for a discussion of exploratory graphical comparison of parametric and non-parametric approaches in R.

The simplest parametric model is the exponential model, under which the leaving rate is constant, defining a stationary process with hazard

$$h(t) = \lambda,$$

survival function $S(t \mid \lambda) = \exp(-\lambda t)$, and density

$$f(t \mid \lambda) = \lambda \exp(-\lambda t).$$

Contributions to the likelihood depend on event status d_i, namely

$$f(t_i \mid \lambda, d_i) = \lambda^{d_i} \exp(-\lambda t_i).$$

With covariates Z_i (excluding an intercept), and assuming proportionality

$$h(t \mid Z_i) = \lambda e^{Z_i\beta}.$$

Alternatively, setting $\lambda = e^{\beta_0}$, exponential hazard regression can be represented as

$$h(t \mid Z_i) = e^{\beta_0 + Z_i\beta}.$$

Equivalently, under the Poisson likelihood approach of Aitkin and Clayton (1980), one has, for event indicators d_i,

$$d_i \sim \text{Po}(\mu_i),$$

where, from (11.4),

$$\log(\mu_i) = \log(\lambda t_i) + Z_i\beta,$$

since $H_0(t) = \lambda t$. Absorbing λ into the regression term, one has

$$\log(\mu_i) = \log(t_i) + \beta_0 + Z_i\beta.$$

This Poisson likelihood device can be used in piecewise exponential models as considered below.

Another commonly used parametric form is the Weibull (e.g. Thamrin et al., 2013), with scale parameter λ and shape κ, namely

$$h(t \mid \lambda, \kappa) = \lambda\kappa t^{\kappa-1},$$

so that

$$S(t \mid \lambda, \kappa) = \exp[-\lambda t^{\kappa}],$$

$$f(t \mid \lambda, \kappa) = \lambda\kappa t^{\kappa-1}\exp[-\lambda t^{\kappa}].$$

The Weibull hazard rate is monotonic, with positive duration dependence if $\kappa > 1$ (and if the 95% credible interval excludes 1), and negative dependence if $\kappa < 1$. Impacts of covariates, including an intercept, on the hazard, are represented via

$$h(t_i \mid \lambda, \kappa, Z_i, \beta) = e^{\beta_0 + Z_i\beta}\kappa t^{\kappa-1}.$$

The generalised gamma density (Stacy, 1962; Cox and Matheson, 2014) is of interest in including the Weibull, gamma, and lognormal as special cases. This has various parameterisations, with BUGS using the representation

$$f(t \mid \alpha, \lambda, \gamma) = \frac{\gamma}{\Gamma(\alpha)}\lambda^{\alpha\gamma}t^{\alpha\gamma-1}\exp\left[-(\lambda t)^{\gamma}\right],$$

as in Morris et al. (1994) (see Example 11.1 below). Instead, one may take $B = \lambda^{\gamma}$, leading to

$$f(t \mid \alpha, \lambda, \gamma) = \frac{\gamma}{\Gamma(\alpha)}B^{\alpha}t^{\alpha\gamma-1}\exp[-t^{\gamma}B],$$

where setting $\gamma = 1$ gives the gamma, $\alpha = 1$ gives the Weibull, and $\alpha \to \infty$ provides the lognormal. The survival function is $1 - I(\gamma, t^{\gamma}B)$, where $I(a, x) = (1/\Gamma(a))\int_0^x u^{a-1}e^{-u}du$. Cox and Matheson (2014) consider a parameterisation involving $\delta = \gamma^{0.5}$, allowed to take both positive and negative values, with the sign of δ leading to different survivor functions.

Preliminary assessment of different parametric hazards can be obtained by density estimation without covariates (e.g. using fitdistr in R), or by graphical means. For example, under the Weibull one has $\log(S(t \mid \lambda,\kappa)) = -\lambda t^{\kappa}$, and hence $\log(-\log(S(t \mid \lambda,\kappa))) = \log\lambda + \kappa\log t$. Therefore, a plot of $\log(-\log(\hat{S}(t|\lambda,\kappa)))$ against $\log(t)$ should be approximately linear when a Weibull is appropriate. An initial assessment can be made using a Kaplan–Meier estimate of $S(t)$ in the *R* package. Assume a dataset named survdat containing variables time (corresponding to observed survival times t_i, whether censored or not), and status (corresponding to d_i). Then the procedure is

```
library(survival)
KMinputs <- Surv(survdat$time,survdat$status)
KM <- survfit(KMinputs)
plot(log(KM$time), log(-log(KM$surv)), type="S").
```

However, many processes exhibit peaks in exit rates; for example, the rate may at first increase, but after reaching a peak, tail off again (Gore et al., 1984; Shao and Zhou, 2004). Parametric models accommodating such a pattern include the log-logistic model and the sickle model (Bennett, 1983; Brüderl and Diekmann, 1995; Diekmann and Mitter, 1983). The log-logistic density has hazard

$$h(t) = \lambda\kappa t^{\kappa-1}\left[1+\lambda t^{\kappa}\right]^{-1},$$

and survivor function

$$S(t) = \left[1+\lambda t^{\kappa}\right]^{-1},$$

where all parameters are positive, and the scale parameter λ can be adapted to model the impact of predictors; see Li (1999) for a Bayesian application to Chapter 11 bankruptcies. An alternative common parameterisation (Florens et al., 1995) sets $\lambda = \nu^{\kappa}$, so that

$$h(t) = \frac{\nu^{\kappa}\kappa t^{\kappa-1}}{[1+(\nu t)^{\kappa}]}. \tag{11.5}$$

The sickle model has corresponding functions

$$h(t) = cte^{-t/\lambda}$$

$$S(t) = \exp\left[-\lambda c\left\{\lambda-(t+\lambda)e^{-t/\lambda}\right\}\right]$$

with both c and λ positive. The sickle model has a permanent survival probability or "cure rate" (Chen et al., 1999) in that $S(\infty) > 0$ (see Section 11.4.1). In general, one may define a cure rate $r = 1-\pi$ as the limit as $t \to \infty$ of the survivor function, namely (Tsodikov et al., 2003)

$$r = \lim_{t\to\infty} S(t) = \exp\left[-\int_0^{\infty} h(u)du\right]$$

with π denoting the proportion of susceptibles.

11.2.3 Accelerated Hazards

In contrast to the proportional hazard model with $h(t_i | Z_i) = h_0(t_i)\exp(Z_i\beta)$, in an accelerated failure time (AFT) model the explanatory variates are assumed to act multiplicatively on time (Wei, 1992; Swindell, 2009; Rivas-López et al., 2014). AFT models focus on the effect of the explanatory variates on the survival function, rather than the hazard function under proportional hazards models. With $B_i = \exp(Z_i\beta)$, one has

$$h(t_i | Z_i) = h_0(t_i B_i) B_i,$$

$$S(t_i | Z_i) = [S_0(tB_i)]^{B_i}$$

and the effect of the predictors Z_i on survival time is more direct, acting to accelerate or decelerate the time to failure. To illustrate this in the case of a treatment comparison, assume Z_i excludes an intercept, and that the baseline hazard includes a scale parameter to model the mean hazard (e.g. the parameter λ in exponential and Weibull models). Also assume a single predictor such as $z_i=1$ for a new treatment and $z_i=0$ for control. Then, with $B_i = e^{\beta z_i} = e^{\beta}(= \phi)$ for a treated subject, one has a hazard $\phi h_0(\phi t_i)$ and a survivor function $S(\phi t_i)$ for a treated subject, but a hazard $h_0(t_i)$ and a survivor function $S(t_i)$ for a control subject. So, the lifetime under the new treatment is ϕ times the lifetime under the control regime.

More inclusive schemes are possible. For example, defining $G_i = \exp(Z_i\gamma)$, one has

$$h(t_i | Z_i) = h_0(t_i G_i) B_i,$$

which includes the AFT and PHM forms as special cases (Chen and Jewell, 2001). For example, for the log-logistic density, this would imply

$$h(t_i | Z_i) = B_i \lambda \kappa (t_i G_i)^{\kappa-1} \left[1 + B_i \lambda (tG_i)^{\kappa}\right]^{-1}.$$

Apart from avoiding the assumption of proportional hazards, the AFT approach has the advantage of a direct regression form which may be useful in modelling nonlinear effects of predictors (Orbe and Nunez-Anton, 2006). Let Z_i be of dimension p, and T_i denote the completed failure time which for censored subjects is unobserved. Then $T_i = t_i$ when $d_i = 1$ but $T_i > t_i$ when $d_i = 0$, so truncated sampling with the censored time as the lower limit is necessary. The regression formulation is then

$$\log(T_i) = Z_i \gamma + \sigma u_i,$$

where σ is a scale parameter, and the errors are defined by the survivor function, namely

$$S(t_i) = Pr(T_i > t_i) = Pr(\log(T_i) > \log(t_i))$$

$$= Pr\left(u_i > \frac{\log(t_i) - \gamma_0 - z_{1i}\gamma_1 - \ldots - z_{pi}\gamma_p}{\sigma}\right).$$

A positive γ_j coefficient means that z_j leads to longer survival or length of stay.

Taking u to be standard normal with variance 1 corresponds to a log-normal density for failure times t_i, under which

$$S(t_i) = 1 - \Phi\left(\frac{\log(t_i) - Z_i\gamma}{\sigma}\right).$$

Taking u to be standard logistic, with density $p(u) = e^u/(1+e^u)^2$, corresponds to a log-logistic failure time density with

$$S(t_i) = \left[1 + \exp\left\{\frac{\log(t_i) - Z_i\gamma}{\sigma}\right\}\right]^{-1}$$

with σ corresponding to the inverse of the shape parameter κ. Finally, consider a Weibull density for failure times with hazard $h(t_i \mid Z_i) = \lambda\kappa t_i^{\kappa-1}\exp(\beta Z_i)$ where Z_i excludes a constant term. Taking u to follow a standard extreme value density, namely $p(u) = \exp(u - e^u)$, the AFT regression takes the form (Keiding et al., 1997)

$$\log(T_i) = -\frac{\log\lambda}{\kappa} - z_{1i}\frac{\beta_1}{\kappa}\ldots - z_{pi}\frac{\beta_p}{\kappa} + \frac{u_i}{\kappa}.$$

so that $\gamma_j = -\beta_j/\kappa$.

Example 11.1 Nursing Home Stays

Morris et al. (1994) analyse lengths of stay t_i for $n = 1601$ nursing home patients, with stay usually terminated by death, with predictors being patient age and gender, their marital status and dependency level, and the type of nursing home assignment (1: receive treatment, 0: control) (www.stats.ox.ac.uk/pub/datasets/csb/). There are 322 censored lengths of stay.

Predictor effects are initially assessed via proportional hazard (PH) Weibull regression. Under the Weibull, one has hazard

$$h(t \mid \lambda_i, \kappa) = \lambda_i\kappa t_i^{\kappa-1}, \quad (11.6)$$

$$\lambda_i = \exp(Z_i\beta),$$

with density

$$f(t \mid \lambda_i, \kappa) = \lambda_i\kappa t_i^{\kappa-1}\exp(-\lambda_i t_i^{a}).$$

This is equivalent to accelerated hazards regression for logged length of stay with error u_i

$$\log(t_i) = Z_i\gamma + \sigma u_i$$

where $\gamma = -\beta/\kappa$. The γ coefficients express influences on length of stay (i.e. survival) while the β coefficients express influences on mortality.

We consider rstan estimation for the PH Weibull model, with both priors and likelihoods represented using the target += option, and an implicit flat prior on the regression coefficients. For the Weibull analysis, the log likelihood for censored cases is provided by the weibull_lccdf function in rstan, namely the log of the Weibull complementary cumulative distribution function for the response t_i. It may be noted that the rstan parameterisation of the Weibull is

$$h(t \mid \lambda_i, \kappa) = \frac{\kappa}{\lambda_i}\left(\frac{t_i}{\lambda_i}\right)^{\kappa-1},$$

differing from that in BUGS and JAGS, and a re-expression of the regression term $\eta_i = Z_i\beta$ is needed to achieve results in line with the parameterisation (11.6) (Buros, 2016). Thus, one obtains the code elements

```
transformed parameters {real eta[n];
real nu[n];
for (i in 1:n) {eta[i]=beta[1]+beta[2]*age[i]/100+beta[3]*trt[i]+
beta[4]*gender[i]
+beta[5]*marstat[i]+beta[6]*hltst3[i]+beta[7]*hltst4[i]+beta[8
]*hltst5[i];
nu[i] = exp(-eta[i]/kappa);}}
model {for (i in 1:n) {if (censored[i] == 0) {target += weibull_
lpdf(time[i] kappa, nu[i]);}
else if (censored[i] == 1) {target += weibull_lccdf(time[i]
kappa, nu[i]);}}}
```

The Weibull shape parameter κ has a posterior 95% credible interval entirely under 1, so mortality is associated with shorter stays (sometimes denoted negative duration dependence). This feature, combined with a negative age effect (albeit not significant), may reflect varying frailty (selection effects). Other regression coefficient estimates (Table 11.1) for the treatment and attribute variables replicate those of Morris et al. (1994). Health status is measured in terms of dependency in activities of daily living; with health=2 if there are four or fewer activities with dependence (reference category), health=3 for five dependencies, health=4 for six dependencies, and health=5 if there were special medical conditions requiring extra care. It can be seen that higher ADL (activities of daily living) dependency is associated with earlier mortality and shorter stays. The LOO-IC for the Weibull model is 16463, with κ estimated as 0.61. Estimation using the AFT form gives the same result.

TABLE 11.1

Nursing Home Stays. Parameter Posterior Summary

	Weibull PH Model			Generalised Gamma AFT			Weibull AFT		
	Mean	2.50%	97.50%	Mean	2.50%	97.50%	Mean	2.50%	97.50%
	Influences on Leaving NH			Influences on Stay Length			Influences on Stay Length		
Age[a]	−0.45	−1.18	0.30	1.27	0.06	2.57	0.71	−0.49	1.93
Treatment	−0.13	−0.24	−0.02	0.11	−0.09	0.28	0.21	0.02	0.39
Male	0.35	0.22	0.48	−0.58	−0.80	−0.34	−0.57	−0.79	−0.35
Married	0.16	0.01	0.31	−0.22	−0.50	0.07	−0.26	−0.51	−0.01
ADL Status 3	−0.03	−0.19	0.13	0.07	−0.10	0.27	0.05	−0.21	0.28
ADL Status 4	0.23	0.08	0.39	−0.44	−0.61	−0.30	−0.38	−0.64	−0.14
ADL Status 5	0.53	0.33	0.74	−0.85	−1.20	−0.44	−0.88	−1.20	−0.54
κ	0.61	0.58	0.64						
α				8.60	7.53	9.25			
γ				0.18	0.17	0.19			
σ							1.64	1.57	1.72
LOO-IC		16463			16364			16463	

[a] Actual age divided by 100.

To assess poorly fitted observations, one may implement different forms of residual. Here the Martingale residual and the normal deviate residual are obtained, with simulation as in Nardi and Schemper (2003) to obtain estimated normal deviate residuals for censored observations. These two residuals have a correlation of 0.94, and high negative values on both highlight subjects (e.g. 1,589 and 1,596) with long lengths of stay despite high ADL dependency.

There is evidence of redundancy among the predictors used above, and covariate selection or shrinkage priors could be applied (e.g. Zhang et al., 2018). The impact of the latter can be demonstrated simply by an application of the BayesMixSurv package, which estimates a two-component discrete mixture of Weibull regressions, but allows a single component option. Lasso shrinkage priors are assumed for regression coefficients in this package. Thus, defining an event indicator (endstay=1-censored) as the complement of the censoring indicator, and defining agec=age/100, one has

C1=bayesmixsurv(Surv(time,endstay)~trt+agec+marstat+gender+hltst3+hltst4+hltst5, D, control=bayesmixsurv.control(iter=1000,single=T)).

This shows an age coefficient much closer to zero (with posterior mean −0.03) than obtained using the flat prior in the rstan code.

A Weibull accelerated failure time regression can be obtained by simply replacing nu[i] = exp(−eta[i]/kappa) by nu[i] = exp(eta[i]) in the preceding code. As in Morris et al. (1994), we compare Weibull and generalised gamma AFT regressions. The latter requires specific functions in the rstan code to define the density and survivor likelihoods.

Convergence issues have been noted for the generalised gamma, even under maximum likelihood, though convergence may be improved by fixing one of the extra generalised gamma parameters (Lawless, 1980). Estimates here are based on a single chain run of 5,000 iterations in rstan, at which point SRFs for the hyperparameters are 1.3 or less. Table 11.1 shows a more pronounced effect of age, and a diminished treatment effect, under the generalised gamma, which produces a lower LOO-IC (16364) than the Weibull AFT. The estimate of α, with 95% CRI (7.5,9.3), suggests the lognormal may be preferred to the Weibull.

11.3 Semiparametric Hazards

In the proportional hazards model

$$h(t_i \mid Z_i) = h_0(t_i)\exp(Z_i\beta),$$

it may be difficult to choose a parametric form for the baseline hazard $h_0(t)$, and semiparametric or non-parametric approaches are often preferable. These have benefits in avoiding possible mis-specification of parametric hazard forms, and in facilitating other aspects of hazard regression, such as time-varying predictor effects (Gamerman, 1991). Such approaches have been applied to the cumulative hazard, and implemented in counting process models (Clayton, 1991). However, they may also be specified for the baseline hazard h_0 itself (e.g. Gamerman, 1991; Sinha and Dey, 1997) and typically use only information about the time intervals in which exit occurs.

Consider a partition of the response time scale into J intervals $(a_0,a_1],\ldots,(a_{J-1},a_J]$, where a_J equals or exceeds the largest observed time, censored or uncensored (Ibrahim et al., 2001, p.106). The partition scheme can be based on distinct values in the profile of observed times $\{t_1,\ldots,t_n\}$, whether censored or not, or by siting knots a_j at selected points in the range (t_{min},t_{max}). Yin and Ibrahim (2006, p.173) propose that the partitioning should ensure an approximately equal number of failures in each of the J intervals, with each interval containing at least one failure. Among alternatives are knots sited at $((j-1)/J)^{\text{th}}$ quantiles of observed times (Gustafson et al., 2000), or evenly spaced along the range of the observed t values. As the number of intervals J tends to infinity, a truly non-parametric model is obtained, but is not likely to be empirically well identified (Lopes et al., 2007).

Different approaches may be based on the assumption that the baseline hazard is constant within each interval. Thus Ibrahim et al. (1999) and Ibrahim et al. (2001, p.55) consider discrete approximation to the gamma process of Dykstra and Laud (1981). This involves a prior on the increments

$$\Delta_j = h_0(a_j) - h_0(a_{j-1}), \quad j = 1,\ldots,J$$

in the baseline hazard, and use of the approximate survival function

$$S(t \mid Z_i) = \exp\left[-B_i \int_0^t h_0(u)du\right] \simeq \exp\left[-B_i \sum_{j=1}^{J} \Delta_j (t-a_{j-1})_+\right],$$

where $B_i = e^{Z_i\beta}$, and $(u)_+ = u$ if $u > 0$ and is zero otherwise. The probability of exit in interval j is then

$$\begin{aligned} q_j &= S(a_{j-1}) - S(a_j) \\ &\simeq \left[\exp\left\{-B_i \sum_{m=1}^{j-1} \Delta_m (a_{j-1} - a_{m-1})\right\}\right]\left[1 - \exp\left\{-B_i (a_j - a_{j-1}) \sum_{m=1}^{j} \Delta_m\right\}\right]. \end{aligned}$$

11.3.1 Piecewise Exponential Priors

Piecewise exponential (PE) priors (Ibrahim et al., 2001; Bender et al., 2018; Sinha et al., 1999; Demarqui et al., 2008; Brezger et al., 2005) are one approach to estimating the hazard function without specifying the hazard parametrically, though semiparametric approaches avoiding the simplifying PE assumptions have been proposed (Murray et al., 2016; Marano et al., 2016). The PE prior specifies a baseline parameter λ_j for each interval, possibly combined with interval-specific regression parameters β_j, so that

$$h(t_i \in (a_{j-1}, a_j] \mid Z_i) = \lambda_j \exp(Z_i \beta_j),$$

where Z_i excludes an intercept. Let $B_{ij} = \exp(Z_i \beta_j)$. For a subject surviving beyond the jth interval, namely with $t_i > a_j$, the likelihood contribution during interval j is

$$\exp(-\lambda_j (a_j - a_{j-1}) B_{ij}).$$

For a subject with $a_{j-1} < t_i \le a_j$, either failing ($d_{ij} = 1$) in interval j, or censored but nevertheless exiting ($d_{ij} = 0$) in the jth interval, the likelihood contribution is

$$\left[\lambda_j B_{ij}\right]^{d_{ij}} \exp\left[-\lambda_k (t_i - a_{j-1}) B_{ij})\right].$$

So, a Poisson likelihood approach may be applied as in (11.2)–(11.4), with responses y_{ij} defined by the event type in each interval, and with offsets Δ_{ij}, defined according to whether the subject survives the interval (see Example 11.2).

The successive baseline parameters λ_j are likely to be correlated, but also possibly to show erratic fluctuations or be imprecisely estimated if treated as fixed effects. Hence, a smoothing prior is indicated. One might assume a parametric model (e.g. polynomial in j) but allowing for additional random variation. Thus, Albert and Chib (2001) and Omori (2003) assume a polynomial for $\alpha_j = \log(\lambda_j)$, whereby

$$\alpha_j = \psi_0 + \psi_1(j-1) + \psi_2(j-1)^2 + u_j,$$

with u_j normal. Pooling strength under autocorrelated priors linking successive λ_j or α_j is also widely applied. These are known as correlated prior processes or Martingale prior processes for the baseline hazard. Possibilities are first or second order random walks (RW) in the α_j, possibly adjusted to reflect unequal width $\delta_j = (a_j - a_{j-1})$ of the intervals. Thus, one might take a 1st order random walk,

$$\alpha_j \sim N(\alpha_{j-1}, \sigma_\alpha^2 \delta_j),$$

with α_1 a separate fixed effect, and with $\tau_\alpha = 1/\sigma_\alpha^2$ following a gamma prior. Alternatively, as in Gustafson et al. (2003), one may take

$$w_j = 0.5(a_j + a_{j+1}),$$

$$\zeta_j = w_j - w_{j-1},$$

$$\alpha_j \sim N(\alpha_{j-1} + (\alpha_{j-1} - \alpha_{j-2})\frac{\zeta_j}{\zeta_{j-1}},\ \sigma_\alpha^2(\zeta_j/\bar{\zeta})^2).$$

Since setting particular partitions of the time scale involves an element of arbitrariness, Sahu and Dey (2004) apply RJMCMC (reversible jump MCMC) techniques in which J is an additional unknown; they specify a sparse precision matrix formulation for the joint prior for the $(\alpha_1, \ldots, \alpha_J)$ under an RW1 prior for particular J values.

Because random walk priors of degree r set a mean level not on the α_j themselves, but on differences of order r (e.g. an RW1 prior specifies a zero mean for $\alpha_j - \alpha_{j-1}$), identifiability may require that a separate regression intercept is omitted or that the α_j are centred to sum to zero at each MCMC iteration, by the operation $\alpha_j' = \alpha_j - \bar{\alpha}$. Alternatives are to set any value, say the hth, to zero (by the operation $\alpha_j' = \alpha_j - \alpha_h$ at each iteration), or set the first effect α_1 to zero (Sahu and Dey, 2004).

A gamma prior in the baseline hazard rates λ_j is also possible (Arjas and Gasbarra, 1994), namely

$$\lambda_j \sim Ga(b, b/\lambda_{j-1}),$$

where λ_1 is a separate positive effect, and larger b values lead to smoother sequences of λ_j. The same identifiability issues obtain as for $\alpha_j = \log(\lambda_j)$ and devices such as normalisation of the λ_j (to value 1) at each iteration may be applied.

Piecewise priors may also be used to model non-constant predictor effects, though typically values of time-varying regression coefficients β_j in successive intervals are expected to be close (Sinha et al., 1999). Sargent (1998) considers alternative gamma priors for the precision $\tau_\beta = 1/\sigma_\beta^2$ of regression coefficients assumed to evolve according to a first order random walk, adjusted for different interval widths,

$$\beta_j \sim N(\beta_{j-1}, \sigma_\beta^2 \delta_j).$$

Prior knowledge in this application (from the Veterans Administration lung cancer trial) suggests that values of time-varying coefficients on successive days would differ by at most 0.001. Taking this as the standard deviation of the normal distribution, the prior mean precision for the gamma is 10^6. This corresponds to quartiles (0.0027, 0.0038, 0.0059) for σ_β. An alternative prior adopted by Sargent has mean precision 10^5. Posterior inferences for the mean precision were different under the alternative priors, but not those for the estimated β_j. Fahrmeir and Knorr-Held (1997) suggest gamma $Ga(1,b)$ priors on precision parameters τ_α on varying log baseline rates, or precisions τ_β on varying predictor effects. Sensitivity is gauged by taking alternative values for b (e.g. $b=0.05$ and $b=0.0005$), since b determines how close to zero the variances are allowed to be a priori.

11.3.2 Cumulative Hazard Specifications

Semiparametric approaches may also be applied to the cumulative baseline hazard H_0 (Kalbfleisch, 1978). Consider a counting process approach with data $(N_i(t), Y_i(t), Z_i(t))$ and independent priors on β and H_0. For an individual i exiting or censored before t, so that $Y_i(s)=0$ for $s>t$, one may apply a Poisson likelihood with binary responses $dN_i(t)$ and means $Y_i(t)\exp(Z_i(t)\beta)dH_0(t)$.

An independent gamma increments prior for dH_0 (Phadia, 2015) may be adopted (assuming a constant baseline hazard in each interval), namely

$$dH_0(t) \sim Ga(c[dH^*(t)], c)$$

where $dH^*(t)$ is a prior estimate of the hazard rate per unit time. Other possibilities include normal priors on $\log(dH_0)$.

Let $J+1$ intervals $(s_0, s_1], \ldots, (s_J, s_{J+1}]$ be defined by the J distinct failure times in a dataset, with s_1 equal to the minimum observed failure time, and s_{J+1} exceeding the largest failure time s_J (Sargent, 1997, p.16). The likelihood for individual i exiting or censored before s_j, so that $Y_{ij}=0$ for $t>s_j$, reduces to a discretised form of Poisson likelihood over all possible intervals j with binary responses dN_{ij} and means $Y_{ij}\exp(Z_{ij}\beta_j)dH_{0j}$. This model may be adapted to allow for unobserved covariates or frailty, as considered in Section 11.4. It also allows for autoregressive dependencies between intervals.

Example 11.2 Veterans Lung Cancer Trial

To illustrate the implementation of semiparametric hazards and the opportunity they offer to model time-varying regression effects, consider data from the Veteran's Administration lung cancer trial (e.g. Bender et al., 2018). In this trial, n=137 male subjects with advanced inoperable lung cancer were randomised to either a standard or a test chemotherapy, with the end point being time to death in days. Only 13 of the 137 survival times are censored. Most analyses find the treatment to be insignificant and consider the remaining predictors, namely:

celltype (1=squamous, 2=smallcell, 3=adeno, 4=large);
the Karnofsky score (KS), reflecting ability to perform common tasks (with scores 0–100, where 100 signifies normal physical abilities);

prior therapy or PT, (0=no, 1=yes);
an interaction between Karnofsky score and PT.

Sargent (1997) considers a counting process version of the Cox model for these data with time-varying effects on the KS predictor.

Here we first consider the piecewise exponential model

$$h(t_i \in (a_{j-1}, a_j] \,|\, Z_i) = \lambda_j \exp(Z_i\beta) = \exp(\alpha_j + Z_i\beta),$$

and a partitioning of the time scale involving $J=20$ intervals, and 21 cut-points a_j at the {0th,5th,10th,...,95th,100th} percentiles of the survival times t_i. With d_i denoting event indicators, the responses y_{ij} and offsets Δ_{ij} are defined in R using the commands

```
for (i in 1:N) {for (j in 1:J){
y[i,j] <- d[i]*(t[i]>=a[j])*(a[j+1] >= t[i]);
# offset terms
del[i,j] <- (min(t[i],a[j+1])-a[j])*(t[i]>=a[j])}}
```

The first model assumes a constant Karnofsky score effect, but a time-varying (log) baseline hazard, namely

$$\alpha_j \sim N(\alpha_{j-1}, \sigma_\alpha^2),$$

with a gamma prior on $1/\sigma_\alpha^2$. To assist identifiability, estimation with the R package rube uses the BUGS car.normal prior, and the re-expression

$$\alpha_j = \beta_0 + \alpha_{0j},$$

where α_{0j} are RW1 centred random effects. With a two-chain run, the posterior density of σ_α is found to be bounded away from zero, with 95% interval (0.05,0.22). The posterior mean α_j show an irregular rise in mortality over the intervals (Figure 11.1). The LOO-IC is 984. This LOO-IC is in fact lower than if more intervals are used (e.g. with $J=50$, and using either quantile cutpoints or equally spaced cutpoints). The coefficient on the Karnofsky score has 95% interval (–0.38,–0.15), while the interaction term has coefficient with 95% interval (–0.57,–0.11). Coefficients for celltypes 1 and 2 are also significantly positive.

A second model additionally takes the Karnofsky score to have a time-varying coefficient, using a random walk prior adjusted for intervals of unequal length. Thus

$$\beta_j \sim N(\beta_{j-1}, \sigma_\beta^2 \delta_j)$$

where the $Ga(1,0.0001)$ prior for $1/\sigma_\beta^2$ has mean and variance 10^5, so supporting large values (Sargent, 1997). The LOO-IC is lower at 971, and there is an upward trend (towards a null value) in the coefficient, with an insignificant effect at higher intervals (Figure 11.2). Topics for further investigation might be the sensitivity of the form of time variation in the KS effect to the partitioning scheme of the durations, or to unobserved heterogeneity between subjects.

We also demonstrate the changing effect of the KS score using a rstan coding of the counting process model [1]. The settings for the gamma increments prior are as for the Winbugs Volume 1 "Leuk: Cox regression" example. Predictors are the KS score, prior

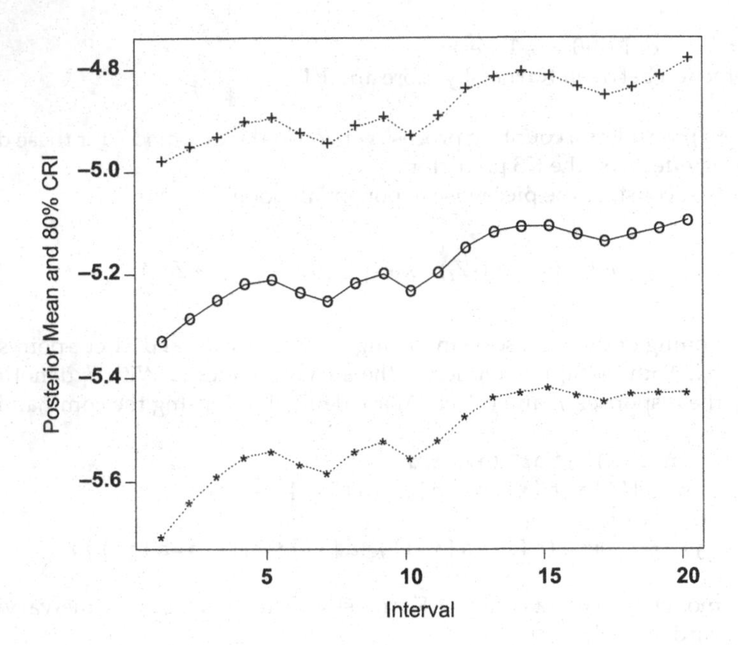

FIGURE 11.1
Trend in α_j by interval.

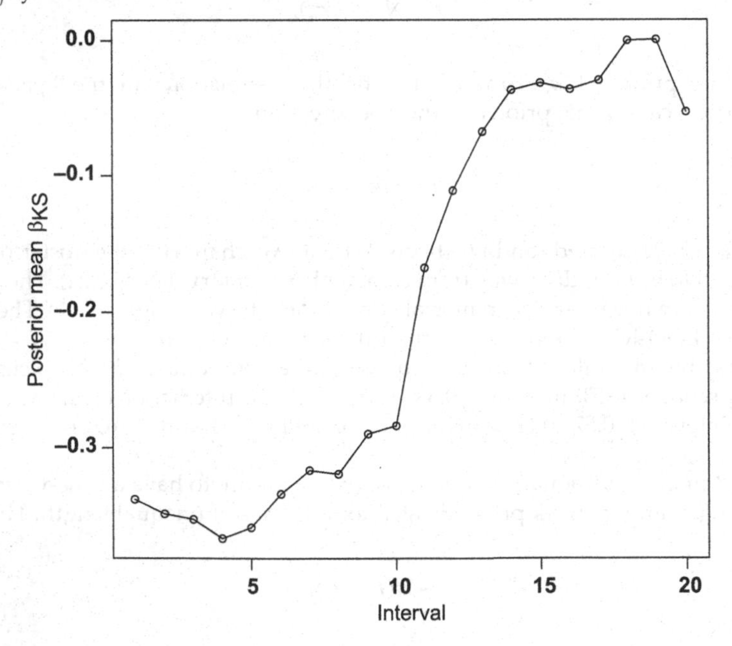

FIGURE 11.2
Trend in beta coefficient.

therapy, and their interaction (the latter two having time-constant effects). A random walk prior is assumed for the varying KS score effect.

Figure 11.3 shows the diminution of the KS effect at higher intervals (based on distinct event times). Figure 11.4 shows differing survival chances according to whether prior therapy was received (upper curve) or not (lower curve), with the KS score set at its upper quartile.

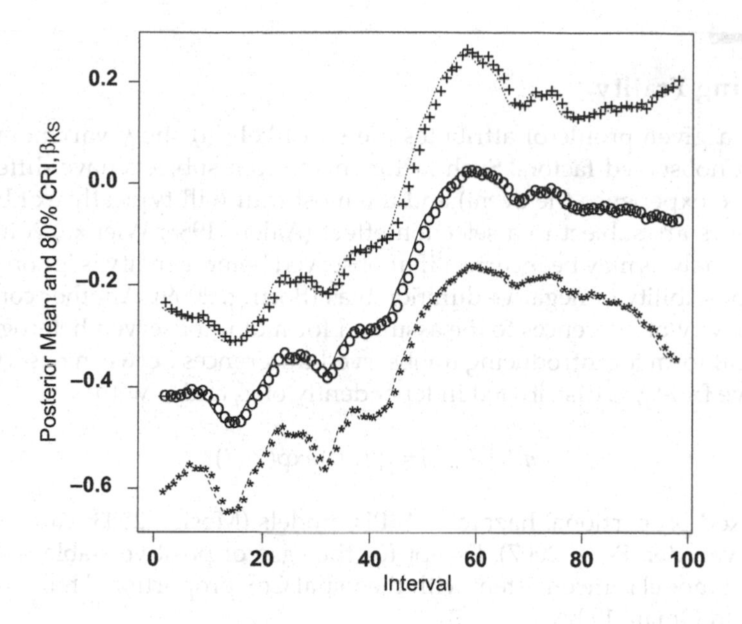

FIGURE 11.3
Trend in KS effect.

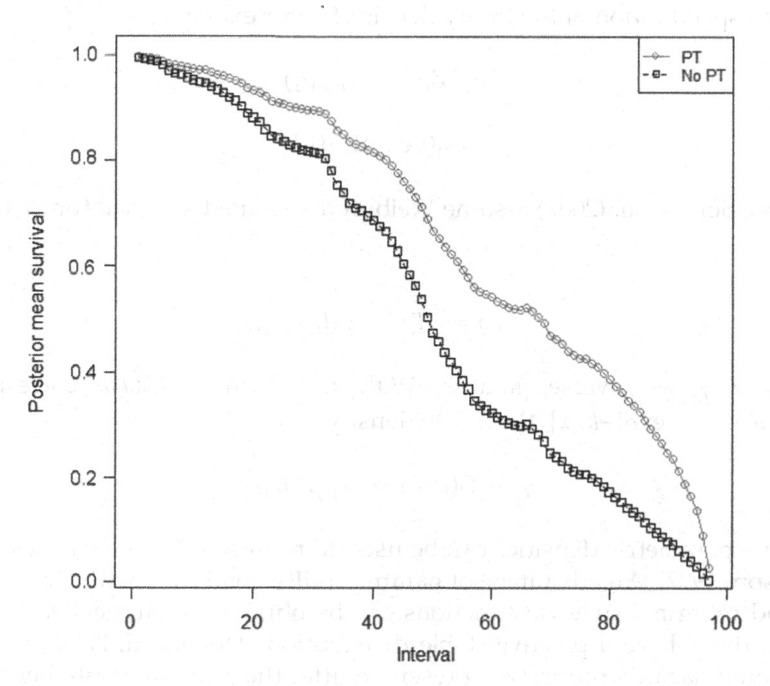

FIGURE 11.4
Survival chances according to therapy.

11.4 Including Frailty

Subjects with a given profile of attributes are still likely to show variations in survival times due to unobserved factors. Such factors mean that subjects have different frailties (i.e. liabilities to experience the event) and the most frail will typically exit before others, so that survivors are subject to a selection effect (Aalen, 1988; Wienke, 2010). Inferences from survival analysis may be incorrect if unobserved heterogeneity is ignored (Lancaster, 1990), with a possibility of negative duration bias (Boring, 2009). Another consideration is possible sensitivity of inferences to the assumed form of unobserved heterogeneity.

The canonical form for introducing unobserved differences between observations is via a multiplicative frailty, γ_i, distributed independently of Z_i and t_i, with

$$h(t_i \mid Z_i, \gamma_i) = \gamma_i h_0(t_i)\exp(Z_i\beta)$$

leading to mixed proportional hazard or MPH models (Mosler, 2003; Van den Berg, 2001; Abbring and van den Berg, 2007). Except for the case of positive stable frailty distributions, the MPH model is inconsistent with the usual Cox proportional hazard formulation (Henderson and Oman, 1999).

A typical assumption for the distribution $p(\gamma_i)$ of multiplicative frailties is that they are gamma distributed (Perperoglou et al., 2006), typically $\gamma_i \sim Ga(k,k)$ where k is unknown. So, the frailties have mean 1, and variance $1/k$, with normalisation to ensure identification when Z_i includes an intercept. Another possibility is to include the regression effect $\exp(Z_i\beta)$ in the specification of the frailty density. So, for example,

$$h(t_i \mid Z_i, \gamma_i) = \gamma_i h_0(t_i)$$

$$\gamma_i \sim Ga(k\exp[Z_i\beta], k).$$

As one option, Sohn et al. (2007) assume Weibull distributed survival times, with density form

$$f(t_i) = \frac{a}{\gamma_i} t_i^{\alpha-1}\exp(t_i^{\alpha}/\gamma_i),$$

and then take γ_i as inverse gamma. With the form $x \sim IG(a,b)$ corresponding to $f(x) = (b^a/\Gamma(a))x^{-(a+1)}\exp[-b/x]$, the frailty density is then

$$\gamma_i \sim IG(\alpha+1, \alpha\exp[Z_i\beta]).$$

Other positive parametric densities can be used to represent frailty, such as the log-normal (Gustafson, 1997). An advantage of gamma frailty combined with Weibull hazard is that joint and marginal survival functions can be obtained analytically. An alternative is to assume the γ_i have a positive stable distribution (Hougaard, 2000), in which case the proportional hazards property is preserved after the γ_i are integrated out (Aalen and Hjort, 2002).

Estimates resulting from the mixed proportional hazard model are often sensitive to the functional form of the heterogeneity distribution, and may be biased if the functional form of the distribution is mis-specified (Baker and Melino, 2000; Keiding et al., 1997). Heckman and Singer (1984) report sensitivity of regression estimates according to different parametric distributions of frailty. They propose discrete mixture models with finite support at a small number K of points, so that

$$h(t_i \mid Z_i, \gamma_i) = \gamma_{G_i} h_0(t_i) \exp(Z_i \beta)$$

where G_i is a multinomial indicator with K categories. Sahu and Dey (2004) compare gamma, stable, and skewed log-t frailty models, and show how the gamma assumption may attenuate covariate effects as compared to the other forms.

Despite such sensitivity, it is important to consider possible heterogeneity. One can show (Lancaster, 1990) that a model neglecting frailty will show spurious duration dependence, and specifically overestimate the extent of negative duration dependence in the true baseline hazard, and underestimate the extent of positive duration dependence. This is a consequence of selection, since in the presence of negative duration dependence, subjects with high values of γ_i exit faster, so survivors at a given survival time are increasingly biased towards relatively low γ_i values, and lower hazard rates. These features can be illustrated with the MPH assumption and particular parametric hazards. Conditional on a particular value of γ_i, the survivor function is

$$S(t_i \mid Z_i, \gamma_i) = \exp\left[-\gamma_i \exp(Z_i \beta) \int_0^{t_i} h_0(u)du\right],$$

or in terms of the cumulative hazard $H_0(t_i)$,

$$S(t_i \mid Z_i, \gamma_i) = \exp[-\gamma_i \exp(Z_i \beta) H_0(t_i)].$$

The unconditional survival function (integrating out the frailties) is therefore

$$S(t_i \mid Z_i) = \int_0^\infty S(t_i \mid Z_i, \gamma_i) p(\gamma_i) d\gamma_i$$

$$= \int_0^\infty p(\gamma_i) \exp\left[-\gamma_i H_0(t_i) e^{Z_i \beta}\right] d\gamma_i.$$

For γ_i following a gamma density, $\gamma_i \sim Ga(a, b)$, the unconditional survivor function is

$$S(t_i \mid Z_i) = b^a \left[b + H_0(t_i) e^{Z_i \beta}\right]^{-a}$$

which for $a = b = k$ (with Z_i including a constant) reduces to

$$S(t_i \mid Z_i) = \left[1 + k^{-1} H_0(t_i) e^{Z_i \beta}\right]^{-k}.$$

Consider exponentially distributed times so that $h_0(t_i) = 1$ and $H_0(t_i) = t_i$. Then

$$S(t_i \mid Z_i) = \left[1 + k^{-1} e^{Z_i \beta} t_i\right]^{-k},$$

$$f(t_i \mid Z_i) = e^{Z_i\beta}\left[1 + k^{-1}e^{Z_i\beta}t_i\right]^{-k-1},$$

$$h(t_i \mid Z_i) = e^{Z_i\beta}\left[1 + k^{-1}e^{Z_i\beta}t_i\right]^{-1}.$$

For a frailty variance $1/k > 0$, the hazard rate is a decreasing function of t, an example of spurious duration dependence. If frailty is present, but ignored, not only will duration effects be mis-stated, but covariate effects will be underestimated (Hougaard et al., 1994; Pickles and Crouchley, 1995). Lancaster (1990) confirmed this analytically for uncensored Weibull survival data.

More general forms of subject-level random variation can be achieved by a general linear mixed model form, where the impact of selected predictors $w_i = (w_{1i}, \ldots, w_{ri})$ is assumed to vary over subjects, or clusters of subjects. Thus

$$h(t_i \mid Z_i, w_i) = h_0(t)\exp(Z_i\beta + w_i b_i).$$

When $r = 1$, and $w_i = 1$, the random effect $b_j \sim N(B, 1/\tau_b)$ is used to represent variability in frailties between subjects. If Z_i contains an intercept, the b_i are constrained to have zero mean, namely $b_j \sim N(0, 1/\tau_b)$.

A general linear mixed model form for frailty may take account of spatial locations of subjects, as in geoadditive hazard regression (Kneib, 2006; Henderson et al., 2002). Suppose subjects ij are nested within J locations, then $\{b_j, j = 1, \ldots, J\}$ would be spatially correlated with local pooling of strength (Zhou et al., 2017). For example, if individuals in neighbouring locations are subject to similar (unobserved) environmental risks, this will affect survival.

In accelerated failure time models (section 11.2.3), frailty is conveniently obtained by discrete mixture modelling of the error term. Following Roeder and Wasserman (1997), a mixture of normals provides a flexible model for estimation of densities. Suppose membership of latent sub-groups is denoted by a categorical variable G_i with K options, and prior $G_i \sim \text{Mult}(1, [\pi_1, \ldots \pi_K])$. Assuming a log-normal density for exit times, one may transform observed failure or censoring times as $r_i = \log(t_i)$, and to account for right censoring, define lower sampling limits $L_i = \log(t_i)$ if $d_i = 0$, and $L_i = 0$ if $d_i = 1$. Then the discrete mixture adopts varying group intercepts and variances in the survivor function

$$S(t_i) = 1 - \sum_{k=1}^{K} \pi_k \Phi\left(\frac{\log t_i - \gamma_{0k} - \gamma_1 z_{1i} - \gamma_2 z_{2i} - \ldots}{\sigma_k}\right).$$

Mixed Dirichlet process and Polya Tree priors for the errors u in an AFT regression are used by Kuo and Mallick (1997) and Walker and Mallick (1999).

11.4.1 Cure Rate Models

A particular form of heterogeneity may arise when permanent survival from an event is possible. Demographic examples are provided by age at first marriage or age at first maternity. The issue is then to identify latent subpopulations in the censored group, namely to distinguish a permanent survival subgroup from a subgroup still liable or susceptible to experience the event, but exhibiting extended survival. Not allowing for permanent survival when it can occur will distort the failure time parameter estimates for the true susceptible population. Herring and Ibrahim (2002) point out – in the context of cancer survival – that improved treatment means that a substantial proportion of patients may now be cured, whereas traditional survival analysis, including the Cox (1972) regression model, assume that no patients are cured, but that all remain at risk of death or relapse. Similarly, in the context of component reliability, Sinha et al. (2003) consider the case where if a unit is free of manufacturing faults, it will never fail in its technological lifetime under usual stress levels.

The most common approach to modelling events with a permanent survival fraction or cure rate assumes the total survival rate is a binary mixture (Ibrahim et al., 2001). The non-susceptible subpopulation has $S_c(t)=1$ with probability $(1-\pi)$, and the other (the non-cured or susceptible subpopulation) follows a conventional survival pattern in which $S_n(t)\rightarrow 0$ as $t\rightarrow\infty$. So the overall survivor function is

$$S^*(t)=(1-\pi)+\pi S_n(t),$$

and the overall distribution function (Bruderl and Diekmann, 1995) is

$$F^*(t)=\pi F_n(t).$$

Ibrahim et al. (2001, p.157) point out that if covariate effects are modelled via binary regression for π_i then the proportional hazard property no longer obtains.

Let R_i be a partially unobserved binary indicator with $R_i=1$ if a subject is susceptible. Schmidt and Witte (1989) and Banerjee and Carlin (2004) follow the standard cure rate model and take R_i to be Bernoulli with $Pr(R_i=1)=\pi_i$ being a propensity to experience the event (e.g. propensity to relapse). For simplicity, omit the subscript n in the survivor function for susceptibles. Then for subjects observed to fail, namely with $d_i=1$, it necessarily follows that $R_i=1$, and so the likelihood contribution from such cases is

$$Pr(R_i=1)f(t_i)=\pi_i f(t_i).$$

Censored subjects may be either susceptibles or non-susceptibles with likelihood contribution

$$Pr(R_i=0)+Pr(R_i=1)Pr(T>t_i)=(1-\pi_i)+\pi_i S(t_i).$$

The total likelihood contribution is then

$$[\pi_i f(t_i)]^{d_i}[(1-\pi_i)+\pi_i S(t_i)]^{1-d_i},$$

which reduces to the usual form $f(t_i)^{d_i}S(t_i)^{1-d_i}$ when $R_i=1$ for all subjects, and so $\pi_i=1$ for all i (i.e. there is no permanent survivor fraction). Any form of binary regression (e.g. logit) may be used for predicting π_i (Schmidt and Witte, 1989). Banerjee and Carlin (2004) carry out a Bayesian analysis with individual level regression in the scale parameter of the failure distribution $f(t)$, but without a regression for the susceptible probability. However, their observations are hierarchical (spatially configured) response times t_{ij} (subjects i within areas j), and they allow spatial variability in the propensities so that $\pi_{ij}=\pi_j$; see also Cooner et al. (2006).

Chen et al. (1999) describe an alternative structure in which there is a latent count of risks C_i, taken to be Poisson with mean θ (for example, tumour cells remaining after treatment that have varying potentials to cause relapse), and unobserved times $U_{i1},\ldots,U_{iC_i}$ associated with each of these risks. The U_{ic} are assumed to follow the same failure distribution $F(t)=1-S(t)$. An observed failure time t_i is the minimum of these times. If $C_i=0$ then a subject survives permanently from the event being modelled (e.g. a form of cancer). In this case the composite survival function is

$$S^*(t_i) = Pr(C_i = 0) + Pr(U_{i1} > t_i,\ldots U_{iC_i} > t_i \mid C_i \geq 1),$$

$$= \exp(-\theta) + \sum_{k=1}^{\infty} S(t)^k \frac{\theta^k}{k!}\exp(-\theta),$$

$$= \exp(-\theta + \theta S(t)) = \exp(-\theta F(t)).$$

and the composite hazard rate is

$$h^*(t_i) = \theta f(t_i).$$

An alternative derivation of this model, not tied to the notion of multiple latent risks, is that the cumulative hazard $H(t) = \int_0^t h(u)du$ tends to a finite positive limit θ as $t \to \infty$ (Tsodikov et al., 2003). Chen et al. (1999) and Ibrahim et al. (2000, p.158) mention that the survivor function of the non-cured subpopulation can be written

$$S_n(t_i) = \frac{\exp(-\theta F(t_i)) - \exp(-\theta)}{1-\exp(-\theta)},$$

so that the composite survival function is in fact also representable as a binary mixture, namely

$$S^*(t_i) = \exp(-\theta) + (1-\exp(-\theta))S_n(t_i).$$

Chen et al. (1999) introduce covariates into a Poisson regression model for subject-specific θ_i. Consider Weibull distributed times with $F(t_i \mid Z_i) = 1-\exp[-\lambda_i t_i^\kappa], \lambda_i = \exp(Z_i\beta)$, and $f(t_i \mid \kappa, Z_i) = \lambda_i \kappa t^{\kappa-1}\exp(-\lambda_i t_i^\kappa)$. The likelihood when predictors are used to explain both θ_i and λ_i, and with d_i being event status indicators, is then

$$\left[h^*(t_i)\right]^{d_i} S^*(t_i) = \left[\theta_i \lambda_i \kappa t_i^{\kappa-1}\exp(-\lambda_i t_i^\kappa)\right]^{d_i} \exp\left(-\theta_i\left\{1-\exp(-\lambda_i t_i^\kappa)\right\}\right).$$

Multiplicative frailty, as in the MPH setup above, can be introduced in cure rate models, but identifiability may be weak because susceptibility responses are partially unobserved themselves. Models for frailty in multivariate cure fraction models are considered by Yin (2005). Thus, for times t_{ij} observed on subjects i and events j, Yin proposes multiplicative frailty at subject level combined with Poisson regression for θ_{ij} in the cure fractions $\exp(-\theta_{ij})$. One option takes

$$S^*(t_{ij}) = \exp(-\theta_{ij}\gamma_i F(t_{ij})).$$

with hazard rates $h^*(t_{ij}) = \theta_{ij}\gamma_i f(t_{ij})$.

Example 11.3 Age at First Maternity

To illustrate frailty modelling in a cure rate model, this example follows Winkelmann and Boes (2005) in analysing ages at first maternity for women in the German General Social Survey for 2002. The subsample considered by Winkelmann and Boes involves 1371 women, comprised of (a) uncensored subjects (event indicator $d_i = 1$) who may have been over 40 at the time of the survey, but whose age at first maternity (AFM) was under 40, and (b) women aged under 40 in 2002, but who had not yet had a child ($d_i = 0$). Here we consider all 1,508 women (including childless women aged over 40).

A log-logistic model with hazard and survivor functions

$$h(t) = \frac{\lambda \kappa t^{\kappa-1}}{[1 + \lambda t^{\kappa}]},$$

$$S(t) = [1 + \lambda t^{\kappa}]^{-1},$$

is appropriate to the non-monotonic form of hazard for first maternity, typically peaking between ages 25 to 35. This model is implemented in rstan using the custom likelihood approach. A standard log-logistic is compared with a log-logistic model with a permanent survivor fraction (PSF), modelled according to the latent count approach (Chen et al., 1999). The PSF log-logistic model is then generalised to allow for unmeasured heterogeneity in the age at first maternity. Permanent survivorship in this case is equivalent to a woman never undergoing a maternity, and at population level is essentially equivalent to the rate of childlessness.

Regression effects are included in the scale parameter of the log-logistic hazard via $\lambda_i = \exp(Z_i \beta)$, with θ assumed constant. However, a Poisson regression for θ_i could be included. Predictors Z_i and regression effects β under the standard log-logistic are as in Table 11.2. Predictors are binary apart from number of siblings and education years. The modal age $\chi = [(\kappa - 1)\exp(-Z_T \beta)]^{1/\kappa}$ reported in Table 11.2 is based on a predictor vector Z_T for a white subject with 13 years of education, and 3 siblings.

The standard log-logistic model gives a LOO-IC of 8,700. Significant coefficients in Table 11.2 show that delayed AFM is associated with longer education, being white, and

TABLE 11.2
Age at First Maternity. Parameter Posterior Summaries

	Standard log-logistic			Log-logistic with Cure Fraction			Cure Fraction and Frailty		
Predictor	Mean	2.50%	97.50%	Mean	2.50%	97.50%	Mean	2.50%	97.50%
Years of education	−0.213	−0.249	−0.184	−0.294	−0.334	−0.26	−0.504	−0.686	−0.377
Number of siblings	0.018	−0.012	0.045	0.017	−0.017	0.046	0.032	−0.03	0.086
White	−0.594	−0.807	−0.407	−0.909	−1.14	−0.706	−1.649	−2.364	−1.133
Immigrant	−0.312	−0.589	−0.083	−0.585	−0.912	−0.317	−0.921	−1.566	−0.45
Low income (age 16)	−0.009	−0.219	0.167	0.122	−0.126	0.325	0.217	−0.244	0.615
Living in city (age 16)	−0.074	−0.256	0.074	−0.007	−0.214	0.157	0.019	−0.343	0.334
Shape parameter	5.04	4.79	5.26	8.79	8.36	9.17	15.9	11.81	20.44
Modal age (typical subject)	33	32.1	33.7	31.7	31.1	32.4	31.5	22.8	38.7
Proportion childless				0.172	0.151	0.19	0.167	0.146	0.185
Frailty SD							2.45	1.44	3.47

immigrant status. Allowing for a permanently childless subpopulation, but without allowing for frailty, reduces the LOO-IC to 7,903. Finally, adding log-normal frailty via

$$\lambda_i = \exp(Z_i\beta + u_i)$$

$$u_i \sim N(0, \sigma_u^2),$$

reduces the LOO-IC to 7,855. This analysis uses a corner constraint on the u_i for identifiability, and this option provides better convergence than (a) excluding the intercept from $Z_i\beta$ and centring the u_i at β_0, namely $u_i \sim N(\beta_0, 1/\tau_u)$, or (b) expressing the random effect as a product of σ_u and $N(0,1)$ terms. The option of centring the u_i is more computationally intensive. The lowest (most negative) frailty values are for subjects with delayed age at first maternity, combined with low education and non-white ethnicity.

Allowing for a childless subpopulation (as a cure rate) is a form of frailty in itself, and enhances (absolutely) the coefficients on significant predictor effects [3]. Formally including frailty in the modelling of the event density further enhances predictor effects. The childless fraction (i.e. the permanent survival fraction), $\exp(-\theta)$, is estimated at around 0.17, regardless of the presence or not of frailty. A standard log-logistic model leads to a significantly later modal age than the extended models. In fact, a better representation of the age at first maternity process may be provided by the generalised log-logistic of Brüderl and Diekmann (1995), as discussed in Congdon (2008).

11.5 Discrete Time Hazard Models

In applications with interval censored times, analysis using a discrete time scale becomes appropriate, and in fact such analysis has certain benefits also for modelling time-varying or lagged predictor effects. Let the time scale be grouped into J intervals $A_1 = [a_0, a_1), \ldots A_J = [a_J, a_{J+1})$, with interval j being $[a_{j-1}, a_j)$, and $a_0 = 0$, $a_{J+1} = \infty$, where a_J denotes either the end of the observation period, or the largest time (censored or failed). The intervals may be of equal length $\delta_j = a_j - a_{j-1}$, but are not necessarily so. Instead of continuous observed failure times, only the discrete times $t_i \in A_j$ are observed. Equivalently, let $t_i = j$ denote that a time of failure or censoring is observed within $[a_{j-1}, a_j)$.

With S_j denoting the probability of surviving to the end of interval j, the unconditional probability of failing in interval j is

$$f_j = Pr(t \in (a_{j-1}, a_j)) = S_{j-1} - S_j,$$

and the hazard function (the conditional probability of failing in interval j given survival till the start of the interval) is

$$q_j = Pr(t \in (a_{j-1}, a_j) \mid t \geq a_{j-1}) = Pr(t = j \mid t \geq j) = f_j / S_{j-1} = \frac{S_{j-1} - S_j}{S_{j-1}}.$$

Alternatively stated, q_j is the proportion of subjects at risk at the beginning of interval j who experience the event sometime during the interval. The survivor function (the probability of surviving beyond interval j) is obtained as

$$S_j = Pr(t > a_j) = \prod_{k=1}^{j} (1 - q_k) = f_{j+1} + f_{j+2} + \ldots + f_J = S_{j-1}(1 - q_j),$$

though an alternative survivor function $\tilde{S}_j = Pr(t > a_{j-1})$ may be defined as the probability of surviving to the start of interval j (Fahrmeir and Tutz, 2001, p.396; Aitkin et al., 2004, p.350).

Let $w_{ij}=1$ if individual i undergoes the event during interval j and w_{ij} otherwise. The likelihood up to interval k for that individual is then (Aitkin et al., 2004, p.351),

$$\begin{aligned} f_{ik}^{w_{ik}} S_{ik}^{1-w_{ik}} &= (q_{ik} S_{i,k-1})^{w_{ik}} [S_{i,k-1}(1-q_{ik})]^{1-w_{ik}} \\ &= S_{i,k-1} q_{ik}{}^{w_{ik}} (1-q_{ik})^{1-w_{ik}} \\ &= q_{ik}^{w_{ik}} (1-q_{ik})^{1-w_{ik}} \prod_{j=1}^{k-1} (1-q_{ij})^{(1-w_{ij})} \\ &= \prod_{j=1}^{k} q_{ij}^{w_{ij}} (1-q_{ij})^{(1-w_{ij})}. \end{aligned}$$

This shows that the likelihood involves binary responses $w_{ij} \sim \text{Bern}(q_{ij})$, where the q_{ij} may vary between time intervals, but are assumed constant within them. So, the hazard probability can be represented as

$$q(j \mid Z_{ij}) = Pr(t = j \mid t \geq j, Z_{ij}) = F(\alpha_j + Z_{ij}\beta_j),$$

where F is a suitable distribution function, and α_j models the baseline hazard (Singer and Willetts, 1993). If the predictors include lagged event status indicators $\{w_{i,j-1}, w_{i,j-2}, \text{etc}\}$, one is led to discrete Markov event histories (e.g. Barmby, 2002). Lagged predictor effects may also be used (Fahrmeir and Tutz, 2001, p.410).

A benefit of the discrete framework is that the baseline hazard can be modelled via polynomial functions of j (Efron, 1988), for example:

$$\alpha_j = \psi_0 + \psi_1(j-1) + \psi_2(j-1)^2 + u_j,$$

where $u_j \sim N(0, \sigma_u^2)$. Parametric time models can also be modelled straightforwardly: a Weibull model is represented in a complementary log-log link for F by taking the log of the time interval as a covariate (Allison, 1997). Non-parametric models for time (e.g. via splines) can also be applied, or a correlated random effect prior assumed, as in Section 11.3.1. Time-varying predictor effects are straightforward to use (Muthen and Masyn, 2005), and non-proportional effects are modelled by including interactions between subject attributes Z_{ij} and j.

Commonly used links for the probabilities q_{ij} are the logit, probit, and complementary log-log. For example, a logit link with time-varying intercepts and predictor effects (where the vector Z_{ij} excludes a constant term) would mean

$$q(j \mid Z_{ij}) = \frac{\exp(\alpha_j + Z_{ij}\beta_j)}{1 + \exp(\alpha_j + Z_{ij}\beta_j)}.$$

Adopting a logit link means the log-odds of the event occurring are modelled as functions of predictors and time (i.e. interval). The complementary log-log link model with

$$q(j \mid Z_{ij}) = 1 - \exp(-\exp(\alpha_j + Z_{ij}\beta_j)),$$

can be derived by assuming an underlying proportional hazard in continuous time, under which

$$S(t_i \mid Z_i) = \exp\left[-\int_0^{t_i} h(u \mid Z_i)du\right] = \exp\left\{-\exp\left[Z_i\beta + \log H_0(t_i)\right]\right\}.$$

Then taking $a_j = \log \int_{a_{j-1}}^{a_j} h_0(t)$ leads to the complementary log-log model, with the same predictor effects as under a PH model (Kalbfleisch and Prentice, 1980; Fahrmeir and Tutz, 2001, p.401).

If correlated priors (e.g. random walks) on the α_j and β_j are adopted, the setting of priors on the hyperparameters (e.g. precisions) follows the same considerations as discussed above in connection with semiparametric models for continuous time hazards (Section 11.3.1). Fahrmeir and Knorr-Held (1997) discuss alternative Hastings sampling schemes for collections of time-varying coefficients $\{a_j, \beta_{j1}, \ldots, \beta_{jp}\}$ in discrete hazard regression.

As for continuous time survival modelling, neglecting unobserved heterogeneity may mean that the estimated baseline hazard parameters are biased downwards, the impact of constant covariates is underestimated, or that spurious time-dependent effects for observed predictors are obtained. For improved identification, frailties may be included at subject level, rather than at subject-interval level, though bilinear schemes are possible. Thus, a log-normal frailty might specify

$$q_{ij} = F(a_j + Z_{ij}\beta + b_i),$$

where $b_i \sim N(0, \sigma_b^2)$. Alternatively, a bilinear scheme might be used

$$q_{ij} = F(a_j + Z_{ij}\beta + \delta_j b_i),$$

where one of the δ_j is set to a fixed value for identification if the variance of b_i is unknown. Muthen and Masyn (2005) use a discrete mixture approach in which $G_i \in (1, \ldots, K)$ are latent groups (e.g. developmental trajectories in educational applications). Then

$$F^{-1}(q_{ij}) = a_{j,G_i} + Z_{ij}\beta_{G_i} + \delta_{j,G_i} b_i,$$

where the probability that $G_i = k$ is defined by predictors U_i in a separate multiple logit regression. The factor scores b_i may be defined by $b_i \sim N(0, \sigma_b^2)$, or by a hierarchical linear regression on the predictors U_i.

11.5.1 Life Tables

Life tables are a particular way of analysing discrete time survival data. They may be applied to situations where permanent survival or withdrawal is possible, such as marital status life tables (Schoen, 2016), or to population mortality. The intervals in such applications refer to age or duration bands, and discretisation may extend beyond that present in the data, as in abridged life tables (Kostaki and Panousis, 2001). The intervals are not necessarily of equal length (Wong, 1977). For example, in one common scheme for human life tables, ages under 1 form the first interval, ages one to four comprise the second interval, ages five to nine, the third interval, and so on for successive five year bands, with the final interval typically open ended, such as ages over 90. Often human life tables are estimated from population deaths data over a specified calendar period, to provide "period" life tables, based on current mortality in individuals born in different periods, as distinct from cohort life tables, based on follow-up studies of mortality in a group of individuals born in the same time period (Richards and Barry, 1998).

Following life table conventions, ages are denoted x and age intervals are denoted $[x, x+n)$, e.g. $n=5$ if intervals are five years in length. Let T denote a random variable for the total lifetime (age of death) of an individual. Also, in line with life table conventions, the probability $Pr(T>x)$ that the age of death T is x or higher (the survivor function) is denoted $l(x)$. The hazard rate – also called the force of mortality in life table applications – is then

$$h(x) = \lim_{\Delta x \to \infty} \frac{l(x) - l(x+\Delta x)}{l(x)\Delta x} = \frac{-l'(x)}{l(x)},$$

with solution

$$l(x) = l(0)\exp\left[-\int_0^x h(u)du\right].$$

With $l(0)=1$, the density of the age at death is $f(x) = h(x)l(x)$. The probability of surviving from age x to age $x+n$, given survival to x, namely $Pr(t > x+n \mid t > x)$, is denoted ${}_np_x$ with

$${}_np_x = l(x+n)/l(x) = \frac{\exp\left[-\int_0^{x+n} h(u)du\right]}{\exp\left[-\int_0^x h(u)du\right]} = \exp\left[-\int_0^n h(x+u)du\right],$$

while the probability of dying before age $x+n$ conditional on reaching age x is

$${}_nq_x = 1 - {}_np_x = 1 - l(x+n)/l(x) = \frac{l(x) - l(x+n)}{l(x)}.$$

Important in linking these functions to estimable quantities is the central rate of mortality, which represents a weighted average of the force of mortality applying over the interval $[x, x+n)$. Let $P(x)$ denote the population of age x. Then the death rate for age interval $[x, x+n)$ is

$${}_nM_x = \int_x^{x+n} h(a)P(a)da \Big/ \int_x^{x+n} P(a)da.$$

Assuming linearity of $l(a)$ in the interval from x to $x+n$, this can be simplified (Namboodiri and Suchindran, 1987, p.36) to

$${}_nM_x = \frac{l(x) - l(x+n)}{0.5n[l(x) + l(x+n)]}.$$

Hence the survivor probability can be written

$$\frac{l(x+n)}{l(x)} = {}_n p_x = \frac{1-0.5n({}_n M_x)}{1+0.5n({}_n M_x)}$$

giving

$${}_n q_x = \frac{n({}_n M_x)}{1+0.5n({}_n M_x)}.$$

To clarify the operations involved, life tables involve hypothetical populations of initial size $l_0 = 100{,}000$ (the radix) with l_x denoting numbers still alive at age x from the initial population. The number dying between age x and $x+n$ is denoted ${}_n d_x = l_x - l_{x+n}$, and from above

$${}_n q_x = 1 - l_{x+n}/l_x = \frac{l_x - l_{x+n}}{l_x} = \frac{{}_n d_x}{l_x}.$$

To develop the life table from observed deaths and populations requires an estimator for the probability ${}_n q_x$. Let D_x denote observed deaths for age band $[x, x+n)$ over a certain period, P_x denote observed mid-period populations at risk (or person-years), and M_x denote age-specific death rates. One estimator of probability of dying in interval $[x, x+n)$ conditional on being alive at the start of the interval is then (Chiang, 1984)

$${}_n q_x = \frac{n_n M_x}{1+n(1-{}_n a_x)_n M_x},$$

where ${}_n a_x$ is the fraction of the interval lived by those dying during it. For most age groups, ${}_n a_x$ is taken as a half, but for infants (ages under one), it can be taken as 0.1, and for the one to four age group as 0.4.

Under conventional life table methods that are usually applied to large populations, the M_x are treated as unrelated fixed effects and estimated by assuming binomial sampling $D_x \sim \text{Bin}(P_x, M_x)$ or Poisson sampling $D_x \sim Po(M_x P_x)$. In a Bayesian version of the fixed effect approach, the M_x would be assigned diffuse beta or gamma priors with known hyperparameters, e.g. $M_x \sim \text{Beta}(1,1)$. Overdispersed versions of binomial or Poisson densities may also be used, involving hierarchical schemes for "borrowing strength" over correlated mortality rates, with a higher stage density for the M_x involving unknown hyperparameters. An example might be when age-specific deaths D_{ix} for a set of areas or hospitals $(i = 1, \ldots, I)$ are to be analysed, and populations at risk are relatively small. Then the conjugate binomial-beta approach would mean taking death rates M_{ix} to be distributed according a hierarchical model, namely

$$D_{ix} \sim \text{Bin}(P_{ix}, M_{ix}),$$

$$M_{ix} \sim \text{Beta}(a, b),$$

where $\{a,b\}$ are unknown parameters. Congdon (2009) adopts a general linear mixed model approach for data involving an additional stratifying group g in which

$$D_{ixg} \sim \text{Bin}(P_{ixg}, M_{ixg}),$$

where i and x denote areas and ages, and a logistic regression with group-specific autoregressive area and age effects has the form

$$\text{logit}(M_{ixg}) = a_g + s_{ig} + h_{xg}.$$

Other options might be to model the impact of age by a parametric function; for example, Neves and Migon (2007) use Makeham's Law, by which

$$D_x \sim Po(M_x P_x),$$

$$M_x = \alpha + \beta\delta^x,$$

and extend this to a time series model for age-specific death rates and times t, namely

$$M_{xt} = \alpha_t + \beta_t \delta_t^x.$$

Example 11.4 Cancer Survival

This example illustrates discrete survival with potential unobserved frailty. It involves survival times in months for 48 participants in a cancer drug trial. Of the 48 patients, 28 receive an experimental drug treatment ($Z_1=1$) and 20 receive a control treatment ($Z_1=0$). The other predictor is patient age at the start of the trial, ranging from 47 to 67 years. The observed times provide the month of death, or the last month the patient was known to be alive.

With a complementary log-log link, Weibull time dependence (model 1) is specified as

$$q(j \mid Z_i) = 1 - \exp(-\exp(Z_i\beta + \kappa \log(j))),$$

with κ a positive parameter, and the regression term Z_i including an intercept. This representation is compared with a semiparametric baseline hazard modelled via a first-order random walk (model 2), namely

$$q(j \mid Z_i) = 1 - \exp(-\exp(\alpha_j + Z_{i1}\beta_1 + Z_{i2}\beta_2 + \kappa \log(j))).$$

Convergence in the latter is assisted by the parameterisation $\alpha_j = \beta_0 + \alpha_{0j}$, where β_0 is the intercept, and the α_{0j} are normal RW1 effects with precision $\tau_\alpha = 1/\sigma_\alpha^2$. The α_{0j} are centred to have mean zero at each MCMC iteration. For numeric stability, age values are divided by 100.

Using rube in R, the semiparametric model gives a WAIC (widely applicable information criterion) of 307 with effective parameter count 33. There is only slight fluctuation about the central value of β_0 which has posterior mean −8.0. The simpler Weibull model has a lower WAIC of 295, with the Weibull parameter having mean (95% CRI) of 0.29 (0.01, 0.76). The treatment and age effects under the Weibull model are −2.1 (−2.9,−1.3) and 9.7 (3.1,16.1). So, mortality declines with time (since $\kappa < 1$) after allowing for the impact of age on mortality, though this decline might be attributable to unmodelled frailty.

A lognormal frailty effect at subject level is accordingly added to the Weibull model, so that

$$q(j \mid Z_i) = 1 - \exp(-\exp(\beta_0 + Z_{i1}\beta_1 + Z_{i2}\beta_2 + \kappa \log(j) + \sigma_b b_i))$$

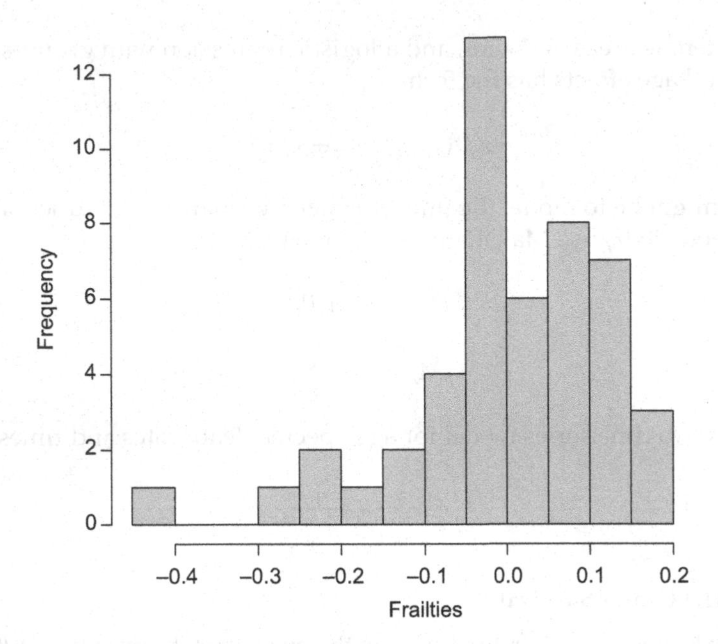

FIGURE 11.5
Posterior mean frailties.

where $b_i \sim N(0,1)$, and the precision $1/\sigma_b^2$ of b_i is assigned a gamma prior. A two-chain run of 10,000 iterations provides a mean (95% CRI) of 0.33 (0.01, 0.81) for the κ coefficient. The treatment and age effects, namely −2.1 (−3,−1.3) and 10.3 (4.0,16.6), are changed only slightly. The WAIC falls slightly to 293, with 31 effective parameters. Figure 11.5 plots out the b_i and shows a negative skew, with negative frailty effects for older subjects still surviving at higher intervals (e.g. subject 34).

Example 11.5 Life Tables and Actuarial Implications

This example considers graduation of mortality data, and illustrates the potential for identifiability issues in models involving multiple random effects. The data are from Neves and Migon (2007) and consist of central numbers exposed to risk (e_x) by age x, and number of deaths observed (d_x) during 1998–2001, collected by insurance companies in Brazil. We consider the male data for ages 0 to 90. Assuming Poisson sampling, forces of mortality μ_x can be estimated using the likelihood $d_x \sim Po(e_x\mu_x)$, with probabilities of death then estimated as $q_x = 1 - \exp(-\mu_x)$ (Brouhns et al., 2002).

Mortality rates by age for different populations often exhibit underlying regularities despite stochastic fluctuations, leading to parametric and non-parametric methods to represent the smooth underlying pattern, under which it is expected that probabilities of death for consecutive ages should be close. Parametric graduation procedures aim to estimate the underlying smooth mortality curve, facilitating actuarial calculation of premiums and reserves. Parametric methods include Makeham's formula, whereby

$$\mu_x = \alpha + \beta\delta^x,$$

with the three parameters all positive. Allowing for variability in these parameters over age groups, one may propose

$$\mu_x = \alpha_x + \beta_x\delta_x^x,$$

$$\log(\alpha_x) = \log(\alpha_{x-1}) + w_{1x},$$

$$\log(\beta_x) = \log(\beta_{x-1}) + w_{2x},$$

$$\log(\delta_x) = \log(\delta_{x-1}) + w_{3x},$$

where the w_{jx} are initially taken as iid normal. Initial conditions ($\log(\alpha_1)$, etc) are taken as $N(0,25)$.

Implementing this model in rstan shows problematic convergence, unless informative priors are assumed for the standard deviations σ_j of the errors w_{jx}. Informative priors can be motivated by an expectation of small changes in death rates between successive ages, and we assume $\sigma_j \sim N^+(0,0.25)$. This model shows impaired convergence in $\log(\beta_x)$ and σ_2. Improved convergence is shown by a model taking w_{2x} and w_{3x} to be Student t with known d.f.=4, and also $\sigma_j \sim t_4^+(0,0.25)$ (a half-t with 4 df). This option has LOO-IC of 605. As one example of the parametric outputs, Figure 11.6 plots out the posterior mean α_x. Predictive checks from this model (comparing replicates and actual d_x) are satisfactory, with no exceedance probability under 0.05 or over 0.95. The minimum exceedance probability is 0.056 for age $x=84$, with a relatively large observed death total as compared to the modelled total.

Neves and Migon (2007) consider the implications of the fitted curve in deriving the monthly whole life annuity-due to a life aged x (Bowers et al., 1986). Figure 11.7 plots out the posterior density of this quantity for a male subject aged 60, at an assumed annual real interest of 6%. This compares closely with Figure 3 in Neves and Migon (2007).

Convergence problems are completely alleviated if a constancy assumption $\delta_x = \delta$ is made, with the specification now

$$\mu_x = \alpha_x + \beta_x \delta^x,$$

$$\log(\alpha_x) = \log(\alpha_{x-1}) + w_{1x},$$

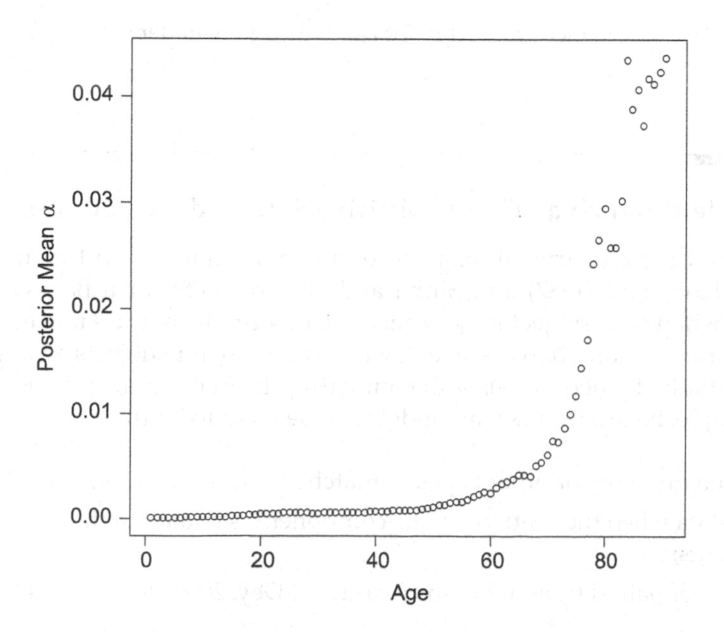

FIGURE 11.6
Alpha by age.

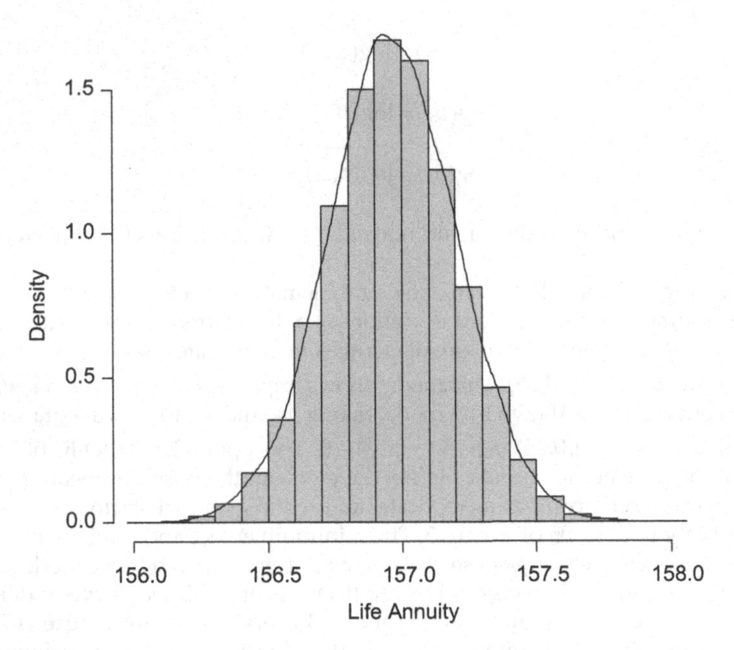

FIGURE 11.7
Posterior density, monthly whole life annuity.

$$\log(\beta_x) = \log(\beta_{x-1}) + w_{2x},$$

with w_{1x} and w_{2x} normal and $\sigma_j \sim N^+(0,0.25)$. However, this raises the LOO-IC to 606. The predictive exceedance probability for age x=84 is now under 0.05.

11.6 Dependent Survival Times: Multivariate and Nested Survival Times

Multivariate and nested survival data can occur in a number of different ways; for discussions, see Hougaard (1987) and Sinha and Ghosh (2005) for a Bayesian perspective. Examples are when each subject may experience repetitions of the same event; when subjects may experience more than one event; when times are for subjects arranged in clusters (including spatially defined units); or in competing risks situations (considered in Section 11.7). For example, bivariate survival models can be used to analyse:

Survival data on twins or other types of matched pair (Anderson et al., 1992);

Reliability data when the lifetime of one component is related to the lifetimes of other components;

Failure times of paired human organs (Sahu and Dey, 2000; Tosch and Holmes, 1980).

Examples of grouped or clustered data are provided by Gustafson (1997) as when several response times are measured for a single patient in a clinical trial, or when responses are for patients categorised according to clinic of treatment. Multivariate perspectives on more specialised survival models are exemplified by Bayesian multivariate cure rate models (Chen et al., 2002; Yin, 2005), and multivariate counting processes (Sinha and Ghosh, 2005).

The statistical model applied to such data needs to account for the intra-cluster or inter-event correlation. It may be possible to model the dependence structure directly, for example, via multivariate versions of widely adopted parametric survival models (Yashin et al., 2001). Thus Sahu and Dey (2000) consider bivariate exponential and Weibull survival models for data on times to visual impairment for paired eyes, while Damien and Muller (1998) provide a Bayesian treatment of a bivariate Gumbel model. The multivariate lognormal is another possibility, which adapts to the situation of conditional multivariate data, when durations on a second event are obtained conditional on the duration in a first event (Henderson and Prince, 2000).

Another approach is to introduce random frailty terms at the cluster level or common frailties across events. The frailty term represents common influences across clusters or events that are neglected or not observed. Responses on members of a cluster (or on correlated events) are typically assumed independent given the value of the cluster effect (or shared frailty factor) (Castro et al., 2014). Sahu and Dey (2004, p.325) describe how different frailty assumptions lead to different correlations between log survival times in a bivariate situation (under the assumption a Weibull baseline hazard).

Let t_{ij} be the failure time for the jth component or outcome $(j = 1,\dots,m_i)$ of the ith subject $(i = 1,\dots n)$. Then the hazard function assuming a common multiplicative frailty takes the form (Sahu et al., 1997; Yin and Ibrahim, 2005)

$$h(t_{ij} \mid Z_{ij}, \gamma_i) = \gamma_i h_0(t_{ij}) \exp(Z_{ij}\beta_j),$$

with the unit frailty effect γ_i distributed independently of Z_{ij} and t_{ij}. If γ_i is high, then all hazards are raised, and so times t_{ij} tend to be low; if γ_i is low then all hazards are lowered and the t_{ij} tend to be relatively extended. In this way, the common frailty induces a positive association between observed times.

In the case of repeated occurrences $r = 1,\dots,R_i$ of the same outcome to the same subject (e.g. multiple occupation shifts or repeat cardiac events), the hazard function conditional on γ_i is independent of the number r of previous occurrences (Sinha, 1993). Unconditionally, however, the hazard for the $(r+1)$th occurrence is

$$h_r(t_{ir} \mid Z_i) = h_0(t_{ir}) \exp(Z_i\beta)(1 + r\text{Var}(\gamma_i)).$$

The same scenario applies when subjects i are nested within clusters j, with cluster effects γ_j shared between the n_j individuals in the same cluster

$$h(t_{ij} \mid Z_{ij}, \gamma_j) = \gamma_j h_0(t_{ij}) \exp(Z_{ij}\beta_j), \quad i = 1,\dots,n_j; \; j = 1,\dots,J.$$

If the γ_j are assumed gamma distributed $\gamma_j \sim Ga(h,h)$ with variance $1/h$, then smaller values of h signify a closer relationship between subjects in the same group and greater heterogeneity between the groups. For models including cure rates, Yin (2005) proposes multiplicative frailty at cluster level combined with Poisson regression for θ_{ij} in the cure fractions $\exp(-\theta_{ij})$. One option takes

$$S^*(t_{ij}) = \exp(-\theta_{ij}\gamma_j F(t_{ij})),$$

with hazard rates $h^*(t_{ij}) = \theta_{ij}\gamma_j f(t_{ij})$.

Survival time data are often highly skewed and this may affect the appropriate form of frailty. Frailty models allowing for fat tails and skewness are obtained under the skew log-normal or skew log-t common frailty approach (Sahu and Dey, 2004). Consider a parametric hazard (e.g. Weibull) for multiple event time data (subjects $i = 1,\ldots,n$ and events $j = 1,\ldots m$) and with subject level scale parameters λ_{ij} for event j, namely

$$h(t_{ij} \mid \lambda_{ij}, \kappa) = \lambda_{ij}\kappa_j t_{ij}^{\kappa_j - 1}.$$

Then a skew log-normal frailty model implies

$$\log(\lambda_{ij}) = Z_i\beta_j + b_i + \delta u_i,$$

where $b_i \sim N(0,\sigma_b^2)$, δ is positive, and $u_i \sim N^+(0,1)$ with u_i independent of b_i. Under the skew log-t model,

$$b_i \sim t(0,\nu,\sigma_b^2),$$

where ν is a degrees of freedom parameter.

In practice, this kind of model may need informative priors for stable identification, bearing in mind that censoring reduces identifiability of complex random effect models, that b_i and u_i are to some extent overlapping in their roles, and that σ_b^2 and δ^2 are confounded in $\text{var}(b_i + \delta u_i) = \sigma_b^2 + \delta^2$. To illustrate relevant strategies for priors on the variance components, uncensored bivariate times (n=100, m=2) are generated with Weibull hazards, and scales $\lambda_{ij} = \exp(\beta_{1j} + \beta_{2j}x_i + b_i + \delta u_i)$ where the x_i are standard normal, with $\beta_1 = (-5,-6), \beta_2 = (0.5,1)$, κ=(1.5,2), δ=0.5, and $1/\sigma_b^2 = \tau_b = 2$, so that $\sigma_b \simeq 0.7$. A U(0,5) prior on δ is adopted in the analysis to re-estimate the parameters (cf. Sahu and Dey, 2004). The re-estimated parameters lead to considerable under-estimation of σ_b, with the second half of a single-chain run of 10,000 iterations leading to posterior mean of 0.054, whereas the posterior mean of δ is 1.6. Assuming instead a U(0,100) prior on $V = \delta^2 + \sigma_b^2$, and a U(0,1) prior on the ratio $\delta^2/(\delta^2 + \sigma_b^2)$, improves the estimation of σ_b with posterior mean 0.55, while the posterior mean of δ is now 1.17.

As for univariate models, flexibility is obtained by adopting a semiparametric hazard, while allowing also for common frailty. An example involves a semiparametric counting process including multiplicative frailty for repeated occurrences of the same event (Sinha, 1993). The semiparametric hazard is based on $J-1$ intervals $A_j = [a_{j-1}, a_j)$ obtained by considering distinct failure times, with a_J equal to the maximum time (censored or failed). Thus, for subject-occurrence index i, subject s, and interval j define an intensity function

$$\lambda(t_{ij} \mid Z_i) = Y(t_{ij})b_0(t_{ij})\exp(Z_{ij}\beta)\gamma_s,$$

where $b_0(t)$ is the baseline intensity function, and the γ_s represent subject level frailty. The integrated baseline intensity $B_0(t) = \int_0^t b_0(u)du$ is assumed to follow an independent increments gamma process, namely

$$dB_0(t) \sim Ga(cdB_0^*(t), c),$$

where $B_0^*(t)$ is an assumed mean intensity. The likelihood kernel for each spell within each subject is Poisson in form [3] with response variables $dN_{ij}=1$ or 0, and means $dB_0(t_{ij})\exp(Z_{ij}\beta)\gamma_s$.

Example 11.6 Clustered Trial of Infection Treatment

This example involves two forms of nesting: repetitions of events within patients, and nesting of patients within hospitals. The data are from Fleming and Harrington (1991) and Yau (2001), and concern a randomised trial of gamma interferon in treating infections among patients with chronic granulomatous disease. The 126 patients were nested in 13 hospitals and patients may experience more than one infection. Of 63 patients in the treatment group, 14 had at least one infection, and 20 infections were recorded in all, whereas in the placebo group, 30 patients had at least one infection, and there were 56 infections in all.

The $n=201$ observations are therefore at three levels: infections at level 1, patients at level 2, and hospitals as level 3 units. Let t_{klm} be times between recurrent infections, with k denoting events within patients, l denoting patients ($l=1,\ldots,L$), and m denoting hospitals ($m=1,\ldots,M$). The analysis seeks to assess the effect of gamma interferon in reducing the rate of infection as well as taking account of the clustering in the data; ignoring such clustering may affect the estimated treatment effect. A piecewise exponential baseline hazard is assumed with $J=20$ intervals $(a_{j-1},a_j]$ based on 5th percentiles of the observed times, with a_J being the maximum time of 389. Then with a single predictor ($z_{lm}=1$ for treated subjects, 0 otherwise)

$$h(t_{klm}\in(a_{j-1},a_j]\,|\,z_{lm})=\lambda_j\exp(z_{lm}\beta+e_{lm}+u_m),$$

with a gamma process prior on the λ_j, and normally distributed patient and hospital effects $e_{lm}\sim N(0,\sigma_e^2)$, and $u_m\sim N(0,\sigma_u^2)$. Because the two sources of variation are confounded, a uniform prior $V\sim U(0,100)$ is adopted on the total variance $V=\sigma_u^2+\sigma_e^2$, and a $U(0,1)$ prior on the ratio $\sigma_u^2/(\sigma_u^2+\sigma_e^2)$.

A two-chain run using jagsUI converges in 10,000 iterations and provides a LOO-IC of 755 and WAIC of 754. Although Yau (2001) reported no significant hospital variation, here posterior means (95% CRI) for σ_e and σ_u of 1.09 (0.59, 1.77) and 0.51 (0.11, 1.12) are obtained, and within hospital correlation estimated as 0.21. The treatment effect is estimated as –1.20 (–1.97,–0.49). Centred hospital effects have posterior means varying from 0.42 (hospital 2) to –0.27 (hospital 10), with the site 2 effect having an 87% probability of being positive.

Q-Q plots of both the posterior mean e_{lm} and u_m suggests a departure from normality, associated with positive skew in both sets of effects. Assuming Student t_4 priors for both sets of effects does not improve fit, and the corresponding Q-Q plots (using the R function TQQPlot) are also not satisfactory.

We then consider a Dirichlet process mixture prior to model the patient random effects (model 3). A gamma(3.5,0.5) prior is adopted on the Dirichlet precision parameter (Dorazio, 2009), with the base density for the random effects taken as normal. This produces a LOO-IC of 749. A skew-normal model also slightly improves fit in terms of the LOO-IC, reducing it to 754. The estimated effective parameters total is unchanged despite the addition of positive normal effects, as in the conditional representation of the skew-normal (Huang and Dagne, 2012; Ghosh et al., 2007).

As a final option, the random patient and hospital effects are represented as multiplicative effects with gamma densities (Glidden and Vittinghoff, 2004). Thus

$$h(t_{klm}\in(a_{j-1},a_j]\,|\,z_{lm})=\lambda_j\alpha_{1lm}\alpha_{2m}\exp(z_{lm}\beta),$$

$$\alpha_{1lm} \sim Ga(\delta_1, \delta_1),$$

$$\alpha_{2m} \sim Ga(\delta_2, \delta_2),$$

$$\delta_j \sim Ga(a_\delta, b_\delta).$$

The prior mean of the gamma effects $\alpha_{1/m}$ and α_{2m} is set at 1 for identification. With the setting $a_\delta = 1, b_\delta = 0.1$, a two-chain run using jagsUI converges in 10,000 iterations. This provides similar fit statistics as the normal errors model, namely a LOO-IC of 755 and WAIC of 753. The estimated gamma parameters (posterior mean and standard deviation) are 1.62 (1.23) and 12.46 (10.29). A plot of actual patient effects as against the implied gamma density, Ga(1.62,1.62), shows a better representation of the skew in the patient effects, albeit with still some discrepancies.

Example 11.7 Bivariate Survival

The Diabetic Retinopathy Study was conducted by the US National Eye Institute to assess the effect of laser photocoagulation in delaying onset of severe visual loss in patients with diabetic retinopathy. One eye of each patient was randomly selected for photocoagulation, and the other was observed without treatment, with patients followed up over several years for the occurrence of blindness in one or other eye. The follow-up time is in months, with 80 patients censored on both eyes at the same time, while 36 patients have onset in both eyes. Censoring is caused by dropout, death, or termination of the study. Following Huster et al. (1989), a subset of the data set containing n=197 high-risk patients is considered here, so there are $i = 1,\ldots,n$ patients and $j = 1,\ldots,m$ events with $m = 2$. The correlation between pairs of uncensored times t_{ij} for patients in both treatment and control groups is 0.28, indicating possible dependence between the two times.

The Weibull is judged to be appropriate for these data, as a plot of the transformed Kaplan-Meier survivor function namely $\log[-\log(S_{KM}(t))]$ on $\log(t)$, is approximately linear when either t_{i1} or t_{i2} are considered. Alternative analyses consider Weibull survival with and without lognormal frailty at patient level. Under the latter

$$t_{ij} \sim \text{Wei}(\kappa_j, \lambda_{ij}),$$

$$\log(\lambda_{ij}) = \beta_0 + \beta_1 \text{Age}_i/10 + \beta_2 Trt_{ij} + \beta_3 \text{Type}_i + b_i,$$

$$b_i \sim N(0, \sigma_b^2),$$

where b_i are random patient effects, Age is age at diagnosis and Type relates to diabetes type. A model (implemented using jagsUI) without frailty produces significant Weibull shape effects: both shape parameters κ_j have 95% intervals below 1, suggesting a lesser chance of impairment at longer follow-ups. The LOO-IC is 2,756, with the worst fitted observation being eye 2 for patient 95. The treatment coefficient β_2 has mean (95% CRI) −0.79 (−1.12, −0.48), and the corresponding hazard ratio $\theta = e^{-\beta_2}$ for untreated eyes averages 2.24, with a 95% credible interval (1.62, 3.06).

In the frailty model, a $U(0,10)$ prior on the standard deviation σ_b of the random effects is adopted, with a posterior mean for σ_b of 0.45 (0.12, 0.82). Under this model, the time effects are attenuated, with the 95% interval for κ_1 now straddling 1. Despite the significant patient heterogeneity, the LOO-IC increases slightly to 2,762. The treatment effect increases, with θ now averaging 2.31 with 95% interval (1.65, 3.22). A histogram and kernel density plot of the posterior mean b_i show a subgroup with high negative values

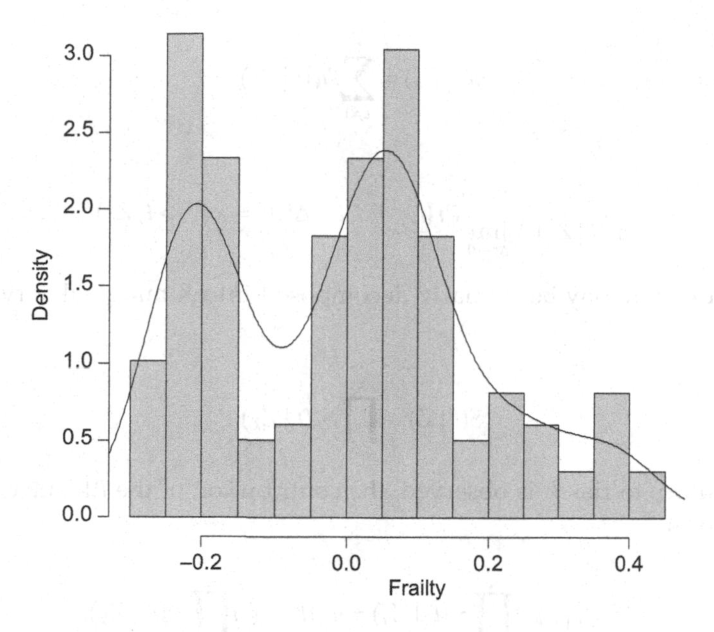

FIGURE 11.8
Posterior mean frailties.

(see Figure 11.8) suggesting that a discrete mixture approach (e.g. a two-group discrete mixture normal) to frailty might be appropriate.

11.7 Competing Risks

Competing risks (CR) models involve the tracking of multiple durations corresponding to different types of exit or transition (Haller et al., 2013). A number of packages in R can estimate competing risks survival regression (Scrucca et al., 2010; Putter, 2018; Scheike and Zhang, 2011). With non-repeatable events, subjects are observed until the first exit and completion of one of the multiple durations, but for repeatable events (e.g. occupational or migration histories), event histories might include repeated transitions between different job or residential destinations. Sometimes the cause of exit may be masked (Sen et al., 2010): exact information on the cause of exit is missing, but information is available that can determine a set of potential causes of exit.

Assume that there are K possible mutually exclusive causes of exit, and C_i be a subject level categorical random variable with K possible outcomes representing observed cause of exit. Under the latent failure time approach (Crowder, 2001; Box-Steffensmeier and Jones, 2004; Kozumi, 2004; Gelfand et al., 2000) with independent risks, there is a latent failure time T_{ik} corresponding to each outcome, but only the minimum time is observed when individual i exits for cause k_i, so that $t_i = \min(T_{i1},\ldots,T_{iK})$ with $k_i = \arg\min(T_{i1},\ldots,T_{iK})$. The remaining times are censored. All times are censored if an individual does not exit for any of the K possible reasons.

With these assumptions, and conditioning on possibly cause-specific predictors Z_k, the total hazard rate may be expressed as a sum of cause-specific hazards,

$$h(t \mid Z_k) = \sum_{k=1}^{K} h_k(t \mid Z_k),$$

where

$$h_k(t \mid Z_k) = \lim_{\Delta t \to 0} \frac{Pr(t < T \le t + \Delta t, C = k \mid T > t, Z_k)}{\Delta t}.$$

The survival function may be similarly decomposed into K marginal survival functions, with

$$S(t \mid Z) = \prod_{k=1}^{K} S_k(t \mid Z_k).$$

Assuming a failure to risk k_i is observed, the contribution of the ith subject to the likelihood has the form

$$f_{k_i}(t_i \mid Z_{ik_i}) \prod_{l \ne k_i}^{K} S_l(t_i \mid Z_{il}) = h_{k_i}(t \mid Z_{ik_i}) \prod_{l=1}^{K} S_l(t_i \mid Z_{il}),$$

while for a subject censored on all risks, the contribution is $\Pi_{l=1}^{K} S_l(t_i \mid Z_{il})$. With event indicators $d_{ik} = 1$ if $C_i = k$, and $d_{ir} = 0$ for $r \ne k$, the likelihood contribution is equivalently

$$\prod_{r=1}^{K} [f_r(t_i \mid Z_{ir})]^{d_{ir}} [S_r(t_i \mid Z_{ir})]^{1-d_{ir}}.$$

For continuous survival times, one may assume parametric forms for the time effect, e.g. a Weibull hazard

$$h_k(t) = \lambda_k \kappa_k t_i^{\kappa_k - 1},$$

or model risk-specific semiparametric hazard sequences that may be correlated over causes. Possible label switching problems under the latent failure approach may require parameter constraints, such as ordering the shape parameters κ_k (Gelfand et al., 2000).

Often competing risks models are applied to repeated transitions between occupational, residential, or marital states. The hazard rate then generalises to reflect moves between the mth observed state and the $(m+1)$th state. If T_{im} denotes the time spent in the mth state, and occupancy of the mth state for subject i is denoted $C_{im} = k$, then

$$h_{kl}(t \mid Z_{ik}) = \lim_{\Delta t \to 0} Pr(t < T_{im} \le t + \Delta t, C_{i,m+1} = l \mid T_{im} > t, C_{im} = k, Z_{ik}) / \Delta t,$$

is the instantaneous risk of moving from state k to state l (with $l \ne k$), given survival in the mth state until t. Under independent risks, the overall hazard for leaving state k is then

$$h_k(t \mid Z_{ik}) = \sum_{l \ne k}^{K} h_{kl}(t \mid Z_{ik}).$$

For discrete time data, the functions described in Section 11.5 similarly generalise to the competing risk case. For non-repeated events, intervals $[a_{j-1}, a_j)$ for $j = 1, \ldots, J+1$, and $C_i \in (1, \ldots, K)$, event probabilities are

$$f_{jk} = Pr(t \in [a_{j-1}, a_j), C = k),$$

with risk specific hazard functions

$$q_{jk} = Pr(t \in [a_{j-1}, a_j), C = k \mid t > a_{j-1}),$$
$$= f_{jk} / S_{j-1},$$

and survivor functions obtained as

$$S_j = \prod_{m=1}^{j} \prod_{h=1}^{K} (1 - q_{mh}).$$

Define event indicators $d_{imh} = 1$ when a non-repeatable event h occurs in interval m, and 0 otherwise. Then for subject i undergoing the kth event in the jth interval, the event indicators are $d_{ijk} = 1$, $\{d_{ijh} = 0, h \neq k\}$ and $d_{i1h} = d_{i2h} = \ldots = d_{i,j-1,h} = 0$ for all h, with likelihood

$$q_{ijk} \left[\prod_{m=1}^{j-1} \prod_{h=1}^{K} (1 - q_{imh}) \right] = q_{ijk} S_{i,j-1}.$$

The response at each interval for discrete competing risks is multinomial, and to model the impact of predictors different links may be used such as the multiple logit, or multiple probit. Consider a multiple logit link with $K+1$ categories (K alternative risks plus an extra category for survival, denoted by $C_i = 0$). Let the reference category be for survival, and define regression coefficients β_k for the kth risk. Assuming the β_r $(r = 1, \ldots, K)$ do not contain an intercept would lead to

$$q(t \in [a_{j-1}, a_j), C_i = 0) = \frac{1}{1 + \sum_{r=1}^{K} \exp(\alpha_{jr} + Z_{ir} \beta_r)},$$

$$q(t \in [a_{j-1}, a_j), C_i = h \mid Z_{ih}) = \frac{\exp(\alpha_{jh} + Z_{ih} \beta_h)}{1 + \sum_{r=1}^{K} \exp(\alpha_{jr} + Z_{ir} \beta_r)}, \quad h = 1, \ldots K.$$

where the parameters α_{jh} describe the baseline hazard for risk. K-dimensional versions of the correlated prior processes discussed in Section 11.3 may be used for the α_{jh}, for example, multivariate normal first- or second-order random walks.

11.7.1 Modelling Frailty

Assuming independent risks, one may introduce unobserved frailties γ_{ik} that impact on each risk, but are uncorrelated across risks, such as independent gamma densities with mean 1 for each possible cause. Under proportionality, the risk specific hazard in a continuous time CR hazard is then

$$h_k(t_i \mid Z_{ik}) = \gamma_{ik} h_{0k}(t_i) \exp(Z_{ik}\beta_k).$$

The assumption of independent risks may not hold in practice because particular groups of subjects may be more likely to experience subsets of the events. Just as it may be unrealistic in multinomial discrete choice situations to assume independence of irrelevant alternatives (i.e. that ratios of choice probabilities of any two alternatives are unaffected by changes in utilities of any other alternatives, or by their removal), so it may be unrealistic in survival analysis that the relative risks of two outcomes will be unaffected by the removal of a third (Gordon, 2002).

To allow for dependent competing risks, especially for multiple spell data, one may assume correlated or dependent frailties. In a generalisation of the MPH scheme, Abbring and van den Berg (2003) mention that the joint distribution of $(T_{i1},\ldots,T_{iK})$, given predictors Z_{ik} and correlated frailties $(\gamma_{i1},\ldots,\gamma_{iK})$, factorises into independent densities $f(T_k \mid Z_{ik},\{\gamma_{i1},\ldots,\gamma_{iK}\})$ which are fully characterised by cause-specific hazard rates

$$h(T_k \mid Z_k,\{\gamma_1,\ldots,\gamma_K\}) = \gamma_k \lambda_k(t) \exp(Z_k\beta_k).$$

Correlated frailties are also obtained by expanding the regression term to a general mixed form, as in Section 11.4, so that in a continuous time analysis,

$$h_k(t_i \mid b_{ik}, Z_{ik}) = \lambda_k(t_i) \exp(\beta_k Z_{ik} + b_{ik}),$$

where b_{ik} are zero mean effects that might be multivariate normal, discrete mixtures of multivariate normal, etc.

Assuming a multivariate normal b_{ik} with covariance matrix Σ_b, dependent risks will be apparent in significant off-diagonal terms. Whether there are significant correlations in the frailty effects over different risks will depend in part on whether observed predictors successfully explain variations in event proneness. Another possibility is a common frailty model with risk specific loadings, so that

$$b_{ik} = \lambda_k b_i,$$

where $\lambda_k > 0$ and $b_i \sim N(0,1)$ for identification.

Example 11.8 Hospital Infection

This example demonstrates parametric competing risk analysis of data on hospital infections from Beyersmann and Scheike (2013), also included as the dataset okiss in the R package compeir. There are 1,000 subjects, and the data concern treatment for hematologic disease by peripheral blood cell stem transplantation. Transplants are either autologous (from the patients' own blood), or allogeneic. After transplantation, patients are deficient in white blood cells (neutropenic), and at risk of bloodstream infection, a severe complication (competing risk 1). Alternatively, competing risk 2 defines survival as until neutropenia ends, or until death without bloodstream infection. Predictors are gender (1 = F, 0 = M), and transplant source (allo = 1 for allogeneic, 0 = autologous).

Figure 11.9 compares Kaplan–Meier, Weibull, and log-logistic survival plots for bloodstream infection, and suggests the Weibull as a working approximation. So

$$h_k(t_i \mid z_i) = \lambda_{ik} \kappa_k t_i^{\kappa_k - 1},$$

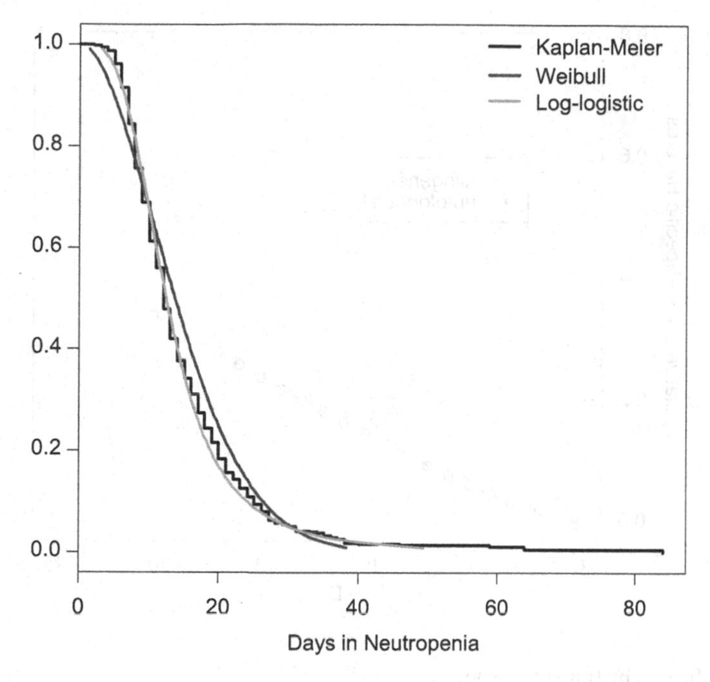

FIGURE 11.9
Survival curve without predictors.

for risks $k=1, 2$, where $\lambda_{ik} = \exp(\beta_k x_i)$. The cause-specific cumulative hazards are

$$\Lambda_k(t_i \mid z_i) = \lambda_{ik} t_i^{\kappa_k}.$$

Estimation uses rstan [4] with early convergence non-problematic. Table 11.3 shows that allogeneic transplants are associated with a lower risk of bloodstream infection, and Figure 11.10 plots out the contrasting cumulative hazard curves for this cause of exit

TABLE 11.3
Posterior Summary. Transplant Data

Weibull Regression		Mean	St Devn	2.5%	97.5%	Standardised Variability
Risk 1	Intercept	-4.20	0.20	-4.58	-3.83	
	Allogeneic Transplant	-0.54	0.14	-0.82	-0.28	0.26
	Female	-0.18	0.14	-0.46	0.09	0.79
	Weibull Shape	1.22	0.07	1.09	1.36	
Risk 2	Intercept	-4.98	0.13	-5.25	-4.72	
	Allogeneic Transplant	-1.19	0.07	-1.33	-1.04	0.06
	Female	0.09	0.07	-0.06	0.23	0.84
	Weibull Shape	2.04	0.04	1.96	2.13	
Cox Regression		**Mean**	**St Devn**	**2.5%**	**97.5%**	**Standardised Variability**
Risk 1	Allogeneic Transplant	-0.25	0.15	-0.54	0.03	0.58
	Female	-0.17	0.14	-0.46	0.11	0.83
Risk 2	Allogeneic Transplant	-1.26	0.08	-1.41	-1.10	0.06
	Female	0.06	0.08	-0.08	0.21	1.17

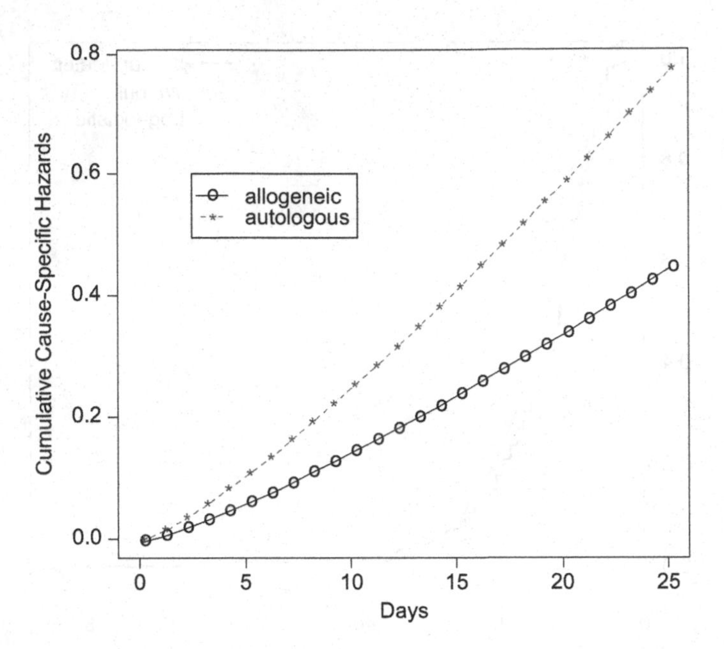

FIGURE 11.10
Cumulative hazard by transplant source.

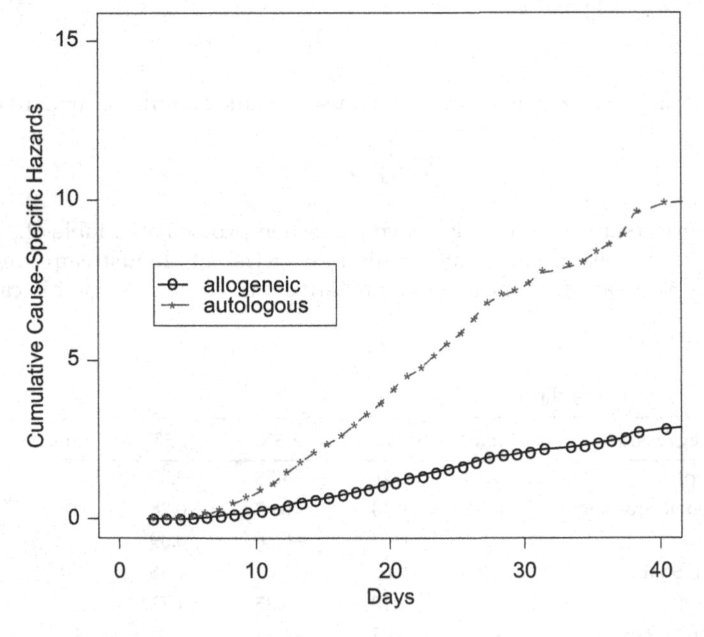

FIGURE 11.11
Cox regression. Cumulative hazard, end of neutropenia, by transplant source.

up to 25 days. The effect of allogeneic transplantation is also negative for the other risk (end of neutropenia), meaning that events of either type are delayed for the allogeneic treatment group.

We also apply Cox regression to these data, based on the distinct event times for each risk. Figure 11.11 shows the resulting cumulative hazard plots for the end of neutropenia.

An issue in comparing Cox and parametric regression is the possibility of differing precision of estimated covariate and treatment effects (Nardi and Schemper, 2003). It can be seen from Table 11.3 that the estimated coefficients for allo and sex from Cox regression have higher standardised variability (i.e. are less efficient). This is especially so for the impact of transplant type on the risk of bloodstream infection. Standardised variability is measured as $sd(\beta)/|\beta|$.

Example 11.9 Follicular Cell Lymphoma, Cause-Specific and Subdistribution Hazards

This example considers the follicular cell lymphoma data from Pintilie (2006), also included in the R package timereg (Scheike and Zhang, 2011). The influence of four covariates is explored on two alternative outcomes: (1) relapse or no response, or (2) death in remission. There are 272 subjects with no response or relapse, 76 deaths without relapse, and 193 censored subjects. Predictors are stage of disease (1=stage II, 0=stage I), treatment (1=radiotherapy and chemotherapy combined, 0=radiotherapy only), age and haemoglobin level, with the latter two predictors divided by 100. Event times are in years since diagnosis. Following discussions such as Latouche et al. (2013), both cause-specific and subdistribution hazard regressions are estimated.

Using the counting process version of the Cox model requires that at risk indicators and increment indicators be set for each of the competing responses, based on distinct event times. So, define a variable icaus in R having values 0 for censoring, 1 for relapse/no response, and 2 for death in remission. With obs_t denoting event (or censoring) times, the command sequence in R to define the two sets of indicators for event 1 is:

```
d1=ifelse(cause==1,1,0)
t.d1=subset(obs_t,cause==1)
# unique event times
t.d1.unique=unique(t.d1)
NT1=length(t.d1.unique)
t1_unique=c(sort(t.d1.unique),max(obs_t)+1)
# define at risk and counting process increments
Y1=dN1=matrix(,N,NT1)
for (i in 1:N) {for (j in 1:NT1) {Y1[i,j]
=ifelse(obs_t[i]>=t1_unique[j],1,0)}}
for (i in 1:N) {for (j in 1:NT1) {dN1[i, j] =Y1[i, j] * (t1_
unique[j + 1] > obs_t[i]) * d1[i]}}
```

This is the usual assignment of at risk and increment indicators in cause-specific hazard regression. The cause-specific hazard is the instantaneous risk of the event (i.e. a specific cause of exit) in subjects currently event-free, namely for cause k,

$$h_k(t) = \lim_{\Delta t \to 0}\left\{\frac{Pr(t \le T < t + \Delta t, K = k \mid T > t)}{\Delta t}\right\}.$$

By contrast, the subdistribution hazard is the instantaneous risk of an event in subjects yet to experience an event of that type. So

$$\tilde{h}_k(t) = \lim_{\Delta t \to 0}\left\{\frac{Pr(t \le T < t + \Delta t, K = k \mid (T > t) \text{ or } (T \le t \text{ and } K \ne k))}{\Delta t}\right\}.$$

TABLE 11.4
Cell Lymphoma. Alternative Hazard Regression Coefficients, Posterior Summaries

		Cause-Specific Hazard			Subdistribution Hazard		
Competing Events	**Predictor**	**Mean**	**2.5%**	**97.5%**	**Mean**	**2.5%**	**97.5%**
Relapse	Stage	0.35	0.09	0.61	0.40	0.16	0.65
	Chemotherapy	0.09	−0.26	0.41	−0.03	−0.39	0.30
	Age/100	4.29	3.35	5.25	1.85	0.97	2.71
	HGB/100	0.79	0.12	1.47	0.60	−0.07	1.26
Death in Remission	Stage	0.11	−0.44	0.63	−0.09	−0.62	0.42
	Chemotherapy	−0.08	−0.85	0.63	−0.38	−1.09	0.23
	Age/100	8.32	6.12	10.52	4.36	2.66	5.95
	HGB/100	−0.01	−1.61	1.56	−0.54	−1.89	0.71

The risk set now includes subjects who have previously experienced a competing cause of exit, as well as subjects currently event-free. For the cause-specific hazard, the risk set reduces every time there is an exit from another cause and is viewed as censored. With the subdistribution hazard subjects that exit for a cause, $j \neq k$ remain in the risk set for cause k and are given a censoring time larger than all event times. The coefficients from a subdistribution hazard model may be interpreted as the impacts of covariates on the incidence of the event (Austin and Fine, 2017).

Table 11.4 summarises the posterior distributions of the covariate effects under these alternative approaches. The effects are not that dissimilar, mainly differing in a lower impact of age on incidence, while the impact of stage on the incidence of relapse is enhanced. The estimated coefficients under classical methods (using coxph from timereg, and crr from cmprsk) are similar to the Bayesian estimates.

11.8 Computational Notes

[1] The rstan code for the counting process model in Example 11.2 includes R calculations to convert time and event indicators (t_i,d_i) into suitable form. Thus

```
obs_t=t
t.d=subset(obs_t,d==1)
# unique event times
t.d.unique=unique(t.d)
NT=length(t.d.unique)
t_unique=c(sort(t.d.unique),max(obs_t)+1)
# define at risk and counting process increments
Y=dN=matrix(,N,NT)
for (i in 1:N) {for (j in 1:NT) {Y[i,j]
=ifelse(obs_t[i]>=t_unique[j],1,0)}}
for (i in 1:N) {for (j in 1:NT) {dN[i, j] =Y[i, j] * (t_unique[j + 1]
> obs_t[i]) * d[i]}}
# centred and scaled Karnosky score
KS.c=(KS-mean(KS))/10
```

```
# dataset
Dstan=list(N=N,NT=NT,t_unique=t_unique,Y=Y,dN=dN,Z=PT,KS=KS.c)
CP.stan ="
data {
int<lower=0> N;
real KS[N];
int<lower=0> NT;
int<lower=0> Y[N,NT];
int<lower=0> dN[N,NT];
int<lower=0> t_unique[NT + 1];
real PT[N];
}
transformed data {
real c;
real r;
c = 0.001;
r = 0.1;
}
parameters {
real beta[2];
real betaKS[NT];
real<lower=0.001> sigmaKS;
real<lower=0> dL0[NT];
}
model {
real dt[NT];
beta ~normal(0, 10);
sigmaKS ~uniform(0,1);
betaKS[1] ~normal(0,1);
//RW prior on KS coefficients
for (j in 2:NT){betaKS[j] ~normal(betaKS[j-1],sigmaKS);}
//gamma increments prior
for (j in 1:NT) {dt[j] = t_unique[j+1] - t_unique[j];
                    dL0[j] ~gamma(r * dt[j] * c, c);
for (i in 1:N) {if (Y[i, j]!= 0)
            target += poisson_lpmf(dN[i, j]
            Y[i, j]*exp(betaKS[j]*KS[i]+beta[1]*PT[i]+beta[2]*PT[i]
            *KS[i]) * dL0[j]);}}}
generated quantities {
real S_noPT[NT];
real S_PT[NT];
for (j in 1:NT) {//Survivor functions by prior therapy, Karnofsky
score set at upper quartile
real s;
s = 0;
for (i in 1:j)
s = s + dL0[i];
S_PT[j] = pow(exp(-s), exp(betaKS[j]*1.64+beta[1]+beta[2]*1.64));
S_noPT[j] = pow(exp(-s), exp(betaKS[j]*1.64));}}
"
# Compilation and Estimation
sm = stan_model(model_code=CP.stan)
fit = sampling(sm,data =Dstan,iter = 1500,warmup=250,chains = 2,seed=
12345)
```

```
print(fit)
betaKS <- extract(fit,"betaKS",permute=F)
```

[2] The code for the cure fraction age of maternity model is

```
loglogistCF.stan ="
functions{
real loglogistCF_lpdf(real t, real kappa, real lambda, real theta) {
return(log(theta)+log(kappa)+log(lambda)+(kappa-1)*log(t)
-2*log(1+lambda*t94kappa)-theta*(1-1/(1+lambda*t94kappa)));}
real loglogistCF_S_lpdf(real t, real kappa, real lambda, real theta) {
return(-theta*(1-1/(1+lambda*t94kappa)));}
}
data {int<lower=1> n;//number of cases
vector[n] t;//response
int<lower=0,upper=1> d[n];//event indicator(1=occurred, 0=censored)
int<lower=0> p;//total regression parameters, incl. intercept
int<lower=0> educ[n];
int<lower=0> sibs[n];
int<lower=0> white[n];
int<lower=0> immig[n];
int<lower=0> lowinc[n];
int<lower=0> city[n];
}
parameters {vector[p] beta;
real<lower=1> kappa;//shape parameter
real<lower=0> theta;//cure fraction parameter
}
transformed parameters {
real eta[n];
real lambda[n];
real lambdaT;
real p_nochild;
real modeT;//modal age first maternity (13 years education, 3
siblings, white)
p_nochild = exp(-theta);//rate of childlessness
lambdaT = exp(beta[1]+beta[2]*13+beta[3]*3+beta[4]);
modeT = ((kappa-1)/lambdaT)94(1/kappa);
for (i in 1:n) {eta[i]= beta[1]+beta[2]*educ[i]+beta[3]*sibs[i]
+beta[4]*white[i]+beta[5]*immig[i]
+beta[6]*lowinc[i]+beta[7]*city[i];
lambda[i] =exp(eta[i]);}}
model {target += gamma_lpdf(kappa 0.01, 0.01);
target += gamma_lpdf(theta 0.01, 0.01);
for (i in 1:n) {
if (d[i] == 1) {target += loglogistCF_lpdf(t[i]kappa,
lambda[i],theta);}
else if (d[i] == 0) {target += loglogistCF_S_lpdf(t[i]kappa,
lambda[i],theta);}}}
generated quantities{real log_lik[n];
for (i in 1:n) {
if (d[i] == 1) {log_lik[i]= loglogistCF_lpdf(t[i] kappa,
lambda[i],theta);}
else if (d[i] == 0) {log_lik[i]= loglogistCF_S_lpdf(t[i] kappa,
lambda[i],theta);}}}
```

[3] An example of the computation involves repeated times being applied to mammary tumour in rats randomly assigned to treatment and control groups (Sinha, 1993). Totals of tumours diagnosed in each rat varying between 0 and 13. So, spell totals for each rat (including possibly censored final spells) range from 1 to 14. There are n=253 spells in all, for K=48 rats, and J=35 distinct times relevant to defining the intervals, with $a_J = t_{\max} = 182$. A BUGS/JAGS code for such an analysis, including gamma frailty for each rat, a treatment covariate, and indicators d[i] of tumour occurrence or censoring, is

```
model {for (j in 1:J) {for(i in 1:n) {# Y indicates whether case
still at risk
Y[i,j] <- step(t[i] - a[j] + eps)
dN[i, j] <- Y[i, j] * step(a[j + 1] - t[i] - eps) * d[i]
dN[i, j] ~dpois(lam[i, j])
lam[i, j] <- Y[i, j] * exp(beta * trt[i]) * dB0[j] * gam[rat[i]]}
# independent increment gamma process
dB0[j] ~dgamma(mu[j], c); mu[j] <- dB0.star[j] * c
dB0.star[j] <- M * (a[j + 1] - a[j])
# Survivorship in two groups
S.tr[j] <- pow(exp(-sum(dB0[1: j])), exp(beta));
S.cntr[j] <- exp(-sum(dB0[1: j]))}
# priors on hyperparameters
c <- 1; M ~dexp(1); beta ~dnorm(0,0.001)
# frailty prior
for (k in 1:K) {gam[k] ~dgamma(h,h)}
h ~dgamma(1,0.001)
var.gam <- 1/h}
```

where eps is a small positive value to ensure at risk and counting indices are correctly defined. The gamma process includes an unknown parameter M defining the mean intensity. The first few records for the spell level data take the form

rat[]	trt[]	t[]	d[]
1	1	182	1
2	1	182	0
3	1	63	1
3	1	68	1
3	1	182	0
4	1	152	1
4	1	182	0

while the other data inputs are list(n=253,J=34,a=c(63,66,68,71,74,77,81,84,85,88,
91,95,98,102,105,108,112,116,119,123,
126,130,134,137,140,145,150,152,157,
161,167,172,174,179,182),eps=0.001,K=48).

[4] The code for this analysis is

```
weibCR.stan ="
data {
int<lower=1> N;//number of cases
int<lower=1> N2;//number of cases
int<lower=1> T;//number of time points for CH profiles
int<lower=1> K;//number of competing causes of exit
vector[N] time;//observed or censored times
```

```
vector[T] timeprof;//time points for CH profiles
int<lower=0,upper=1> cens1[N];//right censoring, cause 1
int<lower=0,upper=1> cens2[N];//right censoring, cause 2
int<lower=0> p;//total regression parameters, including intercept
int<lower=0> allo[N];
int<lower=0> sex[N];
}
parameters {vector[p] beta1;
vector[p] beta2;
real<lower=0> shape[K];//shape parameters
}
transformed parameters {
real eta1[N];
real nu1[N];
real eta2[N];
real nu2[N];
real S1allo[T];//survival functions
real S1auto[T];
real S2allo[T];
real S2auto[T];
real CH1allo[T];//cumulative hazards
real CH1auto[T];
real CH2allo[T];
real CH2auto[T];
for (t in 1:T) {
S1allo[t] = exp(-exp(beta1[1]+beta1[2])*timeprof[t]94shape[1]);
S1auto[t] = exp(-exp(beta1[1])*timeprof[t]94shape[1]);
S2allo[t] = exp(-exp(beta2[1]+beta2[2])*timeprof[t]94shape[2]);
S2auto[t] = exp(-exp(beta2[1])*timeprof[t]94shape[2]);
CH1allo[t] = -log(S1allo[t]);
CH1auto[t] = -log(S1auto[t]);
CH2allo[t] = -log(S2allo[t]);
CH2auto[t] = -log(S2auto[t]);}
for (i in 1:N) {eta1[i]=beta1[1]+beta1[2]*allo[i]+beta1[3]*sex[i];
nu1[i] = exp(-eta1[i]/shape[1]);
eta2[i]=beta2[1]+beta2[2]*allo[i]+beta2[3]*sex[i];
nu2[i] = exp(-eta2[i]/shape[2]);}
}
model {target += gamma_lpdf(shape 0.01, 0.01);
for (i in 1:N) {
if (cens1[i] == 0) {target += weibull_lpdf(time[i] shape[1],
nu1[i]);}
else if (cens1[i] == 1) {target += weibull_lccdf(time[i] shape[1],
nu1[i]);}
if (cens2[i] == 0) {target += weibull_lpdf(time[i] shape[2],
nu2[i]);}
else if (cens2[i] == 1) {target += weibull_lccdf(time[i] shape[2],
nu2[i]);}
}}
generated quantities{real log_lik[N2];
//expanded log-likelihood vector over K=2 causes
for (i in 1:N) {
if (cens1[i] == 0) {log_lik[i]= weibull_lpdf(time[i] shape[1],
nu1[i]);}
```

```
else if (cens1[i] == 1) {log_lik[i]= weibull_lccdf(time[i] shape[1],
nu1[i]);}
if (cens2[i] == 0) {log_lik[i+N]= weibull_lpdf(time[i] shape[2],
nu2[i]);}
else if (cens2[i] == 1) {log_lik[i+N]= weibull_lccdf(time[i]
shape[2], nu2[i]);}
}}
```

References

Aalen O (1988) Heterogeneity in survival analysis. *Statistics in Medicine*, 7, 1121–1137.

Aalen O, Hjort N (2002) Frailty models that yield proportional hazards. *Statistics & Probability Letters*, 58, 335–342.

Abbring J, van den Berg G (2003) The identifiability of the mixed proportional hazards competing risks model. *Journal Royal Statistical Society: Series B*, 65, 701–710.

Abbring J, van den Berg G (2007) The unobserved heterogeneity distribution in duration analysis. *Biometrika*, 94, 87–99.

Aitkin M, Clayton D (1980) The fitting of exponential, Weibull and extreme value distributions to complex censored survival data using GLIM. *Journal of Applied Statistics*, 29, 156–163.

Aitkin MA, Aitkin M, Francis B, Hinde J (2005) *Statistical Modelling in GLIM 4*. OUP, Oxford, UK.

Albert JH, Chib S (2001) Sequential ordinal modeling with applications to survival data. *Biometrics*, 57(3), 829–836.

Allison P (1997) *Survival Analysis Using the SAS System: A Practical Guide*. SAS Institute Inc., Cary, NC.

Anderson JE, Louis TA, Holm NV, Harvald B (1992) Time-dependent association measures for bivariate survival distributions. *Journal of the American Statistical Association*, 87(419), 641–650.

Arjas E, Gasbarra D (1994) Nonparametric Bayesian inference from right censored survival data, using the Gibbs sampler. *Statistica Sinica*, 4, 505–524.

Austin P (2017) A tutorial on multilevel survival analysis: Methods, models and applications. *International Statistical Review*, 85(2), 185–203.

Austin P, Fine J (2017) Practical recommendations for reporting Fine-Gray model analyses for competing risk data. *Statistics in Medicine*, 36(27), 4391–4400.

Baker M, Melino A (2000) Duration dependence and nonparametric heterogeneity: A Monte Carlo study. *Journal of Econometrics*, 96, 357–393.

Banerjee S, Carlin BP (2004) Parametric spatial cure rate models for interval-censored time-to-relapse data. *Biometrics*, 60(1), 268–275.

Barmby T (2002) Worker absenteeism: A discrete hazard model with bivariate heterogeneity. *Labour Economics*, 9, 469–447.

Bender A, Groll A, Scheipl F (2018) A generalized additive model approach to time-to-event analysis. *Statistical Modelling*, 18, 1–23.

Bennett S (1983) Log-logistic regression models for survival data. *Applied Statistics*, 32, 165–171.

Beyersmann J, Scheike T (2013) Classical regression models for competing risks, Chapter 8, pp 157–177, in *Handbook of Survival Analysis*, eds J Klein, H C van Houwelingen, J G Ibrahim, T Scheike. CRC.

Bogaerts K, Komarek A, Lesaffre E (2017) *Survival Analysis with Interval-Censored Data: A Practical Approach with Examples in R, SAS, and BUGS*. CRC Press.

Børing P (2009) Gamma unobserved heterogeneity and duration bias. *Econometric Reviews*, 29(1), 1–19.

Bowers N, Gerber H, Hickman J (1986) *Actuarial Mathematics*, 1st Edition. The Society of Actuaries, Itasca, IL.

Box-Steffensmeier JM, Box-Steffensmeier JM, Jones BS (2004) *Event History Modeling: A Guide for Social Scientists*. Cambridge University Press.

Brard C, Le Teuff G, Le Deley M, Hampson L (2017) Bayesian survival analysis in clinical trials: What methods are used in practice? *Clinical Trials*, 14(1), 78–87.

Brezger A, Kneib T, Lang S (2005) BayesX: Analyzing Bayesian Structured Additive Regression Models , *Journal of Statistical Software September 14 (11)*, 1–22.

Brouhns N, Denuit M, Vermunt J (2002) A Poisson log-bilinear regression approach to the construction of projected lifetables. *Insurance: Mathematics and Economics*, 31(3), 373–393.

Brüderl J, Diekmann A (1995) The log-logistic rate model: two generalizations with an application to demographic data. *Sociological Methods & Research*, 24, 158–186.

Buros J (2016) Model Checking with Simulated Data (Survival Model Example) https://www.bioconductor.org/help/course-materials/2016/BioC2016/ConcurrentWorkshops4/Buros/weibull-survival-model.html

Castro M, Chen M-H, Ibrahim J, Klein J (2014) Bayesian transformation models for multivariate survival data. *Scandinavian Journal of Statistics*, 41(1), 187–199.

Chen Q, Wu H, Ware L B, Koyama T (2014) A Bayesian approach for the cox proportional hazards model with covariates subject to detection limit. *International Journal of Statistics in Medical Research*, 3(1), 32–43.

Chen M-H, Ibrahim J, Sinha D (1999) A new Bayesian model for survival data with a surviving fraction. *Journal of the American Statistical Association*, 94, 909–919.

Chen M-H, Ibrahim J, Sinha D (2002) Bayesian inference for multivariate survival data with a cure fraction. *Journal of Multivariate Analysis*, 80, 101–126.

Chen Y, Jewell N (2001) On a general class of semiparametric hazards regression models. *Biometrika*, 88, 687–702.

Chiang C (1984) *The Life Table and its Applications*. R.E. Krieger, Malabar, FL.

Clayton D (1991) A Monte Carlo method for bayesian inference in frailty models. *Biometrics*, 47, 467–485.

Congdon P (2008) A bivariate frailty model for events with a permanent survivor fraction and non-monotonic hazards; with an application to age at first maternity. *Computational Statistics & Data Analysis*, 52, 4346–4356.

Congdon P (2009) Life expectancies for small areas: A Bayesian random effects methodology. *International Statistical Review*, 77(2), 222–240.

Cooner F, Banerjee S, McBean A (2006) Modelling geographically referenced survival data with a cure fraction. *Statistical Methods in Medical Research*, 15, 307–324.

Cox C, Matheson M (2014) A comparison of the generalized gamma and exponentiated Weibull distributions. *Statistics in Medicine*, 33(21), 3772–3780.

Cox D (1972) Regression models and life-tables. *Journal of the Royal Statistical Society: Series B*, 34, 187–220.

Crippa A (2018) A Not So Short Review on Survival Analysis in R. https://rpubs.com/alecri/258589

Crowder M (2001) *Classical Competing Risks*. CRC Press.

Damien P, Muller P (1998) A Bayesian bivariate failure time regression model. *Computational Statistics & Data Analysis*, 28, 77–85.

Demarqui F, Loschi R, Colosimo E (2008) Estimating the grid of time-points for the piecewise exponential model. *Lifetime Data Analysis*, 14(3), 333–356.

Diekmann A, Mitter P (1983) The "Sickle Hypothesis": A time-dependent Poisson model with applications to deviant behavior and occupational mobility. *The Journal of Mathematical Sociology*, 9, 85–101.

Dorazio R (2009) On selecting a prior for the precision parameter of the Dirichlet process mixture models. *Journal of Statistical Planning and Inference*, 139, 3384–3390.

Dykstra RL, Laud P (1981) A Bayesian nonparametric approach to reliability. *The Annals of Statistics*, 9(2), 356–367.

Efron B (1988) Logistic regression, survival analysis, and the Kaplan-Meier curve. *Journal of the American Statistical Association*, 83(402), 414–425.

Fahrmeir L, Knorr Held L (1997) Dynamic discrete time duration models. *Sociological Methodology*, 27, 417–452.

Fahrmeir L, Tutz G (2001) *Multivariate Statistical Modelling Based on Generalized Linear Models*, 2nd Edition. *Springer Series in Statistics*. Springer Verlag, New-York, Berlin, Heidelberg.

Fleming TR, Harrington DP (1991) *Counting Processes and Survival Analysis*, Vol. 169. John Wiley & Sons.

Florens J, Fougere D, Mouchart M (1995) Duration models, pp 491–534, in *The Econometrics of Panel Data*, eds L Matyas, P Sevestre. Kluwer.

Gamerman D (1991) Dynamic Bayesian models for survival data. *Applied Statistics*, 40, 63–79.

Gelfand A, Ghosh S, Christiansen C, Soumerai S, McLaughlin T (2000) Proportional hazard models: a latent competing risk approach. *Journal of Applied Statistics*, 49, 385–397.

Ghosh P, Branco MD, Chakraborty H (2007) Bivariate random effect model using skew-normal distribution with application to HIV-RNA. *Statistics in Medicine*, 26(6), 1255–1267.

Glidden D, Vittinghoff E (2004) Modelling clustered survival data from multicentre clinical trials. *Statistics in Medicine*, 23(3), 369–388.

Gordon S (2002) Stochastic dependence in competing risks. *American Journal of Political Science*, 46, 200–217.

Gore S, Pocock S, Kerr G (1984) Regression models and non-proportional hazards in the analysis of breast cancer survival. *Journal of Applied Statistics*, 33, 176–195.

Gustafson P (1997) Large hierarchical Bayesian analysis of multivariate survival data. *Biometrics*, 53, 230–242.

Gustafson P (2000) Bayesian regression modeling with interactions and smooth effects. *Journal of the American Statistical Association*, 95(451), 795–806.

Gustafson P, Aeschliman D, Levy A (2003) A simple approach to fitting Bayesian survival models. *Lifetime Data Analysis*, 9, 5–19.

Hagar Y, Dignam J, Dukic V (2017) Flexible modeling of the hazard rate and treatment effects in long-term survival studies. *Statistical Methods in Medical Research*, 26(5), 2455–2480.

Haller B, Schmidt G, Ulm K (2013) Applying competing risks regression models: An overview. *Lifetime Data Analysis*, 19(1), 33–58.

Heckman J, Singer B (1984) A method for minimizing the impact of distributional assumptions in econometric models for duration data. *Econometrica*, 52, 271–320.

Henderson R, Oman P (1999) Effect of frailty on marginal regression estimates in survival analysis. *Journal of the Royal Statistical Society: Series B*, 61, 367–379.

Henderson R, Shimakura S, Gorst D (2002) Modeling spatial variation in leukemia survival data. *Journal of the American Statistical Association*, 97, 965–972.

Henderson R, Prince H (2000) Choice of conditional models in bivariate survival. *Statistics in Medicine*, 19, 563–574.

Herring A, Ibrahim J (2002) Maximum likelihood estimation in random effects cure rate models with nonignorable missing covariates. *Biostatistics*, 3, 387–405.

Hougaard P (1987) Modelling multivariate survival. *Scandinavian Journal of Statistics*, 14(4), 291–304.

Hougaard P (2000) *Analysis of Multivariate Survival Data*. Springer, New York.

Hougaard P, Myglegaard P, Borch-Johnsen K (1994) Heterogeneity models of disease susceptibility, with application to diabetic nephropathy. *Biometrics*, 50, 1178–1188.

Huang Y, Dagne G (2012) Bayesian semiparametric nonlinear mixed-effects joint models for data with skewness, missing responses, and measurement errors in covariates. *Biometrics*, 68(3), 943–953.

Huster W, Brookmeyer R, Self S (1989) Modeling paired survival data with covariates. *Biometrics*, 45, 145–156.

Ibrahim J, Chen M-H, MacEachern S (1999) Bayesian variable selection for proportional hazards models. *The Canadian Journal of Statistics*, 27, 701–717.

Ibrahim J, Chen M-H, Sinha D (2001) *Bayesian Survival Analysis*. Springer-Verlag.

Kalbfleisch JD (1978) Non-parametric Bayesian analysis of survival time data. *Journal of the Royal Statistical Society: Series B (Methodological)*, 40(2), 214–221.

Kalbfleisch JD, Prentice R (1980) *The Statistical Analysis of Failure Time Data*. Wiley, New York.

Keiding N, Andersen P, Klein J (1997) The role of frailty models and accelerated failure time models in describing heterogeneity due to omitted covariates. *Statistics in Medicine*, 16, 215–224.

Kiefer N (1988) Economic duration data and hazard functions. *Journal of Economic Literature*, 26, 646–679.

Kneib T (2006) Mixed model-based inference in geoadditive hazard regression for interval-censored survival times. *Computational Statistics & Data Analysis*, 51, 777–792.

Kostaki A, Panousis V (2001) Expanding an abridged life table. *Demographic Research*, 5, 1.

Kozumi H (2004) Posterior analysis of latent competing risk models by parallel tempering. *Computational Statistics & Data Analysis*, 46, 441–458.

Kuo L, Mallick B (1997) Bayesian semiparametric inference for the accelerated failure-time model. *Canadian Journal of Statistics*, 25, 457–472.

Lambert P (2007) Modeling of the cure fraction in survival studies. *The Stata Journal*, 7(3), 351–375.

Lancaster T (1990) *The Econometric Analysis of Transition Data*. Cambridge University Press.

Latouche A, Allignol A, Beyersmann J, Labopin M, Fine J (2013) A competing risks analysis should report results on all cause-specific hazards and cumulative incidence functions. *Journal of Clinical Epidemiology*, 66(6), 648–653.

Lawless J (1980) Inference in the generalized gamma and log gamma distributions. *Technometrics*, 22(3), 409–419.

Lee K, Chakraborty S, Sun J (2015) Survival prediction and variable selection with simultaneous shrinkage and grouping priors. *Statistical Analysis and Data Mining*, 8(2), 114–127.

Li K (1999) Bayesian analysis of duration models: An application to Chapter 11 bankruptcy. *Economics Letters*, 63(3), 305–312.

Li M (2007) Bayesian proportional hazard analysis of the timing of high school dropout decisions. *Econometric Reviews*, 26, 529–556.

Lopes H, Muller P, Ravishanker N (2007) Bayesian computational methods in biomedical research, in *Computational Methods in Biomedical Research*, eds R Khattree, D Naik.

Manda S, Gilthorpe M, Tu Y, Blance A, Mayhew M (2005) A Bayesian analysis of amalgam restorations in the Royal Air Force using the counting process approach with nested frailty effects. *Statistical Methods in Medical Research*, 14, 567–578.

Marano G, Boracchi P, Biganzoli E (2016) Estimation of the piecewise exponential model by Bayesian P-splines via Gibbs sampling: Robustness and reliability of posterior estimates. *Open Journal of Statistics*, 6, 451–468.

Morris CN, Norton EC, Zhou XH (1994) Parametric duration analysis of nursing home usage, pp 231–248, in *Case Studies in Biometry*, eds N Lange, L Ryan, L Billard, D Brillinger, L Conquest, J Greenhouse. Wiley.

Mosler K (2003) Mixture models in econometric duration analysis. *Applied Stochastic Models in Business and Industry*, 19, 91–104.

Murray T, Hobbs B, Sargent D, Carlin B (2016) Flexible Bayesian survival modeling with semiparametric time-dependent and shape-restricted covariate effects. *Bayesian Analysis*, 11(2), 381–402.

Muthen B, Masyn K (2005) Discrete-time survival mixture analysis. *Journal of Educational and Behavioral Statistics*, 30, 27–58.

Namboodiri K, Suchindran C (1987) *Life Table Techniques and Their Applications*. Academic Press, New York.

Nardi A, Schemper M (2003) Comparing Cox and parametric models in clinical studies. *Statistics in Medicine*, 22(23), 3597–3610.

Neves C, Migon H (2007) Bayesian graduation of mortality rates: An application to reserve evaluation. *Insurance: Mathematics and Economics*, 40, 424–434.

Omori Y (2003) Discrete duration model having autoregressive random effects with application to Japanese diffusion index. *Journal of the Japan Statistical Society*, 33, 1–22.

Orbe J, Núñez-Antón V (2006) Alternative approaches to study lifetime data under different scenarios: From the PH to the modified semiparametric AFT model. *Computational Statistics & Data Analysis*, 50, 1565–1582.

Perperoglou A, van Houwelingen H, Henderson R (2006) A relaxation of the Gamma frailty (Burr) model 2006. *Statistics in Medicine*, 25, 4253–4266.

Phadia E (2015) *Prior Processes and Their Applications; Nonparametric Bayesian Estimation*. Springer.

Pickles A, Crouchley R (1995) A comparison of frailty models for multivariate survival data. *Statistics in Medicine*, 14, 1447–1461.

Pintilie M (2006) *Competing Risks: A Practical Perspective*. John Wiley, West Sussex, UK.

Putter H. (2018) Tutorial in biostatistics: Competing risks and multi-state models Analyses using the mstate package. Leiden University Medical Center, Department of Medical Statistics and Bioinformatics. https://cran.r-project.org/web/packages/mstate/vignettes/Tutorial.pdf

Richards H, Barry R (1998) U.S. Life tables for 1990 by sex, race, and education. *Journal of Forensic Economics*, 11, 9–26.

Rivas-López M, López-Fidalgo J, Campo R (2014) Optimal experimental designs for accelerated failure time with Type I and random censoring. *Biometrical Journal*, 56(5), 819–837.

Roeder K, Wasserman L (1997) Practical Bayesian density estimation using mixtures of normals. *Journal of the American Statistical Association*, 92(439), 894–902.

Sahu S, Dey D (2000) A comparison of frailty and other models for bivariate survival dataata. *Lifetime Data Analysis*, 6, 207–228.

Sahu S, Dey D (2004) On a Bayesian multivariate survival model with skewed frailty, pp 321–338, in *Skew-Elliptical Distributions and Their Applications: A Journey Beyond Normality*, eds M Genton. CRC/Chapman & Hall, Boca Raton, FL.

Sahu S, Dey D, Aslanidou H, Sinha D (1997) A Weibull regression model with gamma frailties for multivariate survival data. *Lifetime Data Analysis*, 3, 123–137.

Sargent DJ (1998) A general framework for random effects survival analysis in the Cox proportional hazards setting. *Biometrics*, 54(4), 1486–1497.

Scheike T, Zhang M (2011) Analyzing competing risk data using the R timereg package. *Journal of Statistical Software*, 38(2), i02.

Schmidt P, Witte A (1989) Predicting criminal recidivism using 'split population' survival time models. *Journal of Econometrics*, 40, 141–159.

Schoen R (2016) The continuing retreat of marriage: Figures from marital status life tables for United States females, 2000–2005 and 2005–2010, pp 203–215, in *Dynamic Demographic Analysis*, ed R Schoen. Springer.

Scrucca L, Santucci A, Aversa F (2010) Regression modeling of competing risk using R: An in depth guide for clinicians. *Bone Marrow Transplantation*, 45(9), 1388–1395.

Sen A, Banerjee M, Li Y, Noone A (2010) A Bayesian approach to competing risks analysis with masked cause of death. *Statistics in Medicine*, 29(16), 1681–1695.

Shao Q, Zhou X (2004) A new parametric model for survival data with long-term survivors. *Statistics in Medicine*, 23, 3525–3543.

Singer JD, Willett JB (1993) It's about time: Using discrete-time survival analysis to study duration and the timing of events. *Journal of Educational Statistics*, 18(2), 155–195.

Sinha D (1993) Semiparametric Bayesian analysis of multiple event time data. *Journal of the American Statistical Association*, 88, 979–983.

Sinha D, Chen M-H, Ghosh S (1999) Bayesian analysis and model selection for interval-censored survival data. *Biometrics*, 55, 585–590.

Sinha D, Dey DK (1997) Semiparametric Bayesian analysis of survival data. *Journal of the American Statistical Association*, 92(439), 1195–1212.

Sinha D, Patra K, Dey DK (2003) Modelling accelerated life test data by using a Bayesian approach. *Journal of the Royal Statistical Society: Series C (Applied Statistics)*, 52(2), 249–259.

Sohn Y, Chang I, Moon T (2007) Random effects Weibull regression model for occupational lifetime. *European Journal of Operational Research*, 179, 124–131.

Stacy EW (1962) A generalization of the gamma distribution. *The Annals of Mathematical Statistics*, 33(3), 1187–1192.

Swindell W (2009) Accelerated failure time models provide a useful statistical framework for aging research. *Experimental Gerontology*, 44(3), 190–200.

Thamrin S, McGree J, Mengersen K (2013) Bayesian Weibull survival model for gene expression data, Chapter 10, pp 171–185, in *Case Studies in Bayesian Statistical Modelling and Analysis*, eds C Alston, K Mengersen, A Pettitt. Wiley.

Tosch TJ, Holmes PT (1980) A bivariate failure model. *Journal of the American Statistical Association*, 75(370), 415–417.

Tsodikov A, Ibrahim J, Yakovlev A (2003) Estimating cure rates from survival data: An alternative to two-component mixture models. *Journal of the American Statistical Association*, 98, 1063–1078.

Umlauf N, Klein N, Zeileis A (2018) BAMLSS: Bayesian additive models for location, scale, and shape (and beyond). *Journal of Computational and Graphical Statistics*, 27(3), 612–627.

Van den Berg G (2001) Duration models: Specification, identification, and multiple durations, in *Handbook of Econometrics 5*, eds J Heckman, E Leamer. North Holland, Amsterdam, Netherlands.

Viswanathan B, Manatunga A (2001) Diagnostic plots for assessing the frailty distribution in multivariate survival data. *Lifetime Data Analysis*, 7, 143–155.

Walker S, Mallick B (1999) A Bayesian semiparametric accelerated failure time model. *Biometrics*, 55, 477–483.

Watson T, Christian C, Mason A, Smith M, Meyer R (2002) Bayesian-based decision support system for water distribution systems. 5th International Conference on Hydroinformatics, Cardiff University, UK.

Wei L (1992) The accelerated failure time model: A useful alternative to the Cox regression model in survival analysis. *Statistics in Medicine*, 11, 1871–1879.

Wienke A (2010) *Frailty Models in Survival Analysis*. Chapman and Hall/CRC.

Winkelmann R, Boes S (2005) *Analysis of Microdata*. Springer-Verlag.

Wong O (1977) A competing-risk model based on the life table procedure in epidemiologic studies*International Journal of Epidemiology*, 6, 153–159.

Yashin A, Iachine I, Begun A, Vaupel J (2001) Hidden frailty: Myths and reality. Research Report 34, Department of Statistics and Demography, SDU - Odense University.

Yau KK (2001) Multilevel models for survival analysis with random effects. *Biometrics*, 57(1), 96–102.

Yin G (2005) Bayesian cure rate frailty models with application to a root canal therapy study. *Biometrics*, 61, 552–558.

Yin G, Ibrahim J (2005) A class of Bayesian shared gamma frailty models with multivariate failure time data. *Biometrics*, 61, 208–216.

Yin G, Ibrahim J (2006) Bayesian transformation hazard models, pp 170–182, in *IMS Monograph Series*, Vol. 49. Institute of Mathematical Statistics.

Zhang Z, Sinha S, Maiti T, Shipp E (2018) Bayesian variable selection in the accelerated failure time model with an application to the surveillance, epidemiology, and end results breast cancer data. *Statistical Methods in Medical Research*, 27(4), 971–990.

Zhou H, Hanson T, Zhang J (2017) Generalized accelerated failure time spatial frailty model for arbitrarily censored data. *Lifetime Data Analysis*, 23(3), 495–515.

Zhou H, Hanson T, Zhang J (2018) spBayesSurv: Fitting Bayesian spatial survival models using R. https://arxiv.org/abs/1705.04584

12

Hierarchical Methods for Nonlinear and Quantile Regression

12.1 Introduction

Standard versions of the normal linear model and general linear models assume additive and linear predictor effects in the regression mean, and a constant variance. While linear regression effects are often suitable, nonlinear predictor effects are common in areas as diverse as economics, hydrology (Qian et al., 2005), and epidemiology (Natario and Knorr-Held, 2003). In some applications, there may be a theoretical basis for a particular form of nonlinearity, though some elements of specification will be uncertain – see Borsuk and Stow (2000) on biochemical oxygen demand, and Meyer and Millar (1998) on models of fishery stock. In other situations, the form of nonlinearity is unknown and to be assessed from the data – hence the term "non-parametric", since a particular form for the mean function is not assumed. Bayesian application of non-parametric smooth regression is facilitated by R libraries such jagam (Wood, 2016) (www.rdocumentation.org/packages/mgcv/versions/1.8-17/topics/jagam), bamlss (Umlauf et al., 2016; https://rdrr.io/rforge/bamlss/), gammSlice (Pham and Wand, 2015), stan_gamm4 within rstanarm (https://cran.rstudio.com/web/packages/rstanarm/index.html), and spikeSlabGAM (Scheipl, 2011).

In many applications, a nonlinear effect is present, or suspected, in only a subset of predictors, leading to partially linear models or semiparametric regression models. Consider outcomes $\{y_i, i=1,\ldots,n\}$ from an exponential density

$$p(y_i \mid \theta_i,\phi) = \exp\left(\frac{y_i\theta_i - a(\theta_i)}{\phi} + c(y_i,\phi)\right),$$

with $E(y_i) = \mu_i = a'(\theta_i)$, and link $g(\mu_i) = \eta_i$ to a regression term η_i. Suppose it is intended that R metric predictors $W_i = (w_{1i}, w_{2i}, \ldots, w_{Ri})$ be modelled non-parametrically via unknown smooth functions $S(w_{ri})$, then

$$g(\mu_i) = \eta_i = \alpha + X_i\beta + S_1(w_{1i}) + \ldots + S_R(w_{Ri}) + u_i,$$

$$u_i \sim N(0,\sigma^2).$$

For instance, Engle et al. (1986) analyse the relationship between temperature and monthly electricity sales (y metric and u normal, and with g an identity link) for four US cities. The impact of electricity price, month (11 dummy variables), and income is modelled parametrically, but an unknown smooth function is adopted to model the impact of monthly temperature.

Residual errors u_i will be present when y_i is metric, and may also be present for over-dispersed discrete outcomes. While an assumption of independent errors with constant variance is standard, non-parametric regression for the regression mean may be extended to modelling heteroscedastic errors (Yau and Kohn, 2003; Krivobokova et al., 2008), or to other distributional features (Mayr et al., 2012). When the observations are observed through time or over space it may also be important to control for correlations in the u. Smith et al. (1998) and Kohn et al. (2000) consider the case when the observations y_t are arranged in time, smooth functions are used for predictor effects, and the u_t are autocorrelated. The estimate of the predictor smooth functions will be adversely affected if independent residuals are incorrectly assumed.

The two major forms of non-parametric regression involve basis functions (e.g. polynomial spline methods) and general additive methods based on smoothness priors. These are considered in Sections 12.2 and 12.5 respectively. Extending non-parametric regression to multiple predictors raises the same issues as multiple linear regression, for example, whether interactions are necessary and how the presence of smooths for other predictors alters the smooth for a given predictor – see Section 12.3. Robustness in non-parametric regression (e.g. to heteroscedastic errors) may be obtained through spatially adaptive methods which allow the level of smoothness to vary over the space of the covariates (Wood et al., 2002; Baladandayuthapani et al., 2005) – see Section 12.4. A major application area for non-parametric regression is in longitudinal settings, as discussed in Section 12.6.

12.2 Non-Parametric Basis Function Models for the Regression Mean

A wide range of methods for non-parametric regression in one or more predictors typically assume linear combinations of basis functions $S_r(w_r)$ of predictors $(w_1,\ldots,w_R)$. Numerous basis functions can be used, including truncated polynomial functions, B-spline functions, radial basis functions (Yau et al., 2003), logistic functions (Hooper, 2001), trigonometric basis functions, and wavelets (Dennison et al., 2002). For exponential family responses y_i with mean μ_i and link g, a truncated polynomial spline (or piecewise polynomial spline) regression on a single predictor w_i has the form (Dennison et al., 2002, p.52)

$$\begin{aligned} g(\mu_i) &= a + S(w_i) + u_i, \qquad (12.1) \\ &= a + \sum_{k=1}^{K} b_k (w_i - \kappa_k)_+^q + u_i, \\ u_i &\sim N(0,\sigma^2), \end{aligned}$$

where q is a known positive integer, and the κ_k are knots placed within the range $[w_{\min}, w_{\max}]$ of w. In (12.1), the piecewise polynomials are fitted in each interval $[\kappa_k,\kappa_{k+1})$ and preferably join smoothly at each knot (e.g. this applies for a cubic spline, as it has continuous 1st and 2nd derivatives at each knot).

An alternative spline specification (e.g. Meyer, 2005; Tutz and Reithinger, 2007, p.2877) matches the degree q of the truncated function $T(w_i) = \Sigma_{k=1}^{K} b_k (w_i - \kappa_k)_+^q$ by a standard polynomial of order q, namely $Q(w_i) = \beta_1 w_i + \ldots + \beta_q w_i^q$. So, the total smooth is $S(w_i) = Q(w_i) + T(w_i)$, and one has

$$g(\mu_i) = \alpha + \beta_1 w_i + \ldots + \beta_q w_i^q + \sum_{k=1}^{K} b_k (w_i - \kappa_k)_+^q + u_i. \tag{12.2}$$

Values $q = 1, 2$, or 3 are most typical, with $q = 1$ often being suitable for reproducing a smooth function given a large enough set of knots (Ruppert et al., 2003, p.68), but also capable of reproducing abrupt changes in the underlying function (Dennison et al., 2002, p.52).

The knots in (12.1) and (12.2) may be known or unknown. If known, then they are typically much less than the sample size in number. They could be sited at percentile points (e.g. deciles) of w, or possibly placed more densely at points where the function is known to be rapidly changing and less densely elsewhere. Choosing too few knots can result in oversmoothing, and choosing too many in overfitting – see the LIDAR data examples discussed by Ruppert et al. (2003, p.63). Coull et al. (2001, p.540) suggest the allocation of one knot for every four to five observations, up to a maximum of about 40 knots. Yau and Kohn (2003) suggest fitting a model with a small number of knots first and gradually increasing their number until estimates and fit stabilise. An alternative procedure known as smoothing splines places a knot at every observed distinct predictor value (Berry et al., 2002; Dias and Gamerman, 2002). The most general model averaging approach takes both the number of knots and their sitings as unknowns, while both Denison et al. (1998) and Biller (2000) assume a large number of potential, but prespecified, candidate knot locations. If knots are taken to have unknown locations within $[w_{min}, w_{max}]$, identification may rely on order constraints such as $\kappa_k > \kappa_{k-1}$, and analysis resembles time series with multiple change points.

If the b_k in (12.1)–(12.3) are modelled as fixed effects, predictor coefficient selection is open as a way of achieving model parsimony, and is especially indicated under the smoothing spline method (Smith and Kohn, 1996). With a large number of preset potential knot sitings, predictor selection involves obtaining posterior probabilities $Pr(\delta_{jk} = 1 | y)$ on binary indicator variables δ_{1k} ($k = 1, \ldots, q$) for retaining coefficients in the Q(w) component, and δ_{2k} ($k = 1, \ldots, K$) in the $T(w)$ component. One then has

$$g(\mu_i) = \alpha + \delta_{11}\beta_1 w_i + \ldots + \delta_{1q}\beta_q w_i^q + \sum_{k=1}^{K} \delta_{2k} b_k (w_i - \kappa_k)_+^q + u_i, \tag{12.3}$$

with coefficients estimated by means of the products $\delta_{1j}\beta_j$ and $\delta_{2k} b_k$.

12.2.1 Mixed Model Splines

In contrast to the fixed effects models (12.1)–(12.3), under a mixed model spline regression, or penalised spline method, the coefficients in $Q(w_i)$ usually remain fixed effects, but those in $T(w_i)$ follow a penalising random effects or P-spline prior (Brumback et al., 1999; Ruppert et al., 2003; Wand, 2003; Yue et al., 2012). Under a P-spline approach the problem of choosing the number and position of the knots is alleviated, since providing enough knots are used, the penalty function should ensure that the resulting fits are very similar (Currie and Durban, 2002). Under the P-spline approach, (12.1) and (12.2) become

$$g(\mu_i) = \alpha + S(w_i) = \alpha + \sum_{k=1}^{K} b_k (w_i - \kappa_k)_+^q + u_i, \tag{12.4}$$

$$g(\mu_i) = \alpha + \beta_1 w_i + \ldots + \beta_q w_i^q + \sum_{k=1}^{K} b_k (w_i - \kappa_k)_+^q + u_i, \tag{12.5}$$

where b_k are a collection of random parameters from a common density with unknown hyperparameters.

Possible priors for the random b_k include an unstructured normal (Ruppert et al., 2003)

$$b_k \sim N(0, \phi), \tag{12.6}$$

which, by comparison with a fixed effects prior, imposes a restriction on the b_k when $\phi < \infty$, and tends to shrink the b_k, leading to a smooth fit (Wand, 2003). A standard approach (e.g. Lang and Brezger, 2004) assumes $\phi \sim IG(g,h)$ with $g=1$ and h small (e.g. 0.001, 0.0001, or 0.00001), though there may be sensitivity to the value of h.

To illustrate equivalence to the broader class of mixed models, define design matrices

$$W = [1, w_i, \ldots, w_i^q]_{1 \le i \le n},$$

$$Z = [(w_i - \kappa_k)_+^q]_{1 \le k \le K, 1 \le i \le n}$$

and vectors $\beta = (\alpha, \beta_1, \ldots, \beta_q)'$ and $b = (b_1, \ldots, b_K)'$. Then, under normal error assumptions, model (12.5) can be written in the mixed model form

$$g = X\beta + Zb + u,$$

$$\begin{bmatrix} b \\ u \end{bmatrix} \sim N\left(\begin{matrix} 0 \\ 0' \end{matrix} \begin{bmatrix} \phi I & 0 \\ 0 & \sigma^2 I \end{bmatrix} \right).$$

Alternatives schemes for b_k are a random walk penalty (Eilers and Marx, 1996), such as $\Delta^d b_k \sim N(0, \phi)$. For instance, taking $d=1$ gives

$$b_k \sim N(b_{k-1}, \phi). \tag{12.7}$$

Another option provides monotonic smooths – in applications where such smooths have a substantive rationalisation (Brezger and Steiner, 2008) – and stipulates $b_k \sim N(0, \phi)$, but subject to

$$b_k \ge b_{k-1}, \ k = 2, \ldots, K,$$

and an increasing function $T(w_i) = \Sigma_{k=1}^K b_k (w_i - \kappa_k)_+^q$, or $b_k \le b_{k-1}$ for a decreasing function.

The function $S(w_i)$ resulting from a fixed effects prior on $\{b_k\}$ in (12.1)–(12.3) may be quite rough, due to the large number of truncated polynomials being fitted, whereas the shrinkage prior under the mixed model approach tends to penalise large coefficients and lead to a smoother fit (Yau et al., 2003; Ngo and Wand, 2004; Meyer, 2005). Under an unstructured prior $b_k \sim N(0, \phi)$, and smoothing or penalty parameter λ, the mode of the posterior density of $\{\beta, b, \phi\}$ is the same as that obtained by maximising a penalised likelihood

$$PL = log[P(y \mid \beta, b, \phi)] - \lambda \sum_{k=1}^{K} b_k^2,$$

where the form of λ (in terms of variance parameters) depends on whether or not there is an unstructured residual term u_i in the regression model. For a metric outcome and $u_i \sim N(0,\sigma^2)$, one has $\lambda = \sigma^2/\phi$ (Fahrmeir and Knorr-Held, 2000). This penalised likelihood is analogous to "ridge" penalties sometimes used with correlated predictors (Eilers and Marx, 2004). For random walk priors of order d, one has (Lang and Brezger, 2004)

$$PL = \log[P(y \mid \beta, b, \phi)] - \lambda \sum_{k=d+1}^{K} (\Delta^d b_k)^2.$$

12.2.2 Basis Functions Other Than Truncated Polynomials

The fixed or random coefficient approaches can equally be applied with other basis functions to represent $T(w_i)$. Truncated polynomial basis functions span the space of degree q polynomials with knots located at $\kappa_1,\ldots\kappa_K$ (Friedman, 1991). This property also holds for radial basis functions (Koop et al., 2003, p.252) based on distances $r_{ik} = |w_i - \kappa_k|$, such as the polyharmonic spline $T(w_i) = \Sigma_{k=1}^{K} H(r_{ik})$, with

$$H(r_{ik}) = r_{ik}^q, \quad q = 1,3,5,\ldots$$

$$H(r_{ik}) = r_{ik}^q \log(r_{ik}), \quad q = 2,4,6,$$

of which the thin-plate spline (Kohn et al., 2001; Koop and Tole, 2004)

$$H(r_{ik}) = r_{ik}^2 \log(r_{ik})$$

is a special case. These are examples of functions which are radially symmetric around knots κ_k, such that the value of the function at w_i depends only on the distance between w_i and the knot location. They have the form $H(u) = H(|w - \kappa_k|)$, where $|v| = \sqrt{v'v}$ is the length of the vector v. Other types of radial basis include Gaussian functions (Konishi et al., 2004) with

$$H_k(w_i) = \exp\left(-\frac{|w - \kappa_k|}{2\nu\eta_k}\right),$$

where ν is the same over different knots. As for truncated splines, smoothing based on radial basis functions may include a parametric polynomial term to degree q to match the degree of the radial function, for example, with $q = 1$

$$g(\mu_i) = \alpha + \beta_1 w_i + \sum_{k=1}^{K} b_k |w_i - \kappa_k| + u_i.$$

Both radial and truncated power splines may be ill-conditioned in terms of broader regression considerations (Eilers and Marx, 2004). An alternative basis less prone to ill-conditioning is provided by B-splines, with health mapping applications exemplified by Silva et al. (2008) and MacNab and Gustafson (2007). B-splines are defined to be non-zero for at most $q+2$ interior knots for a qth degree B-spline (also called a B-spline of order $q+1$), which means the condition number of the design matrix product is relatively low (Eilers and

Marx, 1996, p.90; Biller, 2000; Dennison et al., 2002, p.75). A B-spline of degree q consists of $q+1$ polynomial pieces of degree q and overlaps with $2q$ of its neighbours. For K knots, and so K+1 intervals, in the domain $[w_{min}, w_{max}]$ of a predictor, there will be $K^*=K+1+q$ B-spline schedules, because extra knots are placed outside the domain of w to get q overlapping B-splines in each interval.

Let $B_k(w_i,q)$ be the value at w_i of the kth B-spline of degree q, with $k=1,\ldots,K^*$. Successive B-spline values are defined by the recursion

$$B_k(w_i,0) = I(\kappa_k \leq w_i < \kappa_{k+1}),$$

$$B_k(w_i,q) = \frac{w_i - \kappa_k}{\kappa_{k+q} - \kappa_k} B_k(w_i,q-1) + \frac{\kappa_{k+q+1} - w_i}{\kappa_{k+q+1} - \kappa_{k+1}} B_{k+1}(w_i,q-1).$$

The initial terms in the recursion are simply binary indicators defining a partition of the w values. For equally spaced knots, a simplified B-spline recursion applies involving differences in truncated power splines (Eilers and Marx, 2004). B-spline bases for $T(w)$ can be combined with random or fixed effects priors for the spline coefficients. For example, random b_k in an analysis with a single predictor w_i leads to

$$g(\mu_i) = a + \beta_1 w_i + \ldots + \beta_q w_i^q + \sum_{k=1}^{K^*} b_k B_k(w_i,q) + u_i.$$

In particular, Eilers and Marx (1996) combine a B-spline basis with a penalty on dth order differences in adjacent b_k coefficients. As mentioned above, difference penalties can be achieved by random walk priors under a Bayesian approach (e.g. a second order random walk prior if $d=2$).

Relatively small numbers of knots may be needed to provide an effective smooth, as may be illustrated by drawing on the Stan case study Kharratzadeh (2017), with B-spline schedules defined either by a function, or by using the package splines. Consider the Boston data set (in the R MASS package), and predicting median house values based on the percentage of lower status of the population (lsat) (Figure 12.1). A B-spline of degree 3 is used, and a random walk prior on the coefficients $\{b_k\}$ of the B-spline basis, with

$$b_1 \sim N(0,5),$$
$$b_k \sim N(b_{k-1}, \tau_b),$$
$$\tau_b \sim C^+(0,5).$$

There are 508 observations. For $K=10$ knots located at corresponding quantiles of lsat, we find a LOO-IC (leave-one-out information criterion) of 3117 (Figure 12.2), while $K=5$ knots also gives a LOO-IC of 3117. By contrast, a larger number of knots, $K=20$, shows evidence of undersmoothing (overfitting) with a LOO-IC of 3122.

Bayesian application of spectral basis functions is discussed by Lenk (1999), Fahrmeir and Tutz (2001, Chapter 5), and Kitagawa and Gersch (1996). Here the smooth may be represented by the series

$$T(w_i) = \sum\nolimits_{k=1}^{\infty} b_k H_k(w_i),$$

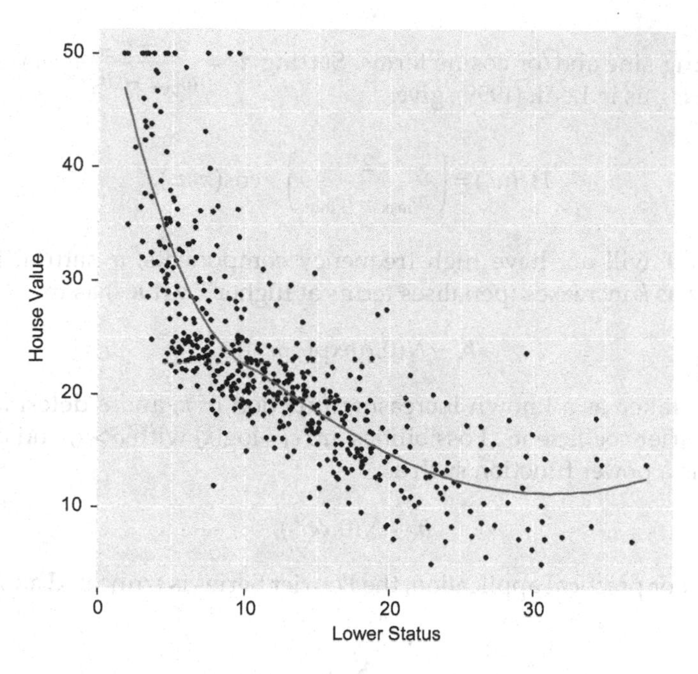

FIGURE 12.1
Median house values and status.

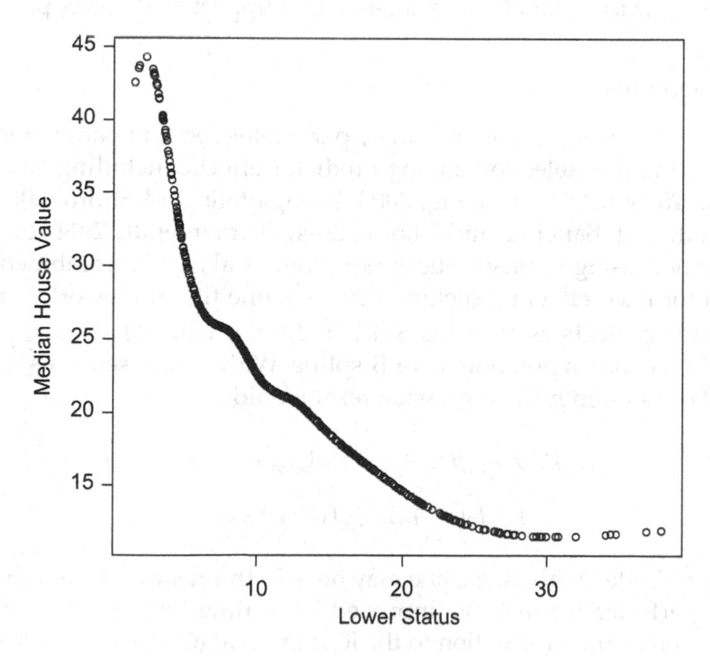

FIGURE 12.2
Smooth for $K = 10$ knots.

with H_k including sine and/or cosine terms. Setting $z_i = \frac{w_i - w_{\min}}{w_{\max} - w_{\min}}$ and including only cosine terms in H_k, as in Lenk (1999), gives

$$H_k(w_i) = \left(\frac{2}{w_{\max} - w_{\min}}\right)^{0.5} \cos(\pi k z_i).$$

Since a smooth T will not have high frequency components, a natural prior on the b_k expresses decay as k increases (penalises terms at higher k values) as in

$$b_k \sim N(0, \phi \exp[-\delta c_k]),$$

where c_k can be taken as a known increasing function of k, and δ determines the rate of decay of the Fourier coefficients. Possibilities are $c_k = \log(k)$ with $\delta > 1$, and $c_k = k$ with $\delta > 0$. An alternative is a power function such as

$$b_k \sim N(0, \phi \delta^k),$$

where $\delta \in (0,1)$. For practical application, the Fourier Series is truncated above at K, namely

$$T(w_i) = \sum_{k=1}^{K} b_k H_k(w_i)$$

where K can be regarded as another parameter (cf. Ruppert et al., 2003, p.86).

12.2.3 Model Selection

Non-parametric regressions are often heavily parameterised and parameter redundancies are likely, indicating that selection among predictor effects, including smooths, is necessary (Yau et al., 2003; Belitz and Lang, 2008; Panagiotelis and Smith, 2008; Wood, 2008; Marra and Wood, 2011; Banerjee and Ghosal, 2014; Gelman et al., 2014). Smooth selection may be approached using binary indicators J_r (Yau et al., 2003), combined with conventional selection for fixed effect predictor terms. Assume the framework in (12.1). Then for $r = 1, \ldots, R$ predictor effects as smooths $S_r(w_{ri}) = T_r(w_{ri})$, where $T_r(w_{ri}) = \Sigma_{j=1}^{K_r} b_{rj} Z_j(w_{ri})$, and $Z_j(w_{ri})$ generically denotes a polynomial or B-spline. With binary selectors γ_j for fixed effect predictors X_i of dimension p, the regression term would be

$$g(\mu_i) = \alpha + \gamma_1 \beta_1 x_{ii} + \ldots + \gamma_p \beta_p x_{pi} + \ldots + J_1 T_1(w_{1i})$$
$$+ J_2 T_2(w_{2i}) \ldots + J_R T_R(w_{Ri}) + u_i.$$

Alternatively (e.g. Cottet et al., 2008), one may have both linear and smooth terms for each w_{ri}. Numerical performance may be improved by scaling both the X and W predictors, e.g. standardisation or transformation to the [0,1] interval (Cottet et al., 2008; Scheipl et al., 2012).

Selection for retention ($J_r = 1$) is influenced by the degree of informativeness of the prior adopted for the variances $(\phi_1, \ldots, \phi_R)$ of $(b_{1k} \ldots, b_{Rk})$. Flat priors will tend to lead to low posterior probabilities $Pr(J_r = 1 | y)$ for retaining random components. One option is to undertake initial runs with diffuse priors to develop an informative data-based prior (Shively et al., 1999).

Related approaches include hierarchical priors (e.g. log-normal) on ϕ_r (Cottet et al., 2008; Panagiotelis and Smith, 2008) as in

$$\log(\phi_r) \sim N(g_r, h_r),$$
$$g_r \sim N(0,100),$$
$$h_r \sim IG(101,10100),$$

independent of those for J_r. Another option involves a multiplicative reparameterisation combined with spike-slab selection on the variances (Scheipl et al., 2012), namely,

$$b_{rk} = c_{rk} d_{rk},$$
$$c_{rk} \sim N(0, \gamma_r \phi_r),$$
$$\gamma_r = h_r + \varepsilon(1-h_r),$$
$$h_r \sim \text{Bern}(\omega_r),$$
$$d_{rk} \sim N(m,1),$$

where the γ_r are scaling factors, ε is a predefined small constant, and the mean m is set to –1 or 1 with equal probability. So for $h_r = 0$ the random effects are effectively excluded, since their variance is near zero. Assuming $\phi_r \sim IG(a_\phi, b_\phi)$, then $\{a_\phi, b_\phi\}$ may be set by default, or set in line with subject matter considerations. Default settings of $\varepsilon = 0.00025$, $a_\phi = 5$, and $b_\phi = 25$ are proposed by Scheipl et al. (2012, p.1525).

Example 12.1 Beta-Carotene Plasma

This example concerns a continuous outcome, the dependence of blood plasma concentrations of beta-carotene on regulatory factors such as age, gender, vitamin use, dietary intake, smoking status, alcohol intake, cholesterol intake, etc. (details at http://lib.stat.cmu.edu/datasets/Plasma_Retinol). There are 315 subjects. Here impacts of $p = 11$ predictors (excluding age and cholesterol) are modelled via linear terms, and the impacts of age and cholesterol intake as smooth terms, S(AGE) and S(CHOL) (Liu et al., 2011; cf. Banerjee and Ghosal, 2014). The initial two analyses compare a linear truncated spline with a cubic B-spline ($q=3$ in $B_k(w_i, q)$). These two analyses use $K=19$ knots sited at the 5th, 10th, ..., 95th percentiles of AGE and CHOL (which are untransformed).

Preliminary coding in R to obtain knots and B-spline values in cholesterol is as follows

```
D <- read.table("betacarotene.txt",header=T)
attach(D)
require(splines)
knots <- quantile(chol,probs=seq(0.05,0.95,0.05))
kap.chol <- as.vector(knots)
bs.chol <- bs(chol, df=NULL,knots, degree = 3, intercept = T,
Boundary.knots = range(chol))
bs.chol <- matrix(as.numeric(bs.chol), nr = nrow(bs.chol))
```

The linear spline model is

$$y_i = \beta_0 + \beta_1 x_{ii} + \ldots + \beta_p x_{pi} + \sum_{k=1}^{19} b_{1k}(\text{AGE}_i - \kappa_{1k})_+$$
$$+ \sum_{k=1}^{19} b_{2k}(\text{CHOL}_i - \kappa_{2k})_+ + u_i,$$

where

$$u_i \sim N(0,\sigma^2),$$
$$b_{rk} \sim N(0,\phi_r).$$

The B-spline model is analogous, namely, with $K^*=23$,

$$y_i = \beta_0 + \beta_1 x_{ii} + \ldots + \beta_p x_{pi} + \sum_{k=1}^{K^*} b_{1k}B_{1k} + \sum_{k=1}^{K^*} b_{2k}B_{2k} + u_i,$$

and assumes a 1st order random walk for b_{rk}, penalising first differences in the b_{rk}. For identification, a corner rather than centring constraint is used to ensure identifiability, namely $b_{r1}=0$. For precisions $1/\sigma^2$ and $\theta_r = 1/\phi_r$, gamma $Ga(1,0.001)$ priors are assumed.

Applying jagsUI to the linear spline, a B-spline model (models 1 and 2) shows similar smooths in AGE and CHOL. Fit values favour the linear spline: a LOO-IC of 673, as against 695 for the B-spline, though both fit values have large SE values. However, convergence is much earlier achieved using the B-spline method, and effective sample sizes are larger.

A third analysis involves predictor and smooth selection in the B-spline model, as in

$$y_i = \beta_0 + \gamma_1\beta_1 x_{ii} + \ldots + \gamma_p\beta_p x_{pi} + J_1\sum_{k=1}^{K^*} b_{1k}B_{1k} + J_2\sum_{k=1}^{K^*} b_{2k}B_{2k} + u_i,$$

In this application, all predictors are standardised and the B-spline coefficients accordingly revised. Diffuse priors for β_j and b_{rk} are likely to lead to an overly parsimonious model, with few predictors retained. Instead, for the β_j, normal N(0,1) priors are adopted (McElreath, 2016). A data-based prior, based on posterior inferences from model 2, is adopted for the precisions $\theta_r = 1/\phi_r$ in the prior on the b_{rk} terms. Specifically, gamma priors Ga(0.265,0.0007) and Ga(0.6815,0.0014) correspond to the posterior mean and variance of θ_r in model 2. In model 3, a four-point discrete prior for the θ_r for CHOL and AGE is accordingly adopted, using values set by the four quintiles of Ga(0.265,0.0007) and Ga(0.6815,0.0014) densities. The 13 retention indicators, $\{\gamma_j, J_r\}$, are assigned Bernoulli priors, with the prior probability ω being an overarching complexity hyperparameter, with prior $\omega \sim \text{Be}(1,1)$.

From a two-chain run of 10,000 iterations, vitamin use (vituse) and BMI among the X predictors have posterior retention probabilities of 0.88 and 1.00 respectively (the indicators J[1] and J[3] in the code), but otherwise retention probabilities are below 0.7. CHOL and AGE have respective retention probabilities of 1 and 0.99 (J[12] and J[13] in the code). Figure 12.3 shows the corresponding smooths. Retention of both smooth terms is also reported by Liu et al. (2011).

A fourth analysis also involves selection, but using a partial adaptation of Scheipl et al. (2012). Thus, it is assumed that

$$b_{rk} \sim N(0,\gamma_r\phi_r),$$
$$\gamma_r = h_r + \varepsilon(1-h_r),$$
$$h_r \sim \text{Bern}(\omega_r),$$
$$1/\phi_r \sim G(a_\phi,b_\phi)$$

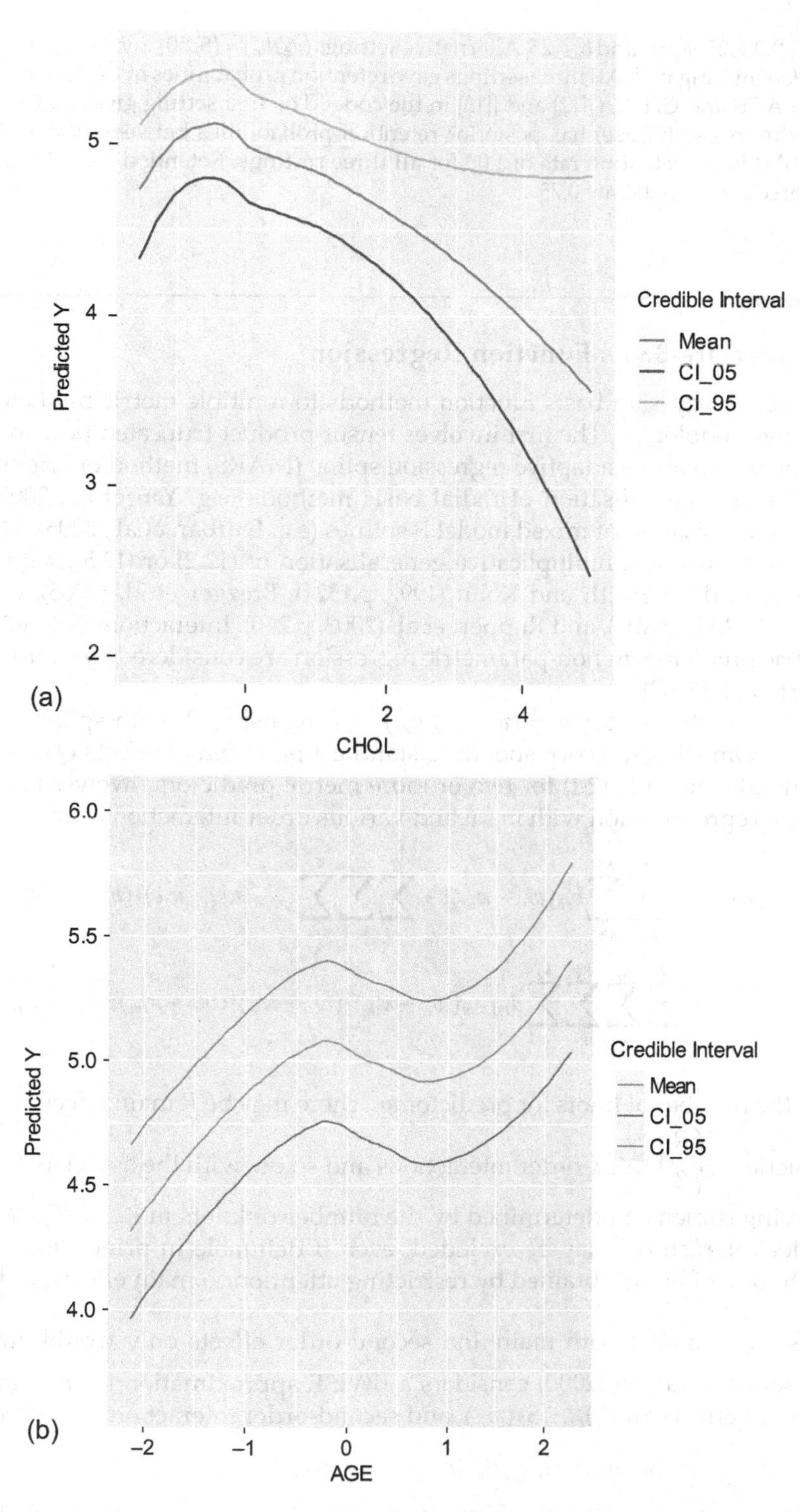

FIGURE 12.3
(a) Beta-carotine smooth in CHOL. (b) Beta-carotine smooth in AGE.

with $\varepsilon = 0.00025, a_\phi = 5$, and $b_\phi = 25$. Alternative settings $(a_\phi, b_\phi) = (5,50)$ and $(a_\phi, b_\phi) = (10,30)$ were also investigated. All three settings gave retention probabilities of 1 for the smooths in both AGE and CHOL (J[12] and J[13] in the code). The first setting gives a LOO-IC of 678. Vitamin use (vituse) has posterior retention probabilities between 0.90 and 0.95, while BMI has a retention rate of 1.00 for all three settings. Retention probabilities for other predictors are below 0.75.

12.3 Multivariate Basis Function Regression

Generalisation of Bayesian basis function methods to multiple metric predictors follows three main methodologies. The first involves tensor product truncated polynomial bases, including the multivariate adaptive regression spline (MARS) method of Friedman (1991); the second is the generalisation of radial basis methods (e.g. Yau et al., 2003); and the third is the generalisation of mixed model P splines (e.g. Durban et al., 2006). The full tensor product approach is a multiplicative generalisation of (12.2) or (12.5), with particular versions discussed by Smith and Kohn (1997, p.1524), Brezger et al. (2005), Chen (1993), Dennison et al. (2002, p.104), and Ruppert et al. (2003, p.240). Interactions between categorical and metric predictors in non-parametric regression are considered by Coull et al. (2001) and Ruppert et al. (2003).

Assume a predictor vector $w_i = (w_{1i},\ldots,w_{Ri})$ of dimension R, with spline degree q for all predictors. Omitting the corresponding standard polynomial effects $Q_r(w_r)$, the tensor product generalisation of (12.1) for two or more metric predictors involves an analysis of variance type representation with main and various order interaction effects,

$$g(\mu_i) = \alpha + \sum_{r=1}^{R}\sum_{k=1}^{K_r} b_{rk}(w_{ri}-\kappa_{rk})_+^q + \sum_{r\neq s}^{R}\sum_{k=1}^{K_r}\sum_{l=1}^{K_s} c_{rs,kl}(w_{ri}-\kappa_{rk})_+^q(w_{si}-\kappa_{sl})_+^q$$

$$+\sum_{r\neq s\neq t}^{R}\sum_{k=1}^{K_r}\sum_{l=1}^{K_s}\sum_{m=1}^{K_t} d_{rst,klm}(w_{ri}-\kappa_{rk})_+^q(w_{si}-\kappa_{sl})_+^q(w_{ti}-\kappa_{tm})_+^q + \ldots + u_i$$

where K_r is the number of knots for predictor w_r. There may be R main effects, $\binom{R}{2}$ second-order interactions, $\binom{R}{3}$ third-order interactions and so on, with the associated parameters $\{b,c,d,\ldots\}$ having dimension determined by the number of knots in K_r, $\{K_r,K_s\}$, $\{K_r,K_s,K_t\}$, etc. Higher order interactions may be excluded, even if definable in principle, as an acceptable smooth may often be obtained by restricting attention to main effects and low order interactions. So, a model with main and second order effects only would have $R+\binom{R}{2}$ parameter sets. Gustafson (2000) considers a BWISE approximation to smooth functions involving main effects $S_1(w_1),\ldots,S_R(w_R)$, and second-order interactions only, namely,

$$S_{12}(w_1,w_2),\ldots,S_{1R}(w_1,w_R),\ldots,S_{(R-1),R}(w_{R-1},w_R).$$

Main effects are assumed to be either conventional linear regression effects or cubic splines. The form of the interaction depends on which form of main effect is selected for predictors w_r and w_s.

As an example, consider a tensor product of truncated polynomials with $q=1$, and $R=3$, so that $w_i=(w_{1i},w_{2i},w_{3i})$ and with $K_1=K_2=K_3=5$ knots. Also just consider linear step functions $(w-\kappa)_+$. Then there may be $R=3$ main effects, $\binom{R}{2}=3$ second-order interactions, and $\binom{R}{3}=1$ third-order interactions. In a model confined to main effects and second-order interactions, the main effects would be terms $\Sigma_{k=1}^{K_1}b_{1k}(w_{1i}-\kappa_{1k})_+$, $\Sigma_{k=1}^{K_2}b_{2k}(w_{2i}-\kappa_{2k})_+$, and $\Sigma_{k=1}^{K_3}b_{3k}(w_{3i}-\kappa_{3k})_+$, involving 15 parameters. The second-order interactions would be terms $\Sigma_{k=1}^{K_1}\Sigma_{l=1}^{K_1}c_{12,kl}(w_{1i}-\kappa_{rk})_+(w_{2i}-\kappa_{2l})_+$, $\Sigma_{k=1}^{K_1}\Sigma_{l=1}^{K_2}c_{13,kl}(w_{1i}-\kappa_{1k})_+(w_{3i}-\kappa_{3l})_+$, and $\Sigma_{k=1}^{K_2}\Sigma_{l=1}^{K_3}c_{23,kl}(w_{2i}-\kappa_{2k})_+(w_{3i}-\kappa_{3l})_+$ involving 75 parameters. If the coefficients $\{b_{rk},c_{rs,kl}\}$ are assumed to be fixed effects, then predictor selection methods are relevant, as in Smith and Kohn (1996) or the RJMCMC (reversible jump MCMC) methods discussed by Dennison et al. (2002, p.105). If the $\{b_{rk},c_{rs,kl}\}$ are assumed to be random effects, smoothness may be achieved by penalising large coefficients, and parsimony achieved by selection between zero and positive variance components $\{\phi_{b_1},\phi_{b_2},\phi_{b_3},\phi_{c_{12}},\phi_{c_{13}},\phi_{c_{23}}\}$.

The tensor product generalisation of (12.2) or (12.5) includes interactions between the terms in $T(w)$ and $Q(w)$ (Smith and Kohn, 1997; Ruppert et al., 2003, p.240). Consider a situation with $R=2$, with K_1 knots in w_{1i} and K_2 knots in w_{2i}. For a linear spline ($q=1$), and random effect spline coefficients $\{b_{rk},d_{rsk},c_{rskm}\}$ one would have

$$\begin{aligned}
g(\mu_i)=\alpha+\beta_1 w_{1i}+\beta_2 w_{2i}+\beta_3 w_{1i}w_{2i}+\sum_{k=1}^{K_1}b_{1k}(w_{1i}-\kappa_{1k})_+\\
+\sum_{k=1}^{K_2}b_{2k}(w_{2i}-\kappa_{2k})_+ +\sum_{k=1}^{K_2}d_{12,k}w_{1i}(w_{2i}-\kappa_{2k})_+\\
+\sum_{k=1}^{K_1}d_{21k}w_{2i}(w_{1i}-\kappa_{1k})_+\\
+\sum_{k=1}^{K_1}\sum_{m=1}^{K_2}c_{12km}(w_{1i}-\kappa_{1k})_+(w_{2i}-\kappa_{2m})_+ +u_i
\end{aligned}$$

where there are six variance components $(\phi_{b_1},\phi_{b_2},\phi_{d_{12}},\phi_{d_{21}},\phi_{c_{12}},\sigma^2)$. In the bivariate example of Smith and Kohn (1997, p.1530), $K_1=K_2=9$ and $q=3$ leading to a (fixed effects) analysis involving 169 coefficients.

A similar scheme applies when interactions between metric and categorical predictors are considered. Thus let $C_i\in(1,\dots L)$ be a categorical predictor, and w_{1i} and w_{2i} be metric predictors. Suppose that only the smooth in w_2 is postulated to vary according to the level of C, and define

$$\begin{aligned}
z_{il}&=1 \quad \text{if } C_i=l\\
&=0 \quad \text{otherwise.}
\end{aligned}$$

Also consider a metric response y_i, and assume that interactions between w_1 and w_2 are not present. Then, with a qth degree truncated polynomial basis in both predictors, one possible representation is

$$y_i = \alpha + Q_1(w_{1i}) + Q_2(w_{2i}) + T_1(w_{1i}) + T_{2,C_i}(w_{2i}) + u_i,$$

$$= \alpha + \beta_{11}w_{1i} + \ldots + \beta_{1q}w_{1i}^q + \beta_{21}w_{2i} + \ldots + \beta_{2q}w_{2i}^q + \sum_{k=1}^{K_1} b_{1k}(w_{1i} - \kappa_{1k})_+^q,$$

$$+ \sum_{k=1}^{K_2} b_{2k}(w_{2i} - \kappa_{2k})_+^q + \sum_{l=2}^{L} z_{il}\{\sum_{k=1}^{K_2} c_{kl}(w_{2i} - \kappa_{2k})_+^q\},$$

where $b_{1k} \sim N(0,\phi_{b1}), b_{1k} \sim N(0,\phi_{b1})$, $c_{kl} \sim N(0,\phi_{cl})$ and $u_i \sim N(0,\sigma^2)$ (Coull et al., 2001). The amount of smoothing under $S_1 = Q_1 + T_1$ and $S_{2,C_i} = Q_2 + T_{2,C_i}$ then depends on the ratios σ^2/ϕ_{b1} and $\sigma^2/[\phi_{b2} + \phi_{cl}]$.

In a multivariate mixed model generalisation of the radial basis, one may consider thin-plate functions with exponents $(2q-d)$ specified by integer combinations (q,d), where d is the dimension of the covariate vectors in the relevant interaction (Yau et al., 2003). So

$$H_k(z) = |z - t_k|^{(2q-d)} \log(|z - t_k|) \text{ for } (2q-d) \text{ even}$$

$$H_k(z) = |z - t_k|^{(2q-d)} \text{ for } (2q-d) \text{ odd}$$

where z are univariate or multivariate vector predictor values, and t_k are univariate or multivariate knots. In applying such functions, heavily parameterised multivariate spline models are often not likely to be well identified, and simpler options involving univariate smooths in each predictor (with $d=1$), and all possible bivariate interactions (with $d=2$), may be considered (Yau and Kohn, 2003). Consider the setting $q=2$, with predictors, w_1 and w_2, and let $z_i = (w_{1i}, w_{2i})$ denote bivariate covariate combinations, with K_{12} bivariate centres $t_k = (t_{1k}, t_{2k})$ that might be provided by an initial cluster analysis. Also denoting distances $h_{ik} = |z_i - t_k|$, the bivariate basis for $2q-d=2$ is of the form $h^2\log(h)$. With linear terms in the parametric component $Q(w)$, this leads to the representation

$$g(\mu_i) = \alpha + S_1(w_{1i}) + S_2(w_{2i}) + S_{12}(w_{1i}, w_{2i}),$$

$$= \alpha + \beta_1 w_{1i} + \sum_{k=1}^{K_1} b_{1k}|w_{1i} - \kappa_{1k}|^3 + \beta_2 w_{2i}$$

$$+ \sum_{k=1}^{K_2} b_{2k}|w_{2i} - \kappa_{2k}|^3 + \sum_{k=1}^{K_{12}} c_k h_{ik}^2 \log(h_{ik}).$$

With K_r knots $\{\kappa_{r1}, \ldots \kappa_{rK_r}\}$ for predictor w_{ri}, the R main effects are

$$S_r(w_{ri}) = \beta_r w_{ri} + \sum_{k=1}^{K_r} b_{rk}|w_{ri} - \kappa_{rk}|^3,$$

where the R sets of coefficients $\{[b_{r1}, b_{r2}, \ldots b_{rK_r}], r = 1, \ldots, R\}$ are assumed random with variances $\phi_{b1}, \ldots \phi_{bR}$. Let the K_{rs} bivariate knots for first order (w_r, w_s) interaction effects be denoted $t_{rs,k} = (t_{rk}, t_{sk})$. Then the interaction bases have the form

$$T_{rs}(w_{ri}, w_{si}) = \sum_{k=1}^{K_{rs}} c_{rs,k} \left|(w_{ri}, w_{si}) - (t_{sk}, t_{rk})\right|^2 \log(\left|(w_{ri}, w_{si}) - (t_{rk}, t_{sk})\right|).$$

The $\binom{R}{2}$ sets of coefficients $c_{rs,k}$ are also assumed to be random.

Lo-rank thin-plate spline smooths as an approximation to the full thin-plate regression spline (TPRS) smoother are considered by Wood (2003, 2006, 2016). Thus, for an R-dimensional predictor vector $w_i = (w_{1i}, \ldots, w_{Ri})$, with linear model

$$y_i = f(w_i) + u_i$$

with u_i random, the full TPS smooth of degree m involves a function g minimising

$$\|y - g\|^2 + \lambda J_{mR}(g),$$

where $J_{mR}(g)$ is a roughness penalty and λ is a smoothing parameter. This penalty is an integral of dimension $M = \binom{m+R-1}{R-1}$ involving all possible terms

$$\frac{m!}{\nu_1! \ldots \nu_R!} \left(\frac{\partial^m g}{\partial w_1^{\nu_1} \ldots \partial w_R^{\nu_R}} \right)^2,$$

where $\nu_1 + \ldots \nu_R = m$. So, for $R = 2$ and $m = 2$,

$$J_{mR}(g) = \iint \left(\frac{\partial^2 g}{\partial w_1^2} \right)^2 + 2\left(\frac{\partial 2g}{\partial w_1 \partial w_2} \right)^2 + \left(\frac{\partial^2 g}{\partial w_2^2} \right)^2 dw_1 dw_2.$$

The function g has the form

$$g(w) = \sum_{i=1}^{n} \delta_i \eta_{mR} \|w - w_i\| + \sum_{j=1}^{M} a_j \phi_j(w),$$

where δ_i and α_j are unknowns. To reduce the number of unknowns, especially for larger samples, a rank k orthonormal basis for the δ parameters is used instead. This approach avoids the knot placement problems of conventional regression spline modelling. Thin-plate regression splines with truncated basis are implemented in the R package mgcv, with the jagam option (Wood, 2016) producing a modifiable rjags code incorporating the TPRS commands (see Example 12.3).

Example 12.2 Fertility, GDP, and Female Education

These data are from the UN Human Development Report for 2015 (http://report.hdr.undp.org/), and relate to a measure of fertility (TFR, the total fertility rate) over 167 countries. The analysis concerns the relation of TFR to GDP per head and average years

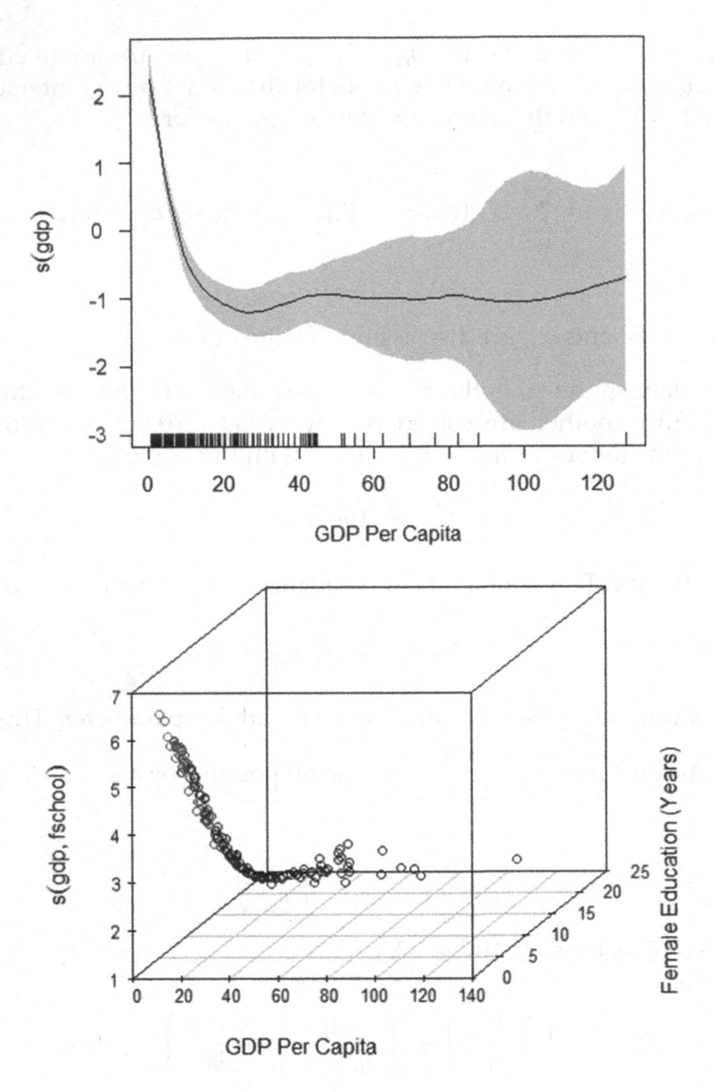

FIGURE 12.4
(a) Smooth for TFR as function of GDP per capita. (b) Smooth for TFR as function of GDP per capita and female education.

of education for females. TPRS models are applied and can be fitted using jagam/mgcv, or the stan_gamm4 option in rtsanarm.

The first model involves a smooth in GDP only, and a truncated TPRS representation with rank $k=20$ and $m=2$. This provides a penalised DIC of 415, with Figure 12.4a showing the resulting centred smooth. Including separate univariate smooths in both GDP and female education improves the pDIC to 380. Both analyses show rapid convergence. The second model is illustrated both by jagam/mgcv and stan_gamm4 codes.

A third model involves a joint smooth s(gdp,fschool) in the predictors, and provides a pDIC of 379. Figure 12.4b shows the resulting three-dimensional scatter plot. Combining both univariate smooths and a joint smooth provides a slightly improved pDIC of 374.

A final analysis modifies the rjags code for this model to include likelihood calculations from which WAIC (widely applicable information criterion) and LOO-IC may be derived, and also includes binary selection indicators, $J_k \sim \text{Bern}(0.5)$, for the three smooths. A penalising complexity prior is adopted for the residual standard deviation, based on an assumed 0.01 probability that this exceeds 2, and exponential, $E(1)$, priors are adopted on the smoothing parameters. This analysis shows a 0.06 probability for retaining the univariate smooth in gdp, and if that smooth is excluded (so that the model consists only of a univariate smooth in fschool and a bivariate smooth), the pDIC falls to 372.

12.4 Heteroscedasticity via Adaptive Non-Parametric Regression

As mentioned above, a random effects spline regression in a predictor w_i typically takes the form

$$g(\mu_i) = \alpha + S(w_i) = \alpha + \beta_1 w_i + \ldots + \beta_q w_i^q + \sum_{k=1}^{K} b_k (w_i - \kappa_k)_+^q + u_i,$$

where $u_i \sim N(0, \sigma^2)$, the κ_k are knots, and the spline coefficients b_k may be taken as normal, for example $b_k \sim N(0, \phi)$. This approach is spatially homogenous (in terms of the predictor space), whereas a spatially adaptive regression may be used to represent heteroscedasticity, which is also related to w values, or possibly to the values of other predictors (Currie and Durban, 2002). Spatial adaptive regression may also be used to allow non-constant variance in the b_k, namely $b_k \sim N(0, \phi_k)$, with $\log(\phi_k)$ determined by a spline regression on the knots (Yue et al., 2012).

For modelling heteroscedasticity, with $u_i \sim N(0, \sigma_i^2)$, a subsidiary spline regression may be applied to the variances $\sigma_i^2 = \exp(h_i)$, with M knots in the same predictor

$$h_i = \gamma_0 + \gamma_1 w_i \ldots + \gamma_q w_i^q + \ldots \sum_{m=1}^{M} c_m (w_i - \psi_m)_+^q,$$

with $c_m \sim N(0, \phi_c)$ (e.g. (Chib and Greenberg, 2013)). Other options (Jerak and Lang, 2005) are random walk priors in h_i, such as an RW1

$$h_i \sim N(h_{i-1}, 1/\tau_h).$$

or discrete mixture over smoothing functions, with mixture probabilities based on multinomial logit regression involving additional covariates x_i. For y metric and M mixture components, one might have

$$p(y_i \mid x_i, w_i) \sim \sum_{m=1}^{M} \pi_m(x_i) N(S_m(w_i, \theta_m), V_m)$$

$$\sum_{m=1}^{M} \pi_m(x_i) = 1$$

where each smooth function $S_m(w, \theta_m)$ has its own parameter set θ_m.

Models allowing for heteroscedasticity via non-parametric regression belong to a broader class of generalised additive models for location, scale, and shape (GAMLSSs). These regress not only the expected mean, but other distribution parameters (e.g. location, scale, and shape) on covariates, leading to what is termed distributional regression (Mayr et al., 2012; Wood et al., 2016). In zero-inflated Poisson or negative binomial (NB) regressions, the additive model framework is used for the Poisson or NB rate, the probability of inflation, and for the NB scale parameter (Klein et al., 2015).

Example 12.3 Elementary School Attainment

The data for this example are a random sample of 400 elementary schools from California Education Department's API datafile for 2000 (www.cde.ca.gov/ta/ac/ap/apidatafiles.asp), which reports school academic performance (y_i) together with school characteristics, such as average class size and the poverty rate among the pupil intake (Chen et al., 2003). A linear regression analysis involves regressing 2000 performance

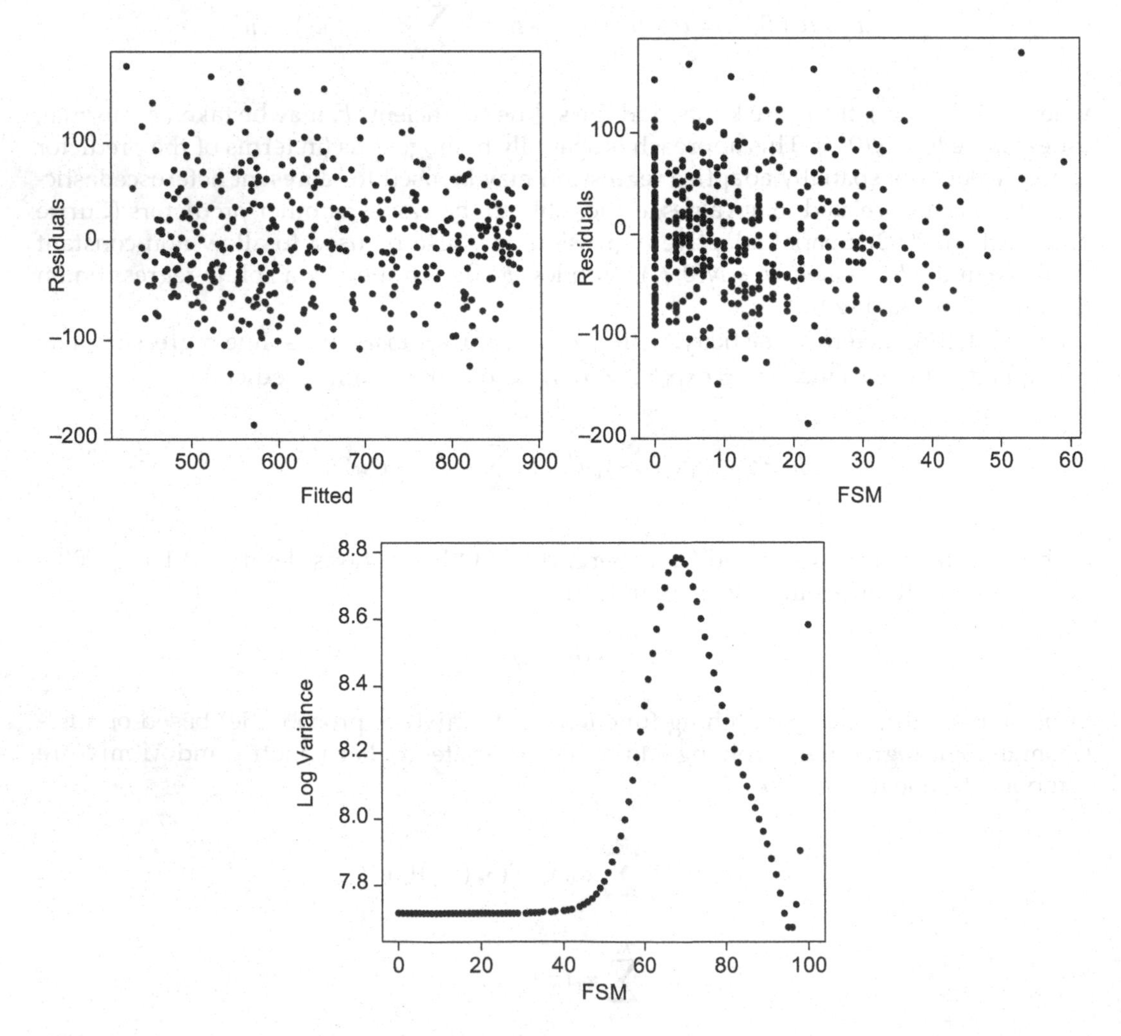

FIGURE 12.5
(a) Residuals against fitted, homoscedastic model. (b) Residuals against free school meals. (c) Plot of log variance

on the percentage of pupils receiving free meals (FSM), percentage of English language pupils (ELP), and percentage of teachers with emergency credentials (EMCRED).

Let $\sigma^2 = \text{Var}(u_i)$, and assume $1/\sigma^2 \sim Ga(1,0.001)$ in a homoscedastic linear regression

$$y_i = \mu_i + u_i = \beta_0 + \beta_1\text{FSM}_i + \beta_2\text{ELP}_i + \beta_3\text{EMCRED}_i + u_i.$$

With computation via jagsUI, this provides a LOO-IC of 4387, with $p_e = 5.1$. However, a plot of the residuals shows residual variation to decrease as fitted attainment increases (Figure 12.5a). All three predictors have significant (negative) effects on attainment, but the highest ratio of posterior mean to standard deviation is for FSM, and a plot of the residuals against FSM (Figure 12.5b) suggests residual variation increases with FSM.

A second model therefore specifies $y_i \sim N(\mu_i, \sigma_i^2)$, where $u_i \sim N(0, \sigma_i^2)$, with $\log(\sigma_i^2)$ modelled by a cubic spline regression

$$\log(\sigma_i^2) = \gamma_0 + \sum_{m=1}^{M} c_m(\text{FSM}_i - \psi_m)_+^3.$$

The spline coefficients are random $c_m \sim N(0, \phi_c)$ with $1/\phi_c \sim Ga(1,1)$. There are $M = 9$ knots, sited at the 10th, 20th, and 90th percentiles of FSM. A corner constraint $c_1 = 0$ is used for identifiability. A two-chain run of 20,000 iterations gives an estimate for $\phi_c^{0.5}$ of 0.52 with 95% interval (0.33,0.84), whereas homoscedasticity would imply $\phi_c^{0.5} = 0$. Figure 12.5c accordingly demonstrates non-constancy in $\log(\sigma_i^2)$ as FSM varies, though there is no consistent monotonic upward or downward trend in variability as FSM increases. The LOO-IC under the second model falls to 4375 ($p_e = 11.3$).

A third model employs a different identification device, namely centring (at each iteration) the observation level smooth $S_i(\text{FSM}) = \Sigma_{m=1}^{M} c_m(FSM_i - \psi_m)_+^3$ around the overall mean of such smooths. The centred smooth is then included in the spline regression for $\log(\sigma_i^2)$. This produces a similar fit (LOO-IC = 4376), and a similar non-monotonic relation between $\log(\sigma_i^2)$ and FSM. The centred c_m (c.cent in the R code) for this implementation have a correlation of 0.99 with those from the corner constraint option.

12.5 General Additive Methods

Consider ranked values of a single predictor $w_1, \ldots, w_n$ such that

$$w_1 < w_2 < \ldots < w_n,$$

and let $S_t = S(w_t)$ be a smooth function representing the locally changing impact of w_t on $g(\mu_t)$ as it varies over its range. Thus

$$g(\mu_t) = a + S(w_t) + u_t,$$

$$u_t \sim N(0, \sigma^2),$$

where depending on identification procedures used, the intercept a may not be present (Koop and Poirier, 2004). Appropriate priors for S_t reflect the ordering and spacing of the w values, and typically follow dynamic linear priors or other time series schemes. Normal or

Student t random walks in the first, second, or higher differences of S_t are one possibility (Knorr-Held, 1999; Fahrmeir and Lang, 2001; Chib and Jeliazkov, 2006). For identifiability, especially when there are smooths $S_{rt} = S(w_{rt})$ in several predictors one may adopt devices such as centring of the S_{rt}, or corner constraints (e.g. $S_{r1} = 0$). Alternatively, to expedite computing speed, one may monitor identified quantities such as the centred series $S_{rt} - \bar{S}_r$ without actually imposing centring constraints within the estimation. Because there is only local smoothing, inferences may also be sensitive to priors assumed for evolution variance τ^2 for the S_t and other aspects of the model.

If the w values are equally spaced and distinct, then 1st and 2nd order random walk priors are just

$$S_t \sim N(S_{t-1}, \tau^2),$$

$$S_t \sim N(2S_{t-1} - S_{t-2}, \tau^2),$$

where smaller values of τ^2 result in a smoother curve. For metric or overdispersed discrete responses, the parameterisation $\tau^2\lambda = \sigma^2$ may be used, allowing for trade-off between the residual variance and the variance of the smooth (Koop and Poirier, 2004).

In ordinary regression applications, values of the w_t are typically unequally spaced, and there may be tied values. To take account of unequal spacing between successive w_t, the prior is modified such that for second and higher order walks, the weighting on lagged values is varied according to how distant they are from the current value (Fahrmeir and Lang, 2001). In all orders of random walk, the precision of S_t is reduced the wider the gap between w_t, and its preceding ordered values. Let gaps between points be denoted $\delta_2 = w_2 - w_1, \delta_3 = w_3 - w_2, \ldots, \delta_n = w_n - w_{n-1}$ (with $\delta_1 = 0$). Then a first-order Normal random walk becomes

$$S_t \sim N(S_{t-1}, \delta_t \tau^2),$$

and a second-order one becomes

$$S_t \sim N([1 + \delta_t / \delta_{t-1}]S_{t-1} - [\delta_t / \delta_{t-1}]S_{t-2}, \delta_t \tau^2).$$

Separate usually fixed effect priors are assumed for the initial values (e.g. S_1 in a first order random walk). A scheme allowing choice between RW1 and RW2 dependence for unequally spaced w is proposed by Berzuini and Larizza (1996), namely

$$s_t \sim N(M_t, \delta_t \tau^2)$$

where

$$M_t = s_{t-1}[1 + (\delta_t / \delta_{t-1})\exp(-\eta\delta_t)) - s_{t-2}[(\delta_t / \delta_{t-1})\exp(-\eta\delta_t)).$$

Larger values of $\eta > 0$, such that $\exp(-\eta\delta_t)$ tends to zero, imply an approximate RW1 prior and less smoothness.

If there are ties in the w values, with only $m < n$ distinct values, denoted $\{w_j^*, j = 1, \ldots, m\}$, then the above priors would be on the differences $\delta_j = w_j^* - w_{j-1}^*$ in the ranked distinct values, and it is necessary to specify a grouping index G_t (ranging between 1 and m) for each observation $t = 1,\ldots,n$ to indicate which distinct value it takes. Assuming an RW1 prior in the smooth of the predictor effects, the regression in w_t can then be written

$$g(\mu_t) = a + S(G_t) + u_t, \quad t = 1,\ldots,n$$
$$S_j \sim N(S_{j-1}, \delta_j \tau^2) \quad j = 1,\ldots,m$$

with $G_t \in (1,\ldots,m)$.

If there is more than one predictor then a semiparametric model might be adopted with smooth functions $S_r(w_r)$ on a subset $r = 1, \ldots, q$ of R predictors, with the remainder modelled by assuming global linearity. So

$$g(\mu_t) = \alpha + S_1(w_{1t}) + S_2(w_{2t}) + \ldots + S_q(w_{qt}) + \beta_1 w_{q+1,t} + \ldots \beta_{R-q} w_{R,t} + u_t.$$

If non-parametric functions are estimated for several regressors $w_{1t}, w_{2t}, \ldots, w_{qt}$, then a unique ordering across all predictors is usually infeasible and grouping indices $G_{1t}, G_{2t}, \ldots, G_{qt}$ for each of q regressors are necessary, even if the regressors have no tied values. In the case of tied values, the indices range between 1 and m_1,1 and m_2,...,1 and m_q (rather than between 1 and n).

Another approach (Wahba, 1983; Biller and Fahrmeir, 1997; Wood and Kohn, 1998) to Bayesian general additive modelling involves the state space version of the polynomial smoothing spline. For a spline of general order $2h-1$, $S_t = S(w_t)$ is generated by a differential equation

$$\frac{d^h S_t}{dt^h} = \tau \frac{dW_t}{dt},$$

with W_t a Weiner process, and τ^2 the evolution variance. The state vector

$$Z_t = \left(S_t, \frac{dS_t}{dt}, \frac{d^2 S_t}{dt^2}, \ldots, \frac{d^{(h-1)} S_t}{dt^{(h-1)}} \right),$$

is then of order h, evolving stochastically according to

$$Z_t = F_t Z_{t-1} + e_t, \tag{12.8}$$

where F_t is an $h \times h$ transition matrix and e_t is a multivariate error. For the cubic spline case with $h = 2$, $Z_t = (S_t, dS_t/dt)$ is bivariate and the transition matrix is

$$F_t = \begin{pmatrix} 1 & \delta_t \\ 0 & 1 \end{pmatrix},$$

where $\delta_t = w_{t+1} - w_t$. The e_t are also bivariate, for example, MVN with zero mean and covariance $\tau^2 E_t$, where

$$E_t = \begin{pmatrix} \delta_t^3/3 & \delta_t^2/2 \\ \delta_t^2/2 & \delta_t \end{pmatrix}.$$

As usual there may be ties in the w values, and the prior (12.8) would be on $j = 1,\ldots,m$ distinct ranked values. Each observation for $t=1, \ldots, n$ would have a grouping index G_t with values between 1 and m.

Example 12.4 Conceptions under 18, RW2 Smooths

This example considers data on conceptions to women aged under 18 (y_i) in 352 English local authorities. Explanatory factors are area deprivation, measured by an Index of Multiple Deprivation (IMD), and the percentage of 15-year-old pupils *not* achieving five or more GCSE subjects at grade C or above. The acronym GCSE refers to the General Certificate of Secondary Education, and educational proficiency is set by the criterion of grade C or above. The model involves additive RW2 priors in w_1=IMD and w_2=GCSE. Let G_{1i} and G_{2i} indicate which of the unique IMD and GCSE values is taken by area i, where such unique values are ranked, with $m_1=352$ and $m_2=351$ unique values (there is a single tie in the GCSE values). Some of these distinct values are, however, very close to each other (a consideration relevant in a BayesX application). With $j=1, \ldots, m_r$ denoting ranked predictor values, $\delta_{rj} = w^*_{rj} - w^*_{r,j-1}$ and assuming RW2 dependence

$$y_i \sim \text{Bin}(n_i, \mu_i),$$

$$\text{logit}(\mu_i) = a + s_1(w_{1,G_{1i}}) + s_2(w_{2,G_{2i}})$$

$$s_{rj} \sim N([1+\delta_{rj}/\delta_{r,j-1}]s_{r,j-1} - [\delta_{rj}/\delta_{r,j-1}]s_{r,j-2}, \tau_r^2\delta_{rj}) \quad r=1,2;\ j=1,\ldots,m_r,$$

where τ_r^2 is the variance for the randomly varying s_{rj}. There is excess dispersion which may be removed by a model also including an unstructured effect

$$\text{logit}(p_i) = a + s_1(w_{1,G_{1i}}) + s_2(w_{2,G_{2i}}) + u_i,$$

where $u_i \sim N(0, \sigma_u^2)$.

A BayesX analysis is applied within R using the BayesXsrc package. Inverse gamma priors $IG(g,h)$ are assumed on variance parameters, with a setting of $\{g=1, h=0.0001\}$. To avoid estimation of a large number of coefficients, BayesX performs internal grouping if a covariate has a large number of distinct values (for first- and second-order random walks), so the actual number of distinct values used will be lower than the observed m_r. Plots of the smooths under $h=0.0001$ are based on 94 distinct IMD values and 89 distinct GCSE values.

Figures 12.6a and 12.6b (which include the intercept) show the resulting smooth functions. The extent of smoothing may depend on prior settings: setting h smaller (e.g. $h=0.001$) produces more short-term variability. A stan_gamm4 code using TPRS smooth functions shows similar results.

12.6 Non-Parametric Regression Methods for Longitudinal Analysis

Two major applications of non-parametric regression to longitudinal datasets are to time-varying regression coefficients and subject-specific curves (James et al., 2000; Wu and Zhang, 2006). Applications to joint models for longitudinal and time-to-event data are also

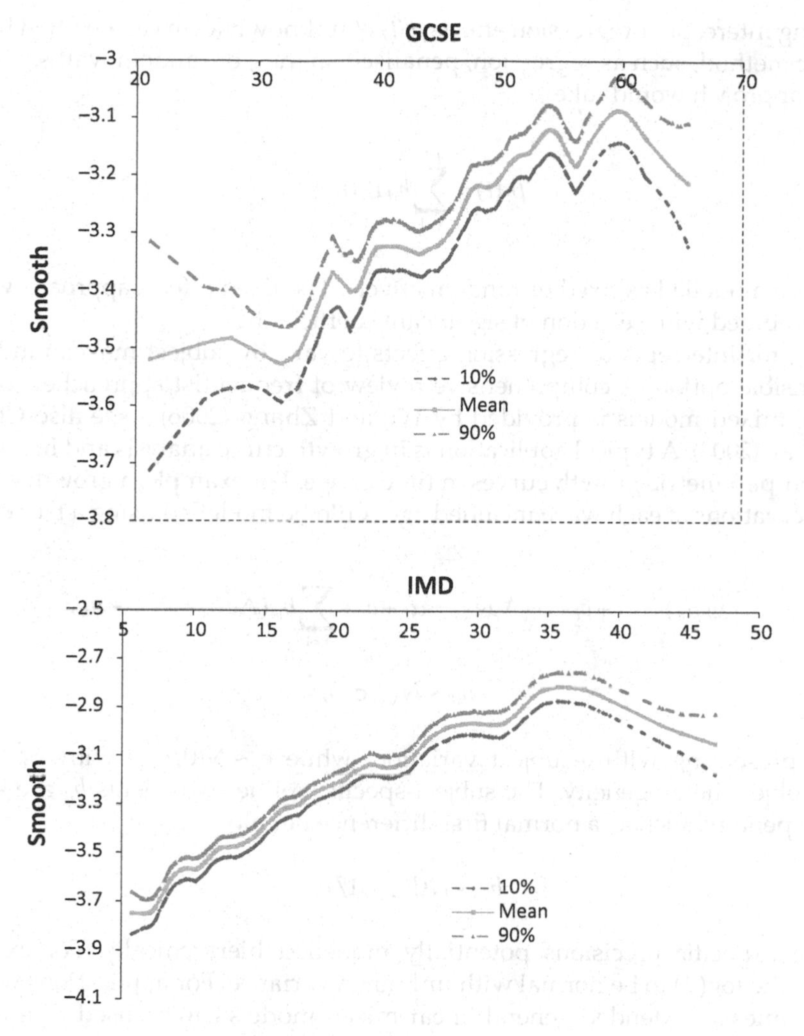

FIGURE 12.6
(a) Smooth in GCSE (80% CRI). (b) Smooth in IMD (80% CRI).

increasing (Kohler et al., 2016). Time-varying regression effects are a special case of the general varying coefficient model of Hastie and Tibshirani (1993), namely

$$g(\mu_i, \upsilon) = \beta_0(\upsilon_0) + w_{1i}\beta_1(\upsilon_1) + \ldots w_{Ri}\beta_R(\upsilon_R),$$

where the effect modifiers $\upsilon = (\upsilon_1, \ldots, \upsilon_R)$ govern the effect of predictors $w = (w_1, \ldots w_R)$. If the modifiers are all the same (e.g. time) with $\upsilon_1 = \upsilon_2 = \ldots = \upsilon_R = t$ then

$$g(\mu_{it}) = \beta_0(t) + w_{1i}\beta_1(t) + \ldots + w_{Ri}\beta_R(t),$$

and the time-varying coefficient model, or dynamic general linear model (West and Harrison, 1997), is obtained. This extends to time-varying predictors w_{rit}, with

$$g(\mu_{it}) = \beta_0(t) + w_{1it}\beta_1(t) + \ldots + w_{Rit}\beta_R(t),$$

Tim-varying intercept or regression effects $\beta_r(t)$ of unknown form can be fitted by any non-parametric method, such as regression, penalised splines, or random walks. For example, a B-spline approach would take

$$\beta_r(t)=\sum_{k=1}^{K^*} b_{rk}B_k(w_{rit},q)$$

where b_{rk} are modelled as fixed or random effects. The fixed effects approach would typically be combined with selection of significant coefficients.

Allowing for intercepts or regression effects to vary by subject makes random effects a more sensible option. A comprehensive review of frequentist approaches to such non-parametric mixed models is provided by Wu and Zhang (2006) – see also Chapter 9 in Ruppert et al. (2003). A typical application is in growth curve analysis and involves subject specific non-parametric growth curves in time or age. For example, a growth curve model where observations at each wave included age could be modelled using a truncated spline

$$g(\mu_{it})=\alpha_t+c_i+S_i(\text{Age}_{it})=\alpha_t+c_i+\sum_{k=1}^{K} b_{ik}(\text{Age}_{it}-\kappa_k)_+^q+u_{it},$$

$$u_{it}\sim N(0,\sigma^2),$$

with σ^2 representing within-subject variation, while $c_i\sim N(0,\sigma_c^2)$, with σ_c^2 measuring between-subject heterogeneity. The subject-specific spline coefficients b_{ik} are subject to a roughness penalty, such as a normal first difference penalty

$$b_{ik}\sim N(b_{i,k-1},1/\theta_i),$$

with subject-specific precisions potentially modelled hierarchically. For example, one might take the $\log(\theta_i)$ to be normal with unknown variance. For applications with distinct recording times a_{it}, extended general linear mixed models can be used (Wu and Zhang, 2006), with

$$g(\mu_{it})=X_{it}\beta+\eta(a_{it})+Z_{it}b_i+S_i(a_{it})+u_{it},$$

where $\eta(a)$ is the population mean function, estimated non-parametrically, and $S_i(a)$ are subject-specific deviation functions. Silva et al. (2008) consider cubic B-spline bases to model region-wide and area-specific trends for health outcomes $y_{it}\sim \text{Bin}(n_{it},\pi_{it})$, namely

$$\text{logit}(\pi_{it})=\alpha+\eta(t)+S_i(t)+d_i=\alpha+\sum_{k=1}^{K^*} b_kB_k(t,3)+\sum_{k=1}^{K^*} c_{ik}B_k(t,3)+d_i,$$

where d_i and c_{ik} are random area effects.

Another possible scheme for allowing variability across subjects is by random "slopes" around the population smooth functions, also sometimes denoted as random scaling of nonlinear functions (Tutz and Reithinger, 2007). For example, consider a longitudinal (e.g. growth curve) application with a single predictor w_{it}, the impact of which is modelled at population level by a smooth function $S(w_{it})$. Then one may wish to allow both for intercept (baseline) variation and for subject level variation around the average function $S(w)$. Thus

$$g(\mu_{it}) = a + b_{1i} + S(w_{it}) + b_{2i}S(w_{it}) + u_{it},$$

$$= a + b_{1i} + S(w_{it})(1 + b_{2i}) + u_{it},$$

with

$$(b_{1i}, b_{2i}) \sim N(0, D),$$

and for identification $\sum_{it} S_{it} = 0$ where $S_{it} = S(w_{it})$. The smooth function $S(w_{it})$ represents the mean effect of predictor w_{it}, but this effect is stronger for subjects with $b_{2i} > 0$, and weaker for subjects with $b_{2i} < 0$. So b_{2i} acts to amplify or attenuate the non-parametric impact of the variable w_{it}. For some subjects, one may even obtain large negative estimates, $b_{2i} < -1$, so that the effect of w_{it} is inverted. This model adapts to cross-sectional data where

$$g(\mu_i) = a + S(w_i) + b_i S(w_i) + u_i,$$

particularly in cases where the units are non-exchangeable, for example, if the units were areas, and b_i followed a spatial prior.

The impact of $(1 + b_{2i})$ on the unknown function $S(w_{it})$ is analogous to (subject specific) factor loadings operating on factor scores, and is subject to identifiability (label switching) issues, since $[-(1 + b_{2i})][-S(w_{it})) = S(x_{it})(1 + b_{2i})$. However, labelling issues should be avoided in practice if the impact of w_{it} represented by $S(w)$ is well-identified by the data. An alternative product scheme is applied by Congdon (2006), based on the Lee and Carter (1992) mortality forecasting model. In this scheme, subject-specific weights q_i that sum to 1 over all subjects operate on $S(w_{it})$, so that for $\Sigma_i q_i = 1$ the product scheme is $q_i S(w_{it})$. The effect of w is stronger for subjects with higher q_i, and weaker for subjects with lower q_i, with the average q_i being $1/n$.

Example 12.5 Progesterone Readings over Menstrual Cycle

This example uses progesterone readings y_{it} (log progesterone) in a study of early pregnancy loss (Brumback and Rice, 1998; Wu and Zhang, 2006). There are $n = 91$ observed cycles of length $T = 24$ days, so the total number of observations is $n \times T = 2184$. The days are coded as −8,−7,...,13,14,15, with 0 as day of ovulation. There are $J = 2$ groups of observations, the first 69 cycles being nonceptive, the last 22 being conceptive. The conceptive group growth paths (model 1), or subject level growth paths (model 2), are modelled non-parametrically. So instead of a linear or polynomial function in the days variable, cubic B-splines are used with knots at (−5,0,5,10).

The $K^* = 8$ basis functions are obtained from the R splines package via the commands:

```
require(splines)
cycval <- seq(-8,15)
bs.cycval <- bs(cycval,df=NULL,knots=c(-5,0,5,10),degree=3,intercep
t=T, Boundary.knots=range(cycval))
bs.cycval <- matrix(as.numeric(bs.cycval), nr = nrow(bs.cycval)).
```

In a baseline group-specific model, the spline coefficients are group-specific random coefficients $\{b_{jk}, j=1,2, k=1,K^*\}$, with group-specific precisions. Let $G_i \in (1,2)$ denote conceptive group, then $y_{it} \sim N(\mu_{it}, 1/\tau)$, with

$$\mu_{it} = a_{G_i} + \sum_{k=1}^{K^*} b_{G_ik} B_k(t,3),$$

$$b_{jk} \sim N(0, 1/\phi_j),$$

$$\phi_j \sim Ga(1, 0.001),$$

$$\tau \sim Ga(1, 0.001).$$

A two-chain run of 20,000 iterations is undertaken, with centring of $c_{jt} = \Sigma_{k=1}^{K^*} b_{G_ik} B_k(t,3)$ within groups for identification. There is a similar path between the two groups, in terms of posterior means of $\{a_j + c_{jt}\}$ up to the week after ovulation, but distinct trends thereafter (Figure 12.7a). The LOO-IC is 6252.

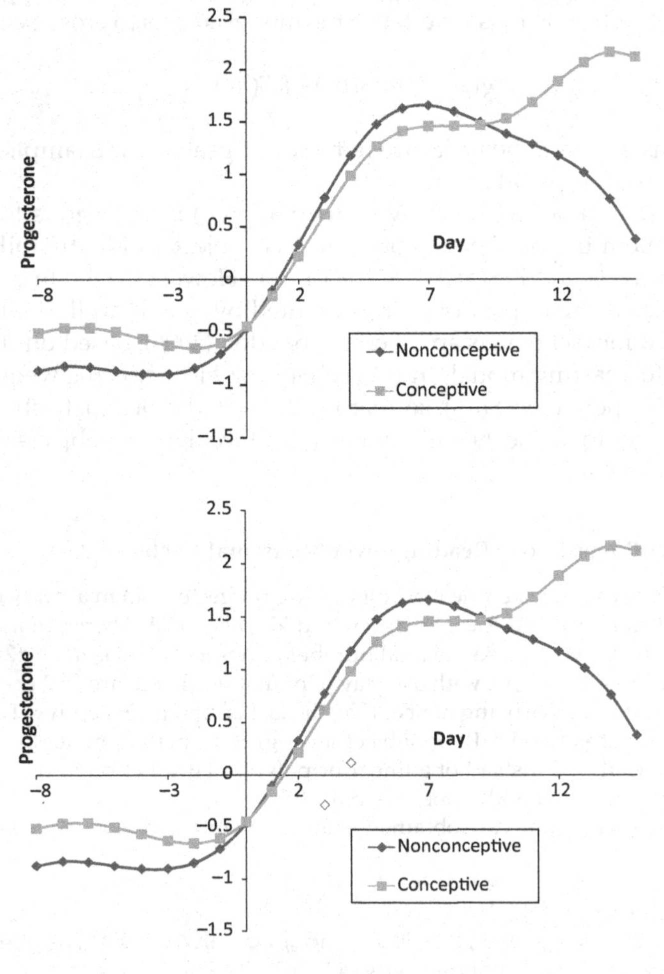

FIGURE 12.7
(a) Growth curve smooths (Model 1). (b) Growth curve smooths (Model 2).

A subject-specific model adds both subject heterogeneity and subject (cycle)-specific growth effects, so that

$$\mu_{it} = \alpha_{G_i} + b_{i0} + \sum_{k=1}^{K^*} b_{ik} B_k(t,3),$$

with

$$b_{i0} \sim N(0, 1/\tau_0), \quad i = 2,\ldots,n,$$
$$b_{i1} = 0,$$
$$b_{ik} \sim N(b_{i,k-1}, 1/\theta_{G_i}), \quad k = 2, K^*,$$
$$\tau_j \sim Ga(1, 0.001), \quad j = 0,1,$$
$$\theta_j \sim Ga(1, 0.001), \quad j = 0,1.$$

The corner constraint $b_{i1} = 0$ aids in identification. Average growth curves are shown in Figure 12.7b. The LOO-IC for this model is 3319.

Example 12.6 Birthweight and maternal age

Neuhaus and McCulloch (2006) consider a subset of data from a more extensive longitudinal study that involves the birthweights of babies born to n = 878 mothers from the state of Georgia, USA, all of whom has at least $T = 5$ babies. The analysis here is focused on the impact on birthweight y_{it} of mother's age at birth w_{it}, and the extent to which there is heterogeneity in the overall smooth $S(w_{it})$, which is based on a second-order random walk.

Thus, for each five birth history for mother i one may stipulate

$$y_{it} = \beta_0 + b_{1i} + S(w_{it}) + b_{2i}S(w_{it}) + u_{it},$$
$$= \beta_0 + b_{1i} + (1 + b_{2i})S(w_{it}) + u_{it},$$

where

$$(b_{1i}, b_{2i}) \sim N(0, D),$$

and D^{-1} follows a Wishart prior with identity scale matrix and 2 degrees of freedom. A second-order random walk smooth is estimated over all (i,t) pairs using a normal prior with a single variance parameter, rather than on the basis of successive ages within each fertility sequence, which would permit distinct variance parameters for each subject. The smooth involves 31 random parameters, namely for maternal ages 12 to 42. Identification is achieved by centring $S(w)$ at each iteration.

A two-chain run of 10,000 iterations using the rube library shows significant heterogeneity around the overall smooth in age, with a posterior mean for var(b_2) of 2.0, and 95% interval {1.2, 3.3}. Figure 12.8a shows the varying non-parametric impact of maternal age w_{it} on birthweight according to b_{2i}, namely for subjects with $b_{2i} = sd(b_2)$, $b_{2i} = 0$, and $b_{2i} = -sd(b_2)$, where the standard deviations are those at particular MCMC iterations. A histogram plot of the posterior mean b_{2i} (Figure 12.8b) indicates normality, though an extreme negative outlier of −4.9 occurs for subject 470, whose fourth and fifth infants weighed under 1kg, whereas the first two exceeded 3kg in weight. To assess outlier status at observation level, one may derive WAIC component scores for individual (mother, infant) pairs: the largest such score (44 out of a total WAIC of 5788) is for the fifth infant to mother 838.

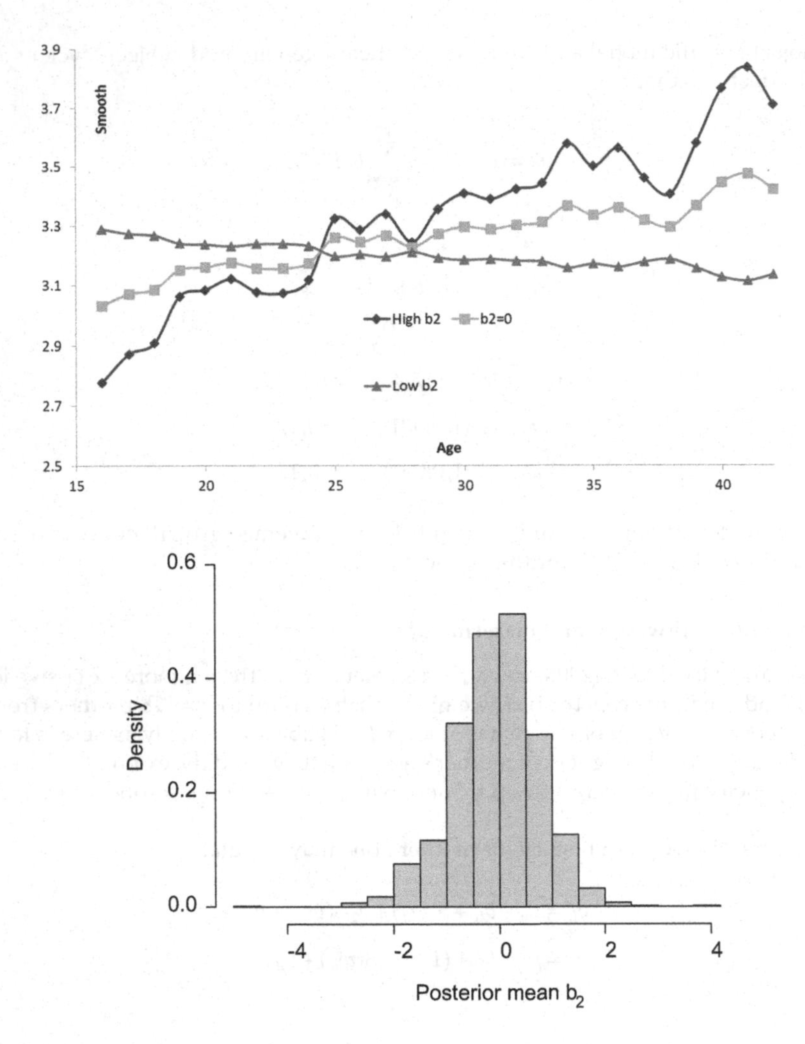

FIGURE 12.8
(a) Smooth impacts of maternal age on birthweight, according to variability in b_2. (b) Histogram of b_2.

12.7 Quantile Regression

Normal linear regression and generalised linear models focus on estimating the conditional mean of the response y_i. Quantile regression (Koenker, 2005) provides a more complete perspective on the conditional density of y_i, and focuses on estimating conditional quantiles (such as the conditional median) of the response. Sometimes, conditional mean regression will show a predictor as having no impact, whereas quantile regression will show a significant impact over at least part of the quantile range (Cade and Noon, 2003), though collinearity between predictors (and hence, predictor selection) may still be an issue (Xi et al., 2016; El Adlouni et al., 2018). With quantiles denoted $q \in [0,1]$, the conditional quantile density is denoted by the quantile (inverse cumulative distribution) function $Q(q \mid X_i)$, defined as $\Pr[y_i < Q(q \mid X_i) = q]$.

For linear regression involving a continuous response, the frequentist quantile regression estimator at quantile q minimises the function

$$Q(q|X_i) = q \sum_{y_i \geqq X_i\beta} |y_i - X_i\beta_q| + (1-q) \sum_{y_i < X_i\beta} |y_i - X_i\beta_q|$$

Equivalently, quantile regression involves minimising $\Sigma_{i=1}^{n}\rho_q(y_i - X_i\beta_q)$, as defined by the loss function (Yu and Moyeed, 2001)

$$\rho_q(u) = u(q - I(u<0) = u(qI(u \geq 0) + (1-q)I(u<0)).$$

This loss function downweights or emphasises absolute errors according to the quantile q. For example, setting $q=0.9$ results in a loss nine times larger for positive residuals with $y_i \geqq X_i\beta$ than for negative residuals with $y_i < X_i\beta$. So, the upper tail of the conditional distribution is emphasised.

A special case is provided by median regression, via minimisation of the absolute deviations:

$$Q(0.5|X_i) = \sum |y_i - X_i\beta|.$$

This reduces the impact of outliers (influential observations) in the response space on estimation, so as to provide a better fit for the majority of observations. Credible intervals (e.g. for observation level predictions) estimated using conditional mean regression by averaging over MCMC samples may also be affected by outliers. By contrast, median regression is more robust to skewness and other departures from normality (Geraci and Bottai, 2006). Thus, Min and Kim (2004) consider different forms of non-Gaussian errors, with asymmetric and long-tailed distributions, and show that median regression outperforms conditional mean regression, since the median is a more suitable centrality measure for data with a skewed response.

Methods for Bayesian quantile regression include asymmetric Laplace likelihood (Yu and Moyeed, 2001), exponentially tilted empirical likelihood (Schennach, 2005), and Dirichlet process mixture median regression (Kottas and Gelfand, 2001). Yu and Moyeed (2001) demonstrate that loss function minimisation is equivalent to estimation using an asymmetric Laplace distribution (ALD), with density function

$$ALD(y|\eta_q,\sigma,q) = \frac{q(1-q)}{\sigma}\exp\left[\rho_q\left(\frac{y-\eta_q}{\sigma}\right)\right].$$

This density can be represented as a scale mixture of normals, thus facilitating Gibbs sampling (Kozumi and Kobayashi, 2011).

Thus, for $y \sim \text{ALD}(\eta_q,\sigma,q)$, one has for quantiles $q=1, \ldots, Q$ the quantile-specific representation

$$y_i = \eta_{iq} + \xi_q W_{iq} + \left[\frac{2\sigma_q W_{iq}}{q(1-q)}\right]^{0.5} Z_{iq},$$

where η_{iq} is the regression term, $\xi_q = (1-2q)/q(1-q)$, $W_{iq} \sim \text{Exp}(\sigma_q)$, and $Z_{iq} \sim N(0,1)$. The practical role of the $\xi_q W_{iq}$ terms is to maintain the model as a satisfactory representation of y, compensating for shifts in η_{iq} between quantiles. The W_{iq} are measures of outlier status. Observations with higher W_{iq} have higher variances and lessened influence on the likelihood. R packages to implement Bayesian quantile linear regression include brq (Alhamzawi, 2012), bayesQR (Benoit and Van den Poel, 2014), and ALDqr (Sanchez et al., 2017).

In practice, it is not necessarily guaranteed that estimated quantile curves will be non-crossing, especially for quantiles not widely separated (e.g. $q=0.05$ compared to $q = 0.10$) (Bondell et al., 2010). Methods to circumvent this, not necessarily fully Bayesian, have been proposed (Cai and Jiang, 2015). An ad hoc approach involves simultaneous estimation of all quantiles of interest, and omitting MCMC samples where the expected ordering of the quantile regression terms η_{iq} is not satisfied.

For longitudinal data (with units i, and times t) (e.g. Geraci and Bottai, 2006; Alhamzawi et al., 2011), the regression term might include quantile-specific unit level random effects b_{iq}. Assuming normal subject effects, the representation would then be

$$y_{it} = X_{it}\beta_q + b_{iq} + \xi_q W_{itq} + \left[\frac{2\sigma_q W_{itq}}{q(1-q)}\right]^{0.5} Z_{itq},$$

with $b_{iq} \sim N(0, \sigma_b^2)$.

12.7.1 Non-Metric Responses

For binary responses, the augmented data method can be applied, combined with the scale mixture version of the ALD (Benoit and Van den Poel, 2012; Benoit and Van den Poel, 2017). Thus, binary responses y_i can be regarded as determined by a continuous latent variable y_i^*. To implement quantile regression for these latent variables, one specifies

$$y_i^* \sim \text{ALD}(\eta_q, \sigma_q = 1, q),$$

with set scale parameter for identifiability and truncated sampling according to the observed value of y_i. Thus

$$y_i^* \sim \text{ALD}(\eta_q, \sigma_q = 1, q) \quad I(,0),\ y_i = 0;$$

$$y_i^* \sim \text{ALD}(\eta_q, \sigma_q = 1, q) \quad I(0,),\ y_i = 1.$$

Yue and Hong (2012) apply quantile tobit regression to highly skewed medical expenditure data, focusing on the latent outcome in combination with the scale mixture ALD, while Rahman (2016) uses the augmented data approach for quantile regression of ordinal data.

To extend quantile regression to count data, Machado and Santos Silva (2005) propose adding uniform noise u to count responses, giving $z_i = y_i + u_i$, where $u_i \sim U(0,1)$, and apply quantile regression of the form

$$Q_{Z_i}(q \mid X_i) = \eta_{qi} = q + \exp(X_i\beta_q).$$

With offsets E_i, the quantile regression is

$$Q_{Z_i}(q \mid X_i) = \eta_{qi} = q + E_i \exp(X_i\beta_q).$$

This can be rearranged (Fuzi et al., 2016) into a linear model

$$Q_{Z_i^*}(q \mid X_i) = \eta_{qi} = X_i\beta_q + \log(E_i),$$

for quantities

$$z_i^* = \log(z_i - q) \quad \text{for } Z_i > q;$$

$$z_i^* = \log(\phi), \qquad \text{for } Z_i \le q \ \ (\text{with } q > \phi > 0).$$

Another approach to quantile regression for overdispersed count data involves a scale mixture version of the ALD (Yu and Moyeed, 2001), within a hierarchical Poisson lognormal representation to account for overdispersion (e.g. Connolly and Thibaut, 2012). The quantile regression is for latent outcomes at the second stage of the hierarchical model, focused on estimating latent incidence rates or relative risks (Congdon, 2017). The Poisson lognormal representation is in itself beneficial, since the tails of the lognormal are heavier than for the gamma distribution, and for data with outliers, the Poisson lognormal model may give a better fit than the negative-binomial model. Thus, for observed counts y_i, one specifies for quantiles $q = 1,\ldots,Q$,

$$y_i \sim \text{Poi}(\mu_{iq}),$$

$$\mu_{iq} = \exp(\nu_{iq}),$$

$$\nu_{iq} \sim N\left(X_i\beta_q + \xi_q W_{iq}, \frac{2W_{iq}\delta_q}{q(1-q)}\right),$$

$$W_{iq} \sim Exp(\delta_q)$$

This approach is less computationally intensive than the uniform noise (jittering) method.

Example 12.7 Trout Density

We consider data on a continuous variable, the density of Lahontan cutthroat trout y (trout numbers per metre of stream) as response, and its varying relationship to stream width-depth (w-d) ratio (x). The code for this example estimates several quantile regressions simultaneously, and illustrates the plots that can be made. There are n=71 observations from 13 streams across 7 years in Nevada (Dunham et al., 2002; Cade and Noon, 2003). Dunham et al. (2002) compare quantile linear regression, and a nonlinear quantile regression $y_i = \exp(\beta_0 + \beta_1 x_i + u_i)$, which can be obtained by taking a log transform of y.

To obtain a plot of the varying influence of the w-d ratio on y, as in Figure 4 of Dunham et al. (2002), regressions are performed at $Q=19$ quantiles, namely 0.05, 0.1, 0.15, etc. through to 0.95. The varying influence is represented by the b2[q] parameters in the code. Because the outcome is necessarily positive, the linear regression is constrained to produce positive values. The b2[q] show no impact of the predictor until significant negative impacts at $q=0.75$ and above for the linear regression, and above $q=0.8$ for the exponential regression (Figure 12.9a). Figure 12.9b shows the predicted relationship between density and w-d ratios (from 0 to 60) for selected quantiles under the exponential model, analogous to Figure 5 in Dunham et al. (2002). This plot uses the posterior median of replicate density values.

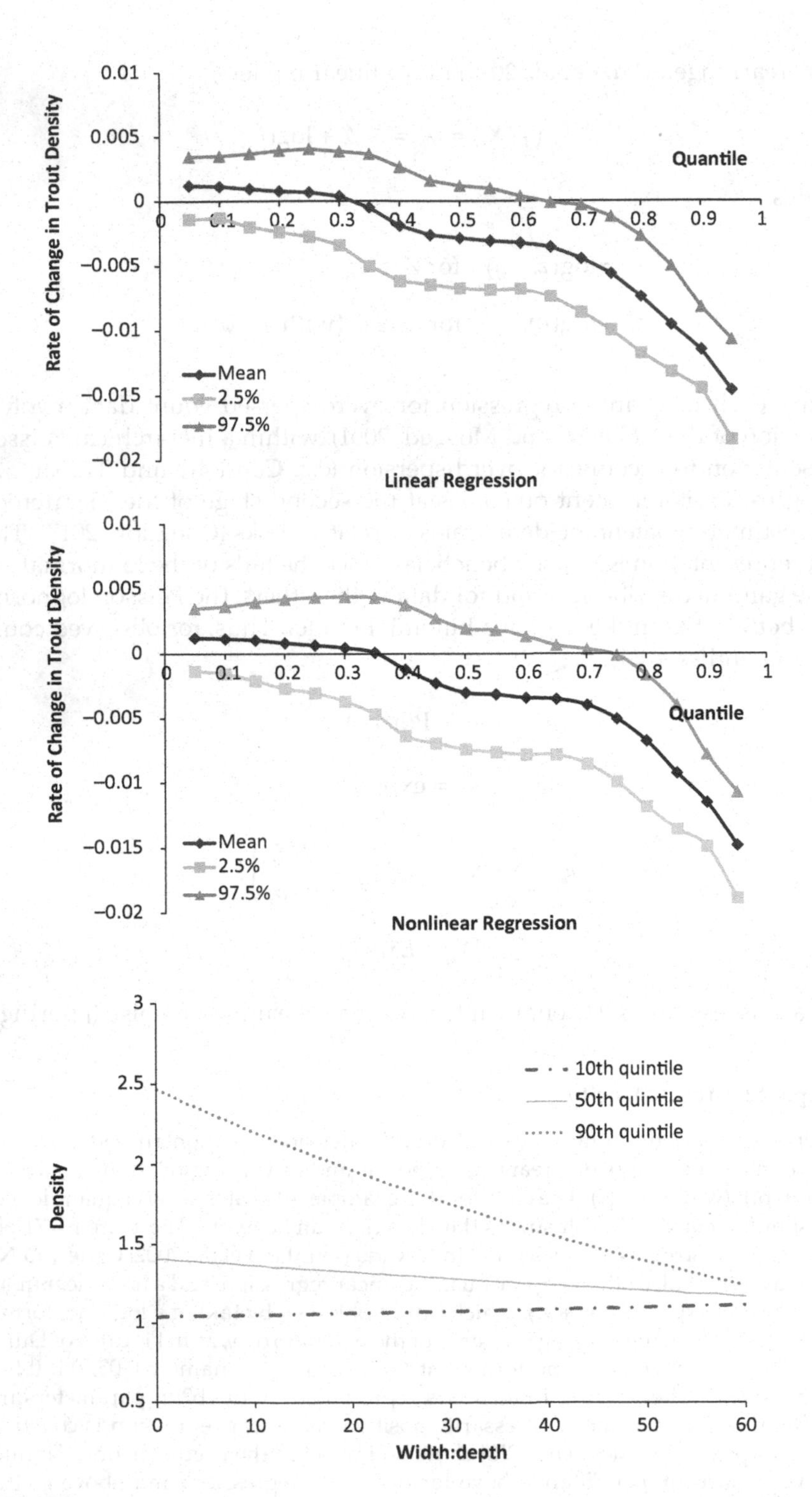

FIGURE 12.9
(a) Trout per meter by width/depth ratio, according to regression quantiles. (b) Relationship between density and W-D ratios.

Posterior predictive p-tests at each quantile are made using total absolute deviations between actual responses (or replicates) and model predictions. These remain between 0.1 and 0.9 under the linear model, though are under 0.1 for middle quantiles (between 0.4 and 0.7). Predictive tests are satisfactory across all quantiles for the exponential model.

An rstan implementation of the linear option confirms the lack of impact of w-d ratio at $q = 0.5$, with β_1 having mean (95% interval) of −0.0029 (−0.0072, 0.0022). However, at $q = 0.9$, the estimate is −0.0114 (−0.0146, −0.0081) (Figures 12.10 and 12.11).

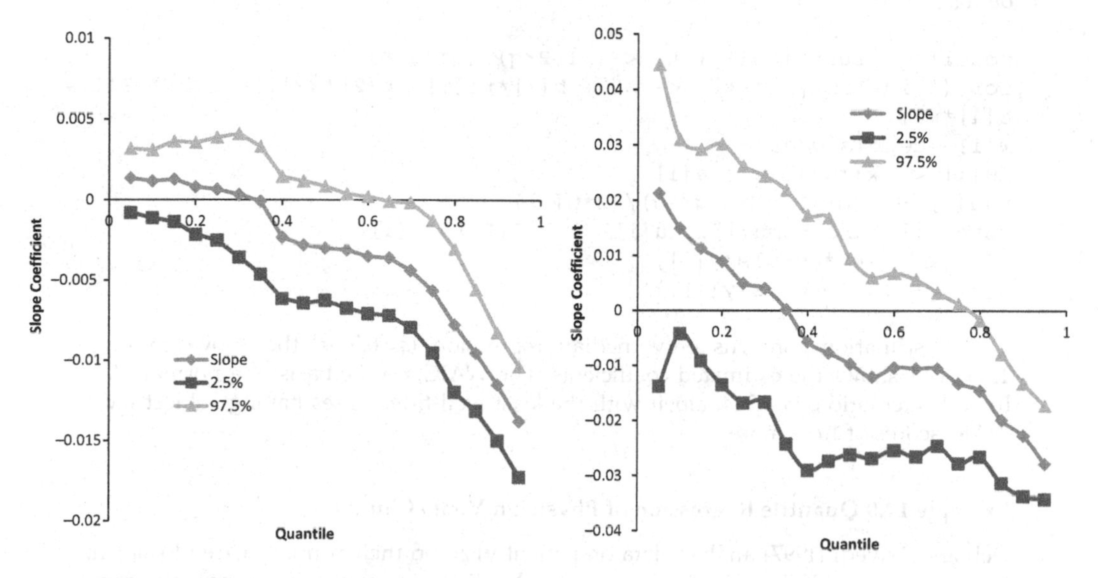

FIGURE 12.10
Quantile Regression Coefficient Plots, Linear (left), Exponential (right), Slope of Density on Width-Depth Ratio.

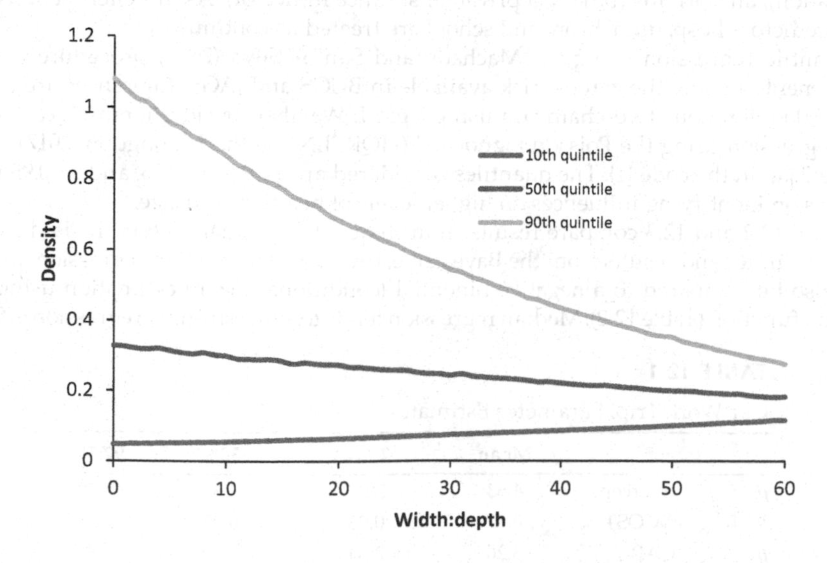

FIGURE 12.11
Conditional predictive profile, trout density against width-depth ratio, 10th, 50th and 90th quantiles, exponential transform model.

Example 12.8 Binary Work Trip Data

This example shows how quantile regression can be applied to binary data using data augmentation. The application is to work trip data, specifically use of car or not as response, with predictors: DCOST (transit fare minus automobile travel cost in cents); CAR (number of cars owned by household); DOVTT (transit out-of-vehicle travel minus automobile outofvehicle travel time in minutes); and DIVTT (transit in-vehicle travel time minus automobile in-vehicle travel time in minutes). For identifiability, the parameter σ_q is set to 1. This model can be fitted using R2OpenBugs, with the relevant coding being:

```
model1 <- function() { xi <- (1-2*q)/(q*(1-q))
for (i in 1:n){ eta[i] <- b0 + b[1]*x1[i]+ b[2]*x2[i]+ b[3]*x3[i]+
b[4]*x4[i]
w[i] ~dexp(sigmaq)
mu[i] <- xi*w[i] + eta[i]
tau[i] <- (q*(1-q)*sigmaq)/(2*w[i])
ystar[i] ~dnorm(mu[i],tau[i]) %_% I(A[i],B[i])
A[i] <- -100*equals(y[i],0)
B[i] <- 100*equals(y[i],1)}
```

The estimation concerns only median regression ($q=0.5$ in the above coding). Table 12.1 shows the estimated coefficients. The WAIC, on the basis of a normal likelihood calculation, is 4,344, albeit with the least well-fitted cases having subject level WAIC scores of 10 or more.

Example 12.9 Quantile Regression of Physician Visits Counts

Deb and Trivedi (1997) analyse data on patient visits to their primary care physician. Data refer to 4,406 individuals covered by Medicare, a public insurance program. Predictors are total hospital stays (hosp), health (binary, excellent self-perceived health status), numchron (number of chronic conditions), gender, school (number of years of education), and privins (binary, a private insurance indicator). As in Zeileis et al (2008), the predictors hosp, numchron, and school are treated as continuous.

Quantile regression using the Machado and Santos Silva (2005) procedure can be implemented using the zeroes trick available in BUGS and JAGS. Inferences are based on a 5,000 iterations two-chain run using jagsUI. We also consider hierarchical quantile regression using the Poisson lognormal (HQRPLN) method (Congdon, 2017), as in model2.jag in the code [1]. The quantiles considered are $q=0.5$, $q=0.75$, and $q=0.95$ with a focus on identifying influences on higher levels of health care usage.

Tables 12.2 and 12.3 compare results from frequentist estimation (via the lqm.counts function in R), and results from the Bayesian estimation. The median regression ($q=0.5$) can also be compared to a negative binomial (conditional mean) estimation using the glm.nb function (Table 12.2). Median regression tends to show stronger regression effects,

TABLE 12.1

Car Work Trip, Parameter Estimates

		Mean	2.5%	50%	97.5%
β_0	Intercept	4.63	3.98	4.63	5.30
β_1	DCOST	0.97	0.53	0.97	1.41
β_2	CAR	3.20	2.64	3.20	3.79
β_3	DOVTT	1.00	0.38	0.99	1.75
β_4	DIVTT	0.24	−0.32	0.23	0.82

TABLE 12.2
Physician Visits, Comparison of Estimates, Median Regression

	Negative Binomial (Conditional Mean)		Median Regression: Machado-Santos Silva (via lqm.counts)		Median Regression: Machado-Santos Silva (Bayesian Estimates)		Median Regression: Hierarchical HQRPLN (Bayesian Estimates)	
	Estimate	SE	Estimate	SE	Mean	Std	Mean	Std
Intercept	1.00	0.05	0.72	0.08	0.43	0.05	0.48	0.06
hosp	0.23	0.02	0.26	0.03	0.28	0.02	0.25	0.02
health	−0.36	0.06	−0.40	0.10	−0.39	0.07	−0.37	0.06
numchron	0.19	0.01	0.22	0.01	0.24	0.01	0.23	0.01
gender	−0.13	0.03	−0.20	0.05	−0.20	0.04	−0.18	0.03
school	0.023	0.004	0.017	0.006	0.029	0.005	0.028	0.005
privins	0.19	0.04	0.22	0.05	0.33	0.05	0.30	0.05

TABLE 12.3
Physician Visits, Comparison of Estimates, Higher Quantiles

	$q=0.75$: Machado-Santos Silva (via lqm.counts)		$q=0.75$: Machado-Santos Silva (Bayesian Estimates)		$q=0.75$: Hierarchical HQRPLN (Bayesian Estimates)	
	Estimate	SE	Mean	Std	Mean	Std
Intercept	1.16	0.07	1.26	0.05	1.34	0.05
Hosp	0.26	0.04	0.26	0.02	0.25	0.02
Health	−0.37	0.06	−0.36	0.06	−0.38	0.06
Numchron	0.21	0.02	0.20	0.01	0.19	0.01
Gender	−0.13	0.04	−0.15	0.03	−0.15	0.03
School	0.026	0.005	0.020	0.003	0.019	0.004
Privins	0.21	0.05	0.21	0.04	0.18	0.04

	$q=0.95$: Machado-Santos Silva (via lqm.counts)		$q=0.95$: Machado-Santos Silva (Bayesian Estimates)		$q=0.95$: Hierarchical HQRPLN (Bayesian Estimates)	
	Estimate	SE	Mean	Std	Mean	Std
Intercept	1.90	0.07	2.32	0.05	2.12	0.05
Hosp	0.21	0.04	0.20	0.02	0.23	0.02
Health	−0.37	0.09	−0.35	0.05	−0.42	0.05
Numchron	0.18	0.02	0.12	0.01	0.15	0.01
Gender	−0.01	0.05	−0.07	0.03	−0.08	0.03
School	0.036	0.006	0.018	0.003	0.020	0.003
Privins	0.20	0.06	0.04	0.05	0.08	0.03

though less precisely estimated, than negative binomial regression. Posterior mean W_{iq} from the HQRPLN estimation show subject 3735 as the most extreme outlier. This subject has no physician visits, despite a high number of hospital stays and chronic conditions.

Estimated regression coefficients for higher quantiles show a diminished influence of gender and insurance status. The Bayesian estimates for $q=0.95$ also show a lessened influence of total chronic conditions.

12.8 Computational Notes

[1] The JAGS code for the HQRPLN model is as follows:

```
cat("model{ xi <- (1-2*q)/(q*(1-q))
for (i in 1:n){ y[i] ~dpois(mu[i])
log(mu[i]) <- nu[i]
eta[i] <- b[1]+b[2]*hosp[i]+b[3]*excelhlth[i]
+b[4]*numchron[i]+b[5]*gender[i]
+b[6]*school[i]+b[7]*privins[i]
w[i] ~dexp(sigmaq)
tau[i] <- (q*(1-q)*sigmaq)/(2*w[i])
nu[i] ~dnorm(xi*w[i] + eta[i],tau[i])
log(L[i]) <- -mu[i]+y[i]*log(mu[i])-logfact(y[i])
LL[i] <- log(L[i])}
sigmaq ~dgamma(1,0.001)
for (j in 1:7) {b[j] ~dnorm(0,0.001) }}
", file="model2.jag")
```

References

Alhamzawi R (2012) R Package 'Brq', Bayesian Analysis of Quantile Regression Models. https://cran.r-project.org/web/packages/Brq/Brq.pdf

Alhamzawi R, Yu K, Pan J (2011) Prior elicitation in Bayesian quantile regression for longitudinal data. *Journal of Biometrics and Biostatistics*, 2, 115.

Baladandayuthapani V, Mallick B, Carroll R (2005) Spatially adaptive Bayesian penalized regression splines (P-splines). *Journal of Computational and Graphical Statistics*, 14, 378–394.

Banerjee S, Ghosal S (2014) Bayesian variable selection in generalized additive partial linear models. *Stat*, 3(1), 363–378.

Belitz C, Lang S (2008) Simultaneous selection of variables and smoothing parameters in structured additive regression models. *Computational Statistics & Data Analysis*, 53, 61–81.

Benoit D, Van den Poel D. (2012) Binary quantile regression: A Bayesian approach based on the asymmetric Laplace distribution. *Journal of Applied Econometrics*, 27(7), 1174–1188.

Benoit D, Van den Poel D (2014) bayesQR: A Bayesian approach to quantile regression. *Journal of Statistical Software*, 76(7). https://www.jstatsoft.org/article/view/v076i07

Benoit D, Van den Poel D (2017) bayesQR: A Bayesian approach to quantile regression. *Journal of Statistical Software*, 76(7). https://www.jstatsoft.org/article/view/v076i07

Berry S, Carroll R, Ruppert D (2002) Bayesian smoothing and regression splines for measurement error problems. *Journal of the American Statistical Association*, 97, 160–169.

Berzuini C, Larizza C (1996) A unified approach for modeling longitudinal and failure time data, with application in medical monitoring. *IEEE Transactions on Pattern Analysis and Machine Intelligence*, 18(2), 109–123.

Biller C (2000) Adaptive Bayesian regression splines in semiparametric generalized linear models. *Journal of Computational and Graphical Statistics*, 9, 122–140.

Biller C, Fahrmeir L (1997) Bayesian spline-type smoothing in generalized regression models. *Computational Statistics*,12, 135–151.

Bondell H, Reich B, Wang H (2010) Noncrossing quantile regression curve estimation. *Biometrika*, 97(4), 825–838.

Borsuk M, Stow C (2000) Bayesian parameter estimation in a mixed-order model of BOD decay. *Water Research*, 34, 1830–1836.

Brezger A, Lang S (2006) Generalized structured additive regression based on Bayesian P-splines. *Computational Statistics and Data Analysis*, 50, 967–991.

Brezger A, Steiner W (2008) Monotonic regression based on Bayesian P-splines: An application to estimating price response functions from store-level scanner data. *Journal of Business & Economic Statistics*, 26, 90–104.

Brumback B, Ruppert D, Wand M (1999) Variable selection and function estimation in additive non-parametric regression using a data-based prior: Comment. *Journal of the American Statistical Association*, 94, 794–797.

Brumback BA, Rice JA (1998) Smoothing spline models for the analysis of nested and crossed samples of curves. *Journal of the American Statistical Association*, 93(443), 961–976.

Cade B, Noon B (2003). A gentle introduction to quantile regression for ecologists. *Frontiers in Ecology and the Environment*, 1(8), 412–420.

Cai Y, Jiang T (2015) Estimation of non-crossing quantile regression curves. *Australian & New Zealand Journal of Statistics*, 57, 139–162.

Chen X, Ender P, Mitchell M, Wells C (2003) Regression with Stata, from http://www.ats.ucla.edu/stat/stata/webbooks/reg/default.htm

Chen Z (1993) Fitting multivariate regression functions by interaction spline models. *Journal of the Royal Statistical Society, Series B*, 55, 473–491.

Chib S, Greenberg E (2013) On conditional variance estimation in nonparametric regression. *Statistics and Computing*, 23(2), 261–270.

Chib S, Jeliazkov I (2006) Inference in semiparametric dynamic models for binary longitudinal data. *Journal of the American Statistical Association*, 101(474), 685–700.

Congdon P (2006) A model framework for mortality and health data classified by age, area, and time. *Biometrics*, 62(1), 269–278.

Congdon P (2017) Quantile regression for overdispersed count data: A hierarchical method. *Journal of Statistical Distributions and Applications*, 4, 18.

Connolly SR, Thibaut LM (2012) A comparative analysis of alternative approaches to fitting species-abundance models. *Journal of Plant Ecology*, 5(1), 32–45.

Cottet R, Kohn R, Nott D (2008) Variable selection and model averaging in semiparametric overdispersed generalized linear models. *Journal of the American Statistical Association*, 103, 661–671.

Coull B, Ruppert D, Wand M (2001) Simple incorporation of interactions into additive models. *Biometrics*, 57, 539–545.

Currie I, Durban M (2002) Flexible smoothing with P-splines: A unified approach. *Statistical Modelling*, 2, 333–349.

Deb P, Trivedi PK (1997) Demand for medical care by the elderly: A finite mixture approach. *Journal of Applied Econometrics*, 12(3), 313–336.

Denison DG, Mallick BK, Smith AF (1998) Bayesian mars. *Statistics and Computing*, 8(4), 337–346.

Dennison D, Holmes C, Mallick B, Smith A (2002) *Bayesian Methods for Non-linear Classification and Regression*. John Wiley, Chichester, UK.

Dias R, Gamerman D (2002) A Bayesian approach to hybrid splines non-parametric regression. *Journal of Statistical Computation and Simulation*, 72, 285–298.

Dunham JB, Cade BS, Terrell JW (2002) Influences of spatial and temporal variation on fish-habitat relationships defined by regression quantiles. *Transactions of the American Fisheries Society*, 131(1), 86–98.

Durban M, Currie I, Eilers P (2006) Multidimensional P-spline mixed models: A unified approach to smoothing on large grids. Working Paper, Department of Statistic, Universidad Carlos III de Madrid, Spain. http://www.unavarra.es/metma3/Papers/Invited/Durban.pdf

Eilers P, Marx B (1996) Flexible smoothing with B-splines and penalties. *Statistical Science*, 11, 89–121.

Eilers P, Marx B (2004) Splines, knots, and penalties. Working Paper. www.stat.lsu.edu/faculty/marx/

El Adlouni S, Salaou G, St-Hilaire A (2018) Regularized Bayesian quantile regression. *Communications in Statistics – Simulation and Computation*, 47(1), 277–293.

Engle R, Granger C, Rice J, Weiss A (1986) Semiparametric estimates of the relation between weather and electricity sales. *Journal of the American Statistical Association*, 81, 310–320.

Fahrmeir L, Knorr-Held L (2000) Dynamic and semiparametric models, pp 513–543, in *Smoothing and Regression: Approaches, Computation and Application*, ed M Schimek. John Wiley.

Fahrmeir L, Lang S (2001) Bayesian inference for generalized additive mixed models based on Markov random field priors. *Journal of the Royal Statistical Society C*, 50, 201–220.

Fahrmeir L, Tutz G (2001) *Multivariate Statistical Modeling Based on Generalized Linear Models*. Springer, Berlin.

Friedman J (1991) Multivariate adaptive regression splines. *Annals of Statistics*, 19, 1–67.

Fuzi M, Jemain A, Ismail N (2016) Bayesian quantile regression model for claim count data. *Insurance: Mathematics and Economics*, 66, 124–137.

Gelman A, Stern H, Carlin J, Dunson D, Vehtari A, Rubin D (2014) *Bayesian Data Analysis*, 3rd Edition. Chapman and Hall/CRC.

Geraci M, Bottai M (2006) Quantile regression for longitudinal data using the asymmetric Laplace distribution. *Biostat*, 8(1), 140–154.

Gustafson P (2000) Bayesian regression modelling with interactions and smooth effects. *Journal of the American Statistical Association*, 95, 795–806.

Hastie T, Tibshirani T (1993) Varying coefficient models. *Journal of the Royal Statistical Society B*, 55, 757–796.

Hooper P (2001) Flexible regression modeling with adaptive logistic basis functions. *Canadian Journal of Statistics*, 29, 343–378.

James G, Hastie T, Sugar C (2000) Principal component models for sparse functional data. *Biometrika* 87, 587–602.

Jerak A, Lang S (2005) Locally adaptive function estimation for binary regression models. *Biometrical Journal*, 47, 151–166.

Kharratzadeh M (2017) Splines in Stan. https://mc-stan.org/users/documentation/case-studies/splines_in_stan.html

Kitagawa G, Gersch W (1996) *Smoothness Priors Analysis of Time Series*. Springer Verlag, New York.

Klein N, Kneib T, Lang S (2015) Bayesian generalized additive models for location, scale, and shape for zero-inflated and overdispersed count data. *Journal of the American Statistical Association*, 110(509), 405–419.

Knorr-Held L (1999) Conditional prior proposals in dynamic models. *Scandinavian Journal of Statistics*, 26, 129–144.

Koenker R (2005) *Quantile Regression*. Cambridge University Press, Cambridge, UK.

Kohler M, Umlauf N, Beyerlein A, Winkler C, Ziegler A-G, Greven S (2016) Flexible Bayesian additive joint models with an application to type 1 diabetes research. arXiv preprint arXiv:1611.01485

Kohn R, Schimek M, Smith M (2000) Spline and kernel regression for dependent data, Chapter 6, pp 135–158, in *Smoothing and Regression Approaches, Computation and Estimation*, ed M Schimek. John Wiley.

Kohn R, Smith M, Chan D (2001) Nonparametric regression using linear combinations of basis functions. *Statistics and Computing*, 11, 313–322.

Konishi S, Ando T, Imoto S (2004) Bayesian information criteria and smoothing parameter selection in radial basis function networks. *Biometrika*, 91, 27–43.

Koop G, Poirier D (2004) Bayesian variants of some classical semiparametric regression techniques. *Journal of Econometrics*, 123, 259–282.

Koop G, Tole L (2004) Measuring the health effects of air pollution: To what extent can we really say that people are dying from bad air? *Journal of Environmental Economics and Management*, 47, 30–54.

Koop GM (2003) *Bayesian Econometrics*. John Wiley & Sons Inc.

Kottas A, Gelfand AE (2001) Bayesian semiparametric median regression modeling. *Journal of the American Statistical Association*, 96(456), 1458–1468.

Kozumi H, Kobayashi G (2011) Gibbs sampling methods for Bayesian quantile regression. *Journal of Statistical Computation and Simulation*, 81(11), 1565–1578.

Krivobokova T, Crainiceanu C M, Kauermann G (2008). Fast adaptive penalized splines. *Journal of Computational and Graphical Statistics*, 17, 1–20.
Lang S, Brezger A (2004) Bayesian P-splines. *Journal of Computational and Graphical Statistics*, 13, 183–212.
Lee R, Carter L (1992) Modeling and forecasting U.S. mortality. *Journal of the American Statistical Association*, 87, 659–675.
Lenk P (1999) Bayesian inference for semiparametric regression using a Fourier representation. *Journal of Royal Statistical Society B*, 61, 863–879.
Liu X, Wang L, Liang H (2011) Estimation and variable selection for semiparametric additive partial linear models. *Statistica Sinica*, 21(3), 1225.
Machado J, Silva J (2005) Quantiles for counts. *Journal of American Statistical Association*, 100(472), 1226–1237.
MacNab Y, Gustafson P (2007) Regression B-spline smoothing in Bayesian disease mapping: with an application to patient safety surveillance. *Statistics in Medicine*, 26, 4455–4474.
Marra G, Wood S (2011) Practical variable selection for generalized additive models. *Computational Statistics & Data Analysis*, 55(7), 2372–2387.
Mayr A, Fenske N, Hofner B, Kneib T, Schmid M (2012) Generalized additive models for location, scale and shape for high dimensional data—A flexible approach based on boosting. *Journal of the Royal Statistical Society: Series C (Applied Statistics)*, 61(3), 403–427.
McElreath R (2016) *Statistical Rethinking: A Bayesian Course with Examples in R and Stan*. Chapman & Hall/CRC.
Meyer K (2005) Random regression analyses using B-splines to model growth of Australian Angus cattle. *Genetics Selection Evolution*, 37, 473–500.
Meyer R, Millar B (1998) Bayesian stock assessment using a nonlinear state-space model, in *Statistical Modeling*, eds B Marx, H Friedl, Proceedings of the 13th International Workshop on Statistical Modelling, New Orleans, pp 284–291.
Min I, Kim I (2004) A Monte Carlo comparison of parametric and nonparametric quantile regressions. *Applied Economics Letters*, 11(2), 71–74.
Natario I, Knorr-Held L (2003) Non-parametric ecological regression and spatial variation. *Biometrical Journal*, 45, 670–688.
Neuhaus JM, McCulloch CE (2006) Separating between-and within-cluster covariate effects by using conditional and partitioning methods. *Journal of the Royal Statistical Society: Series B (Statistical Methodology)*, 68(5), 859–872.
Ngo L, Wand M (2004) Smoothing with mixed model software. *Journal of Statistical Software*, 9(1), 1–54.
Panagiotelis A, Smith M (2008) Bayesian identification, selection and estimation of functions in high-dimensional additive models. *Journal of Econometrics*, 143, 291–316.
Pham T H, Wand MP (2015) Generalized Additive Mixed Model Analysis via gammSlice. http://www.matt-wand.utsacademics.info/PhamWand.pdf
Qian S, Reckhow K, Zhai J, McMahon G (2005), Nonlinear regression modeling of nutrient loads in streams: A Bayesian approach. *Water Resources Research*, 41, W07012. doi:10.1029/2005WR003986
Rahman M (2016). Bayesian quantile regression for ordinal models. *Bayesian Analysis*, 11(1), 1–24.
Ruppert D, Wand M, Carroll R (2003) *Semiparametric Regression*. Cambridge University Press.
Sanchez LB, Galarza CE, Lachos VH (2017) R Package 'ALDqr', Quantile Regression Using Asymmetric Laplace Distribution. https://cran.r-project.org/web/packages/ALDqr/ALDqr.pdf
Scheipl F (2011) spikeSlabGAM: Bayesian variable selection, model choice and regularization for generalized additive mixed models in R. *Journal of Statistical Software*, 43(14), 1–24.
Scheipl F, Fahrmeir L, Kneib T (2012) Spike-and-slab priors for function selection in structured additive regression models. *Journal of the American Statistical Association*, 107(500), 1518–1532.
Schennach SM (2005) Bayesian exponentially tilted empirical likelihood. *Biometrika*, 92, 31–46.
Shively TS, Kohn R, Wood S (1999) Variable selection and function estimation in additive nonparametric regression using a data-based prior. *Journal of the American Statistical Association*, 94(447), 777–794.

Silva G, Dean C, Niyonsenga T, Vanasse A (2008) Hierarchical Bayesian spatiotemporal analysis of revascularization odds using smoothing splines. *Statistics in Medicine* 27, 2381–2401.

Smith M, Kohn R (1996) Nonparametric regression using Bayesian variable selection. *Journal of Econometrics*, 75, 317–344.

Smith M, Kohn R (1997) A Bayesian approach to nonparametric bivariate regression. *Journal of the American Statistical Association*, 92, 1522–1535.

Smith M, Wong C-M, Kohn R (1998) Additive nonparametric regression with autocorrelated errors. *Journal of the Royal Statistical Society, Series B*, 60, 311–331.

Tutz G, Reithinger F (2007) A boosting approach to flexible semiparametric mixed models. *Statistics in Medicine*, 26, 2872–2900.

Umlauf N, Klein N, Zeileis A, Koehler M (2016) BAMLSS: Bayesian additive models for location scale and shape (and beyond). Working Papers in Economics and Statistics, 2017–04, University of Innsbruck.

Wahba G (1983) Bayesian confidence intervals for the cross validated smoothing spline. *Journal of the Royal Statistical Society, Series B*, 45, 133–150.

Wand M (2003) Smoothing and mixed models. *Computational Statistics*, 18, 223–249.

West M, Harrison P (1997) *Bayesian Forecasting and Dynamic Models*, 2nd Edition. Springer-Verlag, New York.

Wood S (2006) *Generalized Additive Models: An Introduction with R*. CRC Press.

Wood S (2008) Fast stable direct fitting and smoothness selection for generalized additive models. *Journal of the Royal Statistical Society, Series B*, 70(3), 495–518.

Wood S (2016) Just another gibbs additive modeller: Interfacing JAGS and mgcv. *Journal of Statistical Software*, 75(7). doi:10.18637/jss.v075.i07

Wood S, Kohn R (1998) A Bayesian approach to robust nonparametric binary regression. *Journal of the American Statistical Association*, 93, 203–213.

Wood S, Pya N, Safken B (2016) Smoothing parameter and model selection for general smooth models. *Journal of the American Statistical Association*, 111(516), 1548–1563.

Wood SN (2003) Thin plate regression splines. *Journal of the Royal Statistical Society: Series B (Statistical Methodology)*, 65(1), 95–114.

Wood SN, Augustin NH (2002) GAMs with integrated model selection using penalized regression splines and applications to environmental modelling. *Ecological Modelling*, 157(2–3), 157–177.

Wu H, Zhang JT (2006) *Nonparametric Regression Methods for Longitudinal Data Analysis: Mixed-Effects Modeling Approaches*, Vol. 515. John Wiley & Sons.

Xi R, Li Y, Hu Y (2016). Bayesian quantile regression based on the empirical likelihood with spike and slab priors. *Bayesian Analysis*, 11(3), 821–855.

Yau P, Kohn R (2003) Estimation and variable selection in nonparametric heteroscedastic regression. *Statistics and Computing* 13, 191–208.

Yau P, Kohn R, Wood S (2003) Bayesian variable selection and model averaging in high dimensional multinomial nonparametric regression. *Journal of Computational and Graphical Statistics*, 12, 23–54.

Yu K, Moyeed RA (2001) Bayesian quantile regression. *Statistics & Probability Letters*, 54(4), 437–447.

Yue Y, Hong H (2012) Bayesian Tobit quantile regression model for medical expenditure panel survey data. *Statistical Modelling*, 12(4), 323–346.

Yue Y, Speckman P, Sun D (2012). Priors for Bayesian adaptive spline smoothing. *Annals of the Institute of Statistical Mathematics*, 64(3), 577–613.

Zeileis A, Kleiber C, Jackman S (2008) Regression models for count data in R. *Journal of Statistical Software*, 27(8), 1–25.

Index